THE CAMBRIDGE HISTORY OF SCIENCE

VOLUME 2

Medieval Science

This volume in the highly respected Cambridge History of Science series is devoted to the history of science in the Middle Ages from the North Atlantic to the Indus Valley. Medieval science was once universally dismissed as nonexistent – and sometimes it still is. This volume reveals the diversity of goals, contexts, and accomplishments in the study of nature during the Middle Ages. Organized by topic and culture, its essays by distinguished scholars offer the most comprehensive and up-to-date history of medieval science currently available. Intended to provide a balanced and inclusive treatment of the medieval world, contributors consider scientific learning and advancement in the cultures associated with the Arabic, Greek, Latin, and Hebrew languages. Scientists, historians, and other curious readers will all gain a new appreciation for the study of nature during an era that is often misunderstood.

David C. Lindberg is Hilldale Professor Emeritus of the History of Science and past director of the Institute for Research in the Humanities at the University of Wisconsin–Madison. He has written or edited a dozen books on topics in the history of medieval and early-modern science, including *The Beginnings of Western Science* (1992). He and Ronald L. Numbers have previously coedited *God and Nature: Historical Essays on the Encounter between Christianity and Science* (1986) and *When Science and Christianity Meet* (2003). A Fellow of the American Academy of Arts and Sciences, he has been a recipient of the Sarton Medal of the History of Science Society, of which he is also past president (1994–5).

Michael H. Shank is Professor of the History of Science at the University of Wisconsin–Madison. He is the author of *"Unless You Believe, You Shall Not Understand": Logic, University, and Society in Late Medieval Vienna* (1988); the editor of *The Scientific Enterprise in Antiquity and the Middle Ages: Readings from Isis* (2000); the coeditor, with Peter Harrison and Ronald L. Numbers, of *Wrestling with Nature: From Omens to Science* (2011); and the author of numerous articles in edited collections and scholarly journals.

THE CAMBRIDGE HISTORY OF SCIENCE

General editors
David C. Lindberg and Ronald L. Numbers

VOLUME 1. *Ancient Science*
Edited by Alexander Jones and Liba Chaia Taub

VOLUME 2. *Medieval Science*
Edited by David C. Lindberg and Michael H. Shank

VOLUME 3. *Early Modern Science*
Edited by Katharine Park and Lorraine Daston

VOLUME 4. *Eighteenth-Century Science*
Edited by Roy Porter

VOLUME 5. *The Modern Physical and Mathematical Sciences*
Edited by Mary Jo Nye

VOLUME 6. *The Modern Biological and Earth Sciences*
Edited by Peter J. Bowler and John V. Pickstone

VOLUME 7. *The Modern Social Sciences*
Edited by Theodore M. Porter and Dorothy Ross

VOLUME 8. *Modern Science in National and International Context*
Edited by David N. Livingstone and Ronald L. Numbers

David C. Lindberg is Hilldale Professor Emeritus of the History of Science and past director of the Institute for Research in the Humanities at the University of Wisconsin–Madison. He has written or edited a dozen books on topics in the history of medieval and early-modern science, including *The Beginnings of Western Science* (1992). He and Ronald L. Numbers have previously coedited *God and Nature: Historical Essays on the Encounter between Christianity and Science* (1986) and *When Science and Christianity Meet* (2003). A Fellow of the American Academy of Arts and Sciences, he has been a recipient of the Sarton Medal of the History of Science Society, of which he is also past president (1994–5).

Ronald L. Numbers is Hilldale Professor of the History of Science and Medicine at the University of Wisconsin–Madison, where he has taught since 1974. A specialist in the history of science and medicine in the United States, he has written or edited more than two dozen books, including *The Creationists* (1992, 2006), *Science and Christianity in Pulpit and Pew* (2007), *Galileo Goes to Jail and Other Myths about Science and Religion* (ed.) (2009), and the forthcoming *Science and the Americans*. A Fellow of the American Academy of Arts and Sciences and a former editor of *Isis*, the flagship journal of the history of science, he has served as the president of the American Society of Church History (1999–2000), the History of Science Society (2000–1), and the International Union of History and Philosophy of Science/Division of History of Science and Technology (2005–9).

THE CAMBRIDGE HISTORY OF SCIENCE

VOLUME 2

Medieval Science

Edited by

DAVID C. LINDBERG

MICHAEL H. SHANK

CAMBRIDGE
UNIVERSITY PRESS

32 Avenue of the Americas, New York NY 10013-2473, USA

Cambridge University Press is part of the University of Cambridge.

It furthers the University's mission by disseminating knowledge in the pursuit of education, learning and research at the highest international levels of excellence.

www.cambridge.org
Information on this title: www.cambridge.org/9781107521643

First published 2013
First paperback edition 2015

A catalogue record for this publication is available from the British Library

Library of Congress Cataloguing in Publication data
(Revised for volume 2)
The Cambridge history of science
p. cm.
Includes bibliographical references and indexes.
Contents: – v. 2. Medieval science / edited by David C. Lindberg and Michael H. Shank
v. 3. Early modern science / edited by Katharine Park and Lorraine Daston
v. 4. Eighteenth-century science / edited by Roy Porter
v. 5. The modern physical and mathematical sciences / edited by Mary Jo Nye
v. 6. The modern biological and earth sciences / edited by Peter J. Bowler and John V. Pickstone
v. 7. The modern social sciences / edited by Theodore H. Porter and Dorothy Ross
1. Science – History. I. Lindberg, David C. II. Numbers, Ronald L.
Q125C32 2001
509–dc21 2001025311

ISBN 978-0-521-59448-6 Hardback
ISBN 978-1-107-52164-3 Paperback

CONTENTS

ILLUSTRATIONS

NOTES ON CONTRIBUTORS

E. JENNIFER ASHWORTH, Distinguished Professor Emerita at the University of Waterloo, was elected Fellow of the Royal Society of Canada in 1991. She has published extensively on medieval and post-medieval logic and philosophy of language, and her first book, *Language and Logic in the Post-Medieval Period*, was published in 1974. Her most recent book, *Les théories de l'analogie du XIIe au XVIe siècle* (2008), is based on the four Pierre Abélard lectures that she delivered at the Sorbonne in 2004. Since her retirement in 2005, she has returned to the United Kingdom.

J. L. BERGGREN received his PhD from the University of Washington in 1966 and is now Emeritus Professor at Simon Fraser University, Canada. He has held visiting positions in the Mathematics Institute at the University of Warwick and the History of Science Departments at Yale and Harvard Universities. He has published numerous papers and books on the history of mathematical sciences of ancient Greece and medieval Islam, among them *Episodes in the Mathematics of Medieval Islam* (1986); Euclid's "Phaenomena" (with Robert Thomas, 1966); and the section on "Islamic Mathematics" in *The Mathematics of Egypt, Mesopotamia, China, India, and Islam: A Source Book* (2007).

CHARLES BURNETT has been Professor of the History of Islamic Influences in Europe at the Warburg Institute, University of London, since 1999. He received his MA and PhD from Cambridge University and has been a Member of the Institute for Advanced Study (Princeton), a Leverhulme Research Fellow at the University of Sheffield, and a Distinguished Visiting Professor in Medieval Studies at the University of California at Berkeley. His work has centered on the transmission of Arabic science and philosophy to Western Europe, which he has documented by editing and translating several texts.

JOAN CADDEN is Professor Emerita of History at the University of California, Davis. Her current research concerns include medieval natural philosophers' explanations of male homosexual desire and the dissemination of medieval natural philosophical and medical learning. She is the author of *Meanings of Sex Difference in the Middle Ages: Medicine, Science, and Culture* (Cambridge University Press, 1993), which was awarded the History of Science Society's Pfizer Prize, as well as articles on the medical and scientific ideas of medieval women, such as Hildegard of Bingen and Christine de Pizan.

BRUCE S. EASTWOOD (PhD, University of Wisconsin) is Professor of History, Emeritus, at the University of Kentucky. His publications include *Astronomy and Cosmology in the Carolingian Renaissance* (2007); *Planetary Diagrams for Roman Astronomy in Medieval Europe, ca. 800–1500* (with Gerd Grasshoff, 2004); *The Revival of Planetary Astronomy in Carolingian and Post-Carolingian Europe* (2002); and an online edition of the ninth-century *Anonymous Commentary on the Astronomy of Martianus Capella*. He has received fellowships from the National Endowment for the Humanities and the Institute for Advanced Study (Princeton), as well as numerous grants from the National Science Foundation and other sources. Among his current projects is a book on Charlemagne and the Christian revival of science.

EDWARD GRANT is Distinguished Professor Emeritus of History and Philosophy of Science at Indiana University, Bloomington. He has published more than ninety articles and twelve books, including one on medieval cosmology titled *Planets, Stars, and Orbs: The Medieval Cosmos 1200–1687* (Cambridge University Press, 1994). During 1985–6, he served as president of the History of Science Society. His honors include the George Sarton Medal of the History of Science Society (1992), Fellow of the American Academy of Arts and Sciences (elected 1984), Fellow of the Medieval Academy of America (1982), and Membre effectif of the Académie Internationale d'Histoire des Sciences, Paris (1969).

DANIELLE JACQUART is full professor at the École Pratique des Hautes Études (Paris 1, Sorbonne, "Section des sciences historiques et philologiques"), where she holds the chair of "History of Science in the Middle Ages." She has written widely on medical thought and practice in the Latin Middle Ages, and on the influence of Arabic medicine on the medieval West. Her major works include *La médecine médiévale dans le cadre parisien (XIVe–XVe siècle)* (1998) and *Le milieu médical en France du XIIe au XVe siècle* (1981). She is corresponding Fellow of the Medieval Academy of America and a member of the Academia Europea.

ELAHEH KHEIRANDISH is a historian of science (PhD, Harvard University, 1991), with a focus on science in Islamic lands. She has taught at Harvard University, received awards from the National Science Foundation, contributed to collaborative projects and major journals, and recently coedited a special issue of *Iranian Studies*. Her publications include the two-volume *The Arabic Version of Euclid's Optics* (1999) and forthcoming books on the Arabic and

Persian traditions of optics and mechanics. She is currently a Fellow at Harvard's Center for Middle Eastern Studies and serves on the advisory boards of *Interpretatio* and the Islamic Scientific Manuscripts Initiative (ISMI).

TOMOMI KINUKAWA received her PhD at the University of Wisconsin. She is now Assistant Professor of History at the University of the Pacific, Stockton, California. Her research has focused on natural history, colonial science, gender, and race. She is currently working on a project on health and citizenship among Korean diaspora communities in Japan in the mid- to late twentieth century.

WALTER ROY LAIRD teaches medieval history and the history of science at Carleton University, Ottawa, Canada. In addition to articles on medieval and renaissance natural philosophy and the mathematical sciences, he is author of *The Unfinished Mechanics of Giuseppe Moletti* (2000) and coeditor of *Mechanics and Natural Philosophy before the Scientific Revolution* (2008).

Y. TZVI LANGERMANN is a professor in the Department of Arabic, Bar Ilan University, Ramat Gan, Israel. His most recent books are *Hebrew Medical Astrology* (coauthored with Gerrit Bos and Charles Burnett) and *Adaptations and Innovations: Studies on the Interaction between Jewish and Islamic Thought and Literature* (coedited with Josef Stern). He is a regular contributor to *Aleph: Historical Studies in Science & Judaism* and has published widely on the history of science and philosophy.

DAVID C. LINDBERG, coeditor of this volume, is Hilldale Professor Emeritus at the University of Wisconsin. He has written or edited more than a dozen books, including editions and translations of medieval Latin texts and a prize-winning survey: *The Beginnings of Western Science*, 2nd ed. (2007). He has been a Guggenheim Fellow, a visiting member of the Institute for Advanced Study (Princeton), and a Fellow of the Medieval Academy of America and the Académie Internationale d'Histoire des Sciences. He has served as president of the History of Science Society and has been awarded its Sarton Medal for lifetime scholarly achievement.

STEPHEN C. McCLUSKEY is Professor Emeritus of History at West Virginia University. His recent work focuses on astronomy and cosmology in the early Middle Ages and the astronomical and religious significance of the orientation of English village churches. Among his publications are *Astronomies and Cultures in Early Medieval Europe* (Cambridge University Press, 1999) and "Boethius's Astronomy and Cosmology," in *A Companion to Boethius in the Middle Ages* (2012), edited by Noel H. Kaylor and Philip E. Phillips.

A. GEORGE MOLLAND (1941–2002) pursued the mathematics tripos at Corpus Christi College, Cambridge, receiving the PhD degree in 1967. He then spent his subsequent academic career at the University of Aberdeen, advancing from Lecturer to Senior Lecturer in History and Philosophy of Science.

Molland's major scholarly contributions were in medieval mathematics and mathematical science (especially the science of motion) and the relationship of medieval mathematical sciences to those of Galileo and the seventeenth century. Toward the end of his career, he returned to studying the Middle Ages, especially Roger Bacon. An edition of a Latin text, with English translation, of Bacon's *Opus Tertium* remains incomplete, owing to Molland's untimely death.

ROBERT G. MORRISON is Associate Professor of Religion at Bowdoin College. His recent book *Islam and Science: The Intellectual Career of Nizam al-Din al-Nisaburi* (2007) won Iran's 2009 World Book Prize for Islamic studies. His research has been funded by NEH and a Graves Award in the Humanities. He is currently studying a Judeo-Arabic text on astronomy and its relation to currents in Islamic science.

WILLIAM R. NEWMAN is Ruth N. Hall Professor and Distinguished Professor in the History and Philosophy of Science at Indiana University. Most of his recent scholarly work has focused on "chymistry" in the early-modern period and on the experimental tradition more broadly. His recent books include *Atoms and Alchemy: Chymistry and the Experimental Origins of the Scientific Revolution* (2006); *Promethean Ambitions: Alchemy and the Quest to Perfect Nature* (2004); and *Alchemy Tried in the Fire: Starkey, Boyle, and the Fate of Helmontian Chymistry* (with Lawrence M. Principe) (2002).

JOHN NORTH (1934–2008) was Professor Emeritus of History of Philosophy and the Exact Sciences at the University of Groningen. A universal scholar, he received his higher education at Merton College, Oxford, where he read mathematics, philosophy, politics, and economics, followed by an external degree in astronomy from the University of London. After earning his doctorate at Oxford, he served as a curator in the Oxford Museum of the History of Science before taking the chair at Groningen. North's many interests included, preeminently, medieval astronomy and astronomical instruments. He was a prolific author, whose major publications included *Chaucer's Universe* (1988); *Richard of Wallingford*, 3 vols. (1976); *Horoscopes and History* (1986); *The Ambassadors' Secret: Holbein and the World of the Renaissance* (2002); and, most recently, *Cosmos* (2008).

VIVIAN NUTTON is Professor of the History of Medicine at the Wellcome Trust Centre for the History of Medicine, University College, London. He has written extensively on the history of medicine from Classical Antiquity to the Renaissance. His books include *Galen, On My Own Opinions* (1999); *Ancient Medicine* (2004); and *Girolamo Mercuriale, De arte gymnastica* (2008). His edition of a forgotten work by Galen, *On Problematical Movements*, will be published by Cambridge University Press. He is a Fellow of the British Academy and of the Deutsche Akademie der Wissenschaften.

GEORGE OVITT received his PhD from the University of Massachusetts. He has taught history at Dean College, Drexel University, and Sidwell Friends School, and is currently at Albuquerque Academy. His scholarly interests include the history of technology and labor and, in particular, the ways in which the material aspects of human life are affected by cultural concerns. He is author of *The Restoration of Perfection: Labor and Technology in Medieval Culture.*

KATHARINE PARK teaches in the Department of the History of Science at Harvard University, where she works on the history of science and medicine in medieval and early-modern Europe and the history of women, gender, and the body. Her most recent books are *Secrets of Women: Gender, Generation, and the Origins of Human Dissection* (2006) and *The Cambridge History of Science*, vol. 3: *Early Modern Science* (2006), the latter coedited with Lorraine Daston.

F. JAMIL RAGEP is Canada Research Chair in the History of Science in Islamic Societies and Director of the Institute of Islamic Studies at McGill University in Montreal, Canada. Educated at the University of Michigan and Harvard University, he has written extensively on the history of astronomy and on science in Islam. He is currently leading an international effort to catalogue all Islamic manuscripts in the exact sciences and is codirecting a project to study the fifteenth-century background to the Copernican revolution.

KAREN MEIER REEDS, of the Princeton Research Forum and Visiting Scholar at Columbia University and the University of Pennsylvania, is an independent historian of science and medicine whose research focuses on the history of botany from antiquity through Linnaeus. She is the author of *Botany in Medieval and Renaissance Universities* (1991) and *A State of Health: New Jersey's Medical Heritage* (2001); coeditor, with Jean Givens and Alain Touwaide, of *Visualizing Medieval Medicine and Natural History, 1200–1550* (2006); and guest curator of "Come into a New World: Linnaeus & America" (2007). She is also a Fellow of the Linnaean Society of London.

EMILIE SAVAGE-SMITH is Professor of the History of Islamic Science at the Oriental Institute, University of Oxford. She has published studies on a variety of medical and divinatory practices in the Islamic world, as well as on celestial globes and mapping. Her most recent book (with Peter E. Pormann) is *Medieval Islamic Medicine* (2007).

MICHAEL H. SHANK (coeditor of this volume) teaches at the University of Wisconsin–Madison, where he is Professor of the History of Science (and Herbert and Evelyn Howe Bascom Professor of Integrated Liberal Studies, 2008–10). A former associate editor of *Isis*, he is the author of *"Unless You Believe, You Shall Not Understand": Logic, University, Society in Late Medieval Vienna* (1988); the editor of *The Scientific Enterprise in Antiquity and the*

Middle Ages (2000); and a coeditor, with Peter Harrison and Ronald L. Numbers, of *Wrestling with Nature: From Omens to Science* (2011) and of Johannes Regiomontanus's *Defensio Theonis contra Georgium Trapezuntium* (Web publication in progress, in association with Richard Kremer).

KATHERINE H. TACHAU earned her PhD from the University of Wisconsin–Madison in 1981. After teaching at Montana State University and Pomona College, she joined the History Department at the University of Iowa in 1985, where she has served as Faculty Senate President. A Guggenheim Fellowship recipient, she studies thirteenth- and fourteenth-century philosophy, science, and art of Paris, Oxford, and other European universities, in publications ranging from *Vision and Certitude in the Age of Ockham: Optics, Epistemology, and the Foundations of Semantics, 1250–1345* (1988) to "God's Compass and *Vana Curiositas*: Scientific Study in the Old French *Bible moralisée*," *Art Bulletin*, 80 (1998).

ANNE TIHON is Doctor in Classical Philology (Université Catholique de Louvain) and also Professor at the Université Catholique de Louvain (Louvain-la-Neuve). Her teaching concerns the history of science in Antiquity and the Middle Ages, Byzantine history and civilization, Greek paleography, Byzantine texts, and methodology of textual editions. She has provided critical editions of the commentaries of Theon of Alexandria on Ptolemy's *Handy Tables (Small Commentary and Great Commentary)* (*Studi e Testi* 282, 315, 340, 390) and several editions of Byzantine astronomical texts. She is the director of the *Corpus des Astronomes Byzantins* (ten volumes published).

DAVID WOODWARD (1942–2004) was Arthur H. Robinson Professor Emeritus of Geography at the University of Wisconsin–Madison. A wide-ranging scholar of the history and the art of cartography, he was founding coeditor (with J. B. Harley) of the award-winning multivolume *History of Cartography*. His essay on "Medieval *Mappaemundi*" for volume one (1987) revitalized the study of cosmographical representations.

GENERAL EDITORS' PREFACE

The idea for *The Cambridge History of Science* originated with Alex Holzman, former editor for the history of science at Cambridge University Press. In 1993, he invited us to submit a proposal for a multivolume history of science that would join the distinguished series of Cambridge histories, launched nearly a century ago with the publication of Lord Acton's fourteen-volume *Cambridge Modern History* (1902–12). Convinced of the need for a comprehensive history of science and believing that the time was auspicious, we accepted the invitation.

Although reflections on the development of what we call "science" date back to antiquity, the history of science did not emerge as a distinctive field of scholarship until well into the twentieth century. In 1912, the Belgian scientist-historian George Sarton (1884–1956), who contributed more than any other single person to the institutionalization of the history of science, began publishing *Isis*, an international review devoted to the history of science and its cultural influences. Twelve years later, he helped to create the History of Science Society, which by the end of the century had attracted some 4,000 individual and institutional members. In 1941, the University of Wisconsin established a department of the history of science, the first of dozens of such programs to appear worldwide.

Since the days of Sarton, historians of science have produced a small library of monographs and essays, but they have generally shied away from writing and editing broad surveys. Sarton himself, inspired in part by the Cambridge histories, planned to produce an eight-volume *History of Science*, but he completed only the first two installments (1952, 1959), which ended with the birth of Christianity. His mammoth three-volume *Introduction to the History of Science* (1927–48), more a reference work than a narrative history, never got beyond the Middle Ages. The closest predecessor to *The Cambridge History of Science* is the three-volume (four-book) *Histoire Générale des Sciences* (1957–64), edited by René Taton, which appeared in an English translation under the title *General History of the Sciences* (1963–4). Edited just before the late-century boom in the history of science, the Taton set quickly became dated.

During the 1990s, Roy Porter began editing the very useful Fontana History of Science (published in the United States as the Norton History of Science), with volumes devoted to a single discipline and written by a single author.

The Cambridge History of Science comprises eight volumes, the first four arranged chronologically from antiquity through the eighteenth century and the latter four organized thematically and covering the nineteenth and twentieth centuries. Eminent scholars from Europe and North America, who together form the editorial board for the series, edit the respective volumes:

Volume 1: *Ancient Science*, edited by Alexander Jones, University of Toronto, and Liba Chaia Taub, University of Cambridge
Volume 2: *Medieval Science*, edited by David C. Lindberg and Michael H. Shank, University of Wisconsin–Madison
Volume 3: *Early Modern Science*, edited by Katharine Park, Harvard University, and Lorraine Daston, Max Planck Institute for the History of Science, Berlin
Volume 4: *Eighteenth-Century Science*, edited by Roy Porter, late of Wellcome Trust Centre for the History of Medicine at University College London
Volume 5: *The Modern Physical and Mathematical Sciences*, edited by Mary Jo Nye, Oregon State University
Volume 6: *The Modern Biological and Earth Sciences*, edited by Peter J. Bowler, Queen's University of Belfast, and John V. Pickstone, University of Manchester
Volume 7: *The Modern Social Sciences*, edited by Theodore M. Porter, University of California, Los Angeles, and Dorothy Ross, Johns Hopkins University
Volume 8: *Modern Science in National and International Context*, edited by David N. Livingstone, Queen's University of Belfast, and Ronald L. Numbers, University of Wisconsin–Madison

Our collective goal is to provide an authoritative, up-to-date account of science – from the earliest literate societies in Mesopotamia and Egypt to the end of the twentieth century – that even nonspecialist readers will find engaging. Written by leading experts from every inhabited continent, the essays in *The Cambridge History of Science* explore the systematic investigation of nature and society, whatever it was called. (The term "science" did not acquire its present meaning until early in the nineteenth century.) Reflecting the ever-expanding range of approaches and topics in the history of science, the contributing authors explore non-Western as well as Western science, applied as well as pure science, popular as well as elite science, scientific practice as well as scientific theory, cultural context as well as intellectual content, and the dissemination and reception as well as the production of scientific knowledge. George Sarton would scarcely recognize this collaborative effort as the history of science, but we hope we have realized his vision.

David C. Lindberg
Ronald L. Numbers

INTRODUCTION

Michael H. Shank and David C. Lindberg

After asking what we do for a living, people often find the answer jarring. "The history of medieval science?" How, indeed, can one use a synonym for "backward" to modify a noun that signifies the best available knowledge of the natural world? Yet the history of medieval science is a recognized and productive area of scholarship, whose practitioners not only use the expression freely but also are acknowledged as significant reinterpreters of medieval history itself.[1] To bridge the chasm between popular and scholarly understandings, we must grapple briefly with the two terms in "medieval science," the way the tension between their meanings arose, and the way the Middle Ages as a general historical category has shaped the framework of medieval science.

THE POSTHUMOUS MIDDLE AGES

When the fourteenth-century poet Petrarch looked back at the Roman Empire, he saw "darkness" separating it from his own day. Although he did not think of this interval as a full-blown historical period, he nevertheless characterized it as contemptible. Barbarians from the misnamed Emperor Charlemagne onward had usurped a title and a dominion that rightly belonged to Romans. The fifteenth century gave this period such names as *media tempestas* or *medium aevum*, the "middle era." For many European intellectuals of that day, the Middle Ages were a useful invention that contrasted the political fragmentation and barbarous degenerate Latin of the recent past with the lost glory and beautiful language of Rome, to which they

[1] Marcia Colish, *Remapping Scholasticism* (The Etienne Gilson Series, 21) (Toronto: Pontifical Institute of Medieval Studies, 2000), pp. 13–15, reprinted in Colish, *Studies in Scholasticism* (Aldershot: Ashgate, 2006).

aspired.[2] To invoke the Middle Ages when discussing empire, language, or art was implicitly to narrate history with the radical discontinuity of a sorry, if not necessarily vacuous, millennium.

By the late seventeenth century, Christoph Cellarius (1638–1707), a Lutheran scholar at the University of Halle, gave the "middle era" a major role in historical periodization. After writing a separate *Historia medii aevi*. . . , he integrated it into his history of the world: *Historia universalis*. . . *in antiquam et medii aevi ac novam divisa* ("Universal history . . . divided into ancient, and of the middle era, and new"). He ended antiquity with Emperor Constantine (d. 337), and the "Middle Era" with the Ottoman conquest of Constantinople (1453). His 200-odd pages of medieval history included some culture. The Middle Ages ended with the resurgence of "Latin letters out of darkness," the invention of printing, the foundation of new universities, and medieval theologians' foreshadowing of the Reformation, which would begin the New Era.[3] The medieval period and modernity each opened with landmarks in the history of Christianity: Constantine, who legalized it, and Luther, who reformed it. The confessional, even parochial, character of Cellarius's divisions did not undermine their universal reach. They applied not merely to Europe but to the world.

This schema has had an astonishing career. In a few centuries, a slur born from Petrarch's nostalgia for lost Roman power grew into the central hinge of the European past. By the nineteenth century, a layering of humanist, Protestant, and Enlightenment sensibilities had transformed Cellarius's tripartite division into the framework that historians from the European colonizing nations routinely used to structure their understanding of the globe. The Middle Ages thus became a standard period of world history, which even the critical outlook of Marxist historiography not only left untouched but also helped to entrench and to export.[4]

The threefold division of global history remains firmly anchored in our conceptualization of the past, despite long-standing criticisms.[5] It may take

[2] Jean Dagenais and Margaret Greer, "Decolonizing the Middle Ages: Introduction," *Journal of Medieval and Early Modern Studies*, 30 (2000), 431–48, and the literature cited therein; Theodor E. Mommsen, "Petrarch's Conception of the 'Dark Ages,'" in Theodor E. Mommsen, *Medieval and Renaissance Studies*, ed. Eugene F. Rice, Jr. (Ithaca, N.Y.: Cornell University Press, 1959), pp. 106–21 (original 1942); and Paul Lehmann, "Mittelalter und Küchenlatein," *Historische Zeitschrift*, 137 (1928), 196–213s, especially pp. 206–13.

[3] Lehmann, "Mittelalter und Küchenlatein," p. 201; Ilja Mieck, "Die Frühe Neuzeit: Definitionsprobleme, Methodendiskussion, Forschungstendenzen," in *Die frühe Neuzeit in der Geschichtswissenschaft: Forschungstendenzen und Forschungserträge*, ed. Nada Boškovska Leimgruber (Paderborn: Ferdinand Schöningh, 1997), pp. 17–38, especially pp. 18–19. I cited here Cellarius, who was not the first to use it with positive historical content, from the general preface, pp. 11–12 of the seventh edition (Jena, 1728). See Christophorus Cellarius, *Historia medii aevi a temporibus Constantini Magni ad Constantinopolim a Turcis captam deducta* . . . (1688; 7th ed.: Jena: Bielke, 1724), pp. 214–18.

[4] Timothy Reuter, "Medieval: Another Tyrannous Construct?" *Medieval History Journal*, 1 (1998), 25–45.

[5] For example, Oswald Spengler, *Der Untergang des Abendlandes: Umrisse einer Morphologie der Weltgeschichte* [1923 ed.] (Munich: Deutscher Taschenbuch Verlag, 1972), Introduction, especially

the future dominance of Asian historiography to dislodge it. Upon hearing the expressions "medieval China" or "medieval India," few wince, but everyone should who understands that for these civilizations the withering of the Western Roman Empire and the fall of Constantinople mean little. Meanwhile, "medieval Islam" designates its earliest centuries; that is, ancient or classical Islam. Such universalized usages conveniently avoid verbal "time warps" between cultures, but the price of calling everything between 400 and 1450 "medieval" is that the deeply entrenched unflattering connotations associated with the European Middle Ages automatically color other civilizations.

Significantly, the problems of the category "Middle Ages" are also severe when applied to Europe, where it originated. For historians, the most pernicious trait associated with the thousand-year label is the implication that the period shares a fundamental unity rooted in the "medieval mentality" and its many ramifications, all of which change as one approaches 1450–1500. As Jacob Burckhardt, one of the most influential creators of the Renaissance, evocatively characterized the contrast in 1860: "In the Middle Ages both sides of human consciousness – that which was turned within as that which was turned without – lay dreaming or half awake beneath a common veil. The veil was woven of faith, illusion and childish prepossession, through which the world and history were seen clad in strange hues."[6] More recent quotations in a similar vein could be cited for medieval religiosity, the medieval cosmos, and other aspects of medieval thought and life. Newspaper editorials and ordinary language continue to cast a pall of negativity on the period and its image.[7]

Once upon a time, the modern world immediately followed the Middle Ages. Later, it was buffered by "Renaissance and Reformation"; nowadays, more cautiously, by the "early-modern" world.[8] The further we recede from 1500, the more qualifications "modern" is likely to receive. The crux of the problem is that, in any guise, modernity and early modernity are fundamentally European categories, the universalization of which is far more than academic; they implicitly turn the idiosyncrasies of European/Western

pp. 21–30; Rainer Thurnher, "Oswald Spengler: Sa théorie de l'histoire et ses implications politico-idéologiques," in *Expansions, ruptures et continuités de l'idée européenne*, ed. Daniel Minary, 3 vols. (Paris: Diffusion Les Belles Lettres, 1993–1997), (Annales Littéraires de l'Université de Besançon, 562), vol. 2, pp. 155–70, especially pp. 161–2.

6 Jakob Burckhardt, *The Civilization of the Renaissance in Italy*, trans. S. G. C. Middelmore (London: Penguin Books, 1990), Part II, chap. 1, pp. 97–8.

7 Skeptics should browse the World Wide Web or watch the film "Pulp Fiction" (1994) to see astonishing uses of the expression "going medieval."

8 Charles H. Parker, "Introduction: Individual and Community in the Early Modern World," and Jerry H. Bentley, "Early Modern Europe and the Early Modern World," in *Between the Middle Ages and Modernity: Individual and Community in the Early Modern World*, ed. Charles H. Parker and Jerry H. Bentley (Lanham, Md.: Rowan and Littlefield, 2007), pp. 1–9, 13–31, respectively, especially pp. 1–3; Lorraine Daston and Katharine Park, "Introduction: The Age of the New," in *The Cambridge History of Science*, vol. 3: *Early Modern Science*, ed. Katharine Park and Lorraine Daston (Cambridge: Cambridge University Press, 2006), pp. 1–20.

history into a normative developmental pattern and impose these expectations on vast areas of the world that are still wrestling with the legacies of colonialism. To be "not yet modern" is implicitly to be medieval.[9]

Most of our knowledge of the vast medieval period has been uncovered since Burckhardt, whose seductive oversimplifications historians now overwhelmingly reject. In political and institutional history, the expressions "medieval" and "the Middle Ages" have long enjoyed more neutral connotations than they do in either ordinary language or the history of high culture, including science. After all, general historians could discuss *something* rather than nothing – rulers, battles, even economic and social practices and institutions. In these domains, once the expression "medieval civilization" ceased to be a contradiction, scholars of the period embraced the former slur and called themselves "medievalists." By the late twentieth century, one distinguished medievalist went so far as to proclaim the Middle Ages "a true period."[10] It could now partake of the real and the good.

Assigning a positive affect to the old monolithic periodization adds nothing to our critical understanding, however. More helpful are the scholarly challenges to it embodied in such journals as *Mediaeval and Renaissance Studies* (1941–) and the *Journal of Medieval and Early Modern Studies* (1971–). A few generations ago, such titles would have met with disbelief. Despite their use of "medieval," they obviously reject a sharp break and reflect a principled skepticism about the traditional periodization of European history. Other historians go further and want altogether to eliminate "medieval" and "Middle Ages" from our lexicon. One, who has called these "the worst terms that have ever found their way into the vocabulary of historians," has promoted "Old Europe" as a more coherent unit spanning roughly 1000–1800. Some historians of law and science, among other fields, have tried spans that avoid making 1450–1500 the central hinge of recent history.[11] Such approaches perform an invaluable service by usefully correcting two of the most misleading aspects of the traditional periodization: the false unity implied by lumping a very diverse millennium under a single heading and the false impression that everything changed circa 1450–1500. The preceding millennium was also a period of great change in almost every facet of human life, from the movements of peoples to the creation of institutions and the

[9] See the suggestive remarks of Dagenais and Greer, "Decolonizing the Middle Ages," pp. 435–7.

[10] Cited in Wilhelm Kamlah, "'Zeitalter' überhaupt, 'Neuzeit,' und 'Frühneuzeit'," *Saeculum*, 8 (1957), 313–32, especially pp. 319–22.

[11] Dietrich Gerhard, *Old Europe: A Study of Continuity, 1000–1800* (New York: Academic Press, 1981), p. 3; Howard Kaminsky, "From Lateness to Waning to Crisis: The Burden of the Later Middle Ages," *Journal of Early Modern History*, 4 (2000), 85–125, especially pp. 123–5; Harold Berman, *Law and Revolution: The Formation of the Western Legal Tradition* (Cambridge, Mass.: Harvard University Press, 1988) goes to the twentieth century; A. C. Crombie, *Augustine to Galileo: The History of Science, A.D. 400–1650* (London: Falcon Press, 1952) and later editions; and Stephen Gaukroger, *The Emergence of a Scientific Culture: Science and the Shaping of Modernity, 1210–1685* (Oxford: Oxford University Press, 2006).

transformation of ideas. As Marcia Colish phrased it, "diversity, inconsistency, and contradiction, and the ability to live with them and to thrive on them, are features of medieval culture that militate against any monolithic or schematic understanding of it."[12]

These alternative periodizations allow us to avoid the easy contrasts in which the traditional schema imprisons us. The old tripartite schema will, however, die very hard. It is entrenched in the lexicon, readers expect it, it defines academic positions and curricula, and many historians use it un–self-consciously. Under the circumstances, our best hope is to treat the Middle Ages as a conventional name and subvert the stereotype associated with it, notably by using jarring expressions like "medieval science" and publishing books about it.

THE SCIENCE IN MEDIEVAL SCIENCE

Early-twenty-first-century English typically uses the word "science" to denote the systematic study of natural phenomena. This dictionary-like definition is purposefully very general. (In other modern languages, the equivalent word is even more general, encompassing all systematic knowledge, whether of nature or not). When we are talking about science as it is practiced today, we implicitly modify this general definition with such specific connotations as professionalization, governmental funding, large laboratories, and experimental activity, sometimes on a grand scale.

Clearly, science in 1300 or even 1800 did not involve white lab coats or Nobel prizes. The fact that the meanings of "science" today are not precisely what they were in the past is no reason to ban the term from speech about the past, as some want to do. Let us not forget that, as used throughout the world today, the meanings of "science" and its cognates are far from unified. Many people disagree about them now.

If historians of science were to investigate only those past practices and beliefs about the study of nature that most resemble the latest science, the result would be both thin and seriously distorted. We would not be responding to the richness and variety of the past as it existed but filtering it through a modern grid. To be as fair as possible to the past in its own terms, we must refrain from scouring it only for examples or precursors of the latest science. We must respect the various ways in which earlier generations investigated nature, acknowledging that they are of great interest even though many of their approaches differed from the modern ones. Most obviously, some belong to our immediate intellectual ancestry

[12] Marcia Colish, "When Did the Middle Ages End? Reflections of an Intellectual Historian," in *Schooling and Society: The Ordering and Reordering of Knowledge in the Western Middle Ages*, ed. Alasdair McDonald and Michael Twomey (Leuven: Peeters, 2004), pp. 213–23, especially pp. 222–3.

(e.g., mathematical astronomy) and can help us to understand how modern science became what it is. But other related activities with no counterpart in modern science (e.g., medical astrology) were also important to our predecessors and deserve our attention for that reason alone. Historians, then, require a very broad working definition of "science" – one that will permit investigation of the vast range of practices and beliefs about the operations of nature that preceded the modern scientific enterprise and that can help us to understand how the latter came about. We need to be broad and inclusive – even broader, the farther back we go – rather than narrow and exclusive.[13]

English, a relatively new language, adopted the word "science" from the much older Latin. This history means that translation is at the heart of what the historian of early science must do. Obviously, many meanings of "science" before 1800 differ from some of the many meanings of the word in 1850 or today (many other words face the same problem). But these meanings have evolved in describable ways and for the most part have not changed into their opposites. Thankfully, adjectives allow us to qualify nouns and specify the range of their meaning. Using the expressions "Babylonian science" or "medieval science" does not presuppose that science has an unchanging essence, only that general terms are useful in communicating family resemblances and can be qualified as needed.[14] Our readers will grasp at once that "medieval science" means something different from both "ancient science" and "contemporary science" while sharing some similarities with them.

Long before 1500, we encounter languages for describing nature, the systematic collection and analysis of data about it, methods for exploring or investigating it (including some experiments), factual and theoretical claims (sometimes stated mathematically) that derive from such explorations and lead to new ones, and criteria for judging the validity of these claims. Moreover, in the planetary astronomy, geometrical optics, natural history, and some aspects of medicine of the Middle Ages, we clearly recognize a close kinship with what we now call science. This is not to deny significant differences – in motivation, instrumentation, institutional support, methodological preferences, mechanisms for the dissemination of theoretical results, economic importance, and social function. Despite such differences, terms like "science" or "natural science" were used in the various contexts of the Middle Ages for goals and activities that bear a family resemblance with those of modern scientific disciplines, to whose history they therefore squarely belong. It is the burden of this volume to illustrate the similarities as well as the differences.

13 See David Pingree's definition: "Science is a systematic explanation of perceived or imaginary phenomena, or else is based on such an explanation." See David Pingree, "Hellenophilia versus the History of Science," *Isis*, 83 (1992), 554–63, especially p. 559.

14 See the wise words of David Hull, "The Professionalization of Science Studies: Cutting Some Slack," *Biology and Philosophy*, 15 (2000), 61–91, especially p. 64.

The chapters that follow will use "science" for purposes as broad as those of the historical actors whose intellectual efforts they seek to understand. Additional distinctions and unusual terminology will also appear. At their most general, many medieval theoretical efforts fell under the general rubric of what scholars in the Byzantine, Arabic, and Latin worlds called "philosophy." Much, but by no means all, medieval scientific activity fell under the rubric of "natural philosophy." It is this expression that ancient and medieval scholars in the Greek tradition generally applied to investigations involving the causes of change in nature. In the medieval civilizations that interacted with the Greek heritage, natural philosophy was closely identified with, but not restricted to, topics covered by the large corpus of Aristotle's writings devoted to nature – the elements and transformations of them, the growth, decay, and classification of living things, motion, the heavens, and so forth.[15] Long after Aristotle's thought ceased to structure the natural world, the expression "natural philosophy" was still a synonym for "physics" in the late nineteenth century. The category was not all-encompassing, however. Another broad area of activity, albeit one with fewer practitioners, drew heavily on mathematical analysis and grouped its disciplines into various "mathematical sciences," most notably astronomy, optics, and the "science of weights." Significantly, much intriguing work occurred in areas in which natural philosophy and the mathematical sciences overlapped, shared questions, and contested each other's boundaries. Not least, practitioners developed a technical vocabulary for identifying subdisciplines with their own specific foci, notably astronomy, optics, meteorology, metallurgy, the science of motion, the science of weights, geography, the natural history of both plants and animals, medicine, and others. In the chapters that follow, then, "science" and its cognates signify the attempts to acquire, to evaluate, or to create systematic knowledge of the natural world, and occasionally to exploit it. The reader's close attention to context should in every case make the meaning clear.

In the medieval period as today, most literate individuals interested in such questions about nature were already drawing on a long tradition of past activity. Accordingly, many of their efforts were "bookish" or textual in nature. They occurred in a study or a library or a disputation hall, and proceeded by reading arguments in the books of predecessors and contemporaries, reflecting critically on their contents, and scrutinizing or debating their conclusions, in disputations and in writing. (Contemporary scientists do a lot of this, too.) But medieval science also had an empirical component, especially in what we now call biology and the biomedical sciences (botany, zoology, and medicine) but also in such physical sciences as

[15] Edward Grant, *A History of Natural Philosophy* (Cambridge: Cambridge University Press, 2007), pp. 234–8.

astronomy, optics, and alchemy, which used tools of measurement and sometimes performed experiments.

In the millennium between roughly the fifth and the fifteenth centuries, investigations of nature were not homogeneous across time or space. They underwent significant changes and developed notable local emphases across the vast lands from the Atlantic to Central Asia covered by this volume.

THE HISTORY OF MEDIEVAL SCIENCE AS A FIELD

Although a few individual scholars had worked on aspects of the subject since the late eighteenth century, the history of medieval science came into being as a field of study in the early twentieth century. The catalyst was a series of provocative claims about the role of fourteenth-century Parisian natural philosophy in explaining the science of the sixteenth and seventeenth centuries. Medieval science thus represented a new milestone in the sketchy linear narrative connecting classical Greece to Isaac Newton. Although the impetus for the new field began with an almost exclusively European focus, it has now expanded vastly beyond those limitations, chronologically and geographically. The newest growth area is the scientific enterprise in Islamic civilization, which is drawing more of the attention it richly deserves. The pioneering work of Joseph Needham's *Science and Civilization in China* and of David Pingree in the exact sciences in India have laid the groundwork for new fundamental scholarship in these areas. The field seems poised to "go global."[16]

Before the history of medieval science could become a field, it had to overcome several centuries of contempt for its subject matter. For Cellarius, as we have seen, medieval political history already had some content; for his immediate predecessors, however, medieval scientific efforts were null. In *The Advancement of Learning*, the English philosopher Francis Bacon (1561–1626) wrote of the "degenerate learning" of the schoolmen, who like spiders "did out of no great quantity of matter, and infinite agitation of wit, spin out unto us those laborious webs of learning which are extant in their books." "In the inquisition of nature," Bacon continued, the schoolmen "ever left the oracle of God's works and adored the deceiving and deformed images which the unequal mirror of their own minds or a few received authors or principles did represent unto them."[17] Similar themes were ubiquitous in the eighteenth

[16] Nathan Sivin, *Granting the Seasons: The Chinese Astronomical Reform of 1280, with a Study of its Many Dimensions and an Annotated Translation of its Records* (Berlin: Springer, 2009).

[17] Francis Bacon, "Of the Interpretation of Nature," in *The Works of Francis Bacon*, ed. Basil Montagu, 3 vols. (Philadelphia: A. Hart, 1852), vol. 1, p. 84; Francis Bacon, "Of the Advancement of Learning, I," in *The Works of Francis Bacon*, ed. James Spedding, Robert Leslie Ellis, and Douglas Denon Heath, 7 vols. (London: Longmans, 1857–1874), vol. 3, pp. 285, 287; and David C. Lindberg, "Conceptions of the Scientific Revolution from Bacon to Butterfield: A Preliminary Sketch," in

and nineteenth centuries. The French *philosophe* Voltaire (1694–1778) wrote of the degeneracy of the human spirit after the fall of Rome, illustrated by the scholastic theology of the Middle Ages, "the bastard offspring of the Aristotelian philosophy, badly translated, and as ill understood, did more injury to understanding and polite studies than ever the Huns and Vandals had done."[18] These witty diatribes articulated a broad intellectual consensus that associated the medieval universities with idolatrous pre-Reformation Christianity, debased barbarous intellects, and bad philology.

More serious because they were more cogently argued were the views of the Cambridge scholar William Whewell, one of the most prolific and influential historians and philosophers of science in the nineteenth century. Whewell saw the medieval period as a

> long and barren period, which intervened between the scientific activity of ancient Greece, and that of modern Europe; and which we may, therefore, call the Stationary Period of Science... men's Ideas were obscured, their disposition to bring their general views into accordance with Facts was enfeebled. They were thus led to employ themselves unprofitably, among indistinct and unreal notions. And the evil of these tendencies was further inflamed by moral peculiarities in the character of those times – by an abjectness of thought on the one hand, which could not help looking towards some intellectual superior, and by an impatience of dissent on the other.[19]

Whewell's outlook was consistent with the triumphalist narratives of "rebirth" of such contemporaries as Jules Michelet (*La Renaissance*, 1855) and Jacob Burckhardt (*Die Cultur der Renaissance in Italien*, 1860). Since rebirth presumes prior death, invoking the Renaissance presupposes a dead or dying antecedent.

Many otherwise well-educated people have long taken this picture for granted. No one has diffused it more widely than astronomer Carl Sagan (1934–1996), whose television series *Cosmos* drew an audience estimated at half a billion. In his 1980 book by the same name, a timeline of astronomy from Greek antiquity to the present left between the fifth and the late fifteenth centuries a familiar thousand-year blank labeled as a "poignant

Reappraisals of the Scientific Revolution, ed. David C. Lindberg and Robert S. Westman (Cambridge: Cambridge University Press, 1990), pp. 3–5.

18 Voltaire, "Ancient and Modern History in Seven Volumes," in *The Works of Voltaire: A Contemporary Version*, trans. William F. Fleming, 43 vols. (Akron, Ohio: Werner, 1905), vol. 26, p. 54; Eugenio Garin, "Medio Evo e tempi bui: concetto e polemiche nella storia del pensiero dal XV al XVIII secolo," in *Concetto, storia, miti et immagini del medio evo*, ed. Vittore Branca (Florence: Sansoni, 1973), pp. 199–224, especially p. 208; and Voltaire, *Works*, trans. Tobias Smollett and others, 39 vols. (London: J. Newbery et al., 1761–1774), vol. 1, pp. 41–42.

19 William Whewell, *History of the Inductive Sciences*, 3rd ed., 3 vols. (London: John W. Parker, 1857), vol. 1, p. 181.

lost opportunity for mankind."[20] The timeline reflected not the state of knowledge in 1980 but Sagan's own "poignant lost opportunity" to consult the library of Cornell University, where he taught. In it, Sagan would have discovered large volumes devoted to the medieval history of his own field, some of them two hundred years old.[21] He would also have learned that the alleged medieval vacuum spawned the two institutions in which he spent his life: the observatory as a research institution (Islamic civilization) and the university (Latin Europe).

PIERRE DUHEM (1861–1916)

Sagan also overlooked the ten volumes of Pierre Duhem's *Le Système du monde* (1916–54), subtitled "a history of cosmological doctrines from Plato to Copernicus" and devoted mostly to the Middle Ages. Duhem was a French physicist who made fundamental contributions to physics and the history and philosophy of science. Late in his career (1903–4), Duhem became very excited upon learning that the "science of weights" of the thirteenth-century Parisian master Jordanus Nemorarius anticipated views associated with Leonardo da Vinci and Galileo.[22]

When Duhem's research turned up more precursors, he became convinced that late-medieval Parisian criticisms of, and alternatives to, Aristotle's views marked the origins of modern science, which had wrongly been ascribed to the sixteenth and seventeenth centuries. Paradoxically, Duhem argued that the bishop of Paris's condemnation in 1277 of 219 propositions defended by university masters freed their successors to think outside the Aristotelian box and to propose views about motion, for example, that eventually led to the theories of Copernicus and Galileo. Duhem developed this startling

[20] Carl Sagan, *Cosmos* (New York: Random House, 1980), p. 335. Sagan's outlook recently regained currency thanks to Alejandro Amenábar's spectacular and spectacularly anachronistic film "Agora" (2009), which portrays Hypatia (d. 415) as on the verge of discovering the law of free fall and heliocentric planetary ellipses before she is murdered by fanatical monks.

[21] Jean-Etienne Montucla, *Histoire des mathématiques*, 2 vols. (Paris: Charles Antoine Jombert, 1758) (2nd ed. in 4 vols., Paris: Henri Agasse, 1799–1802; repr. Paris: Blanchard, 1968); Jean-Baptiste J. Delambre, *Histoire de l'astronomie au moyen âge* (Paris: V. Courcier, 1819); and Guillaume Libri, *Histoire des sciences mathématiques en Italie depuis la renaissance des lettres jusqu'à la fin du dix-septième siècle*, 4 vols. (Paris: Jules Renouard et Compagnie, 1838; repr. New York: Elibron Classics, 2003), especially vols. 1–2. Outstanding scholarship also includes C. A. Nallino, *Al-Battānī sive Albategni Opus astronomicum* (1904); Baldassare Boncompagni, *Bullettino di bibliografia e di storia delle scienze matematiche e fisiche*, 20 vols. (Rome, 1868–1887); Maximilian Curtze, *Urkunden zur Geschichte der Mathematik im Mittelalter und der Renaissance*, 2 parts (Leipzig: Teubner, 1902); and Paul Tannery, *Les Sciences exactes au moyen âge*, vol. 5 of his *Mémoires scientifiques* (Toulouse: E. Privat, 1922).

[22] A few specialists had been aware of Jordanus's work since Montucla, *Histoire des mathématiques*. See Tannery, *Sciences exactes au moyen âge*, p. 316; and John E. Murdoch, "Pierre Duhem (1861–1916)," in *Medieval Scholarship: Biographical Studies on the Formation of a Discipline*, ed. Helen Damico, 3 vols. (New York: Garland, 2000), vol. 3 (Philosophy and the Arts), pp. 23–42, especially p. 26.

thesis in three multivolume works that soon stimulated much new research on medieval Latin science.[23]

Duhem had the missionary fervor of a new convert and the rhetoric to back it up. His preface to *Les Origines de la statique* (1905–6) states: "the mechanics and physics of which the modern world is rightly proud proceed, by an uninterrupted series of barely perceptible improvements, from doctrines taught at the heart of the medieval schools. The alleged intellectual revolutions were most often merely slow and long prepared evolutions, the so-called renaissances were merely unfair and sterile reactions."[24] Leading themes of Duhem's outlook appear here.[25] Clearly, he believed that there was such a thing as modern (i.e., anti-Aristotelian) science, but it had emerged much earlier than previously thought. His advocacy of incremental change was not a generalized historical position but a specific rejection of the misplaced rebirths associated with the "Scientific Revolution."

Note that Duhem did not promote the scientific importance of the Middle Ages generally. Rather, he highlighted the anti-Aristotelian achievements of fourteenth-century Paris, stressing their conceptual connections with sixteenth- and seventeenth-century physics and astronomy (e.g., from the possible rotation of the Earth to Copernicus). About universities other than Paris, he at first had little to say, except to blame Oxford logic for the subsequent decline of Parisian natural philosophy.[26] Although he gave short shrift to the early Middle Ages and painted the fifteenth century as regressive, Duhem's interpretations were exciting. Based on many new empirical findings, they effectively established late-medieval Latin science as a new frontier. By commission, omission, and reaction, his writings deeply shaped the subsequent historiography of medieval Latin and Arabic science, and even the Scientific Revolution.

On the Latin side, Duhem's refutation of medieval scientific vacuity challenged the periodization of European science. Although modest, his inroad into the discontinuity of a millennium of nothingness was sufficient to shatter its unity. He had spanned a once-unbridgeable medieval–modern gap in

[23] Pierre Duhem, *Les Origines de la statique*, 2 vols. (Paris: Hermann, 1905–1906); Pierre Duhem, *Études sur Léonard de Vinci: Ceux qu'il a lus et ceux qui l'ont lu*, 3 vols. (Paris: Hermann, 1906–1913); and Pierre Duhem, *Le Système du monde: Histoire des doctrines cosmologiques de Platon à Copernic*, 10 vols. (Paris: Hermann, 1913–1959).

[24] Duhem, *Origines de la statique*, vol. 1, p. iv; Roger Ariew and Peter Barker, eds., "Pierre Duhem: Historian and Philosopher of Science," *Synthèse: International Journal for Epistemology, Methodology and Philosophy of Science*, 83 (1990); Murdoch, "Pierre Duhem (1861–1916)," pp. 23–42 and the literature cited therein.

[25] Floris Cohen, *The Scientific Revolution: A Historiographical Inquiry* (Chicago: University of Chicago Press, 1994), pp. 45–53.

[26] John E. Murdoch, "Pierre Duhem and the History of Late Medieval Science and Philosophy in the Latin West," in *Gli studi di filosofia medievale fra otto e novecento: contributo a un bilancio storiografico*, ed. Ruedi Imbach and Alfonso Maierù (Rome: Edizioni di Storia e Letteratura, 1991), pp. 253–302, especially pp. 266–71.

the history of *science*, the hallmark of modernity. But Duhem was an advocate of continuity in a very limited sense. The anti-Aristotelian *conceptual* continuity between the fourteenth and the sixteenth and seventeenth centuries leapt over the temporal *discontinuity* of a fifteenth-century Aristotelian relapse. Despite its narrow scope, Duhem's research seeded a "continuity–discontinuity debate" about the relations of late-medieval science and the "Scientific Revolution" (early-modern science) that flourished into the 1960s and still flares up occasionally.

Although Duhem's thesis was thoroughly Parisian, his *To Save the Phenomena* and his *Système du monde* touched, sometimes jarringly, on areas beyond Europe. In treating Arabic science, Duhem judged Islamic contributions to astronomy, for example, to be vitiated by mechanical models that led nowhere and illustrated the failure of the Semitic mind to think either abstractly or logically.[27] His judgment meshed smoothly with the long-standing view of Islamic civilization as an Arabic repository of Greek texts that patiently awaited Latin translators. Since Duhem, the history of Arabic science has undergone exceptional valorization both in relation to Latin science and even more so on its own terms.

THE GENERATION OF THE EARLY 1920S

Thanks to the impetus of Duhem, the history of medieval science by the early 1920s was showing its vitality in research and reaping the benefits of institutionalization, especially in the United States. In 1924, the History of Science Society was organized as an international source of financial support for the struggling history of science journal *Isis* (1913–). Its founder and editor was George Sarton, a Belgian mathematician who promoted the history of science as the common patrimony of the human race. Accordingly, to escape the divisions of World War I, he had moved to the still neutral United States in 1915. Unlike Duhem, Sarton was an internationalist whose journal covered worldwide scholarship, including medieval science in China, India, and Arabic civilization. He learned Arabic later in life and, despite his precarious position at Harvard, supervised the first PhD degree under the new Committee on Higher Degrees in the History of Science and Learning (1942). The thesis, by the Turkish scholar Aydin Sayili, was later published as *The Observatory in Islam* (1960), a groundbreaking institutional history of Arabic astronomy that remains a fundamental contribution to science in Islamic civilization and to medieval science generally.[28]

[27] Duhem, *Système du monde*, vol. 2, pp. 117–18; F. Jamil Ragep, "Duhem, the Arabs, and the History of Cosmology," *Synthèse: International Journal for Epistemology, Methodology and Philosophy of Science*, 83 (1990), 201–14, especially pp. 208–11.

[28] I. Bernard Cohen, "A Harvard Education," *Isis*, 75 (1984), 13–21, especially pp. 14–15; Aydin Sayili, *The Observatory in Islam and Its Place in the General History of the Observatory* (Ankara: Türk Tarih

The early 1920s also witnessed four seminal publications, each of which sketched a different path for medieval science. Three of them were aware of Duhem. Their authors – Charles Homer Haskins (1870–1937), Lynn Thorndike (1882–1965), E. A. Burtt (1892–1989), and E. J. Dijksterhuis (1892–1965) – form an impressive if motley quartet whose influence, direct and indirect, both shaped the history of medieval science and radiated far beyond it.

Unlike Sarton and most of his early colleagues, Haskins and Thorndike were trained as historians and became leading figures in the institutionalization of both medieval studies and the history of science. Their outlooks were more fundamentally contextual than those of other early historians of science, whose training in the sciences or philosophy left them comfortable with the relations of ideas alone. Both Haskins and Thorndike appreciated the central significance of Arabic science for their main quarry, the "pre-Duhem" twelfth and thirteenth centuries in the Latin world.[29]

In 1923, Thorndike published *A History of Magic and Experimental Science* in two volumes that ended in "the course of the fourteenth century" because "by then the medieval revival had spent its force."[30] After discovering Duhem's work, he must have winced, but the shock led to two more volumes on the fourteenth and fifteenth centuries (1934), then to four more, which concluded with Newton. Thorndike promoted continuity by juxtaposition, broadening Duhem's claims about the medieval understanding of nature but also stressing the importance of earlier Arabic materials for his subject. Contemporaries like Sarton hated Thorndike's focus on magic and astrology (pseudo-science!), which he loved to find lurking in the early-modern heroes of science, but Thorndike's eclectic contextual vision won the day posthumously.

In 1924, the title of Haskins's *Studies in the History of Mediaeval Science* named the emerging field, whereas its content self-consciously focused on the twelfth and thirteenth centuries in Europe. Haskins had read Duhem but, unlike him, gave prominent attention to "science among the Arabs," emphasizing their contributions to the "revival" that became the central focus of his classic *Renaissance of the Twelfth Century* (1927). His title was subversive, for it characterized the heart of the Middle Ages with a term

Kurumu Basımevi, 1960; repr. New York: Arno Press, 1981); and Gül A. Russell, "Eloge: Aydin Sayili, 1913–1993," *Isis*, 87 (1996), 672–5.

29 Sally Vaughn, "Charles Homer Haskins (1870–1937)" and Michael H. Shank, "Lynn Thorndike (1882–1965)," in *Medieval Scholarship: Biographical Studies on the Formation of a Discipline*, ed. Helen Damico and Joseph Zadavcil, 3 vols. (New York: Garland, 1995), vol. 1 (History), pp. 169–84, 185–204, respectively. See the appreciation of Thorndike in Maria Mavroudi, "Occult Science and Society in Byzantium: Considerations for Future Research," in *The Occult Sciences in Byzantium*, ed. Paul Magdalino and Maria Mavroudi (Geneva: La Pomme d'Or, 2006), pp. 39–95, especially pp. 39–44, 59–63.

30 Lynn Thorndike, *A History of Magic and Experimental Science*, 2 vols. (New York: Macmillan, 1923), vol. 1, p. 3.

coined to mean "post-medieval." Significantly, Haskins's account of twelfth- and thirteenth-century science had a notable impact on the wider community of historians.[31] When added to Duhem's, Haskins's work produces a much longer continuity argument, one that not only extends from the twelfth century into early-modern Europe but also stretches much further back once the narrative takes seriously contacts with Islamic civilization.

In that same year of 1924, Duhem also figured prominently in the work of E. J. Dijksterhuis, a Dutch mathematics teacher who became a university professor at Utrecht and Leiden in his sixties. Still available only in Dutch, his *Val en Worp* (1924) was a history of free fall and projectile motion from Aristotle to Newton that summarized Duhem in the eighty pages he devoted to the "scholastic" period. Despite its chronological leaps, the book implicitly offered a conceptual continuity argument – one topic arranged in three progressive phases: ancient, medieval, and early modern. The wider history of science community did not assimilate Dijksterhuis's work until the late 1950s and 1960s, when a new Dutch work appeared in English as *The Mechanization of the World Picture*, an influential synthesis extending from antiquity to the Scientific Revolution.[32]

A very different reading of Duhem, and one with profound consequences, appears in E. A. Burtt's *Metaphysical Foundations of Modern Physical Science* (1924). Although aware of Duhem's multivolume works, Burtt drew especially on his *To Save the Phenomena*, which sharply contrasts metaphysical realists (who believe that their theories map the real world) with instrumentalists (who use theories hypothetically, with no commitment to their reality). Duhem had presented Copernicus, Galileo, and Kepler as misguided metaphysical realists who, despite abandoning the clear-headed instrumentalism of their most successful predecessors, nevertheless contributed to the (instrumentalist) mathematical worldview of Newtonian mechanics.[33] Burtt liked Duhem's framework. Studying the metaphysical shift from medieval to modern physical thought allowed him to explore at one remove the one looming in his day, from Newtonian physics to relativity theory.[34] The ironies here are deep. Building on Duhem's metaphysical discontinuity, Burtt's work

31 *Renaissance and Renewal in the Twelfth Century*, ed. Robert Benson and Giles Constable (Cambridge, Mass.: Harvard University Press, 1982), pp. xvii–xxx. See also the remarks on Michelet in Johan Huizinga, "The Problem of the Renaissance," in his *Men and Ideas: History, the Middle Ages, the Renaissance; Essays by Johan Huizinga*, trans. James Holmes and Hans van Marle (New York: Meridian Books, 1959), pp. 243–87.

32 E. J. Dijksterhuis, *Val en Worp: Een Bijdrage tot de Geschiedenis der Mechanica van Aristoteles tot Newton* (Groningen: Noordhoff, 1924), pp. v–vi, 57–137; and E. J. Dijksterhuis, *The Mechanization of the World Picture*, trans. C. Dikshoorn (Oxford: Oxford University Press, 1961). On Dijksterhuis, see Cohen, *Scientific Revolution*, pp. 59–73.

33 Pierre Duhem, *To Save the Phenomena: An Essay on the Idea of Physical Theory from Plato to Galileo*, trans. E. Dolan and C. Maschler (Chicago: University of Chicago Press, 1969, 1985), especially p. 3.

34 Diane E. D. Villemaire, *E. A. Burtt, Historian and Philosopher: A Study of the Author of The Metaphysical Foundations of Modern Physical Science* (Boston Studies in the Philosophy of Science, 226) (Boston: Kluwer, 2002), pp. 48, 52–54, 185.

would inspire Alexandre Koyré (1892–1964), a Russian émigré to Paris who became the staunchest defender of the Scientific Revolution as metaphysically discontinuous with the medieval past.[35]

THE POSTWAR YEARS

In the postwar generation, Koyré's discontinuity arguments targeted not Duhem but Alistair C. Crombie (1915–1996), an Australian zoologist who had turned to the history of science and spent his career at Oxford. Crombie saw the origins of the Scientific Revolution's experimentalism in medieval variants of Aristotle's methodology, especially in the work of Robert Grosseteste (d. 1253), more than a generation before Duhem's beloved Condemnations of 1277. Crombie's work was therefore fundamentally non-Duhemian. The Aristotelian tradition was not a fossil that had endured; despite its bad press, it was alive and evolved to fertilize even Galileo in the seventeenth century.[36]

Nowadays, historians of medieval science are often suspected of defending continuity, perhaps because the widely read synthetic treatments of medieval science by Crombie and Dijksterhuis promoted continuity theses. It is therefore important to note that other colleagues in the field were wary even of Duhem's limited claims of conceptual continuity between fourteenth- and seventeenth-century science.[37] Two of the most prominent argued that the scientific theories and practices of these centuries were more discontinuous than Duhem had implied. Anneliese Maier (1905–1971), trained as a Kantian philosopher, spent most of her career as an independent scholar working in Rome, where she made fundamental contributions to our understanding of fourteenth-century natural philosophy; and Marshall Clagett (1916–2005), a student of Thorndike's at Columbia who turned to medieval mathematics and motion, helped to establish the first autonomous department of the history of science at the University of Wisconsin.[38] Both scholars

35 Cohen, *Scientific Revolution*, pp. 100–2.

36 A. C. Crombie, *Robert Grosseteste and the Origins of Experimental Science* (Oxford: Clarendon Press, 1953); and Bruce S. Eastwood, "On the Continuity of Western Science from the Middle Ages: A. C. Crombie's *Augustine to Galileo*," *Isis*, 83 (1992), 84–99.

37 On the continuity debate, see Alexandre Koyré, "The Origins of Modern Science: A New Interpretation," *Diogenes*, 16 (Winter 1956), 1–22; Ernan McMullin, "Medieval and Modern Science: Continuity or Discontinuity?" *International Philosophical Quarterly*, 5 (1965), 103–29; and David C. Lindberg, *The Beginnings of Western Science: The European Scientific Tradition in Philosophical, Religious, and Institutional Context, 600 B.C. to A.D. 1450*, 2nd ed. (Chicago: University of Chicago Press, 2008), pp. 357–67.

38 Anneliese Maier, *Die Vorläufer Galileis im 14. Jahrhundert* (Rome: Edizioni di Storia e Letteratura, 1949); Anneliese Maier, *An der Grenze von Scholastik und Naturwissenschaft* (Rome: Edizioni di Storia e Letteratura, 1952); Anneliese Maier, *Zwischen Philosophie und Mechanik* (Rome: Edizioni di Storia e Letteratura, 1958); Crombie, *Augustine to Galileo*; Crombie, *Robert Grosseteste and the Origins of Experimental Science*; and Marshall Clagett, *The Science of Mechanics in the Middle Ages* (Madison: University of Wisconsin Press, 1959).

criticized Duhem's tendency to characterize as "barely perceptible" the differences between sixteenth- and seventeenth-century themes (e.g., inertia) and the fourteenth-century Parisian achievements he had championed (i.e., impetus).

Institutionally, the most successful of these scholars was surely Clagett. During the postwar explosion of American higher education, he trained at the University of Wisconsin a "school" of historians of medieval science that has now ramified severalfold (including the editors of and many contributors to this volume). Like his mentor Thorndike, Clagett committed himself and his students to editing important manuscript sources. Together, they produced editions, translations, and interpretations of leading mechanical, mathematical, and cosmological works of the thirteenth and fourteenth centuries. Such efforts made many medieval scientific texts more accessible, some in still unpublished dissertations, others in Clagett's monograph series at the University of Wisconsin Press – mostly in Latin, but also in Arabic, Persian, Hebrew, and Middle French.

The vagaries of world politics combined with peculiar institutional circumstances led to a burst of activity in North American history of science around and after World War II in particular. Émigrés such as Otto Neugebauer (1899–1990) and his "school" at Brown made extraordinary contributions to the exact sciences from antiquity to Kepler, including work on medieval Greek, Latin, and Arabic, extended to Sanskrit with the arrival of David Pingree (1933–2005). A. I. Sabra's arrival in the United States from Alexandria via London made a lasting impact on scholarship in medieval Islamic science through his own work and that of a cohort of students at Harvard.

Thankfully, the aggregate effort to advance the study of medieval science, however fragile, has always been international, as it must continue to be if it is to survive. For many years, Josep María Millàs Vallicrosa (1897–1970) held the chair in Hebrew and Arabic at the University of Barcelona. From the 1920s to the 1960s, he devoted his research to medieval Spanish and Catalonian science. His efforts and those of his students made Barcelona a thriving center of multicultural medieval science, with a strong emphasis on the Arabic tradition. In the Soviet Union, A. A. P. Yushkevich (1906–1993), who headed the department of the history of mathematics at the Institute for the History of the Natural Sciences and Technology in Moscow, made leading contributions to medieval mathematics, notably in the Arabic tradition.[39] At the École Pratique des Hautes Études in Paris, Guy Beaujouan (1935–2007) was a superb codicologist who branched into science in medieval Spain.[40] Willy Hartner (1905–1981), an astronomer with a PhD in physics, built the

[39] I. Bashmakova et al., "In Memoriam: Adolph Andrei Pavlovich Yushkevich (1906–1993)," *Historia Mathematica*, 22 (1995), 113–18.

[40] Nicolas Weill-Parot, "Guy Beaujouan et l'histoire des sciences au moyen-âge," *Revue de Synthèse*, 129 (2008), 625–34.

Goethe University (Frankfurt am Main) into a leading center of the history of science from 1943 on. His passionately cross-cultural approach to medieval science worked with primary sources in Chinese, Arabic, and Hebrew as well as the traditional Latin and Greek. Nowadays, the greatest concentration of historians of medieval science is certainly in Europe, with special strength in France. The career of Roshdi Rashed (Centre National de la Recherche Scientifique and University of Paris-7) illustrates the transformation of the field in the last century. With several dozen books and many more articles devoted primarily to the history of Arabic mathematics and sciences, the founding of a journal, and the establishment of a leading department and research team, Rashed's scholarly and institutional efforts have both surpassed those of Duhem and changed the direction of research.

Like Duhem, many early historians of medieval science had concentrated on aspects of mathematics and the physical sciences, especially mechanics and optics. All engaged in the "conceptual analysis" of texts, and all grappled in some form with the question of continuity between medieval science and the Scientific Revolution.[41] More than happenstance probably links the temporal coincidence of the high prestige of physics in the twentieth century with the professionalization of the history of science and its institutionalization in the 1940s–1960s. In recent years, scholars have devoted greater attention to the diversity of the medieval scientific enterprise, including such areas as natural history, psychology, matter theory, geography, and sexuality, topics that sometimes eluded the gazes of the pioneers.[42] If the relations of natural philosophy, the mathematical sciences, and medicine with theology have drawn renewed attention, those with the all-important discipline of law remain understudied.

As a community of scholarship, the history of medieval science first developed within the framework of European historiography, from which it has remained slow to branch out. Recent research beyond these traditional confines, however, offers the hope of recasting linguistic boundaries, intellectual geography, and the periodization of history in ways that stretch beyond the standard pseudo-national and regional frameworks and make it reflect more faithfully the sometimes astonishing cross-cultural interactions that characterize the period.

Long-term projects are excellent gauges of change, and this one is no exception. Conceived late in the last millennium, the structure of this volume bears the imprint of its day. Without rethinking the periodization of the history of science on a global scale, the following brief reflections from the

[41] Clagett, *Science of Mechanics in the Middle Ages*, p. xxii; and Crombie, *Augustine to Galileo*, p. 273.

[42] Joan Cadden, *Meanings of Sex Difference in the Middle Ages: Medicine, Science, and Culture* (Cambridge: Cambridge University Press, 1993); William Newman, *The Summa perfectionis of pseudo-Geber* (Leiden: Brill, 1991); and J. Brian Harley and David Woodward, eds., *History of Cartography: Cartography in Prehistoric, Ancient, and Medieval Europe and the Mediterranean* (Chicago: University of Chicago Press, 1987).

angle that we know best may illuminate the direction in which research in medieval science is developing.

Accordingly, this introduction concludes with a brief examination of the language of decline, which played a formative role in the emergence of the category "Middle Ages," invented – as noted earlier – to name a millennium-long decline. In many instances, the imagery of decline is useful and appropriate, but it has had an overbearing influence on our understanding of the relations between ancient and medieval science. It is ironic that many historians of science, who quickly spot the first hint of a "Whig interpretation" (i.e., progressive narrative), think nothing of using its complement, decline. Like the language of progress, the language of decline conveys affects and assumptions that stifle questions and hinder inquiry. These vexing problems are especially salient in the history of early science. In particular, they have touched deeply, and continue to touch, the traditional post-Roman civilizations – those of the Latin world, Islamic civilization, and the Byzantine Empire – at their beginnings, in their middles, and at their ends.

BEYOND THE DECLINE OF ROME

The Petrarch who coined the term "Dark Ages" also wrote, rather myopically: "What else, indeed, is all of history but the praise of Rome?"[43] The coinage and the question are linked, and it is decline that links them. Traditionally bounded by the falls of Rome and the "New Rome," the millennial wait for the other shoe to drop points to the complexity of the links, historical and affective, between our understanding of the Middle Ages and the fate of the Roman Empire. The medieval decline is implicit in every form of the narrative, and is often conjoined with the aspiration of emulating, if not resurrecting, the (western) Roman Empire. These hopes have motivated an amazing array of often misbegotten efforts – not only from Constantine through the Byzantine emperors and Charlemagne but also from Peter the Great and the three Napoleons to fascism (with its Roman symbols and aspirations) and the Third Reich/Empire.[44] The Roman Empire still stimulates the political imagination as analysts continue to draw on images of its pitfalls and accomplishments when thinking about twenty-first-century politics.[45]

43 "Quid est enim aliud omnis historia, quam romana laus?" Petrarch, *Invectiva contra eum qui maledixit Italie* (1373).

44 Herwig Wolfram, *The Roman Empire and Its Germanic Peoples*, trans. Thomas Dunlap (Berkeley: University of California Press, 1997), pp. 313–14; Patrick Geary, *The Myth of Nations: The Medieval Origins of Europe* (Princeton, N.J.: Princeton University Press, 2002), pp. 4–8, 16–40; Yuri M. Lotman, *Universe of the Mind: A Semiotic Theory of Culture*, trans. Ann Shukman (London: Tauris, 2001), pp. 191–94; and Solomon Volkov, *St. Petersburg: A Cultural History*, trans. Antonina Boris (New York: Free Press, 1995), pp. 6–7.

45 Anthony Pagden, *Worlds at War: The 2500-Year Struggle between East and West* (New York: Random House, 2008), and others.

Like Petrarch's, however, our image of the Roman Empire is still too monolithic, too Western, too Latin, and too narrowly identified with the city of Rome. We tend to forget the Roman Empire's enormous cultural and human diversity, a fact that deeply shaped the history that followed, including the facets of its science. The case of language nicely illustrates the point. Latin was once the very local language of the Latium, near the center of the Italian peninsula. By the age of Augustus (first century), however, Rome was the capital of an empire that ringed the Mediterranean and extended deep inland for hundreds of leagues. Much of this territory was effectively a vast colony on which the occupying power imposed Latin as its administrative language. Rome's cultural politics, by contrast, had to be subtler. The impact of Latin in the occupied territories thus had highly asymmetrical cultural consequences as a function of geography. These regional variations profoundly marked the history of science in these regions.

In the largely Greek eastern empire, Latin was militarily and politically correct, but it was barbarous.[46] Greek was the language for almost everything else, including Hellenistic philosophy and science, domains in which neither Rome nor Latin could compete. On the contrary, as Horace famously put it, "Captive Greece took captive her fierce conqueror, and introduced her arts into rough Latium" (*Epistles*, II.1).

This reverse conquest was very uneven, however. Whereas the alphabet, poetry, and religion of Greece captivated the Romans, the latter left most of Greek philosophy, natural philosophy, the mathematical sciences, and medicine untranslated. In Rome itself, the practitioners of these disciplines were so overwhelmingly foreigners that the Romans stereotyped them. The astrologers were "Chaldeans," whereas physicians were Greeks, portrayed with ambivalence or worse. As the case of Rome pointedly reminds us, a vibrant scientific tradition is a highly contingent phenomenon; not every major civilization develops one of its own. Overall, Roman civilization produced only a handful of works concerned with the workings of nature, most of them encyclopedic. In retrospect, we can see that it neither appropriated nor extended the Greek scientific tradition in the ways that the medieval Arabic and Latin worlds would.[47]

[46] Raymond Van Dam, *The Roman Revolution of Constantine* (Cambridge: Cambridge University Press, 2007), pp. 184–93.

[47] Heinrich von Staden, "Liminal Perils: Early Roman Reception of Greek Medicine," in *Tradition, Transmission, Transformation*, ed. F. Jamil Ragep and Sally Ragep, with Steven Livesey (Leiden: Brill, 1996), pp. 369–418. Leading Latin works include Lucretius's *On the Nature of Things*, Seneca's *Natural Questions*, Manilius's *Astronomicon*, and Celsus's *De medicina*, the chief surviving medical work in classical Latin (one of his six treatises on the "arts"). See William Stahl, *Roman Science: Origins, Development, and Influence to the Later Middle Ages* (Madison: University of Wisconsin Press, 1962), pp. 66, 71, 79, 96; Lionel Casson, *Libraries in the Ancient World* (New Haven, Conn.: Yale University Press, 2001), pp. 80–5; Jean Irigoin, "Les textes grecs circulant dans le nord de l'Italie aux Ve et VIe siècles: Attestations littéraires et témoignages paléographiques," in *Teodorico il Grande e i Goti tra oriente e occidente*, ed. Antonio Carile (Ravenna: Longo, 1995), pp. 391–400; and Catherine Salles, *Lire à Rome* (Paris: Les Belles Lettres, 1992).

Unlike the Greek East, much of the western Roman Empire away from the Mediterranean first encountered literacy in tandem with subjugation. From Britain to such Roman colonial outposts as "Colonia" (Cologne), Trier, and Vienna, Latin was not only the conquerors' administrative language but also the language of learning, the vehicle for Rome's cultural interests. The population centers in these occupied regions underwent extensive Roman cultural colonization. This process did not include most of Greek science, which was largely unavailable in Latin, as Rome at its height had shown only a low-level interest in it.

Rome had passed its prime long before Alaric sacked the city in 415. Already a century earlier, Emperor Constantine had made the extraordinary decision of moving the imperial capital to Constantinople, the "New Rome" that he founded in the Greek East. The administrative heart of the Roman Empire was now much closer to its greatest perceived threat – not the Germanic tribes along the Rhine but the Persian Sasanian Empire (224–651).[48] Equally significant, Constantine ended the persecutions of Christianity and favored the new religion. He funded St. Peter's basilica and, when he abandoned Rome, he left it to the local bishop, who partially filled the power vacuum, landholdings and the imperial Lateran Palace. The pope still cares for the palace today.[49]

The political importance of the papacy grew steadily after the fourth century. The Church of Rome would play a long-range cultural role in the survival and diffusion of Latin. Radiating from Monte Cassino, Benedictine monasticism became a missionary endeavor that not only spread the rites and beliefs of the Roman Church throughout the western Roman Empire but also kept alive the language and secular writings of Rome.

This background suggests that one should treat with considerable skepticism the stereotype of a widespread precipitous early-medieval decline of learning without attention to local contexts.[50] At any given time, gauging "what the most able knew" may serve as a crude measure of progress or decline for an entire regime or era. On these grounds, the Roman Empire can "get credit" for Ptolemy's Greek work in Alexandria. But should it? Should not the historian of science be more interested in a regional rather than a universal criterion? In considering natural philosophy, medicine, and the mathematical sciences, we often implicitly evaluate early-medieval Europe against the apex of the Hellenistic achievement. Since Rome at its height did

48 Peter Heather, *The Fall of the Roman Empire: A New History of Rome and the Barbarians* (Oxford: Oxford University Press, 2006), pp. 96–7, 110, 385–6.

49 Richard Fletcher, *The Barbarian Conversion: From Paganism to Christianity* (Berkeley: University of California Press, 1997), pp. 18–22.

50 "The intellectual crisis of the fifth to the seventh centuries was deepest in Italy and Gaul." See Paul Leo Butzer and Dietrich Lohrmann, eds., *Science in Western and Eastern Civilization in Carolingian Times* (Basel: Birkhäuser Verlag, 1993), p. vii.

not translate this work, its cultural colonies had no access to it. A fortiori, they could scarcely be expected to master it, let alone emulate it.

Seen from the other end of the stick, the inhabitants of these territories under Roman occupation at first had to deal with an alien language, alien laws, and a foreign political system in which they eventually developed a significant stake. By the late empire, Rome granted citizenship to everyone within its borders, making it a multiethnic, multilingual honor shared by many former "barbarians." After Constantine moved the imperial administration to Constantinople, the city of Rome faded as a political center, and the peoples in the occupied territories eventually found themselves in a post-colonial situation.[51] They cobbled together political, institutional, and legal structures that drew eclectically on elements from tribal traditions, the former Roman colonizers, immigrants, new invaders, and their own innovations. Like other post-colonial societies, those of early-medieval Europe cultivated the tongue of the former metropolis as their literate language, which provided their main access to the modest scientific interests of Rome, represented mostly in encyclopedias, handbooks, and didactic poetry.

To call such a complex process an early-medieval scientific decline is to mischaracterize it fundamentally, notwithstanding the fact that historians of late-medieval science like ourselves have used and taught this narrative. By eagerly embracing the story of decline, we, too, could hitch the period from the twelfth century onward to the bandwagon of modernity, in effect replicating the traditional tripartite narrative structure that requires a decline so that a triumphal rebirth can occur around 1100.[52]

SCIENCE IN ISLAMIC CIVILIZATION: RISE AND DECLINE?

In Constantinople, Constantine's successors saw the Persian threat disappear, but not as he had imagined. From the Arabian peninsula, on the edges of both the Hellenistic world and the Roman Empire, emerged the new and vigorous religion of Islam, which reshaped Eastern and Southern Mediterranean politics and culture from the Straits of Gibraltar into Central Asia and Northern India. The conquests of the Umayyad dynasty carried the Qurʾān and the Arabic language far outside the Arabian peninsula to cities and territories that had been deeply Hellenized in both the Byzantine and Persian empires. Damascus, which the Umayyads made their capital, was

[51] Peter S. Wells, *The Barbarians Speak: How the Conquered Peoples Shaped Roman Europe* (Princeton, N.J.: Princeton University Press, 1999), pp. 20–2; and Frederick Cooper, *Colonialism in Question: Theory, Knowledge, History* (Berkeley: University of California Press, 2005), pp. 158–61.

[52] For the parallel case in history tout court, see Chris Wickham, *The Inheritance of Rome: Illuminating the Dark Ages, 400–1000* (New York: Viking, 2009), pp. 5–6.

so Greek that the conquerors retained Greek as the administrative language until the late seventh century.[53]

Access to language and books does not imply engagement with them, of course. Indeed, the Byzantine elite in Damascus were evidently unsympathetic to the secular learning of Hellenism. After overthrowing the Umayyads in the mid-eighth century, however, the new 'Abbāsid dynasty founded a new capital in Baghdad, in multiethnic, multireligious territory unshackled by this Byzantine outlook. The 'Abbāsids and their entourage promoted a vigorous translation movement of Greek and other scientific works, out of which Islamic civilization developed a thriving and original scientific culture that would last for many centuries.[54]

How many? Historians of Islamic civilization have also adopted the language of scientific decline. They have variously linked the beginning of the end with the eleventh century for internal reasons, or the Mongol invasions in the mid-thirteenth for external ones. One recent work postpones the decline to the sixteenth century, pointing to shifting trade patterns associated with the European exploitation of the New World.[55] Since science in the Ottoman Empire is still a vast underexplored field, one might well suspect that even this generalization will not survive a confrontation with future research.[56]

Even in this streamlined form, debates about the decline of Islamic science clearly illustrate a notable general point. Behind most decline narratives stands the implicit expectation of ongoing progress. Once science is in motion, we tend to assume that it should keep accelerating or, at the very least, behave inertially – that is, remain in motion unless impeded by some resistance. If, by whatever measure, science "declines" (or even remains on a plateau), historians search for obstacles, hindrances, and enemies. The pattern is clear in classic accounts of Greek science, which rose and then declined because of Goths, Huns, Christians, and other aliens.[57] Most human activities, including science, do not take place in a frictionless vacuum, however. If left to themselves, if the positive interests of their movers – intellectuals, patrons, or institutions – turn elsewhere, benign neglect may be sufficient to explain the change.

53 Dimitri Gutas, *Greek Thought, Arabic Culture: The Graeco-Arabic Translation Movement in Baghdad and Early 'Abbāsid Society (2nd–4th/8th–10th Centuries)* (London: Routledge, 1998), pp. 17–18; and L. E. Goodman, "The Greek Impact on Arabic Literature," in *Arabic Literature to the End of the Umayyad Period*, ed. A. F. L. Beeston et al. (Cambridge: Cambridge University Press, 1983), pp. 462–82.

54 A. I. Sabra, "The Appropriation and Subsequent Naturalization of Greek Science in Medieval Islam: A Preliminary Statement," *History of Science*, 25 (1987), 223–43; Gutas, *Greek Thought, Arabic Culture*, especially chaps. 1–4; and George Saliba, *Islamic Science and the Making of the European Renaissance* (Cambridge, Mass.: MIT Press, 2007).

55 Saliba, *Islamic Science and the Making of the European Renaissance*, chap. 7.

56 Sonja Brentjes, "Between Doubts and Certainties: On the Place of History of Science in Islamic Societies within the Field of History of Science," *NTM Zeitschrift für Geschichte der Wissenschaften, Technik und Medizin*, 11 (2003), 65–79.

57 For a classic statement, see G. E. R. Lloyd, *Greek Science after Aristotle* (New York: Norton, 1973), chap. 10.

However this may be, science in Islamic civilization recently became an established field of exceptional interest on its own terms, quite apart from its relation to Latin science. The many discoveries of recent research in this area remain underappreciated, when not ignored, in the history of science at large. Institutionally, the increasing neglect of the history of all early science has been particularly hard on Islamic science, a field in which the scholars are few and the number of professorships is smaller still and shrinking. Yet the amount of fundamental research that remains to be done is staggering. In astronomy alone, Arabic manuscripts outnumber those in Greek and Latin together.[58] Despite these handicaps, recent studies of the textual riches on this vast frontier have yielded exciting results, which are summarized in this volume. Indeed, a proper account of the scientific enterprises of medieval Islamic civilization, like those of Chinese and Indian civilizations, will probably require a separate volume in the next iteration of the Cambridge History of Science.

BEYOND THE FIFTEENTH-CENTURY DECLINE

As the most important hinge of the traditional tripartite schema of world periodization, the fifteenth century has long presented problems for historians. Cellarius's "universal" history had already emphasized the "unhappiness of the century"[59] and left in limbo the half-century between 1453 (the end of the Middle Ages) and the Reformation (modernity). Mid-nineteenth-century historians such as Jules Michelet and Jacob Burckhardt were happy to turn the fifteenth century into the first breath of modernity (i.e., the Renaissance). For Michelet, Christianity had suppressed science until the fifteenth century, when it, too, would have a "rebirth" associated with "the discovery of the world and of man," phrases that Burckhardt borrowed and popularized.[60] Both historians thought that the period had something to do with science, but neither developed the theme. Once the Middle Ages became a legitimate area of investigation in intellectual history, the thirteenth century soon emerged as its apex (the universities, Thomas Aquinas). Everyone now knew that the Renaissance was coming; the Middle Ages therefore had to be ending. The fourteenth century and parts of the fifteenth were ripe for the images and language of decline (the Black Death, scholasticism as logic-chopping gone to seed, etc.). This picture was reinforced by

58 Noel Swerdlow and Otto Neugebauer, *Mathematical Astronomy in Copernicus's De Revolutionibus* (Berlin: Springer-Verlag, 1984), p. 41.

59 Cellarius, *Historia medii aevi a temporibus Constantini Magni ad Constantinopolim a Turcis captam deducta*, p. 182.

60 Jules Michelet, *La Renaissance* (the subtitle of his *Histoire de France*, bk. 7), quotation in his *History of France*, vol. 2 (bk. 7), trans. G. H. Smith (New York: Appleton, 1847), p. 18; Burckhardt, *Civilization of the Renaissance in Italy*; Peter Burke, *The Renaissance* (London: Longman, 1964); and Huizinga, "Problem of the Renaissance."

the seemingly innocuous metaphors of the "late" or "waning" Middle Ages, with their connotations of aging, crisis, and impending death.[61] Far from being harmless verbal games, these images drive away potential researchers, for decline is a self-fulfilling trope that generates its own downward spiral. Why should eager graduate students choose to bury themselves in a dying period when the glorious origins of modernity beckon?

Ironically, the historical neglect of fifteenth-century Latin science was partly a problem of the founding medievalist fathers' own making. Although Duhem's research changed the image of fourteenth-century science, he saw the fifteenth century as regressive – enamored of the very antiquity (notably Aristotle) that the critical Parisian natural philosophers had begun to abandon. In effect, Duhem created "mini-Middle Ages," a decline separating the highlights of fourteenth-century Paris from Copernicus. He was not alone. George Sarton referred to the "antiscientific tendencies of the humanists," whereas Thorndike attacked the very notion of a Renaissance ("nothing is ever reborn").[62] Medievalists and many others took the antagonism of science and humanism for granted until a more nuanced picture emerged in the last generation.[63]

Just as, for Cellarius, Luther had inaugurated modernity in religion, so has his contemporary Copernicus traditionally ushered in the Scientific Revolution, the canonical beginning of modern science. Cast as a sharp break with the medieval past, this enticing narrative has offered historians of science little incentive to undermine its dramatic story line with shades of gray better suited to its complexities.[64] Even recent syntheses that have ostensibly backed away from the classic story of the Scientific Revolution have left its language, its termini, and the larger periodization intact.[65] This narrative, whatever its disguise or its name, remains a cosmogony that recounts the birth of the modern world. Like all creation myths, it necessarily must begin with either chaos or nothing, a starting point that the Middle Ages has ably offered. Eliminating that starting point demands a completely different plot.

Given the central role of Copernicus in this classic narrative, research in the history of astronomy has done much to shake up the old creationist

[61] Kaminsky, "From Lateness to Waning to Crisis," pp. 85–6, 123–5.

[62] George Sarton, "Science in the Renaissance," in *The Civilization of the Renaissance*, ed. James Westfall Thompson et al. (Chicago: University of Chicago Press, 1929), pp. 75–98, especially pp. 78–9.

[63] See Brian W. Ogilvie, "Science," in *Palgrave Advances in Renaissance Historiography*, ed. Jonathan Woolfson (New York: Palgrave Macmillan, 2005), pp. 241–69; Pamela O. Long, "Humanism and Science," in *Renaissance Humanism: Foundations, Forms, and Legacy*, ed. Albert Rabil, Jr., 3 vols. (Philadelphia: University of Pennsylvania Press, 1988), vol. 3, pp. 486–512; and Ann Blair and Anthony Grafton, "Reassessing Humanism and Science," *Journal of the History of Ideas*, 53 (1992), 535–40.

[64] Marie Boas, *The Scientific Renaissance, 1450–1630*, is a notable exception.

[65] Steven Shapin, *The Scientific Revolution* (Chicago: University of Chicago Press, 1998). The continuity of medieval and early-modern science, which Shapin takes for granted (pp. 3–4), has left no trace on his narrative or its structure.

story. The more intriguing emergent picture is fundamentally cross-cultural in orientation and contextual at its core. A half-century since the discoveries by Otto Neugebauer, E. S. Kennedy, and their students, it now seems likely that Copernicus put to heliocentric use several of the newer thirteenth- and fourteenth-century geocentric planetary models invented in Islamic civilization. Like Duhem's discoveries circa 1903, these have opened a growing breach in a Scientific Revolution that had been safely buffered from Islamic influence by three late-medieval Latin centuries.[66]

Earlier insular accounts are becoming untenable as new research documents the increasing number of contacts among scholars in Islamic, Byzantine, and Latin civilizations of the Middle Ages – to say nothing of Mughal India and China. These interactions – whether personal, textual, or diplomatic – are introducing much complexity into the large-scale narrative of the history of science, which has often cast the history of medieval science as a monofilament line leading to the telos of modern science. This framework follows unidirectional translation movements – from Greek to Arabic (perhaps via Syriac) in the eighth to ninth centuries, from Arabic to Latin in the twelfth, from more Greek into Latin in the fifteenth. But such a schema overlooks the dialogic, multistranded, and often personal character of many intellectual interactions,[67] to say nothing of cross-cultural ones.

Despite the efforts of pioneering scholars, Byzantine science remains both marginal to mainstream Byzantine history and poorly integrated into history of science surveys. Science in Byzantium is often portrayed as akin to a repository awaiting later use, whether by Islamic civilization or Latin Europe.[68] One recognizes at once, in lightly edited form, the familiar narrative structure that once framed early accounts of Latin and Islamic science, back when historians could more plausibly claim to know nothing about them. Surprisingly, it was marriage negotiations between daughters of Byzantine emperors and Ilkhanid rulers in Persia that brought news of the astronomical work at the recent Maragha observatory to the notice of Greek scholarly circles. In the fourteenth century, these contacts mediated astronomy with Arabic and Persian content into the highest Byzantine circles. Between the eleventh and fifteenth centuries, Byzantine scholars also translated into Greek works that were lost in Greek but available in Arabic or Latin, especially in medicine and astronomy.[69] Some of this material arrived in Italy with Italian "study-abroad" students returning from Constantinople, and especially with Byzantine diplomats and émigrés, from the 1390s to the 1460s. Conversely,

[66] For the lively francophone debates on the subject, see Philippe Büttgen et al., eds., *Les Grecs, les Arabes, et nous: Enquête sur l'islamophobie savante* (Paris: Fayard, 2009).

[67] Kenneth Caneva, "Objectivity, Relativism, and the Individual: A Role for Post-Kuhnian History of Science," *Studies in History of Philosophy of Science*, 29 (1998), 327–44.

[68] Mavroudi, "Occult Science and Society in Byzantium," pp. 47–53.

[69] Maria Mavroudi, "Exchanges with Arabic Writers during the Late Byzantine Period," in *Byzantium: Faith and Power (1261–1557), Perspectives on Late Byzantine Art and Culture*, ed. Sarah T. Brooks (New Haven, Conn.: Yale University Press, 2006), pp. 62–75, especially pp. 63–7.

Arabic works that were well known in Latin translation, such as Averroes's commentaries on Aristotle, drew the attention of fifteenth-century Byzantine scholars.

In the fifteenth century, Ottoman pressure on the Byzantine Empire and political turmoil in Central Asia also played crucial roles in these interactions. The Timurid ruler-scholar Ulug Begh turned fifteenth-century Samarqand (in present-day Uzbekistan) into an exceptional center of learning, which hosted leading figures in mathematics and astronomy. After Ulug Begh was assassinated, several of these individuals made their way to Istanbul, where Mehmed II the Conqueror was gathering to his new capital texts from Maragha and elsewhere.[70] For the Byzantine and Latin worlds that mourned the end of Constantine's "New Rome," Mehmed II was both the agent and the symbol of an apocalyptic decline, a point of view that the Conqueror himself no doubt had great difficulty appreciating. Both of these contrasting narratives belong in a cross-cultural history of "late-medieval" science whereas one of them now monopolizes the stage for the sake of a single story line. Once again, one should suspect that such a pattern reflects not the deep structure of history itself but rather the narrative veneer that covers our ignorance of the documentary evidence and drives away potential researchers, steering them away from groundbreaking work and toward traditional topics that the narrative itself has defined as more obviously upbeat.

The growing awareness of cross-cultural interaction in the history of medieval science offers a powerful heuristic for framing future research with a principled disregard for the constraints of traditional periodization and the confines of one culture, however it is defined. Among the chapters that follow, alert readers will find some of the fruits of such an outlook.

[70] F. Jamil Ragep, "ʿAlī Qushjī and Regiomontanus: Eccentric Transformations and Copernican Revolutions," *Journal for the History of Astronomy*, 36 (2005), 359–371, especially p. 360; and Ihsan Fazlioglu, "The Samarqand Mathematical-Astronomical School: A Basis for Ottoman Philosophy and Science," *Journal of the History of Arabic Science*, 14 (2008), 3–68.

I

ISLAMIC CULTURE AND THE NATURAL SCIENCES

F. Jamil Ragep

The civilization of Islam arose from a religious revelation in the seventh century that formed, for Muslims, their core of belief about God, His creation of the universe, the individual's relation to the Supreme Being and the created world, and each person's personal and social obligations. Although there was (and is) broad unanimity about fundamental Islamic beliefs and values, interpretations of what these beliefs and values imply have varied greatly. It should not surprise us, therefore, that there is no single Islamic view of nature and that Muslim scholars have defended very different, at times conflicting, notions of the physical and metaphysical worlds. Generally speaking, we can identify two major approaches to the physical world in Islam.[1] The first arose in the religious context of understanding the Qur'ānic revelation and the traditions of the Prophet Muḥammad, while the other developed as the result of the recovery and appropriation of the ancient Hellenistic tradition of natural philosophy. How these approaches evolved and interacted during the premodern period of Islamic history is the subject of this chapter.

Some readers may be surprised to learn that Islamic civilization accommodated such a diversity of views. Even more surprising is the claim that both religious and secular viewpoints coexisted in Islamic intellectual history. Here we need to be careful not to think of this dichotomy in modern terms – the secularism of modern societies, including the "separation of church and state," would have been as alien to the premodern Islamic world as to the premodern European world. Nevertheless, nonreligious ("secular") knowledge was recognized as a distinct category in Islamic thought. As early as the ninth century, Islamic classifications of knowledge made a fundamental distinction between the "religious sciences" (*al-ʿulūm al-sharʿiyya*) and what was usually referred to as the "sciences of the ancients" (*ʿulūm al-awāʾil*). Although the

[1] Technically, it is inappropriate to speak of "physics" or "nature" with respect to the writings of the Islamic theologians since they denied both of these Greek notions, as we shall see further. Although imperfect, "physical world" is employed here as the best alternative.

latter encompassed topics that were metaphysical or religious – the nature of God and so forth – these subjects employed a rational discourse whose origins went back to the Greeks rather than a religious revelation based upon faith. This was precisely how Islamic thinkers depicted the difference – religious sciences based on "transmitted" (*manqūl*) knowledge, sciences of the ancients a product of "rationalized" (*maʿqūl*) knowledge.[2] Knowledge of the physical world occupied an important place within both traditions. Before turning to a detailed exposition of that knowledge, we need to examine these traditions themselves and the place they assigned to this knowledge of the physical world.

The religious sciences emerged early in Islamic history as a way of understanding and interpreting the revelation (the Qur'ān) and systematizing and authenticating the sayings and actions of the Prophet (the *ḥadīth*). These sciences included grammar (*naḥw*), which was necessary for understanding difficult passages and expressions; Qur'ānic hermeneutics (*tafsīr*), which was crucial for determining the correct interpretations as a basis of law and dogma; *ḥadīth* studies, which were important not only for interpreting the Prophet's life and sayings but also for authenticating them; law (*fiqh* and *sharīʿa*), which regulated most aspects of life in Islamic societies; and *kalām* (theology), which explored rationally the implications of the Islamic revelation in all its aspects.[3] Although the religious "sciences" may seem to us less than scientific since their starting points were an unquestioned revelation, it should be emphasized that within that framework rational methods were developed for religious purposes. Among these was the desire to explore God's creation, a hallmark of Abrahamic monotheism, which many theologians took to be a religious obligation. Thus, within the context of *kalām*, one finds extensive discussions that rely exclusively on a rationalist approach to understanding the physical world.

The other major source of information regarding the physical world was the "ancient sciences," particularly the natural philosophy and mathematical sciences of Greek antiquity. These entered the Islamic world by means of a massive translation movement starting in the eighth century and continuing into the tenth; there was also some relatively minor continuity of Greek learning in a few Middle Eastern centers, such as Antioch in Syria and Edessa in southeastern Anatolia, that had fallen under Islamic control. It is noteworthy that natural philosophy and mathematics did not arrive in Islam independently of other branches of philosophy, such as metaphysics

[2] On the classification of the sciences in Islam, see Franz Rosenthal, *The Classical Heritage in Islam* (The Islamic World Series) (Berkeley: University of California Press, 1975), pp. 52–73; and Osman Bakar, *Classification of Knowledge in Islam: A Study in Islamic Philosophies of Science* (Cambridge: Islamic Texts Society, 1998).

[3] A good way to learn about these disciplines is by looking them up (under their Arabic names) in the *Encyclopaedia of Islam*, 2nd ed., ed. P. Bearman, Th. Bianquis, C. E. Bosworth, E. van Donzel, and W. P. Heinrichs, 12 vols. (Leiden: Brill, 1960–2004).

and ethics, but as part of a larger philosophical system. Almost the entire Aristotelian philosophical corpus was translated into Arabic, and major parts of the works of other great Hellenistic philosophers were known, either directly or indirectly. Islamic civilization appropriated this material to such an extent that an Islamic scholar of the Middle Ages, whether trained in the philosophical sciences or not, would be expected to know something of the various Greek philosophical doctrines (even heretical ones), which often provided the starting point for the development and refinement of what came to be accepted as "Islamic" doctrines. Indeed, an eminent modern scholar has proclaimed that "Islamic civilization as we know it would simply not have existed without the Greek heritage."[4]

Islamic scholars of various persuasions were particularly well disposed toward the ancient mathematical sciences, which appeared to be less contentious than the philosophical ones, inasmuch as they were perceived to be less implicated in metaphysical and physical doctrines that might be taken as "un-Islamic." By "mathematical sciences" we mean not only pure mathematics, such as geometry, number theory, arithmetic, algebra, and trigonometry (the last two being systematized for the first time in Islam), but also such sciences as astronomy, geometrical optics, and music theory, which applied mathematics to natural phenomena. Indeed, in many ways the mathematical sciences held a privileged position; one finds numerous references, in both the religious and the secular literature, to the incontestability of mathematical truths. No doubt, this status allowed the mathematical sciences to flourish more extensively than the less precise physical sciences.[5]

THE HISTORICAL AND CULTURAL BACKGROUND

Before we turn to a fuller exploration of various Islamic approaches to the physical world and their interactions, we need to understand the historical and cultural background in which these traditions would develop and flourish. But first a note about terminology. Products of Islamic civilization will often be referred to in this essay as "Islamic" – whether these were produced by Muslims, Jews, Christians, or others – to indicate a shared intellectual heritage. If a particular religious aspect is being emphasized, then the terms "Muslim" (or "Jewish" or "Christian") will be used. Although Arabic was the main language of intellectual life (including science and philosophy) in all Islamic countries during the medieval period, there were substantial numbers of works in Persian and Turkish. Therefore we will use the expression "Islamic science" (or "philosophy") rather than "Arabic science" to refer to

[4] Rosenthal, *Classical Heritage in Islam*, p. 14.

[5] See F. Jamil Ragep, "Freeing Astronomy from Philosophy: An Aspect of Islamic Influence on Science," *Osiris*, 16 (2001), 49–71. See Berggren, Chapter 2, and Kheirandish, Chapter 3, this volume.

the tradition as a whole. "Arabic," "Persian," and "Turkish" will be reserved to indicate language or ethnic background.

Islamic history is long, complex, and multifaceted, and for most people (even Arabs and Muslims) keeping its many names and dates straight represents a formidable challenge. Fortunately one does not need to know all of the details of Islamic history in order to understand its scientific and philosophical traditions, but it will help if we are aware of a few key political and cultural details. Above all, it is crucial to keep in mind that Islamic civilization fostered numerous intellectual traditions over a 1,500-year period, which often emerged as the result of local social, political, and religious developments.[6] Nevertheless, until recently, many of these traditions were sustained, argued for and against, and generally remained vital owing to the remarkably free circulation of people, books, and goods over much of the Islamic world. Thus the "unity" of Islamic civilization was due to the maintenance of a variety of vibrant intellectual traditions rather than the sterile uniformity of a single sustaining ideal or worldview.

According to the traditional reckoning, Islamic history[7] began in the year 622 (= *hijra* year 1), the year in which the Prophet Muḥammad was invited to lead a community of believers in the Arabian city of Medina, some 300 miles to the north of his native Mecca. It is significant that this event, at once religious and political, rather than the Prophet's birthday (ca. 570) or the date of his first revelation (610), marks the beginning of Islam. From that time on, the political and the religious in Islam were intertwined, and we must remain aware of the religious background if we are to understand the many political and intellectual events in Islamic history. At the same time, it would be a mistake to assume that everything in Islamic history can be reduced to religion or religious causes; as we have already seen, Muslim intellectuals themselves readily distinguished between religious and nonreligious spheres of knowledge.

With the death of the Prophet Muḥammad in the year 632, the community of believers, which by then had become extensive owing not only to his religious but also his political, military, and diplomatic skills, faced a crisis. Muḥammad left no surviving male heir; consequently, the community, centered by then in Mecca, chose one of the Prophet's earliest converts as caliph (or successor). He was followed by three others, who collectively came to be called the four orthodox caliphs. This period (632–661) was a time of enormous expansion, with large parts of the Byzantine and Sasanian

[6] A. I. Sabra, "Situating Arabic Science: Locality Versus Essence," *Isis*, 87 (1996), 654–70.

[7] A good short survey of Islamic history is Karen Armstrong, *Islam: A Short History* (New York: Modern Library, 2002). A more elaborated version can be found in Ira M. Lapidus, *A History of Islamic Societies*, 2nd ed. (Cambridge: Cambridge University Press, 2002). For a quick survey of the many Islamic dynasties, a good source is Clifford Edmund Bosworth, *The New Islamic Dynasties: A Chronological and Genealogical Manual*, enlarged and updated edition (New York: Columbia University Press, 1996). Information on dynasties, realms, and political movements can also be found in the *Encyclopaedia of Islam* (2nd edition).

(Persian) empires falling to the Arab/Muslim armies; it was also a time of political unrest in Mecca. Among the many parties vying for control of this expanding empire, two in particular stand out: the traditional Meccan Arab aristocracy and the followers of the Prophet's son-in-law ʿAlī. The latter, who came to be called the partisans (*shīʿa*) of ʿAlī, held that succession should have been within the family of the Prophet. Since ʿAlī was not only married to Muḥammad's daughter Fāṭima but was also a cousin, he seemed a likely successor. But he had to wait his turn, becoming the fourth orthodox caliph. His accession was not universally accepted, however, and the ensuing civil war resulted in the defeat of the Shīʿa, who were forced underground. Their religious/political rivals came to be known as Sunnis because of their claim to follow the practice (*sunna*) of the Prophet rather than an enlightened successor/leader (imam) from his family. The Sunnis would become the majority in Islam and thus define Islamic orthodoxy.

The defeat of ʿAlī and his immediate family led to the ascendancy of the Umayyad dynasty (661–750), which moved the capital to the more strategically located city of Damascus in Syria. During this period, the empire expanded to southern France in the west and to the borders of China in the east. There were, however, many disaffected elements within this empire, including the Shīʿa, who managed to mount a successful insurgency that resulted in the rise of the ʿAbbāsid dynasty (750–1258) – a Sunni dynasty, despite early Shīʿa support.

The early ʿAbbāsid period (750–945) marked a time of great economic and political power for the Islamic empire, one of the largest in human history. It also witnessed a remarkable intellectual and cultural achievement, only partially reflected in the *Thousand and One Nights*, the fabulous work that helped make this an age legendary in the popular imagination of many peoples. The important point to bear in mind is that during this time a remarkable artistic, literary, and cultural humanism came into being, one of whose fruits was the appropriation of the ancient scientific and philosophical traditions not only from the Hellenistic world but also from India and pre-Islamic Persia.

The cultural and political conditions that made this possible are worth reflecting upon. As we have seen, the ʿAbbāsid revolutionaries appealed to disaffected Umayyad subjects, many of whom had found that despite converting to Islam and belonging to an empire that purportedly protected minority rights, they felt inferior to the Arabs of Damascus and the Arabian peninsula. The ʿAbbāsids, once in power, sought to strengthen their broad-based appeal in a number of ways. By moving the capital to the newly created Mesopotamian city of Baghdad, closer to the ancient lands of the Persians, the ʿAbbāsid court attracted many people with cultural and political ties to the old Persian order and regime. Christians, Jews, Buddhists, Zoroastrians – even pagans – found a welcoming atmosphere in the new capital. People whose native tongues were Greek, Arabic, Persian,

Khwarizmian (from Central Asia), Syriac, Coptic, and Turkish (and varieties thereof) rubbed shoulders on the street and in their daily work. It is not surprising that such a multicultural, multireligious, and multilingual environment would stimulate cultural creativity and help foster the notion that the ʿAbbāsids were the rightful custodians of the precious heritage of the ancient world.

The unity of the Islamic empire lasted less than two centuries. The difficulties of maintaining so far-flung an empire – centrifugal forces that promoted the establishment of independent governors, the need to import "slave soldiers" from Central Asia to protect the caliph, and the resurgence of various Shīʿa groups – all this and more contributed to frequent breakdowns of the early unity. From about the middle of the tenth century until modern times, diverse regional centers emerged with their own distinctive intellectual and cultural styles.

One of these, the Fatimid dynasty (lasting from the tenth through twelfth centuries), ruled in Egypt and parts of North Africa and adhered to Ismāʿīlism, a branch of Shīʿism. Like other Ismāʿīlī courts, the Fatimids promoted the ancient sciences and philosophy, in part because both Shīʿa and Platonists aspired to the rule of enlightened – and inspired – leaders.[8] Another Shīʿī dynasty, the Buyids (932–1062), controlled large areas of Persia and Iraq, including Baghdad, generally pursuing an enlightened policy in which both Sunnis and Shīʿīs participated in a great flourishing of the arts, the sciences, and philosophy.[9] The Saljūq Turks (1038–1194), originally from Central Asia, displaced the Buyids in this region, including Baghdad, and embraced a strongly orthodox Sunnism that led to their establishing religious schools (*madrasas*).

Lest it appear that scientific and philosophical work was undertaken only in the Islamic heartland of the Middle East, we should mention some of the many regional centers where science and philosophy flourished. Since the Alexandrian conquests of the fourth century B.C., Afghanistan and the lands beyond the Oxus River in Central Asia (Transoxiana) had been important centers of learning and culture, and this continued into the Islamic period. During the tenth through twelfth centuries, the Khwārizm Shāhs and the Ghaznawids supported some of the most important scientists and philosophers in all of Islamic history.[10] Yemen was a significant center of astronomical and agricultural science during the Rasūlid period

[8] In his *Republic*, Plato discusses the need for enlightened rulers trained in the mathematical sciences. The Ismāʿīlīs picked up on this theme, connecting it to the inspired rule of ʿAlī and his successors.

[9] The Buyid period has been called the "renaissance of Islam" by several authors. See Adam Mez, *The Renaissance of Islam* (New York: AMS Press, 1975; original German, 1922); Joel L. Kraemer, *Humanism in the Renaissance of Islam: The Cultural Revival During the Buyid Age*, 2nd rev. ed. (Leiden: Brill, 1992); and Joel L. Kraemer, *Philosophy in the Renaissance of Islam: Abū Sulaymān Al-Sijistānī and His Circle* (Leiden: Brill, 1986).

[10] Foremost among these were Ibn Sīnā (Avicenna) and al-Bīrūnī.

(1229–1454).[11] And, finally, we should not fail to note the major cultural and intellectual achievements in far western Islam, specifically Islamic Spain. From the early eighth century, when Muslim armies first entered Spain, until the late fifteenth century, when Muslims were expelled, Islamic Spain (al-Andalus) was the scene of distinctive Islamic intellectual traditions.[12] Of particular note is the strong defense of a "pure" Aristotelianism, often associated with Ibn Rushd (Averroes), which took place under Almohad rule during the twelfth century.

The later period of Islamic history (after 1200) is characterized by a number of strong, centralized authorities. Despite the traditional view that this was a period of precipitous intellectual decline, recent scholarship has shown that significant work in the sciences and philosophy continued. Under the Mamlukes (1250–1517), activity in a number of fields, especially astronomy and medicine, was maintained at a high level in Egypt and Syria.[13] Mongol invasions from Central Asia in the thirteenth century, which brought an end to both the ʿAbbāsids and the Ismāʿīlīs as political forces, and further invasions by the Tīmūrids in the fourteenth and fifteenth centuries, are often portrayed as disastrous for the Islamic world. Nevertheless, both Mongol and Tīmūrid rulers supported a variety of scientific and philosophical work in Persia and Central Asia. Their patronage brought into being the observatory as a full-fledged scientific research institution in such centers as Marāgha (thirteenth century, northwest Persia) and Samarqand (fifteenth century).[14]

The Ottomans (1281–1924) and the Safavids (1501–1732) were, roughly speaking, geographically coterminous with the Byzantine and Sasanian empires, respectively. The two rival empires – the former Sunni, the latter Shīʿī – attained magnificence in literature and the arts, but their attainments in scientific and philosophical matters are still not sufficiently known or appreciated. Recent historical work has shown that a number of intellectual traditions continued, perhaps even flourished, during this time.[15] There is no doubt, though, that the attainments of the post-1600 period did not rival the major transformations going on contemporaneously in Europe.[16]

11 See Daniel M. Varisco, *Medieval Agriculture and Islamic Science: The Almanac of a Yemeni Sultan* (Seattle: University of Washington Press, 1994).

12 See Salma Jayyusi, ed., *The Legacy of Muslim Spain*, 2nd ed. (Leiden: Brill, 1994).

13 See David A. King, "The Astronomy of the Mamluks," *Isis*, 74 (1983), 531–55.

14 See Aydın Sayılı, *The Observatory in Islam and Its Place in the General History of the Observatory* (Ankara: The Turkish Historical Society, 1960); and E. S. Kennedy, "The Heritage of Ulugh Beg," in his *Astronomy and Astrology in the Medieval Islamic World* (Aldershot: Ashgate Variorum Reprints, 1998), p. xi.

15 See Ekmeleddin Ihsanoğlu, "Ottoman Educational and Scholarly-Scientific Institutions" and "The Ottoman Scientific-Scholarly Literature," in *History of the Ottoman State, Society & Civilisation*, ed. Ekmeleddin Ihsanoğlu, 2 vols. (Istanbul: Research Center for Islamic History, Art and Culture, 2002), vol. 2, pp. 361–515, 517–603; and Seyyed Hossein Nasr and Mehdi Amin Razavi, *The Islamic Intellectual Tradition in Persia* (Richmond, U.K.: Curzon Press, 1996).

16 This later period of the history of science in Islamic societies has not been well studied; for the time being, the reader is referred to Ekmeleddin Ihsanoğlu, ed., *Transfer of Modern Science & Technology to the Muslim World: Proceedings of the International Symposium on "Modern Sciences and the Muslim*

THE TRANSLATION OF GREEK NATURAL PHILOSOPHY INTO ARABIC: BACKGROUND AND MOTIVATIONS

The Islamic appropriation of Hellenistic natural philosophy, mathematical sciences, medicine, and philosophical teachings is one of the most remarkable events in the history of learning. How did the early adherents to Islam, mainly cut off from the great intellectual traditions of antiquity, evolve in a few generations into a cultural polity that would give rise to an intensive and relatively organized translation movement that would make this possible?[17]

Part of the answer lies in the rapid development of the Islamic community into a sophisticated, multicultural empire under the early ʿAbbāsid caliphs, at which time (the late eighth century) the translation movement began. These caliphs, in addition to being defenders of the Islamic faith, often saw themselves as successors to the great pre-Islamic Persian kings and encouraged the translation into Arabic of not only Hellenistic science and philosophy but also a substantial amount of pre-Islamic Persian and Indian material (both scientific and literary). Islamic intellectuals thus gained an extraordinary degree of access to these ancient traditions – rivaling, and in some cases surpassing, our modern knowledge of them.

The scope of this transfer of knowledge was unprecedented. Although there had been a steady transmission of knowledge between peoples in the Mediterranean basin since at least the second millennium B.C. (if not before), there had never been such a conscious attempt by one culture to appropriate and incorporate the entire scientific and philosophical corpus of another. Even in the case of the Romans, who greatly admired Greek culture and accomplishments, very little of the Hellenistic scientific and philosophical textual tradition was translated into Latin during the period of Roman hegemony.[18] And, before Islam, the examples we have of transmission tend to be mainly piecemeal and of a practical sort. Thus, in addition to discussing

World": Science and Technology Transfer from the West to the Muslim World from the Renaissance to the Beginning of the XXth Century (Istanbul, 2–4 September 1987) (Studies and Sources on the History of Science Series, No. 5) (Istanbul: Research Centre for Islamic History, Art and Culture, 1992); Feza Günergun and Shigehisa Kuriyama, *International Symposium, the Introduction of Modern Science and Technology to Turkey and Japan* (Kyoto-shi: International Research Center for Japanese Studies, 1998); and Donald M. Reid, *Cairo University and the Making of Modern Egypt* (Cambridge: Cambridge University Press, 1990). For recent discussions of science in the contemporary Islamic world, see *The Arab Human Development Report 2002: Creating Opportunities for Future Generations* (New York: United Nations Development Programme Regional Bureau for Arab States, 2002); and Pervez Hoodbhoy, *Islam and Science: Religious Orthodoxy and the Battle for Rationality* (London: Zed Books, 1991).

17 One might contrast such an imported intellectual tradition with the religious sciences, which were, despite influences from the outside, mainly developed internally within an Islamic context.

18 Heinrich von Staden provides a fascinating depiction of the ambivalence the Romans felt toward Greek science, especially medicine, in his "Liminal Perils: Early Roman Receptions of Greek Medicine," in *Tradition, Transmission, Transformation: Proceedings of Two Conferences on Pre-modern Science Held at the University of Oklahoma*, ed. F. Jamil Ragep and Sally P. Ragep (Leiden: Brill, 1996), pp. 369–418.

the details of the transmission movement itself, we will need to take into account the contexts in which such an appropriation of knowledge could occur.

We must first examine the precedents for the ʿAbbāsid translation movement, for such antecedents can serve both as historical background and as contrasting episodes. Prior to the translation movement under the early ʿAbbāsids, an indigenous intellectual tradition was pursued by an international and multilingual group of scholars. Nestorian and Syrian Orthodox Christians produced works in Syriac, a form of Aramaic, dealing with Aristotelian logic, Ptolemaic astronomy and cosmology, and Hippocratic and later Greek medicine. Deserving mention here is a Syrian bishop, Severus Sebokht (d. 666/667), who flourished after the appearance of Islam, knew Persian, Greek, and Syriac, was an expert on logic, wrote on the astrolabe, and knew of Indian (later to be called Arabic) numerals, which he highly praised.[19] In pre-Islamic Persia, we find scholars at the time of the Sasanid ruler Khusraw Anūshirvān (r. 531–78) writing works in Pahlavi Persian on Aristotelian logic and developing new astronomical tables based on both Greek and Indian sources.[20] A Greek medical tradition flourished in the Persian town of Gondēshāpūr, and there were Pahlavi translations of Indian medical works.[21] The preponderance of Persian astrologers such as Nawbakht, Māshā'allāh, and al-Ṭabarī at the early ʿAbbāsid court suggests a long-standing tradition of work on astrology in Pahlavi.[22]

These pre-Islamic translation efforts served as a precedent, and in some cases a model, for subsequent translation. Royal patronage of translations into Pahlavi presented a useful historical model for the early ʿAbbāsid caliphs, who wished to solidify their base in a Persian heartland that remained loyal for some time to the Sasanian past, while translations from Greek into Syriac provided an important guide for translating from Greek into Arabic, a sister Semitic language. These earlier translation efforts, however, can go only so far in helping us understand the Arabic translation movement. It appears that the pre-Islamic Pahlavi and Syriac translations may have been directed at practical knowledge, and there is little evidence, during this early period, of a sustained or coherent effort to translate the entire Greek philosophical and scientific heritage.[23]

The Umayyads (661–750), the immediate predecessors of the ʿAbbāsids, did not sponsor any widespread or sustained translation activity of scientific or philosophical texts, despite the fact that their capital of Damascus was

[19] Dimitri Gutas, *Greek Thought, Arabic Culture: The Graeco-Arabic Translation Movement in Baghdad and Early ʿAbbāsid Society (2nd–4th/8th–10th centuries)* (London: Routledge, 1998), pp. 20–2.

[20] Ibid., pp. 25–7.

[21] Manfred Ullmann, *Islamic Medicine* (Edinburgh: Edinburgh University Press, 1978), pp. 15ff.

[22] David Pingree, "Astrology," in *Religion, Learning, and Science in the ʿAbbāsid Period*, ed. M. J. L. Young, J. D. Latham, and R. B. Serjeant (Cambridge: Cambridge University Press, 1990), pp. 293–5.

[23] Gutas, *Greek Thought, Arabic Culture*, p. 22.

closer to the Greek heartland than would be the ʿAbbāsid capital of Baghdad, and despite the many multilingual scholars within its boundaries.[24] On the other hand, it was under the Umayyads that we find the beginnings of an intellectual tradition of great distinction in the religious sciences, Arabic linguistic and literary studies, and humanist literature (*adab*).[25] It would thus appear that the foundations for a vigorous intellectual tradition had already been laid by the time of the ʿAbbāsid revolution, but without any inclination to incorporate the teachings of other cultures.[26] Why did the ʿAbbāsids seek such incorporation?

It has been argued that the early ʿAbbāsid caliphs, particularly al-Manṣūr (r. 754–75), sought to consolidate their hold on power by promoting the notion that they were the successors of the ancient Near Eastern dynasties, which had culminated with the Sasanids. Included in this campaign was the conscious imitation of the Sasanid Zoroastrian imperial figure of the king as patron of philosophy and science, represented now by a program of cultural integration under the banner of Islam. There may also have been an aspect of competition with Byzantine emperors, who were viewed as having allowed the glorious Hellenistic intellectual tradition to deteriorate to the point of nonexistence. The use of astrological cycles to make the ʿAbbāsid role in world history seem inevitable, something the Sasanids had previously used for their own imperial strategy, provided an important incentive for the translation of certain scientific texts.[27]

This political context helps us to understand one of the incentives that would lead the caliphs of Islam, the "commanders of the faithful," to support the translation of materials often held to be contrary to Islamic doctrine. Astrology in particular seemed to give to the stars powers competitive with God's. It also helps us to understand the breadth and depth of the translation movement. The central importance of the role of the caliphs ensured that the administrative weight of the empire would be brought to bear on the endeavor and fostered the establishment of an infrastructure of learning. In its tangible aspects, this infrastructure included the creation of libraries, hospitals, and nascent research institutions as well as direct caliphal support for the translations. More intangibly, such support from the top emphasized

[24] Ibid., pp. 23–4.

[25] The major fields of *adab*/humanism and their origins are discussed in George Makdisi, *The Rise of Humanism in Classical Islam and the Christian West, with Special Reference to Scholasticism* (Edinburgh: Edinburgh University Press, 1990), pp. 119–200.

[26] The Egyptian scholar Aḥmad Amīn (1886–1954) strongly argued, contrary to many historians, for the continuity between the Umayyad and ʿAbbāsid dynasties in his *Ḍuḥā al-Islām*, 3 vols. (Cairo: Maktabat al-nahḍa al-Miṣriyya, 1964), especially vol. 1, pp. 1–4.

[27] Dimitri Gutas has forcefully and coherently articulated the views in this paragraph in his *Greek Thought, Arabic Culture*, especially chap. 2; cf. his article in the *Encyclopaedia of Islam* ("Tardjama (section 2)"). David Pingree, in numerous books and articles, has emphasized the role of Sasanid astrological history in early Islamic science and society; see, for example, his "Astrology."

the ideal of the cultivated man of learning among the upper classes, thus stimulating their own system of patronage for translators and researchers.

The political explanation, however, can carry us only so far. The Sasanids, though imbued with the belief that it was their sacred duty to recover the lost scientific and philosophical heritage that had been allegedly plundered by the Greeks at the time of Alexander the Great, undertook what seems at best to have been a modest amount of translation from Greek for the purpose of "recovery" – translating activity that fell far short of that promoted by the ʿAbbāsids.[28] It is here that the intellectual context of early Islam may help us understand the translation movement. The growth of the religious sciences and the humanist (*adab*) literary genres under the Umayyads and early ʿAbbāsids provided a critical mass of scholars and interested laymen who could promote and criticize the intellectual products of other cultures within the context of developing Islam, both as a religion and as a literary civilization. It is within this culture that the value of knowledge – whether religious, practical, or theoretical – was promoted, partly to deal with this burgeoning quantity of texts and ideas from a variety of sources and partly to conform to various Qurʾānic and prophetic injunctions to seek knowledge from whatever source.

An important indication of this promotion of knowledge comes from a purported dream of the caliph al-Maʾmūn, employed to explain why the Commander of the Faithful would support the translation movement. According to various sources, the caliph was visited one night by Aristotle, who granted him a series of questions.[29] The caliph began with the all-important one, "What is good?" to which the immediate answer was "That which is good in the mind." "And what then?" The answer: "That which is good in law (*sharʿ*)." By placing rational knowledge first, the dream as reported indicates that the caliph had adopted a view held by the early theologians (the Muʿtazilites) as well as Aristotle – namely, that the mind (rationality) is the arbiter of all knowledge, even religious knowledge. The result of this acknowledgment, as the story was told, was enthusiastic support for the rational sciences of the Greeks. The dream may not, in fact, be the reason for the translation movement, which, at any rate, was well under way before al-Maʾmūn. It does, however, offer a window on the cultural and intellectual context within which the translation movement was being promoted. This context also helps us to understand the ever-increasing interest

[28] Gutas (*Greek Thought, Arabic Culture*, pp. 40–5) emphasizes the Sasanid imperial ideology that promoted translation and provides considerable information regarding possible and actual translations from Greek into Pahlavi. Concrete evidence for a massive translation movement, however, remains scant.

[29] One prominent report is in Ibn al-Nadīm, for which see Bayard Dodge, ed. and trans., *The Fihrist of al-Nadīm: A Tenth-Century Survey of Muslim Culture*, 2 vols. (New York: Columbia University Press, 1970), vol. 2, p. 583.

in theoretical, nonutilitarian matters throughout the ninth century – a move away from the practical approaches of Syriac, Persian, and Indian science.[30]

TRANSLATORS AND THEIR PATRONS

As one of the most important intellectual movements in human history, the translation movement in Islam resulted from the happy coincidence of a number of factors. Of fundamental significance was the new urban, multicultural, multireligious, and multilingual milieu of Baghdad. The new ʿAbbāsid capital exhibited a cosmopolitanism and tolerance in which members of its diverse population routinely exchanged ideas, techniques of learning, and technology. Caliphs, as well as prominent figures of the bureaucracy, often attended and supported gatherings of Christians, Jews, pagans, and Muslims in which there were intellectual discussions and study of texts. It is notable that non-Arab converts to Islam assumed a prominence denied them under the Umayyads.

Another factor was the institutionalization of Arabic as the language of administration and daily life. This trend, already begun under the Umayyads, continued and expanded under the ʿAbbāsids, now encompassing scholarly works that were rendered into Arabic from Greek, Syriac, Pahlavi, and Sanskrit. No other civilization had ever supported the translation of so many works into a single language, and this would make Arabic the principal medium for studying and doing science and philosophy for centuries to come.

Many of those with the requisite language skills were non-Muslims. It was therefore of great importance that Islam was generally tolerant toward members of other religious groups, especially people of the book – Jews and Christians – but often also other groups with a sacred text. Ibn al-Nadīm, the tenth-century bibliophile and biobibliographer, lists some sixty-five translators.[31] More often than not, they were Nestorian Christians, whose native language was Syriac and whose liturgical language was Greek; others were Jewish, Zoroastrian, Hindi, and even pagan.

Two well-known and representative translators were Ḥunayn ibn Isḥāq (808–873) and Thābit ibn Qurra (d. 901). Ḥunayn was a Nestorian Christian of Arab extraction from al-Ḥīrā in Iraq. Probably bilingual in Syriac and Arabic, he learned Greek later in life and became remarkably proficient in its use. Credited with the translation of well over 100 titles (mostly medical and philosophical works), he stands as one of the greatest translators of all time. These works were first translated from Greek into either Arabic or, more

30 This is most clearly seen in the move away from the practical astronomy of the Persians and Indians and the almost universal adoption of a Ptolemaic/Aristotelian theoretical framework for astronomy by the end of the ninth century.

31 Dodge, *Fihrist of al-Nadīm*, vol. 2, pp. 586–90.

commonly, Syriac; later either Ḥunayn or one of his students, who included his son Isḥāq ibn Ḥunayn and a nephew Ḥubaysh, would often translate into the final target language. (Although the Arabic translations would form the basis for a 1,000-year scientific and philosophical tradition, Ḥunayn's Syriac translations have mostly disappeared.)[32]

Thābit ibn Qurra was part of an ancient pagan, Semitic-speaking group from Ḥarrān in upper Mesopotamia (now in present-day Turkey) that had been Hellenized after the conquests of Alexander the Great. Although, in theory, pagans were not allowed within the Islamic polity, the Ḥarrānians claimed to be "Sabians," who were mentioned in the Qur'ān, and this expedient was readily accepted. Given his background, it is not surprising that Thābit was proficient in languages and became a fixture at the court in Baghdad. He is credited with numerous translations or revised translations of major works, including Euclid's *Elements*, Ptolemy's *Almagest*, and several works by Archimedes and Apollonius. Like Ḥunayn, Thābit was a prolific writer as well as a translator, with some seventy original works ascribed to him.[33]

The existence of tolerated non-Muslims with the requisite linguistic skills, however, might not have led anywhere without patronage. The ʿAbbāsid upper classes, with their newly acquired wealth and power, had the desire and necessary means for acquiring knowledge and the social status that it conveyed.[34] The list of wealthy patrons of Ḥunayn ibn Isḥāq's group of translators included Abū al-Ḥasan ʿAlī (d. 888), secretary to the caliph al-Mutawakkil and commander of the Muslim forces on the Byzantine border, whose magnificent library diverted the famous astrologer Abū Maʿshar from his pilgrimage to Mecca; Muḥammad ibn ʿAbd al-Malik, vizier to the caliph al-Muʿtaṣim, who is said to have spent 2,000 dinars[35] a month for translations of Galen's works by Ḥunayn and his school; and the Banū Mūsā, a wealthy family of mathematicians and physicists in their own right, who spent their fortune procuring Greek manuscripts from the former Byzantine provinces and directing the intellectual careers of protégés like the translator-mathematician-astronomer Thābit ibn Qurra.[36] Another major source of patronage was the *Bayt al-ḥikma* (House of Wisdom), a famous scientific institution in Baghdad, which functioned also as a major library. Probably founded in the eighth century, it seems to have been given additional support by the caliph al-Ma'mūn. Whether or not it was a site for translations is open to dispute.[37]

32 G. Strohmaier, "Ḥunayn b. Isḥāḳ al-ʿIbādī," in *Encyclopaedia of Islam*, 2nd ed., vol. 3, pp. 578b–581a.

33 B. A. Rosenfeld and A. T. Grigorian, "Thābit ibn Qurra," *Dictionary of Scientific Biography*, XIII, 288–95.

34 Kraemer, *Humanism in the Renaissance of Islam*, p. 4.

35 One dinar was approximately 4.25 grams of gold.

36 Max Meyerhof, "New Light on Ḥunain Ibn Isḥāq and His Period," *Isis*, 8 (1926), 685–724 at pp. 713ff; and Gutas, *Greek Thought, Arabic Culture*, pp. 138–9, 185.

37 Gutas, *Greek Thought, Arabic Culture*, pp. 53–60.

An important by-product of this elite patronage was an insistence that the translated texts be, as far as possible, in a flowing and easily comprehensible style.[38] The translators displayed remarkable ingenuity and creativity as they fashioned a philosophical and scientific Arabic whose vocabulary and syntax were the result of considerable experimentation that took place over a long period of time. Certain translators, especially those associated with the school of Ḥunayn ibn Isḥāq, developed a semantic style of translation that emphasized meaning over the word-for-word renderings prevalent in earlier translations. They also coined new words based upon Arabic grammatical forms rather than using transcriptions of Greek technical terms.[39] Ḥunayn also developed sophisticated techniques for editing and translation, which included the collection and collation of as many Greek manuscripts as possible for a given text.[40] Specialization also occurred: Ḥunayn was noted for his translations of Galen, whereas his son Isḥāq seems to have been given the task of translating mathematical and astronomical works, many of which he sent to Thābit ibn Qurra for revision. This was all in marked contrast to previous, rather limited, translations, such as one might find in the Syriac schools, where the audience consisted mainly of church scholars rather than an educated lay public.[41]

THE NATURAL PHILOSOPHY TRADITION IN ISLAM

Against this background, we now examine how the physical world was dealt with and understood in Islam. The reader is reminded that we employ the expression "physical world," rather than "nature," because the very existence of a fixed nature that adhered to unchanging rules was a point of contention among Islamic intellectuals. Those Islamic philosophers who defended this idea of nature were the inheritors of a worldview rooted in classical Greek antiquity that, for the most part, entered Islam through the translation movement. Consequently, in addition to discussing the details of natural philosophy in this section, we will need to inquire how such an "alien"

38 An indication of this is in a report we have that individuals in Ma'mūn's court asked the translator of Euclid's *Elements* to make it more accessible to members of the court. See Sonja Brentjes, "The Relevance of Non-primary Sources for the Recovery of the Primary Transmission of Euclid's *Elements* into Arabic," in Ragep and Ragep, *Tradition, Transmission, Transformation*, pp. 201–25 at p. 222.

39 One might contrast this with English, whose scientific vocabulary tends to be derived from Latin and Greek.

40 Rosenthal, *Classical Heritage in Islam*, pp. 17–18.

41 J. L. Berggren, "Islamic Acquisition of the Foreign Sciences: A Cultural Perspective," in Ragep and Ragep, *Tradition, Transmission, Transformation*, pp. 263–83 at p. 268. The limited nature of the Syriac translations is discussed in Gutas, *Greek Thought, Arabic Culture*, p. 22. Hogendijk notes that many of the Syriac translations were by-products, or intermediaries, produced as part of the translation from Greek into Arabic; see Jan P. Hogendijk, "Transmission, Transformation, and Originality: The Relation of Arabic to Greek Geometry," in Ragep and Ragep, *Tradition, Transmission, Transformation*, pp. 31–64 at p. 33, fn. 7.

tradition came to occupy such an important place in Islamic intellectual life. We will also need to learn who its defenders were.

It is important to keep in mind that the natural philosophy tradition in Islam was part of a larger tradition of philosophy. By "philosophy" we mean here the Neoplatonized Aristotelianism that was known in Arabic as *falsafa*, or sometimes *ḥikma* (wisdom); it had been a hallmark of the schools of late antiquity and was therefore well known and easily available to translators and teachers during the early centuries of Islam. In its broadest manifestation, this tradition would provide the philosophers of Islam with a coherent body of literature, allowing them to employ natural philosophy for exploring the physical world, make use of mathematics for studying abstract entities based upon that physical world, and apply metaphysics for investigating those higher beings that exist without physical bodies; it would also afford them the rational tools of logic for studying all of the above.

This tradition owes much of its basic content and many of its doctrines to Aristotle. However, intellectual developments during the twelve or thirteen centuries intervening between the fourth century B.C., when Aristotle reformulated the ideas of his teacher Plato, and the ninth century A.D., when philosophy was reborn in the Islamic context, ensured that Islamic philosophers would be the beneficiaries of various reinterpretations and reformulations of Aristotle's thoughts. Three of these developments deserve special mention. First, Aristotelianism was coupled during late Hellenism with ideas that are collectively known as Neoplatonism. Neoplatonists advocated an emanation of being from the highest celestial region through a hierarchy of spheres (the Aristotelian/Ptolemaic celestial orbs) down to the sphere of the Moon, which is closest to the Earth. This lowest level was associated both with Aristotle's "agent intellect" and Plato's world of ideas that were accessible to those people who purified their souls through the intense study of philosophy. Second, a number of prominent Greek philosophers wrote after Aristotle and either recast various of his doctrines or even directly criticized them. These include luminaries of the millennium running from the fourth century B.C. to the sixth century A.D. such as Theophrastus, Alexander of Aphrodisias, Ptolemy, Olympiodorus, Themistius, Simplicius, and John Philoponus, all of whom were known, either directly or indirectly, to Islamic natural philosophers. Philoponus (ca. 490–570), in particular, was severely critical of Aristotle, so that Islam also inherited some decidedly non-Aristotelian views from late antiquity. Third, mathematical philosophers such as Ptolemy (second century A.D.) held, contra Aristotle, that physical principles could be established by the mathematical sciences. This found a following among certain Islamic mathematicians, and even Islamic natural philosophers who disagreed with Ptolemy were deeply influenced by such views.

This worldview inherited from the Greeks gave the philosophers of Islam a mandate, indeed a moral imperative, for the study of science and nature. Except for prophets, who had a direct channel to the higher world, human beings wishing to achieve true happiness and salvation are required to exert great efforts, employing their powers of reasoning.[42] According to Plato, training in the mathematical arts was one way of rising above the debased material world to the world of Ideas;[43] without disputing the importance of the mathematical sciences, Aristotle also advocated study of the natural world as a way of understanding its wonderful design and purpose.[44] The study of nature was further sanctioned by the religiously inspired notion of the need to study God's creation. What was distinctive about *falsafa* was that it regarded rational means – logic, mathematics, and empirical study – as the road to truth. It followed that divine revelations regarding the natural world could be – indeed must be – subjected to interpretation by those capable of this higher rationality. Clearly this outlook was at the root of the conflict between *falsafa* and those who insisted that revelation alone was a sufficient basis for knowledge. The fact that such an argument could occur within Islam, and that it would last for almost a millennium, tells us something important about the diversity of that civilization.

The tradition of Hellenistic natural philosophy, along with philosophy in general, took root within the Islamic intellectual domain in the aftermath of the translation movement. The range of natural philosophy (*al-ṭabīʿiyyāt*) was considerably broader than our physics or even physical sciences, taking into account, as it did, all sensible bodies and studying them in terms of their capacity for change and motion.[45] We can gather something of its content by looking at the eight chapters on natural philosophy included by the preeminent philosopher Ibn Sīnā (Latin: Avicenna; d. 1037) in his monumental encyclopedic work *al-Shifāʾ*: (1) "Received Physics," covering principles common to all natural philosophy; (2) The "Sky and the World," dealing with both the heavens and the sublunar world; (3) "Generation and Corruption," concerning the processes of coming into and passing out of existence (such as birth and death); (4) "Actions and Passions," covering the processes of acting upon something and being acted upon by it (the radiation of light and its reception are good examples); (5) "Minerals and Meteorology"; (6) "The Soul," which is the source of change and motion in all living things; (7) "Plants"; and (8) "Animals."

The physical world of the Islamic philosophers was, first of all, the familiar plenum of Aristotle's finite, spherical cosmos centered on an immobile,

42 A good summary of the relationship of this philosophical worldview to the pursuit of science and philosophy in Islam can be found in A. I. Sabra, "Some Remarks on Al-kindi as a Founder of Arabic Science and Philosophy," in *Dr. Mohammad Abdulhādi Abū Ridah Festschrift*, ed. Abdullah O. Al-Omar (Kuwait: Kuwait University Press, 1993), pp. 607–601 (sic).

43 See especially Plato, *The Republic*, bk. VII.

44 Aristotle nicely argues the point in the introduction to his *Parts of Animals*.

45 On the other hand, mathematics and metaphysics were assigned the realms of the insensible, the unchanging, and the unmoving.

spherical Earth.[46] This universe was subdivided into parts: at the most basic level between the celestial realm and the terrestrial or sublunar realm. The celestial region was usually divided into nine spherical orbs (or shells).[47] The first seven (moving from bottom to top) were further subdivided into the spherical bodies that moved the seven visible planets (including, in ancient and medieval astronomy, the Moon and Sun); the eighth contained the fixed stars; and an upper, ninth orb was responsible for the daily rotation of the heavens. The terrestrial realm, which began below the lunar orb, consisted of the spherical levels (also shells), each of which was predominantly composed of one of the four classical elements: a level of fire followed by one of air (these two comprising the atmosphere), a level of water, and finally a level of earth. These two lowest levels were, of course, not perfectly spherical, because part of the earth (i.e., the continents) protruded above the level of water, thereby providing a place for land creatures, including humankind, to exist; this feature was sometimes taken as an indication of divine providence.

The celestial realm, including the planets and stars, is composed of a perfect substance called the "quintessence" or "aether," which is eternally unchanging; bodies made of aether rotate with uniform circular motion and can undergo no other change of state. By contrast, substances in the sublunar world, including human beings, can undergo all manner of changes owing to the imperfection and finiteness of straight-line motion, for in the sublunar realm, natural motion (that which occurs if there is no outside interference – thus according to the nature of the body) is in a straight line toward its natural place – upward for fire and air, downward toward the center of the Earth for water and earth. If these heavy or light bodies reach their natural places, their motion ceases. But since most substances that we see around us are not simple elements but are instead made up of two or more, the predominant element in the compound will determine the direction of motion. Even the air we breathe and the earth we walk on are considered composed of multiple elements.

The dynamics of the motion of sublunar substances are generally determined according to whether the substances are inanimate or animate. Natural, noncompulsory motion of inanimate objects is considered to occur as a result of a nature (*ṭabʿ*)[48] inherent in the body. Earth and water have a heavy nature, whereas the nature of air and fire is lightness. Thus a falling body is said to move downward because of its heavy nature; because this motion increases in speed as it approaches the center of the Earth, it cannot be said

[46] This spherical cosmos had already been upheld by the pre-Socratic philosopher Parmenides (fl. 480 B.C.).

[47] These nine spherical orbs or shells can be thought of as hollowed-out spheres that fit inside one another.

[48] On use of the terms *ṭabʿ*, *ṭabīʿa*, and so on to render the Greek concept of nature (φύσις), see D. E. Pingree and S. N. Haq, "Ṭabīʿa," *Encyclopaedia of Islam*, 2nd ed., vol. 10, pp. 25a–28b.

to be uniform (as is the case with the celestial motions). Nevertheless, many Islamic natural philosophers held that this accelerating motion is like the celestial motions in that both are the result of a "nature" and both move "in a single way" (*ʿalā nahj wāḥid*).[49]

Animate beings (i.e., beings with a soul) will be affected as well by these natural tendencies (heavy or light) of the elements of which they are composed, but they can also move as a result of action by their souls. In the case of plants, this is due to a vegetative soul, which is responsible for nutrition, growth, and reproduction. Animals also have a vegetative soul but in addition an animal soul, which accounts for the motor and sensory faculties. Human beings, who have a vegetative and animal soul, also possess a rational soul, which is responsible for their higher faculties, such as rational activity.[50] Note that most Islamic natural philosophers followed Aristotle in making study of the soul a part of physics rather than of metaphysics, though the latter usually included the study of the souls of the celestial orbs.

Other important Aristotelian doctrines widely held in Islam included the four causes (the material, the formal, the efficient, and the purposeful) as a way of explaining natural things; the concept of natural places for the elements; time as a measure of change, the daily celestial motion determining unit-time; magnitudes, motion, and time as continuous and infinitely divisible (in order to counter atomism and avoid the threat of Zeno's paradoxes); and denial of an actual infinite and of any void.[51]

Natural philosophy included the study of meteorology, which treated not only weather phenomena but also any atmospheric[52] (or allegedly atmospheric) changes such as halos, rainbows, meteors, and (according to Aristotle) the Milky Way and comets.[53] Earthquakes, sources of rivers, the salinity of the seas, the habitability of regions of the Earth, and even the creation of metals and minerals could count as part of meteorology. The reason for such a wide-ranging array of topics is that each of these was considered to be affected or caused by one of the two exhalations: one vaporous, moist and cold from the water on the surface of the Earth, the other windy, smoky, hot, and dry from the Earth itself.[54]

49 See F. J. Ragep, *Naṣīr al-Dīn al-Ṭūsī's Memoir on Astronomy*, 2 vols. (New York: Springer-Verlag, 1993), vol. 1, pp. 44–6, 100, and vol. 2, p. 380.

50 The three levels of the soul ultimately go back to Plato but also have an important role to play in Galenic medicine.

51 Rosenthal, *Classical Heritage in Islam*, pp. 162–81; and Paul Lettinck, *Aristotle's Physics and Its Reception in the Arabic World: With an Edition of the Unpublished Parts of Ibn Bājja's Commentary on the Physics* (Aristoteles Semitico-Latinus, 7) (Leiden: Brill, 1994). Physical principles are often presented in a clear, summary fashion at the beginning of theoretical works on astronomy; see, for example, Ragep, *Naṣīr al-Dīn al-Ṭūsī's Memoir on Astronomy*, vol. 1, pp. 41–6, 98–101.

52 Here, by atmosphere is meant the region below the lunar (i.e., the lowest celestial) orb.

53 Some Islamic natural philosophers claimed that the Milky Way and comets were celestial, not atmospheric, thus challenging Aristotelian doctrine.

54 An excellent survey of Islamic meteorology is Paul Lettinck, *Aristotle's Meteorology and Its Reception in the Arab World: With an Edition and Translation of Ibn Suwār's Treatise on Meteorological Phenomena*

Natural philosophy might also include what are now regarded as the "occult sciences" – astrology, alchemy, and magic – which were thought to operate by obscure or arcane principles not easily accessible to reason and the senses.[55] Although usually considered spurious today, they did encourage empirical activity and indicate a desire to understand and control hidden aspects of nature. Astrology required astronomical observations that would permit the casting of horoscopes, making of judgments, and so forth. Alchemy involved a good bit of simple metallurgy but also accepted the possibility (predictable within Aristotelian matter theory) that metals could be converted one into another. Finally, magic is a catch-all category dealing with surprising phenomena not easily explained in the light of accepted natural philosophy and attributed to incomprehensible hidden forces. The claims of some adherents of the occult sciences, especially that they could lead to an understanding of the inner workings of the universe, would inevitably lead to opposition. Religious scholars could object that the occult sciences sought insights and powers properly reserved to God.[56] Hellenized natural philosophers, such as Ibn Sīnā, also cast doubt on the veracity of the claims made by occult practitioners and indeed sought to differentiate mathematical astronomy from astrology.[57]

DEFENDERS AND PRACTITIONERS OF NATURAL PHILOSOPHY

Having presented a broad overview of the natural philosophy tradition in Islam, we now need to examine the rather remarkable ways in which many committed Muslims, from caliphs on down, came to appropriate, accept, and Islamize this seemingly foreign body of knowledge. It has sometimes been held that the Hellenistic philosophical tradition was opposed to the essential nature of Islamic religion and Islamic civilization and therefore inevitably able to occupy only a marginal place within that civilization. But, as we have seen, there was ample patronage and intellectual justification for the translation movement from within the heart of that civilization, and a careful study of the next thousand years of this tradition reveals its intellectual power throughout all levels of Islamic society and culture.[58] Far from being

and Ibn Bājja's Commentary on the Meteorology (Aristoteles Semitico-Latinus, 10) (Leiden: Brill, 1999).

55 See "Nudjūm, Aḥkām al-" [astrology], "Siḥr" [magic], and "Kīmīyā'" [alchemy] in the *Encyclopaedia of Islam*, 2nd ed.; and Pingree, "Astrology," pp. 290–300. Ibn Khaldūn provides a good overview of the occult sciences in Islam (before launching into a strong refutation of them) in his *The Muqaddimah: An Introduction to History*, trans. F. Rosenthal, 2nd ed., 3 vols. (Princeton, N.J.: Princeton University Press, 1967), vol. 3, pp. 156–246, 258–80.

56 Ibn Khaldūn, *Muqaddimah*, vol. 3, pp. 272–80.

57 Ragep, *Naṣīr al-Dīn al-Ṭūsī's Memoir on Astronomy*, vol. 1, pp. 34–5.

58 A well-argued case against the view that the Greek tradition was marginal to Islam can be found in A. I. Sabra, "The Appropriation and Subsequent Naturalization of Greek Science in Medieval

marginal, the Hellenistic tradition was at the center of debates and intellectual life, compelling even its opponents to deal with it, sometimes by borrowing its concepts and terminology. One way of understanding the situation is to see Islamic civilization as a kind of marketplace of ideas in which it proved impossible to impose a single dogma. A wide range of ideas was tolerated, including criticism of aspects of Islam itself, and the one attempt to impose a single doctrine (about the createdness of the Qur'ān, under the caliph al-Ma'mūn) was the exception rather than the rule.[59] Widespread literacy – religiously mandated because of the need for Muslims to read the Qur'ān – and the use of cheap paper (its manufacture learned from Chinese technicians captured during the early ʿAbbāsid period)[60] led to a vigorous literary culture that allowed the Hellenistic tradition to thrive and develop.

One of the first Hellenized philosophers was Abū Yaʿqūb al-Kindī (d. ca. 870), credited with being the first Muslim philosopher and often called the "Philosopher of the Arabs." The scion of a prominent Arab tribe patronized by the caliphs al-Ma'mūn (r. 813–33) and al-Muʿtaṣim (r. 833–42), whose son al-Kindī tutored, he sought to provide legitimacy for *falsafa* among both Arabs and Muslims. A polymath who wrote on astrology, astronomy, arithmetic, geometry, medicine, and philosophy, Kindī's intellectual development was supported by the translation activity of a number of people, who became part of what has been called the "Kindī Circle." In the course of the ninth century, this circle produced translations of Aristotle's *On the Heavens*, *On the Soul*, *Meteorology*, the nineteen books of zoology, and the *Metaphysics*, and Plato's *Timaeus* (in Galen's synopsis), along with representative texts of Hermetic gnostic Platonism and Neoplatonism such as pseudo-Aristotle's *Theology* (a paraphrase of parts of Plotinus's *Enneads*), the *Liber de Causis* (influenced by Proclus's *Elements of Theology*), and the Christian Neoplatonist John Philoponus's commentary on Aristotle's *On the Soul*.

A number of notable features of the Kindī Circle are discernible in the writings of Kindī himself. Unlike most other Islamic philosophers, Kindī advocated the doctrine of creation ex nihilo, which put him at odds with Plato, Aristotle, and virtually all other Greek philosophers. There is a clear affinity here with Islamic theology (*kalām*), which sought to affirm God's ultimate and absolute power of creation. But we should not read too much into this agreement since Kindī in other ways distanced himself from the

Islam: A Preliminary Statement," *History of Science*, 25 (1987), 223–43, reprinted in A. I. Sabra, *Optics, Astronomy and Logic: Studies in Arabic Science and Philosophy*, no. 1 (Aldershot: Variorum, 1994), and in Ragep and Ragep, *Tradition, Transmission, Transformation*, pp. 3–27.

59 Here it is useful to recall that historically Islam did not have an equivalent of a pope (or even a "church") that could enforce dogma. Rulings were by way of the learned (*ʿUlamā'*), and these could only be upheld by a ruler who himself was constrained by various social and political factors. Needless to say, such a consensus changed over time, and even during a given period different dynasties and religious groups could well disagree on any of a number of matters, some fundamental.

60 Jonathan M. Bloom, *Paper before Print: The History and Impact of Paper in the Islamic World* (New Haven, Conn.: Yale University Press, 2001).

theologians, with whom he had profound disagreements; and they, of course, found in him much to disapprove of as well.[61] The Kindī Circle seems also to have developed what would become a staple of later *falsafa* – namely explanation of prophecy within the context of Aristotelian psychology. According to this explanation, prophets are able to attain truth without the theoretical inquiry of philosophy through the imaginative faculty of the soul and direct inspiration from the Active Intellect, which inhabits the lunar sphere. But that is *their* knowledge, and other humans must employ the often-plodding method of rational thought and argument. Kindī and other philosophers claimed this was what the theologians did not do since they assumed direct revelation and used rhetoric rather than logic to make their case. Only the latter, Kindī contended, offered true knowledge and understanding.[62]

Another aspect of Kindī's worldview was his defense of astrology, conceived as God's way of dispensing divine providence through the action of the celestial orbs. Although the emanationist view underlying Kindī's astral science would play a major role in subsequent articulations of Islamic philosophy, astrology itself would meet with increasing disdain, not only from theologians but also from philosophers and, in some cases, astronomers.[63]

Finally, Kindī and his circle were instrumental in establishing a role, in some ways primary, for mathematics and the mathematical sciences within Islamic investigations of the physical world. As we will see, this marked a departure from trends in the ancient world, especially late antiquity. What seems to have developed was an interest in mathematics not only for its own sake but also as a means of doing philosophy.[64] This seems to have had the further consequence of inducing Kindī to attempt a reconciliation of mathematical and physical approaches to the so-called mixed sciences. This is most clearly visible in his optical work, where his predilection for a mathematical approach did not deter him from criticizing Euclid and Ptolemy on the physics of vision or from injecting physical considerations into a "mathematical" field.[65] It is tempting to see this as part of a larger trend (beginning in the ninth century) that would seek to reconcile the mathematical and physical aspects of astronomy.

Was this a hidden, surreptitious enterprise? Certainly not. Kindī advocated an audacious program of education for his coreligionists in the philosophical sciences and wrote a very large number of expository treatises (many

61 See the discussion of *kalām* later in this chapter.

62 Fritz W. Zimmermann, "Al-Kindi," in Young, Latham, and Serjeant, *Religion, Learning and Science in the ʿAbbasid Period*, pp. 364–9, especially p. 367.

63 S. Fazzo and H. Wiesner, "Alexander of Aphrodisias in the Kindī-Circle and in al-Kindī's Cosmology," *Arabic Sciences and Philosophy*, 3 (1993), 119–53.

64 Roshdi Rashed, "Al-Kindī's Commentary on Archimedes' 'The Measurement of the Circle,'" *Arabic Sciences and Philosophy*, 3 (1993), 7–53.

65 A. I. Sabra, "Optics, Islamic," in *Dictionary of the Middle Ages*, ed. Joseph R. Strayer et al., 12 vols. (New York: Charles Scribner's Sons, 1987), vol. 9, pp. 240–7; and David Lindberg, *Theories of Vision from Al-Kindi to Kepler* (Chicago: University of Chicago Press, 1976), pp. 18–32.

quite short) to accomplish this goal.[66] Thus, far from being marginal or conceiving of themselves as persecuted outsiders, Kindī and his circle were boldly attempting to set the agenda for Islamic civilization, not only in its intellectual pursuits but also in its religious understanding of itself. In his own day, official sanction for his philosophical viewpoint, or even general acceptance, never came. Kindī was later swept up in the measures taken by the caliph al-Mutawakkil against both rationalist philosophers and theologians. Moreover, he had few immediate followers to carry on his mission; but his program was an unequivocal success in setting the stage and agenda for the philosophical and scientific traditions of Islam, which were to continue for almost a millennium after his death.

Al-Fārābī (d. 950), known as the "second teacher" (after Aristotle), marks both a continuation and a departure from al-Kindī. Associated both as a student and teacher with a group of Hellenized philosophers drawn from Christian monasteries in Syria and Mesopotamia, al-Fārābī could trace his intellectual lineage back to a branch of the school of Alexandria, whose influence, especially in logic, seems to have continued even after the Islamic conquests. Like al-Kindī, his own metaphysical works incorporate a Neoplatonic emanationist scheme, along with the Ptolemaic planetary system, to explain the relationship between the First Principle and the sublunary world. But Fārābī's system, unlike that of the Kindī Circle, displays no concern for reconciling the natural theology of the ancients (which generally rejected creation ex nihilo) with the creationist requirements of Islamic theology. This departure from Kindī's earlier attempts at accommodation does not seem accidental; Fārābī, who for most of his life was neither a part of the scribal class nor associated with a court, held philosophy to be above the various religions, whose messages were necessarily meant for the masses rather than the elite; truth for the latter came from rational demonstration, not revelation. This does not mean that Fārābī forsook Islam; it does mean that in his view the Islamic message was ultimately to be understood by philosophers rather than theologians, and it was philosophy that should inform the ideal state.

Fārābī's strictly scientific writings (as distinct from his writings on metaphysics, logic, and politics) were not as extensive as those of some of his successors. But his views on science broadly conceived, as well as the individual sciences, were to have considerable influence. In his famous *Enumeration of the Sciences,* Fārābī excludes medicine from the sciences because of its lack of demonstrative principles, distinguishes astrology ("judgments of the stars") from mathematical astronomy, and reduces *kalām* to a mere defense of prescriptive religion. His treatises on Euclid's *Elements* and Ptolemy's *Almagest* are extant, as are works critical of astrology, atomism, the vacuum, the idea of a created universe, and Galen's understanding of Aristotle's *Parts*

[66] Ibn al-Nadīm lists a staggering 242 titles; see Dodge, *Fihrist of al-Nadīm*, vol. 2, pp. 615–26.

of Animals. In all of this, we see a clear predilection for theoretical sciences that can be based on rational and unassailable First Principles. Fārābī no doubt had Platonist inclinations, but he also wrote a work entitled *Harmony between the Views of Plato and Aristotle*, in which he demonstrates high regard for both the mathematical tradition of the Platonists and the empiricism of Aristotle. These sentiments are applied in his major work on music, which combines extensive fieldwork on musical practice with an abstract, mathematical approach to the subject.[67]

Another strand of the natural philosophical tradition in Islam is represented by the unorthodox writings of Abū Bakr Muḥammad ibn Zakariyyā' al-Rāzī (ca. 854–ca. 930). Born in the Persian city of Rayy (whence his name, al-Rāzī), he became most famous as an alchemist and physician. As an alchemist, he tended to be much more empirical and naturalistic than his predecessors; as a physician, he was noted for his clinical observations and criticism of predecessors, especially Galen. In all of his work, he exhibited an extraordinary independence of mind. This is nowhere better displayed than in his philosophical writings. Breaking with the Aristotelian tradition, he argued that both absolute time and absolute space existed coeternally, along with God, soul, and matter. According to Rāzī, God is the creator of our current universe, but the latter is neither eternal, as held by Aristotle, nor created out of nothing, as advocated by most Islamic theologians. Like Plato's Creator God in the *Timaeus*, Rāzī's is constrained by the necessities of his coeternals – soul and matter. Indeed, it was the soul's insistence on experiencing the pleasures of the body that led, in Rāzī's mythopoeic account, to creation. But God also endowed humans with reason, which could impel the soul, through philosophical study, to return to its previous beatific state. Rāzī combined these Platonic ideas with atomism, a void, and self-moving bodies, all of which were required by his absolute space and time that existed independently of any corporeal body. Rāzī's revolutionary ideas, though, were never mathematized – curious for someone often described as a Platonist; indeed Rāzī seems to have had some aversion to mathematics.[68]

Rāzī was as iconoclastic in religion as in philosophy. His attack on the prophets, who, he felt, unfairly took upon themselves the role of arbiters between man and God, won him few adherents. In particular, he clashed with the Ismāʿīlīs, a powerful Shīʿa group of the time whose leader (imam)

[67] Deborah L. Black, "Al-Fārābī," in *History of Islamic Philosophy*, ed. Seyyed Hossein Nasr and Oliver Leaman (London: Routledge, 2001), pp. 178–97; Dimitri Gutas, "Fārābī and Greek Philosophy," in *Encyclopaedia Iranica*, ed. Ehsan Yarshater (New York: Bibliotheca Persica, 1999), vol. 4, pp. 219–22; and Alfred Ivry, "Al-Fārābī," in Young, Latham, and Serjeant, *Religion, Learning, and Science in the ʿAbbasid Period*, pp. 378–88. See also M. Fakhry, *A History of Islamic Philosophy*, 2nd ed. (New York: Columbia University Press, 1983), pp. 107–28.

[68] L. E. Goodman, "Al-Rāzī, Abū Bakr Muḥammad b. Zakariyyā'," in *Encyclopaedia of Islam*, 2nd ed., vol. 8, pp. 474a–477b; and L. E. Goodman, "Muḥammad ibn Zakariyyā' al-Rāzī," in Nasr and Leaman, *History of Islamic Philosophy*, pp. 198–215. Compare Fakhry, *History of Islamic Philosophy*, pp. 97–106.

was held to be the conduit of God's continuing revelations. This fit in nicely with Plato's notion of the philosopher-king, and several Ismāʿīlī groups promoted Neoplatonist viewpoints. Among these were the Ikhwān al-Ṣafāʾ (the Brethren of Purity), who lived sometime in the tenth century. Their major encyclopedic work, divided into some fifty *Epistles*, seems to have been written to further the political goals of the Ismāʿīlīs, who attained political power during that century in North Africa and Egypt. In addition to Neoplatonic philosophical and political views, the Brethren blended a curious variety of other elements – Pythagorean numerology, Hermetic mysticism and magic, and Hindu, Persian, and Christian beliefs – all situated within the framework of the Islamic religion.

According to the Brethren, God first created the celestial orbs through a series of emanations. Creation then continued, ascending from the lowest level of the four elements through the plant and animal kingdoms. This hierarchy of being had both a cosmological and a moral significance. Mankind stood at the cusp between the lower and higher worlds; lower human forms were condemned to the prison of the flesh, whereas a select few, through knowledge and moral behavior, were able to liberate themselves from the constraints of matter. Astrological considerations and numerology played an important role in the *Epistles*, predicting the rise of the prophets and imams who guide and sustain mankind. But in addition to Pythagorean mysticism, the Brethren incorporated a substantial amount of the Hellenistic mathematical tradition from Euclid to Ptolemy and were quite influential in legitimizing the Greek tradition within Shīʿa circles and perhaps among other Muslims as well; it is no coincidence that one of the greatest scientists of medieval Islam, Ibn al-Haytham, would find patronage among the Ismāʿīlī rulers of Egypt.[69]

It would be a mistake, however, to conclude that Hellenism found a home only among the Shīʿa. In fact, a number of prominent Islamic philosophers and scientists claimed that philosophy should transcend such sectarianism. Although brought up in a home with strong Ismāʿīlī leanings, Abū ʿAlī ibn Sīnā (980–1037), known as Avicenna in the Latin West, rejected these for what he considered the higher truths of philosophy. Strong claims have been made for Ibn Sīnā's importance to the history of philosophy, in particular metaphysics. Ibn Sīnā was a child prodigy who claimed understanding (if not complete mastery) of most human knowledge by the age of eighteen. In metaphysics, he formulated an approach, akin to that of the *mutakallimūn* and unlike Aristotle's, whereby the universe was contingent upon God. However, unlike the *mutakallimūn*, he claimed that the universe adhered to natural necessity and order along the lines laid out by Aristotle. This

[69] Fakhry, *History of Islamic Philosophy*, pp. 163–81; and Y. Marquet, "Ikhwān al-Ṣafāʾ," in *Encyclopaedia of Islam*, 2nd ed., vol. 3, pp. 1071a–1076b. Compare Ian Richard Netton, "The Brethren of Purity (Ikhwān al-Ṣafāʾ)," in Nasr and Leaman, *History of Islamic Philosophy*, pp. 222–30.

reconciliation of science and religion would have significant influence in Europe, where Avicenna's writings would influence such diverse figures as Aquinas and Descartes, whose famous dictum "*cogito ergo sum*" may have owed something to Ibn Sīnā's "Floating Man" argument. For our purposes, it is important to note that Ibn Sīnā marks a melding and a culmination of a number of trends in both Islam and the ancient world. As a philosopher, he carried on the Neoplatonic cosmological reworking of Aristotle that we have already noted in al-Kindī. Also following al-Kindī, Ibn Sīnā takes account of the physical and mathematical sciences in ways that fly in the face of the "Athenian–Alexandrian" split of antiquity. Although criticized for being a mediocre mathematician and astronomer by his contemporary and rival al-Bīrūnī, Ibn Sīnā saw fit to include competent and appreciative summaries of both Euclid's *Elements* and Ptolemy's *Almagest* (the latter totaling 689 pages in a modern edition) in his philosophical encyclopedia, the *Shifā'* – an act that has no true parallel among the philosophers of antiquity.[70] It is also important to recall his fame as a physician and medical writer. Ibn Sīnā's broad interest in the details of the human body, the physical world, and the mathematical sciences sets him apart from both ancient philosophers and Islamic theologians. Here perhaps we can see the importance of a metaphysics that proclaimed God's glory through his created works and the moral imperative to learn as much as possible about those works.[71]

Ibn Sīnā had an enormous influence on the subsequent course of philosophy and theology in Islam. Among the Ashʿarites, his work provoked a strong counterreaction. Specifically, during the conservative "Sunni revival" that occurred under the Saljūqs in the twelfth century, his fellow Persian Abū Ḥāmid al-Ghazālī took Ibn Sīnā to task for allegedly anti-Muslim views. These included a belief in the eternity of the world, God's lack of knowledge of particulars, denial of bodily resurrection, and the Aristotelian doctrine of an unchanging natural order and fixed causality. But although attacking Ibn Sīnā in some of his works, Ghazālī paid him the ultimate compliment by writing a book that offered a clear and objective presentation of Ibn Sīnā's philosophical views.[72]

Ibn Sīnā was also well known, and criticized, by the "Spanish Aristotelians," who spanned the twelfth century in al-Andalus (Islamic Spain). Beginning with the Saragossan philosopher Ibn Bājja (Avempace; d. 1139),

[70] One might contrast here Proclus's critical and brief summary of astronomy in his *Hypotyposis*.

[71] Lenn Goodman, *Avicenna* (New York: Routledge, 1992); and Fakhry, *History of Islamic Philosophy*, pp. 128–62. For a translation of Ibn Sīnā's autobiography, continued after his death by one of his associates, see William E. Gohlman, *The Life of Ibn Sīnā: A Critical Edition and Annotated Translation* (Albany: State University of New York Press, 1974).

[72] This is his book *The Aims of the Philosophers* (*Maqāṣid al-falāsifa*). That al-Ghazālī would seek to gain a deep knowledge of philosophy and use its very own tools of logic and rational discourse against it tells us something important about the *mutakallims* (and other religious scholars as well), who did not simply use religious authority and divine revelation to make their points. This is emphasized in Massimo Campanini, "Al-Ghazzālī," in Nasr and Leaman, *History of Islamic Philosophy*, pp. 258–74.

continuing through Ibn Ṭufayl (d. 1185/1186), and culminating with his protégé, the redoubtable judge and Aristotelian commentator Ibn Rushd (Averroes; 1126–1198), these Andalusian scholars sought to define a philosophical position that was often in opposition to both the philosophers and the theologians of eastern Islam. Generally speaking, the tendency was to return to a purer Aristotelian philosophy, stripped of Neoplatonic and Avicennian accretions. This is particularly true of Ibn Rushd; rejecting the transcendent Ineffable One of the Neoplatonists as well as Avicenna's Necessary Being, he insisted that God should be considered, following Aristotle, part as well as cause of the universe.[73]

This "return to Aristotle" extended to the sciences as well. Ibn Bājja, Ibn Ṭufayl, Ibn Rushd, and the Jewish philosopher-theologian Maimonides (who was born in Cordoba but later moved to Egypt) all objected to Ptolemaic astronomy, with its eccentrics and epicycles, which they saw as having moved away from the pure Aristotelian astronomy of concentric spheres carrying the celestial bodies about the central Earth. Although these philosophers lamented their lack of time required to set astronomy back on its true foundations, another Andalusian, al-Biṭrūjī (fl. 1185–92), undertook the research project they set forth.[74]

The Spanish Aristotelians were also noteworthy for attempts to deal with the tensions between Islam and the science and philosophy of the ancients. Both Ibn Bājja and Ibn Ṭufayl wrote works exploring the relationship of the philosopher to the Islamic state. The latter did this within a fascinating romance in which a child, growing up alone on an island, discovers on his own the truths of ancient philosophy. Given the chance to spread these on a neighboring island, which strongly resembles an Islamic society, he finds it impossible to convince the inhabitants to accept the harmony of their religion and philosophical truth. This pessimistic assessment echoes that of Ibn Bājja, who compared philosophers to "weeds."[75]

Ibn Rushd, however, as a judge and follower of Aristotle, believed that he could and should show how the Islamic religion and Greek philosophy could be reconciled. In several remarkable works, he confronted Ghazālī's charges of heresy leveled against the philosophers and upheld the right and duty of philosophers, but not the general public, to explore philosophical questions and seek to reinterpret the religious texts when they were in conflict with philosophical conclusions. He also sought to undermine the contentions of Ghazālī and other theologians who objected to the Aristotelian notions

73 R. Arnaldez, "Ibn Rushd," in *Encyclopaedia of Islam*, 2nd ed., vol. 3, pp. 909b–920a; and Dominique Urvoy, "Ibn Rushd," in Nasr and Leaman, *History of Islamic Philosophy*, pp. 330–45.

74 For details, see Morrison, Chapter 4, this volume. Compare A. I. Sabra, "The Andalusian Revolt Against Ptolemaic Astronomy: Averroes and al-Biṭrūjī," in *Transformation and Tradition in the Sciences*, ed. Everett Mendelsohn (Cambridge: Cambridge University Press, 1984), pp. 133–53.

75 Lenn E. Goodman, "Ibn Bājjah" and "Ibn Ṭufayl," in Nasr and Leaman, *History of Islamic Philosophy*, pp. 294–312, 313–29.

of an independent natural order and fixed causality. However, his defense of Aristotle was not simply a matter of answering the objections of the theologians. He also undertook a vast project, probably with the encouragement of the Almohad rulers, of writing summaries as well as commentaries (both "medium" and "large") on the entire Aristotelian corpus, including the works on natural philosophy. Subsequently translated into Latin, these had an enormous influence on European scholarship and teaching for some five centuries, but, interestingly, little impact on Islamic philosophy.[76] The later Tunisian historian Ibn Khaldūn (1332–1406), for example, was not swayed by Ibn Rushd's arguments and adopted a negative stance toward ancient philosophy and some of the sciences in his monumental work *al-Muqaddima*.[77]

In the millennium after Ibn Sīnā's death, the eastern regions of Islam saw a different philosophical turn. Although denounced in some quarters, the philosophical tradition that Ibn Sīnā had been so instrumental in establishing was championed by others. Several important scholars in Syria, Iraq, and Iran defended his work, including Naṣīr al-Dīn al-Ṭūsī (thirteenth century) and others associated with the famous Marāgha observatory in northwestern Persia. Although more research needs to be undertaken, it is clear that the terminology and tools of philosophy (if not its doctrines) were increasingly accepted during this period, as we see from the teaching of Aristotelian logic in almost all the major *madrasas* and the incorporation of aspects of philosophy into both *kalām* and Sufism. Another important tradition includes "illuminationism," a mystically inspired philosophy that reached its zenith during the Safavid period in Persia. The relationship of these developments to natural philosophy remains to be investigated.[78]

THE THEOLOGICAL (*KALĀM*) APPROACH TO THE PHENOMENAL WORLD

The natural philosophy movement inherited from the Hellenistic tradition represents only one Islamic approach for investigating the physical world. Islamic religious scholars, theologians (*mutakallims*) in particular,[79]

[76] Arnaldez, "Ibn Rushd"; and Urvoy, "Ibn Rushd." For his influence, or lack thereof, see S. Harvey, "Conspicuous by His Absence: Averroes' Place Today as an Interpreter of Aristotle," in *Averroes and the Aristotelian Tradition*, ed. G. Endress and J. A. Aertsen (Leiden: Brill, 1999), pp. 32–49.

[77] Abderrahmane Lakhsassi, "Ibn Khaldūn," in Nasr and Leaman, *History of Islamic Philosophy*, pp. 350–64.

[78] Dimitri Gutas, "The Heritage of Avicenna: The Golden Age of Arabic Philosophy, 1000–ca. 1350," in *Avicenna and His Heritage: Acts of the International Colloquium, Leuven – Louvain-La-Neuve September 8–September 11, 1999*, ed. Jules Janssens and Daniel De Smet (Leuven: Leuven University Press, 2002), pp. 81–97. Compare the various articles in Sections IV and V ("Philosophy and the Mystical Tradition" and "Later Islamic Philosophy") in Nasr and Leaman, *History of Islamic Philosophy*, pp. 365–670, especially the articles by Hossein Ziai.

[79] There is much controversy about whether one should translate *kalām* as theology and *mutakallim* as theologian, in part because *kalām* deals with matters not often taken to be theological (such as the

also undertook to understand the world in which we live. But their insistence that God had created the world from nothing, that God had the power to create or destroy this world at will, and that consequently the world was totally contingent upon His will generally set them apart from their natural philosophy coreligionists who held that nature represented an independent realm in which events are governed by a uniform and unbreakable set of rules. This view, which was an invention of Greek thinkers, having evolved over several centuries from the pre-Socratics to Aristotle, was challenged by Islamic theologians as well as by their Christian and Jewish counterparts. For those in the Abrahamic tradition, the independent "Nature" of Hellenistic philosophy seemed to limit the action of the omnipotent God of the monotheistic religions, whose freedom to perform miracles was incompatible with an unchanging and unchangeable natural order. It was also challenged because it was often (though not necessarily) tied to Aristotle's eternity of the world – thus seeming to deny the ultimate miracle of God's creation – and also because philosophy seemed to sanction astrology, which gave to the stars power that was beyond divine control.

The earliest Islamic theologians known to us are the Muʿtazilites. Although their origins remain obscure, the Muʿtazilites found a social function as spokesmen for the egalitarian and proselytizing Islam of the ʿAbbāsid state. The egalitarian ideology of both the state and the theologians seems to be tied to the strongly rationalist tendencies of the Muʿtazilites. According to them, Muslims were individually responsible for establishing rationally their own religious beliefs, which could not be accepted simply on faith. By the same token, Muslims were called upon to use rational means to understand the physical world created by God. This line of argument, as well as the early intellectual exchanges and interfaith dialogues promoted by the ʿAbbāsid family, gave rise to widely divergent views toward cosmological as well as more strictly religious matters.[80]

At a later stage, beginning sometime late in the ninth century and continuing until the middle of the tenth, Muʿtazilism developed a coherent body of doctrine regarding the physical world, much of which continued to exercise its influence on Islamic theologians even after the waning of Muʿtazilism in the eleventh century. Included among these doctrines were the following: God is the Creator of the world; the world is composed of atoms and their inherent qualities; accidental qualities (such as perceptible colors, imperceptible motions, or psychological states) combine with atoms

physical world) and in part because it insists, at least in its earliest stages, on using rational methods to establish religious truths. But neither was the Christian theology of the Middle Ages limited to discussions of God, and many of its practitioners insisted upon rational methods. Given that there is also good evidence of the influence of *kalām*, either directly or indirectly, upon a number of medieval Christian theologians, theology and theologian will be used interchangeably with *kalām* or *mutakallim*.

80 D. Gimaret, "Muʿtazila," in *Encyclopaedia of Islam*, 2nd ed., vol. 7, pp. 783a–793b.

to form inanimate bodies and animate beings; causation results only from volitional beings, God's volition being unlimited, whereas man's, though circumscribed, is critical since it allows for individual accountability; a vacuum exists in which the atoms can move; and time is also atomistic in character.[81]

What was the significance of these Muʿtazilite doctrines? First, the primary role that the Muʿtazilites assigned to reason, to establish both religious truth and knowledge about the physical world, marks their endeavor as religious philosophy rather than simple apologetics. The continuing debates, carrying over into subsequent centuries, are sufficient to reveal that no single enforceable dogma emerged, despite widespread agreement over fundamental principles. Second, the use of reason to establish religious truth was the *obligation* not just of the theologians or the elite but of all Muslims. Thus, as we have seen, Muʿtazalism emerged as a political and social movement as well as a religious one. Third, Muʿtazilite atomism, with its discontinuities in both space and time, provided a way for God to intervene in the world (so that miracles could occur) and thus preserve His omnipotence and absolute volition. This use of atomism represented a way to escape from its ancient association with materialist and atheistic doctrines that led both Aristotle and the early church fathers to oppose it.[82] Finally, it is clear that Muʿtazilite views regarding the physical world were substantially opposed to Aristotle's conception of "natures," which exist independently and function as "causes."

The Muʿtazilite doctrines would not go unchallenged, however, and would encounter substantial opposition, both politically and intellectually, that would lead to their demise by the eleventh century. One challenge came from the "traditionist" opponents of the Muʿtazilites such as Aḥmad ibn Ḥanbal (d. 855), who took a more literal approach to the Qurʾān and *ḥadith* (the "traditions") and looked askance at the rationalizing imperatives of the Muʿtazilites. Another challenge to Muʿtazilism came from an internal schism initiated by Abū al-Ḥasan al-Ashʿarī (d. 935). Al-Ashʿarī may be viewed as representative of an attempt to reconcile rationalist theology with its traditionist opposition. In particular, he retained much of the subject matter, rationalist methodology, and cosmology of the Muʿtazilites, while renouncing many of their characteristic positions. In particular, al-Ashʿarī affirmed that the Qurʾān is eternal, abandoning the Muʾtazilite view that it was created; insisted on God's absolute freedom of action, thereby opposing the Muʿtazilite view that God always acts according to human rationalist ideas of justice; denied the independence of human volition, preferring instead to see God as giving man the power to commit each individual action; and asserted, more forcefully than the Muʿtazilites had, that God was the

[81] Alnoor Dhanani, *The Physical Theory of Kalām: Atoms, Space, and Void in Basrian Muʿtazilī Cosmology* (Leiden: Brill, 1994).

[82] For the early church opposition to atomism and its surprising adoption in early *kalām*, see Alnoor Dhanani, "Kalām Atoms and Epicurean Minimal Parts," in Ragep and Ragep, *Tradition, Transmission, Transformation*, pp. 157–71.

direct cause of *every* action in the world. More than any other doctrine, this denial of natural causality (often called occasionalism) would distinguish the Ashʿarite position from that of the Islamic Hellenistic philosophers. Although the positions of the Ashʿarites might seem congenial to the traditionists, their continuing program of explaining points of doctrine through the terminology and methodological argumentation of rationalist *kalām* was met with hostile suspicion by conservative groups such as the Ḥanbalites.[83]

By the eleventh century, the Ashʿarites gained political and intellectual influence under the Saljūqs and found a home in the newly established *madrasas*. These schools were originally set up for the teaching of religious subjects and for countering the doctrines of political enemies of the Saljūqs, especially the Ismāʿīlī Shīʿa, who were well established in Egypt under the Fatimids and making inroads in Syria, Iraq, and Iran.[84]

One of the most important figures in the history of Islamic thought, Abū Ḥāmid al-Ghazālī (1058–1111) was a law professor in one of these *madrasas*. Ghazālī was also a major Ashʿarite theologian, who at one point gave up his high position to live the life of a poor Sufi mystic. His importance in the development of *kalām* is twofold and, at first sight, contradictory. On the one hand, he forcefully denounced various aspects of Hellenistic philosophy, especially the views of his fellow Persian Ibn Sīnā (d. 1037), in his *Tahāfut al-falāsifa* (*Disintegration of the Philosophers*). On the other, he explicitly endorsed the legitimation of certain parts of the ancient heritage, such as logic, medicine, and arithmetic, that might benefit the religious sciences and the community. Ghazālī's opposition to Aristotelian philosophy was directed toward the eternity of the world, corporeal resurrection, God's knowledge of particulars, prophecy, and natural causation.[85] But no less important than this opposition to specific views of the philosophers was his role in furthering the acceptance of certain aspects of Aristotelianism within *kalām*, including the absorption of dialectical syllogism as the methodological mainstay of legal and theological argumentation, and the adoption of philosophical terminology in theological treatises.[86]

The majority of later Ashʿarī treatises, beginning with the rise of the theological "manuals" of al-Ījī (d. 1355) and al-Taftazānī (d. 1389) and continuing with commentaries, glosses, and superglosses on these manuals, paid due attention to the epistemology, methodology, and topics of Aristotelian philosophy as it developed in the Islamic tradition. Such manuals, composed in

83 James Pavlin, "Sunni *Kalām* and Theological Controversies," in Nasr and Leaman, *History of Islamic Philosophy*, pp. 105–18; and George Makdisi, "The Sunnī Revival," in *Islamic Civilization, 950–1150*, ed. D. H. Richards (Oxford: Bruno Cassirer, 1973), pp. 155–68. For Islamic occasionalism, see Fakhry, *History of Islamic Philosophy*, pp. 209–33.

84 George Makdisi, *The Rise of Colleges: Institutions of Learning in Islam and the West* (Edinburgh: Edinburgh University Press, 1981), especially pp. 23–4, 41, 54, 301–4, 306–7, 311.

85 This is also true in the writings of later theologians such as Tāj al-Dīn al-Shahrastānī (d. 1153) and Fakhr al-Dīn al-Rāzī (d. 1209).

86 Campanini, "Al-Ghazzālī," pp. 258–74.

the course of the next five centuries, included introductory essays on natural philosophy (and even expositions on Ptolemaic astronomy) that adopted much terminology and methodology from the philosophers while seeking to refute them. This is a strong indication of the influence of ancient science and philosophy upon mainstream Islamic thought – a point to which we will return.[87]

To conclude this discussion of *kalām*, it is worth noting that although Muʿtazilite and Ashʿarite *kalām* originally arose within a Sunni milieu, it would later cut across sectarian divisions and even move beyond Islam. Among the Shīʿa, the Zaydīs adopted Muʿtazilite *kalām*, as did the early Twelver (Imāmī) Shīʿa.[88] Later, the eminent thirteenth-century philosopher-scientist Naṣīr al-Dīn al-Ṭūsī would establish Twelver *kalām* on the basis of Hellenistic philosophy. *Kalām* also had a profound effect upon Jewish intellectual history; the Karaites, a Jewish sect, were influenced by the Muʿtazilites, whereas Maimonides sought to discredit the *mutakallims* in his *Guide for the Perplexed*.

TRANSFORMATIONS AND INNOVATIONS IN ISLAMIC NATURAL PHILOSOPHY

The natural philosophy tradition in Islam underwent a number of important changes over the course of a millennium. The extent of the Islamic empire made it natural that in areas of medicine, pharmacology, zoology, botany, and so forth, new knowledge from a variety of civilizations would be added to the inheritance from Greek sources. Innovation also occurred in the mathematical, astronomical, and optical traditions (matters dealt with elsewhere in this volume). Although research into these developments is still in its infancy, it is fairly clear that many of them were the result of the transplantation of ancient ideas into a new environment, as well as interactions among the various groups seeking to explain the physical world. This concluding section will discuss some of the important developments and innovations that occurred within this scientific tradition, viewing them within the rich diversity of Islamic civilization.

An important theme running through these innovations is what we might call novelty out of diversity. For example, the development of trigonometry in its modern form occurred when Islamic mathematicians replaced the clumsy chord function of Greek astronomy with the Indian sine function, subsequently developing the five other trigonometric functions and their

87 See A. I. Sabra's discussion of "naturalization" in his "Appropriation and Subsequent Naturalization of Greek Science in Medieval Islam."

88 The Twelver, or Imāmī, Shīʿa are today a majority in Iran and represent the largest of the Shīʿa groups worldwide.

relationships from this model. In the end, one had something new, the discipline of trigonometry, in which one had not only handier tools for solving a variety of astronomical (and other) problems but also a new mathematical field that could be developed in its own right. What makes this example particularly interesting is that the impetus behind these developments was religious – the need to find the direction of the Kaaba in Mecca (which must be faced during prayer) and tables for the five astronomically determined prayer times.

Many developments in natural philosophy can be understood as the result of reactions to pre-Islamic ideas, as well as interactions among different groups of Islamic thinkers as they dealt with this ancient heritage. One example is Ibn Sīnā's advocacy of the notion of *mayl*, or impressed force, to replace Aristotle's explanation of projectile motion after the moving object leaves the hand of the thrower. Aristotle, who held that motion needed a mover in direct contact with the thing moved, had to resort to the less than satisfactory explanation that the ambient air somehow kept the object moving. The pre-Islamic Christian philosopher-theologian John Philoponus (ca. 490–570) opposed this view with his theory of impressed force, by which an incorporeal moving force, impressed on the projectile by the mover, moves the projectile until the impressed force is exhausted over the course of the movement.[89] Philoponus also claimed that the natural rectilinear motions of the elements, as well as the uniform circular motions in the heavens, were the result of an impressed force. These views clearly diverged from Aristotle in a number of ways, in that motion could occur without direct contact between mover and moved, the causes of natural motion in the heavens and below the Moon's orb were of the same kind, and the exhaustibility of the impressed force of celestial motion implied that the world was not eternal.[90]

Ibn Sīnā's theory of *mayl* is close to Philoponus's impressed force, but refined in a number of ways. Ibn Sīnā makes it dependent upon the "mass" (quantity of matter) of a body – a stone can receive more *mayl* than a feather. More importantly, *mayl* does not weaken unless there are resisting counterforces; thus, in theory, a body would continue to move indefinitely in a void (if such existed, which Ibn Sīnā denies) or in the resistance-free celestial realm. Unlike Philoponus, Ibn Sīnā does not seem bothered by an eternally moving celestial realm; as we have seen, his metaphysics allowed for both God's creation and an eternal world based upon natural principles. Among those influenced by Ibn Sīnā on this point was the Jewish philosopher (later a convert to Islam) Abū al-Barakāt al-Baghdādī (d. after 1164–1165),

89 See R. R. K. Sorabji, ed., *Philoponus and the Rejection of Aristotelian Science* (London: Duckworth, 1987); and C. Wildberg, "Impetus Theory and the Hermeneutics of Science in Simplicius and Philoponus," *Hyperboreus*, 5 (1999), 107–24. Compare C. Wildberg, "John Philoponus," in *The Stanford Encyclopedia of Philosophy* (fall 2003 ed.), ed. Edward N. Zalta, http://plato.stanford.edu/archives/fall2003/entries/philoponus.

90 Lettinck, *Aristotle's Physics and Its Reception in the Arabic World*, pp. 337, 665.

who also held to a version of *mayl* in which, like Philoponus's, the impressed force is depleted in the process of motion.

Abū al-Barakāt is also noted for a number of other innovative physical ideas. He used the idea of *mayl qasrī* (violent *mayl* or violent impressed force) to explain the acceleration of falling bodies. Since weight causes natural *mayl* according to Abū al-Barakāt, it continuously adds *mayl* as a body falls, thus accounting for its acceleration. He advocated, as had Muḥammad ibn Zakariyyā' al-Rāzī, the notion, contra Aristotle, of a three-dimensional empty, infinite space. He also rejected the widely held view that time is a measure of movement, holding instead that it is prior to and independent of movement.[91]

Anti-Aristotelian physical theories were also put forth by Thābit ibn Qurra, who rejected the notion of "natural place" and instead advocated the remarkable idea that similar bodies come together because of mutual attraction.[92] The twelfth-century Andalusian philosopher Ibn Bājja also challenged Aristotle, disputing his view that the velocity of a moving body is inversely proportional to the resistance of the medium in which motion occurs – an Aristotelian idea that ruled out motion in a void since, in the absence of resistance, velocity must be infinitely large. Instead, Ibn Bājja maintained that motion in a void, if it existed, would be finite, and that the resistance of the medium would simply reduce that base velocity.[93] These physical ideas of Ibn Bājja, Ibn Sīnā, and other Islamic thinkers would exert a major influence on Latin scholastic authors such as Thomas Aquinas and Jean Buridan.[94] Most historians now agree that such innovations were an important step between Greek physical views and the new physical theories (such as those put forth by Galileo) in the early-modern period.

Important reforms of ancient theories grew out of the desire to reconcile the doctrines of natural philosophy with mathematical approaches to natural phenomena. In the ancient world, mathematicians and natural philosophers frequently adopted distinctive approaches. For example, Aristotle's cosmological views were based upon a priori physical and metaphysical principles from which his universe could be deduced. By contrast, Ptolemy claimed to

91 Shlomo Pines, "Abu'l-Barakāt al-Baghdādī, Hibat Allāh," *Dictionary of Scientific Biography*, I, 26–8.

92 A. I. Sabra, "Thābit ibn Qurra on the Infinite and other Puzzles: Edition and Translation of His Discussions with Ibn Usayyid," *Zeitschrift für Geschichte der Arabisch-Islamischen Wissenschaften*, 11 (1997), 1–33.

93 Lettinck, *Aristotle's Physics and Its Reception in the Arabic World*, p. 8.

94 Ibid., Epilogue ("The Influence of Arabic Philosophers on the Development of Dynamics in the Middle Ages"), pp. 665–73. The question of whether the scholastics knew of impetus theory through Islamic sources has not been settled but seems to be a reasonable assumption given that Philoponus's commentaries were unknown and there are strong parallels in Buridan with passages from Ibn Sīnā. Compare A. Sayılı, "Ibn Sīnā and Buridan on the Motion of the Projectile," in *From Deferent to Equant: A Volume of Studies in the History of Science in the Ancient and Medieval Near East in Honor of E. S. Kennedy*, ed. David A. King and George Saliba (New York: New York Academy of Sciences, 1987), pp. 477–82. Ibn Bājja's views were not known directly but were disseminated through the writings of his countryman Ibn Rushd, who was critical of many of Ibn Bājja's anti-Aristotelian views.

base his cosmology upon mathematical principles and observations. That, roughly speaking, they came to similar conclusions about the structure of the universe was no doubt comforting. In other cases, such as the question of how vision occurs, the "intromissionist" views of the physicists were quite distinct from the "extramissionist" theories of the mathematicians. This split can conveniently be given a geographical designation since many of the mathematicians worked in Alexandria, whereas the natural philosophers found homes in the schools of Athens. That Islam inherited both made the project of reconciliation irresistible, and some of the results are impressive indeed. Ibn al-Haytham's revolution in optics and visual theory, the attempts to redefine the idea of number, the development of a long-term project to reform the Ptolemaic system, and more came out of the process of dealing with the many contradictions and unsolved problems of physics versus mathematics inherited from the ancients.

One can also detect cases in which the Islamic religion played a role in the transformation of ancient science. We have already seen the example of trigonometry. There were also other, more subtle, ways in which religion affected the course of science in Islam. The theologians, as noted, objected to any number of Aristotelian metaphysical and physical doctrines. In particular, they opposed any ideas that seemed to undermine God's absolute power (such as natural causes, whether terrestrial or celestial). Thus they condemned astrology, and it is interesting that, beginning with Ibn Sīnā and almost universally after him, Islamic astronomers and philosophers of science, unlike their ancient predecessors, separated astrology from mathematical astronomy. There was also a trend toward making mathematical astronomy as free as possible from Aristotelian natural philosophy. We see this in theologians such as al-Ghazālī, who rejected the philosophical doctrines of Aristotle but held that the mathematical sciences, based upon true principles, were unassailable. Some mathematical scientists, especially astronomers, began to search for ways of establishing their discipline without recourse to the philosophers. This radical notion even led one fifteenth-century scientist, ʿAlī al-Qūshjī, to propose that astronomy could dispense with Aristotelian natural philosophy altogether, including the doctrine that the Earth is at rest.[95]

That Qūshjī's proposal was made within the context of a theological commentary should alert us to the ways in which the sciences and theology came to be intertwined in later centuries. It was not unusual to find, especially after the twelfth century, theological texts with long discussions of natural philosophy and even Ptolemaic astronomy. There is also strong evidence that the teaching of arithmetic, astronomy, logic, and perhaps other "ancient sciences" came to be staples in the curricula of the *madrasas*. It may seem odd

95 F. Jamil Ragep, "Ṭūsī and Copernicus: The Earth's Motion in Context," *Science in Context*, 14 (2001), 145–63.

that institutions originally established to propagate Sunni orthodoxy against various Hellenistic and Shīʿī doctrines should succumb to teaching some of the very subjects it was established to counteract. The key to understanding this is twofold. On the one hand, many of these sciences had found a useful handmaiden role within the orthodox Islamic world. Science in the service of Islam included algebra for inheritance problems, astronomy for religious ritual, and logic to counter heresy.[96] But, more significantly, in the later centuries scientists who often combined the roles of Hellenistic thinker and Islamic theologian found in science a means to glorify God.[97]

96 A. I. Sabra has called this "naturalization." See Sabra, "Appropriation and Subsequent Naturalization of Greek Science in Medieval Islam."

97 See Ragep, "Freeing Astronomy from Philosophy."

2

ISLAMIC MATHEMATICS

J. L. Berggren

SOURCES OF ISLAMIC MATHEMATICS

When Muḥammad fled from Mecca to Medina in 622, the beginning of the Muslim (*hijri*) era, the Arabs were neither highly numerate nor aware of the mathematical traditions around them. Yet, within two centuries of that date, some of their descendants, along with those of peoples they had converted or conquered, had begun the appropriation of ancient mathematics and laid the foundation of a medieval Islamic mathematical tradition.

Nor was the appropriation of the ancient mathematical heritage a marginal activity, but rather one supported by rulers and encouraged by wealthy patrons. During the reign of al-Mansūr (who founded Baghdad in 762, on a date determined by astrologers), an astronomical work by Brahmagupta was, along with other Indian astronomical material, translated from Sanskrit to Arabic and circulated as the *Sindhind*. This handbook contained the use of the sine function and introduced this fundamental feature of trigonometry to Arabic mathematical literature.

Al-Mansūr's son, Harūn al-Rashīd, whose reign was immortalized in the *Thousand and One Nights*, built a library in Baghdad to contain the growing production of the translators, which included an Arabic version of Euclid's *Elements*, and the third ʿAbbāsid caliph, al-Ma'mūn (r. 813–33), built the House of Wisdom in Baghdad and staffed it with translators and scribes, who produced translations from Greek, Pahlavi, Sanskrit, and Syriac. Al-Ma'mūn was an adherent of the Muʿtazilite sect in Islam, a rationalizing group friendly to the foreign sciences (or "sciences of the ancients," as the sciences taken over from the Greeks were called). Furthering these efforts were wealthy families, such as the Banū Mūsa and the Barmakids, who provided personal involvement and/or financial support for mathematical sciences.

Thus, the Banū Mūsa, three brothers who were active between 850 and 900, found, on one of their trips in search of Greek manuscripts, an able translator in Harran (the modern Turkish Diyar bakir) named Thābit ibn Qurra (836–901). Working with these patrons, Thābit either commenced or improved on translations of basic works of Euclid, Archimedes, Apollonius, and other Greek mathematicians.

The early translations were often difficult to understand, partly because Arabic began with virtually no technical mathematical vocabulary, and the earliest translations sometimes had recourse simply to transliterating obscure Greek words and letting the reader struggle with the meaning. Gradually, however, the resources of Arabic word-formation came into play, and an Arabic mathematical vocabulary was created in the course of translation. Another reason for the initial difficulties was that a good translation demands a thorough understanding of the material, and so there was a dynamical interplay between the development of good translations and mathematical understanding. For example, Euclid's *Elements* was twice translated into Arabic, first in the early ninth century by al-Hajjaj and later by Ishaq ibn Hunayn (d. about 910), the latter a translation that Thābit revised. Euclid's *Elements* was of fundamental importance for the history of medieval geometry, and its centuries-long tradition of study notes, commentary, critique, improvement, and extension testifies to medieval use of that work as a basis both of instruction for the study and investigation of higher geometry or astronomy and for raising metamathematical issues.

Of course, neither Euclid's *Elements* nor other parts of mathematics beyond elementary reckoning were taught in the elementary schools associated with the mosque, where instruction was virtually entirely preparatory to the training of pious Muslims in Arabic grammar and the basics of Qur'ān reading. Nor were they generally taught in *madrasas*, institutions founded for the advanced study of Qur'ānic law and exegesis.[1] To learn advanced mathematics, one hired a tutor. Ibn Sīna tells how, in the late tenth century, his father hired a tutor to teach him arithmetic, and how, in Bukhara, evidently around the age of fourteen, he studied the early part of Euclid with a tutor and the later part of the work by himself. Al-Samaw'al names the tutors with whom he studied the early books of the *Elements* in twelfth-century Baghdad but goes on to tell us that he could find no one able to tutor him in the later books. The use of tutors for mathematics was sufficiently common to give rise to stories like that of an itinerant scholar who heard a tutor making mistakes in his instruction on Euclid. When the itinerant scholar interrupted the lesson to correct the errors, he was hired instead!

Having learned the *Elements*, one could then move on to a corpus known as the Middle Books, which were studied after Euclid's *Elements* and before Ptolemy's great second-century astronomical text, the *Almagest*. This corpus included Euclid's *Data* and his *Phaenomena*, the former intended to train

[1] See the remarks on mathematical instruction in the *madrasas*.

the student in the solution of advanced geometrical problems and the latter to provide instruction on the geometry of the sphere.

Another part of this corpus was Archimedes' *Measurement of the Circle*, which, together with his *Sphere and Cylinder*, Books I and II, constituted the only part of the present Archimedean canon known in medieval Islam. But these two works provided mathematicians of the late ninth and tenth centuries with sufficient inspiration and instruction to author a number of original studies of areas and volumes. Additional, at least partially pseudonymous, material bearing Archimedes' name, such as *The Division of the Circle into Seven Equal Parts*, was also translated and inspired a number of medieval studies.

In the case of Apollonius, the translation of his *Conics*,[2] with its extensive treatment of the parabola, ellipse, and hyperbola, provided Muslim geometers with their basic text on the geometry of the conic sections. The same geometers found further sources for their study of advanced geometry in the translations of most of his other geometrical works. Many of them, unlike the *Conics*, were included among the Middle Books.

All of this ninth-century activity took place within a learned environment in which one could find Christians, Jews, and Zoroastrians able to explain aspects of various ancient texts to their Muslim patrons and collaborators. For example, Thābit belonged to a star-worshipping cult that the Muslims called Sabians; the engineer/algebraist Sanad ibn ʾAlī was a Jew; and Qusta ibn Lūqa, who translated the *Arithmetica* of Diophantus, was a Christian.

MATHEMATICS AND ISLAMIC SOCIETY

Like any other society with complex administrative and commercial institutions, the kingdoms of medieval Islam needed the mathematics of computation and measurement, and three authors wrote works that represent literary versions of this material at a high level. The first was al-Khwārizmī, a member of al-Maʾmūn's House of Wisdom in the early ninth century, who wrote a work entitled *The Book of Addition and Subtraction According to Hindu Calculation*[3] that treated arithmetical computation. In this book, al-Khwārizmī explains the base-10 place-value system, including the extraction of square roots, and, since decimals (e.g., .17, 3.14, 1.412) had not yet

2 The Banū Mūsa relate, in their preface, how they struggled with the translation until they found the commentary of Eutocius, which aided them immensely. As with so many of the translations, theirs of the *Conics* was improved by the work of Thābit ibn Qurra. Details of this fascinating story may be found in Gerald J. Toomer, *Apollonius, Conics Books V–VII: The Arabic Translation of the Lost Greek Original in the Version of the Banū Mūsā* (New York: Springer, 1990). See also Jan P. Hogendijk, "Arabic Traces of Lost Works of Apollonius," *Archive for History of Exact Sciences*, 35 (1986), 187–253.

3 For translations of material from this work, and from many others cited in this chapter, see V. Katz, ed., *Mathematics of Egypt, Mesopotamia, China, India and Islam: A Source Book* (Princeton, N.J.: Princeton University Press, 2007).

been invented, he also explains the use of fractions based on sixtieths, exactly as we use minutes and seconds for fractions of units of time and angles. Al-Khwārizmī's book, in its Latin translation, was an important source for the West's knowledge of Hindu-Arabic arithmetic, and from a Latin version of his name ("*algorismi*") comes the modern word "algorithm."

Quite a different system of computation was explained by Abu'l-Wafa' al-Būzjanī, the author of the second pair of practically oriented books. One is the *Book of Arithmetic Needed by the Secretary and Official*, based on a traditional system of mental arithmetic in which the practitioner stored the results of intermediate calculations on his fingers. In this system, the mind and the hand were the essential elements, and writing intervened, if at all, only to record the final results. Its fractions were based on either the above-mentioned base-60 fractions or sums of products of unit fractions (so that $^4/_7$ would be represented as $^1/_2 + {}^1/_2 \cdot {}^1/_7$). Abu'l-Wafa's second work aimed at practitioners was the *Geometric Constructions Necessary for the Artisan*, which included, in addition to much mensurational material, a variety of construction problems, including those of constructing figures using a compass with fixed, rather than variable, radius, now referred to as the "rusty compass."

Al-Khwārizmī wrote around 825, and Abu'l-Wafa' around 975. About 450 years later, one finds the crown of Islamic arithmetics in *The Calculators' Key*, written by the Persian Jamshīd al-Kāshī and dedicated to the Timurid sultan Ulug Begh of Samarqand. Here al-Kāshī explains computation not only with whole numbers but also with decimal fractions, which Muslim mathematicians had invented in the tenth century.

However, beyond the ordinary needs of civil and military activity typical of any complex society, Islamic society posed other challenging mathematical problems, notably that of finding the *qibla* (direction) of Mecca, toward which all Muslims were required to face when saying their five daily prayers. To find the exact direction of a locality not visible on one's horizon is a difficult mathematical problem, and from the time of al-Khwārizmī to that of al-Kāshī it elicited a remarkable number of solutions.

One type consisted of geometrical solutions, which ranged from approximate constructions involving only a plane right triangle to exact solutions employing descriptive geometry or spherical trigonometry. Such geometrical solutions served as a basis for producing a second type of solution: numerical tables that could contain as many as 3,000 entries, of which certain entries were computed exactly, whereas the remaining entries were interpolated according to schemes of varying sophistication.[4]

[4] A useful survey of such schemes may be found in Javad Hamadanizadeh, "A Survey of Medieval Islamic Interpolation Schemes," in *From Deferent to Equant: A Volume of Studies in the History of Science in the Ancient and Medieval Near East in Honor of E. S. Kennedy*, ed. David A. King and George Saliba (Annals of The New York Academy of Sciences, 500) (New York: New York Academy of Sciences, 1987), pp. 143–52.

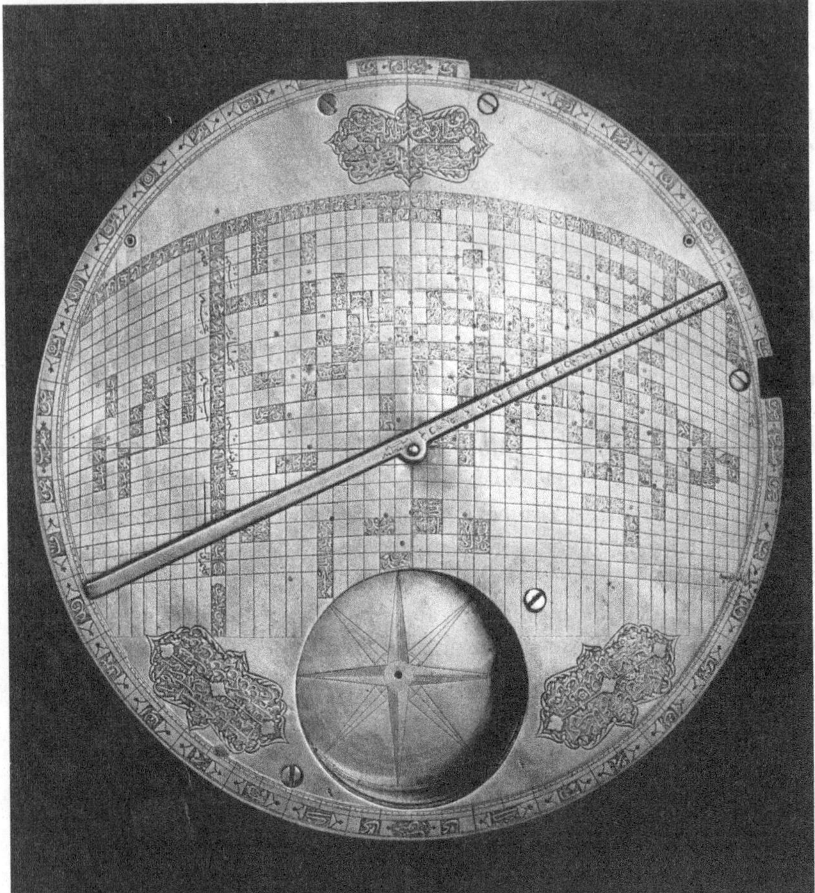

Figure 2.1. A map of the medieval Islamic world, centered on Mecca. When the ruler is rotated to pass through a given cell in the grid, the direction of Mecca from the locality named in the cell can be read off from the rim of the disk. The author thanks Prof. David King for permission to use this photo from his collection.

A third type of solution that has come to light is embodied in a mathematically constructed world map based on Mecca and incised on a large disk with a rotatable ruler (Figure 2.1). The vertical straight lines represent longitudes, and the circular arcs crossing them represent circles of latitude. Names of localities are incised on the appropriate cells of the grid that these lines form. When one rotates the ruler so that it passes through a given locality on the disk, one can read from the numbers incised around the rim the direction of Mecca from that locality, and from the scale along one-half of the ruler the distance between Mecca and that locality. The two known copies of this instrument are from seventeenth-century Iran, but there are good arguments for believing that it was a medieval discovery.[5]

Islam, in addition to prescribing the direction of prayer, also requires that the five obligatory daily prayers be said within certain specified time intervals. These came to be defined in terms of dawn, noon, twilight, and the lengths of shadows, all of which could be (and were) reduced to calculations involving the altitude of the Sun. Here again, methods including simple numerical schemes, geometric constructions, and elaborate trigonometric calculations produced an extensive corpus of material known under the general heading of "the science of timekeeping."

However, the religious requirements of Islam posed problems not only to the geometers but also to its specialists in the science of calculation. Some of these problems arose from the calculation of "*zakat*," a tax reflecting the community's share of an individual's wealth. Other problems arose from the division of inheritances according to the requirements of Muslim law and the rules of algebra. Found in al-Khwārizmī's *Algebra* in the early ninth century, such problems no doubt had been treated mathematically well before his time.

Evidently, the problems of inheritance sometimes developed well beyond simple practical rules into an art form for virtuosi, for Ibn Khaldūn, the great Tunisian historian of the fourteenth century, wrote:

> Religious scholars in the Muslim cities have paid much attention to it [inheritance calculation]. Some authors are inclined to exaggerate the mathematical side of the discipline and to pose problems requiring for their solution various branches of arithmetic, such as algebra, the use of roots, and similar things. It [exaggeration of the mathematical side] is of no practical use in inheritance matters because it deals with unusual and rare cases.[6]

THE SOCIAL SETTING OF MATHEMATICS IN MEDIEVAL ISLAM

Ibn Khaldūn was certainly no obscurantist, opposed to all things new or scientific, but the view he expressed represents a significant sector of Islamic opinion, a sector ranging from Ibn Khaldūn's skeptical view of arithmetical niceties to others' denunciation of the mathematical sciences because they were not Arab, or even Islamic, in origin but came from heathen foreign

[5] See David A. King, *World Maps for Finding the Direction and Distance of Mecca: Innovation and Tradition in Islamic Science* (Leiden: Brill, 1999).

[6] Ibn Khaldūn, *The Muqaddimah: An Introduction to History*, trans. Franz Rosenthal, ed. and abridged by N. J. Dawood (Princeton, N.J.: Princeton University Press, 1981), p. 347.

sources. These sciences, one argument went, created pride in human accomplishments and drew the practitioner's attention away from God; moreover, they had little relevance to the needs of an Islamic community, and that little could equally well be obtained from indigenous folk sciences. Thus the tenth-century mathematician al-Sijzī says that in his region people hold it lawful to kill geometers, and al-Ghazalī, in the late eleventh century, asks, caustically, just how many mathematicians a community needs.[7]

There has been considerable debate, since Goldziher published his study in 1915,[8] on the relation between science and orthodoxy in Islam. What is generally agreed is that medieval Islam was not a monolith but an amalgam of peoples, places, and powers that found, over a great interval of time and space, many different answers to the Muslim version of the question "What has Athens to do with Jerusalem?"[9]

We have already mentioned the support the early ʿAbbāsid caliphs, al-Mansūr, al-Rashīd, and al-Maʾmūn, gave to the efforts to acquire the best of ancient mathematics. The proliferation of kingdoms during the tenth century provided further venues for patronage, prominent among them being the courts of the various Būyid kings, who ruled parts of Iraq and Iran. Among the intellectual ornaments of these courts were the best mathematicians of the tenth century, such as the geometer Abū Sahl al-Kūhī, who wrote more than thirty works on topics arising from his deep study of Euclid, Apollonius, and Archimedes.[10] On the other hand, Mahmūd of Ghazna's patronage of the eminent mathematician and astronomer al-Bīrūnī in the first part of the eleventh century was not always clearly distinguishable from captivity.

With the distinctively Islamic development of the astronomical observatory, a social institution for the support of science, including mathematics, came into being. ʾUmar al-Khayyamī worked for eighteen years at an observatory supported by Malikshah in Isfahan. The eminent philosopher, theologian, astronomer, and mathematician Nasīr al-Dīn al-Tūsī was for many years director of the observatory at Maragha in Iran. We earlier mentioned al-Kāshī, who was but one – albeit the best – of a group of scientists at the observatory built by Ulug Begh in Samarqand in the 1420s.

7 See Sijzī, in the preface to his treatise on Menelaus's theorem, published in J. L. Berggren, "Al-Sijzī's Treatise on the Transversal Figure," *Journal for the History of Arabic Science*, 5 (1981), 23–36.

8 English translation in Ignaz Goldziher, "The Attitude of Orthodox Islam toward the Ancient Sciences," in *Studies on Islam*, trans. and ed. Merlin Swartz (Oxford: Oxford University Press, 1981), pp. 185–215.

9 See J. L. Berggren, "Islamic Acquisition of the Foreign Sciences: A Cultural Perspective," in *Tradition, Transmission, Transformation: Proceedings of Two Conferences on Pre-modern Science Held at the University of Oklahoma*, ed. F. Jamil Ragep and Sally P. Ragep (Leiden: Brill, 1996), pp. 263–83.

10 For a study of al-Kūhī, see "Tenth-Century Geometry through the Eyes of al-Kūhī," in *On the Enterprise of Science in Medieval Islam*, ed. Jan P. Hogendijk and A. I. Sabra (Cambridge, Mass.: MIT Press, 2003). Other geometers of this period were Abuʾl-Wafaʾ, an astronomer whose accounts of geometry and arithmetic for the practical person we have already cited, ʾAbd al-Jalīl al-Sizjī, and Ahmad al-Saghanī, who invented a new astronomical instrument.

Institutions such as observatories virtually demand royal patronage, and such patronage was often motivated by hopes of gain from astrology. In his pioneering study of the Islamic observatory, Aydin Sayili wrote, "In Islam astrology, with all its branches, enjoyed great popularity also with the kings and thrived well in the royal courts." He went on to state that this encouraged the growth of mathematics because the kind of astrology that many kings encouraged required elaborate mathematical treatment and observations.[11]

In the latter part of the three centuries separating 'Umar al-Khayyamī from Ulug Begh, there emerged in Egypt and Syria another mathematically based profession – that of the timekeeper. The practitioners of this profession were attached to mosques and specialized in the science of astronomical timekeeping to which we referred earlier. This association of mathematical sciences with the core institution of Islamic society, the mosque, provided powerful arguments for the utility of mathematics to the community. It also stimulated new accomplishments in mathematics, such as those of Ibn al-Shāṭir, a fourteenth-century timekeeper at the Umayyad Mosque in Damascus.[12]

Another institution that offered some support for the mathematical sciences was the *madrasa*, an institution for educating the legal and spiritual leaders of Islamic communities, which began in eleventh-century Baghdad but soon spread throughout the Islamic lands. Both arithmetic (especially as it applied to inheritance problems) and algebra, as well as some aspects of astronomy, were taught in these schools, even though their primary purpose was instruction in the religious sciences.[13] Ibn Munʿim (d. 1228) of Marrakesh, who excelled in the theory of numbers and the study of combinations and permutations, taught in such an institution. On the other hand, al-Samaw'al tells us that in twelfth-century Baghdad, where the institution of the *madrasa* had been born, he could find no one to teach him the later books of Euclid.

ARITHMETIC

"Calculation" (*hisab*), in medieval mathematics, included both arithmetic and algebra, and even an astronomer could be given the nickname "the Calculator" (*al-Hasib*). Medieval Islam knew at least three different systems

[11] See Aydin Sayili's discussion of the popularity (and criticisms) of astrology in medieval Islam in his *The Observatory in Islam*, 2nd ed. (Ankara: Turkish Historical Society, 1988), pp. 30–49.

[12] For a brief survey of Islamic timekeeping and other uses of mathematics for religious purposes, see David A. King, "Astronomy and Islamic Society: Qibla, Gnomonics, and Timekeeping," in *Encyclopedia of the History of Arabic Science*, ed. Roshdi Rashed and Régis Morelon, 3 vols. (London: Routledge, 1996), vol. 1, pp. 331–48. On Ibn al-Shāṭir, see E. S. Kennedy and I. Ghanem, eds., *The Life and Work of Ibn al-Shāṭir: An Arab Astronomer of the Fourteenth Century* (Aleppo: Institute for the History of Arabic Science, 1976).

[13] Sonja Brentjes sets out some of this evidence in "Reflections on the Role of the Exact Sciences in Islamic Culture and Higher Education between the 12th and 15th centuries," unpublished manuscript.

of arithmetic. We have already described mental arithmetic, known as the astronomers' arithmetic and done with a base of 60 (following Ptolemy). This was useful principally because it provided an efficient way of treating fractions in a place-value system. This system used fourteen letters of the Arabic alphabet to represent the numbers 1–10, 20, . . . , 50, and, by juxtaposition, all numbers from 1 to 59. It was ubiquitous in astronomical texts, and its arithmetic was discussed in a variety of works, including those of al-Khwārizmī, Kūshyar ibn Labban's *Hindu Reckoning*, and al-Kāshī's *Reckoners' Key* – representing a span of 600 years.

The third arithmetic system was decimal arithmetic. Although it was known to the Syriac Christian abbot, Severus Sebokht, who wrote in praise of the system in the mid-seventh century, it seems that no written account of it circulated in the Islamic world until that of al-Khwārizmī in the early ninth century. In general, the decimal system seems to have been slow in making its way through Islamic society. Ibn Sīna, in an autobiographical essay, tells how, over 150 years after al-Khwārizmī's work, he learned Hindu arithmetic not in school but from a greengrocer.[14] Lemay has argued that these numerals appear preponderantly in divinatory and magical texts.[15] Certainly Abu'l-Hasan al-Uqlīdisī, who in 952 (341 in Islamic chronology) wrote a comprehensive work on Hindu arithmetic,[16] refers to one of the reasons for changing from the dust board to pen and paper algorithms in this science – namely, the need to distance oneself from the astrologers who used the dust board. (The early arithmetics, such as that of Kūshyar, make it clear that the procedures were to be carried out in limited space, such as one might have on the commonly used dust board. The introduction of paper from China in about 730 led to fundamental changes in the practice of arithmetic computation by al-Uqlīdisī.)

The first work known to have used decimals is that of al-Uqlīdisī, his "decimal point" being a vertical stroke above the unit's place. Thus he expressed 2.34 as 2|34, so that the ciphers to the right of the decimal point denote tenths, hundredths, etc. Although al-Uqlīdisī did not use decimals consistently, it is not unusual for a fruitful idea to appear first as an ad hoc device in a special context. Decimals appear in a number of subsequent writers, and, over two centuries after al-Uqlīdisī, we find them used extensively in the works of al-Samaw'al (a Jew who grew up in Baghdad and subsequently converted to Islam), who used decimals extensively in division, root extraction, and approximating solutions to equations. Slightly more than

14 This information comes from an autobiographical essay of Ibn Sīna in which the great philosopher also explains that geometry and Indian arithmetic figured in the religious propaganda of Ismaili missionaries from Egypt who had converted his father and brother. See A. J. Arberry, *Aspects of Islamic Civilization: As Depicted in the Original Texts* (New York: A. S. Barnes, 1964), pp. 136–40.

15 Richard Lemay, "Arabic Numerals," in *Dictionary of the Middle Ages*, ed. Joseph R. Strayer et al., 12 vols. (New York: Charles Scribner's Sons, 1982), vol. 1, pp. 382–98, especially p. 384.

16 Abu'l-Hasan al-Uqlīdisī, *Book of Chapters on Hindu Arithmetic*, trans. A. S. Saidan as *The Arithmetic of al-Uqlīdisī* (Dordrecht: Reidel, 1978).

two centuries after that, al-Kāshī, in his *Reckoners' Key*, developed decimals as a consistent, closed system for doing arithmetic.

The modern common fractions were also used, but only in the western part of the Islamic world. The earliest known use of the horizontal bar to separate the numerator and denominator of a fraction is in the *Book of Proof and Recall* by the twelfth-century writer al-Hassar, who achieved fame as a specialist in inheritance law.

Just as the development of decimals and the spread of the Hindu-Arabic system of arithmetic were gradual processes, so was the discovery of methods for finding fourth, fifth, and higher roots of numbers. The earliest treatises contain no references to finding roots higher than the third. The cases of square and cube roots were, according to al-Khayyamī, taken from the Indians, but he says of himself that he has

> written a treatise on the proof of the validity of those [Indian] methods and that they satisfy the conditions. In addition we have increased their types, namely in the form of the determination of the fourth, fifth, sixth roots up to any desired degree. No one preceded us in this, and those proofs are purely arithmetical, founded on the arithmetic of the *Elements*.

Al-Khayyamī's treatise has not survived, but we may suppose that its methods must be the basis for those in al-Kāshī's *Reckoners' Key*, where one finds a detailed example of the extraction of the fifth root of a number on the order of trillions, namely 44,240,899,506,197. Al-Kāshī's methods depend on knowledge of the binomial theorem for the expansion of $(a + b)^n$, for $n = 2, 3, 4$, etc. This fundamental theorem was discovered by Abū Bakr al-Karajī, who was active in Baghdad in the late tenth century and wrote a work containing the so-called Pascal Triangle of coefficients needed in the sum of terms making up the expansion of $(a + b)^n$.

ALGEBRA

To find the nth root of a number, r, is to find a root of the equation $x^n = r$, which is simply a problem in algebra, another part of the science of calculation. Solutions to a variety of problems that we now express by the single quadratic equation $ax^2 + bx + c = 0$, for various possible coefficients a, b, and c, were common knowledge to Babylonian scribes in the early second millennium B.C. and to Greek mathematicians such as Heron of Alexandria in the first century of our era.

The earliest Muslim contribution to the study of quadratic equations was al-Khwārizmī's organization of ancient material into a new discipline in *The Condensed Book on the Calculation of Algebra*, and from the Arabic word "*al-jabr*" in its title comes our word "algebra." Al-Khwārizmī says that he has confined his book to "what is easiest and most useful in calculation"; in it he explains how to find the unknown in first- and second-degree equations and how to find the portions due to the heirs of a deceased relative according to Muslim law and the rules of algebra.[17] The scientific character of this work is revealed in its systematically organized rules and its arguments, based on an informal geometry of squares and rectangles, for the correctness of his solutions. Al-Khwārizmī's demonstrations referred to specific positive values of the coefficients a, b, and c, such as one-half, ten, or twenty-one, but not long after the *Algebra* was written Thābit ibn Qurra proved the correctness of the ancient methods for arbitrary coefficients using theorems from Euclid's *Elements*.

One must remember that Islamic mathematicians achieved their greatest results in algebra without the use of algebraic symbolism. Although, as indicated, such symbolism did arise by the twelfth century in the western part of the Islamic world in the context of teaching activity, the achievements we have described took place in the eastern caliphate, where algebraic symbolism was not used. The unknown was called "root" or "thing," its square "property" or "capital" (*māl*), its cube "cube," and higher powers were referred to by terms compounded of these, so that, for example, the fifth power was referred to as "square cube." Their reciprocals (our $1/x$, $1/x^2$, etc.) were expressed by prefixing the word "part" to the expressions for the respective powers. Equations were written out in words, and even the numerical coefficients were written out!

However, the Arabs of Morocco, Algeria, and Tunisia (al-maghrib) possessed a concise algebraic symbolism in which initial letters or other parts of appropriate Arabic words, written above the respective numerical coefficients, represented the unknown, its powers, and the square root. There was also a special symbol for the equals sign, which we shall represent by "*k*."[18] Thus, if we substitute for the Arabic letters the initial letters of the corresponding words in English (so "r" stands for "root," "n" for "numbers," "s" for "*x*-squared," and "c" for "*x*-cubed"), one would write "2^{r} *k* $7^{1}/_{2}{}^{\mathrm{n}}$ less 2^{s} 4^{c}" for what we now write as $\sqrt{2} = 7^1/_2 x - 2x^2 - 4x^3$.

In his *Algebra*, the Egyptian mathematician Abū Kāmil (850–930) stated in general form rules that al-Khwārizmī had stated in terms of examples. Abū Kāmil worked with the powers of the unknown up to the eighth. He also provided proofs of rules for manipulating algebraic quantities, including rules for the multiplication of expressions like $a - b$ and $c - d$, and he could simplify complex algebraic expressions. These developments climaxed in

[17] On al-Khwārizmī's mensurational work, see the discussion later in this chapter.

[18] F. Woepcke, "Recherches sur l'histoire des sciences mathématiques chez les Orientaux d'après des traités inédits arabes et persans. Premier Article. Notice sur des notations algébriques employées par les Arabes," *Comptes rendus hebdomadaires des séances de l'Académie des Sciences*, 39 (1854), 162–5. Woepcke's source was a treatise of al-Qalasadī. See also Ahmed Djebbar, *Une histoire de la science arabe* (Paris: Editions du Seuil, 2001), especially pp. 229–30. Djebbar has shown that this symbolism can be traced back at least to Ibn al-Yasamīn, who worked in Marrakesh and died in 1204.

The Marvelous [Book] by the aforementioned al-Karajī, who worked with expressions equivalent to arbitrary positive or negative powers of the unknown and stated a general rule for their multiplication, equivalent to our $x^n \cdot x^m = x^{n+m}$. In his extension of al-Karajī's work, al-Samaw'al displays a mastery of algebra, expressing, for example, the quotient of two algebraic expressions (in his terms "20 mal and 30 things" by "60 mal and 12") in terms of a series in "parts of thing," "parts of mal," etc., which he realized could be continued to arbitrarily many terms.

Cubic equations, which we write as $ax^3 + bx^2 + cx + d = 0$, arose in a variety of problems, including one from Proposition 4 of Book II of Archimedes' *Sphere and Cylinder*. Despite their advances in algebra, no medieval mathematician appears to have been able to find a formula for solving these equations in the manner of the formula for quadratics. It was instead al-Khayyamī who, in a work dedicated to the Chief Judge of Samarqand, approached the problem geometrically as one of constructing a segment representing the root of the equation (given a segment representing the unit). To do this, he interpreted the terms of the cubic as solids and used intersecting conic sections and, as he warns his reader, a knowledge of Euclid's *Elements* and *Data* and Apollonius's *Conics*. Al-Khayyamī was justifiably proud of having discovered solutions to certain equations that earlier writers had thought unsolvable, but he missed some cases and gave deficient arguments about the intersection of conics in other cases. It fell to Sharaf al-Dīn al-Tūsī,[19] in his *On Equations* (from the late twelfth century), to fill in the gaps in al-Khayyamī's arguments.[20] He also added a discussion of a computational method for accurately approximating a root in each possible case.

Sharaf al-Dīn's approximations were, however, just examples of the range of methods that Islamic mathematicians found for approximating solutions of equations. Many of these were iterative methods, an example being that of Habash al-Hasib, who, in the ninth century, found a method for solving an equation that we would write as $\vartheta(t) = t + 24\sin\vartheta(t)$ and is now known as Kepler's equation. In the early fifteenth century, al-Kāshī described a rapidly converging iterative method[21] of solving the cubic equation that arises from the problem of approximating the sine of 1°. Since the accuracy of astronomical calculations depends on the accuracy of trigonometric tables, such methods are of considerable importance, although as a practical matter no observational results in the ancient or medieval worlds were so accurate as to demand the accuracy in computation that al-Kāshī achieved.

[19] Not to be confused with Nasīr al-Dīn al-Tūsī, who lived in the thirteenth century, but on Sharaf al-Din's methods, see Jan P. Hogendijk, "Sharaf al-Dīn al-Tūsī on the Number of Positive Roots of Cubic Equations," *Historia Mathematica*, 16, no. 1 (1989), 69–85.

[20] Sharaf al-Dīn al-Tūsī, *Oeuvres Mathématiques*, ed. and trans. Roshdi Rashed, 2 vols. (Paris: Les Belles Lettres, 1985).

[21] See Asger Aaboe, "Al-Kashī's Iteration Method for Sin(1°)," *Scripta Mathematica*, 20 (1954), 24–9.

INDETERMINATE EQUATIONS

Islamic algebraists contributed importantly to several other disciplines, including the study of Diophantine equations, which demand integer or fractional solutions to a single equation in more than one unknown. Islamic mathematicians would have met one famous example in the equation that we would write as $x^2 + y^2 = z^2$, an equation in three unknowns x, y, and z, whose integer solutions for $n = 2$ (such as $x = 3$, $y = 4$, and $z = 5$) had been described by Euclid in Lemma 1 to *Elements* X.29.

Abū Kāmil, in his *Algebra*, discussed these equations, known to the Arabic mathematicians as "fluid" (*sayyala*), and presented about forty such problems, of both the second degree and underdetermined systems of linear equations. The systematic style of his exposition, however, suggests that he was not the one who introduced the subject.

Following Abū Kāmil's discussion, Qusta ibn Lūqa in Baghdad translated Diophantus's *Arithmetica*, which was devoted to the study of such equations, probably at the beginning of the tenth century. However, he viewed the subject, as did Abū Kāmil, as a part of algebra and, accordingly, retitled Diophantus's work *Algebra*. Diophantus's book strongly influenced several of the leading mathematicians in the following centuries, including the algebraist al-Karajī. Al-Karajī approached the subject in two works, *Fakhrī* and *The Marvelous*, by grouping the equations systematically by the number of terms and the differences between their powers. His work, in turn, stimulated studies by a series of later figures, including al-Samaw'al, who, with only modest success, tried to extend al-Karajī's work to cubic indeterminate equations. Several medieval Islamic mathematicians tried unsuccessfully to prove that one example of this latter type, $x^3 + y^3 = z^3$, had no integer or fractional solutions.

NUMBER THEORY

Investigations of equations of the Diophantine type and the theory of numbers are closely related subjects. For example, the tenth-century mathematician Abū Ja'far al-Khazin generalized the equation $x^2 + y^2 = z^2$ to the problem of expressing the square of a whole number as the sum of three or more such squares, a problem in number theory. No doubt, study of the theory of numbers in medieval Islam was stimulated by the requirements of Diophantine equations and the fact that Euclid's *Elements* had three books devoted to number theory.

Number theory, the Greek "arithmetic," benefited from two ancient traditions. One was a logical investigation of the mathematical properties of the whole numbers (2, 3, 4, etc.) found in Books VII–IX of the *Elements*. The other was a mystical Pythagorean tradition typified by Nicomachus of

Figure 2.2. The three triangular arrays of points represent the numbers 3, 6, and 10.

Gerasa's *Introduction to Arithmetic* (first century). No rigid division between topics treated in these two areas existed, however, and one finds in *Elements* IX.36, a theorem about what the Pythagoreans called "perfect numbers"– that is, numbers equal to the sum of their proper factors (thus $6 = 3 + 2 + 1$).

Similarly, in some Arabic writings on number theory, one found rigorous proofs for problems about amicable numbers, which are generalizations of perfect numbers. Pythagoreans defined amicable numbers as two positive whole numbers, such as 220 and 284, each of which is the sum of the proper factors of the other; Thābit established a sufficient condition for two numbers to be amicable. This subject led Kamal al-Dīn al-Farisī, searching for a new proof, to study both the number of factors and their sums for an arbitrary integer, and to prove that every positive integer greater than 1 is a product of prime factors.[22]

Another problem related to the sums of factors of a number that one finds in Arabic sources is that of finding equiponderant numbers (i.e., any set of whole numbers the factors of which all have the same sum). Among these writers was the jurist-theologian 'Abd al-Qāhir al-Baghdadī (d. 1037), who also stated rules for general terms of series of figured numbers[23] (called, in Arabic, "numerical figures") as well as for the general terms of the series whose terms are sums of the figured numbers.

Nicomachus discussed figured numbers, a Pythagorean specialty, in his *Introduction*. Readers will be familiar with "square numbers," such as 1, 4, 9, 16, and so on, each of which can be represented by a square grid of points. The Pythagoreans also studied numbers representing other polygonal shapes; thus the triangular numbers 3, 6, and 10 record the numbers of points in the figures comprising Figure 2.2.

The Greeks had also begun the study of series whose terms are the sums of the terms of the series, the so-called pyramidal numbers, the bases of the pyramids being triangles, squares, pentagons, and so on.

The mathematicians of medieval Islam carried the study of such numbers considerably further, and both the aforementioned al-Baghdadī and

[22] A number is prime if its only proper divisor is 1. Kamal al-Dīn's result is not the same as proving the Fundamental Theorem of Arithmetic – that every whole number has a *unique* expression as a product of prime numbers. On this point, see Ahmet G. Agargün and Colin R. Fletcher, "Al-Farisī and the Fundamental Theorem of Arithmetic," *Historia Mathematica*, 21 (1994), 162–73.

[23] In modern terms, these are numbers formed by summing the terms of arithmetic series, beginning with 1, having constant differences of 1, 2, 3, etc. See the chapter by Ahmad S. Saidan, "Numeration and Arithmetic," in Rashed and Morelon, *Encyclopedia of the History of Arabic Science*, vol. 2, pp. 331–48.

1	14	11	8
12	7	2	13
6	9	16	3
15	4	5	10

Figure 2.3. In this four-by-four magic square, the sum of each row and column and of each of the two diagonals is equal to 34.

al-Umawī explicitly stated rules for the general terms of the series of figured and pyramidal numbers as well as for the general terms of the series whose terms are sums of the terms of the previous series.[24] Finally, Islamic mathematicians provided rules for summing the sequences of successive integers, their squares, their cubes, and their fourth powers. All of this material was treated in a systematic fashion and was made the subject of rigorous proofs.

Ibn al-Haytham, who was born in Iraq in 965 and died in Egypt in about 1040, established the values of the sums of the sequence of fourth powers (e.g., $1 + 16 + 81 + \cdots + n^4$), and he proved a result now known as Wilson's theorem, that a whole number n is a prime exactly when it divides $(n - 1)! + 1$.[25] This work shows that Islamic mathematicians were familiar with techniques for solving special problems of the "Chinese remainder" type. At about the same time, al-Baghdadī, in the context of recreational problems, had dealt with general problems of this type, showing how to find a "hidden number" N that leaves given remainders, a, b, and c when divided by 3, 5, and 7.

Also falling within the general area of number theory is the problem of constructing what are now called magic squares.[26] A magic square of order n is an $n \times n$ array of the whole numbers from 1 to n^2 such that the sums of the elements in each row, column, and diagonal are equal. For example, Figure 2.3 shows a magic square of order four. Muslim writers had long studied these "harmonies of numbers," as they were called, and had worked out methods for generating squares of any odd order or any order divisible by 4. By the end of the twelfth century, they had found methods for generating squares of all orders. Among those participating in these developments were

24 Ibid.

25 Here the expression $(n - 1)!$ represents the product of all whole numbers from 2 to $n - 1$. Thus $(4 - 1)! = 6$, $(5 - 1)! = 24$, and so forth. Note that 4, which is not prime, does not divide $(4 - 1)! + 1$, but 5, which is prime, does divide $(5 - 1)! + 1$.

26 For a further discussion, see Jacques Sesiano, *Un Traité médiéval sur les carrés magiques: De l'arrangement harmonieux des nombres* (Lausanne: Presses Polytechniques et Universitaires Romandes, 1996), which includes an informative English language summary.

such distinguished mathematicians as the tenth-century writers Abu'l-Wafa' al-Būzjanī and Ibn al-Haytham.

COMBINATORICS

Another set of problems requiring determination of an unknown number, and therefore related to algebra, involved what we now call combinatorics.[27] Such problems first appeared as isolated investigations in areas such as lexicography, linguistics, music, and astrology. Similar problems had been solved in the Greek and Hindu civilizations, but during the Islamic period the results were systematized as subjects of instruction and were made the subject of proofs. Among the problems solved by the thirteenth-century Moroccan Ibn Mun'im, who figured in our discussion of *madrasas*, is that of counting the number of pom-poms one can make using at most ten colors of silk when no color is repeated. Another, indicative of the roots of the subject in linguistic questions, requires finding "the number of words such that a human being cannot express himself except by the aid of one of them."[28]

THE TRADITION OF GEOMETRY

Among the earliest Arabic treatments of geometry (*al-handasa*) were works by al-Khwārizmī and the Banū Mūsa, the latter being the *Book of Measurement of Plane and Spherical Figures*. Both works combine material from ancient sources in ways that were widely influential, and both were eventually translated into Latin. Al-Khwārizmī's treatment of geometry in his *Algebra* provides procedures for calculating the height of a triangle given its sides, the areas of circles and sectors of circles, and volumes of cylinders with polygonal and circular bases. To measure the circumference of a circle, the rule in practical life, according to al-Khwārizmī, is to take 3 1/7 as the value that, multiplied by the diameter, produces the circumference, "though it is not quite exact."[29] Geometers, he informs us, take the circumference either as $\sqrt{(10d \times d)}$ or, if they are astronomers, as $\frac{62{,}832 \times d}{20{,}000}$. Since both of these latter values are found in Indian writings predating the Islamic period, it appears

[27] In works by medieval authors, this is the study of permutations and combinations. See the articles by Ahmed Djebbar "Combinatorics in Islamic Mathematics" and "Ibn Mun'im" in *Encyclopedia of the History of Science, Technology and Medicine in Non-Western Cultures*, ed. H. Selin (Dordrecht: Kluwer, 1997), pp. 230–32 and 427–28, respectively.

[28] A. Djebbar, "Mathématique et linguistique dans le Moyen Âge arabe: L'exemple de l'analyse combinatoire au Maghreb," in *Le Moyen Âge et la Science: Approche de quelques disciplines et personalités scientifiques médiévales*, ed. Bernard Ribémont (Paris: Klincksieck, 1991), pp. 15–29.

[29] F. Rosen, *The Algebra of Mohammed ben Musa* (London: Oriental Translation Fund, 1831; repr. Hildesheim: Georg Olms Verlag, 1986), pp. 71–2.

that al-Khwārizmī is dipping into a fund of knowledge of measurements well known to mathematical practitioners over a large geographical region.

It is striking how mensuration and other aspects of geometry developed over the centuries in Islam in response to practical needs. In the second half of the tenth century, Abu'l-Wafa' devoted the third part of his *Book of Arithmetic Needed by the Secretary and Official* to mensuration, noting that the rules for measuring such abstract solids as spheres, cylinders, and cones can be applied to the measurement of domes. At that time, the geometry of these characteristic shapes of Islamic architecture had not developed beyond the above-mentioned three basic shapes. By the time al-Kāshī wrote his *Reckoners' Key* in the early fifteenth century, however, the geometry of architecture had become more sophisticated, a variety of shapes being used for *muqarnas* and domes.[30] Al-Kāshī analyzes these complex structures in terms of simpler shapes and provides accurate coefficients for calculating their areas from measurable dimensions, as well as an explanation of how these coefficients were derived.

In their mensurational work, the Banū Mūsa relied on Greek sources of a high level. For example, their work contains the first proof that the ratio of the circumference of a circle to its diameter is the same for all circles and also a procedure for calculating the volume of a sphere, the surface of a hemisphere, and other results from Archimedes' *Sphere and Cylinder* (a subject al-Khwārizmī does not treat). They also borrowed other Archimedean material, such as the so-called Heron's Formula for the area of a triangle in terms of its sides and material from Eutocius's commentary on *Sphere and Cylinder* related to the classical problems of trisecting the angle and duplicating the cube.

Other early Arabic geometrical investigations were those of Thābit ibn Qurra, who read in the preface to Archimedes' *Sphere and Cylinder* of the great Greek mathematician's work on the area of a segment of a parabola but could not find Archimedes' treatise on the subject with its proof. He therefore set out to prove it for himself, but the resulting proof, although successful, was so long that many geometers criticized it. (Thābit's grandson, Ibrahīm ibn Sinan, later decided to rescue the family honor by providing a shorter proof based on a different approach.) Thābit also successfully established the volumes of the solid (the paraboloid of revolution) obtained by rotating a segment of a parabola around its axis, a topic to which both al-Kūhī and Ibn al-Haytham returned about a century later. And, in his treatise *The Sections of*

30 *Muqarnas* are honeycombs of squinches that gave an aesthetic solution to the problem of putting round domes on the square bases found in so many Islamic shrines and other buildings. For a study of the construction of this feature of Islamic architecture by Mohammad al-Asad, as well as a geometric analysis of many features of Islamic architectural decoration, see the recent work by Gülru Necipoğlu, *The Topkapi Scroll: Geometry and Ornament in Islamic Architecture* (Santa Monica, Calif.: The Getty Center for the History of Art and the Humanities, 1995).

a Cylinder and Their Surfaces, he establishes, among many beautiful results, the areas of an ellipse and a segment of an ellipse.

Archimedes' *Measurement of the Circle* and its medieval tradition also inspired one of the great medieval computational feats: al-Kāshī's *Treatise on the Circumference*. In it, al-Kāshī computed the value of π so exactly that using it to calculate the circumference of the cosmos (whose radius was taken to be 600,000 times that of the Earth) would, as he put it, yield a value not differing from the exact value by more than the width of a horse's hair.

At the same time, Eutocius's commentary on the problems of trisecting the angle and duplicating the cube,[31] as well as the Arabic translation of a work attributed to Archimedes, *On Dividing the Circle into Seven Equal Parts*, inspired medieval geometers to justify or complete classical solutions. Classical problems sometimes demanded constructions that went beyond the use of a compass and straightedge, and although these were used without comment by geometers like Archimedes and Apollonius, they were not regarded as satisfactory by tenth-century Islamic geometers. One of them, Abu'l Jūd, said of the construction required by Archimedes' solution to the construction of a regular heptagon that it was perhaps more difficult than the original problem! The advanced constructions used by these authors inspired a lively debate among tenth-century geometers over what sorts of geometric constructions were valid.[32]

In addition to bringing constructions such as that of Archimedes for the heptagon into the theory of conic sections, medieval Islamic geometers also found constructions of the regular nonagon and two possible ways of inscribing a pentagon with equal sides within a given square. One by Abū Kāmil, noteworthy for its application of algebra to a geometrical problem, treats the case in which the pentagon shares a vertex with the square. The other case, which Abū Sahl al-Kūhī considered, is more difficult, for it involves constructions equivalent to solving an equation of the fourth degree; al-Kūhī solved it by purely geometrical methods.

Problems involving conic sections also exercised geometers' ingenuity. For example, Ibrahīm ibn Sinan found constructions that allowed a geometer to construct an arbitrary number of points on any of the conic sections, and al-Kūhī invented his "complete compass" for drawing conic sections. Such an instrument would have been of at least theoretical interest because of the importance of the conics, with their foci, in the design of burning mirrors and burning lenses, their importance in the design of sundials, and their frequent use in solutions to the advanced construction problems mentioned previously. They also entered into Abu Jaʿfar al-Khazin's investigations of

[31] Islamic geometers could also have learned about these problems from the work of the Banū Mūsa; see n. 2.

[32] Jan P. Hogendijk, "Greek and Arabic Constructions of the Regular Heptagon," *Archive for History of Exact Sciences*, 30 (1984), 197–330.

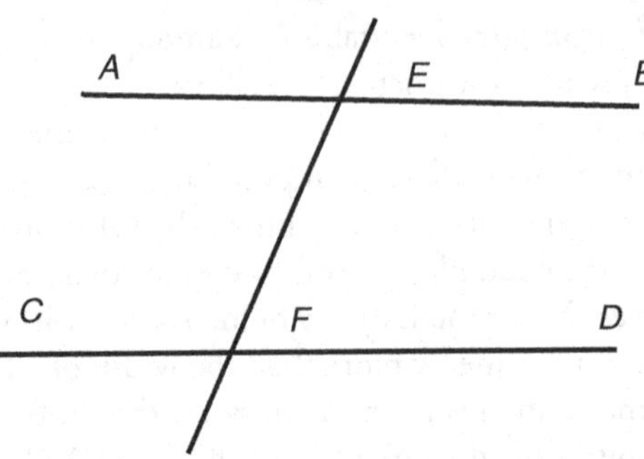

Figure 2.4. Whether Euclid's famous parallel postulate was truly a postulate or could be proved on the basis of his other postulates was one of the questions on the foundations of mathematics that medieval Islamic mathematicians investigated. In the situation illustrated here, lines *AB* and *CD* are not parallel.

Archimedes' solution to his problem of using a plane to divide a sphere into segments of a given ratio. Later in the tenth century, al-Kūhī modified Archimedes' problem into one of constructing a sphere and a segment thereof having both a given area and a given volume. These and other achievements stimulated, as al-Khayyamī tells us, his own systematic solution of cubic equations using conic sections.

FOUNDATIONS OF GEOMETRY

Another medieval Islamic tradition was the discussion of metamathematical issues arising in geometrical problems, one of these being the question of the necessity for Euclid's Postulate 5 of *Elements* Book I (the "parallel postulate"), which states that if two lines *AB* and *CD* (Figure 2.4) are cut by a straight line *EF*, and if the two angles *BEF* and *EFD* are together less than two right angles, then lines *AB* and *CD* meet in the direction of *B* and *D*. Many later writers, Greek and Islamic, felt that the postulate could be proved as a theorem, and Greek writings on this topic, such as those of Simplicius and Heron, stimulated considerable work on the topic from Thābit in the ninth century down to Nasīr al-Dīn in the thirteenth.[33]

Al-Khayyamī was the first to introduce into the study of parallels the so-called Saccheri quadrilateral *ABCD* (Figure 2.5), in which angles *C* and *D* are right and sides *AC* and *BD* are equal. He avoided treatments of the topic using motion, including that of Ibn al-Haytham, and he used a principle that he attributed to "the Philosopher" (Aristotle) to argue that in this case, independently of Euclid's fifth postulate, angles *A* and *B* are right.

[33] For details on the medieval Islamic tradition, see Boris A. Rosenfeld and Adolf P. Youschkevitch, "Geometry," in Rashed and Morelon, *Encyclopedia of the History of Arabic Science*, vol. 2, pp. 447–94, especially pp. 463–70.

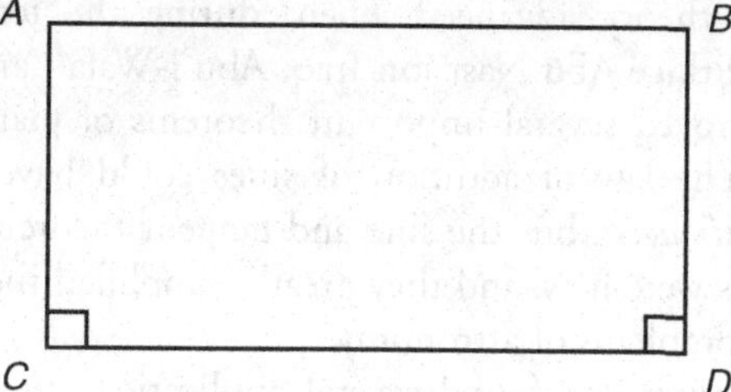

Figure 2.5. Saccheri's quadrilateral, in which the angles at *C* and *D* are assumed to be right and sides *AC* and *BD* are assumed to be equal.

(This assumption is, however, equivalent to the parallel postulate.) ʾUmar's views influenced Nasīr al-Dīn's discussion, published by Nasīr al-Dīn's son in his version of Euclid's *Elements*. Some three centuries later, in 1594, this Arabic version of the *Elements* was published in Rome. In the seventeenth century, the English mathematician John Wallis included a Latin translation of a proof of the fifth postulate from this work, and Wallis's discussion was quoted by G. Saccheri in his famous *Euclid Freed from Every Fault* (1733). By such circuitous routes did Arabic learning come westward to influence European thinking.

TRIGONOMETRY

Despite these impressive developments, it would be a mistake to claim that medieval Islam transformed classical geometry. One area, however, that medieval Islam did transform was trigonometry. Here their starting materials were the chord table and ancillary results (including Menelaus's theorem) in Ptolemy's *Almagest* and, from India, sine tables for steps of $7\frac{1}{2}°$. On this base, Muslim mathematicians built all six trigonometric functions, though not in modern terms as ratios but rather as absolute lengths.[34] They refined the methods of calculating tables from the early tables of Habash, accurate to three sexagesimal places, to al-Bīrūnī's (ca. 1000) four-place tables and Ulug Begh's (ca. 1400) five-place (equivalent to nine-place decimal) tables – far more than the astronomy of the day called for. These tables required ingenious methods of interpolation and approximation of the roots of cubic equations to obtain the fundamental value, sin 1°, from that of sin 3°; such methods included the beautiful iterative procedure used to great effect by al-Kāshī, as well as methods of interpolation.

[34] Thus the medieval sine of a central angle in a circle was the length of half the chord of its double rather than that length divided by the radius. Abu'l-Wafaʾ did, however, suggest taking the radius of the circle as equal to 1 and in this way defined the modern trigonometric functions.

Trigonometric theory advanced when, during the tenth century, al-Bīrūnī's teacher, Prince Abū Nasr ibn Iraq, Abu'l-Wafa', and other geometers stated and proved several important theorems of plane and spherical trigonometry.[35] The law of addition of sines could have been extracted from Ptolemy's *Almagest*, but the sine and tangent theorems for plane and spherical triangles were new, and they greatly simplified the applications of trigonometry to problems of astronomy.

This new trigonometry found several applications to geodesy as well, notably in al-Bīrūnī's *Determination of Coordinates of Cities*,[36] a work devoted to finding the coordinates of Ghazna, a city in present-day Afghanistan and the capital of his patron, Mahmūd. His calculated difference of longitudes between Baghdad (whose position was known) and Ghazna was in error by only about 1/3° (one part in 72), though this was to some extent due to the good fortune of errors in opposite directions canceling each other out.

Nasīr al-Dīn al-Tūsī's creation of trigonometry as a mathematical discipline independent of astronomy, in his work *The Complete Quadrilateral*, climaxed these developments. In it, he treated the theory of plane and spherical trigonometry and systematically explained methods for the solution of both types of triangles.[37]

THE ASTROLABE

Despite the important use of trigonometry in the solution of problems in spherical astronomy, medieval Islamic astronomers had inherited from ancient Greece a special instrument, devised for the solution of such problems, whose use requires no knowledge of trigonometry. The planispheric astrolabe uses stereographic projection of the sphere from one of its poles onto a circular plate representing the plane of its equator; the ninth-century astronomer al-Farghānī gave the first proof that this technique projects circles on the sphere as circles or straight lines in the plane.[38]

To the Greek astrolabe, Islam added the azimuth circles, circles corresponding to those passing through the user's overhead point, or zenith, and

35 For a more detailed discussion of the various advances, see Marie-Thérèse Debarnot, "Trigonometry," in Rashed and Morelon, *Encyclopedia of the History of Arabic Science*, vol. 2, pp. 495–538.

36 Jamil Ali, *The Determination of the Coordinates of Positions for the Correction of Distances between Cities: A Translation from the Arabic of al-Bīrūnī's Kitāb Tahdīd Nihāyāt al-Amākin litashīh masāfāt al-masākin* (Beirut: American University of Beirut Press, 1967). See also E. S. Kennedy, *A Commentary upon Bīrūnī's Kitāb Tahdīd Nihāyāt al-Amākin, an 11th Century Treatise on Mathematical Geography* (Beirut: American University of Beirut Press, 1973).

37 For an extensive account of the development of trigonometry in medieval Islam, consult Glen Van Brummelen, *The Mathematics of the Heavens and the Earth: The Early History of Trigonometry* (Princeton, N.J.: Princeton University Press, 2009).

38 Although medieval Islamic mathematicians must have been aware that this projection also preserves angles between circles on the sphere (for several medieval applications use this property), no medieval proof of the angle-preserving property of the projection has been found.

his antipodal point, or nadir. Such circles were of special importance to users of the instrument for finding the azimuth of Mecca – the local direction of prayer. The construction of these circles offers a geometrical challenge, and over a half-dozen mathematicians suggested ways of solving the problem thus posed, including Habash al-Hasib in the ninth century, al-Kūhī in the tenth, and al-Bīrūnī in the eleventh.

The planispheric astrolabe needs a different plate for each latitude, but a universal astrolabe, which serves for all latitudes, was invented in al-Andalus. It was reinvented by the fourteenth-century Syrian Ibn al-Sarraj. The resulting magnificent instrument is now in the Benaki Museum in Athens.

CONCLUSION

By the time al-Kāshī, the last major figure in medieval Islamic mathematics, died in 1429, the Muslims had brought mathematics a long way from the Greek and Hindu elements with which they had begun. Algebra had been systematized and could deal with polynomials of any positive or negative integer degree; the decimal arithmetic of the Hindus had been adapted to pen-and-paper algorithms and extended to decimal fractions; in geometry, the configuration whose study was to lead to non-Euclidean geometry had begun its journey from al-Khayyamī to Sacchieri; and trigonometry had gained a panoply of functions, theorems, and tables. These achievements had an impact on mathematics that is still felt today, but there were other such developments no less remarkable for being of apparently more limited impact. These included new mappings of the sphere and instruments based on them, geometrical and numerical methods for solving equations, and studies of permutations and combinations. Additionally, the mathematical sciences had revealed an ability to serve not only the needs of commerce, kingdom, and empire, as they had in the ancient world, but those of a religious community as well.[39]

[39] For more on Islamic mathematics, see J. L. Berggren, *Episodes in the Mathematics of Medieval Islam* (New York: Springer-Verlag, 1986); J. L. Berggren, "Mathematics and Her Sisters in Medieval Islam," *Historia Mathematica*, 24 (1997), 407–40; and A. P. Youshchkevitch, *Les mathématiques arabes (VIIIe–XVe siècles)*, trans. M. Cazenave and K. Jaouiche (Paris: Vrin, 1976).

3

THE MIXED MATHEMATICAL SCIENCES

Optics and Mechanics in the Islamic Middle Ages

Elaheh Kheirandish

Of the scientific traditions of medieval Islamic civilization, optics and mechanics especially stand out in their decisive roles in the transformation of the relations and applications of the mathematical and natural sciences. In the ancient Greek tradition, long before they became parts of classical physics, both disciplines belonged to the mathematical sciences. But not long after the Arabic translations of ancient, primarily Greek, scientific works in the third Islamic century (ninth century), optics and mechanics moved beyond their ancient conceptions as respective "subordinates to plane and solid geometry" and "mathematical sciences closer to physical/natural sciences," as expressed in the *Posterior Analytics* and *Physics* of Aristotle (d. ca. 322 B.C.).[1] The explicit "co-operation," and later "combination," of mathematical and physical sciences in the evolving Arabic optical and mechanical traditions, however differently formulated and applied, occasioned new and constructive developments in both fields. As "mixed sciences," a modern expression first applied to Arabic astronomy,[2] the optics and mechanics of the Islamic Middle Ages shared close conceptual links. But they had distinct historical developments and fates, including the nature of their contributions to the most vital interval of scientific activity in medieval Islam. That interval extended temporally from at least the ninth through the thirteenth centuries, and spatially from Mesopotamia and Persia to Central Asia in the

[1] See, respectively, Aristotle, *Posterior Analytics*, I.13.78b36–39, and *Physics*, II.2.194a7, in *Aristotle in Twenty-Three Volumes*, ed. G. P. Goold et al., trans. Hugh Tredennick (Loeb Classical Library) (Cambridge, Mass.: Harvard University Press; London: Heinemann, 1960), vol. 2, pp. 89–91, and ibid., trans. Philip H. Wicksteed and Francis M. Cornford (Loeb Classical Library) (Cambridge, Mass.: Harvard University Press; London: Heinemann, 1957, 1980 printing), vol. 4, pp. 120–1, especially the latter, for the expression *ta physikōtera tōn mathēmatōn*.

[2] On the "mixed sciences" and some modern uses of the expression, see Elaheh Kheirandish, "Organizing Scientific Knowledge: 'Mixed Sciences' and Early Classifications," in *Organizing Knowledge: Encyclopaedic Activities in the Pre-Eighteenth Century Islamic World*, ed. Gerhard Endress (Leiden: Brill, 2006), pp. 135–54.

east and Andalusia in the west.[3] Linguistically, optics and mechanics also comprised a vast corpus of works written first in Arabic and later also in Persian, an inversion of the order in astronomy.[4]

Like their Greek prototypes, the Arabic and Persian traditions of optics and mechanics each had two disciplinary subdivisions. Optics treated direct vision or the science of "aspects" (optics proper, or vision through air); indirect vision, or the science of "mirrors," treated vision through media beyond polished surfaces (in Greek, "*optika*" and "*catoptrika*," respectively). Mechanics encompassed the science of "devices" (mechanics proper, corresponding to the Greek "*mēchanē*") and a separate "science of weights," involving measures and motions (see the sections on "Developments"). These four endeavors had no fewer activities or achievements than the mainstream mathematical sciences of the classical quadrivium (arithmetic, geometry, astronomy, and music).[5] The main traditions of optics (aspects) and mechanics (devices) were both primarily Alexandrian, deriving directly from works devoted to optics by Euclid (ca. 300 B.C.) and Ptolemy (second century), and to mechanics by Heron (first century) and Pappus of Alexandria (fourth century).[6] The parallel traditions of "mediated vision" and "weights and balances" were derived from similarly authoritative works now mostly considered anonymous or pseudonymous.

Although they were the two closest Aristotelian "mixed" sciences, the optics and mechanics of the Islamic Middle Ages nevertheless combined their mathematical and physical components in different ways. This disparity underlay the methodological distinction between optics (initially a "demonstrative" and later also an "experimental" science) and mechanics

[3] A. I. Sabra, "The Scientific Enterprise: Islamic Contributions to the Development of Science," in *The World of Islam: Faith, People, and Culture*, ed. Bernard Lewis (London: Thames and Hudson, 1976), pp. 181–7; and A. I. Sabra, "Situating Arabic Science: Locality versus Essence," *Isis*, 87 (1996), 654–70.

[4] See Elaheh Kheirandish, "Optics: Highlights from Islamic Lands," in *The Different Aspects of Islamic Culture*, vol. 4: *Science and Technology in Islam*, part I, ed. A. Y. al-Hassan, M. Ahmed, and A. Z. Iskandar (Paris: UNESCO, 2001), pp. 337–57; and Robert E. Hall, "Mechanics," in *Different Aspects of Islamic Culture*, pp. 297–336. On astronomy, see F. Jamil Ragep, *Naṣīr al-Dīn al-Ṭūsī's Memoir on Astronomy*, 2 vols. (Berlin: Springer-Verlag, 1993); and George Saliba, "Persian Scientists in the Islamic World: Astronomy from Maragha to Samarqand," in *The Persian Presence in the Islamic World*, ed. Richard G. Hovannisian and Georges Sabagh (Cambridge: Cambridge University Press, 1998), pp. 126–46.

[5] See the second edition of the *Encyclopedia of Islam* under the following entries: "Manāẓir, or ʿIlm al-Manāẓir" (optics or science of aspects), Ḥiyal (mechanics, mechanical devices), and "Mīzān" (balance), respectively; also ʿIlm al-ḥisāb (arithmetic, science of reckoning), ʿIlm al-handasa (science of geometry), ʿIlm al-hayʾa (theoretical astronomy, cosmography), Nujūm (astronomy, astrology), and Mūsīqī (music, harmonics).

[6] On Euclid and Ptolemy, see Elaheh Kheirandish, *The Arabic Version of Euclid's Optics*, 2 vols. (Berlin: Springer-Verlag, 1999). On Heron and Pappus, see "Introduction of Pappus to the Science of Mechanics," in D. E. P. Jackson, *The Arabic Version of the Mathematical Collection of Pappus Alexandrinus Book VIII* (unpublished PhD thesis, University of Cambridge, 1970). A critical edition and English translation is under preparation by the present author.

(largely a "productive" and "operative" craft).[7] The distinctions went beyond classification or method. The prefaces and dedications of both translations and original compositions in these fields suggest that mechanics involved more commissions from patrons, and possibly more competition among practitioners, than did optics. But the most marked contrast between these fields is that, unlike mechanics, optics had a remarkably continuous and influential life in medieval and early-modern Europe.

HIGHLIGHTS

The Islamic Middle Ages witnessed breakthroughs in all the Aristotelian "mixed sciences," a point that, along with their comparative rates of success, was not lost on one early observer. Even before most of the major achievements in these fields, the polymath Qusṭā ibn Lūqā (d. ca. 912), who wrote on optics, mechanics, and astronomy, among many other subjects, remarked prophetically not only that "the best of the demonstrative sciences is that in which the natural and the geometrical sciences both participate" but that "nowhere is this better and more perfect than in the science of rays."[8] The participation of natural philosophy and mathematics noted in Ibn Lūqā's first statement produced different results within the combined sciences themselves. In optics, "the best and most perfect" of the mixed sciences, the mathematical and the physical produced a particularly successful mix.

The theoretical, disciplinary, and methodological contributions of optics, which mark the height of scientific activity in Islamic regions, are primarily associated with the monumental *Optics* of Ibn al-Haytham (d. ca. 1040), the great scholar of Baṣra (and later Cairo), best known in Europe as "Alhacen" or "Alhazen" after his first name, Al-Ḥasan.[9] But as early as the ninth century,

7 On "demonstrative," "experimental," "productive," and "operative" traditions, see Elaheh Kheirandish, "Science and Mithāl: Demonstrations in Arabic and Persian Scientific Traditions," in *Sciences, Crafts, and the Production of Knowledge: Iran and Eastern Islamic Lands (ca. 184–1153 AH/800–1740 CE)*, special issue of *Iranian Studies*, 41, no. 4 (2008), 465–89; and Elaheh Kheirandish, "Footprints of Experiment in Early Arabic Optics," in *Evidence and Interpretation: Studies on Early Science and Medicine in Honor of John E. Murdoch*, ed. William Newman and Edith D. Sylla, special issue of *Early Science and Medicine*, 14 (2009), 79–104.

8 Elaheh Kheirandish, "Qusṭā Ibn Lūqā," in *Biographical Encyclopedia of Astronomers*, ed. Thomas Hockey et al., 2 vols. (New York: Springer Reference, 2007), vol. 2, pp. 948–9. A transcription and French translation of Qusṭā's "The Book on the Reasons for What Occurs in Mirrors on Account of the Difference of Aspects" is included in Roshdi Rashed, *Oeuvres philosophiques et scientifique d'al-Kindī* (Leiden: Brill, 1997), vol. 1, pp. 571–645.

9 See A. I. Sabra, *Al-Ḥasan Ibn al-Haytham, Kitāb al-Manāẓir: Books I-II-III* (Kuwait: National Council for Culture, Arts, and Letters, 1983); Books IV–V (1983–2002); A. I. Sabra, *The Optics of Ibn al-Haytham*, Books I–III: *On Direct Vision*, 2 vols. (London: The Warburg Institute, 1989); and A. I. Sabra, "Ibn al-Haytham's Revolutionary Project in Optics: The Achievement and the Obstacle," in *The Enterprise of Science in Islam: New Perspectives*, ed. Jan Hogendijk and A. I. Sabra (Cambridge, Mass.: MIT Press, 2003), pp. 85–118.

the Greco-Arabic translations had begun to transform the optical tradition, with key developments that extended into the next few centuries.[10] Optics proper (the study of vision as opposed to light or mirrors) witnessed an initial merger of sense perception and geometrical demonstration in the works of Aḥmad ibn ʿĪsā (fl. ca. ninth to tenth centuries?) and Yaʿqūb al-Kindī (d. ca. 870), which extended demonstrative reasoning and geometrical proofs to illustrations involving sense and physical setups.[11]

Al-Kindī has been credited with the earliest important breakthroughs. Among these is the formulation and use of the optical principle known as the "punctiform analysis of light radiation," a principle based on a linear correspondence between multipoint radiating surfaces and visual objects. The one-to-one correspondence between points on the visible object and the sensitive area within the eye was an ancient conception that al-Kindī explicitly stated and that Ibn al-Haytham raised to the status of an optical principle; it would serve as the foundation of the geometry of sight for Latin perspectivists as late as Kepler.[12] An early conception of the "sine law of refraction" as a fixed ratio between the angles of incident and refracted rays moving from one medium to another has been associated with another early author, Abū Saʿd al-ʿAlāʾ Ibn Sahl (d. ca. 984), long before Hariot, Snell, and Descartes.[13] The apex of the study of vision, however, which is what optics was at this time, is the eleventh-century *Optics* of Ibn al-Haytham, an acknowledged and influential author in Europe, who placed the field on a new foundation.[14]

[10] See Elaheh Kheirandish, "The Arabic 'Version' of Euclidean Optics: Transformations as Linguistic Problems in Transmission," in *Tradition, Transmission, Transformation*, ed. F. Jamil Ragep and Sally Ragep (Leiden: Brill, 1996), pp. 227–43; and Kheirandish, *Arabic Version of Euclid's Optics*. On the Greco-Arabic translation movement, see Dimitri Gutas, *Greek Thought, Arabic Culture: The Graeco-Arabic Translation Movement in Baghdad and Early ʿAbbāsid Society* (London: Routledge, 1998); and Roshdi Rashed, "Greek into Arabic: Transmission and Translation," in *Arabic Theology, Arabic Philosophy, From the Many to the One: Essays in Celebration of Richard M. Frank*, ed. James E. Montgomery (Leuven: Peeters, 2006), pp. 157–96.

[11] A partial transcription and French translation of Aḥmad ibn ʿĪsā's *Optics and Burning Mirrors (Kitāb al-Manāẓir wa al-marāyā al-muḥriqah)* appears in Rashed, *Oeuvres philosophiques et scientifique d'al-Kindī*, vol. 1, pp. 571–645; an edition and English translation of the optics section with historical commentary is under preparation by the present author. For additional details, see Kheirandish, *Arabic Version of Euclid's Optics*; and Elaheh Kheirandish, "The Many Aspects of Appearances: Arabic Optics to 950 AD," in Hogendijk and Sabra, *Enterprise of Science in Islam*, pp. 55–83.

[12] David Lindberg, *Theories of Vision from al-Kindi to Kepler* (Chicago: University of Chicago Press, 1976), pp. 26–30, passim; and Sabra, *Optics of Ibn al-Haytham*, vol. 2, pp. 23–4, 28.

[13] On the relevant works of al-Kindī and Ibn Sahl, see, respectively, Rashed, *Oeuvres philosophiques et scientifique d'al-Kindī*; and Roshdi Rashed, *Géométrie et dioptrique au Xe siècle: Ibn Sahl, al-Qūhī et Ibn al-Haytham* (Paris: Les Belles Lettres, 1993). On the sine law of refraction, see the review of the preceding work and alternative arguments in A. I. Sabra, "Book Review: Géometrie et dioptrique au Xe siècle: Ibn Sahl, al-Quhi, Ibn al-Haytham Roshdi Rashed (Paris: Les Belles Lettres, 1993)," *Isis*, 85, no. 4 (1994), 685–6. On the "punctiform analysis of light radiation," see Lindberg, *Theories of Vision from al-Kindi to Kepler*, pp. 26–30.

[14] On the medieval Latin tradition, see Lindberg and Tachau, Chapter 20, this volume. On the Arabo-Latin transmission of Ibn al-Haytham's *Optics*, see Sabra, *Optics of Ibn al-Haytham*; A. Mark Smith, "Alhacen's Theory of Visual Perception (Books 1–3)," *Transactions of the American Philosophical Society*, 91, nos. 4 and 5 (2001); A. Mark Smith, "Alhacen on the Principles of Reflection (Books

This voluminous *Optics* integrated comprehensive theories of light (Book I), visual perception (Book II), and visual illusions (Book III), as well as vision through reflection and refraction (Books IV–VII), into a new theory of vision that influenced European epistemology, linear perspective, and the study of light and vision as late as the early-modern period. Ibn al-Haytham's systematic and interdisciplinary combination of the natural and mathematical sciences stood in clear contrast to the limiting hierarchical subordination schemes of Aristotle and his most thorough commentator, Ibn Rushd (Averroes; d. ca. 1198).[15] The success of the optical program was not limited to its ambitious combinations of the theory of forms (natural philosophy) with the visual models of geometrical optics (mathematics). With these, Ibn al-Haytham definitively solved the intromission–extramission puzzle (as will be discussed later) in favor of intromission (radiation moving from the object to the eye). His conscious combination of mathematical demonstrations and physical setups with empirical "tests" gave optics an increasingly experimental character.[16]

Later optical authors such as Naṣīr al-Dīn al-Ṭūsī (d. ca. 1274) and Kamāl al-Dīn al-Fārisī (d. ca. 1318) also made valuable contributions to the studies of vision and the rainbow. Ṭūsī's inquiry into the accuracy of vision continued to advance that study based on radiation distribution and regional intensity. Fārisī's study of the rainbow involved methodological as well as theoretical leaps; glass flasks filled with water reproduced the effect of sunlight on raindrops approximating what is now called "controlled" experimentation.[17]

The mechanics of the Islamic Middle Ages had its own share of achievements in both theory and practice. Most immediately, Arabic mechanics created a specific "science of weights" independent from, and more productive than, the activities corresponding to Greek "mechanics."[18] Whereas the

4–5)," *Transactions of the American Philosophical Society*, 96, nos. 2 and 3 (2006); and A. Mark Smith, "Alhacen on Image-Formation and Distortion in Mirrors," *Transactions of the American Philosophical Society*, 98, no. 1 (2008).

15 Sabra, *Optics of Ibn al-Haytham*; A. I. Sabra, "Ibn al-Haytham," *Dictionary of Scientific Biography*, VI, 189–210, at pp. 192–4; and A. I. Sabra, "The Physical and the Mathematical in Ibn al-Haytham's Theory of Light and Vision," in *The Commemoration Volume of Bīrūnī International Congress in Tehran*, publication no. 38 (Tehran: High Council of Culture and Arts, 1976), pp. 439–78 at p. 67.

16 See David Lindberg, "The Intromission–Extramission Controversy in Islamic Visual Theory: Alkindi versus Avicenna," in *Studies in Perception: Interrelations in the History of Philosophy and Science*, ed. Peter K. Machamer and Robert G. Turnbull (Columbus: Ohio State University Press, 1978); A. I. Sabra, "Ibn al-Haytham and the Visual-Ray Hypothesis," in *Ismāʿīlī Contributions to Islamic Culture*, ed. Seyyed Hossein Nasr (Teheran: Imperial Iranian Academy, 1977), pp. 189–205; and A. I. Sabra, "The Astronomical Origin of Ibn al-Haytham's Concept of Experiment," in *Actes du XIIe Congrès International d'Histoire des Sciences* (Paris: Albert Blanchard, 1971), pp. 133–6.

17 On the optical works of Ṭūsī and Fārisī, see Elaheh Kheirandish, "Mathematical Sciences through Persian Sources: The Puzzle of Ṭūsī's Optical Works," in *Sciences, techniques et instruments dans le monde iranien, X^e–XIXe siècle*, ed. Nasrollah Pourjavady and Živa Vesel (Teheran: Institut Français de Recherche en Iran/Presses Universitaires d'Iran, 2004), pp. 197–213; and A. I. Sabra, "The 'Commentary' That Saved the Text: The Hazardous Journey of Ibn al-Haytham's Arabic 'Optics'," *Early Science and Medicine*, 12 (2007), 117–33.

18 On the Arabic "science of weights," see Mohammed Abattouy, Jürgen Renn, and Paul Weinig, Introduction to "Intercultural Transmission of Scientific Knowledge in the Middle Ages," *Science*

latter primarily involved theoretical explanations, constructions, and operations of machines such as levers and pulleys, the former extended both the theoretical and practical aspects of mechanical knowledge to the study of equilibria and balances. The historical scope of the discipline corresponds to the medieval Latin "scientia de ponderibus," and later to the statics and hydrostatics that, along with kinematics and dynamics, formed an integral part of the science of mechanics in Europe after the sixteenth and seventeenth centuries.[19]

The theoretical and methodological advances of the Arabic mechanical traditions are relatively well documented for their contributions to laws governing simple machines, weighing instruments, specific weights, and projectile motions, most within the first few centuries of their development. Early examples include the theory and construction of mechanical devices and lifting machines such as the lever, treated extensively by the three Mūsā brothers (the Banū Mūsā, fl. ca. 860s) and later Badīʿ al-Zamān Ibn Razzāz al-Jazarī (fl. ca. 1200); theory and construction of balances and other weighing instruments involving Thābit ibn Qurra al-Ḥarrānī (d. ca. 901) long before ʿAbd al-Raḥmān al-Khāzinī (fl. 1115–30) and his contemporary, Muẓaffar al-Isfizārī (fl. ca. 1050–1116), himself associated with the initial stages of weight calculations and verification of precious metals, among other objects. Among other examples, Abū Rayḥān al-Bīrūnī (d. ca. 1050) and ʿUmar Khayyām (d. ca. 1137) carried out measurements of specific weights and volumes, while the Persian Ibn Sīnā (Avicenna; d. ca. 1037) and the Spaniard Ibn Bājja (Avempace; d. ca. 1138) advanced theories of projectile motion.[20] Figures directly associated with the sciences of the balance and motion, therefore, stretch in period, from the earlier to the later centuries, and in region, from the eastern to the western parts of Islamic lands. But the historical marker of mechanics as preeminently a science of "devices," rather than one of balances

in Context, 14 (2001), 1–12; Mohammed Abattouy, "Greek Mechanics in Arabic Context: Thābit ibn Qurra, Isfizārī, and the Arabic Tradition of Aristotelian and Euclidean Mechanics," *Science in Context*, 14 (2001), 179–247; Mohammed Abattouy, "The Arabic Science of Weights: A Report on an Ongoing Research Project," *Bulletin of the Royal Institute for Inter-Faith Studies*, 4 (2002), 90–130; and Hall, "Mechanics."

19 Hall, "Mechanics," p. 297. On medieval and early-modern mechanics, see Joseph E. Brown, "The Science of Weights," and John E. Murdoch and Edith D. Sylla, "The Science of Motion," in *Science in the Middle Ages*, ed. David C. Lindberg (Chicago: University of Chicago Press, 1978), pp. 179–205 and 206–64, respectively.

20 See Banū Mūsā, *Book of Mechanics (Kitāb al-Ḥiyal)*, ed. A. Y. al-Ḥassan (Aleppo: the Banū (sons of) Mūsā bin Shākir, 1981); Donald R. Hill, trans., *The Book of Ingenious Devices (Kitāb al-Ḥiyal) by the Banū Mūsā bin Shākir* (Dordrecht: Reidel, 1979); Al-Jazarī, *A Compendium of Theory and Practice in the Mechanical Arts*, ed. A. Y. al-Ḥassan (Aleppo: Maʿhad al-Turāth al-ʿIlmī al-ʿArabī, Jāmiʿat Ḥalab, 1979); and al-Jazarī, *The Book of Knowledge of Ingenious Mechanical Devices*, trans. Donald K. Hill (Dordrecht: Reidel, 1974). On the relevant works of Bīrūnī, Khayyām, Khāzinī, Isfizārī, Ibn Sīnā, and Ibn Bājja, see Hall, "Mechanics." On Khāzinī, see also Robert E. Hall, "Al-Khāzinī, Abu'l-Fatḥ ʿAbd al-Raḥmān," *Dictionary of Scientific Biography*, VII, 335–51 at pp. 340–3; and Nikolai Khanikoff, "Analysis and Extracts of Kitāb al-Mizān al-Ḥikma: Book of the Balance of Wisdom, an Arabic Work on the Water-Balance, written by al-Khāzinī in the Twelfth Century," *Journal of the American Oriental Society*, 6 (1860), 1–128.

or motion, is its function as a combined "operative–productive" craft as well as a theoretical–practical science, as distinguished in sources as early as the classification work of Ibn Lūqā.[21] The nature of transmission, disciplinary combinations, and natural or physical manifestations of mechanics in the Islamic Middle Ages are all features that distinguish it from both earlier optics and later mechanics.

HERITAGE

Optics and mechanics in the Islamic Middle Ages both belonged to more than one classical tradition. Optics drew on at least three textual traditions: (1) the mathematical visual models of Euclid (third to fourth centuries B.C.) and Ptolemy (second century), which applied geometrical demonstration to the behavior of rays; (2) the very different natural-philosophical traditions of Aristotle and the atomists (fifth to fourth centuries B.C.); and (3) the anatomical and medical tradition of Galen (second century) and his followers. Mechanics followed the tradition of mechanical devices by Heron and Pappus of Alexandria (first and fourth centuries, respectively), of balances by pseudo-Aristotle and Archimedes (third century B.C.), and of gadgets by Philo of Byzantium (second century B.C.).

The near-simultaneous reception of these sources in Arabic culture in the ninth century led to subtraditions and overlaps. In optics, visual-ray theories following authorities as different as Euclid and Galen produced overlaps between the medical and the mathematical traditions.[22] In mechanics, despite their diverse foci, the aforementioned works by Heron, Pappus, pseudo-Aristotle, Archimedes, and Philo of Byzantium all shared subjects such as weights.[23] The traditions of optics and mechanics proper were, however, both primarily mathematical, with a transparently Greek legacy. Both were subdivisions of geometry (plane and solid, respectively) within the classical quartet of "mathematical" sciences (arithmetic, geometry, astronomy, and music). And both had been considered "intermediate" sciences since

[21] For the distinctions and original expressions, see Hans Daiber, "Qosṭā ibn Lūqā (9. Jh.) über die Einteilung der Wissenschaften," *Zeitschrift für Geschichte der arabisch-islamischen Wissenschaften*, 6 (1990), 93–129; the Four Classes of Crafts (*ṣināʿāt*) are at pp. 102–4. (The English translations of the title and terms are mine.)

[22] On the Euclidean, Aristotelian, and Galenic traditions, see Lindberg, *Theories of Vision from al-Kindi to Kepler*. On the cross-overlaps, see Lindberg, "The Intromission–Extramission Controversy in Islamic Visual Theory." On the mathematical models, see Sabra, "Ibn al-Haytham and the Visual-Ray Hypothesis," pp. 189–205. On the relevance of the works of Alexander of Aphrodisias (ca. second to third centuries) and John Philoponus (ca. fifth to sixth centuries), see, most recently, Peter Adamson, "Vision, Light and Color in al-Kindī, Ptolemy and the Ancient Commentators," *Arabic Sciences and Philosophy*, 16 (2006), 207–36.

[23] The term "balances" (Mawāzīn) was included in the early Arabic title of the Aristotelian *Mechanica*, itself partially preserved through the *Book of the Balance of Wisdom (Kitāb al-Mizān al-Hikmah)* of Khāzinī (twelfth century); see Mohammed Abattouy, "Nutaf min al-Ḥiyal: A Partial Arabic Version of Pseudo-Aristotle's *Problemata Mechanica*," *Early Science and Medicine*, 6 (2001), 96–122.

antiquity because of the ontological position they held between the lower, natural or physical sciences and the higher, metaphysical sciences. While the Islamic scientific tradition continued to treat optics and mechanics as "intermediary" as part of the "Intermediate" or "Middle" Books, studied after Euclid's *Elements* and before Ptolemy's *Almagest*,[24] the two fields shared, as "mixed" sciences, the combination of mathematical and natural entities that their Arabic traditions advanced in both content and method more explicitly and effectively than their classical prototypes.[25]

The historical parallels are striking. The science of optics was initially a study of vision; its classification a subdivision of plane geometry; its focus sight rather than light; its method geometrical rather than experimental; and its subject the sciences of direct and indirect vision. The science of mechanics was initially a science of devices; its classification a subdivision of solid geometry; its focus on machines and powers, not just motion; its method operative and productive more than demonstrative; and its subject the science of devices and later the "science of weights." No less striking, however, are the ways in which optics and mechanics differed from one another, beginning with transmission.

TRANSMISSION

As in most Islamic scientific traditions, the early transmission of optics and mechanics involved Greco-Arabic and Arabo-Latin phases on a cross-cultural level and Arabo-Persian phases on an intracultural level. The differences between the external and internal transmissions of the two disciplines deeply affected later transformations and applications.

In optics, the works of several important Greek authors associated with ancient Alexandria were translated into Arabic. Euclid and Ptolemy had each written independent works entitled *Optics*. In addition, the fourth-century commentators Theon and Pappus and other Alexandrian commentators of antiquity are associated with optical texts. The Arabic versions of Euclid's *Optics* and pseudo-Euclid's *De speculis* were translated early enough for use in original ninth-century optical compositions.[26] The best known of these are

[24] On the "Intermediate Works," also known as "Middle Books" or the "Little Astronomy" (consisting of the *Elements, Optika, Data,* and *Phaenomena* by Euclid, *Spherics* of Theodosius and of Menelaus, and the *Moving Sphere* of Autolycus), see Moritz Steinschneider, "Die 'mittleren' Bücher der Araber und ihre Bearbeiter," *Zeitschrift für Mathematik und Physik*, 10 (1865), 456–98.

[25] For the original works on optics and mechanics proper, see Kheirandish, "Organizing Scientific Knowledge."

[26] On the *Optics* of Euclid and pseudo-Euclid in Arabic, see Elaheh Kheirandish, "What Euclid Said to His Arabic Readers: The Case of the Optics," in *Optics and Astronomy, Proceedings of the XXth International Congress of History of Science in Liège (20–26 July 1997)*, ed. Gérard Simon and Suzanne Débarbat (*Diversis Artibus*, Collection of Studies from the International Academy of the History of Science, 55, no. 18) (Turnhout: Brepols, 2001), pp. 17–28; and Kheirandish, *Arabic Version of Euclid's Optics*. On the Greek tradition, see Wilbur R. Knorr, "Pseudo-Euclidean Reflections in Ancient

works by Ya'qūb al-Kindī, the celebrated scholar of Kufa and Baghdad. His two major works on optics proper are *De aspectibus*, surviving only in a Latin version, and a recently discovered "Rectification" (*Taqwīm*) of problems in Euclid's *Optics*, known through a single Arabic manuscript presumably written after the widely circulated *De aspectibus*.[27]

In some cases, the same author produced both Arabic translations and original compositions. The prominent scholar of Nestorian Christian origin Ḥunayn ibn Isḥāq (d. ca. 877), for example, translated into Arabic Galen's *On the Usefulness of the Parts of the Body* and *Doctrines of Plato and Hippocrates*, and composed related works such as his *Ten Treatises on the Eye* and the *Treatise on Light*.[28] There were also a few combined optical works. Ibn Lūqā wrote, in addition to his translations, a joint work on optics and catoptrics. Likewise, Ibn 'Īsā combined optics and burning mirrors in another single composition, unusual for its specific references to ancient Greek authors (Euclid, Anthemius, Archimedes, Aristotle, Hippocrates, and Galen), as well as its transmission through a Hebrew transcription.[29]

The material aspects of the transmission of Greek scientific works, both in the original and through the intermediary of Syriac and middle-Persian, are well documented since rulers, translators, and authors often played direct roles in commissioning and collecting manuscripts of those works. The mechanism and order of transcriptions, dictations, and illustrations have also been recorded sufficiently to provide a picture of both textual production and nontextual transmission in the early Islamic period in particular.[30] But many puzzling questions remain for further research, ranging from the limited

Optics: A Re-examination of Textual Issues Pertaining to the Euclidean *Optica* and *Catoptrica*," *Physis*, 31, no. 1 (1994), 1–45; and Alexander Jones, "Peripatetic and Euclidean Theories of the Visual Ray," *Physis*, 31, no. 1 (1994), 47–76. On late Greek antiquity, see Adamson, "Vision, Light and Color in al-Kindī, Ptolemy and the Ancient Commentators."

27 Al-Kindī, *De Aspectibus*, in *Alkindi, Tideus und Pseudo-Euklid: Drei optische Werke*, ed. Axel A. Björnbo and Sebastian Vogl (*Abhandlungen zur Geschichte der mathematischen Wissenschaften*, 26, no. 3) (Leipzig: Teubner, 1912), pp. 1–176; editions and French translations of *De aspectibus* and *Rectification* are included in Rashed, *Oeuvres philosophiques et scientifique d'al-Kindī*, vol. 1, pp. 437–534 (ed. J. Jolivet, H. Sinaceur, and H. Hugonnard-Roche) and pp. 161–335, respectively. On the Latin transmission, see David Lindberg, *A Catalogue of Medieval and Renaissance Optical Manuscripts* (Toronto: Pontifical Institute of Mediaeval Studies, 1975), pp. 21–2; and Pinella Travaglia, *Magic, Causality, and Intentionality: The Doctrine of Rays in al-Kindī* (Florence: SISMEL/Edizioni del Galluzzo, 1999), pp. 118–20.

28 See, respectively, Margaret T. May, *Galen On the Usefulness of the Parts of the Body*, 2 vols. (Ithaca, N.Y.: Cornell University Press, 1968–1969); Phillip De Lacy, *On the Doctrines of Hippocrates and Plato: Galen's De placitis Hippocratis et Platonis, edition, translation and commentary* (Berlin: Akademie-Verlag, 1978–1984); and Max Meyerhof, *The Book of the Ten Treatises of the Eye Ascribed to Hunain ibn Ishâq (809–877 A.D.)* (Cairo: Government Press, 1928). See also Lindberg, *Theories of Vision from al-Kindi to Kepler*, pp. 33–57.

29 Kheirandish, *Arabic Version of Euclid's Optics*. On Ibn Lūqā and Ibn 'Īsā's works, see, respectively, Kheirandish, "Science and Mithāl"; and Kheirandish, "Footprints of Experiment in Early Arabic Optics."

30 See Johannes Pedersen, *The Arabic Book*, trans. Geoffrey French (Princeton, N.J.: Princeton University Press, 1984), pp. 20–36; and Jonathan Bloom, *Paper before Print: The History and Impact of Paper in the Islamic World* (New Haven, Conn.: Yale University Press, 2001).

transmission of the *Optics* of both Ptolemy and Ibn al-Haytham in Islamic versus European lands to the impact of the defective nature of the opening books of both sources on later concepts and developments.[31]

The wide diffusion in Europe of the seven books of the *Optics* of Ibn al-Haytham, owing to well-circulated Latin translations and a printed edition,[32] contrasts even more sharply with their limited circulation in Islamic lands. Remaining unknown to people as prominent as Naṣīr al-Dīn al-Ṭūsī, the "teacher of mankind," the great volumes became the subject of Kamāl al-Dīn al-Fārisī's impressive commentary, *Tanqīḥ al-Manāẓir*, only after his teacher Quṭb al-Dīn Shīrāzī (d. 1311), himself a student of Ṭūsī, obtained them from a "distant land." The work of Fārisī, best known for a theory of the rainbow put forward simultaneously with that of Theodoric of Freiberg (d. ca. 1310) in Europe, opens with a revealing preface. It provides explicit evidence for the discontinuous transmission in optics, identifying problematic treatments of such subjects as reflection and refraction in the works of "leading philosophers" as a primary motive for his own study, one imbued with novelties from double refraction to artificially produced conditions for observation.[33]

Despite a few similarities, the transmission of mechanics presents a sharp contrast to this picture. As in optics, the main Greek texts of mechanics proper were available in Arabic early on: Heron's *Mechanics*, through a direct Arabic translation by Ibn Lūqā, and Pappus's *Mechanics*, as part of his *Mathematical Collection*, apparently sponsored by the three sons of Mūsā ibn Shākir al-Munajjim (the Banū Mūsā), who wrote their own *Book of Mechanics* (*Kitāb al-Ḥiyal*). The books of Heron and Pappus also stimulated early mechanical works by Sīnān ibn Thābīt (d. 943) and, reportedly, Abū al-wafā' al-Buzjānī (d. 997) and Abū Alī Sinā.[34] But in contrast to the transmission of optics, neither the *Mechanics* of Heron and Pappus nor their Arabic versions and dependent compositions were as well known in medieval

[31] On the limited circulation of Ptolemy's *Optics*, see Sabra, *Optics of Ibn al-Haytham*, vol. 2, pp. lii–liii; for the case of Ibn Sahl, see pp. lix–lx; on the internal transmission of Ibn al-Haytham's own *Optics*, see pp. lxiv–lxxiii; and on the nontransmission of the first three books and its effects, see pp. lxxvi–lxxvii.

[32] Sabra, *Optics of Ibn al-Haytham*, vol. 2, pp. lxxiii–lxxix; Friedrich Risner, ed., *Opticae thesaurus* (Basel: n.p., 1572), reprinted with Introduction by David C. Lindberg (New York: Johnson Reprint, 1972); Smith, *Alhacen's Theory of Visual Perception*; and A. Mark Smith, "The Latin Source of the Fourteenth Century Italian Translation of Alhacen's *De aspectibus*," *Arabic Sciences and Philosophy*, 11 (2001), 27–44.

[33] For relevant references to Ibn al-Haytham, Ṭūsī, and Fārisī, see Sabra, *Optics of Ibn al-Haytham*, especially vol. 2, p. lxx; and Sabra, "The 'Commentary' That Saved the Text." On the optical works of Ṭūsī, see Kheirandish, "Mathematical Sciences through Persian Sources."

[34] Sīnān's *Principles of the Craft of Mechanics* corresponds to Book II of Heron's *Mechanics* (manuscript copy and bibliographical information provided by Mohammed Abattouy). Buzjānī's link to Pappus's *Mechanics* is mentioned by Jackson, the editor and translator of the Arabic Pappus (unpublished); see Elaheh Kheirandish, "The 'Fluctuating Fortunes of Scholarship': A Very Late Review Occasioned by a Fallen Book," *Early Science and Medicine*, 11 (2006), 207–22. For the mechanics attributed to Ibn Sinā, see *Miʿyār al-ʿuqūl*, ed. Jalāl al-Dīn Humāyī (Teheran: Anjuman-i Āthār-I Millī, 1952).

Europe as Euclidean and Ptolemaic optics.[35] Only much later did Pappus reach Guidobaldo del Monte (d. 1609), a patron of Galileo's, who called the "union of geometry and physics" in mechanics the "noblest art." Heron's text also bloomed very late, this time mostly through the transmission of its second book, including a Persian-French version, on the five mechanical devices or "powers" (axle, lever, pulley, wedge, and screw).[36]

DEVELOPMENTS: CONTEXT

Most scientific work produced in various Islamic regions was sponsored by patrons. As institutions of religious learning, the *madrasa* (school) and the mosque did not directly sponsor the teaching or practice of the "sciences of the ancients."[37] Patronage, however, stimulated a large number and wide range of scientific works, including those in optics and mechanics, however different in form. Court support was strong from the earliest centuries. The substantial body of work produced in both the sciences and crafts throughout the classical period range, in the case of texts, from theoretical tracts to practical manuals, and for instruments from constructed models to marvelous devices.[38] Of the many preserved or reported texts and instruments, those dedicated to or associated with patrons or companions provide important clues for the relations between the application, commission, or transmission of a given work.

The earliest scientific works on optics and mechanics were largely associated with ninth-century Baghdad. The best-known early optical author is Yaʿqūb al-Kindī, a prominent Arab scholar and ʿAbbāsid courtier under caliphs al-Maʾmūn (r. 813–33) and his brother and successor al-Muʿtaṣim

35 For the Latin versions of Euclid's *Optics*, see Wilfred R. Theisen, "The Mediaeval Tradition of Euclid's *Optics*" (unpublished PhD thesis, University of Wisconsin, 1972); and Wilfred R. Theisen, "Liber de visu: The Greco-Latin Translation of Euclid's *Optics*," *Mediaeval Studies*, 41 (1979), 44–105. For the pseudo-Euclidean *De Speculis*, see Björnbo and Vogl, *Alkindi, Tideus und Pseudo-Euklid*, pp. 97–158. For Ptolemy's *Optics*, see Albert Lejeune, *L'Optique de Claude Ptolémée dans la version latine d'après l'arabe de l'émir Eugène de Sicile: Édition critique et exégétique augmentée d'une traduction française et de compléments* (Leiden: Brill, 1989; first published 1956).

36 On the Latin tradition of Heron's *Mechanics* and its five simple machines, see Stillman Drake and I. E. Drabkin, *Mechanics in Sixteenth-Century Italy: Selections from Tartaglia, Benedetti, Guido Ubaldo and Galileo* (Madison: University of Wisconsin Press, 1969), pp. 259–328. On a late Persian source, see Edgar Blochet, *Catalogue des manuscrits persans de la Bibliothèque Nationale*, 4 vols. (Paris: Imprimerie Nationale, 1905–1934), vol. 2, no. 803, pp. 73–4.

37 See Fazlur Rahman, *Islam*, 2nd ed. (Chicago: University of Chicago Press, 1979), chap. 11; and Sabra, "Situating Arabic Science."

38 On the numbers, see David King, "Some Remarks on Islamic Scientific Manuscripts and Instruments and Past, Present, and Future Research," in *The Significance of Islamic Manuscripts*, ed. John Cooper (London: Al-Furqān Islamic Heritage Foundation, 1992), pp. 115–43 at p. 115. On some constructions, see George Saliba, "The Function of Mechanical Devices in Medieval Islamic Society," in *Science and Technology in Medieval Society*, ed. Pamela O. Lang (Annals of the New York Academy of Sciences, 441) (New York: New York Academy of Sciences, 1985), pp. 141–51.

(r. 833–42), an author of many scientific works, who reportedly fell out of favor and had his library confiscated.[39] A good instance of private support is his own "Rectification" (*Taqwīm*) of Euclid's *Optics*, which refers to "companions" requesting the composition of the work in the preface. But ruler support still dominated, especially for subjects with practical applications. The combined optical and catoptrical composition of Ibn Lūqā, which focuses on appearances in mirrors and other media, is dedicated to al-Muwaffaq (ca. 870–891), the ʿAbbāsid commander and proxy ruler.[40] The works of slightly later optical authors such as Ibn Sahl and Ibn al-Haytham, respectively associated with the courts of Būyīd Persia (r. 945–1055) and Fātimīd Egypt (r. 909–1171), similarly include such practically oriented areas as the construction of burning instruments and the regulation of water flow.[41] Some courtly figures were, in turn, personally engaged with theoretical works. The king of Saragossa, Mu'taman ibn Hūd (r. 1081–5), was apparently among the few who had access to Ibn al-Haytham's *Optics* before the thirteenth century, and wrote on the subject himself.[42] Other forms of patronage include the case of Naṣīr al-Dīn al-Ṭūsī (d. 1274), the prolific author and director of the Maragha observatory, who had a close association with Ismāʿīlī and Mongol rulers through astronomical works in both Arabic and Persian that provide interesting contrasts to their optical counterparts in that the latter lack any trace of dedications or commissions. The same observation holds for the much more important optical works of Ibn al-Haytham and Kamāl al-Dīn Fārisī and their dominant theoretical discourses.[43]

Not surprisingly, practical interests and direct links with specific patrons are more frequent in the case of mechanics. The dedication of Qusṭā ibn Lūqā's Arabic translation of Heron's *Mechanics* is to "Aḥmad ibn [Muḥammad ibn] Muʿtaṣim" (the prince or later caliph, depending on the exact reading); the translation of Pappus's *Mechanics* is associated with the

[39] See George N. Atiyeh, *Al-Kindi: The Philosopher of the Arabs*, Publication no. 6 (Rawalpindi: Islamic Research Institute, 1966), which lists 264 titles; and Gerhard Endress, "The Circle of Al-Kindī: Early Arabic Translations from the Greek and the Rise of Islamic Philosophy," in *The Ancient Tradition in Christian and Islamic Hellenism*, ed. Gerhard Endress and Remke Kruk (Leiden: Research School CNWS, 1997), pp. 43–76.

[40] For "On the Reasons for What Occurs in Mirrors on Account of the Difference of Aspects," see Kheirandish, "Qusṭā Ibn Lūqā," which includes references to relevant texts and dates.

[41] On the related works of these authors, see Roshdi Rashed, *Optique et mathématiques: Recherches sur l'histoire de la pensée scientifique en arabe* (Aldershot: Ashgate Variorum Reprints, 1992); Rashed, *Géométrie et dioptrique au Xᵉ siècle*, pp. 53–6; and Sabra, *Optics of Ibn al-Haytham*, vol. 2, pp. lii–liii, lix–lx.

[42] Jan P. Hogendijk, "Discovery of an Eleventh-Century Geometrical Compilation: The Istikmāl of Yūsuf ibn Hūd, King of Saragossa," *Historia Mathematica*, 13 (1986), 43–52. On the king's apparent access to Ibn al-Haytham's *Optics*, see Sabra, *Optics of Ibn al-Haytham*, vol. 2, p. lxiv and n. 94.

[43] On Ṭūsī's optical and catoptrical works, see Kheirandish, "Mathematical Sciences through Persian Sources." On the astronomical dedications, see F. Jamil Ragep, "The Persian Context of the Ṭūsī Couple," in *Naṣīr al-Dīn al-Ṭūsī: Philosophe et savant du XIIIᵉ siècle*, ed. Nasrollah Pourjavady and Živa Vesel (Teheran: Institut Français de Recherche en Iran, 2000), pp. 113–30, especially p. 116.

three "Banū Mūsā" brothers, owners of a transcribed copy by ʿAbd al-Jalīl al-Sijzī (from ca. 969–70),[44] authors of a *Mechanics* of their own for both function and entertainment,[45] and courtiers of the ʿAbbāsid caliph Mutawwakil (r. 847–61), identified with construction and irrigation projects in Baghdad and Samarrah. Sīnān, son of the contemporary translator-scholar Thābit ibn Qurra and reported author of a treatise on the five simple machines, was apparently well supported as the personal physician of three early-tenth-century ʿAbbāsid caliphs. And the slightly later Ibn Razzāz al-Jazarī (ca. 1200) reports that his patron, the Artuqid prince of Diyārbakr, "asked him to commit to writing a description of the things he had made," a "Compendium of Theory and Practice in Mechanical Arts," that came to attract patrons like the sponsor of a late Persian translation and continued to generate impressive descriptions and illustrations of mechanical models well known to art historians and manuscript collectors.[46]

Analogous patronage connections appear in the mechanical tradition of weights and balances. Thābit ibn Qurra, the author credited with the important "Book of the Steelyard" (*Kitāb al-Qarasṭūn*), was a prominent scholar of Ṣabīʾan (star-worship) origin and well known for commissions by the Banū Mūsā, in particular.[47] Al-Khāzinī, the famous author of the "Book on the Balance of Wisdom," or "Philosophical Balance" (ca. 1121/1122), and his older contemporary, al-Isfizārī, were well connected to the Seljūq court, where they worked extensively on the theory and construction of balances and other weighing instruments. The court also had its risks: al-Isfizārī reportedly died of grief after the court treasurer destroyed his hydrostatic balance.[48]

These and other sources show that optics and mechanics proper, as well as the neighboring traditions of catoptrics and weights, followed comparable courses in involving various doses of practice and theory. In optics and catoptrics, they covered in addition to reflection and refraction burning mirrors and instruments, and in the sciences of mechanics and weights, they ranged from balances and steelyards to centers of gravity and inclination.

44 On the identification problems, see Gutas, *Greek Thought, Arabic Culture*, pp. 125–6; and Kheirandish, "Fluctuating Fortunes of Scholarship."

45 Banū Mūsā, *Book of Mechanics*; Hill, *Book of Ingenious Devices*.

46 Al-Jazarī, *Compendium of Theory and Practice in the Mechanical Arts*; and al-Jazarī, *Book of Knowledge of Ingenious Mechanical Devices*. See Saliba, "Function of Mechanical Devices in Medieval Islamic Society," especially pp. 147–51. On the patrons of Al-Jazarī's book in Arabic and Persian, see, respectively, Sabra, "Scientific Enterprise"; and Hasan Taromi-Rad, "The Persian Translation of Badīʿ al-Zamān al-Jazarī's Treatise in Mechanics," in *La science dans le monde iranien à l'époque islamique*, ed. Živa Vesel, H. Beikbaghban, and Bertrand Thierry de Crussol des Epesse (Bibliothèque Iranienne, 50) (Teheran: Institut Français de Recherche en Iran, 2004), pp. 199–204.

47 On *Kitāb al-Qarasṭūn* and authorship problems, see Wilbur R. Knorr, *Ancient Sources of the Medieval Tradition of Mechanics: Greek, Arabic and Latin Studies of the Balance* (Florence: Istituto e Museo di Storia della Scienza, 1982).

48 On al-Khāzinī, al-Isfizārī, and the Seljūq court, see Hall, "Al-Khāzinī, Abuʾl-Fatḥ ʿAbd al-Raḥmān."

Optical works of the postclassical period have the distinction of not only being independent compositions (as opposed to translated versions of original works) but also largely lacking the stamp of a ruler or patron, as in the optical and catoptrical works of Ghiyāth al-Dīn Manṣūr al-Dashtakī (d. ca. 1541), after whom the "Manṣūrīyah School" in Shiraz was named, and Qāsim-ʿAlī al-Qāyinī (ca. 1661), who belonged to an important circle of instrument makers.[49] Late works associated with patrons come from beyond Arabic and Persian regions even when written in those languages.[50]

The range of sources in optics and mechanics, in both Arabic and Persian, provides further clues to their respective applications, commissions, transmissions, and contributions. That optics was more continuously and effectively transmitted owing to its internal problems and applications is indicated by the prefaces to optical works of people like Ibn Lūqā and Ibn al-Haytham. Mechanics, on the other hand, was naturally driven through external applications and commissions as well, and both textual and material evidence are needed to identify patrons beyond ʿAbbāsid and Seljūq rulers. The critically important role of transmission and communication for both scientific traditions becomes much more transparent with the identification of several scholarly circles, in the case of optics around al-Kindī and Ibn Lūqā in ninth-century Baghdad and Ṭūsī and Shirāzī in thirteenth-century Maragha, and in mechanics around Ibn Qurra and Banū Mūsā in ninth- and tenth-century Baghdad and Khāzinī and Isfizārī in twelfth-century Khurāsān.

DEVELOPMENTS: OPTICS

The well-known classification work of Abū Naṣr al-Fārābī (d. ca. 950), written in Arabic sometime before the mid-tenth century, depicts optics as a science of vision devoted to both direct and mediated appearances (*manāẓir* and *marāyā*). Of these subdivisions (corresponding to Greek *optika* and *catoptrika*), optics proper (literally "the science of aspects") underwent particularly outstanding development during the Islamic Middle Ages.[51] After centuries of competing visual hypotheses and models, a decisive and lasting theory was established about the nature and manner of vision, the main focus

[49] See Kheirandish, "The Manāẓir Tradition Through Persian Sources," in Vesel, Beikbaghban, and Thierry de Crussol des Epesse, *La science dans le monde iranien*, pp. 125–45; and Kheirandish, *Arabic Version of Euclid's Optics*, vol. 1, pp. lviii–lix.

[50] Ibid.; and Kheirandish, "Science and Mithāl."

[51] Al-Fārābī, *Enumeration of the Sciences*, ed. Osman Amine (Cairo: Maktabat al-Anjalū al-Miṣrīyah, 1968), pp. 79–84. An English translation of the optics section is included in Sabra, *Optics of Ibn al-Haytham*, vol. 2, pp. lvi–lvii; and Kheirandish, "Many Aspects of Appearances."

of early optics.[52] The theoretical developments of the first few centuries of the Arabic optical tradition also coincided with considerable advances in the disciplinary boundaries of the field, from a study of vision (*manāẓir*) that at best included vision involving media (*marāyā*) and measurements (*misāḥa*) to one that integrated much broader areas from theories of light, shadows, and rainbows to the psychology and physiology of visual perception. With increasing disciplinary exchanges with natural philosophy as well as astronomy, the field's methodology also shifted from strictly geometrical modes to more physical and experimental realms. The disciplinary, theoretical, and methodological developments of optics as a study of vision have been treated elsewhere, for both the Arabic and Persian traditions and for direct and indirect forms of vision.[53] An updated outline and time frame for the more outstanding developments is the subject of the present section.

A major theoretical development in optics was a comprehensive and definitive theory of vision and perception that combined visual models advocated primarily by the mathematical, natural-philosophical, and medical traditions after authorities such as Euclid, Aristotle, and Galen.[54] Authors with self-contained optical works whose dominant traditions and positions were often reflected in their very titles were by no means the only ones to write on the subject. Al-Fārābī, the famous philosopher and "second teacher" (after Aristotle), for example, wrote about optics and not on it, but he is nevertheless associated with a position on the subject. Of the several visual models available at the time – the visual-ray theory (mostly Euclidean and Galenic), the theory of forms (both Aristotelian and atomist), and the theory of coalescence (mainly Platonic) – Fārābī apparently favored the latter. And, with further variations, options for later authors came to include, besides the visual models of mathematicians and natural philosophers, those of rational theologians and illuminationist philosophers.[55]

Before an inclusive combination of mathematical and physical criteria created a new foundation for a new theory of vision and science of optics, optical authors commonly presented alternative models before taking a stance on the subject, often without reference to specific authorities or traditions. Al-Kindī, writing as he did within the Euclidean tradition, presented four alternative models of vision – as a power proceeding from the

52 See Lindberg, "Intromission–Extramission Controversy in Islamic Visual Theory"; and Katherine Tachau, *Vision and Certitude in the Age of Ockham: Optics, Epistemology, and the Foundations of Semantics, 1250–1345* (Leiden: Brill, 1988).

53 On these developments, see A. I. Sabra, "Optics, Islamic," in *Dictionary of the Middle Ages*, ed. J. R. S. Strayer (New York: Charles Scribner's Sons, 1987), vol. 9, pp. 240–7; and "Manāẓir, or 'Ilm al-Manāẓir," in *Encyclopedia of Islam*, 2nd ed., vol. 6, fasc. 103–4, pp. 376–7. See also Elaheh Kheirandish, "Optics in the Islamic World," in *Encyclopaedia of the History of Science, Technology, and Medicine in Non-Western Cultures* (Dordrecht: Kluwer, 1997), pp. 795–99.

54 See Lindberg, *Theories of Vision from al-Kindi to Kepler*; and Sabra, *Optics of Ibn al-Haytham*.

55 See Franz Rosenthal, "On the Knowledge of Plato's Philosophy in the Islamic World," *Islamic Culture*, 14 (1940), 412–16.

eye, as forms traveling to the eye, as the two occurring simultaneously, or as forms being stamped in the eye through air – without mentioning the corresponding Euclidean, Aristotelian, Platonic, and atomistic traditions.[56] His contemporary, Ḥunayn Ibn Isḥāq, associated mostly with the Galenic tradition, did likewise before offering his variant visual model. And Ibn Sīnā, the famous eleventh-century philosopher, writing mostly in the Aristotelian tradition, presented his own version of the models but stopped short of offering a positive theory.

Ibn al-Haytham, by contrast, not only pledged in his *Optics* to combine the natural and mathematical forms of inquiry and the theories of vision but also offered introductory chapters that critically reviewed and extensively analyzed several existing visual models.[57] His proposed visual theory was a combined model drawing at once from the Aristotelian (and atomist) intromission theory of forms and the Euclidean (and Ptolemaic) extramission theory of visual cones. In this clever synthetic model, the incoming forms of the natural philosophers reached the cognitive faculties along the conic visual lines of the mathematicians, themselves selectively restricted to imaginary perpendiculars forming the outward extension of a visual cone.[58] The celebrated principle of the "punctiform analysis of light radiation" (the assumption of a linear correspondence between illuminating and illuminated points) was critical to the conception of that model. Closely associated with visual models as early as al-Kindī, this principle was a key component of Ibn al-Haytham's combination of aspects of extramission and intromission theories that would later be adopted by Western perspectivists as late as Kepler, the first to use inverted images in the model.[59] Significantly, the formulation of the principle itself was largely dependent on both a combined conception of mathematical and physical lines and the analogous behavior of visual and luminous radiation (Figure 3.1).

Major theoretical developments during this period coincided with methodological and disciplinary upgrades. As Aristotle's *Physics* had explicitly described optics as a more "natural/physical" mathematical science and "visual rays" as more physical than mathematical lines (in contrast to geometry's treatment of physical lines as mathematical), the topical and theoretical domains of the field took their cues from the tradition of Aristotelian physics.

[56] On the relevant Greek traditions, see Lindberg, *Theories of Vision from al-Kindi to Kepler*. For an extended scope of al-Kindī's ancient sources, see Rüdiger Arnzen, *Aristoteles' De anima: eine verlorene spätantike Paraphrase in arabischer und persischer Überlieferung: arabischer Text nebst Kommentar quellengeschichtlichen Studien und Glossaren* (Leiden: Brill, 1998); and Adamson, "Vision, Light and Color in al-Kindī, Ptolemy and the Ancient Commentators."

[57] Sabra, "Ibn al-Haytham," p. 191; Sabra, "The Physical and the Mathematical in Ibn al-Haytham's Theory of Light and Vision"; and Sabra, *Al-Ḥasan Ibn al-Haytham, Kitāb al-Manāẓir*, p. 60.

[58] The transition from the Euclidean visual model to that of Ibn al-Haytham is illustrated in Kheirandish, "The Arabic 'Version' of Euclidean Optics." See also n. 59.

[59] Lindberg, *Theories of Vision from al-Kindi to Kepler*, pp. 30, 63, 193, and 241 n. 5. For the case of Ibn al-Haytham and the Latin perspectivists, see Sabra, *Optics of Ibn al-Haytham*, vol. 2, pp. 23–4, 28. With reference to Kepler, see Sabra, "Ibn al-Haytham's Revolutionary Project in Optics."

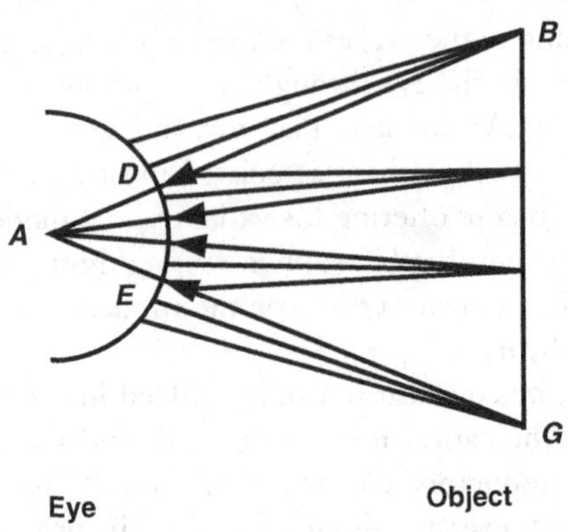

Figure 3.1. Combined extramission-intromission optical model. The incoming rays (intromission) along a set of perpendicular lines (*BDA*, *GEA*) with one-to-one correspondence with sensitive visual points (*D*, *E*) are geometrically equivalent to the imaginary visual cone *BAG* formed by rays assumed to issue from the eye (extramission).

But so long as the theorems of optics were made subordinate to those of geometry (and those of rainbows to optics itself), the methodological guidelines of the field were initially set by Aristotle's *Posterior Analytics*.[60] For authors as early as al-Kindī, Ibn Lūqā, and al-Fārābī, all of whom discussed the methodology of optics, both Aristotelian works acted as indispensable sources before other models and criteria provided extended methodological contexts.

In a key passage of the *Posterior Analytics*, Aristotle had reserved the privileged knowledge of the "cause" or reason for the mathematician and, with respect to subordinated sciences, the optician. This passage legitimated a method of applying geometrical proofs to propositions on vision, a method that found its way early into optical texts along with the terminology of "reasoning" (*'illa*) that entered some early arguments and titles. As other influential sources like Galen's *On the Usefulness of the Parts of the Body* licensed the flow of natural philosophy into mathematical sciences from physiological and other angles,[61] alternative methods increasingly integrated the natural philosophical criteria of direct experience and physical apparatus in the earliest optical works. In his *De aspectibus* (extant only in Latin), al-Kindī qualified optics as one of the "propaedeutic sciences [*artes doctrinales*] in which geometrical demonstrations would proceed in accordance with the requirements of physical things" and supplemented geometrical proofs with evidence from the world of experience.

60 Aristotle, *Physics*, 194a6–11, *Posterior Analytics*, 75b15, Loeb ed., pp. 120–1 and 60–3, respectively.
61 Aristotle, *Posterior Analytics*, 79a3–15; Galen, *De usu partium*, in May, *On the Usefulness of the Parts of the Body*, pp. 496–9.

It was at almost the same time that Ibn Lūqā, describing optics as the "perfect" mixed mathematical science, spoke about the incomparable excellence of the "coming together of sense perception and geometrical demonstrations" in the science of optics. Even in an optical text such as that of Ibn ʿĪsā, the full title of which placed it "in the tradition of Euclid," the author supported the geometrical "illustration" (such as the "conical spreading out" of rays) that was part of a Euclidean proposition with a physical illustration involving devices such as "tubes" for "comparative testing."[62]

Ibn al-Haytham continued the tradition of his predecessors, especially Ptolemy's *Optics*,[63] in regarding demonstrations involving experience as part of a mathematical inquiry. His own *Optics* placed both the theoretical and methodological dimensions of the field on a new platform. His arguments and physical setups refuted the hypothesis of extramitted visual rays and settled an age-old deadlock in the visual theories of the two main camps, the mathematical extramission and the natural-philosophical intromission traditions. Each was typically challenged for its inability to explain specific phenomena: in the former case, the appearance of objects as far away as stars (extramitted visual rays being arguably too weak by the time they reach the stars); and, in the latter case, the reception of visible forms as large as mountains (the eye being too small to receive their full form). Ibn al-Haytham also made direct contributions to the study of light through the conception of a finite but imperceptible interval of time for the movement of light.[64] Ibn al-Haytham's understanding of experimentation as a distinct form of argumentation appropriate for optical inquiries was among his most significant contributions. His adoption and appropriation of the astronomically inspired language of "testing" (*iʿtibār*) from Ptolemy's *Almagest*, along with the concept of "test or proof" (*peira*) used in astronomical records, led to the lasting conceptual and terminological apparatus of *experimentare* and *experiri* in the Latin tradition.[65]

For Ibn al-Haytham, the recurring theme of synthesis between mathematics and natural philosophy extended beyond his studies of light and vision. In his *Optics*, he proposed the combination of natural and mathematical inquiry according to subject matter (light, ray, transparency, air, and sense of vision for the natural, and shapes, magnitudes, and positions for the mathematical), and his later *Discourse on Light* discussed the physical "nature" of light and the mathematical "manner" of its propagation independently.

[62] For the early optical works of al-Kindī, Ibn Lūqā, and Ibn ʿĪsā and their full titles, see Kheirandish, *Arabic Version of Euclid's Optics*, vol. 1, p. liv; and Sabra, *Optics of Ibn al-Haytham*, vol. 2, pp. 25–6.

[63] Sabra, *Optics of Ibn al-Haytham*, vol. 2, pp. lii–liii, lviii–lx.

[64] Sabra, *Theories of Light from Descartes to Newton* (London: Oldbourne, 1967; Cambridge: Cambridge University Press, 1981), p. 47.

[65] Sabra, "Astronomical Origin of Ibn al-Haytham's Concept of Experiment," p. 133. See also Roshdi Rashed, "Experimentation," in *Geometry and Dioptrics in Classical Islam* (London: al-Furqān Islamic Heritage Foundation, 2005), pp. 1039–45; and Kheirandish, "Footprints of Experiment in Early Arabic Optics."

But the fine lines of Ibn al-Haytham's great "synthesis," in method, not just subject, lay in his subtle, but momentous, remark in the *Halo and Rainbow*, written toward the end of his life, on the appropriate application of "composite inquiry for composite matter." This was the same sort of reasoning behind his earlier combinations of observation-based justifications of astronomy (natural) with the logically drawn propositions on light and vision (mathematical).[66]

There was much more to this methodological transformation than experimental verifications of assumptions such as the rectilinear propagation of light, already attempted by Ptolemy and al-Kindī, transformations that continued through the optical works of European perspectivists dependent on them, from Roger Bacon, John Pecham, and Witelo to early-modern scholars such as Kepler.[67] Kamāl al-Dīn al-Fārisī, one of the few heirs of Ibn al-Haytham's *Optics* in Islamic lands, refined the method of controlled experimentation to the point of extending it to phenomena such as rainbows, which he explicitly claimed as "a subject properly belonging to optics." Diverse subjects such as "dark chambers," involving classical "forms" rather than modern "pictures" in the more advanced treatments of the camera obscura, still excluded subjects partial to the presence of the eye – a condition that largely defined the subject matter of optics at the time. Whereas the study of vision itself had long ago moved beyond the appearance of heavenly bodies and earthly objects to the determination of heights, widths, and depths insofar as it involved vision, a subject such as burning mirrors largely continued to be treated independently, not only in works as early as the treatise of Ibn Sahl on the subject (ca. 984) but as late as the optical works of al-Fārisī himself.[68]

The methodological breakthroughs accompanying disciplinary expansions of optics still focused on verification rather than discovery, as would be the case in a much later period in Europe. These occurred as the Aristotelian notion of the subalternation of disciplines was replaced, first by their "co-operation" (Ibn Lūqā) and later their full-fledged "combination" (Ibn al-Haytham). The shift in the hierarchy and authority of the sciences was itself brought to optics mainly through a critical mix with astronomy,

66 Sabra, "The Physical and the Mathematical in Ibn al-Haytham's Theory of Light and Vision."

67 Sabra, *Optics of Ibn al-Haytham*, vol. 2, pp. lxxiii–lxxix; and David C. Lindberg, *Roger Bacon and the Origins of Perspectiva in the Middle Ages: A Critical Edition and English Translation of Bacon's "Perspectiva" with Introduction and Notes* (Oxford: Oxford University Press, 1996).

68 See Roshdi Rashed, "Kamāl al-Dīn al-Fārisī," *Dictionary of Scientific Biography*, VI, 212–19; Sabra, "Optics, Islamic," vol. 9, pp. 242 and 246; Sabra, *Optics of Ibn al-Haytham*, vol. 2, p. lxii, nn. 91–92; S. H. Nasr, "Quṭb al-Dīn Shīrāzī," *Dictionary of Scientific Biography*, XI, 250; and Kheirandish, "Optics," especially p. 347. On the *Optics* of Ibn al-Haytham and Fārisī and their respective treatments of camera obscura phenomena, see Sabra, *Optics of Ibn al-Haytham*, vol. 2, pp. li–lii and n. 72. On Ibn Sahl's work on burning mirrors, see Roshdi Rashed, "A Pioneer in Anaclastics: Ibn Sahl on Burning Mirrors and Lenses," *Isis*, 81 (1990), 464–91.

a field for which optics had acted since antiquity as a preparatory study in the mathematical curriculum. What astronomy brought to optics was more than scientific content – namely, the means of accommodating elements of observation and experimental "testing" to the study of light and vision. This was a privilege that optics, in turn, exercised with respect to the study of rainbows, the one science explicitly and repeatedly subordinated to it. It was, however, a privilege that did not similarly extend to the other mixed mathematical sciences, including mechanics, the one historically closest to it, despite its many advances in both mechanical knowledge and practice.

DEVELOPMENTS: MECHANICS

In contrast to the science of optics, in which direct and indirect vision (*manāẓir, marāyā*) formed two aspects of the same discipline, the sciences of mechanics and weights (*ḥiyal, athqāl*), presented as such in Fārābī's *Catalogue*,[69] represented distinct traditions and developments throughout the Islamic Middle Ages. The ancient tradition of pseudo-Aristotle's *Mechanical Problems*, which was clearly distinguished from "physical problems" and endowed with a "share in mathematical and natural disciplines," was, however, foundational to both traditions. On the one hand, the Aristotelian "mechanical problems" were defined in the science of devices or contrivances (*mēchanē*), involving "lesser weights mastering greater weights," with special reference to the lever (one of the five mechanical devices central to the *Mechanics* of both Heron and Pappus of Alexandria).[70] On the other hand, they were also closely tied to the sciences of weights through both the standard and nonstandard forms of the balance. The former was a weighted beam suspended from its midpoint, with scale pans at both ends (equal-arm balance); the latter, known as a steelyard or unequal-arm balance, involved a beam suspended from a point close to one end, with weights on the shorter arm and counterweights on the longer arms.[71] The force–distance

[69] Al-Fārābī, *Enumeration of the Sciences*, pp. 88–9. An English translation of the "science of mechanics" is included in Saliba, "Function of Mechanical Devices in Medieval Islamic Society," pp. 145–6; and that of the "science of weights" in Marshall Clagett, "Some General Aspects of Physics in the Middle Ages," *Isis*, 39 (1948), 29–44 at p. 32.

[70] Pseudo-Aristotle, *Mechanical Problems*, 847a24–847b, in Aristotle, *Minor Works*, trans. W. S. Hett (Cambridge: Cambridge University Press, 1936), p. 331. On the Greek tradition, see A. G. Drachmann, *The Mechanical Technology of Greek and Roman Antiquity* (Copenhagen: Munksgaard, 1963). On the Arabic tradition, see Abattouy, "Nutaf min al-Ḥiyal"; and Saliba, "Function of Mechanical Devices in Medieval Islamic Society."

[71] On the traditions of balances and steelyards, and the original Greek, Arabic, and Latin terms (*mīzān = zugon, libra,* or *bilanx; qarasṭūn = charistiôn, statera, trutina, falagges*), see Knorr, *Ancient Sources of the Medieval Tradition of Mechanics*; K. Jaouiche, "Karasṭūn," in *Encyclopedia of Islam*, 2nd ed., vol. 4; and Abattouy, "Arabic Science of Weights," pp. 112–13. On Ibn Lūqā's *Kitāb fī al-Qarasṭūn*, see Hall, "Mechanics," p. 302.

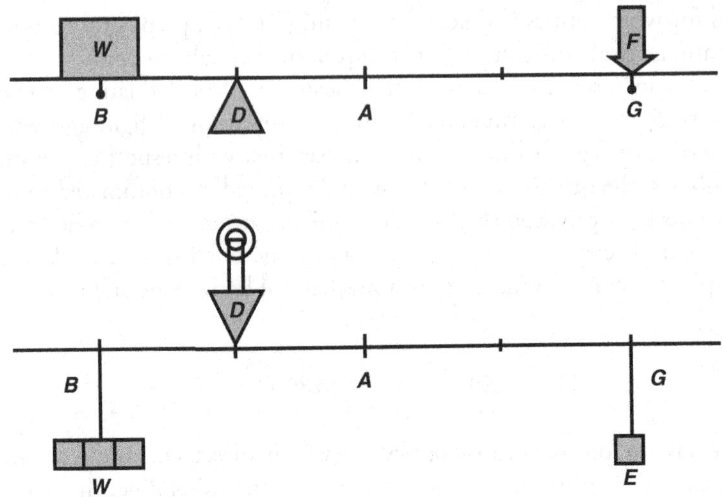

Figure 3.2. Combined lever-balance mechanical model. Lever model (top): When beam *BG* has its fulcrum at *D* (where $BD = \frac{1}{3}DG$), upper weight *W* (3 units) at *B* is lifted with force *F* (1 unit $= \frac{1}{3}W$) since $3 \times 1 = \frac{1}{3}(3) \times 3$. Balance model (bottom): When beam *BG* has its suspension point at *D* (where $BD = 1$ and $DG = 3$), lower weight *W* (3 units) is balanced with lower weight *E* (1 unit $= \frac{1}{3}W$) since $3 \times 1 = \frac{1}{3}(3) \times 3$.

proportionality of the *Mechanical Problems*, according to which the same force moves the endpoints on the radius of a circle faster than its midpoint, was close to the model of the lever commonly associated with Archimedes, for whom the same force would counter greater weights the longer their distance from the lever's fulcrum. The principle behind the function of the lever was, in turn, close to that of an unequal-arm balance, in which distance and equilibrium both played significant roles[72] (Figure 3.2).

Arabic texts in the tradition of weights included works as early as the important ninth-century *Kitāb al-Qarasṭūn*, attributed to Thābit ibn Qurra, a text with a complex Latin tradition through both the *Liber Karastonis* and *Liber de Canonio* and with an impact in Europe long before the tradition of mechanical "devices" (*Ḥiyal*).[73] The tradition of "weights" is also well represented by the famous book on the "Balance of Wisdom" of the twelfth-century author al-Khāzinī, a work that preserves many earlier treatments, including the Arabic version of the pseudo-Aristotelian *Mechanical Problems*. Developments within the same tradition included the construction of

72 See Jürgen Renn, Peter Damerow, and Peter McLaughlin, "Aristotle, Archimedes and the Origin of Mechanics: the Perspective of Historical Epistemology," in *Symposium Arquímedes*, ed. J. L. M. Sierra (Berlin: Max Planck Institut für Wissenschaftsgeschichte, 2003), preprint 239, pp. 43–59.

73 Knorr, *Ancient Sources of the Medieval Tradition of Mechanics*, p. 106.

balances by predecessors of Khāzinī, such as Zakariyā al-Rāzī (d. ca. 987), as well as his own contemporaries, Khayyām and Isfizārī. Although Bīrūnī in the eleventh century had already calculated specific weights and volumes, al-Khāzinī's work brought the field not only to a high point in the theory of construction of mechanical instruments but also to the culmination of mechanical work in the entire Islamic Middle Ages.[74]

Developing alongside the Arabic "science of weights," which continued its impressive course in Europe through the "scientia de ponderibus," was the science of motion. This subject belonged to Aristotelian natural philosophy, which remained a part of the science of weights in Arabic mechanics in accordance with al-Fārābī's description of the science of weights as investigating "motion" and "measure." Concepts in dynamics, based on Aristotle's *Physics* and *De caelo*, were elaborated by the works of Ibn Sīnā in the East and Ibn Ṭufayl (d. ca. 1130) and Ibn Rushd in the West, as were theories of projectile motion.[75] But what was historically conceived as "mechanics" was a different tradition that included independent works with such a title following Heron and Pappus of Alexandria. In contrast to modern treatments of early mechanics that encompass the sciences of weights, motion, statics, dynamics, and their philosophical frameworks, the present account concerns the historically defined "science of mechanics" and their earlier conceptions and perceptions, especially as a combined science or craft.

The historical dominance of the science of mechanics or devices in the early Islamic period is well reflected by early sources such as the *Key of the Sciences* (*Mafātiḥ al-ulūm*) of al-Kātib al-Khwārizmī (d. ca. 987), an encyclopedia of technical terms with two chapters on "mechanics" (*ḥiyal*) next to mainstream mathematical sciences,[76] and with no mention of the science of weights, like the contemporary *Enumeration of the Sciences* of al-Fārābī (d. ca. 950). The distinct nature and functions of the discipline are further indicated by later classifications in both Arabic and Persian, where people like Ibn Sīnā, al-Ṭūsī, and al-Shīrāzī classified under "geometry" various forms of "mechanical" subjects, from "devices in motion" to "traction by weights" and "transfer of water."[77]

[74] For details, see Hall, "Mechanics," pp. 310–14.

[75] On the Arabic tradition of Aristotelian physics, see Paul Lettinck, *Aristotle's Physics and Its Reception in the Arabic World, with an Edition of the Unpublished Parts of Ibn Bājja's Commentary on the Physics* (Aristoteles Semitico-Latinus, 7) (Leiden: Brill, 1994). On the Latin tradition of the science of motion, see Murdoch and Sylla, "Science of Motion."

[76] G. van Vloten, ed., *Liber Mafâtîh al-olûm, explicans vocabula technica scientiarum tam Arabum quam peregrinorum, auctore Abû Abdallah Mohammed ibn Ahmed ibn Jûsof al-Kâtib al-Khowarezmi* (Leiden: Brill, 1895, 1968). See also C. E. Bosworth, "A Pioneer Arabic Encyclopedia of the Sciences: Al-Khwārizmī's Key of the Sciences," *Isis*, 54 (1963), 341–5.

[77] See Mahdi Muhaqqiq, "The Classification of the Sciences," in *Different Aspects of Islamic Culture*, vol. 4, part I, pp. 116–19. See also Peters, *Aristotle and the Arabs: the Aristotelian Tradition in Islam* (New York: New York University Press; London: University of London Press, 1968), pp. 105–9; Naṣir al-Dīn al-Ṭūsī, *Akhlāqi Nāṣirī*, trans. G. M. Wickens (London: Allen and Unwin, 1964);

The applications of mathematics to natural bodies in both branches of mechanics were largely limited to geometrical models or proofs, in contrast to the observational methods of both optics and astronomy. Mechanics had no explicit counterpart to the concept of "testing" that entered optics from the "trials" and "comparisons" of Ptolemaic astronomy.[78] Following al-Fārābī's description of the "mechanical sciences" characterized under that entry as "applying mathematical statements and proofs to natural bodies through a craft," al-Khāzinī's *Balance of Wisdom* describes the methods of a subject such as the balance as "founded upon geometrical demonstrations deduced from physical causes."[79] Every craft is said to be "a combination [*murakkab*] of geometrical and natural/physical crafts through the two complementary [*jāmiʿ*] entities of quantity and quality." Within a chiefly geometrical approach, the mechanical tradition of the Islamic Middle Ages still extended the methods of the Greek *mēchanē*. Going beyond the theoretical and practical components of the "sciences" and "arts" (the *epistemai* and *technai* of Aristotelian works) and the professed and practiced aspects of mechanical works (the theory and construction of lifting devices of Heron and Pappus), it also stressed the "operative" and "productive" components of "combined" (*mushtarak*) crafts.[80]

The notably broader methods of mechanics proper, applied to various devices and machines of the Greek *Mechanics*,[81] are represented in a range of later texts. Two important examples will suffice. The Persian mechanical work *Miʿyār al-ʿuqūl*, attributed to Ibn Sīnā (d. 1039),[82] a rarely noted source with a problematic authorship, included, as part of the "science of devices" and its five simple machines, instructions not only for their combination (*tarkīb*) but also their actualization (*fiʿl*) through production. The Arabic "Compendium" of Badīʿ al-zamān al-Jazarī (fl. 1200),[83] with its five categories of water clocks, vessels, basins, fountains, and water lifts, involved along with

J. Stephenson, "The Classification of the Sciences According to Nasiruddin Tusi," *Isis*, 5 (1923), 329–38; Quṭb al-Dīn Shīrāzī, *Durrat al-tāj*, part 1 (Teheran: Majlis, 1938–1941), sec. 3, pp. 74–5; and Shams al-Dīn Āmulī, *Nafā'is al-funūn* (Teheran: Islamiyah, 1957), 3, 4, p. 557.

78 Sabra, "Astronomical Origin of Ibn al-Haytham's Concept of Experiment," pp. 133–6 (includes the relevant traditions and terms). On earlier and later developments, see Kheirandish, "Footprints of Experiment in Early Arabic Optics."

79 Khanikoff, "Analysis and Extracts of Kitāb al-Mizān al-Ḥikma," pp. 1–128 at pp. 10–11. See also Hall, "Al-Khāzinī, Abu'l-Fatḥ ʿAbd al-Raḥmān," pp. 340–3.

80 Daiber, "Qosṭā ibn Lūqā (9. Jh.) über die Einteilung der Wissenschaften," pp. 102–5; Al-Fārābī, *Enumeration of the Sciences*, pp. 108–9, English translation in Saliba, "Function of Mechanical Devices in Medieval Islamic Society," pp. 145–6.

81 Drachmann, *Mechanical Technology of Greek and Roman Antiquity*, includes sections on the *Mechanics* of (pseudo)-Aristotle and of Heron, pp. 13–18 and pp. 19–140, respectively.

82 Humāyī, *Miʿyār al-ʿuqūl*. Reference to "Miʿyār al-aql" (sic) is made in Maryam Rozhanskaya (in collaboration with I. S. Levinova), "Statics," in *Encyclopedia of the History of Arabic Science*, ed. Roshdi Rashed, 3 vols. (London: Routledge, 1996), vol. 2, pp. 614–42.

83 Al-Jazarī, *Compendium of Theory and Practice in the Mechanical Arts*. See Saliba, "Function of Mechanical Devices in Medieval Islamic Society," especially pp. 147–51.

marvelous designs and useful mechanisms their construction, illustration, and especially reproduction. None of these methodological progressions, however, is quite comparable to either the "controlled experimentation" of the Islamic and European Middle Ages or the "artificial production" of Renaissance Europe.

CONCLUSION

Optics and mechanics of the Islamic Middle Ages represent crucial transformations in the understanding of the relations and applications of mathematical and natural sciences. This era marks a significant departure from classical emphases, one to be associated not only with new theoretical frameworks and practical solutions but especially with new methodological advances. Optics embodies the clearest and most successful example of a "science" with a continuous transmission of ancient Greek traditions, a creative cross-fertilization with astronomy, and the clear transformation of a critically important field of endeavor with increasingly more experimental approaches. Mechanics, on the other hand, makes its historical landmark mostly through its continued function as an explanatory and productive "craft" with a different transmission and combination of mathematical and natural entities, directed more toward geometrical models and natural bodies than toward the physical demonstrations or natural observations of both earlier optics and later European mechanics.

The close historical overlap of the optical and mechanical traditions throughout their development within Islamic lands must not eclipse the distinctions in their parallel transmissions, contributions, and indeed modern states of research. In mechanics, research and publication have given rise to pioneering works that are still concerned more with the theoretical than with the methodological aspects of their subjects, and more with the subjects of weights and balances or statics and dynamics than with mechanics proper and the science of devices, often grouped under fine technology or engineering.[84] This scholarly development stands in clear contrast to the case of both the Greek traditions of mechanics and harmonics and the Arabic traditions of astronomy and optics; in each of these areas, monographic studies

[84] On mechanical technology, see Donald Hill, *Studies in Medieval Islamic Technology: From Philo to al-Jazarī, from Alexandria to Diyār Bakr*, ed. David A. King (Aldershot: Ashgate Variorum, 1998); and Donald Hill, "Mechanical Technology," in *Different Aspects of Islamic Culture*, pp. 165–92. See also Aḥmad Yūsuf al-Hassan, *Islamic Technology: An Illustrated History* (Cambridge: Cambridge University Press, 1986); and Aḥmad Yūsuf al-Ḥassan, "Engineers and Artisans," in *Different Aspects of Islamic Culture*, vol. 4, pt. II, pp. 271–97. On the tradition of mechanical devices, see Saliba, "Function of Mechanical Devices in Medieval Islamic Society," and forthcoming publications by the present author.

and editions are further enriched by fundamental scholarship directed toward methodological issues. The present chapter is a step in that same direction by including the methodological context of developments within the mixed mathematical sciences of optics and mechanics, the fields historically described, respectively, as the "most perfect" science and the "noblest arts."

4

ISLAMIC ASTRONOMY

Robert G. Morrison

In this chapter, "Islamic astronomy" designates the astronomy of Islamic civilization, the civilization of regions where Islam was the religion of either the rulers (Muslims were a minority in their own empire at the beginning of Islamic history) or the majority of the populace (some rulers did not convert to Islam). The expression "Islamic" thus includes scientists, patrons, and students from diverse religious and ethnic backgrounds and scientific texts in a variety of languages.

Several features distinguish Islamic astronomy from the astronomies that preceded it. Above all, there was a profound concern with the physical consistency of models and with the relationship between these models and observations. At times, this attention manifested itself through an awareness of how faulty instrumentation or observational techniques could produce poor observations. At other times, when astronomers of Islamic civilization confined themselves to critiques of physical models, they made certain that their critiques and proposed improvements did not detract from the models' predictive accuracy. At all times, Islamic astronomers unrelentingly pursued increasing accuracy in their description of the universe. Finally, certain applications of astronomy, such as the determination of the direction of prayer, the *qibla*, can be understood to have arisen from within the context of Islamic civilization, and are often specific to it.

The theoretical advances of Islamic astronomy, particularly its innovative solutions to the contradictions of Ptolemy's *Almagest* and *Planetary Hypotheses*, have deservedly attracted much attention.[1] The practice of astronomy was not attested only in texts, however: Theoretical and practical activities

[1] See, for example, Ahmad Dallal, ed., trans., and comment., *An Islamic Response to Greek Astronomy: Kitāb Taʿdīl al-aflāk of Ṣadr al-Sharīʿa* (Leiden: Brill, 1995); F. J. Ragep, ed., trans., and comment., *Naṣīr al-Dīn al-Ṭūsī's Memoir on Astronomy (al-Tadhkira fī ʿilm al-hayʾa)*, 2 vols. (New York: Springer-Verlag, 1993); and George Saliba, *A History of Arabic Astronomy: Planetary Theories during the Golden Age of Islam* (New York: New York University Press, 1994).

were often simultaneous. In the past decade, significant work on Islamic astronomical instruments has shown that the latter deserve more scholarly attention, even if the history of instrumentation does not fit a neat narrative of progress.[2] Whether or not new observations spawned new theories, developments in each area often occurred in the same milieu. To take a famous example, the patron of the renowned observatory at Marāgha, the non-Muslim Ilkhanid sultan Hülegü Khan (d. 1265), was probably motivated by an interest in astrology when he charged astronomers with the production of new tables of planetary positions.[3] Impressive instruments and important new observations were the initial result. Yet the Marāgha observatory became most famous for attracting figures responsible for some of the greatest theoretical breakthroughs in Islamic astronomy.

Some recent scholarship has taken Pierre Duhem's *To Save the Phenomena* and his *Système du Monde* as a starting point for a reappraisal of methodological approaches to the study of Islamic astronomy.[4] Duhem contrasted what he understood to be the naive concern of Islamic (in his words "Arabic") astronomy for mirroring the physical world with the sophisticated concern of Greek astronomy for models that emphasized good predictions (whether or not such models mirrored the physical world). An additional century of research, however, has demonstrated that astronomers writing in Greek based their theories on physical postulates and cared about the physical implications of their astronomical theories.[5] And though the Islamic astronomers were occupied with the reality of the physical cosmos, they rarely allowed physical concerns to prevent their proposed models from accounting for available observations. Rather, astronomers attempted to resolve the contradictions between the physical principles and the mathematical models.[6] Moreover, Duhem's facile characterizations should not obscure the fact that Ptolemy, for all his successes, did not save all of the phenomena.[7] Islamic astronomers understood profoundly the difference between the mathematical and physical components of their science.[8]

2 See, for example, the works of David King, François Charette, and Richard Lorch cited herein.

3 A. I. Sabra, "Situating Arabic Science: Locality Versus Essence," *Isis*, 86 (1996), 654–79 at pp. 666–7.

4 F. Jamil Ragep, "Duhem, the Arabs, and the History of Cosmology," *Synthèse*, 83 (1990), 201–14.

5 See G. E. R. Lloyd, citing the rejection of heliocentrism, in "Saving the Appearances," *Classical Quarterly*, n.s. 28 (1978), 202–22 at pp. 219–22.

6 See Ragep, "Duhem, the Arabs, and the History of Cosmology," p. 213. "In accepting that astronomy was based on both mathematical and physical principles, Arab astronomers reached a rather simple conclusion – the mathematical models had to be consistent with the physical principles." A recent article questions the suitability of the categories of "realist" and "fictionalist" for studies of premodern astronomy; see Gad Freudenthal, "'Instrumentalism' and 'Realism' as Categories in the History of Astronomy: Duhem vs. Popper, Maimonides vs. Gersonides," *Centaurus*, 45 (2003), 227–48.

7 Ptolemy's solar model predicted variations in the solar diameter that had not yet been observed, and his lunar model entailed massive 2:1 variations in the lunar diameter that should have been observed.

8 Compare G. E. R. Lloyd, *Methods and Problems in Greek Science* (Cambridge: Cambridge University Press, 1991), pp. 249–50.

THE APPLICATIONS OF ASTRONOMY: TIME, PRAYER, AND ASTROLOGY

Inasmuch as observational activity, and often the patronage that supported it, were central to the enterprise of Islamic astronomy, we must acquaint ourselves with the practical applications of astronomy that led scientists to such exacting levels of precision. Judging by the quantity of primary and secondary sources, timekeeping was a central preoccupation.[9] The recent convention of standardized time zones makes it easy to forget that local (solar) time varies with longitude. The foundation of timekeeping was the hour angle, the measure of the arc between the Sun's maximum local altitude on a given day and its current position that day. The Sun's maximum angle above the horizon, its altitude, varies from one latitude to the next. The astronomer approximates the local latitude from observing the local altitude of the North Star. Or, knowing the date and the local latitude, he can calculate the Sun's maximum altitude, its zenith. After observing the Sun's instantaneous position, he in turn can calculate the hour angle. A determination of the Sun's zenith was also necessary for observing or calculating the times of the midday and afternoon prayers. The times of the sunrise, sunset, and nightfall prayers depended on measuring the Sun's first appearance and its depression below the horizon.[10] Such timekeeping methods based on spherical astronomy presumed knowledge of the Sun's daily motions, including its extremes (at the solstices) and the mean (at the equinoxes, where the ecliptic, the Sun's apparent yearly path through the zodiacal constellations, intersects the equator); see Figure 4.1. To ensure accurate timekeeping, local observations of the Sun and Moon were a necessity, and later Islamic astronomers were often dissatisfied with the results of their predecessors.

Al-Khwārizmī (fl. ca. 830) computed the earliest known tables of the five daily prayer times, expressing them in terms of the Sun's shadow at the latitude of Baghdad, for six-degree increments of the Sun's longitude.[11] Ultimately, tables for timekeeping, employing a variety of approaches, proliferated in Islamic civilization.[12] Timekeeping at night depended on observations of the fixed stars (i.e., not the planets). Stars near the equinoxes were particularly well suited for this application. The drive for ever more precise

[9] David King, *In Synchrony with the Heavens: Studies in Astronomical Timekeeping and Instrumentation in Medieval Islamic Civilization*, 2 vols. (Leiden: Brill, 2004–2005); David King, "Mīḳāt," in *Encyclopaedia of Islam*, new edition, ed. C. E. Bosworth et al., 12 vols. (Leiden: Brill, 1993), vol. 7, pp. 27–32.

[10] For an example of this time reckoning, see David King, "Ibn Yūnus' Very Useful Tables for Reckoning Time by the Sun," *Archive for History of the Exact Sciences*, 10 (1973), 342–94.

[11] On al-Khwārizmī's work, see Bernard Goldstein, *Ibn al-Muthannā's Commentary on the Astronomical Tables of al-Khwārizmī* (New Haven, Conn.: Yale University Press, 1967); and Otto Neugebauer, *The Astronomical Tables of al-Khwārizmī: Translation with Commentaries of the Latin Version edited by H. Suter* (Copenhagen: Munksgaard, 1962).

[12] See King, *In Synchrony with the Heavens*, vol. 1.

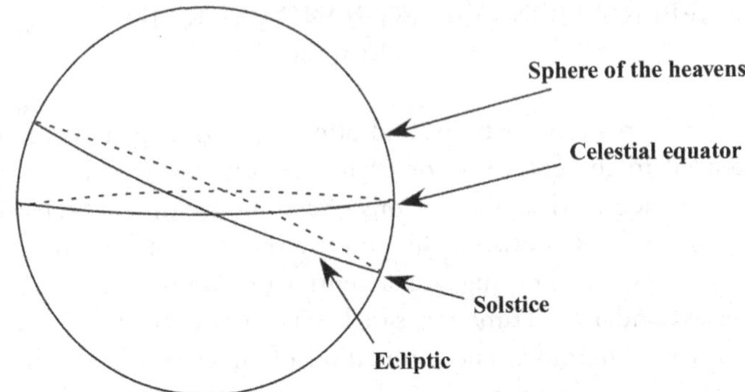

Figure 4.1. The ecliptic, the Sun's path through the sphere of the heavens.

timekeeping entailed continued observations of the stars, Moon, and Sun, and constant attention to instrument construction.

In addition to tables, sundials were another way to determine the time. Sundials could be either vertical or horizontal. In either case, drawing the curves onto which the gnomon, the part of a sundial perpendicular to the dial, cast its shadow, as well as aligning the sundial, were the key problems. Sundials could be calibrated for twenty-four equal hours (like ours), unequal (seasonal) hours, when the unequal night and day are each divided into twelve hours, or both. Khwārizmī also wrote the earliest Arabic treatise on sundial construction, treating the problem of how variations in latitude affected the curves drawn on the sundial to mark the passage of time.[13] Thābit ibn Qurra (d. 901) explained how to construct a sundial that was functional at all latitudes. Although Thābit was not Muslim, he lived in a Muslim culture, and his work on sundials, and particularly on lunar crescent visibility, illustrates how topics with Islamic applications could nevertheless interest non-Muslims, given the history of the lunar calendar in the Near East.[14] Subsequently, Thābit's grandson Ibrāhīm ibn Sinān (d. 946) applied some of the same mathematics from sundials to the problems of lenses and burning mirrors.[15] One famous *muwaqqit* (mosque timekeeper), Ibn al-Shāṭir (d. 1375), constructed a sundial in the Umayyad Mosque in Damascus. Other examples of mosque sundials from two centuries later also survive.[16]

The *qibla* is the direction of Mecca from any location, relevant for the orientation of daily prayer as well as mosque construction. Attempts to improve *qibla* computation significantly aided the broader development of Islamic astronomy. The earliest methods for *qibla* determination were

13 Fuat Sezgin, *Geschichte des arabischen Schritftums*, 13 vols. (Leiden: Brill, 1978), vol. 6, p. 143.

14 Régis Morelon, "Thābit ibn Qurra and Arabic Astronomy in the Ninth Century," *Arabic Sciences and Philosophy*, 4 (1994), 111–39.

15 Roshdi Rashed and Hélène Bellosta, *Ibrāhīm ibn Sīnān: Logique et géométrie au Xe siècle* (Leiden: Brill, 2000).

16 David King, "Mizwala," in Bosworth et al., *Encyclopaedia of Islam*, new edition, vol. 7, pp. 210–11.

nonscientific.[17] Pre-Islamic astronomy had connected meteorological phenomena, such as wind, with the appearances and disappearances of certain stars (*anwāʾ*); the *Kaʿba*, Islam's geographical center, was oriented toward the south wind.[18] In some cities, the legal schools (*madhāhib*, singular *madhhab*) had their own *qiblas*; one such *madhhab*, the Shāfiʿī *madhhab*, used due south, following the practice of Muḥammad in Medina.[19] These nonscientific early solutions endured after scientific alternatives appeared.

By the ninth century, precise scientific solutions to the *qibla* problem had emerged. These scientific solutions connected the *qibla* to the great circle arc on the celestial sphere between the longitude and latitude of the observer and Mecca's coordinates.[20] The intersection of that arc with the local meridian created the *qibla* angle itself, measured from the north or south. The *qibla* problem was one application for the Islamic astronomers' expansion of the rudimentary spherical trigonometry that they inherited from Greek texts. Scientists such as Ḥabash al-Ḥāsib (d. 864–74) also produced solutions to the *qibla* problem based on analemmas, the projection of the celestial sphere into a plane (simpler planar approximations were also available).[21] Later astronomers devised universal solutions of the *qibla* problem and accompanying tables,[22] along with new instruments that provided graphical solutions.[23] Eventually, astronomers recognized formal similarities between timekeeping and *qibla* determination.[24] All of these theoretical refinements reflected the astronomers' active engagement with these practical problems.

Although the observations of the Sun, Moon, and fixed stars were sufficient for timekeeping, astrological forecasts required more: good tables based on repeated observations of the planets and a thorough knowledge of astronomy. There were different types of astrological forecasts. The horoscope was defined as the part of the zodiac[25] rising over the horizon at the time of one's birth. Horoscopic astrology made predictions on the basis of

17 Monica Rius, "La Orientación de las Mezquitas según el *Kitāb Dalāʾil al-qibla* de al-Mattīyī (s. XII)," in *From Baghdad to Barcelona: Studies in the Islamic Exact Sciences in Honour of Prof. Juan Vernet*, ed. Josep Casulleras and Julio Samsó (Barcelona: Universidad de Barcelona, 1996), pp. 781–830. See also David King, "The Sacred Direction in Islam: A Study of the Interaction of Religion and Science in the Middle Ages," *Interdisciplinary Science Reviews*, 10 (1985), 315–28 at p. 319.

18 Daniel Varisco, "The Origin of the Anwāʾ in Arab Tradition," *Studia Islamica*, 74 (1991), 5–28 at p. 23; and King, "Sacred Direction in Islam," p. 320.

19 King, "Sacred Direction in Islam," p. 325.

20 David King, "Ḳibla," in *Encyclopaedia of Islam*, new edition, ed. C. E. Bosworth et al. (Leiden: Brill, 1986), vol. 5, pp. 83–8.

21 On Ḥabash al-Ḥāsib's analemma, see Yūsuf ʿĪd and E. S. Kennedy, "Ḥabash al-Ḥāsib's Analemma for the Qibla," *Historia Mathematica*, 1 (1974), 3–11.

22 Ahmad S. Dallal, "Ibn al-Haytham's Universal Solution for Finding the Direction of the *Qibla* by Calculation," *Arabic Sciences and Philosophy*, 5 (1995), 145–93.

23 David King, *World-maps for Finding the Direction and Distance of Mecca: Examples of Innovation and Tradition in Islamic Science* (Boston: Brill, 1999).

24 King, "Ḳibla," in Bosworth et al., *Encyclopaedia of Islam*, new edition, vol. 5.

25 The zodiac is the belt of twelve constellations that is the background for the Sun's apparent yearly path against the fixed stars.

the other planets' positions with respect to the horoscope. Interrogations aimed to determine the best time, based on auspicious configurations of the planets, for a given undertaking, such as marriage, war, or laying a foundation.[26] Astrologers believed that the conjunctions of planets had an impact on political events. In the earliest texts, *ʿilm al-nujūm* (the science of the stars) could mean either astrology or astronomy. But as early as the ninth century, the term *ʿilm al-hayʾa* [the science of the configuration (of the stars)] appeared as a possible translation for astronomy, and by the eleventh century *ʿilm al-hayʾa* came to denote the entire discipline of astronomy.[27] Astrology came to be known as *ʿilm al-aḥkām* (the science of the judgments; i.e., *ʿilm aḥkām al-nujūm*). Still, information about astrological forecasts was included in most Islamic handbooks of astronomy (*azyāj*) throughout the history of Islamic science.[28] At the above-mentioned Marāgha observatory, planetary observations were part of the program, and astrological applications could have been a selling point for the patrons.[29] At Marāgha, too, astronomers used observations of the planets to explore in greater detail the subtle intertwining of the theoretical models and the parameters derived from observations.[30]

THE ASTROLABE

The astrolabe was the instrument best suited to all of these applications of observational astronomy in timekeeping, astrology, and sacred geography (Figures 4.2 and 4.3; see also North, Chapter 19, this volume). Although it originated in classical antiquity, the astrolabe has pride of place in the history of Islamic scientific instruments owing to the elegance, sophistication, and versatility of its design. Its face is a stereographic projection of the celestial sphere onto the plane with latitude and longitude lines. Originally, these projections were for a single latitude, so some astrolabes came with a set of interchangeable plates for different latitudes. In the eleventh century, astronomers worked on universal astrolabe plates that functioned

26 T. Fahd, "Nudjūm (Aḥkām al-)," in *Encyclopaedia of Islam*, new edition, ed. P. J. Bearman et al. (Leiden: Brill, 1997), vol. 8, pp. 105–8.

27 George Saliba, *Islamic Science and the Making of the European Renaissance* (Cambridge, Mass.: MIT Press, 2007), p. 18.

28 The astrological material in the fifteenth-century *Zīj-i Khāqānī* of Jamshīd al-Kāshī constitutes a late example; see E. S. Kennedy, "A Survey of Islamic Astronomical Tables," *Transactions of the American Philosophical Society*, n.s. 46, no. 2 (1956), 121–77 at pp. 165–6. Kennedy's publication should be supplemented with David King et al., "Astronomical Handbooks and Tables from the Islamic World (750–1900), an Interim Report," *Suhayl*, 2 (2001), 9–105.

29 See George Saliba, "Horoscopes and Planetary Theory: Ilkhanid Patronage of Astronomers," in *Beyond the Legacy of Genghis Khan*, ed. Linda Komaroff (Leiden: Brill, 2006), pp. 357–68.

30 George Saliba, "The Determination of New Planetary Parameters at the Maragha Observatory," *Centaurus*, 29 (1986), 187–207.

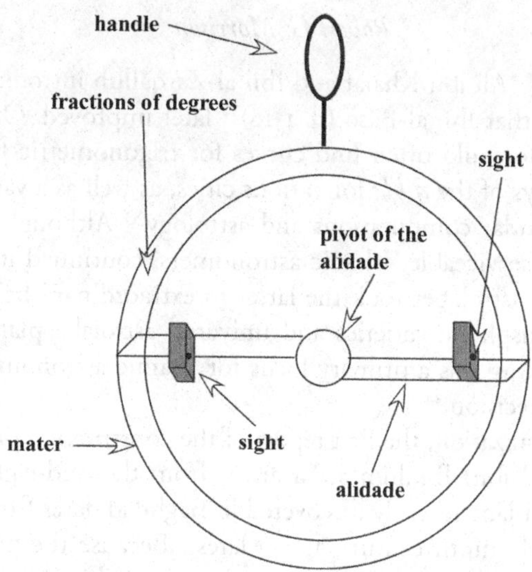

Figure 4.2. Schematic planispheric astrolabe.

Figure 4.3. Rare spherical astrolabe (Eastern Islamic, c. 1480–1). Courtesy of the Museum of the History of Science, Oxford.

at any latitude.[31] ʿAlī ibn Khalaf and Ibn al-Zarqālluh introduced universal astrolabe plates that Ibn al-Bāṣo (d. 1316)[32] later improved. On the back of an astrolabe, one could often find curves for trigonometric functions and graphical displays of the *qibla* for various cities, as well as a variety of information for calendar computations and astrology.[33] Although early Islamic astrolabes were serviceable, Islamic astronomers' continued interest in and development of astrolabes took the latter to extraordinary heights. Sophisticated nonplanispheric varieties and universal astrolabe plates prove that instrument making was a primary locus for Islamic astronomers' creativity and quest for precision.[34]

In Islamic civilization, the first report of the construction of an astrolabe, by (Muḥammad ibn) Ibrāhīm al-Fazārī, is from the mid-eighth century.[35] At least one astrolabe recently discovered in Baghdad dates from that period or from the early ninth century at the latest. Because the first substantial description of the astrolabe is from the ninth century, this discovery makes the point that astronomical knowledge could be transmitted through instrument construction.[36]

TRANSMISSION AND TRANSLATIONS

Traditionally, scholars have focused their attention on how translations of texts on astronomy (and astrology) from Greek, Sanskrit, and Pahlavi flourished under the ʿAbbāsid caliphs (~750–1258) owing in some part to their desire to establish an identity apart from those of their predecessors, the Umayyads (661–750).[37] But there is evidence of translations under the Umayyads. Indeed, the ʿAbbāsid caliph al-Manṣūr's public consultation of astrologers regarding the foundation of their new capital, Baghdad, in 762 strongly suggests that their technical competence came from a prior transmission of astronomy.[38] There were no ʿAbbāsid era translations before

31 Ibn al-Zarqālluh (ed., trans., and comment. Roser Puig), *Al-Shakkāziyya* (Barcelona: Universidad de Barcelona, Instituto Millás Vallicrosa de Historia de la Ciencia Árabe, 1986).

32 Emilia Calvo, *Ibn Bāṣuh: Risālat al-ṣafīḥa al-jāmiʿa li-jamīʿ al-ʿurūḍ: tratado sobre la lamina generál para todas las latitudes* (Madrid: Consejo Superior de Investigaciones Científicas, Instituto de Cooperación con el Mundo Árabe, 1993).

33 Willy Hartner, "Aṣṭurlāb," in *Encyclopaedia of Islam*, new edition, ed. H. A. R. Gibb et al. (Leiden: Brill, 1960), vol. 1, pp. 722–8.

34 Al-Fazārī, like Ḥabash al-Ḥāsib, wrote a book on the melon-shaped (*mubaṭṭaḥ*) astrolabe. See Boris A. Rosenfeld, "A Supplement to Mathematicians, Astronomers, and other Scholars of Islamic Civilisation and Their Works (7th–9th c.)," *Suhayl*, 4 (2004), 87–140. For more information on astrolabes, see John D. North, "The Astrolabe," *Scientific American*, 230, no. 1 (January 1974), 96–106.

35 Al-Farghānī (ed., trans., comment. Richard Lorch), *On the Astrolabe* (Stuttgart: Franz Steiner Verlag, 2005), p. 3.

36 King, *In Synchrony with the Heavens*, vol. 2, pp. 407ff and 588–9. On the date of al-Farghānī's description of the astrolabe, see al-Farghānī, *On the Astrolabe*.

37 Dimitri Gutas, *Greek Thought, Arabic Culture* (London: Routledge, 1998), pp. 28–60, especially pp. 28–9.

38 Saliba, *Islamic Science and the Making of the European Renaissance*, pp. 15–19, 45–72.

the foundation of Baghdad. The astrologers, among them Muḥammad ibn Ibrāhīm al-Fazārī, whom al-Manṣūr consulted regarding the construction of the capital, may have used the Pahlavi original of the *Zīj al-Shāh*, which would later be translated into Arabic.[39] George Saliba has argued that in the late seventh and early eighth century, the decision of the Umayyad caliph ʿAbd al-Malik (d. 705), or perhaps Hishām (d. 724), to change the language of administration to Arabic created an atmosphere of competition in which a knowledge of science would have been an asset, enhancing one's ability to solve practical problems in architecture, engineering, arithmetic, and other areas.[40] Although no translations from this period survive, the Arabization of the administration must have entailed the translation of the elementary scientific texts necessary for governing.[41] These socioeconomic and political explanations for the translation movement assume that the translations awakened a spirit of scientific discovery, which in turn accounted for the outstanding achievements of Islamic astronomy long after the translation movement had ended. Older accounts of the translation movement had ascribed a more central role to purely intellectual motives.[42]

Whatever the root causes of the translation movement, the earliest type of astronomy text to be composed in Arabic was the *zīj* (plural *azyāj*), a handbook of astronomy that included tables of planetary positions. The earliest known exemplar was the *Zīj al-Arkand*, composed in Sind (in South Asia) in 735. The *Zīj al-Arkand* was an Arabic version of a Sanskrit original that itself drew on a Pahlavi text. Another *zīj* from the Umayyad period was the *Zīj al-Harqan*, with an epoch date of 742. Under the ʿAbbāsid al-Manṣūr, an embassy from Sind brought a text that al-Fazārī translated as *Zīj al-Sindhind al-Kabīr* in the early 770s.[43] Theoretical components of the *Zīj al-Sindhind al-Kabīr* (e.g., the size of the epicycles) also relied heavily on Sanskrit texts, but included Ptolemaic and Pahlavi material as well.[44] All of these early *azyāj* survive only as fragments recorded in later works.[45] The fact that these *azyāj*, whatever their date, were composed before the mass of translations under the ʿAbbāsids reminds us that the translation movement was intentional. Interest in existing scientific work created the need for more and better translations. Al-Khwārizmī's *Zīj al-Sindhind* (different from the

39 David Pingree, "The Greek Influence on Early Islamic Mathematical Astronomy," *Journal of the American Oriental Society*, 93 (1973), 37.

40 Saliba, *Islamic Science and the Making of the European Renaissance*, pp. 45–72.

41 Ibid., pp. 55–6.

42 For an explanation of the translation movement that favors intellectual motives, see Franz Rosenthal, *The Classical Heritage in Islam*, trans. Émile and Jenny Marmorstein (Berkeley: University of California Press, 1975), pp. 1–12. See also A. I. Sabra, "The Appropriation and Subsequent Naturalization of Greek Science in Medieval Islam: A Preliminary Statement," *History of Science*, 25 (1987), 223–38.

43 David Pingree, "The Fragments of the Works of Yaʿqūb ibn Ṭāriq," *Journal of Near Eastern Studies*, 27 (1968), 97; and Pingree, "Greek Influence on Early Islamic Mathematical Astronomy," pp. 35–8.

44 Pingree, "Fragments of the Works of Yaʿqūb ibn Ṭāriq," p. 119.

45 See, e.g., Pingree, "Fragments of the Works of Yaʿqūb ibn Ṭāriq"; and David Pingree, "The Fragments of the Works of al-Fazārī," *The Journal of Near Eastern Studies*, 29 (1970), 103–23.

above-mentioned *Zīj al-Sindhind al-Kabīr*) was the first original astronomical composition in Arabic to survive (only in Latin, however).[46]

After the ʿAbbāsid Revolution (747–50), the birth of an independent Umayyad caliphate in Andalusia coincided with the progress of astronomy in the Iberian Peninsula beyond its level under the Visigoths. Astrology again played a central role, beginning with the independent Umayyad Hishām (d. 796), who summoned the astrologer al-Ḍabbī to predict the length of his reign.[47] In the ninth century, either ʿAbbās ibn Firnās or ʿAbbās ibn Nāṣiḥ, both poet-astrologers, introduced tables ascribed to Khwārizmī into Andalusia.[48] By the ninth or tenth century, Spanish timekeepers were using astronomical methods.[49]

Probably by the end of the Umayyad caliphate in the Islamic East, and certainly by the time of the ninth-century embassy from Sind, Islamic astronomers had already encountered Ptolemaic astronomy indirectly. But the first of Ptolemy's books to be translated into Arabic was the *Almagest*.[50] From copious observations, the *Almagest* derived, and presented in geometric terms, models that could account for those observations with a remarkable degree of accuracy. We have reports of a total of four translations of the *Almagest* during the ʿAbbāsid era, of which only two survive. The earliest (829–30) is that of al-Ḥajjāj ibn Maṭar, completed during the reign of the ʿAbbāsid caliph al-Maʾmūn (d. 833).[51] The second surviving translation was produced by Isḥāq ibn Ḥunayn ibn Isḥāq (d. 910–911) and corrected by Thābit ibn Qurra in 892.[52]

OBSERVATIONAL ASTRONOMY

The need to correct parameters and answer questions posed by the translations of the *Almagest* brought the first documented program of systematic astronomical observation. Soon after the al-Ḥajjāj *Almagest* translation,

46 For al-Khwārizmī's *zīj*, see Goldstein, *Ibn al-Muthannā's Commentary on the Astronomical Tables of al-Khwārizmī*; and Neugebauer, *Astronomical Tables of al-Khwārizmī*.

47 On al-Ḍabbī, see Julio Samsó, "La primitiva version árabe del *Libro de las Cruces*," in *Nuevos Estudios sobre Astronomía española en el siglo de Alfonso X*, ed. Juan Vernet Ginés (Barcelona: Instituto de Filología, Institución "Milá y Fontanals," Consejo Superior de Investigaciones Científicas, 1983), pp. 149–61.

48 Juan Vernet and Julio Samsó, "The Development of Arabic Science in Andalusia," in *Encyclopedia of the History of Arabic Science*, ed. Roshdi Rashed, 3 vols. (London: Routledge, 1996), vol. 1, pp. 243–75 at p. 248.

49 Roser Puig, "La astronomía en al-Andalus : Aproximacíon historiográfica," *Arbor*, 142 (1992), 167–84 at p. 171.

50 Sezgin, *Geschichte des arabischen Schriftums*, vol. 6, pp. 88–95.

51 Today, there are eleven known Arabic *Almagest* manuscripts. See Paul Kunitzsch, "A Hitherto Unknown Arabic Manuscript of the *Almagest*," *Zeitschrift für Geschichte der arabisch-islamischen Wissenschaften*, 14 (2001), 31–7.

52 The classic work on the Arabic translations of the *Almagest* is Paul Kunitzsch, *Der Almagest: Die Syntaxis mathematica des Claudius Ptolemäus in arab.-latein. Überlieferung* (Wiesbaden: Harrassowitz, 1974).

astronomers under al-Ma'mūn (d. 833) conducted observations on Mount Qāsiyyūn near Damascus and at Shammāsiyya at Baghdad, reportedly using gnomons, an armillary sphere, and a mural quadrant.[53] Both the nature of the observations and accounts of the activity indicate a minimum period of observation of between one and two years for each observatory. In 831–2, there was a full year of solar observations, including work on the length of the solar year, the motion in precession, and the obliquity of the ecliptic. The precession of the equinoxes is the movement of the location of the equinoxes in the opposite direction of the zodiacal signs. Because the spring equinox was the reference point for measuring celestial positions, its precession affected the longitudes of the fixed stars. In turn, the positions of planets measured with respect to the zodiac would have to be adjusted for the rate of precession. The obliquity of the ecliptic is the angle (between 23° and 24°) between the ecliptic and the celestial equator. As an outgrowth of attempts to measure the length of a solar year, these astronomers also determined the location of the solar apogee and its motion, which would affect the length of the solar year. Measuring the solar apogee's motion, of which Ptolemy had been unaware, required precise observations of the daily motions of the Sun and the fixed stars.[54] The observations of 828–9 that al-Ma'mūn ordered led to Yaḥyā ibn Abī Manṣūr's (d. 832) *Al-Zīj al-Mumtaḥan*, a *zīj* that included applications in astrology.[55] This *zīj* reported that the solar apogee moved and posited that its movement was connected to the motion in precession.[56]

The career of Ḥabash al-Ḥāsib (d. 864–874) marked the ascent of Ptolemaic astronomy in Islamic civilization. Ḥabash's planetary theory generally followed or improved upon that of Ptolemy, though aspects of his solar theory and his method for determining parallax are found in Indian astronomy. Building on the work of al-Khwārizmī's pioneering sine table, Ḥabash compiled a table of tangents. In addition, he defined various formulas for spherical astronomy, and his *al-Zīj al-Dimashqī* contained tables of first visibilities of the lunar crescent, the marker of the new month.[57] Earlier tables of crescent visibility had drawn heavily on Indian astronomy.[58] From this same period, we have the *Treatise on the Solar Year*, attributed to Thābit but

53 Aydın Sayılı, *The Observatory in Islam* (Ankara: Türk Tarih Basımevi, 1960), pp. 56–63. The following sentences rely on these pages.

54 Otto Neugebauer, "Thābit ben Qurra 'On the Solar Year' and 'On the Motion of the Eighth Sphere'," *Proceedings of the American Philosophical Society*, 16 (1962), 264–98 at p. 287.

55 Benno van Dalen, "A Second Manuscript of the *Mumtaḥan Zīj*," *Suhayl*, 4 (2004), 9–44, especially pp. 28–30.

56 Régis Morelon, "Eastern Arabic Astronomy," in Rashed, *Encyclopedia of the History of Arabic Science*, vol. 1, pp. 20–57 at p. 26.

57 Sevim Tekeli, "Ḥabash al-Ḥāsib," *Dictionary of Scientific Biography*, V, 612–20. See also Marie-Thérèse Debarnot, "The Zīj of Ḥabash al-Ḥāsib: A Survey of MS Istanbul Yeni Cami 784/2," in *From Deferent to Equant*, ed. David King and George Saliba (New York: New York Academy of Sciences, 1987), pp. 35–69.

58 E. S. Kennedy and Mardiros Janjanian, "The Crescent Visibility Table in Al-Khwārizmī's Zīj," *Centaurus*, 20 (1965–1967), 73–8.

probably due to an earlier author.[59] That book rejected Ptolemy's observational work as outdated but used his models as the framework for moving forward.[60]

The composition of al-Battānī's (d. 929) *Zīj al-Ṣābi'* marked the supremacy of Ptolemaic astronomy in Islamic civilization. Battānī himself was a gifted observer who detected a variation in the angular diameter of the Sun, something that Ptolemy's models predicted but that Ptolemy was unable to observe. Those variations meant that annular eclipses were possible.[61] Battānī also played a central role in the development of models for trepidations (*al-iqbāl wa-'l-idbār*), oscillations in the rate of precession. Many models for precession also implied changes in the obliquity of the ecliptic. These models show that Islamic astronomers were reluctant to separate physical concerns from the production of tables for planetary positions.[62] The motion of the fixed stars or a moving ecliptic meant that measurements of celestial positions would have to be adjusted further. In addition to their practical applications, observations had theoretical implications. It is therefore not surprising that the quest for better observations led to improvements in instrumentation, notably in the tenth century under the Buyids, a dynasty that wrested effective control from the ʿAbbāsid caliphs. The Buyid ruler ʿAḍud al-Dawla (d. 982) patronized the work of ʿAbd al-Raḥmān al-Ṣūfī (d. 986), who used a ring, probably 2.5 meters in diameter, to make a careful measurement of the obliquity of the ecliptic.[63] Al-Ṣūfī produced a famous text on the constellations that went beyond the star catalogues in the *Almagest*. He also reobserved the magnitudes and longitudes of many stars. His measurements would figure into debates about whether variations in the obliquity of the ecliptic in fact existed and, if so, whether it was necessary to posit a system of orbs to account for them.

The observatory of the Buyid ruler Sharaf al-Dawla (d. 989) featured a section of a sphere with a radius of 12.5 meters that could be used to trace the Sun's day circles.[64] Abū Sahl al-Qūhī, who built that instrument, was also a gifted mathematician who performed an experiment to disprove Aristotle's position that there had to be an instant of rest between rising and falling when a heavy projectile is thrown upward.[65] Abū al-Wafā' al-Būzajānī (d. 998) was another astronomer, one particularly interested in mathematical methods, associated with the observatory of Sharaf al-Dawla. In 994, al-Khujandī,

59 Régis Morelon, ed. and trans., *Thābit ibn Qurra: Oeuvres d'astronomie* (Paris: Les Belles Lettres, 1987), p. LII.

60 Morelon, *Thābit ibn Qurra*, pp. LV, LXXII–IV. See also Neugebauer, "Thābit ben Qurra 'On the Solar Year' and 'On the Motion of the Eighth Sphere'," p. 274.

61 For a summary of observations of the solar and lunar diameters in Islamic astronomy, see Ragep, *Naṣīr al-Dīn al-Ṭūsī's Memoir on Astronomy (al-Tadhkira fī ʿilm al-hay'a)*, pp. 460–1.

62 F. Jamil Ragep, "Al-Battānī, Cosmology, and the Early History of Trepidation in Islam," in Casulleras and Samsó, *From Baghdad to Barcelona*, pp. 267–303 at p. 292.

63 Sayılı, *The Observatory in Islam*, pp. 105–6.

64 Ibid., p. 116.

65 Roshdi Rashed, "Al-Qūhī Versus Aristotle on Motion," *Arabic Sciences and Philosophy*, 9 (1999), 7–24 at pp. 17–18.

under the patronage of the Buyid Fakhr al-Dawla (d. 997), measured the obliquity using a 60° meridian arc, faced with wood, with a radius of 20 meters, constructed between two walls. It surpassed the precision of previous instruments used for this measurement. With these scientists, one finds a real concern for the accuracy of instruments and plentiful evidence for their observational acumen. For instance, al-Khujandī determined the amount by which the center of the arc had sunk, in turn displacing the upper part of its arc.[66] The matter of skill was related to the reliability of astrological forecasts and to the theoretical implications of observations of, for example, a moving ecliptic, oscillations in the rate of precession, and annular eclipses. Such concerns about the accuracy of earlier observations gave impetus to the production of a new *zīj* for a given locale even when, like that of Ibn Yūnus in Cairo, it was not the product of systematic new observations.

By the tenth century, Islamic astronomers had completely revamped the parameters found in Ptolemy's (fl. 125–50) *Almagest*, which nevertheless continued to wield the greatest influence through its models for the celestial motions. Ptolemy had also produced a work entitled *Planetary Hypotheses*, which was translated into Arabic after the *Almagest* and summarized the *Almagest*'s models in terms of three-dimensional orbs or cross sections of orbs (though Islamic astronomers tended to write about orbs, not cross sections).[67] A careful assessment of the relationship between the *Almagest* and *Planetary Hypotheses* became the springboard for Islamic astronomy's subsequent theoretical insights. Although criticisms of the *Almagest*'s models may have come as early as the ninth century,[68] they intensified in the eleventh century. Meanwhile, in tenth-century Andalusia, we see the first evidence of acquaintance with Ptolemaic planetary theory, in the *Kitāb al-Hay'a* of Qāsim ibn Muṭarrif al-Qaṭṭān. Subsequently, Abū al-Qāsim Maslama al-Majrīṭī (d. 1007) and his school became known for their serious observational work and study of Ptolemy.[69]

PTOLEMY'S MODELS AND ENSUING CRITICISMS OF THE PTOLEMAIC EQUANT HYPOTHESIS

To give some background, Ptolemy's planetary models relied on two devices, the eccentric and the epicycle, that were known as early as the time of Apollonius (third century B.C.). In the *Almagest*, these hypotheses first

[66] Sayılı, *The Observatory in Islam*, pp. 119, 125.

[67] For the Arabic translation of the *Planetary Hypotheses*, see Bernard R. Goldstein, "The Arabic Version of Ptolemy's Planetary Hypotheses," *Transactions of the American Philosophical Society*, 57, no. 4 (1967), 1–55. For an edition and French translation of the first book, see Régis Morelon, "La version arabe du *Livre des Hypothèses* de Ptolémée," *Mélanges de l'Institut Dominicain des Études Orientales du Caire*, 21 (1993), 7–85.

[68] George Saliba, "Early Arabic Critique of Ptolemaic Cosmology: A Ninth-Century Text on the Motion of the Celestial Spheres," *Journal for the History of Astronomy*, 25 (1994), 115–41.

[69] Juan Vernet and Julio Samsó, "The Development," in Rashed, *Encyclopedia of the History of Arabic Science*, vol. 1, p. 254.

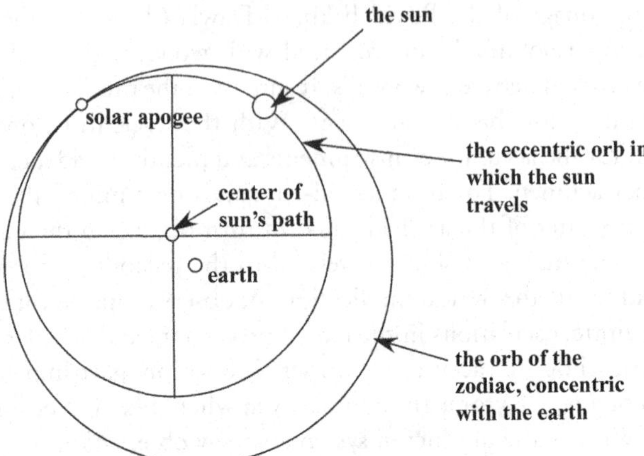

Figure 4.4. Ptolemy's eccentric model for the Sun.

appeared in the solar model. Pre-Ptolemaic observations had shown that the seasons do not all have the same length, implying that the Sun's apparent annual motion varied slightly from day to day. The eccentric hypothesis entailed that a circle (or orb) could rotate uniformly about its center and yet produce nonuniform motions from the perspective of an observer removed from that center (Figure 4.4). In the case of the Sun, Ptolemy used the data about the inequality of the seasons to compute the distance and direction of the Earth's displacement from the center of the eccentric.[70]

In the *Almagest*, the epicyclic hypothesis placed the planet on a small circle, the epicycle (perhaps representing an orb), the center of which was carried by a larger circle, the deferent (again perhaps representing an orb). Then, as the center of the epicycle rotates about the center of the deferent, the epicycle rotates about its own center. Apollonius had demonstrated that if the epicycle and the deferent rotate at the same rate but in opposite directions, then the hypothesis of the epicycle is equivalent to that of the eccentric.[71] Indeed, Ptolemy noted that a model with either an eccentric or an epicycle could account for the Sun's observed motions (Figure 4.5). With Venus, Mars, Jupiter, and Saturn, Ptolemy had the epicycle and the deferent moving in the same direction with different velocities so that the motion on the epicycle would help account for the planets' retrograde motion (the planets' brief apparent motion against their usual west-to-east motion against the background of the zodiac). In addition, by analyzing observations of these planetary motions, Ptolemy concluded not only that the Earth was removed from the center of the deferent (i.e., that the deferent was eccentric) but also that the deferent's motion was uniform about a point other than

70 Gerald J. Toomer, *Ptolemy's Almagest* (London: Duckworth, 1984), pp. 153–7.

71 Ibid., pp. 148–53.

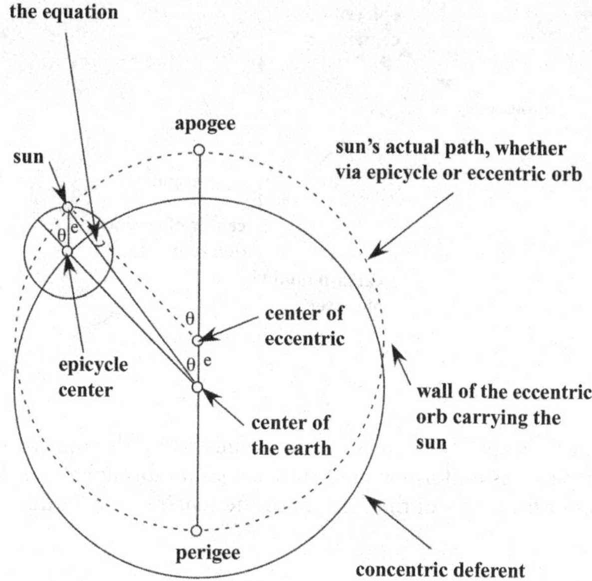

Figure 4.5. Ptolemy's epicyclic model for the Sun (solid lines) superimposed upon the eccentric model (dotted lines), illustrating their equivalence.

its own center. The point about which the deferent's motion was uniform would come to be called the equant; its distance from the Earth was twice the eccentricity (Figure 4.6).[72] Ptolemy himself did not like the equant compromise, which the data forced him to make. In Book XIII of the *Almagest*, Ptolemy also registered his own dissatisfaction with his latitude model, which suffered from inconsistencies analogous to those of the equant hypothesis.[73]

Ptolemy's *Planetary Hypotheses* presented the eccentric hypothesis as consisting of an orb or torus in the wall of which the planet (e.g., the Sun) was embedded; the orb in turn rotated uniformly about its center, whereas the Earth was slightly removed from the center of the orb. The *Planetary Hypotheses* also physicalized the epicyclic hypothesis, which held that the planet was fixed on a small orb, the epicycle, that was in turn embedded in the wall of a larger orb (or torus), the deferent. The *Planetary Hypotheses*' emphasis on physical models led to a tension with the *Almagest.* The predictive advantages of the eccentric orbs rotating uniformly about their equant points clashed with the requirement that the orbs rotate uniformly in place about axes passing through their centers.

[72] Otto Neugebauer, *A History of Ancient Mathematical Astronomy*, 3 vols. (with continuous pagination) (New York: Springer-Verlag, 1975), p. 155.

[73] Toomer, *Ptolemy's Almagest*, pp. 600–1.

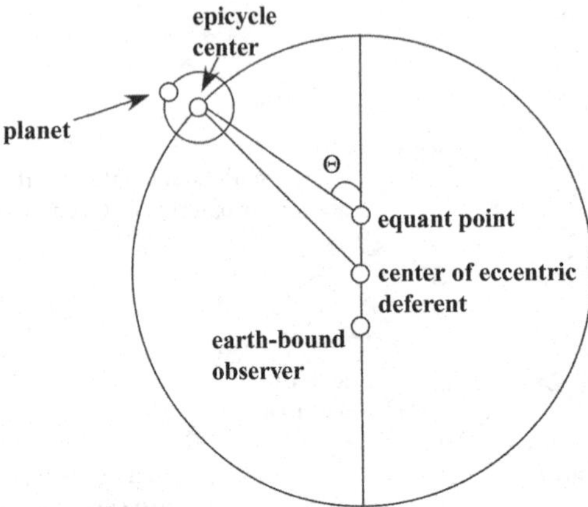

Figure 4.6. Generic epicyclic planetary model illustrating the equant point. The epicycle center rotates uniformly (angle Θ changes uniformly) about the equant point, not about the center of the deferent or the center of the Earth.

Astronomers had become aware of some inconsistencies in Ptolemy's models by the ninth century. A text composed by the ninth-century Banū Mūsā, three sons of Mūsā ibn Shākir named Muḥammad, Aḥmad, and al-Ḥasan, bucks the trend of most ninth-century astronomy, being more focused on theory than on observation.[74] The Banū Mūsā raised questions about how one concentric orb can move another without creating friction, and these concerns are connected to the systematic criticisms of Ptolemy that emerged later, in the eleventh century. But even at this early stage, the Banū Mūsā criticized Ptolemy after thinking carefully about how orbs could rotate, a consideration shared by natural philosophy. The Banū Mūsā favored models that could account for the observations without sacrificing a physically consistent description of the orbs' rotations.

ASTRONOMY AND NATURAL PHILOSOPHY

In addition, the eleventh century saw a reassessment of astronomy's relationship to natural philosophy, one result of which was that an astronomy that assumed the reality of the physical orbs came to mean something more than an astronomy in accord with Aristotelian philosophy. The starting point for our examination of the eleventh century is the career of al-Bīrūnī (d. ca. 1048), a scholar of perhaps unparalleled talents in premodern Islamic civilization. For al-Bīrūnī, the motion of the solar apogee had been an opportunity for an

[74] Saliba, "Early Arabic Critique of Ptolemaic Cosmology," pp. 115–41.

incisive mathematical analysis of instantaneous rates of change.[75] For later astronomers, such as Quṭb al-Dīn al-Shīrāzī (d. 1311), the need for the solar apogee to have its own mover recommended the solar model based on the eccentric hypothesis (which had been Ptolemy's choice), as it could account for the solar apogee's motion with fewer orbs. Bīrūnī's *Kitāb Maqālīd ʿilm al-hayʾa* was a book of trigonometric methods that, like his *Kitāb al-Tafhīm fī ʿilm al-tanjīm* and his treatise on transits, paid attention to astronomy's applications in astrology.[76] Still, Bīrūnī distanced himself from astrology, a science that had been attacked even in Hellenistic civilization for its theoretical foundations and for its questionable predictive accuracy. Bīrūnī's most important work in astronomy was his *al-Qānūn al-Masʿūdī*, a massive *zīj* featuring his own observations, earlier observations, and perceptive mathematical analyses of celestial motions.[77] He cooperated on simultaneous eclipse observations with Abū al-Wafāʾ al-Būzajānī (d. 998), a key figure in the further development of spherical trigonometry. Bīrūnī's work stands out as well for its perceptive view of the weaknesses of proofs based on the conclusions of natural philosophy.[78] For instance, he disagreed with Ibn Sīnā's (Avicenna; d. 1037) stance, based on natural philosophy, that a void was impossible.[79] Famously, Bīrūnī argued that one could use observational tests to determine whether the Earth moved or not.[80] Was Bīrūnī saying that astronomy could make statements about the reality of the celestial orbs without having to rely on natural philosophy?[81] Or was he saying simply that he refused to confine his search for truth to either the philosophical or the mathematical approach?[82] Islamic astronomy after Bīrūnī reconsidered its relationship with natural philosophy.

The eleventh-century critique of natural philosophy's role in astronomy coincided with a critique of astrology. The astrology of Abū Maʿshar

[75] Willy Hartner and Matthias Schramm, "Al-Bīrūnī and the Theory of the Solar Apogee, an Example of Originality in Arabic Science," in *Scientific Change*, ed. Alistair C. Crombie (London: Heinemann, 1963), pp. 206–18. Thābit ibn Qurra had been the first to raise the question of a planet's instantaneous velocity.

[76] Al-Bīrūnī (ed. and trans. Marie Thérèse Debarnot), *Kitāb Maqālīd ʿilm al-hayʾa: La Trigonométrie sphérique chez les Arabes de l'Est à la fin du Xe siècle* (Damascus: Institut Français de Damas, 1985). For al-Bīrūnī's work on transits, see al-Bīrūnī (trans. Mohammad Saffour and Adnan Ifram, comment. E. S. Kennedy), *Al-Bīrūnī On Transits* (Beirut: American University of Beirut, 1959).

[77] Abū al-Rayḥān al-Bīrūnī, *al-Qānūn al-Masʿūdī*, 3 vols. (Hyderabad: Dāʾirat al-maʿārif al-ʿuthmāniyya, 1954–1956).

[78] F. Jamil Ragep, "Ṭūsī and Copernicus: The Earth's Motion in Context," *Science in Context*, 14 (2001), 145–64, especially p. 155.

[79] Seyyed Hossein Nasr and Mahdi Mohaghegh, eds., *Al-Biruni and Ibn Sina: al-Asʾilah waʾl-ajwibah* (Tehran: High Council of Culture and Art, Centre of Research and Cultural Coordination, 1972), p. 58.

[80] Al-Bīrūnī, *al-Qānūn al-Masʿūdī*, vol. 1, pp. 54–6.

[81] F. Jamil Ragep, "Freeing Astronomy from Philosophy: An Aspect of Islamic Influence on Science," *Osiris*, 16 (2001), 49–71, especially pp. 58–60.

[82] Anton Heinen, "Al-Bīrūnī and Al-Haytham: A Comparative Study of Scientific Methodology," in *Al-Bīrūnī Commemorative Volume*, ed. Hakim Mohammed Said (Karachi: The Times Press, 1979), pp. 501–13 at p. 508.

(d. 886) already had been a vehicle for the introduction of Aristotelian philosophy into Islamic civilization.[83] This connection between astrology and Aristotelian philosophy did not go unnoticed. Like al-Bīrūnī, Ibn Sīnā denigrated astrology. Whereas Ibn Sīnā was the philosopher directly targeted by Islam's dialectical theologians, such as al-Ghazālī (d. 1111), his philosophy had represented the greatest attempt to that point to reconcile the Hellenistic intellectual heritage with Islam. Ibn Sīnā's criticism of astrology began not from astrology's foundations but rather from astrology's inability to perform as advertised. Certainly, none of these arguments proved sufficient to eliminate astrology as an application of astronomy.[84] But the reevaluation of astronomy's relationship with both astrology and natural philosophy led to more nuanced arguments that would lead to a flowering of astronomy in a tradition of religious scholarship.

Ghazālī's famous argument against natural philosophy, found in his *Tahāfut al-falāsifa* (*The Incoherence of the Philosophers*), rebutted the philosophers' belief in a necessary connection between cause and effect.[85] The import of Ghazālī's position was that celestial motions could not be understood as caused by the natures of the celestial bodies. Rather, the movements were the direct result of God's will. The response of astronomers was not to cease studying the orbs as physical bodies but instead to do so in such a way as to render their conclusions as independent as possible of a given position on metaphysics. Conversely, some of the tenets of astrology, such as the heavens being a means for God's control over the earth, became part of religious scholarship. One should overlook neither the fact that the *Incoherence* pointed out the folly of denying astronomy's predictive abilities nor that Ghazālī's arguments, whatever their motivations, were philosophical in style.[86] Most of all, Ghazālī was influential because he attacked the philosophers on their own terms, via a mastery of the philosophy of Ibn Sīnā, who had deliberately tried to make philosophy coexist with Islam. The *Incoherence* was preceded by *Maqāṣid al-falāsifa* (*Intentions of the Philosophers*), an informed summary of Ibn Sīnā's philosophy. By any standard, the presence of science and philosophy in religious texts increased after Ghazālī's career.

Religious scholars such as Ghazālī were not the only ones to take a careful look at the relationship between astronomy and natural philosophy in Islamic civilization. The aforementioned Ibn al-Haytham (d. ca. 1040)[87]

83 Abū Maʿshar, *Kitāb al-Madkhal al-kabīr ilā ʿilm al-nujūm*, ed. Richard Lemay (Naples: Istituto Universitario Orientale, 1995). See also George Saliba, "Islamic Astronomy in Context: Attacks on Astrology and the Rise of the Hay'a Tradition," *Bulletin of the Institute for Inter-Faith Studies*, 4 (2002), 73–96.

84 George Saliba, "The Role of the Astrologer in Medieval Islamic Civilization," *Bulletin d'Études Orientales*, 44 (1992), 45–68.

85 Al-Ghazālī, *The Incoherence of the Philosophers*, trans., intro., and annot. Michael Marmura (Chicago: Brigham Young University Press, 1997 and 2000), pp. 166–77.

86 See Frank Griffel, *Al-Ghazali's Philosophical Theology* (Oxford: Oxford University Press, 2009).

87 On Ibn al-Haytham's identity, see Roshdi Rashed, *Les Mathématiques infinitésimales du IXe au XIe siècle*, vol. 2: *Ibn al-Haytham* (London: Al-Furqan Islamic Heritage Foundation, 1993); A. I. Sabra,

took Ptolemy's astronomy to task in his book *al-Shukūk ʿalā Baṭlamyūs* (*Doubts Against Ptolemy*).[88] Ibn al-Haytham's most famous "doubt" was that the center of the epicycle of the upper planets moved uniformly about a point other than the center of the epicycle's deferent (the equant problem).[89] Not only was that arrangement a clear violation of Aristotle's principle that celestial motions had to be uniform, but Ptolemy had contradicted his own stated principles. In three dimensions, the hypothesis of the equant point would entail that the orbs would rotate uniformly, and in place, about an axis that did not pass through the orb's center.[90] Such a state of affairs would be physically and geometrically impossible. Ibn al-Haytham's text had an immediate impact. A student of his contemporary Ibn Sīnā, Abū ʿUbayd al-Jūzjānī, attempted the first (albeit unsatisfactory) solution of the equant problem.[91]

PLANETARY THEORY IN THE ISLAMIC WEST

Advances in models were not confined to the Islamic East; the eleventh century was also a creative time for astronomy in Andalusia. Ibn al-Zarqālluh (d. 1100), an astronomer closely involved with the Toledan Tables and a keen observer of the fixed stars, developed three different models that accounted for trepidation (oscillations in the rate of precession of the equinoxes), with combinations of uniformly rotating orbs.[92] His model for variations in the obliquity of the ecliptic (Figure 4.7) posited that the pole of the ecliptic moved on an epicycle whose center moved on a circle centered on the north celestial pole.[93] Ibn al-Hāʾim also developed a sophisticated model for trepidation.[94] Ibn al-Hāʾim's *zīj* contained a new method (perhaps Ibn

"One Ibn al-Haytham or Two? An Exercise in Reading the Bio-bibliographical Sources," *Zeitschrift für Geschichte der arabisch-islamischen Wissenschaften*, 12 (1998), 1–50; and A. I. Sabra, "One Ibn Al-Haytham or Two? Conclusion," *Zeitschrift für Geschichte der arabisch-islamischen Wissenschaften*, 15 (2002–2003), 95–108.

88 For a recent summary of Ibn al-Haytham's position on the equant, and an ensuing debate, see A. I. Sabra, "Configuring the Universe: Aporetic, Problem Solving, and Kinematic Modeling as Themes of Arabic Astronomy," *Perspectives on Science*, 6 (1998), 288–330; George Saliba, "Arabic versus Greek Astronomy: A Debate over the Foundations of Science," *Perspectives on Science*, 8 (2000), 328–41; and A. I. Sabra, "Reply to Saliba," *Perspectives on Science*, 8 (2000), 342–5.

89 Ibn al-Haytham, *al-Shukūk ʿalā Baṭlamyūs*, ed. A. I. Sabra and Nabil Shehaby (Cairo: National Library Press, 1996; original printing 1971), p. 26.

90 Saliba, "Arabic versus Greek Astronomy," pp. 331–2.

91 George Saliba, "Ibn Sīnā and Abū ʿUbayd al-Jūzjānī: The Problem of the Ptolemaic Equant," *Journal for the History of Arabic Science*, 4 (1980), 376–403.

92 Julio Samsó, *Islamic Astronomy and Medieval Spain* (Aldershot: Variorum, 1994), p. VIII. See also José Maria Millás Vallicrosa, *Estudios sobre Azarquiel* (Madrid: Consejo Superior de Investigaciones Científicas, Instituto "Miguel Asín," Escuelas de Estudios Arabes de Madrid y Granada, 1943–1950), pp. 245–343.

93 Mercè Comes, "Accession and Recession in al-Andalus and the North of Africa," in Casulleras and Samsó, *From Baghdad to Barcelona*, pp. 349–64 at pp. 353–4.

94 Mercè Comes, "Ibn al-Hāʾim's Trepidation Model," *Suhayl*, 2 (2001), 291–408.

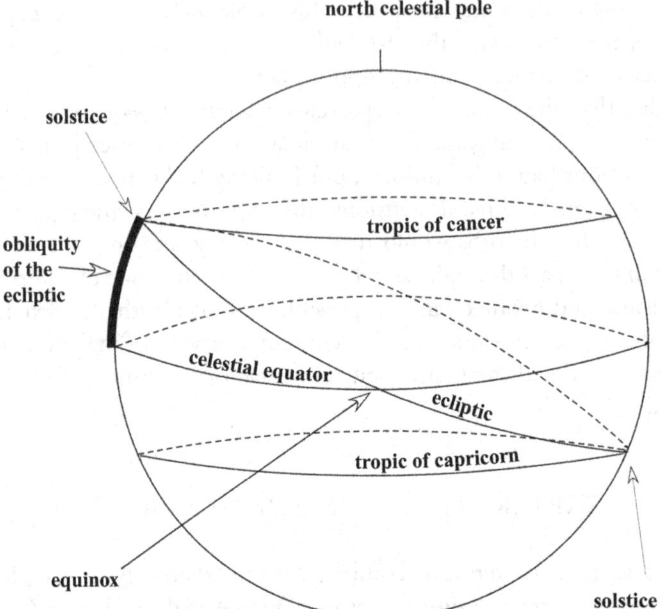

Figure 4.7. Obliquity of the ecliptic (in relation to the celestial equator).

al-Zarqālluh's) for producing improved calculations of the lunar longitude.[95] Ibn al-Zarqālluh was also the first Islamic astronomer to demonstrate that the separate motion of the solar apogee was not equal to the precession of the equinoxes. That discovery was a result of twenty-five years spent on solar observations.[96] Ibn al-Zarqālluh's interest in accounting for his own observations led to interesting theoretical developments such as homocentric planetary models from Andalusia.

In Andalusia, for a brief period, Aristotelian philosophy played an extremely dominant role in astronomy even as the connection between theory and observation was deemphasized. In the twelfth century, such prominent philosophers as Ibn Bājja (Avempace; d. 1138), Ibn Rushd (Averroes; d. 1198), and Maimonides (d. 1204) considered the epicyclic and eccentric orbs unacceptable to Aristotelian philosophy.[97] Inspired by Ibn Bājja's allusions to his own work on astronomy, now lost, al-Biṭrūjī (fl. ca. 1200) endeavored

95 Roser Puig, "The Theory of the Moon in the Al-Zīj al-Kāmil fi'l-Ta'ālīm of Ibn al-Hā'im (ca. 1205)," *Suhayl*, 1 (2000), 71–99. Mercè Comes has attributed the prevalence of work on trepidation in al-Andalus and North Africa to the strong impact of Indian astronomy in the region. See Mercè Comes, "Some New Maghribī Sources Dealing with Trepidation," in *Science and Technology in the Islamic World*, ed. S. M. Razaullah Ansari (Turnhout: Brepols, 2002), pp. 121–41 at pp. 121–2.

96 Emilia Calvo, "Astronomical Theories Related to the Sun in Ibn al-Hā'im's al-Zīj al-Kāmil fi'l-Ta'ālīm," *Zeitschrift für Geschichte der arabisch-islamischen Wissenschaften*, 12 (1998), 51–111 at p. 55. See also Millás Vallicrosa, *Estudios sobre Azarquiel*, p. 241.

97 The classic account of this movement in astronomy remains A. I. Sabra, "The Andalusian Revolt Against Ptolemaic Astronomy: Averroes and al-Biṭrūjī," in *Transformation and Tradition in the*

to construct a system of astronomical models that had neither eccentrics nor epicycles. Poles moving on small circles about the pole of the celestial orb, reminiscent of the models for trepidation, drove the motions of planets near the plane of the zodiac. In addition, all of the motions were in the same direction, as Biṭrūjī deemed seemingly opposite motions of the orbs to be philosophically undesirable. Al-Biṭrūjī's efforts were most successful with the solar model. Even there, though, his model predicted that the sun would stray as much as 1.5° from its observed path through the zodiac, a significant error. With the upper planets, the divergences in latitude were much greater.[98] Despite al-Biṭrūjī's lack of success with predictive accuracy, his text *Kitāb al-Hay'a* enjoyed wide circulation via the Latin translation by Michael Scot. Homocentric astronomy was not confined completely to the Iberian Peninsula. Ibn al-Haytham (d. ca. 1040) used homocentric orbs to construct a model for some of the planets' motions in latitude.[99]

One astronomer writing on the Iberian Peninsula did try to improve Biṭrūjī's models. Ibn Naḥmias's (fl. ca. 1400) *Nūr al-ʿālam* (*Light of the World*) is the only known original composition on theoretical astronomy in Judeo-Arabic. By discarding the need for all motions to be in a single direction, it improved upon Biṭrūjī's predictive accuracy for planetary longitudes and latitudes. All the planets moved on small circles with centers on a great circle inclined to the ecliptic. Even with Ibn Naḥmias's changes, some discrepancies remained.[100] Indeed, any system of astronomy premised on homocentric orbs would have a tough time coping with any variation in planetary distances. These astronomers and philosophers from Andalusia went far beyond holding Ptolemy to his own principles.[101] Theirs was a rare theoretical enterprise divorced from any work on observational astronomy.

THE MARĀGHA OBSERVATORY: PLANETARY THEORY AND OBSERVATIONAL ASTRONOMY

By contrast, both theoretical and observational advances occurred at the institution associated with the most famous figures of Islamic astronomy, the

Sciences: Essays in Honour of I. Bernard Cohen, ed. Everett Mendelsohn (Cambridge: Cambridge University Press, 1984; repr. 2003), pp. 133–53.

98 Al-Biṭrūjī, *On the Principles of Astronomy*, ed., trans., and intro. Bernard R. Goldstein, 2 vols. (New Haven, Conn.: Yale University Press, 1971).

99 F. J. Ragep, "Ibn al-Haytham and Eudoxus: The Revival of Homocentric Modelling in Islam," in *Studies in the History of the Exact Sciences in Honour of David Pingree*, ed. Charles Burnett et al. (Leiden: Brill, 2004), pp. 786–809 at pp. 788–9.

100 Robert Morrison, "The Solar Model of Joseph ibn Naḥmias," *Arabic Sciences and Philosophy*, 15 (2005), 57–108.

101 See Gerhard Endress, "Mathematics and Philosophy in Medieval Islam," in *The Enterprise of Science in Islam: New Perspectives*, ed. Jan P. Hogendijk and A. I. Sabra (Cambridge, Mass.: MIT Press, 2003), pp. 121–76.

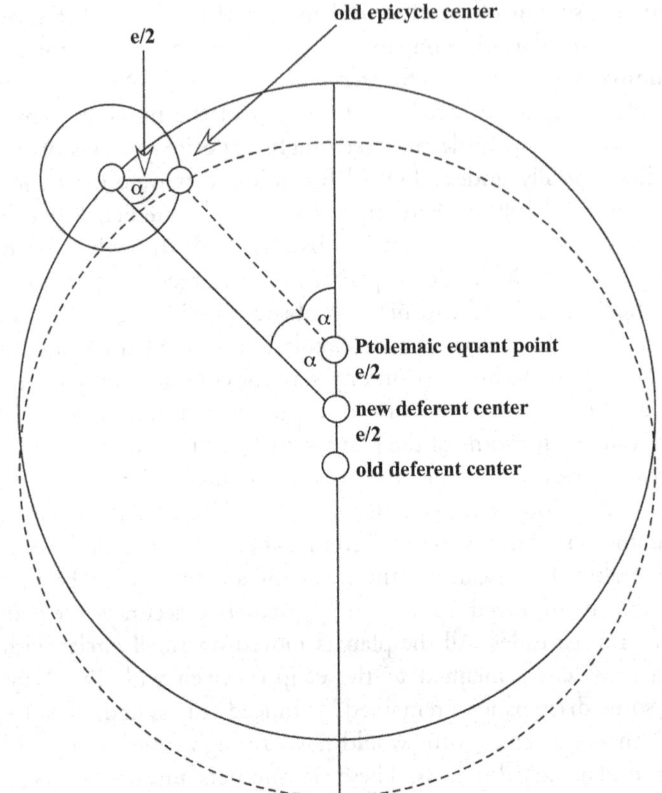

Figure 4.8. ʿUrḍī's planetary model (incorporating the ʿUrḍī lemma).

great observatory at Marāgha (in modern Iran's East Azerbaijan province). The observatory's construction, ordered by either the Ilkhanid sultan Hülegü or Naṣīr al-Dīn al-Ṭūsī (d. 1274), began in 1259.[102] As opposed to the above-mentioned astronomers on the Iberian Peninsula, the astronomers associated with Marāgha aimed to retain Ptolemy's predictive accuracy while removing the inconsistencies that they noticed in the *Almagest.* Ṭūsī and Muʾayyad al-Dīn al-ʿUrḍī (d. 1266), the engineer responsible for the construction of Marāgha's instruments, produced two hypotheses that were the building blocks for their non-Ptolemaic models. The first of these hypotheses to appear at Marāgha is known to historians of astronomy as the ʿUrḍī lemma (Figure 4.8).[103] The ʿUrḍī lemma itself proved that if any two lines of equal length create equal angles on the same side of a third line, then a fourth

[102] On the origins of this observatory, see Ragep, *Naṣīr al-Dīn al-Ṭūsī's Memoir on Astronomy (al-Tadhkira fī ʿilm al-hayʾa)*, pp. 13–15. See also Sayılı, *Observatory in Islam*, pp. 189–223.

[103] For the first identification of the ʿUrḍī lemma, see George Saliba, "The First Non-Ptolemaic Astronomy at the Maraghah School," *Isis*, 70 (1979), 571–6; and George Saliba, "The Original

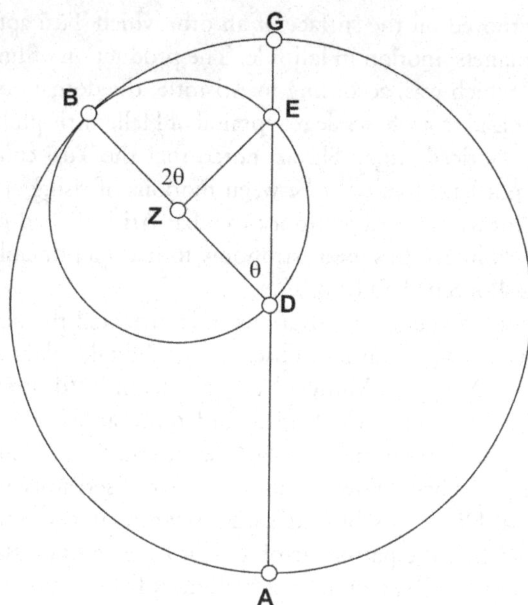

Figure 4.9. The Ṭūsī couple.

line connecting the two equal lines is parallel to the third. On that basis, ʿUrḍī's models for many planets incorporated an eccentric deferent whose center was halfway between the equant and the center of the old Ptolemaic deferent.[104] The result is that the epicycle's center traces a motion that is uniform about the equant's center.

The Ṭūsī couple was another hypothesis that answered some of Ibn al-Haytham's doubts about Ptolemy. It was based on the following geometrical lemma (see Figure 4.9). Given two circles arranged such that the diameter of one is a radius of the other ($DB = DG$), if the smaller rotates at twice the velocity of the larger but in the opposite direction, then a given point E on the diameter of the small circle will oscillate along the diameter GA of the larger circle.[105] Then, Ṭūsī replaced the two circles with two orbs and incorporated that device into models. For all planets except Mercury, Ṭūsī proposed a model with a new deferent, whose center was the old equant point. The inclusion of a Ṭūsī couple in those models meant that the motions of the planet virtually mirrored those that the Ptolemaic models predicted. Ṭūsī's *Tadhkira* contained a curvilinear version of the Ṭūsī couple in which

Source of Quṭb al-Dīn al-Shīrāzī's Planetary Model," *Journal for the History of Arabic Science*, 3 (1979), 376–403.

104 Animations of these models can be found at https://people.scs.fsu.edu/~dduke/arabmars.html.

105 Ragep, *Naṣīr al-Dīn al-Ṭūsī's Memoir on Astronomy (al-Tadhkira fī ʿilm al-hayʾa)*, pp. 194–223.

the two circles moved on the surface of an orb, which Ṭūsī applied to the model for the planets' motion in latitude. The production of linear motion in the heavens, which was, according to Aristotle, the domain of rotational motion, did not lead to a wholesale reappraisal of Hellenistic philosophy, but it did not go unnoticed either. Shīrāzī noted that the Ṭūsī couple showed how there did not have to be rest between motions of rising and falling of a heavy body thrown upward, contrary to what Aristotle had propounded (*Physics* 262a).[106] Shīrāzī thus used astronomy to reject a principle of natural philosophy that Ibn Sīnā had held.[107]

Although these theoretical innovations have attracted the lion's share of scholarly attention, important new observations did take place at Marāgha. An astronomer at Marāgha, Muḥyī al-Dīn al-Maghribī (d. 1283), used, for the first time, a mechanical clock to record more accurately the time of observations.[108] The astronomers at Marāgha produced new parameters for the solar model. The best-known product of that observatory was the *Zīj-i Īlkhānī*. Niẓām al-Dīn al-Nīsābūrī (d. ca. 1330) wrote, in his commentary on the *Zīj-i Īlkhānī*, that the patronage of a great king was necessary for such an enterprise, and we do know that astronomers from China were present at the Marāgha observatory. He and Ṭūsī remarked on the importance of astrological applications for a thorough observational program. In addition, Maghribī wrote his own *zīj* based on his observations at Marāgha and, like ʿUrḍī, recomputed the solar eccentricity using a method that involved three randomly chosen observations, with two of them in opposition.[109] Maghribī also evinced, through his recomputation of the parameters for the planetary models, a deep understanding of the equant's connection to observations.[110] That is, the position of the center of the planet's mean motion (the equant) had to be adjusted as the parameters of the planets' motions were updated. Thanks to their wide-ranging skills, the Marāgha astronomers were well aware of the connection between instrument construction, observational technique, and the resulting implications for planetary distances and physical models of planetary motion.

Whereas Ṭūsī had argued that the difference between the positions predicted by his models and those predicted by Ptolemy's models was negligible, ʿUrḍī and Quṭb al-Dīn al-Shīrāzī (d. 1311), a student of Ṭūsī, pointed out that their own model for the upper planets fit the observations just as

106 Robert Morrison, "Quṭb al-Dīn al-Shīrāzī's Hypotheses for Celestial Motions," *Journal for the History of Arabic Science*, 13 (2005), 21–140 at pp. 58 and 91–2.

107 Abū al-Barakāt al-Baghdādī (d. 1152) had made a similar point; see Saliba, *Islamic Science and the Making of the European Renaissance*, pp. 182–4.

108 Saliba, "Determination of New Planetary Parameters at the Maragha Observatory," pp. 268–9.

109 George Saliba, "Solar Observations at the Maraghah Observatory Before 1275," *Journal for the History of Astronomy*, 16 (1985), 113–22. See also Sayılı, *The Observatory in Islam*, p. 33.

110 Saliba, "Determination of New Planetary Parameters at the Maragha Observatory," p. 255.

well as Ptolemy's.[111] Like Maghribī, Urḍī and then Shīrāzī had a profound understanding of where the connection between observations and the *Almagest*'s models was strong and where it was weak. Specifically, the observations upon which Ptolemy relied did not necessitate a circular path for the epicycle's center. A subsequent theoretical innovation emerging from close attention to observations was Ibn al-Shāṭir's (d. 1375) solar model. That model was the first to account for newly observed variations in the diameter of the Sun.[112] Such observations had led the Marāgha astronomers to accept the existence of annular eclipses, solar eclipses in which the Moon leaves an illuminated ring from the solar disk unobscured. [113] Ibn al-Shāṭir's planetary models not only retained predictive accuracy and eliminated the contradiction of the equant point but also eliminated the need for eccentric orbs.[114] Although based in Damascus, Ibn al-Shāṭir is considered to be one of the Marāgha astronomers owing to the way in which his theories built on the foundation of astronomers such as ʿUrḍī and Ṭūsī. And of all the Marāgha astronomers, he is the one whose Earth-centered models bear the closest resemblance to Copernicus's Sun-centered ones.[115]

The most intractable problem for the Marāgha astronomers was answering the objections to Ptolemy's Mercury model. Shīrāzī was the first to succeed in that respect.[116] Two texts by Shīrāzī on theoretical astronomy, *al-Tuḥfa al-shāhiyya* (*The Royal Gift*) and *Nihāyat al-idrāk fī dirāyat al-aflāk* (*The Highest Attainment in Comprehending the Orbs*), each contain chapters that list the available hypotheses (*uṣūl*), the building blocks of planetary models. Subsequent writers did not invent new *uṣūl*; rather, they applied the ones that Shīrāzī had described in increasingly ingenious ways. In Shīrāzī, as in ʿUrḍī, one detects a refinement of the underlying foundations of planetary theory that parallels the refinements in instrument construction and observational astronomy.[117] ʿAlāʾ al-Dīn al-Qushjī (d. 1474) presented another

111 Morrison, "Quṭb al-Dīn al-Shīrāzī's Hypotheses for Celestial Motions," p. 85. See also George Saliba, *The Astronomical Work of Mu'ayyad al-Dīn al-ʿUrḍī (Kitāb al-Hay'a): A Thirteenth-Century Reform of Ptolemaic Astronomy* (Beirut: Markaz Dirāsāt al-Waḥdah al-ʿArabiyyah, 1990), p. 223.

112 George Saliba, "Theory and Observation in Islamic Astronomy: The Work of Ibn al-Shāṭir of Damascus," *Journal for the History of Astronomy*, 18 (1987), 35–43 at p. 36.

113 Ragep, *Naṣīr al-Dīn al-Ṭūsī's Memoir on Astronomy (al-Tadhkira fī ʿilm al-hay'a)*, p. 461.

114 George Saliba, "Arabic Planetary Theories after the Eleventh Century AD," in Rashed, *Encyclopedia of the History of Arabic Science*, vol. 1, pp. 58–127 at pp. 110–13.

115 Saliba, *Islamic Science and the Making of the European Renaissance*, pp. 193–232. See also Noel M. Swerdlow and Otto Neugebauer, *Mathematical Astronomy in Copernicus' De Revolutionibus*, 2 vols. (New York: Springer-Verlag, 1984), vol. 1, pp. 41–64; vol. 2, pp. 564–73; and Willy Hartner, "The Man, the Work, and Its History," *Proceedings of the American Philosophical Society*, 117 (1973), 413–22 at p. 421.

116 E. S. Kennedy, "Late Medieval Planetary Theory," *Isis*, 57 (1966), 365–78. For a more updated and complete presentation, see George Saliba, "Arabic Planetary Theories," in Rashed, *Encyclopedia of the History of Arabic Science*, vol. 1, pp. 58–127 at pp. 115–25.

117 See King, *In Synchrony with the Heavens*, vol. 2, pp. 153–61. King draws on François Charette, *Mathematical Instrumentation in Fourteenth-Century Egypt and Syria: The Illustrated Treatise of Najm al-Dīn al-Miṣrī* (Leiden: Brill, 2003).

non-Ptolemaic Mercury model in his *Treatise Regarding the Solution of the Equant Problem*. And Shams al-Dīn al-Khafrī (fl. ca. 1525), in *al-Takmila fī sharḥ al-Tadhkira* (*The Complement to the Commentary on the Tadhkira*), provided multiple non-Ptolemaic Mercury models, one of which was Qushjī's and all of which were mathematically equivalent.[118]

One observational problem that the Marāgha astronomers were unable to resolve fully was that of planetary sizes and distances. The models of the *Planetary Hypotheses* yielded values for the sizes and distances of the planets in terms of Earth radii. Given the order of the planets, the principle that the apogee of one planet was the perigee of the next, and the distances of the Sun and Moon determined from parallax and eclipses, one could determine planetary sizes and distances. But sometimes the greatest distance found for one planet was not the least distance found for the planet placed above it. In particular, the distances that astronomers found for Venus did not easily support its being below the Sun. At best, the greatest distance for Venus was slightly greater than the Sun's nearest distance.[119] There was also the issue of the eccentricity of Mars being so large as to encroach on the Sun's path at times. Indeed, a few generations earlier, Maimonides' (d. 1204) *Guide for the Perplexed* had cited the contradictions of the astronomers on the topic of planetary sizes and distances as a reason for his perplexity.[120] 'Urḍī's extensive research into this problem led him to conclude, despite Ibn Sīnā's (d. 1037) report that Venus had been observed passing in front of the Sun, that Venus was above the Sun.[121] Other inaccuracies resulted, however.[122] And the distances of Mars, Jupiter, and Saturn would have to be increased.[123]

Another way of approaching the question of the locations of Mercury and Venus was through transit observations, and in this area astronomers had greater success.[124] Nevertheless, astronomers of the thirteenth and fourteenth

118 On Qushjī, see George Saliba, "Al-Qushjī's Reform of the Ptolemaic Model for Mercury," *Arabic Sciences and Philosophy*, 3 (1993), 161–203. On Khafrī, see George Saliba, "A Redeployment of Mathematics in a Sixteenth-Century Arabic Critique of Ptolemaic Astronomy," in *Perspectives arabes et médiévales sur la tradition scientifique et philosophique grecque*, ed. A. Hasnawi, A. Elamrani-Jamal, and M. Aouad (Leuven: Peeters, 1997), pp. 105–22.

119 Noel M. Swerdlow, *Ptolemy's Theory of the Distances and Sizes of the Planets: A Study of the Scientific Foundations of Medieval Cosmology* (diss., Yale University, 1968), pp. 123–5. See also Ragep, *Naṣīr al-Dīn al-Ṭūsī's Memoir on Astronomy (al-Tadhkira fī 'ilm al-hay'a)*, pp. 519–20.

120 See Moses Maimonides, *The Guide of the Perplexed*, trans. and intro. Shlomo Pines (Chicago: University of Chicago Press, 1963), pp. 322–7.

121 Bernard Goldstein, "Theory and Observation in Medieval Astronomy," *Isis*, 63 (1972), 39–47 at p. 44.

122 Swerdlow, *Ptolemy's Theory of the Distances and Sizes of the Planets*, pp. 194–5.

123 Swerdlow, *Ptolemy's Theory of the Distances and Sizes of the Planets*, p. 195; and Bernard R. Goldstein and Noel Swerdlow, "Planetary Distances and Sizes in an Anonymous Arabic Treatise Preserved in Bodleian Ms. Marsh 621," *Centaurus*, 15 (1970), 135–70 at p. 137. George Saliba found that the author of this anonymous treatise was Mu'ayyad al-Dīn al-'Urḍī (d. 1266); see George Saliba, "The First Non-Ptolemaic Astronomy at the Maraghah School," *Isis*, 70 (1979), 571–6.

124 Bernard R. Goldstein, "Some Medieval Reports of Venus and Mercury Transits," *Centaurus*, 14 (1969), 49–59 at p. 55.

centuries who situated Venus and Mercury below the Sun continued to resort to arguments from symmetry in order to do so.

ASTRONOMY IN RELIGIOUS SCHOLARSHIP

The religious applications of observational astronomy remained prominent. The Mamlūk period (1250–1517) was characterized by a dissemination of astronomy indicated by the proliferation of basic textbooks during that period and by remarkable advances in instrumentation. In the Mamlūk lands of Egypt and the Fertile Crescent, the office of *muwaqqit* (religious timekeeper), which Ibn al-Shāṭir held, emerged, and *ʿilm al-mīqāt* (the science of timekeeping) became a recognized division of astronomy. Not only were Ibn al-Shāṭir's theoretical achievements coincident with this heightened attention to practical astronomy but, paralleling Khafrī's research on the Mercury model, scholars developed numerous equivalent but different ways to construct instruments.[125] The study of astronomy was also widespread in Mamlūk *madrasas*.[126] The increased popularity of *mīqāt* in Mamlūk society led to new developments in instrument construction.[127] The astonishing variety of astrolabe designs, some never constructed, in Najm al-Dīn al-Miṣrī's treatise on instrument building, indicates a complete mastery of the theoretical underpinnings of that instrument.

Even before Ibn al-Shāṭir's career, astronomy had become a component of religious texts, and many of the astronomers this chapter has mentioned, including Ibn al-Shāṭir and Shīrāzī (a judge in Anatolia), served in a religious capacity. Ṭūsī, for instance, was a noted Shiite theologian, and Ṣadr al-Dīn al-Sharīʿa (d. 1347) wrote a three-part encyclopedia on astronomy, *kalām* (a speculative investigation into God's nature), and Islamic law. [128] Because natural philosophy had now been subjected to a thorough critical examination and because many of natural philosophy's conclusions were not objectionable, texts on *kalām* incorporated, while still examining critically, material from *falsafa* (Hellenistic natural philosophy) and sciences such as astronomy.[129] The *kalām* texts and Qurʾān commentary of Fakhr al-Dīn al-Rāzī (d. 1210) present much scientific and philosophical material, occasionally in a critical light.[130] Moreover, statements scientists made in

125 François Charette, "The Locales of Islamic Astronomical Instrumentation," *History of Science*, 44 (2006), 123–38 at pp. 129–31.

126 Ibid., pp. 128–31.

127 Charette, *Mathematical Instrumentation in Fourteenth-Century Egypt and Syria*, p. 9.

128 On Ṣadr al-Sharīʿa, see Ahmad Dallal, *Islamic Response to Greek Astronomy*. See also Ahmad Dallal, *Islam, Science, and the Challenge of History* (New Haven, Conn.: Yale University Press, 2010), pp. 135–38.

129 A. I. Sabra, "Science and Philosophy in Medieval Islamic Theology," *Zeitschrift für Geschichte der arabisch-islamischen Wissenschaften*, 9 (1994), 1–42 at p. 12.

130 Ayman Shihadeh, "From al-Ghazālī to al-Rāzī: 6th/12th Century Developments in Muslim Philosophical Theology," *Arabic Sciences and Philosophy*, 15 (2005), 141–79.

their religious texts lend insight into astronomers' views on both the reality of their models and their science. This is because religious considerations created a tension between understanding the orbs as physical reality and considering the orbs simply as a device for calculating planetary positions, a tension found in Ghazālī's writings themselves. On the one hand, Ghazālī's *Incoherence* stressed the contingency of all causal processes. Since one could not be sure of chains of causes and effects, astronomical theories would remain mathematical constructs. On the other hand, Ghazālī's position on God's eternal foreknowledge meant that natural processes would be open to profitable study, despite the contingency of natural causes, because the outcomes of those causes were fixed.[131] Thus there was a value in studying the orbs, the causes of celestial motions, with regard to their physical reality.

The final stage in the incorporation of natural philosophy and science, including astronomy, into *kalām* texts was the *Kitāb al-Mawāqif* (*The Book of Stations*) of ʿAḍud al-Dīn al-Ījī (d. 1355). *Kitāb al-Mawāqif*, with its informed précis of Ptolemaic astronomy, has remained a staple of the *madrasa* curriculum, and thus evidence for the study of astronomy within the *madrasa*. In a famous passage from a commentary on *Kitāb al-Mawāqif*, al-Sayyid al-Sharīf al-Jurjānī (d. 1413), also a key commentator on Ṭūsī's *Tadhkira*, rebutted Ījī's allegation that the astronomers' models were as tenuous as the threads of a spider's web. Jurjānī conceded that astronomers made assumptions, but argued that by means of those assumptions "the conditions of (celestial) movements are regulated in regard to speed and direction, as perceived (directly) or observed with the aid of instruments. [By means of these notions also] discovery is made of the characteristics of the celestial orbs and the earth, and of what they reveal of subtle wisdom and wondrous creation."[132] Thus, although a fictionalist perspective might best defend a strict interpretation of God's omnipotence, an approach that recognized the orbs' physical reality would give scholars of astronomy greater insight into God's glory. In certain cases, there was cross-fertilization between scientific and religious scholarship.[133] As was the case with *qibla* determination, a nonscientific astronomy based on the Qur'ān and *Ḥadīth* existed. A very important example was the *hay'a sunniyya* (traditional astronomy) of Jalāl al-Dīn al-Suyūṭī (d. 1505), who developed his astronomy and cosmography entirely from revealed texts.[134] Suyūṭī's astronomy did not aspire to or succeed in predicting planetary positions.

131 Richard M. Frank, *Creation and the Cosmic System* (Heidelberg: Carl Winter Universitätsverlag, 1992).

132 Sabra, "Science and Philosophy in Medieval Islamic Theology," p. 39. On the reference to a spider's web, see p. 37. For background about *Kitāb al-Mawāqif*, see p. 13.

133 Robert Morrison, *Islam and Science: The Intellectual Career of Niẓām al-Dīn al-Nīsābūrī* (London: Routledge, 2007).

134 Anton Heinen, *Islamic Cosmology: A Study of As-Suyūṭī's al-Hay'a as-sanīya fī l-hay'a as-sunnīya, with Critical Edition, Translation, and Commentary* (Beirut: Franz Steiner Verlag, 1982).

DEVELOPMENTS IN THE FIFTEENTH CENTURY AND THEREAFTER

The most remarkable institution of the fifteenth century was the observatory in Samarqand, which flourished under the leadership of Ulug Begh (d. 1449), the first governor of parts of Khurasan and Mazandaran, and eventually the occupant of the Timurid throne.[135] The letters of Ghiyāth al-Dīn al-Kāshī (d. ca. 1430) tell us of a *madrasa* with organized scientific instruction and of a methodical program of observation.[136] The most famous theoretician associated with the Samarqand observatory was ʿAlāʾ al-Dīn al-Qushjī (d. 1474), who in addition to his non-Ptolemaic Mercury model perhaps produced a non-Ptolemaic lunar model.[137] In addition, despite skilled predecessors such as Shīrāzī, who denied that it could be done, Qushjī demonstrated the equivalence of the epicyclic and eccentric hypotheses for the retrograde motions of Venus and Mercury.[138] In Renaissance Europe, that equivalence was a key to Copernicus's transformation from a geocentric to a heliocentric cosmos. The observational activity of the Samarqand observatory led to the *Zīj-i Sulṭānī* (a.k.a. *Zīj-i Ulug Begh*). The best-known instrument of the Samarqand observatory was its meridian sextant dug into the ground, which had a radius of over 40 meters. Kāshī himself wrote a treatise on instruments. The instruments at Samarqand inspired the eighteenth-century observatory constructed at Jaipur. Qushjī, like others associated with the Samarqand observatory and *madrasa*, wrote significant works of religious scholarship related to his work on astronomy.[139] Qushjī's work on *kalām* raised new questions about the relationship of physical models to observations. Specifically, he advocated an astronomy founded entirely on observations. He also proposed severing *hayʾa* from the ground rules adopted from Peripatetic natural philosophy.[140] In other words, astronomers should construct their models using their own principles, not those borrowed from natural philosophy.

Taqī al-Dīn Ibn Maʿrūf, the chief of the Istanbul observatory, predicted military success for the Ottomans against the Safavids on the basis of the

135 Sayılı, *The Observatory in Islam*, p. 260.

136 E. S. Kennedy, "A Letter of Jamshīd al-Kāshī to his Father," *Orientalia*, 29 (1960), 191–213. On Kāshī's experiences at the *madrasa* in Samarqand, see Aydın Sayılı, *Ghiyāth al-Dīn al-Kāshī's Letter on Ulugh Bey and the Scientific Activity in Samarqand* (Ankara: Türk Tarih Basımevi, 1960); and M. Bagheri, "A Newly Found Letter of Al-Kāshī on Scientific Life in Samarkand," *Historia Mathematica*, 24 (1997), 241–56.

137 Saliba, "Al-Qushjī's Reform of the Ptolemaic Model for Mercury," pp. 161–203.

138 F. Jamil Ragep, "ʿAlī Qushjī and Regiomontanus: Eccentric Transformations and Copernican Revolutions," *Journal for the History of Astronomy*, 36 (2005), 359–71.

139 On Samarqand, see İhsan Fazlıoğlu, "The Samarqand Mathematical–Astronomical School: A Basis for Ottoman Philosophy and Science," *Journal for the History of Arabic Science*, 14 (2008), 3–68.

140 Ragep, "Freeing Astronomy," *Osiris*, 16 (2001), 49–71.

comet of 1577.[141] Even if religious scruples prevented astrological forecasts from explicitly motivating observational astronomy, the increased prominence of science in religious scholarship meant that the cosmos, as a sign of God's majesty, deserved continuous, precise study. For instance, eclipses, apart from any role in an astrological prediction, could be studied as a sign of God's power.[142] Thus, the inclination of the paths of the Sun and the Moon, contrived so that eclipses occur periodically, reflects divine design.

The development of a heliocentric astronomy in Europe did not, on its own, mean the end of Islamic science. Interesting texts on Islamic astronomy continued to be written through the sixteenth century. Ghiyāth al-Dīn al-Dashtaghī (d. 1542/1544) and Ghars al-Dīn al-Ḥalabī (d. 1563), in addition to Khafrī (q. v.), wrote important works of astronomy that deserve examination. Commentaries on the introductory astronomy text *Tashrīḥ al-aflāk* of Bahā' al-Dīn al-ʿĀmilī (d. 1622) continued to draw attention to the faults of Ptolemaic astronomy and to the available solutions.[143] By the seventeenth century, intellectuals in the Ottoman Empire, such as Katib Çelebi (d. 1657) and Ibrahim Müteferrika (d. 1745), composed summaries of early-modern European (i.e., Copernican) astronomy.[144] Educational reforms in the Ottoman Empire in the eighteenth century led to the incorporation of early-modern European science into the curriculum of government engineering schools, the *mühendishanes*.[145] The study of Islamic astronomy nevertheless continued in religious institutions, both in the Ottoman Empire and elsewhere.[146] A nineteenth-century commentary on *Tashrīḥ al-aflāk* from the Ottoman Empire written in Arabic presents achievements of early-modern European astronomy such as the Copernican system and the observations of Uranus and Neptune.[147] Indeed, in some locales, instruction in Islamic astronomy continued into the twentieth century.[148]

141 Sayılı, *The Observatory in Islam*, pp. 290–1.

142 Morrison, *Islam and Science*, p. 30.

143 Saliba, *Islamic Science and the Making of the European Renaissance*, pp. 114–16.

144 Ekmeleddin İhsanoğlu, "Introduction of Western Science to the Ottoman World: A Case Study of Modern Astronomy (1660–1860)," in *Transfer of Modern Science and Technology to the Muslim World*, ed. Ekmeleddin İhsanoğlu (Istanbul: Research Centre for Islamic History, Art, and Culture, 1992), especially pp. 67–96.

145 İhsanoğlu, "Introduction of Western Science to the Ottoman World," pp. 99–103.

146 For the study of astronomy in Iranian *madrasas* in the twentieth century, see Roy Mottahedeh, *The Mantle of the Prophet: Religion and Politics in Iran* (New York: Pantheon Books, 1985), pp. 100, 103. To be sure, some religious scholars criticized the motion of the Earth. See Juan R. I. Cole, *Roots of North Indian Shīʿism in Iran and Iraq* (Berkeley: University of California Press, 1988), p. 264.

147 Robert Morrison, "The Reception of Early-Modern European Astronomy by Ottoman Religious Scholars," *Archivum Ottomanicum*, 21 (2003), 187–96.

148 Ḥasan Taqīzādeh, *Yādnāmeh* (Tehran: Anjuman-i Āsār-i Millī, 1971; original date 1349), p. 290.

5

MEDICINE IN MEDIEVAL ISLAM

Emilie Savage-Smith

The geographical contours of the medieval Islamic world extended from North Africa and Spain in the west to India and Central Asia in the east, with the central lands of Egypt, Syria, Iraq, and Persia playing a pivotal role. In temporal terms, it covered a period of roughly eleven centuries – from the middle of the seventh to the middle of the eighteenth. Over such a vast area and time span the nature of medical care varied greatly.[1]

The everyday medical practices and the general health of the Islamic community were influenced by many factors: the dietary and fasting laws and the general rules for hygiene and burying the dead of the various religious communities of Muslims, Jews, Christians, Zoroastrians, and others; the climatic conditions of the desert, marsh, mountain, and littoral communities; the different living conditions of nomadic, rural, and urban populations; local

[1] This chapter was first written in 1996 and then slightly revised in 2000 and again in 2006 to allow citation of recent publications. For further elaboration of material and interpretations presented here, see Peter E. Pormann and Emilie Savage-Smith, *Medieval Islamic Medicine* (The New Edinburgh Islamic Surveys) (Edinburgh: Edinburgh University Press, 2007). For other general surveys of the topic, see Manfred Ullmann, *Islamic Medicine* (Edinburgh: University of Edinburgh Press, 1978); Lawrence I. Conrad, "The Arab-Islamic Medical Tradition," in *The Western Medical Tradition, 800 BC to AD 1800*, ed. Lawrence I. Conrad et al. (Cambridge: Cambridge University Press, 1995), pp. 93–138; Emilie Savage-Smith, "Ṭibb [Medicine]," in *The Encyclopaedia of Islam*, new ed., ed. H. A. R. Gibb et al. (Leiden: Brill, 1960–2005), vol. 10, pp. 452–60; Emilie Savage-Smith, "Medicine," in *Encyclopedia of the History of Arabic Science*, ed. Roshdi Rashed, 3 vols. (London: Routledge, 1996), vol. 3, pp. 903–62; H. D. Haskell, "Arabic Medical Literature," in *Religion, Learning, and Science in the ʿAbbasid Period*, ed. M. J. L. Young, J. D. Latham, and R. B. Serjent (Cambridge: Cambridge University Press, 1990), pp. 342–63; G. Strohmaier, "Reception and Traditions: Medicine in the Byzantine and Arab World," in *Western Medical Thought from Antiquity to the Middle Ages*, ed. Mirko D. Grmek (Cambridge, Mass.: Harvard University Press, 1998), pp. 139–69; Lutz Richter-Bernberg, "Iran's Contribution to Medicine and Veterinary Science in Islam AH 100–900/AD 700–1500," in *The Diffusion of Greco-Roman Medicine into the Middle East and the Caucasus*, ed. J. A. C. Greppin, E. Savage-Smith, and J. L. Gueriguian (Delmar, N.Y.: Caravan Books, 1999), pp. 139–68; and the introduction to Michael W. Dols, *Medieval Islamic Medicine: Ibn Riḍwān's Treatise 'On the Prevention of Bodily Ills in Egypt'* (Berkeley: University of California Press, 1984). Still useful are Edward G. Browne, *Arabian Medicine* (Cambridge: Cambridge University Press, 1921) and (used with caution) Cyril Elgood, *A Medical History of Persia and the Eastern Caliphate* (Cambridge: Cambridge University Press, 1951).

economic conditions and agricultural successes or failures; the amount of travel undertaken for commerce, for attendance at courts, or as a pilgrimage; the maintenance of a slave class and slave trade; the injuries and diseases attendant upon army camps and during battles; and the incidence of plague and other epidemics, as well as the occurrence of endemic conditions such as various dysenteries and certain eye diseases. The institutions and policies responsible for dispensing medical care were, moreover, subject to political and social fluctuations.

The medical practices of the society varied not only according to time and place but according to the various strata comprising the society. The economic and social levels of the patient determined to a large extent the type of care sought, and the expectations of the patients varied along with the approaches of the practitioners. Throughout medieval Islamic society, a medical pluralism existed that may be viewed as a continuum running from the formal theories and practices of learned medicine to those of local custom and magic. The medieval Islamic community itself comprised Muslims and non-Muslims speaking many languages – Arabic, Persian, Syriac, Hebrew, and Turkish – though Arabic became the lingua franca and Islam the dominant religion. The resulting medical care involved a rich mixture of religions and cultures to be seen in both the physicians and the patients, for which reason in this context the terms Islamic culture or Islamic medicine are not to be interpreted as applying only to the religion of Islam.

For learned medicine, the medical theories inherited from the Hellenistic world supplied a thread of continuity and uniformity. The formal treatises, however, provide only a limited view of the actual medical practices of the day, and on occasion they are not corroborated by other available evidence. The following discussion will begin with what little is known of medicine in the Middle East prior to the rise of Islam and then proceed to a brief chronological and typological overview of early Islamic medicine. Topics that particularly interested Islamic writers on learned medicine – ophthalmology, pharmacology, and anatomy – will be given separate attention. The final two sections will be concerned with evidence for the everyday practice of medicine and the occasional discrepancies between theory and practice.

PRE-ISLAMIC MEDICINE

In the pre-Islamic Near East, as in any society, there were medical concerns and practitioners who tended the needs of the sick and injured. The primary sources for this period are the accounts of the sayings of the Prophet Muḥammad and others in the early Islamic community. It is evident that the digestive system and its disorders were the central concern, and diet the focus of both prevention and therapy, with much use of locally grown herbs and foodstuffs such as honey. Wise women and elders were the main dispensers of medical care. Protection from the evil eye and the use of talismans and amulets to aid in childbirth or ward off disease figured prominently. Bleeding and cauterization were performed, but surgery was not part of therapeutics.

A genre of Arabic medical writing that arose in the ninth century reflected pre-Islamic practices. These treatises are usually referred to as "prophetic medicine" (*al-ṭibb al-nabawī*) and were intended as an alternative to the exclusively Greek-based medical systems that were at that time being introduced. The authors were generally clerics rather than physicians, and they advocated the traditional medical practices of the Prophet's day over the medical ideas assimilated from Hellenistic society, sometimes blending the two approaches. One of the earliest examples is the ninth-century Arabic treatise called *Medicine of the Imams* compiled in the eastern provinces by clerics of the Shīʿī branch of Islam. At about the same time, Ibn Ḥabīb al-Andalūsī, working in Muslim Spain, also composed a treatise on prophetic medicine. In the thirteenth and fourteenth centuries, the genre became quite popular, and it remains so today. The treatises by the historian al-Dhahabī (d. 1348), the legal scholar Ibn Qayyim al-Jawzīyah (d. 1350), and theologian Jalāl al-Dīn al-Suyūṭī (d. 1505) are still available in modern printings. Treatises on prophetic medicine, advocating many folkloric and magical remedies, flourished for centuries alongside those of the Greek-based humoral tradition. Although we know of a considerable number of treatises on prophetic medicine, we do not have the names of any practitioners. The reason may be that our sources are for the most part skewed toward the Greek-based system that came to dominate the learned medical tradition.[2]

EARLY ISLAMIC MEDICINE

There is little evidence available regarding the medicine practiced during the first century and a half of Islam. The sources tell us virtually nothing about the medical care extended to the first four caliphs – that is, the first four successors to the Prophet Muḥammad, known as the Orthodox caliphs (632–661). There is a story of an Arab named al-Ḥārith ibn Kaladah who is said to have studied medicine at Gondēshāpūr in Persia. He was said to have held learned discussions with the Persian ruler Khusraw Anūshirwān (d. 579) and to have been consulted during the final illnesses of the last two of the Orthodox caliphs nearly a hundred years later. The therapy associated

[2] For the "prophetic medicine" literature and related plague tracts, see Irmeli Perho, *The Prophet's Medicine: A Creation of the Muslim Traditionalist Scholars* (Studia Orientalia, 74) (Helsinki: Finnish Oriental Society, 1995); Michael W. Dols, *Majnūn: The Madman in Medieval Islamic Society* (Oxford: Clarendon Press, 1992), pp. 211–60; Michael W. Dols, *The Black Death in the Middle East* (Princeton, N.J.: Princeton University Press, 1977); and Lawrence I. Conrad, "Epidemic Disease in Formal and Popular Thought in Early Islamic Society," in *Epidemic and Ideas: Essays on the Historical Perception of Pestilence*, ed. Terence Ranger and Paul Slack (Cambridge: Cambridge University Press, 1992), pp. 77–99.

with his name reflects traditional practices of using locally available plants, but the accounts of al-Ḥārith ibn Kaladah were elaborated over time to the point where they now include conflicting elements that make it difficult to assess the historical figure. For similar reasons, questions arise regarding the authenticity of reports regarding Ibn Abī Rimthah, who was supposed to have been a contemporary of the Prophet and to have practiced minor surgery. Evidently a need was later felt to justify and defend the use of medicine by appealing to accounts that showed the Prophet and early members of the Muslim community having recourse to doctors.[3]

Only a few meager details emerge regarding the physicians serving the caliphs of the early Umayyad dynasty, which acquired the caliphate after the four immediate successors to the Prophet. The first Umayyad caliph, Muʿāwiyah (r. 661–80), is said to have employed at the court in Damascus a Christian physician, Ibn Uthāl. The physician to the caliph ʿUmar ibn ʿAbd al-ʿAzīz (r. 717–20) is said to have been one ʿAbd al-Malik ibn Abjar al-Kinānī, a convert to Islam who reportedly studied at the surviving medical school in Alexandria. One of the few Umayyad physicians known by extant writings, and possibly the first to translate a medical treatise into Arabic, is Māsarjawayh, sometimes called Māsarjīs, a Judeo-Persian physician living in Basra near Baghdad. Whereas some accounts have Māsarjawayh living at the end of the eighth or the beginning of the ninth century, others state that, for either the caliph Marwān I (r. 684–5) or ʿUmar ibn ʿAbd al-ʿAzīz (r. 717–20), he translated into Arabic from Syriac a medical handbook by Ahrun, a seventh-century physician of the Alexandrian medical school.

Our sources for the entire period are fragmentary and conflicting. The motivation for recording most accounts was not to delineate the available medical care but rather to illustrate a pious or literary theme. For example, there is a report of an amputation carried out just prior to the reign of the Umayyad caliph al-Walīd I (r. 705–15). The ninth-century poet Ibn Qutaybah recorded that in 704, the year before Walīd was made caliph, the legal authority ʿUrwah ibn al-Zubayr of Medina "suffered from gangrene in his foot while in Syria with Walīd. They cut off his foot while Walīd was present, but ʿUrwah was completely still, and Walīd did not realize that it had been cut off until it was cauterized and the odour of the cautery was evident. After that ʿUrwah lived eight years."[4] By the next century, the account had been elaborated, and it was said that ʿUrwah was offered a drug to drink so he would not feel the pain but that he refused it as being beneath his dignity. By the twelfth century, it was related that ʿUrwah was not only offered a

[3] G. R. Hawking, "The Development of the Biography of al-Ḥārith ibn Kalada and the Relationship between Medicine and Islam," in *The Islamic World from Classical to Modern Times: Essays in Honor of Bernard Lewis*, ed. C. E. Bosworth et al. (Princeton, N.J.: Darwin Press, 1989), pp. 127–40.

[4] Ibn Qutaybah, *Kitāb al-Maʿārif*, ed. Tharwat ʿUkāshah (Cairo: Mabaʿ at Dār al-Kutub, 1960), p. 222.

soporific to drink but that after the amputation he took the leg from the hands of the physicians and, addressing the leg, said: "What makes me feel good about you is that I never moved you to disobey God."[5]

These changing accounts of the amputation of ʿUrwah's leg illustrate the difficulties for the medical historian. Because the accounts were recorded to illustrate dignity and piety rather than medical care, the only evidence for therapeutics to be extracted from them is that later generations viewed the Umayyad court as supporting physicians capable of performing amputations and administering soporifics to relieve pain. Little can be said on the basis of these reports regarding what actually transpired in Syria in the year 704.

For several centuries prior to the rise of Islam, Alexandria, as well as Rome, Constantinople, Antioch, Edessa, and Amida, had flourished as centers of scientific and medical activity. A combination of political and religious events caused many Greek- and Syriac-speaking scholars to move eastward to Persia and to establish centers of learning there. By the sixth century, the city of Gondēshāpūr, near the present-day village of Dizfūl in southwest Persia, had become an outpost of Hellenism. The Umayyad court in Damascus may well have supported people who had some connection with Hellenistic medicine as practiced in Alexandria or Gondēshāpūr. It is likely, however, that the pre-Islamic customs of treatment prevailed even in the emerging urban society well into the ninth century.

In the year 750, the ʿAbbāsids overthrew the Umayyad caliphate and moved the center of power from Damascus to Baghdad. From the mid-eighth century to the mid-thirteenth, the ʿAbbāsid dynasty of caliphs greatly influenced the development of Islamic medicine through their extensive patronage.

The influence of Gondēshāpūr upon early ʿAbbāsid medicine is evident in the prominent role given a family of Nestorian Christian physicians from that city. For eight generations, from the middle of the eighth century until well into the second half of the eleventh, twelve members of the Bukhtīshūʿ family served the caliphs in Baghdad as physicians and advisers, often sponsoring the translation of texts and composing original treatises. In 765, the caliph al-Manṣūr, suffering from a stomach complaint, called Jurjīs ibn Jibrīl ibn Bukhtīshūʿ to Baghdad from Gondēshāpūr, where he had been the leading physician and author of a Syriac medical handbook. He eventually returned to Gondēshāpūr, where he died sometime after 768, but his son was called to Baghdad in 787, where he remained until his death in 801, serving as physician to the caliph Hārūn al-Rashīd (r. 786–809). The subsequent generations of Bukhtīshūʿ remained in Baghdad. The Christian physician Sābūr ibn Sahl,

[5] Ibn al-Jawzī, *Dhamm al-hawá*, ed. Muṣṭafá ʿAbd al-Wāḥid (Cairo: Dār al-Kutub al-Hadīthah, 1962), pp. 221–2; and Savage-Smith, "Medicine," pp. 909–10.

court physician to the caliph al-Mutawakkil (r. 847–61), was also said to have practiced medicine in Gondēshāpūr before coming to Baghdad.[6]

Historians have customarily asserted that Gondēshāpūr had an important hospital and medical school that supported the translation of Greek and possibly Sanskrit texts into Middle Persian and Syriac, but this interpretation has been challenged recently. There seems to be no evidence that there was a hospital or a formal medical school in Gondēshāpūr. There may have been a modest infirmary where Greco-Roman medicine was practiced and a forum where medical texts could be read, as was the case in other towns such as Susa, nearby to the west. The alleged prominence of Gondēshāpūr as a medical center with a hospital was possibly a result of the dominance of Nestorian Christians among the early physicians at the court in Baghdad who wished to claim the hospital as their idea and to establish a history to support their medical authority. Certainly the Nestorian Christian monopolization of medicine in Baghdad during the eighth and ninth centuries meant that the medicine they advocated, based upon Greek texts, was promoted over the rival practices of Zoroastrians and Indians or other pre-Islamic practices.[7]

Baghdad was the venue for most of the translations into Arabic of Greek medical writings and for the earliest medical treatises composed in Arabic. The Greco-Arabic translation movement that took place in Baghdad from the mid-eighth through the tenth century was the result of a sustained program subsidized publicly and privately by the elite of ʿAbbāsid society. Although the court had shifted from a Greek-speaking area (Damascus) to a non–Greek-speaking area (Baghdad), nonetheless this remarkable intellectual venture produced translations of virtually all the Greek medical, philosophical, and scientific writings, many of which had ceased to interest Byzantium.[8]

The most influential of the Greek medical writings to be translated into Arabic included the compendium on materia medica by Dioscorides (d. ca. 90), various treatises by Rufus of Ephesus (fl. ca. 100), the Greek encyclopedia by Paul of Aegina, working in Alexandria in the seventh century, and especially the voluminous medical writings by Galen (d. ca. 216). The Hippocratic writings, although familiar to Islamic physicians, were not as direct a formative influence as the Galenic writings. By the end of the ninth century, the Hellenistic medical theories and the humoral system

6 Lutz Richter-Bernburg, "Boktīšūʿ," in *Encyclopedia Iranica*, ed. Ehsan Yarshater (London: Routledge; Costa Mesa, Calif.: Mazda Publications, 1985–), vol. 4, pp. 333–6; and Sābūr ibn Sahl, *The Small Dispensatory*, trans. Otto Kahl (Leiden: Brill, 2004).

7 Lawrence I. Conrad, "Did al-Walīd I Found the First Islamic Hospital?" *ARAM Society for Syro-Mesopotamian Studies (Majallat Ārām)*, 6 (1994), 225–44; and Michael W. Dols, "The Origins of the Islamic Hospital: Myth and Reality," *Bulletin of the History of Medicine*, 61 (1987), 367–90.

8 Dimitri Gutas, *Greek Thought, Arabic Culture: The Graeco-Arabic Translation Movement in Baghdad and Early ʿAbbāsid Society (2nd–4th/8th–10th Centuries)* (London: Routledge, 1998), especially pp. 53–60.

of pathology formed the basis of nearly all the learned Arabic medical discourses.[9]

The most productive translator was Ḥunayn ibn Isḥāq (d. 873 or 877), another Nestorian Christian but originally from southern Iraq. He translated into both Syriac and Arabic, often working in collaboration with others. Ten years before his death, Ḥunayn recorded that he had made 95 Syriac and 34 Arabic versions of Galen's works. Ḥunayn (known to Europeans as Johannitius) also composed original medical writings, including ophthalmological treatises and the influential *Introduction on Medicine* (*Kitāb al-Mudhkal*), which later in Europe, through a Latin version called the *Isagoge*, established the basic conceptual framework of medieval medicine.

There are a number of early physicians who are not known to have made translations themselves but whose writings reflect the very early period of adaptation of foreign material. Foremost among this group was another Nestorian Christian, Yūḥannā ibn Māsawayh (d. 857, known in the West as Mesue), whose father had been a physician in Gondēshāpūr before coming to Baghdad. Ibn Māsawayh composed a considerable number of Arabic medical monographs on topics including fevers, leprosy, melancholy, dietetics, eye diseases, and medical aphorisms, and the name Mesue was associated with several influential Latin treatises, only some of which were actually written by Ibn Māsawayh. It was reported that Ibn Māsawayh regularly held an assembly of some sort, where he consulted with patients and discussed subjects with pupils, among them Ḥunayn ibn Isḥāq. At times Ibn Māsawayh apparently attracted considerable audiences, having acquired a reputation for repartee. Another important figure was ʿAlī ibn Rabban al-Ṭabarī, who died not long after 855. He not only summarized Greco-Roman practices in his compendium *The Paradise of Wisdom*, dedicated in 850 to the caliph al-Mutawakkil, but devoted a separate chapter to Indian medicine. Neither he nor subsequent writers, however, really tried to integrate the Indian material with Hellenistic medicine.[10]

THE LEARNED MEDICAL TRADITION

Following the rapid appropriation of Hellenistic medicine (with a few Persian and Indian elements) in the ninth century, the organization of this knowledge into a logical and accessible format became a primary concern.

9 M. M. Sadek, *The Arabic Materia Medica of Dioscorides* (Quebec City: Les Éditions du Sphinx, 1983); and P. E. Pormann, *The Oriental Tradition of Paul of Aegina's Pragmateia* (Leiden: Brill, 2004).

10 For individual figures as well as topics, see the comprehensive bibliography by Manfred Ullmann, *Die Medizin im Islam* (Handbuch der Orientalistik, Abteilung I, Ergänzungsband vi, Abschnitt 1) (Leiden: Brill, 1970). For those living before 1038, see Fuat Sezgin, *Geschichte des arabischen Schrifttums*, vol. 3: *Medizin – Pharmazie – Zoologie – Tierheilkunde bis ca. 430 H.* (Leiden: Brill, 1970).

In the tenth and early eleventh centuries, several Arabic medical encyclopedias were composed that proved to be particularly influential in the learned medical tradition.

Two of these fundamental works were written by Abū Bakr Muḥammad ibn Zakarīyā al-Rāzī (d. 925), known to Europeans as Rhazes or Rasis. The *Book of Medicine for al-Manṣūr* (*al-Kitāb al-Manṣūrī fī al-ṭibb*) was dedicated in 903 to a local Iranian prince named Abū Ṣāliḥ al-Manṣūr ibn Isḥāq, governor of the town of Rayy near present-day Tehran. This medical compendium proved very influential throughout the Islamic world as well as in Europe, where it was known in Latin as *Liber ad Almansorem*. The second work was assembled posthumously from Rāzī's working files of readings and personal observations and was called *al-Kitāb al-Ḥāwī fī al-ṭibb* (*The Comprehensive Book on Medicine*), later translated into Latin as *Continens*. The *Ḥāwī* is a unique type of work in the history of medicine, for it contains a lifetime of reading notes interspersed with case histories and personal observations. Even though it was so enormous that few could afford copies and though not as tightly structured as most medieval encyclopedias, it was highly valued by later physicians.[11]

Rāzī's *Ḥāwī* did have its critics, however, for the *Complete Book of the Medical Art* (*Kitāb Kāmil al-ṣināʿah al-ṭibbīyah*) by ʿAlī ibn al-ʿAbbās al-Majūsī (Haly Abbas, fl. ca. 983) was written in part as an attempt to redress the lack of proper organization and insufficient attention to anatomy and surgery in the *Ḥāwī*. Majūsī, as his name suggests, was from a Persian family with Zoroastrian forebears, though he was a Muslim. He dedicated his *Complete Book of the Medical Art* to ʿAḍud al-Dawlah Fanā-Khusraw, the ruler of Persia and Iraq from 949 to 983 and founder of hospitals in both Baghdad and Shiraz. This encyclopedia, translated into Latin as the *Pantegni* and as *Regalis dispositio*, is one of the most comprehensive and well-organized medical compendia of early medical literature. Its division into two discrete parts, theoretical and practical, established a format common to later medieval medical writings.[12]

In about 1000, Abū al-Qāsim al-Zahrāwī (known to Europe as Albucasis) composed at Córdoba another large encyclopedia of thirty books, with the lengthy title *The Arrangement of Medical Knowledge for One Who Is Not Able to Compile a Book for Himself*. Its final book, devoted to surgery, was widely influential throughout the Islamic world and in Europe.[13]

The most influential of all Arabic medical encyclopedias was *The Canon of Medicine* (*Kitāb al-Qānūn fī al-ṭibb*) by Ibn Sīnā (d. 1037), known to Europeans as Avicenna. Composed in Persia before 1015, the compendium

[11] A. Z. Iskandar, "al-Rāzī," in Young et al., *Religion, Learning, and Science in the ʿAbbāsid Period*, pp. 370–7.

[12] Lutz Richter-Bernburg, "ʿAlī b. ʿAbbās Maǧūsī," in Yarshater, *Encyclopedia Iranica*, vol. 1, pp. 837–8.

[13] Zahrāwī, *Albucasis, On Surgery and Instruments*, ed. and trans. M. S. Spink and G. L. Lewis (Berkeley: University of California Press, 1973).

consisted of five books: (1) general medical principles, (2) materia medica, (3) diseases occurring in a particular part of the body, (4) diseases such as fevers that are not specific to one bodily part, and (5) recipes for compound drugs. The first book sometimes circulated separately as *al-Kullīyāt* (*General Principles*), or *Colliget* in Latin.[14]

An enormous medical compendium undertaken in the thirteenth century by the Syrian physician Ibn al-Nafīs had a title, *Complete Book of the Medical Art* (*Kitāb Shāmil fī al-ṣināʿah al-ṭibbīyah*), nearly identical to that written three centuries earlier by Majūsī. Ibn al-Nafīs intended the encyclopedia to extend to 300 volumes, but he completed only 80 before he died in 1288 in Cairo, where he spent much of his working life, bequeathing his house and library to a recently constructed hospital there.[15]

These attempts at systematizing and synthesizing the Hellenistic medical literature were very successful in producing a coherent and orderly medical system, essentially Galenic in nature but much modified and elaborated. Their sheer size, reinforced by titles such as *The Canon*, gave them an aura of authority that was to prove stultifying rather than invigorating. Furthermore, as the medical historian Michael McVaugh has aptly noted: "however much [such treatises] might emphasize the importance of practical knowledge, or the nature of medicine as 'art', the effect of their very structure was to favor the logical element over the clinical."[16]

Ibn Sīnā's *Canon* was not greeted everywhere with praise. In Spain, when Abū al-ʿAlāʾ Zuhr (d. 1131) was presented with a copy, he so disliked it that he refused to put it in his library, preferring to cut off its margins for use in writing prescriptions for patients. He also composed a treatise criticizing the materia medica in the *Canon*. This hostile reaction on the part of the progenitor of a five-generation family of prominent Andalusian physicians raises the question (yet to be investigated by historians) of whether medicine in Islamic Spain developed with less dependence upon the ideas of Ibn Sīnā than elsewhere. His son, Abū Marwān ibn Zuhr (Avenzoar to Europeans), wrote several important works, including *The Easy Guide to Therapy and Dietetics*, translated into Latin as *Theisir*, while his compatriot Ibn Rushd (Averroes; d. 1198) wrote the widely read *Book of General Principles* (*Kitāb al-Kullīyāt*), known in Europe as *Colliget* (the same name by which the first book of the *Canon* was known).[17]

[14] Muhsin Mahdi et al., "Avicenna," in Yarshater, *Encyclopedia Iranica*, vol. 3, pp. 66–110. Two English translations of the first book only have been published, one made from the Latin version and the other from an Urdu version, but neither adequately represents the Arabic original. See O. C. Gruner, *A Treatise on the Canon of Medicine of Avicenna* (London: Luzac, 1930); and O. C. Gruner, *The General Principles of Avicenna's Canon of Medicine*, trans. Mazhar H. Shah (Karachi: Naveed Clinic, 1966).

[15] A. Z. Iskandar, "Ibn al-Nafīs," *Dictionary of Scientific Biography*, IX, 602–6.

[16] Michael McVaugh in Edward Grant, ed., *A Source Book of Medieval Science* (Cambridge, Mass.: Harvard University Press, 1974), p. 715 n. 1.

[17] Roger Arnaldez, "Ibn Zuhr," in Gibb et al., *Encyclopaedia of Islam*, vol. 3, pp. 976–9. For the anecdote regarding the copy of the *Canon*, see A. Z. Iskandar, *A Catalogue of Arabic Manuscripts*

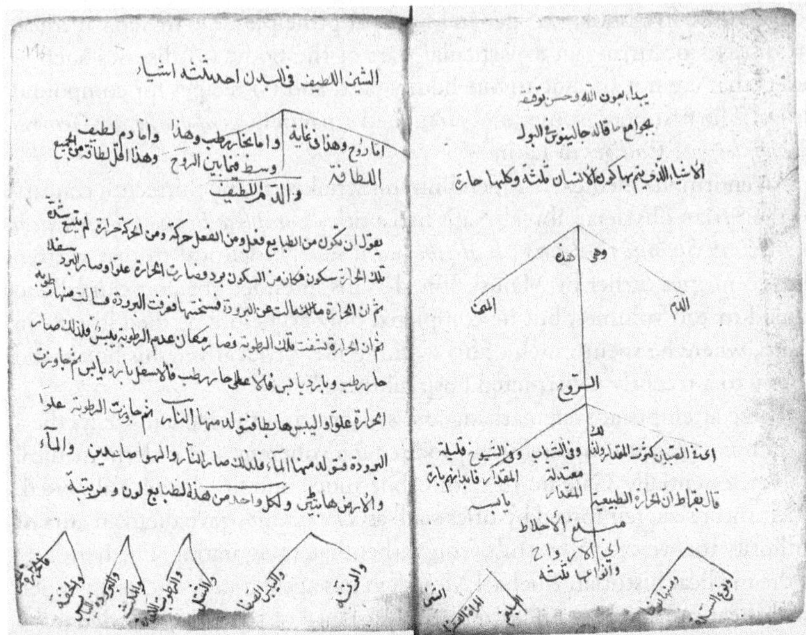

Figure 5.1. Branch diagram in an early Arabic summary of Galen's treatise on diagnosis by urine. By permission of the Bodleian Library, Oxford, MS. Turk. e. 33, folios 195b–196a.

It appears that, having achieved a high level of exhaustive systematization, an awareness set in that these compendia were too large to be really useful for ready reference. After the middle of the eleventh century, epitomes were produced, particularly of the *Canon*, to make the ideas more readily accessible; commentaries were made to clarify the contents and argue different interpretations; and the number of treatises on single topics increased. Few writers thereafter undertook to produce an encyclopedia on the scale of Majūsī or Ibn Sīnā (Ibn al-Nafīs being a notable exception).

In addition to monographs on single topics, such as fevers or how to treat stomach disorders, various new formats were devised for presenting medical information. As early as the ninth century, branch diagrams were used to illustrate the relationship between related diseases (Figure 5.1). Ibn Māsawayh appears to have been among the earliest to employ them, though branch diagrams are also found in some Arabic copies of summaries of Galenic treatises. These branch diagrams, called in Arabic *tashjīr* ("ramification," from a root meaning "to plant with trees"), organized information into a diagrammatic format clearly delineating categories, divisions, and subdivisions as an aide-mémoire. This technique of diagramming was later

on Medicine and Science in the Wellcome Historical Medical Library (London: Wellcome Historical Library, 1967), pp. 36–7.

introduced into the Latin West, where from the twelfth century it was frequently employed in discourses on the classification of the sciences and other scientific and philosophical topics. Modern scholars have called the Latin branch diagrams "dichotomies" to distinguish them from "arbores," or tree diagrams, in which material was written in small cells arranged within the outline of a tree having a large trunk with a root at the bottom. The earliest instance of a tree diagram occurs in a ninth-century copy of the *Etymologies* written in the seventh century by Isidore of Seville. Although it is evident that Arabic treatises employed branch diagrams in the ninth century (possibly continuing a now lost Alexandrian tradition) and that tree diagrams were being used simultaneously in the Latin West, the two techniques of diagramming are sufficiently different to suggest independent traditions. The Arabic method transferred to Europe, whereas the Latin one remained restricted to Western compositions.[18]

Another popular format for medical discourse was that of questions and answers. Ḥunayn ibn Isḥāq employed the technique in his *Questions on Medicine for Beginners* and also in an ophthalmological tract, *Questions Concerning the Eye*.[19]

Others followed suit, such as Saʿīd ibn Abī al-Khayr al-Masīḥī (d. 1193), court physician in Baghdad, in his introductory guide to medicine called *The Brevity*. Didactic medical poetry was also a popular device, with rhyming quatrains that enabled the student or practitioner to easily remember the basic ideas. Ibn Sīnā wrote one that was especially popular, judging from the large number of manuscripts and commentaries preserved today.[20]

Some essays record debates that occurred between physicians. A particularly acrimonious one took place in Cairo between two prominent physicians, Ibn Riḍwān (d. 1068) and Ibn Buṭlān (d. 1066). The former was a self-taught physician, with an enormous ego and a quick temper, who had been appointed chief physician in Egypt, where he attained great political power. When Ibn Buṭlān, a Nestorian Christian from Baghdad educated under the leading physician of the day, arrived in Old Cairo in 1049 and challenged Ibn Riḍwān's position, an exchange of ten increasingly vitriolic essays took place. The debate ostensibly centered upon an issue in Aristotelian biology (which is warmer, the adult chicken or the newborn chick?) but was probably motivated by the desire to acquire or protect social status. In the debate, both displayed their learning and criticized the education of

18 For the Latin tradition, see John E. Murdoch, *Album of Science: Antiquity and the Middle Ages* (New York: Charles Scribner's Sons, 1984), pp. 38–51. For the Arabic tradition, see Emilie Savage-Smith, "Galen's Lost Ophthalmology and the *Summaria Alexandrinorum*," in *The Unknown Galen*, ed. Vivian Nutton (Bulletin of the Institute of Classical Studies, Supplement 77) (London: Institute of Classical Studies, University of London, 2002), pp. 121–38.

19 Ḥunayn ibn Isḥāq, *Questions on Medicine for Scholars*, trans. Paul Ghalioungui (Cairo: Al-Ahram Center for Scientific Translations, 1980).

20 Henri Jahier and Abdelkader Noureddine, *Avicenne, Poème de la médecine, Urǧūza fī ʿṭ-ṭibb, Cantica Avicennae* (Paris: Les Belles Lettres, 1956).

the other (though never, interestingly enough, the opponent's religion); little attention was actually devoted to medical issues. Ultimately, Ibn Buṭlān was forced to leave, but rather than return to Baghdad he went to Constantinople and subsequently to Syria, where in 1063 he supervised the building of a hospital in Antioch, thereafter retiring to a nearby monastery. He composed a medical manual for the use of monks, a tract on how to detect illnesses in slaves who were for sale, a satirical piece exposing the shortcomings of a physician and other medics, and the extremely popular *Almanac of Health* (*Kitāb Taqwīm al-ṣiḥḥah*), which in the course of 40 tables presents 210 plants and animals and 70 other items and procedures useful for maintaining good health. The Latin version of this manual of dietetics and regimen, the *Tacuinum sanitatis*, was immensely popular in Europe, where its author was known as Elbuchasem Elimithar as well as Ibn Botlan.[21]

Perhaps with Ibn Buṭlān's *Almanac* as a model, synoptic tables became a common didactic element in Arabic medical literature. They were employed, for example, in the widely read therapeutic handbooks of Ibn Jazlah (d. 1100) as well as in treatises on materia medica composed by Ibn Biklārish al-Isrā'īlī (fl. ca. 1100) in Saragossa, by Ibrāhīm ibn Abī Sa'īd al-'Alā'ī al-Maghribī (fl. mid-twelfth century) in Anatolia, and by Ḥubaysh ibn Ibrāhīm al-Tiflīsī (fl. end of the twelfth century).[22]

Numerous abridgements and explanatory commentaries on parts of the *Canon* of Ibn Sīnā were composed. Although the earliest abridgement of the *Canon* seems to have been that by al-Īlāqī (fl. 1068), a pupil of Ibn Sīnā, it was not until the late twelfth century that a serious need was perceived for aids to understanding it. The Egyptian Jewish physician Ibn Jumay', who died in 1198, composed possibly the earliest commentary on the *Canon*. In the next two centuries, commentaries and epitomes followed in rapid succession, and it was this industry of glossing and condensing the *Canon* that assured the encyclopedia its preeminent position in medieval medicine.[23]

The most widely read of all abridgements of Ibn Sīnā's *Canon* was that titled *al-Mūjiz* (*The Epitome*), written by the Syrian physician Ibn al-Nafīs,

[21] Lawrence I. Conrad, "Scholarship and Social Context: A Medical Case from the Eleventh-Century Near East," in *Knowledge and the Scholarly Medical Traditions*, ed. Don Bates (Cambridge: Cambridge University Press, 1995), pp. 84–100; Dols, *Medieval Islamic Medicine*; and Ibn Buṭlān, *Le "Taqwīm al-ṣḥḥa" (Tacuini sanitatis) d'Ibn Buṭlān*, ed. and trans. H. Elkhadem (Académie Royale de Belgique, Classe des Lettres, 7) (Louvain: Peeters, 1990).

[22] For the synoptic tables of these physicians, see E. Savage-Smith, "Ibn Baklarish in the Arabic Tradition of Synonymatic Texts and Tabular Presentations," in *Ibn Baklarish's Book of Simples: Medical Remedies between Three Faiths in Twelfth-Century Spain*, ed. Charles Burnett (Studies in the Arcadian Library, 3) (London: The Arcadian Library in association with Oxford University Press, 2008), pp. 113–31.

[23] For commentaries on the *Canon*, see Iskandar, *Catalogue of Arabic Manuscripts on Medicine and Science in the Wellcome Historical Medical Library*, pp. 33–64; E. Savage-Smith, *Islamic Medical Manuscripts at the National Library of Medicine* (http://www.nlm.nih.gov/hmd/arabic); and E. Savage-Smith, *A New Catalogue of Arabic Manuscripts in the Bodleian Library, University of Oxford*, vol. 1: *Medicine* (Oxford: Oxford University Press, 2011), pp. 242–318.

mentioned earlier in connection with his own encyclopedia. He also composed a commentary on the *Canon* that became an authoritative work in its own right. Both *The Epitome* and the commentary by Ibn al-Nafīs continued to be copied and read over the following five centuries, often generating further commentaries and supercommentaries.

Another genre of medical writing was the collection of biographies of physicians, with lists of their writings and anecdotes illustrating their character. One of the most important of those preserved today was composed by the thirteenth-century physician Ibn Abī Uṣaybiʿah (d. 1270), who was born into a family of Damascene physicians. In his *Sources of Information about the Classes of Physicians,* he gave biographies of over 380 physicians and scholars.

OPHTHALMOLOGY

Ophthalmology was the subject of specialized treatises and a topic in which medieval Islamic writers displayed considerable originality. Early in the ninth century, Ibn Māsawayh and Ḥunayn ibn Isḥāq wrote influential monographs on the subject. Although largely based upon Greek sources, they already show considerable advancement in knowledge over that in the extant Greek writings, including knowledge of some previously unrecognized pathological conditions such as pannus (*sabal* in Arabic), a complication of trachoma for which intricate surgical procedures soon developed. One of the most highly regarded ophthalmological manuals was written by ʿAlī ibn ʿĪsá (Jesu Haly in Latin), who practiced in Baghdad in the tenth century. A near contemporary of his was ʿAmmār ibn ʿAlī al-Mawṣilī, who was originally from Iraq but moved to Egypt, where he dedicated his only writing, a treatise on eye diseases, to the Fāṭimid ruler of North Africa and Egypt, al-Ḥākim (r. 996–1021). ʿAmmār's treatise discusses only forty-eight diseases but contains some interesting clinical cases and a claim to have designed a hollow cataract needle for the removal of a cataract from the eye by suction.[24]

During the twelfth and thirteenth centuries, there was an unprecedented outpouring of treatises on ophthalmology. In Spain, Muḥammad ibn Qassūm ibn Aslam al-Ghāfiqī, of whom essentially nothing is known, wrote a comprehensive textbook on the treatment of ocular disorders. In Egypt, Fatḥ al-Dīn al-Qaysī (d. 1258) composed a manual covering the

[24] C. A. Wood, *Memorandum Book of a Tenth-Century Oculist* (Evanston, Ill.: Northwestern University Press, 1936; repr. Birmingham, Ala.: The Classics of Ophthalmology Library, division of Gryphon Editions, 1985); Max Meyerhof, *The Cataract Operations of ʿAmmār ibn ʿAlī al-Mawsilī, Oculist of Cairo* (Barcelona: Laboratorios del Norte de España, 1937); and Frederick C. Blodi et al., trans., *The Arabian Ophthalmologists, compiled from original texts by J. Hirschberg, J. Lippert and E. Mittwoch and translated into English,* ed. M. Zafer Wafai (Riyadh: King Abdulaziz City for Science and Technology, 1993), pp. 7–184.

treatment of 124 eye conditions, some apparently described for the first time. Shortly thereafter, another comprehensive ocular manual was composed in Syria by Khalīfah ibn Abī al-Maḥāsin al-Ḥalabī, who meticulously cited the previous writers on the subject from whom he drew material. The manual by Khalīfah includes a considerable amount of novel material, including diagrammatic charts of ophthalmological instruments and the first recorded instance of the use of a magnet to remove a foreign object from the eye – in this case a piece of a needle that had broken while couching a cataractous eye. Also working in Syria, Ibn al-Nafīs presented an extremely thorough and systematic explication of the ophthalmological knowledge of the day in *The Perfected Book on Ophthalmology*.[25]

Another thirteenth-century ophthalmological treatise is unusual in examining at length the geometrical explanation of vision. This treatise, composed by Abū Zakarīyā᾿ Yaḥyá ibn Abī al-Rajā᾿ (often erroneously attributed to Ṣalāḥ al-Dīn ibn Yūsuf al-Kaḥḥāl al-Ḥamawī), also has illustrations of instruments and an interesting diagram showing a quarter section of an eye along two different planes.[26]

In the following century, the Egyptian scholar Ibn al-Akfānī composed an ophthalmological text in two forms: a full treatise and an abridgement. Later in the same century, Ṣadaqah ibn Ibrāhīm al-Shādhilī wrote yet another treatise on the topic that contains some interesting evidence of the level and frequency of ocular surgery in his day. The decline in originality characteristic of the majority of medical compositions after the fourteenth century can be seen in a manual written in the fifteenth century by the Egyptian oculist Nūr al-Dīn ʿAlī ibn al-Munāwī, for it merely consists of the abridgement that al-Akfānī wrote of his own treatise, alongside the text of al-Akfānī's longer manual as well as the relevant passages from Ibn al-Nafīs's treatise on ophthalmology, all of the sources unacknowledged.

PHARMACOLOGY

The field of pharmacology also generated its own specialized literature. There were two major Greek sources for medicinal substances available through Arabic translations: Dioscorides' treatise on materia medica, which described approximately one thousand substances, and Galen's treatise on simple remedies. The latter set the format for most Arabic writings on materia

[25] al-Qaysī, *Das Ergebnis des Nachdenkens über die Behandlung der Augenkrankheiten von Fatḥ al-Dīn al-Qaisī*, trans. Hans-Dieter Bischoff (Europäische Hochschulschriften; Asiatische und Afrikanische Studien, ser. 27, vol. 21) (Frankfurt am Main: Peter Lang, 1988); Blodi et al., *Arabian Ophthalmologists*, pp. 185–231; and Emilie Savage-Smith, "Ibn al-Nafīs's 'Perfected Book on Ophthalmology'," *Journal for the History of Arabic Science*, 4 (1980), 147–206.

[26] Gregor Schoeler, "Der Verfasser der Augenheilkunde K. Nūr al-ʿuyūn und das Schema der 8 Präliminarien im 1. Kapitel des Werkes," *Der Islam*, 66 (1987), 87–97; Blodi et al., *Arabian Ophthalmologists*, pp. 235–302.

medica by giving an alphabetical listing of medicinal substances rather than the confusing classification given by Dioscorides.

The Arabic literature on pharmacology, however, quickly assumed a different form from that inherited from the Hellenistic world. A class of literature arose that explained unfamiliar foreign terms for drugs and compiled synonym lists giving equivalent terms in different languages.[27] A large proportion of the plants described by Dioscorides and Galen would not have been known in various regions of the Middle East. The differing climatic conditions of the desert, marsh, mountain, and coastal communities meant that the species of medicinal plants, as well as animal species and mineral resources, varied greatly from one region to another. Sometimes related local species and varieties could be identified as similar to those described by Dioscorides or Galen, but in other instances the substances described in the Greek sources meant little to an Arab practitioner. Conversely, the broader and different geographic horizons of Islamic writers brought them into contact with new drugs. Traders and travelers played as important a role in the knowledge and development of medicinal substances in the Islamic world as the treatises of Dioscorides and Galen in their Arabic dress.

From the formularies of Sābūr ibn Sahl (d. 869) and al-Kindī (d. after 870), it is evident that by the ninth century many medicaments were being used that were unknown in Hellenistic medicine, including camphor, musk, and sal ammoniac, as well as commodities previously unknown to Europe, such as cotton. Most influential of all the Arabic treatises on materia medica was the manual by Ibn al-Bayṭār (d. 1248), originally from Malaga in the kingdom of Granada. This was an alphabetical guide to over 1,400 medicaments in 2,324 separate entries, taken from his own observations as well as over 260 written sources that he quoted.[28]

ANATOMY

In contrast to ophthalmology and pharmaceutics, the Islamic writings concerned with anatomy remained quite conservative, deviating little from their Hellenistic models. The subject did not generate a specialized body of

[27] Martin Levey, *Early Arabic Pharmacology: An Introduction Based on Ancient and Medieval Sources* (Leiden: Brill, 1973); and Martin Levey, *Substitute Drugs in Early Arabic Medicine, with Special Reference to the Texts of Masarjawaih, al-Rāzī, and Pythagoras* (Veröffentlichungen der Internationalen Gesellschaft für Geschichte der Pharmazie, n.f. 37) (Stuttgart: Wissenschaftliche Verlagsgesellschaft, 1971).

[28] Remke Kruk, "Nabāt," in Gibb et al., *Encyclopaedia of Islam*, vol. 7, pp. 831–4; Sābūr ibn Sahl, *The Small Dispensatory; al-Kindī, The Medical Formulary or Aqrābādhīn of al-Kindī*, trans. Martin Levey (Madison: University of Wisconsin Press, 1966); Ibn al-Baiṭār, *Die Dioskurides-Erklärung des Ibn al-Baiṭār*, ed. and trans. Albrecht Dietrich (*Abhandlungen der Akademie der Wissenschaften in Göttingen*, phil.-hist. Kl., fol. 3, no. 191) (Göttingen: Vandenhoeck and Ruprecht, 1991); and Selma Tibi, *The Medicinal Use of Opium in Ninth-Century Baghdad* (Sir Henry Welcome Asian Series, 5) (Leiden: Brill, 2006).

literature until the end of the fourteenth century, by which time some additional European influences may have been at work. Despite this conservative approach, medieval Islamic scholars made two important anatomical observations.

All the major Arabic and Persian medical encyclopedias had sections on anatomy describing the bones, muscles, nerves, arteries, veins, and the compound organs, which included the eye, the liver, the heart, and the brain, as well as chapters on embryological theories. Such works summarized Aristotelian biological ideas and Galenic anatomical concepts, and were occasionally illustrated with schematic diagrams of the cranial sutures and the bones of the upper jaw. Debates arose regarding several issues, such as the total number of bones and muscles in the human body, the male and female roles in generation, and the length of fetal development in humans and animals. Ocular anatomy was also discussed in treatises concerned with ophthalmology or optics, often with diagrams of the eye or the visual system.[29]

No anatomical illustrations of the entire body are known to have been produced in the Islamic world before those that accompany copies of *Manṣūr's Anatomy*, by Manṣūr ibn Muḥammad ibn Aḥmad ibn Yūsuf ibn Ilyās. This Persian-language treatise was dedicated to a ruler of a Persian province from 1394 to 1409, probably a grandson of Tīmūr, known to Europeans as Tamerlane. It consists of an introduction followed by five chapters on the "systems" of the body – bones, nerves, muscles, veins, and arteries – each illustrated by a full-page diagram with numerous labels. A concluding section on compound organs and the formation of the fetus is usually illustrated with a diagram showing a pregnant woman.

A similarity has been noted between the first five illustrations accompanying *Manṣūr's Anatomy* and a set of anatomical illustrations that appeared in earlier, apparently twelfth-century, Latin medical treatises. All the figures are in a distinctive squatting posture, but the similarity is particularly evident in the diagram of the skeleton, which in both the Latin and Islamic versions is viewed from behind, with the head hyperextended so that the face looks upward.[30] The origin of this anatomical series remains a puzzle, but as it

[29] See Emilie Savage-Smith, "Tashrīh [Anatomy]," in Gibb et al., *Encyclopaedia of Islam*, vol. 10, pp. 354–6; E. Savage-Smith, "Anatomical Illustration in Arabic Manuscripts," in *Arab Painting: Text and Image in Illustrated Arabic Manuscripts*, ed. Anna Contadini (Handbook of Oriental Studies, Section 1, vol. 90) (Leiden: Brill, 2007), pp. 147–59; Gül Russell, "The Anatomy of the Eye in ʿAlī ibn al-ʿAbbās al-Mağūsī: A Textbook Case," in *Constantine the African and ʿAlī ibn al-ʿAbbās al-Mağūsī: The Pantegni and Related Texts*, ed. Charles Burnett and Danielle Jacquart (Leiden: Brill, 1994), pp. 247–65; and Basim Musallam, "The Human Embryo in Arabic Scientific and Religious Thought," in *The Human Embryo: Aristotle and the Arabic and European Traditions*, ed. Gordon R. Dunstan (Exeter: University of Exeter Press, 1990), pp. 32–46.

[30] See Ynez V. O'Neill, "The Fünfbilderserie: A Bridge to the Unknown," *Bulletin of the History of Medicine*, 51 (1977), 538–49; Roger French, "An Origin for the Bone Text of the 'Five-Figure Series'," *Sudhoffs Archiv*, 68 (1984), 143–58; Andrew Newman, "*Tašrīḥ-i Manṣūrī*: Human Anatomy between the Galenic and Prophetical Medical Traditions," in *La science dans le monde iranien à l'époque islamique*, ed. Živa Vesel, H. Beikbaghban, and Bertrand Thierry de Crussol des Epesse (Tehran:

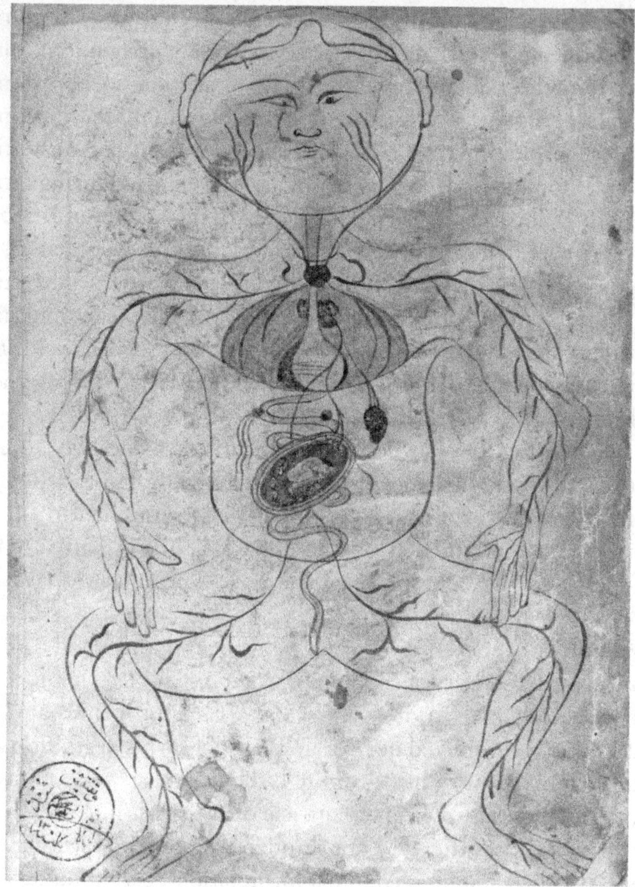

Figure 5.2. The figure of a pregnant woman. This is essentially the arterial figure on which a gravid uterus with the fetus in a transverse position has been superimposed. From *The Anatomy of the Human Body* (*Tashrihi-i badan-i insān*), written in Persian at the end of the fourteenth century by Manṣūr ibn Ilyās. By permission of the National Library of Medicine, Bethesda, Maryland, MS. P 18, folio 39b.

clearly predates the Persian treatise by Manṣūr ibn Ilyās by at least two centuries, it is possible that it originated in the Latin West, from which it was then transmitted to Persian intellectual circles without passing through an Arabic medium. The sixth figure in the Islamic series, the pregnant woman (Figure 5.2), has no parallel in the earlier Latin series and was probably a contribution by Ibn Ilyās himself. It was constructed by removing the labels from the arterial figure and superimposing an oval gravid uterus having the fetus in a breech or transverse position.

Institut Français de Recherche en Iran, 1998; 2nd ed. 2004), pp. 253–71; and Gül Russell, 'Ebn Ilyās,' in Yarshater, *Encyclopaedia Iranica*, vol. 8, pp. 16–20.

Systematic human anatomical dissection was no more a pursuit of medieval Islamic society than it was of medieval Christendom, although it is clear from the available evidence that in neither society were there explicit legal or religious strictures banning it. Indeed, many Muslim scholars lauded the study of anatomy, primarily as a way of demonstrating the design and wisdom of God. Typical of such sentiments is a saying attributed to Ibn Rushd: "Whoever has been occupied with the science of anatomy has increased his belief in God." What is meant by the "science of anatomy" in such statements is not the dissection of an animal in order to determine its structure but rather the elaboration of the ideas of Galen regarding structure and function. There are, however, some references in scholarly and medical writings to dissection, though to what extent these reflect actual practice it is difficult to say.[31]

What is certain is that medieval Islamic writers made two noteworthy contributions to the knowledge of human anatomy. One was the result of chance observation, for following the discovery of some skeletons during a famine in Egypt in 1200, the scholar and physician ʿAbd al-Laṭīf al-Baghdādī (d. 1231) improved the description of the bones of the lower jaw and the sacrum.

The second was the description of the movement of the blood through the pulmonary transit given by the Syrian physician Ibn al-Nafīs in his commentary on the anatomical portions of Ibn Sīnā's *Canon*. In this commentary, preserved in a copy completed in 1242, 46 years before his death, Ibn al-Nafīs described the movement of blood through the pulmonary transit, explicitly stating that the blood in the right ventricle of the heart must reach the left ventricle by way of the lungs and not through a passage connecting the ventricles, as Galen had maintained. This formulation of a pulmonary transit for the movement of blood, sometimes called the "lesser" circulation, was made three centuries before those of Michael Servetus (d. 1553) and Realdo Colombo (d. 1559), the first Europeans to describe the pulmonary transit. Within the Islamic world, Ibn al-Nafīs's commentary on Ibn Sīnā's anatomy remained not as widely known as his commentary on the complete *Canon*. Yet two fourteenth-century Arabic physicians knew of his anatomical commentary and the theory of a pulmonary transit, a sixteenth-century Persian physician (ʿImād al-Dīn Maḥmūd Shīrāzī) wrote two anatomical treatises dependent upon it, and the passages from it are not infrequently encountered as marginalia in Arabic manuscripts.[32]

[31] Quotation given in Ibn Abī Uṣaybiʿah, *ʿUyūn al-anbāʾ fī ṭabaqāt al-aṭibbāʾ*, ed. A. Müller, 2 vols. (Cairo: al-Maṭbaʿah al-Wahbīyah, 1882–1884), vol. 2, p. 77. See Emilie Savage-Smith, "Attitudes Toward Dissection in Medieval Islam," *Journal of the History of Medicine*, 50 (1995), 68–111.

[32] Iskandar, "Ibn al-Nafīs," pp. 603–4; and, for marginal quotations, see Savage-Smith, *Islamic Medical Manuscripts*. Historians have debated whether Ibn al-Nafīs's commentary on the anatomy might have been available to European physicians through translation. It is known that Ibn al-Nafīs's commentary on the last part of the *Canon*, concerned with compound remedies, was translated into

THE PRACTICE OF MEDICINE

The development of urban hospitals was a major achievement of medieval Islamic society. It is evident that the medieval Islamic hospital was a more elaborate institution with a wider range of functions than the earlier poor and sick relief facilities offered by some Christian monasteries, though no systematic study and comparison of Islamic hospitals with Byzantine ones has yet been undertaken. The Islamic hospital served several purposes: as a center of medical treatment, a convalescent home for those recovering from illness or accidents, an insane asylum, and a retirement home giving basic maintenance for the aged and infirm who lacked a family to care for them. It is unlikely that any truly wealthy person would have gone to a hospital unless they became ill while traveling. Otherwise, all the medical needs of the wealthy and powerful would have been administered in the home. Although Jewish and Christian doctors working in hospitals were not uncommon, we do not know what proportion of the patients would have been non-Muslim. An association with a hospital seems to have been highly desirable for a physician, and some teaching occurred in hospitals, especially in Baghdad and later in Damascus and Cairo. Most medical instruction, however, was probably acquired through private tutoring and apprenticeship, though some physicians, such as Ibn Riḍwān, were self-taught. Many physicians had other occupations as well, with their fame as philologists, historians, or jurists, for example, sometimes eclipsing their medical reputations.[33]

The establishment of hospitals in medieval Islam is one aspect of medical care for which there is considerable evidence outside the context of formal medical treatises, for a number of documents relating to the endowment of hospitals and their administrative costs have been preserved. The association of the Umayyad caliph al-Walīd I (r. 705–15) with the establishment of the first hospital in Islam has been demonstrated to be unjustified, and the formative role of Gondēshāpūr in their development has been overemphasized. Evidence suggests that the first Islamic hospital was founded in Baghdad by order of the caliph Hārūn al-Rashīd (r. 786–809). The most important of the Baghdad hospitals was that established in 982 by local ruler ʿAḍud al-Dawla Fanā-Khusraw. When it was founded, it had twenty-five doctors, including oculists, surgeons, and bonesetters, and two centuries later, in 1184, it was described by the traveler Ibn Jubayr as being a large structure having many rooms and all the appurtenances of a palace. We possess the

Latin by the Renaissance physician Andrea Alpago (d. 1522), who had also prepared a new translation of Ibn Sīnā's *Canon*. The translation of Ibn al-Nafīs's commentary on compound remedies was not published until 1547, when it was printed at Venice, but the possibility remains that other parts of Ibn al-Nafīs's commentary were transmitted through translations that never made it into print.

33 D. S. Richards, "A Doctor's Petition for a Salaried Post in Saladin's Hospital," *Social History of Medicine*, 5 (1992), 297–306; Gary Leiser, "Medical Education in Islamic Lands from the Seventh to the Fourteenth Century," *Journal of the History of Medicine*, 38 (1983), 43–75; and Pormann and Savage-Smith, *Medieval Islamic Medicine*, pp. 80–5.

fullest information about the great Syro-Egyptian hospitals of the twelfth and thirteenth centuries. In Damascus, Nūr al-Dīn Maḥmūd ibn Zangī, the Ayyūbid ruler of Syria (r. 1146–74), founded a hospital that was named after him (the Nūrī hospital), and in 1171 Saladin, whose full name was al-Malik al-Nāṣir I Ṣalāḥ al-Dīn Yūsuf ibn Ayyūb, followed his example by founding a hospital in Cairo called the Nāṣirī hospital. The latter was surpassed in size and importance by the Manṣūrī hospital, completed in 1284, which remained the primary medical center in Cairo through the fifteenth century.[34]

The medical treatises themselves occasionally mention hospitals, but give no details of their organization and functioning. The vast majority of recorded clinical cases do not mention a hospital, suggesting that most treatment was undertaken in another setting. There are also some formularies composed for use in a hospital, such as that by Ibn Abī al-Bayān al-Isrā'īlī (d. 1240) compiled for the Nāṣirī hospital in Cairo. In general, however, few details regarding the activities in a medieval hospital can be ascertained from the medical literature, and no records are preserved providing evidence regarding the admittance of patients or the success of the care.

Valuable sources for assessing the practice of medicine, as opposed to the learned theories on the subject, are the preserved collections of case histories. Rāzī, for example, scattered case histories throughout the *Ḥāwī*, with thirty-three lengthy accounts grouped together in one section. His students also recorded and assembled almost 900 of his case notes under the title *Kitāb al-Tajārib*, or *Book of Experiences*. A similar title (*Kitāb al-Mujarrabāt*) was used later by the students of Abū al-ʿAlāʾ Zuhr, who, following his death in Seville in 1131, compiled his therapeutic procedures. Case histories are also scattered throughout the formal literature, including the surgical chapter from the encyclopedia written in Spain around the year 1000 by Zahrāwī (Albucasis). A late Egyptian physician, Dāwūd al-Anṭākī (d. 1599), prefaced the essay on his therapeutic techniques (*Risālat al-Mujarrabāt*) with the statement: "I wished to summarize in this essay what I have experienced personally, and if I have omitted a disease from my discussion, it is because I lacked experience with it."[35]

The practice of surgery appears to have been learned to a large extent by apprenticeship, in the same way that other technologies and applied sciences were transmitted. Surgery (and especially ophthalmological surgery) did, of

[34] Conrad, "Did Walīd I Found the First Islamic Hospital?"; Dols, "Origins of the Islamic Hospital"; D. M. Dunlop, G. S. Colin, and B. N. Şehsuvaroğlu, "Bīmāristān," in Gibb et al., *Encyclopaedia of Islam*, vol. 1, pp. 1222–6; P. E. Pormann, "Theory and Practice in the Early Hospitals in Baghdad – al-Kaškarī on Rabies and Melancholy," *Zeitschrift für Geschichte der Arabisch-Islamischen Wissenschaften*, 15 (2002–2003), 197–248; and P. E. Pormann, "Islamic Hospitals in the Time of al-Muqtadir," in *Abbasid Studies II: Occasional Papers of the School of ʿAbbasid Studies*, Leuven, 28 June – 1 July 2004, ed. John Nawas (Orientalia Lovaniensia Analecta, 117) (Leuven: Uitgeverij Peeters en Departement Oosterse Studies, 2010), pp. 337–82.

[35] For Anṭākī and the quotation given here, see Oxford, Bodleian Library, MS. Hunt. 427, folios 108b–114a; and Savage-Smith, *New Catalogue of Arabic Manuscripts in the Bodleian Library, Oxford. I: Medicine*, p. 551, Entry no. 152. For case histories, see Max Meyerhof, "Thirty-Three Clinical

course, have written literature, some of it quite extensive, and in this respect it differed from most other technologies and applied sciences, which are the subject of very few preserved treatises.[36]

Surgery included bloodletting and cauterization, the latter employing caustics or a heated metal rod not just to stop bleeding but as a treatment in itself. Both of these procedures were very old techniques indigenous to the pre-Islamic Near East as well as ancient Greece. In the Islamic world, these practices were to a large extent conducted by barbers and cuppers and others outside the sphere of the learned physicians who composed treatises. Every medieval *ḥammām*, or steam bath, had a barber and a cupper or bloodletter in attendance, and often the barber served dual roles. The *ḥammām* was a vital center for the maintenance of health and regimen in Islamic society, and every town had one or more of them.[37] Yet the medical literature tells us little about this aspect of medical care.

Surgical procedures were frequently described, sometimes in great detail, in the formal learned medical encyclopedias. The most widely read discussions of the subject were those in the tenth-century encyclopedias by Majūsī and Zahrāwī, which, like all such Islamic discourses, were heavily indebted to the chapter on surgery from the Greek encyclopedia by Paul of Aegina, available through translation. An apparent Islamic innovation in the history of surgical literature was the introduction by Zahrāwī of illustrations of instruments.[38]

It is generally recognized that Islamic writers on surgery modified some of the earlier instrumentation and designed some new instruments and techniques, sometimes to accommodate newly recognized conditions. This is particularly true for eye surgery, where, for example, a technique for surgically removing pannus was developed by the tenth century and apparently not infrequently carried out. Zahrāwī appears to have invented a concealed knife for opening abscesses in a manner that would not alarm the nervous patient. He also designed obstetrical forceps (though not for use in live births), new variations in specula or dilators, and a scissor-like instrument for use in tonsillectomies that had transverse blades that apparently both cut the gland and held it for removal from the throat. In addition to bonesetting, surgeons routinely undertook the removal of abscesses and growths, including swollen

Observations by Rhazes (circa 900 AD)," *Isis*, 23 (1935), 321–56; Abū al-ʿAlāʾ Zuhr, *Kitāb al-Muŷarrabāt (Libro de las experiencias médicas)*, ed. and trans. C. Álvarez-Millán (Fuentes Arábico-Hispanas, 17) (Madrid: Consejo Superior de Investigaciones Científicas, 1994); and C. Álvarez-Millán, "Practice versus Theory: Tenth-Century Case Histories from the Islamic Middle East," *Social History of Medicine*, 13 (2000), 293–306.

[36] Timekeeping would be the technology with the next largest body of literature.

[37] See J. Sourdel-Thomine and A. Louis, "Ḥammām," in Gibb et al., *Encyclopaedia of Islam*, vol. 3, pp. 139–46; and M. A. J. Beg, "Faṣṣād, ḥadjdjām," in Gibb et al., *Encyclopaedia of Islam*, suppl., pp. 303–4.

[38] See Zahrāwī, *Albucasis, On Surgery and Instruments*; E. Savage-Smith, "Zahrāwī, Abū ʾl-Qāsim," in Gibb et al., *Encyclopaedia of Islam*, vol. 11, pp. 398–9. For Paul of Aegina, see Pormann, *Oriental Tradition of Paul of Aegina's Pragmateia.*

tonsils, hemorrhoids, and certain ocular disorders. Following treatment, the area was cleansed or dressed with vinegar and water, saltwater, wine, or oil of roses, which have some antiseptic properties. Ointments also contained some items with antiseptic qualities, including frankincense, myrrh, cassia, lead and copper salts, alum, mercury, and borax. Other operations usually followed accidents or battle wounds, in which infection may have developed.

A method of cataract treatment undoubtedly widely practiced was that of couching, an ancient technique in which the opaque lens (or "crystalline humour") is not removed but rather pushed to one side. There is evidence that in some locales there were people who did nothing but couch cataracts; they were probably itinerant and not highly trained in other medical matters. For example, Quṭb al-Dīn al-Shīrāzī wrote in 1283 in his commentary on Ibn Sīnā's *Canon* that he carried out procedures such as bloodletting, suturing, and the surgical removal of the eye conditions pannus and pterygium, but would not undertake couching of cataracts as the operation did not befit him.[39]

Magical and folkloric practices also formed part of the medical pluralism. One of the most obvious uses of charms and incantations was to protect against epidemics, but they were also used to protect against every sort of disease and misfortune, as well as the evil eye. Certain verses of the Qur'ān were considered especially beneficial, and magical alphabets and other *sigla* were combined to form amulets, which (after the twelfth century) might include magic squares. God's blessing and protection was sought on all occasions and by every available means, sometimes by wearing an amulet, sometimes by employing magical equipment or a talismanic chart, and sometimes by placing a talismanic or benedictory inscription on a very utilitarian object, such as a spoon or a mortar and pestle. That such practices were not the sole domain of the poor is evident in the magic-medicinal bowls made for rulers of Egypt and Syria in the twelfth and thirteenth centuries. Still preserved today are large numbers of magic-medicinal bowls engraved with magical symbols and Qur'ānic verses (Figure 5.3). According to inscriptions on the early specimens, when water was consumed from them, they would assist in childbirth or benefit a variety of ailments, including stomach complaints, headaches, nosebleeds, scorpion stings, and bites of snakes or mad dogs.[40]

Magic-medicinal bowls, amulets, and folkloric use of herbs represent medical care at a more popular level than the formal, learned face of medicine

39 Iskandar, *Catalogue of Arabic Manuscripts on Medicine and Science in the Wellcome Historical Medical Library*, p. 43.

40 F. Maddison and E. Savage-Smith, *Science, Tools & Magic* (Nasser D. Khalili Collection of Islamic Art, 12), 2 pts. (London: Azimuth Editions; Oxford: Oxford University Press, 1997), pt. 1, pp. 72–105; Jacques Sesiano, *Un traité médiéval sur les carrés magiques: De l'arrangement harmonieux des nombres* (Lausanne: Presses Polytechniques et Universitaires Romandes, 1996); and E. Savage-Smith, "Safavid Magic Bowls," in *Hunt for Paradise: Court Arts of Safavid Iran, 1505–1576*, ed. J. Thompson and S. R. Canby (Milan: Skira, 2003), pp. 240–7.

Figure 5.3. Magic-medicinal bowl, made in Syria in 1167–8 (563 H) for Nūr al-Dīn Maḥmūd ibn Zangī, founder of the Nūrī hospital in Damascus. On the outside of the bowl, beneath the rim, a circular inscription reads: "This blessed cup is for every poison. In it have been gathered proven uses, and these are for the sting of the serpent, scorpion and fever, for a woman in labor, the abdominal pain of a horse caused by eating earth, and for the bites of a rabid dog, for abdominal pain and colic, for migraine and throbbing pain, for hepatic and splenic fever, for increasing strength, for stopping hemorrhage, for chest pain, for the eye and vision, for driving out spirits, for releasing the bewitched, and for all diseases and afflictions. If one drinks water or oil or milk from it, one will be cured, by the help of God Almighty." By permission of the Khalili Collection of Islamic Art, Inv. no. MTW1443.

displayed by most treatises. The writings of learned physicians, such as Rāzī, although never mentioning magic-medicinal bowls and seldom referring to an amulet, are not entirely devoid of sympathetic magic, for occasional references are found to remedies involving magical principles. Divinatory techniques were also apparently used by many people for the prognosis and diagnosis of mental and physical illnesses, to find the appropriate time for treatment, and to determine the well-being of someone who was absent or in jail.[41]

41 Maddison and Savage-Smith, *Science, Tools & Magic*, pt. 1, pp. 59–71, 106–63; Dols, *Majnūn*, pp. 261–310; Felix Klein-Franke, *Iatromathematics in Islam: A Study on Yuḥannā Ibn aṣ-Ṣalt's Book on 'Astrological Medicine'* (Texte und Studien zur Orientalistik, 3) (Hildesheim: Georg Olms, 1984); Charles Burnett, *Magic and Divination in the Middle Ages: Texts and Techniques in the Islamic and Christian Worlds* (London: Variorum, 1997); and Emilie Savage-Smith, *Magic and Divination in Early Islam* (The Formation of the Classical Islamic World, 42) (London: Ashgate, 2004).

Throughout the society, there was room for popular explanations and cures alongside the more learned approaches, and it is likely that a larger proportion of medieval society used divination and magic than turned to the Greek humoral medicine of the learned elite physicians. The sophisticated learned medical texts represent only one facet of the actual medical practice of the society, for medical care in the medieval Islamic world was pluralistic and fluid, with various practices serving different needs.

THEORY VERSUS PRACTICE

The formal discourses on medicine and surgery are not necessarily reliable guides to what procedures were actually undertaken and are certainly not to be taken as evidence of the incidence or success of such undertakings. As the seventeenth-century Ottoman historian and bibliographer Ḥājjī Khalīfah expressed it, "the usefulness of surgery [*ʿilm al-jirāḥah*] is very great, but the practice of it is less certain than its theory."[42]

The case histories that have been studied, for example, reveal a considerable discrepancy between the therapeutics practiced and those advocated in the theoretical tracts.[43] The medicinal substances named in these case histories are surprisingly limited in scope and number, the recipes given are not particularly elaborate, and their application does not appear to be as determined by the humoral theory of disease as might be expected from the formal, highly theoretical medical literature. There appears to be a discontinuity between the complex drug treatises, with their humoral and pharmacological theory, and the actual practice of medieval drug therapy. Everyday drug therapy, moreover, was greatly influenced by the traditional, pre-Islamic practices employing local herbs and various foodstuffs. Such ancient customs were a vital part of medical diversity.

It is important to note also that surgical procedures, apart from bloodletting and cauterization, are noticeably absent in these collections of clinical cases, including those of Rāzī, who repeats in his more formal medical writings many of the complex procedures given in encyclopedias. This lack of surgery in the recorded clinical literature suggests that some of the surgical literature was not serving as a practical guide to therapy.

The surgical writings in both the Hellenistic and Islamic worlds appear, in many respects, to have had a literary life of their own, separate from

42 Ḥājjī Khalīfah, *Kashf al-Ẓunūn: Lexicon bibliographicum et encyclopaedicum*, ed. G. Flügel, 7 vols. (Leipzig: Oriental Translation Fund of Great Britain and Ireland, 1835–1858), vol. 2, p. 589, no. 4001.

43 See, for example, Álvarez-Millán, "Practice versus Theory"; and E. Savage-Smith, "The Practice of Surgery in Islamic Lands: Myth and Reality," *Social History of Medicine*, 13 (2002), 307–21. The author wishes to thank Dr. Álvarez-Millán for showing her the preliminary results of her studies on the case notes of al-Rāzī compiled under the title *Kitāb al-Tajārib*.

the actual practice of surgery. There is considerable evidence that some of the more elaborate procedures that were repeated from ancient sources were never intended to be actually performed, possibly not even by the first Greek writers who recorded them. A substantial proportion of the surgical literature can be seen as thought experiments outlining procedures one might conceivably undertake if the situation were to arise. In these instances, they do not reflect actual experience, much less successful achievements.

Some Islamic physicians, while repeating the surgical procedures in full detail as given in the ancient sources, did occasionally make comments regarding the fact that no one actually performed them. When such statements are not explicitly made, a writer's failure to mention an example of its use, or to modify in any way the procedure or instrumentation from that handed down from their authorities, encourages the interpretation that many of the procedures described represented a literary tradition unrelated to surgical practice.

For example, abdominal surgery was said to have been performed for an umbilical hernia, treated by ligaturing it with a thread or silk cord, after which the tumor was to be opened above the ligature. If the intestines were found to be in it, the ligature was to be released and the intestines pushed inward. After tightening the ligature, the tumor was then to be cut off, the vessels ligatured, and the incision stitched closed. The accounts given in Arabic, however, are simply repetitions of those in Greek treatises. No Islamic physician mentions seeing it done or modifies the procedure in any way, and the very nature of the description is so imprecise that its applicability is dubious. On the other hand, somewhat similar surgical techniques are described for the treatment of wounds of the abdomen and the protrusion of the intestines, but in the latter context Zahrāwī (Albucasis) extends earlier discourses on methods of suturing and also presents a case history of a man wounded in the abdomen with a knife. Consequently, we might conclude that the operation for an umbilical hernia was not performed, whereas physicians on occasion did attempt to treat abdominal wounds surgically.[44]

Another operation described in the manuals, and of great potential risk to the patient, is a tracheotomy or laryngotomy (an incision in the windpipe for relief of an obstruction to breathing). This procedure was described by Greco-Roman physicians, but not advocated or approved by many. Rāzī repeats the description, drawing upon the account given by Antyllus in the second century, but adds nothing new and displays no experience with it. Zahrāwī, working in Spain around 1000, stated that he had not seen the operation performed in his day, but he did recount his experience with a slave girl who had been wounded in the throat with a knife. His successful

[44] Zahrāwī, *Albucasis, On Surgery and Instruments*, pp. 376–9, 536–51; and Paul of Aegina, *The Seven Books of Paulus Aegineta*, trans. Francis Adams, 3 vols. (London: Sydenham Society, 1844–1847), vol. 2, pp. 340–2. See also F. Sanagustin, "La chirurgie dans le *Canon* de la médecine (*al-Qānūn fi-ṭ-ṭibb*) d'Avicenne (Ibn Sīnā)," *Arabica*, 33 (1986), 84–122.

treatment of her suggested to him that a tracheotomy or laryngotomy might be possible. In the twelfth century, Abū Marwān ibn Zuhr (Avenzoar) said that he practiced the procedure on goats, in case he ever had to perform it, but that he had never seen it performed on humans.[45]

Arabic writers on surgery also described abdominal surgery for dropsy (an accumulation of fluid in the abdominal cavity or ascites). In the Hellenistic literature, it was said to have been treated by making an incision in the abdominal wall and then inserting a metal tube, or cannula, for drawing off the liquid. Galen, however, stated that he had seldom seen anyone recover from such an operation. This procedure was repeated in Islamic manuals, but the tenth-century physician Majūsī (Haly Abbas) did not approve of this treatment and said that he saw it attempted only once, in which case the patient died, while both Ibn Sīnā and Zahrāwī warned against trying to perform it. It is possible that some did attempt the procedure, but clearly it was not performed with any frequency.[46]

In chapters on gynecological matters, there was much concern shown for extracting a dead fetus from the womb, and instruments were described in the Greek and Arabic literature for cutting up the fetus into parts that could then be removed.[47] It has again been questioned whether such techniques were ever actually used. Pertinent to the argument is a fourteenth-century comment made by the Egyptian oculist Ṣadaqah ibn Ibrāhīm al-Shādhilī:

> We possess written accounts of various procedures which cannot be performed nowadays because there is no one who has actually seen them performed; an example is the instrument designed to cut up a dead fetus in the womb in order to save the mother's life. There are many such procedures: they are described in books, but in our own time [i.e., the fourteenth century] we have never seen anyone perform them because the practical knowledge has been lost, and nothing remains but the written accounts.[48]

From the same source, Shādhilī, we also have an important critique of a technique for removing a cataract (an opaque lens) from the eye. Rāzī had briefly described the procedure and the hollow instrument, attributing its invention to the Greek physician Antyllus of the second century. The author of an eleventh-century ophthalmological treatise, *The Book of Vision and Perception* (incorrectly attributed to Thābit ibn Qurrah, a near contemporary

45 Zahrāwī, *Albucasis, On Surgery and Instruments*, pp. 336–9; Paul of Aegina, *Seven Books of Paulus Aegineta*, vol. 2, pp. 301–3; and F. S. Haddad, "Ibn Zuhr's Contributions to Surgery," *Journal for the History of Arabic Science*, 10 (1994), 69–79, especially p. 76. See also Sanagustin, "La chirurgie dans le *Canon* de la médecine (*al-Qānūn fī-ṭ-ṭibb*) d'Avicenne (Ibn Sīnā)," pp. 116–17.

46 Paul of Aegina, *Seven Books of Paulus Aegineta*, vol. 2, pp. 337–40; Zahrāwī, *Albucasis, On Surgery and Instruments*, pp. 382–87; and Sanagustin, "La chirurgie dans le *Canon* de la médecine (*al-Qānūn fī-ṭ-ṭibb*) d'Avicenne (Ibn Sīnā)," p. 116.

47 Zahrāwī, *Albucasis, On Surgery and Instruments*, pp. 484–94; and Paul of Aegina, *Seven Books of Paulus Aegineta*, vol. 2, pp. 387–92.

48 Bethesda, Maryland, National Library of Medicine, MS. A 29.1, folio 118b, and Munich, Bayerische Staatsbibliothek, cod. arab. 834, folio 78b.

of Rāzī), said the operation with a hollow needle was an illusion, that it could not be performed, and that a person should avoid anyone who says he can do it, even if he calls himself an oculist. The technique, however, was supposedly reinvented in Cairo in the late tenth century by ʿAmmār ibn ʿAlī al-Mawṣilī, who claimed much success with it. The technique that was described involved a rather large incision in the eye, a hollow needle, and an assistant with an extraordinary lung capacity, who provided the suction. Meanwhile, in Spain, his contemporary Zahrāwī said that he had heard of the removal of a cataract by suction with a hollow needle and that he understood it to be in vogue in Iraq but had not seen it used. In the thirteenth century, the Syrian oculist and historian Ibn Abī Uṣaybiʿah said that a predecessor of his in the Nūrī hospital in Damascus often undertook surgery for eye diseases and couched cataracts by using a cataract needle that was hollow and curved so as to permit the suction of the cataract during the couching.[49]

In the fourteenth century, Shādhilī gave us an interesting account of these hollow instruments. He stated that one could still buy in the markets in Cairo two types of hollow cataract needles, one like a thick sewing needle and the other with a screw to replace suction by the mouth. He also said that he had never seen them used or heard of anyone using them. Shādhilī then tested some of these hollow needles in water and in a mucilaginous solution and found that they would draw up only pure water but not slightly thickened water, which he argued was nearer the consistency of a cataract. A friend of his had traveled to Russia, where he met a Christian oculist who had tried to use a hollow needle on a female patient and found it merely loosened but did not remove the cataract. Shādhilī then concluded his discussion with ten logical reasons why such a hollow needle could not remove a cataract successfully.[50]

Such conflicting evidence is difficult to interpret, but at the very least it can be said that the operation was very rarely, if ever, undertaken. Available evidence suggests that Shādhilī was probably correct in his evaluation of these hollow needles and that the author of the tract falsely attributed to Thābit ibn Qurrah may well have been correct in categorizing the technique as one employed by illusionists and medical charlatans. Yet if that interpretation is accepted as the most likely, then we must dismiss as credible evidence the "case histories" provided by ʿAmmār ibn ʿAlī al-Mawṣilī testifying to its successful performance.

[49] M. Feugère, E. Künzel, and U. Weisser, "Les aiguilles à cataracte de Montbellet (Saône-et-Loire): Contribution à l'étude de l'ophtalmologie antique et islamique," *Jahrbuch des römisch-germanischen Zentralmuseums Mainz*, 32 (1985), 436–508, especially pp. 482–508; Zahrāwī, *Albucasis, On Surgery and Instruments*, pp. 256–7; and Ibn Abī Uṣaybiʿah, *ʿUyūn al-anbāʾ fī ṭabaqāt al-aṭibbāʾ*, vol. 2, p. 220.

[50] For a study of Shādhilī's chapters concerned with cataracts, see E. Savage-Smith and B. Inksetter, "Could Medieval Islamic Oculists Remove Cataracts? – The Views of a Fourteenth-Century Sceptic" (forthcoming).

A false view of the capabilities of medieval Islamic surgeons has also arisen from certain manuscript illustrations. Because some Islamic miniatures depict births by abdominal delivery, people have sometimes been misled into thinking that Caesarian sections were performed by medieval Islamic surgeons. One miniature painted in 1307 in a copy of the world history by Abū al-Rayhān al-Bīrūnī (d. 1048) illustrates the abdominal birth of Augustus, whom al-Bīrūnī mistakenly thought to be the first of the Caesars (confusing him with Gaius Julius Caesar, whose mother was said to have died in childbirth).[51] Other miniatures, accompanying copies of the *Shāhnāmah* (*The Book of Kings*), written in Persian at the end of the tenth century by Firdawsī, illustrate the birth of the mythical hero Rustam.[52] The latter have been used to support the assertion that Caesarian sections on living women were performed in medieval Islam, for in the course of the poem it is said that the mother was given a drug to stupefy her and that the operation was performed successfully with her full recovery. Such depictions of abdominal births from a living woman, however, are merely illustrations of a legend attributing to its hero a miraculous birth, a common attribute in antiquity for great men. There is no mention in the surgical literature of such a procedure ever being attempted, even as a postmortem effort to save the fetus after the mother had died, and none of these illustrations should be taken as evidence that Caesarian sections, postmortem or live, were performed in the medieval Islamic world.

When considering what the actual practice of medicine might have entailed, evidence provided by written treatises must be supplemented by artifacts and other corroborating data, such as administrative and legal documents. Unfortunately, few documents, such as hospital records, that might provide pertinent evidence are available, and even fewer artifacts remain. In contrast to classical antiquity and medieval Europe, almost no surgical instruments are preserved that can be assigned with certainty to the medieval Islamic world. For those artifacts that are preserved, there are often discrepancies between object and written text, as in the case of some glass utensils usually described by historians as medieval Islamic cupping glasses.[53] The surviving evidence is sufficiently fragmentary and problematic that it cannot provide a complete or entirely reliable picture of medical practice in medieval Islam. It is evident, however, that the everyday medical care of the educated and the illiterate, the affluent and the poor, encompassed a wide spectrum of

51 Edinburgh, Edinburgh University Library, MS. Or. 161, folio 6b, copied in 707 A.H. [1307–8 A.D.].

52 For examples, see Maddison and Savage-Smith, *Science, Tools & Magic*, pt. 1, pp. 27–8.

53 For the so-called cupping glasses, see Maddison and Savage-Smith, *Science, Tools & Magic*, pt. 1, pp. 42–7, and for other medical artifacts, pp. 48–57 and pt. 2, pp. 290–319. For instruments said to have been found at Fusṭāt (Old Cairo), see Sami Khalaf Hamarneh and H. A. Awad, "Medical Instruments," in *Fustat Finds: Coins, Medical Instruments, Textiles, and Other Artefacts from the Awad Collection* (Cairo: AUC Press, 2002), pp. 176–83. Most of these items could have multiple purposes, many of them nonmedical.

practices not necessarily reflected in the medieval medical literature preserved today.

Through the translations made in ninth-century Baghdad, a continuity of ideas was maintained between Greco-Roman and Byzantine medicine on the one hand and that of the medieval Islamic world on the other. Yet Islamic culture did not simply provide custodial care for classical medicine, serving as a mere transmitter to medieval Europe of ancient Greek medicine and learning. Learned Islamic physicians produced a vast quantity of medical literature of their own, in which they imposed a logical and coherent structure on the earlier Greek medicine while developing a more precise scientific vocabulary. They also added an extensive pharmacology, more elaborate notions of medical pathology, knowledge of certain new diseases and disorders, new therapies, and some new surgical techniques and instrumentation. The development of large hospitals in nearly every major city throughout the medieval Islamic world was a major achievement in terms of public health and welfare, even though such urban foundations had little impact upon the large rural population. Available evidence suggests that there were, however, discrepancies between some of the ideas put forward in the learned literature and the actualities of medical practice, just as between the ideal of a large hospital offering a range of medical services and the actual benefits provided by such institutions. When the Arabic medical literature was translated into Latin, a basic change in European learned medicine occurred, and the European acquaintance with previously unknown medicinal substances such as camphor, new medical equipment such as the albarello, and elaborate hospital structures such as those seen in the crusader states profoundly affected the subsequent development of Western medical practices.

6

SCIENCE IN THE JEWISH COMMUNITIES

Y. Tzvi Langermann

The medieval period (500–1500) saw the creation of a body of Hebrew scientific literature. Jews participated actively in the sciences, often in collaboration with non-Jews or, at least, in the awareness of the universality of the scientific enterprise and the achievements of other nations. It was, moreover, a period of reckoning, tense and intense, as thinkers strove to formulate the basic tenets of Judaism and to reconcile them with the claims of the natural sciences. The postures toward the scientific enterprise that emerged from these deliberations have guided Jewish attitudes ever since. The medieval period corresponds more or less with what traditional Jewish historiography identifies as the epochs of the religious authorities called "geonim" and "rishonim." In Jewish historical consciousness, this was a golden age of scientific accomplishment, associated with Moses Maimonides (1138–1204), Abraham ibn Ezra (1089–1164), and other illustrious names.[1]

The lower chronological bound of the period is clear enough. However, the upper bound is much more problematic, and there is a certain amount of arbitrariness in fixing it at the beginning of the sixteenth century. That date is relevant insofar as the Spanish Expulsion of 1492 is one of the traumatic events of Jewish history and Spain was the most important center of Jewish scientific activity. By the sixteenth century, there was a detectable decrease in the writing of Hebrew scientific treatises, though many medieval treatises continued to be copied. There was also less awareness of recent scientific advances, Italy being the most important exception to this rule and, to a lesser extent, cultural centers such as Cracow and Prague. Jews living in

[1] A good but somewhat outdated survey, very thoroughly annotated, is available in Salo W. Baron, *A Social and Religious History of the Jews*, 2nd ed., 16 vols. (New York: Columbia University Press, 1952–1983), vol. 8. Many of the key texts are described, sometimes in great detail, in Moritz Steinschneider, *Die hebraeische Uebersetzungen des Mittelalters und die Juden als Dolmetscher* (Berlin: Kommissionsverlag des Bibliographischen Bureaus, 1893). Fortunately, a comprehensive and up-to-date volume of studies is now in press: *Science in the Medieval Jewish Communities*, edited by Gad Freudenthal (Cambridge: Cambridge University Press, forthcoming).

North Africa and the Near East generally shared in the "decline" of the Islamic world; this meant, among other things, that they continued to study medieval texts into the twentieth century.

THE EMERGENCE OF A HEBREW SCIENTIFIC LITERATURE

Early rabbinic texts (Talmud and Midrash) give evidence of early encounters with science. Several pages of the tractate *Giṭṭin* are given over to medicine, much of which is Babylonian in origin. Some discussions in the tractate *Pesaḥim*, at the beginning of *Bereishit Rabba*, and other places, display cosmological and cosmogonic speculations, often spurred by challenges from non-Jewish philosophers. These passages, however, are all subsumed within larger works whose main purpose is legal or exegetical. True scientific literature – that is, a body of written works whose chief or exclusive topic may be called scientific – emerged only in the murky centuries between the compilation of the Talmud sometime in the fifth century and the proliferation of Arabic scientific literature in the ninth. Each of the writings I describe here presents a bundle of problems for the specialist; only one can be dated with any precision. Nonetheless, taken as a whole, they indicate that Hebrew writers, living within the Islamic world under the Umayyad (661–750) and early ʿAbbāsid (750–1258) dynasties, began to produce Hebrew scientific texts. The literary model was the *Mishnah*, the earliest postbiblical canonical text, composed of relatively short chapters, each of which is made up of small groups of terse legal statements. The language is also Mishnaic Hebrew, with some neologisms and some terminology borrowed from Greek and other languages. All incorporate earlier Hebrew materials, some of which may be traced back to the first centuries of the common era. However, most of these texts were assembled much later, even as late as the ninth century; by that time Arabic had so thoroughly established itself as the language of scientific and philosophical discourse within the Islamic orbit that Jews wishing to disseminate their contributions felt compelled to do so in that language.

The one treatise that can be dated is *Baraitha di-Shmu'el*, which displays as its epoch the year 776. This work of elementary astronomy and astrology deals with such topics as eclipse calculations, the use of gnomons, and planetary distances. The methods employed, at least in the first half of the book, have been shown to be Babylonian. In the second half, some Greek terms are employed: *stirigmos* (planetary station) and *trigon* (astrological triplicity). Several chapters of *Pirqei di-Rebbe Eliezer* are devoted to cosmography and cosmology, as are several sections of *Midrash Konen*. These works all address in one way or another problems related to the Jewish calendar; the earliest Hebrew scientific writings probably included some calendric texts within the corpus. It is noteworthy that a monograph on the Jewish calendar written by the Muslim scientist al-Khwārizmī (fl. 830), which certainly drew upon

Jewish sources, displays, in addition to the rules for fixing the calendar and some interesting translations of biblical passages, calculations for the planets at epochs labeled "Adam" and the "Construction of the Temple." The computations testify clearly to an early interest in planetary astronomy, if only as an adjunct to calendrics.[2]

Mishnat ha-Middot is a work on geometry and mensuration, some parts of which (especially the solution of quadratic equations) strongly resemble passages in the *Algebra* of al-Khwārizmī. The late Solomon Gandz labored to prove that the Hebrew work came first; and, as we have just seen, al-Khwārizmī did have access to Jewish literature, most likely by way of the library of the caliph al-Ma'mūn (r. 813–33), where he is known to have worked.[3]

The earliest Hebrew medical treatises were compiled about the same time. *Sefer Asaf* is a lengthy medical compendium that utilizes many diverse sources. It certainly antedates the efflorescence of the Arabic medical literature, but estimations of its date vary widely. Another quasimedical work that may belong to this period is *Yeṣirat ha-Welad*, which describes the stages in the formation of a fetus, prescribes the sexual hygiene appropriate for the different stages of pregnancy, and, in one version at least, offers some decidedly misogynic exhortations. Parts of the treatise are included in the Talmud; much of it is found in the medieval Yemeni compilation *Midrash ha-Gadol*; and one Yemeni manuscript exhibits the text in full Mishnaic structure, along with a Judeo-Arabic commentary.[4]

The most influential writing of this corpus of early Hebrew texts, in terms of its impact on later Jewish thought, is certainly *Sefer Yeṣira* (*The Book of Creation*). On account of its later appropriation by the kabbalists, its identification as a work of mysticism remains deeply entrenched. However, a slew of commentaries on the work written during the tenth and eleventh centuries in places ranging from Baghdad to southern Italy, Tunisia, and Spain all read *Sefer Yeṣira* as a book on natural science. Indeed, the text

2 On Baraitha di-Shmu'el, see Eliyahu Beller, "Ancient Jewish Mathematical Astronomy," *Archive for History of Exact Sciences*, 38 (1988), 51–66; and on the work of al-Khwārizmī, E. S. Kennedy, "Al-Khwārizmī on the Jewish Calendar," *Scripta Mathematica*, 27 (1964), 55–9, reprinted in E. S. Kennedy, *Studies in the Islamic Exact Sciences* (Beirut: American University of Beirut Press, 1983), pp. 661–5.

3 Solomon Gandz, "The Mishnat ha-Middot," *Proceedings of the American Academy for Jewish Research*, 4 (1933), reprinted in Solomon Gandz, *Studies in Hebrew Astronomy and Mathematics*, ed. Shlomo Sternberg (New York: Ktav, 1970), pp. 295–400. Gad B. Sarfatti, "Mishnat ha-Middot," in *Ḥeqer 'Ever ve-'Arav mugashim li-Yehoshua Blau*, ed. H. Ben-Shammi (Tel Aviv: Tel Aviv University Press, 1993), pp. 463–90 (in Hebrew), has challenged Gandz's claims about the dating of the treatise. In any event, the work cannot have been written much later than the mid-ninth century.

4 Suessman Muntner, "Asaph ha-Rofe," in *Encyclopaedia Judaica*, 16 vols. (Jerusalem, 1972), vol. 3, cols. 673–6. The *Encyclopaedia* remains the only source for information concerning many of the individuals whose activity we shall be discussing; the articles are generally reliable and the bibliographies adequate. See also Y. Tzvi Langermann, "Manuscript Moscow Guenzburg 1020: An Important New Yemeni Codex of Jewish Philosophy," *Journal of the American Oriental Society*, 115 (1995), 373–87 at pp. 386–7.

describes in systematic fashion the formation of the universe by a series of emanations, as in Neoplatonism; it recognizes three elements (air, water, and fire); and it lists the correspondences between the stars and zodiacal signs on the one hand and the organs of the body on the other. The author was also committed to a belief in the role of letters and numbers as building blocks of the universe, as some Neoplatonic writings maintain. A considerable portion of the text is devoted to classification and permutation of the letters of the Hebrew alphabet and the implications of all this for the structure of the cosmos.[5]

There is a certain chronological overlap between the production of this corpus of Hebrew writings and the first instances of Jewish writers composing scientific treatises in Arabic. Mash'allāh and Sahl ibn Bishr, two Jewish astrologers, worked in Baghdad in the early ninth century and participated in its fervent scientific culture. Their works, which were written in Arabic and directed to a general rather than a specifically Jewish audience, were, unlike the Hebrew writings already discussed, fully assimilated into the Arabic scientific tradition.

SURVEY BY COMMUNITY

Jews who contributed to the scientific enterprise during the main period of concern (ca. 900–1500) resided in communities ranging from the Atlantic seaboard to Mesopotamia, reaching as far north as England and as far south as Yemen. Most of the activity, however, was concentrated along the Mediterranean littoral. These communities can be grouped into larger entities according to their common linguistic, cultural, or political bonds.

The largest and most important communities were those farthest west, in present-day Spain and Morocco. The period of early Muslim rule in Andalusia (ca. 960–1100) has earned for itself the rubric "Golden Age" in Jewish historical consciousness. Indeed, we know from the writings of expatriates such as Moses Maimonides (1135–1204) and Abraham ibn Ezra (1089–1164) that Jews hailing from Andalusia retained a strong and proud identity with their place of birth. The Iberian Peninsula was also the site of a momentous revival of Hebrew prose and poetry, both of which were

[5] An English translation and commentary is available in Aryeh Kaplan, *Sefer Yezirah: The Book of Creation* (Northvale: Jason Aronson, 1995). Kaplan relates the kabbalists' interpretation but, ironically, brings to bear his knowledge of twentieth-century physics, much like the early-medieval commentators who read the book in the light of the science of their day. See also Raphael Jospe, "Early Philosophical Commentaries on Sefer Yezira: Some Comments," *Revue des études juives*, 149 (1990), 369–415; Steven M. Wasserstrom, "Sefer Yesira and Early Islam: A Reappraisal," *Jewish Thought and Philosophy*, 3 (1993), 1–30; and Y. Tzvi Langermann, "On the Beginnings of Hebrew Scientific Literature and On Studying History Through Maqbilot (Parallels)," *Aleph: Historical Studies in Science & Judaism*, 2 (2002), 169–89. Langermann discusses there *Sefer Yeṣira* as well as some other early texts mentioned in this chapter.

put to use for scientific writing. Jews participated in some large-scale (by medieval standards) research projects sponsored by non-Jews. We know of two such efforts, both in astronomy and both located in Toledo. One was sponsored by the Muslim historian and astronomer Ṣāʾid al-Andalusī (1029–1070) and resulted in the Toledan Tables. The other, promoted by Alfonso X (1252–1284) of Castille, led to the Alfonsine Tables. The Iberian Peninsula also presented an important interface between the Muslim and Christian worlds; Jews played a significant role in transferring science across boundaries. They exported Arabic science to coreligionists, especially in Provence, and, together with or in the service of Christians, they had a hand in preparing Latin translations from the Arabic.[6]

Closely linked to northern Spain, both geographically and culturally, were the communities of Provence. Gad Freudenthal's researches into the role of the sciences in these communities, undertaken from a sociological perspective, are the only studies of their kind for medieval Jewry.[7] The most significant difference between the communities of Spain and Provence was linguistic. Relations between the two communities were close, and ideas were exchanged. Provençal Jews realized that their Andalusian coreligionists, because of their fluency in Arabic, had been able to thoroughly absorb the corpus of Arabic scientific writings – texts that had important implications for religion. This factor, more than any other, was responsible for the creation of a huge body of Hebrew scientific literature beginning in the twelfth century. Included are original Hebrew compositions, such as those of Abraham bar Ḥiyya (d. ca. 1136) and Abraham ibn Ezra, and the translations executed by the members of the Ibn Tibbon family. In northern Spain and Provence, Jews interacted with Latin culture as well. The opportunity afforded Jews to study at the medical college at Montpellier – a rare case of tolerance – led to the translation of some Latin medical writings into Hebrew. The achievements of Levi ben Gerson (1288–1344) were transmitted in both Latin and Provençal.[8]

Italy, of course, was not at the time a political or cultural unit. One of the earliest Jewish writers on science, Shabettai Donollo (913–ca. 982), lived in Byzantine Italy. Although he received Hebrew books from the Orient, his education was in Greek and Latin sources. Jews living in Sicily and southern Italy were exposed to Arabic culture and were also in close contact with

[6] Y. Tzvi Langermann, "Science in the Jewish Communities of Medieval Iberia: An Interim Report," in Y. Tzvi Langermann, *The Jews and the Sciences in the Middle Ages* (London: Variorum, 1999).

[7] Gad Freudenthal, "Les sciences dans les communautés juives médiévales de Provence: leur appropriation, leur rôle," *Revue des études juives*, 152 (1993), 30–136; and Gad Freudenthal, "Science in the Medieval Jewish Culture of Southern France," *History of Science*, 23 (1995), 23–58.

[8] Luis García-Ballester, Lola Ferre, and Eduard Feliu, "Jewish Appreciation of Fourteenth Century Scholastic Medicine, "*Osiris*, 2nd ser., 6 (1990), 85–117; J. L. Mancha, "The Latin Translation of Levi ben Gerson's Astronomy," in Gad Freudenthal, *Studies on Gersonides, a Fourteenth-Century Jewish Philosopher-Scientist* (Leiden: Brill, 1992), pp. 21–46; and Gad Freudenthal, "The Provençal Version of Levi ben Gerson's Tables for Eclipses," *Archives Internationales d'Histoire des Sciences*, 48 (1998), 269–353.

Hispano-Jewish communities, and much of the scientific activity there was the work of Spanish emigrés and their descendants, such as the astronomer and poet Yiṣḥaq al-Ḥadib (fl. ca. 1396). A considerable amount of translation was also done in Sicily, some of it as part of scholarly interchanges sponsored by royal courts, most notably that of Emperor Frederick II (d. 1250). Central and northern Italy were less lively. Abraham ibn Ezra lived for a while in Lucca; his Pisan Astronomical Tables were well known to Christians, though there is no trace of them in Hebrew. In the late Middle Ages, scientific activity in northern Italy intensified considerably, as seen, for example, in the work of Mordecai Finzi of Mantua (late fifteenth century). Around the same time, there was a short-lived Jewish school for nonreligious studies in Sicily, under royal license.[9]

It now appears that Byzantium and its environs were the scene of much more intellectual activity than has hitherto been recognized. True, only a few individuals, such as Mordecai Comteano (fifteenth century), wrote original treatises in the sciences. However, recent advances in paleography that allow the identification of Byzantine hands reveal that Byzantium was a center for the intense copying of manuscripts. It was also notable for scientific collaboration between members of the Rabbanite and Karaite communities.[10]

The Middle East was, as we have seen, the birthplace of Hebrew scientific literature and the zone in which Jews also first began to share in the efflorescence of Arabic science. By the tenth century, Jews wishing to write on the sciences did so in Arabic, the one exception being some medical and astronomical matters that Maimonides included in his Hebrew law code. For the sake of those who could not read Arabic script, such treatises were often transcribed into Hebrew letters. (This took place on a smaller scale in Spain and North Africa as well.)[11] Although Jews throughout the Islamic orbit participated in the scientific enterprise, it is difficult to formulate any generalizations that adequately characterize their activity other than to note a strong preference for medicine. For example, Abū Manṣūr Sulaymān's abridgement of Avicenna's *Canon of Medicine* is one work that enjoyed wide circulation.[12]

[9] Suessman Muntner, "Donnolo, Shabbetai," in *Encyclopaedia Judaica*, vol. 6, cols. 168–9; Bernard R. Goldstein, "Descriptions of Astronomical Instruments in Hebrew," in *From Deferent to Equant: A Volume of Studies in the History of Science in the Ancient and Medieval Near East in Honor of E. S. Kennedy*, ed. David A. King and George Saliba (Annals of the New York Academy of Sciences, 500) (New York: New York Academy of Sciences, 1987), pp. 105–41; and Y. Tzvi Langermann, "The Scientific Writings of Mordekhai Finzi," *Italia: Studi e ricerche sulla storia, la cultura e la letteratura degli ebrei d'Italia*, 7 (1988), 7–44.

[10] The Karaites were a sect, defined principally by their rejection of postbiblical legislation, who, though small in number, were very active in many small spheres of intellectual activity.

[11] Y. Tzvi Langermann, "Arabic Writings in Hebrew Manuscripts," *Arabic Sciences and Philosophy*, 6 (1996), 137–60.

[12] Moritz Steinschneider, *Die arabische Literatur der Juden* (Frankfurt am Main: J. Kauffmann, 1902), pp. 233–4. The Jews of the Yemen, however, maintained throughout a very strong interest in astronomy.

Finally, turning to Northern Europe, Abraham ibn Ezra did visit some of these communities, and he wrote one of his books in London. There is an extant Hebrew scientific treatise written somewhere in present-day Belgium during the fourteenth century, and there are scattered references in Ashkenazic literature to the mathematics of bar Ḥiyya and other sources.[13] Some medical and calendrical writings are also found in early Ashkenazic manuscripts. It appears that some magic may have reached Spain by way of Germany early enough to have played a formative role in the kabbalah (ca. 1200). On the whole, however, the Ashkenazic communities were a backwater as far as science is concerned. It is worth noting, however, that the analytical methods of the Ashkenazi commentators on the Talmud (known as the Tosafists), who knew no Aristotelian logic, completely overwhelmed the sophisticated intellectual elites of Spain and North Africa.[14]

SURVEY BY DISCIPLINE

Medieval Jews were in wide agreement that, of all the sciences, astronomy was a Jewish specialty. The passage from the Babylonian Talmud that sees, in the words of Deuteronomy 4:6 ("your wisdom and sagacity in the eyes of the nations"), a reference to Jewish distinction in astronomical computations was cited repeatedly.[15] The level of interest in astronomy, measured by the number of original compositions, translations, and manuscript copies, is matched only by medicine. But, unlike medicine, astronomy could not generally provide a livelihood; consequently, economic motivations alone cannot explain the intensity of interest. Nor can the need to compute the calendar; although this is the justification given by some medieval writers (and overemphasized by some modern scholars), by the tenth century at the very latest the rules for the calendar were firmly fixed. Computing the calendar required no understanding of the motions of the Sun and the Moon, not to mention the planets and the disposition of their orbs; nor did anyone have the authority to make any changes in the calendar were he to conclude that modification was necessary. Rather, the explanation for the high level of interest involves a combination of intellectual curiosity, belief that astronomy was a traditional Jewish science, and, especially after

13 "Ashkenaz" is the common Hebrew designation for northern France and Germany.

14 One manuscript is described and others are listed in Y. Tzvi Langermann, "An Unknown Ashkenazic Treatise on Natural Science," *Kiryat Sefer*, 62 (1988–1989), 448–9 (in Hebrew); see now Y. Tzvi Langermann, "Was There no Science in Ashkenaz?" in *Jahrbuch des Simon-Dubnow-Instituts/Simon Dubnow Institute Yearbook*, 8 (2009), 67–92. On Byzantium, see Y. Tzvi Langermann, "Science in the Jewish Communities of the Byzantine Cultural Orbit: New Perspectives," in *Science in the Medieval Jewish Communities*, ed. Gad Freudenthal (Cambridge: Cambridge University Press, forthcoming). However, there are no surveys of science in the Near Eastern, Byzantine, or Ashkenazic communities. Nonetheless, Freudenthal's in-depth study of Provence is still the only published study of its kind.

15 *Shabbat*, 75a.

the spread of Maimonides' philosophy, the sense that a knowledge of the workings of the heavens was an indispensable step on the path to religious fulfillment.

A number of treatises were written in order to explain the luni-solar calendar employed in Judaism, with its intercalations, "postponements," and other features. The practical rules and tables necessary for the computation take up only a small portion of the treatises of Abraham bar Ḥiyya, Abraham ibn Ezra, and Isaac Israeli of Toledo (fourteenth century, not to be confused with the tenth-century physician bearing the same name). All of these writers used the calendar as a pretext for a whole range of astronomical and historical investigations, including comparative studies between the rabbinic and other (sectarian and non-Jewish) calendars. The longest and most popular of these works is Isaac Israeli's *Yesod ʿOlam,* which covers a lot of material relating to cosmology, Andalusian astronomy (which is an important source for the history of the Toledan and Alfonsine Tables), spherical geometry, and astronomy. Isaac Israeli dedicated this treatise to Asher ben Yeḥiel, a refugee from Germany who assumed the rabbinate in Toledo in 1305, believing clearly that there could be no better way to impress upon the northern legist the cultural achievements of Spanish Jewry than to present him with a work of fine Hebrew prose in which the exigencies of the calendar serve as a springboard for a comprehensive investigation of many scientific questions.

By far the most widely read astronomical treatises were nonmathematical in character. These include translations from the Arabic of the books of al-Farghānī (fl. ca. 830) and Ibn al-Haytham (965–ca. 1040), Shlomo Avigdor's Hebrew version of John of Sacrobosco's *Sphere* (written in Latin in the thirteenth century), and Abraham bar Ḥiyya's *Ṣurat ha-Aretz.* Works of this sort allocate the greatest amount of space to a description of the systems of orbs that account for the observed motions of the stars and planets, but they also usually contain rudimentary treatments of terrestrial physics and geography as well. Demand for works of this sort was created by the assertion of Jewish philosophers, first and foremost Moses Maimonides, that the surest (if not the only) path to knowledge of God lay in contemplation of his works – meaning the sober and serious contemplation that science alone can provide. Maimonides develops this point in his *Guide for the Perplexed*; the same point appears as a summary cosmological exposition in the opening chapters of his monumental and enormously influential law code, *Mishneh Torah.*

Levi ben Gerson (1288–1344) was the outstanding astronomical theoretician and observer of medieval Judaism. He invented an instrument, later known as the Jacob's staff, for determining the angular separation between two stars; designed a transversal scale to reduce errors in measurement; and undertook a long series of observations, paying close attention not only to the positions of the planets but their brightnesses as well. He was one of a mere handful of astronomers to make use of his own observations in his theoretical work. Levi devised a model for the Moon in which, among other things, the Moon would not experience variations in its observed diameter as great as those predicted by Ptolemy's model. He then tested the model against his own observations, which included measurements taken at octants in addition to the usual apsides and quadratures. His estimates of the cosmic distances are greater than Ptolemy's by more than ten orders of magnitude. This new scale for the size of the universe, which Bernard Goldstein has called "one of Levi's most amazing innovations," is just one of a number of revisions that Levi introduced into cosmology. A number of writers, for example Maimonides in Book II, Chapter 24, of his *Guide*, had discussed the apparent contradiction between Aristotelian philosophy and the Ptolemaic models, as had Averroes (1126–1198), Levi's chief source for natural philosophy. Levi proposed a complex system of layered shells, separated by an elastic substance that would help to isolate the motion of each planet from that of its neighbors. Levi also computed a set of tables.[16] Closely connected with astronomy, and also of great interest to Jewish scholars, was the art of astrology. Abraham ibn Ezra wrote a large number of short Hebrew astrological tracts, most of which were issued in two somewhat different versions, one written in 1146 and the other in 1148. *The Beginning of Wisdom* is a general introduction to the art; the other treatises are monographs on such topics as the astral cycles that govern history (*The Book of the World*) and astrological interrogations. Ibn Ezra's writings were translated very early into Latin and a number of vernacular languages, giving them a major impact on Christians as well as Jews. He drew upon a variety of Arabic sources, some of which are no longer extant in the original. Astrology, along with arithmology and Neoplatonism, was a basic component of Ibn Ezra's religious philosophy, and it figured prominently in most of his writings. Astrological references and interpretations are invoked in some key parts of his biblical commentary, and these played a crucial role in legitimizing astrology in much of Jewish thought.[17]

Opposition to astrology had been long-standing, though always a minority view; it drew upon belief in human freedom of action, as well as the

[16] Bernard R. Goldstein, *The Astronomical Tables of Levi ben Gerson* (Hamden, Conn.: Archon, 1974); Bernard R. Goldstein, *The Astronomy of Levi ben Gerson (1288–1344)* (New York: Springer, 1985); and Bernard R. Goldstein, "Levi ben Gerson's Contributions to Astronomy," in Freudenthal, *Studies on Gersonides*, pp. 3–20. Many of Goldstein's studies, which comprise the most important contribution to the history of astronomy in the Jewish communities, have been collected in Bernard R. Goldstein, *Theory and Observation in Ancient and Medieval Astronomy* (London: Variorum, 1985).

[17] Y. Tzvi Langermann, "Some Astrological Themes in the Thought of Abraham ibn Ezra," in *Rabbi Abraham ibn Ezra: Studies in the Writings of a Twelfth-Century Jewish Polymath*, ed. Isadore Twersky and Jay M. Harris (Cambridge, Mass.: Harvard University Press, 1993), pp. 28–85; and Bernard R. Goldstein, "Astronomy and Astrology in the Works of Abraham ibn Ezra," *Arabic Sciences and Philosophy*, 6 (1996), 9–21. Over the past decade, Shlomo Sela has made a telling contribution toward the study of Ibn Ezra's astrological writings; see, for example, Shlomo Sela, "*Sefer ha-Tequfah*: An Unknown Treatise on Anniversary Horoscopy by Abraham Ibn Ezra," *Aleph: Historical Studies in Science and Judaism*, 9, no. 2 (2009), 241–54, and the other publications cited therein.

conviction that astrology represented, if only in vestigial form, paganism itself. Abraham bar Ḥiyya once ordered the postponement of a wedding because an unexpected delay pushed the ceremony beyond the auspicious hour; in response to the ensuing controversy, he wrote a defense of astrology, in which he distinguished between the forbidden "Chaldean" divination, which claims to reveal the future in detail, and the legitimate science of astrology, which provides only general indications. The most outspoken opponent of astrology, Moses Maimonides, acknowledged a gross effect of the celestial motions, as well as the light of the Sun, on terrestrial processes – almost no one denied this – but took every opportunity to rail against the practice of consulting astrologers. His vigorous protestations notwithstanding, a curious amalgamation of the thought of Ibn Ezra and Maimonides swept the Jewish communities in the fourteenth century, combining, in one form or another, Maimonides' call for rationality and scientific investigation with Ibn Ezra's claim that the stars are the agents through which the deity controls the world.[18] Jews were avid consumers of logical treatises. Aristotle's *Organon* was widely read, as was Jacob Anatoli's (fl. ca. 1230) translation of Averroes's shorter commentaries on the individual works that make up this collection of logical treatises. Anatoli tells us that he was pressed into undertaking the translations by fellow Jews who felt that they were ill equipped to answer "the sharp-minded people of other nations, who disagree with us" – that is, to hold their own in interreligious polemics.[19] However, Anatoli reveals that there was some opposition within the Jewish community to the study of logic – an opposition that seems not to have had any serious influence. The logical section of al-Ghazālī's *Intentions of the Philosophers*, treatises of al-Fārābī and Maimonides, and some works of Peter of Spain – all available in Hebrew translations – were the most popular works. As in so many other fields, Levi ben Gerson demonstrated an independence of thought in his commentaries on Averroes and other authors.[20] The study of geometry was for the most part limited to Euclid's *Elements* and some short works on spherics. The existence of about half a dozen codices containing the treatises of Theodosios and Menelaus (the two most important Hellenistic authorities on the sphere) and Jābir ibn Aflaḥ (fl. ca. 1120) of Seville, all bearing numerous Hebrew marginalia, testifies to interest in the geometry of the sphere independent of the study of astronomy. *Meyasher ʿAqov*, which survives only in an incomplete copy, is the most strikingly original treatise.

18 Gad Freudenthal, "Epistémologie, astronomie, et astrologie chez Gersonide," *Revue des études juives*, 146 (1987), 357–65; Y. Tzvi Langermann, "Science in the Jewish Communities of Medieval Iberia"; and Y. Tzvi Langermann, "Maimonides' Repudiation of Astrology," *Maimonidean Studies*, 2 (1991), 123–58.

19 Steinschneider, *Hebräischen Übersetzungen des Mittelalters und die Juden als Dolmetscher*, pp. 57–62.

20 Recently, Charles Manekin has begun a thorough and systematic study of this literature. See Charles Manekin, *The Logic of Gersonides* (Dordrecht: Kluwer, 1992); and Charles Manekin, "When the Jews Learned Logic from the Pope: Three Medieval Hebrew Translations of the Tractatus of Peter of Spain," *Science in Context*, 10 (1997), 395–430.

The author identifies himself simply as "Alfonso"; it has been suggested that he is none other than Abner of Burgos, the fourteenth-century philosopher and polemicist who, after his conversion to Christianity, assumed the name Alfonso of Valladolid. The expressed purpose of the book is the squaring of the circle, and this was indeed a topos among the scholastics; however, it is unlike anything in the Latin tradition. A very wide range of sources is cited, including Simplicius (an important commentator on Aristotle), the early geometer Nicomedes, and an anonymous description of the so-called Ṭūsī couple. "Alfonso" is quite aware of the fact that the circle cannot be squared by means of conventional geometry; his aim is to discover a geometry that will reveal the "higher conformity" in which there does exist a square that perfectly equals the circle.[21]

There seems not to have been much demand for arithmetical works, and even less for those dealing with algebra. Basic arithmetical operations were learned from Abraham ibn Ezra's *Sefer ha-Mispar*, the *ʿIr Siḥon* of Yosef ben Moshe Ṣarfatti (fl. ca. 1422), and other works; these invariably discuss other mathematical topics and, occasionally, philosophical ones as well. Levi ben Gerson's *Ma'aseh Ḥoshev*, a comprehensive mathematical treatise, has some interesting treatments of combinations. In the mid-fifteenth century, Mordecai Finzi traveled throughout northern Italy, translating and compiling works on a number of different subjects, including algebra (the book of Abū Kāmil) and mensuration.

Jews were very well represented in the medical profession, in both Islamic and Christian lands. Preliminary statistics for a number of towns in Spain and Provence indicate that Jews constituted roughly 40 or 50 percent of the medical profession, although they formed less than 10 percent of the general population. Similarly, a high proportion of households within the Jewish communities were headed by medical practitioners.[22] There seems little reason to doubt that in Islamic lands as well, Jewish involvement in medicine was much higher than the Jewish share of the general population. These figures of course do not and cannot take into account folk healers. Physicians were a significant component of the intelligentsia – major consumers of philosophical literature and key propagators of philosophic doctrines within their communities. Finally, in no other field did texts authored by Jews have such a strong impact outside of their communities.[23]

21 Tony Lévy, "Hebrew Mathematics in the Middle Ages: An Assessment," in *Tradition, Transmission, Transformation*, ed. F. Jamil Ragep and Sally P. Ragep (Leiden: Brill, 1996), pp. 71–88; and Y. Tzvi Langermann, "Medieval Hebrew Texts on the Quadrature of the Lune," *Historia Mathematica*, 23 (1996), 31–53.

22 Joseph Shatzmiller, *Jews, Medicine, and Medieval Society* (Berkeley: University of California Press, 1994), chap. 6.

23 Ibid.; see also Samuel S. Kottek and Luis García-Ballester, eds., *Medicine and Medieval Ethics in Medieval and Early Modern Spain: An Intercultural Approach* (Jerusalem: Magnes, 1996).

Communal leaders all over the diaspora were likely to be medical practitioners. Physicians, especially those employed by rulers, nobles, and dignitaries, were often better situated than others to represent the interests of the Jewish community, whose vulnerable position as a religious minority must be emphasized in any honest history. However, we must be careful not to insist on too close a causal connection between these two claims. Individuals such as Yehudah ha-Levy (d. 1140), Moses Maimonides, and Moses Naḥmanides (1194–1270), all of whom practiced medicine, rose to the leadership of their communities through a variety of talents, first and foremost their mastery of Jewish law. Even when they did not hold any official appointment, Jewish medical students and professionals had more opportunity than others to meet with adherents of other religious traditions and thus to represent their own creed, so to speak, in the marketplace of ideas. Potentates often held several physicians, representing different religions, on retainer. Maimonides reports a case in which four physicians, two Jews and two Muslims, held a joint consultation concerning the ruler of Marrakesh. Elsewhere Maimonides informs us that he had to visit his patron's court every day and spend (in his view, waste) a good number of hours there, whether or not anyone was ill. It seems that in a setting of this sort he met the Muslim polymath and physician ʿAbd al-Laṭīf al-Baghdādī.[24]

Some notable conversions of Jews to the dominant religions may also be connected to medicine. Abū al-Barakāt al Baghdādī (twelfth century) was initially refused instruction in medicine because he was Jewish. A number of other affronts, suffered in social settings for which his calling may have been responsible, prompted his conversion to Islam late in life.[25] One of the earliest Hebrew medical texts, *Bedeq ha-Bayit* (*The Maintenance of the House*), written between 1197 and 1199, was issued under a pseudonym ("Doeg the Edomite") by a former Jew who, late in life, wished to atone for abandoning his ancestral faith by contributing to Hebrew medical literature.[26]

Moses Maimonides, himself a physician and philosopher, scarcely ever missed an opportunity to lambast dilettante philosophers, who, he asserts, are by and large physicians.[27] Indeed, it was in large measure physicians who mediated philosophy (as they understood it) to a wide Jewish public, thereby playing an important role in inducing the tensions that were later to emerge between science and tradition.

Isaac Israeli of Qairou'an (ca. 850–950) and Maimonides were the two most important Jewish medical writers. Both wrote in Arabic, and the works

[24] Y. Tzvi Langermann, "L'oeuvre médicale de Maïmonide: Un aperçu général," in *Maïmonide: Philosophe et Savant (1138–1204)*, ed. Tony Lévy and Roshdi Rashed (Leuven: Peeters, 2004), pp. 275–302.

[25] Y. Tzvi Langermann, "Abū al-Barakāt al-Baghdādī," in *Routledge History of Philosophy*, vol. 1: *From the Beginning to Plato*, ed. C. C. W. Taylor (London: Routledge, 1997), pp. 636–8.

[26] Steinschneider, *Hebräischen Übersetzungen des Mittelalters und die Juden als Dolmetscher*, pp. 712–13.

[27] Isadore Twersky, *A Maimonides Reader* (New York: Behrman, 1972), p. 408.

of both were translated into Hebrew and Latin. Many Arabic and Latin works were also translated into Hebrew. The fundamental text was Avicenna's *Canon*; dozens of Hebrew-letter copies are extant, including both translations and transcriptions. The Judeo-Arabic commentary of Shlomo ibn Yaʿish is an extensive, highly scholastic work that resolves contradictions within the *Canon*, between the *Canon* and other writings of Avicenna, and between Avicenna and Galen. It displays, in a fourteenth-century Spanish – and, for all we know, noninstitutional – setting, the same type of approach that later took hold in Italian universities, where the Latin text was studied. Nathan Falaquera (thirteenth century) and Moses Narboni (fourteenth century) wrote large compendia; these are probably the most important original Hebrew contributions to the genre.[28]

The preceding account pertains only to the Hellenistic medical tradition, as elaborated and modified in medieval culture, including the integration of some Indian and other non-Western medicine. For a fuller picture, we must note two additional currents: medical knowledge and related pronouncements found in rabbinic literature, especially the Talmud; and folk medicine, including amulets and other magical devices. Owing to the problematic status of medicine as a science, these three traditions overlap considerably. For example, even Maimonides, the champion of logical rigor and nemesis of superstition, recognized – as did all of his medical colleagues – that the medicinal properties of some plants and other materials cannot be proven demonstratively. Maimonides was thus more tolerant of folk remedies than he was of astrology. He tried to discount the folk remedies described in the Talmud in a respectful sort of way. However, the efficacy and validity of the Talmudic prescriptions were staunchly defended by others, most notably Rabbi Shlomo ben Adret of Barcelona, whose long vindication of the Talmudic remedies, in which he attacks the epistemological foundations of Hellenistic science, is one of the important documents for the traditionalist reaction against Maimonides.[29]

Outside of medicine there was little interest in biology. Encyclopedias, such as those written by Judah ben Solomon of Toledo, Gerson ben Solomon of Arles, and Shem Ṭov ibn Falaquera, all of whom were active in the mid-thirteenth century, were the main sources of information for the Hebrew reading public.[30] Levi ben Gerson was the only writer to comment upon Averroes's *De animalibus*.

28 Narboni's book has been studied by Gerrit Bos, "R. Moshe Narboni: Philosopher and Physician, a Critical Analysis of His Orah Hayyim," *Medieval Encounters*, 1 (1995), 219–51. On the whole, the Hebrew medical literature has not been given the attention it deserves.

29 Langermann, "Science in the Jewish Communities of Medieval Iberia."

30 Colette Sirat, "Juda b. Salomon ha-Cohen, philosophe, astronome et peut-être kabbaliste de la première moitié du XIIIe siècle," *Italia*, 1 (1978), 39–61; Rabbi Gershom ben Shlomoh d'Arles, *The Gate of Heaven*, trans. F. S. Bodenheimer (Jerusalem: Kiryath Sepher, 1953); and Raphael Jospe, *Torah and Sophia: The Life and Thought of Shem Tov Ibn Falaquera* (Cincinnati: Hebrew Union College Press, 1988), pp. 48–61.

Maimonides' impact on natural philosophy was enormous. He studied the Peripatetic corpus closely, but it is not clear how much of Aristotle he read directly; he surely relied heavily on the writings of Alexander of Aphrodisias, al-Fārābī, and Avicenna.[31] In this context, the two most resonant disquisitions in his *Guide* are the rebuttal of atomism (at the end of Book I) and the twenty-five principles (at the beginning of Book II). Muslim theologians, whose views are known collectively as *kalām*, had developed a natural philosophy based on the atomistic division of matter, space, and time. This allowed them to integrate their physical teachings into their religious precepts, especially the concept of a God who exercises direct, unmitigated, and unlimited control over each event in the universe. Maimonides recognized the convenience of this approach for theology, but he felt that the theory was wrong on purely scientific grounds. He grounded his own religious philosophy on Aristotelian principles, such as the denial of the existence of an actual infinite, the concepts of actuality and potentiality, the related notions of change and motion, and so forth. These he formulated in twenty-five clear and concise principles.

Many Jewish thinkers learned their natural philosophy from the *Guide*, even if they preferred the view rejected by Maimonides. For example, Moses Isserles, rabbi at Cracow in the early sixteenth century, developed an atomistic physics of his own based on what he had learned from Maimonides' refutation of the *kalām*. The twenty-five principles were excerpted from the *Guide* and studied as a separate text, and as such they reached non-Jewish audiences. The Muslim scholar al-Tabrīzī (thirteenth century) wrote a commentary on them, which was consulted by Jews both in his Arabic original and in Hebrew translation.[32]

Jews living in the Christian orbit, and very likely many or most living in Islamic lands as well, learned Aristotelian physics through their reading of Averroes. Many manuscripts of Hebrew versions of Averroes are extant, and a number of Jews, most notably Levi ben Gerson, wrote commentaries on them. Averroes also served as a conduit for the views of other commentators, most notably Ibn Bājja (Avempace). Some Latin writings were available in Hebrew. An anonymous Jewish scholar produced a compendium that draws on both the Arabic and Latin traditions and comprises two versions of Aristotle's text, one translated from Arabic and the other from Latin, Averroes's middle commentary, and an anthology of pertinent writings of Avicenna, al-Ghazālī, Giles of Rome, Robert of Lincoln, Walter Burley,

[31] Shlomo Pines, "Translator's Introduction: The Philosophic Sources of the Guide of the Perplexed," in *The Guide of the Perplexed*, trans. Shlomo Pines, 2 vols. (Chicago: University of Chicago Press, 1963), vol. 1, pp. lvii–cxxxiv, reprinted in Shlomo Pines, *Studies in the History of Jewish Thought* (The Collected Works of Shlomo Pines, 5) (Jerusalem: Magnes, 1997).

[32] Colette Sirat, *A History of Jewish Philosophy in the Middle Ages* (Cambridge: Cambridge University Press, 1985), passim; and Shlomo Pines, "Scholasticism after Thomas Aquinas and the Teachings of Ḥasdai Crescas and His Predecessors," in Pines, *Studies in the History of Jewish Thought*, pp. 489–589.

William of Ockham, Albertus Magnus, and others, including some remarks by the editor.[33]

Generally speaking, interest in questions of the infinite and the nature of time or place were theologically motivated and discussed in theological works. This is certainly true with regard to Ḥasdai Crescas (d. ca. 1412), author of one of the most thorough critiques of Aristotle in any language.[34] In his *Light of the Lord*, Crescas examines some fundamental claims of Aristotelian physics and finds them wanting. In particular, he denies the essential distinction between celestial and terrestrial matter, as well as the distinction between their motions – earthly elements do not seek their natural place, nor is the circular motion of the spheres a voluntary movement performed by rational beings. He allows the existence of an actual infinite, rejects the Aristotelian definition of place, and asserts that time is something present in the soul rather than an accident of motion. However, it must be borne in mind that Crescas's critique was defined strictly by his firm opposition to the employment of philosophical proofs for theological claims rather than any independent interest in physics.

The interpretation of late-medieval Jewish natural philosophy involves two major problems. The first is essentially historical and concerns the influence of Christian scholasticism. It is generally true that Jews living in Islamic lands freely cited non-Jewish sources, whereas those living in Christian lands (and hence writing in Hebrew) as a rule did not. (Jewish scholars working in Italy, and also some who worked in Spain late in the fifteenth century, are exceptions.) The question is how to interpret this silence. Were these authors reticent about citing authorities from a rival religion? Or was the Hebrew literature sufficient, so that in fact Jewish thinkers had no cause to look elsewhere? Although numerous analogies can be found between some theories of Gersonides, Crescas, and others and a number of Latin sources, it cannot yet be demonstrated that those Jews actually read Latin works. The matter is complicated by the fact that some Latin deviations from Aristotle may be traced to Ibn Bājja, and Jewish thinkers may have learned of the same ideas by way of citations in Averroes.

The other question concerns the implications of Crescas's critique for Jewish attitudes toward science. Crescas labored to repudiate some basic components of the Aristotelian system, maintaining all the while – in stark contrast to Maimonides – that Jewish beliefs need not ground themselves in Aristotle's natural philosophy. It ought to be noted that neither Crescas nor his students seem to have been interested in a research program that would pursue their radical doctrines with the aim of uncovering scientific knowledge for its own sake. Those who judge the intimate connections

33 Steinschneider, *Hebräische Übersetzungen des Mittelalters und die Juden als Dolmetscher*, pp. 123–4, describes one manuscript (now at Moscow, Ginzburg 396). His description is only partially correct, and some additional manuscripts have since been uncovered.

34 Harry A. Wolfson, *Crescas' Critique of Aristotle* (Cambridge, Mass.: Harvard University Press, 1929).

between physics and theology (of the sort that are prominent in the writings of Maimonides, for example) to be an impediment to science will find that Crescas's work marks a major watershed, for Crescas's expressed purpose was to disengage physics from theology. However, the causal connection between (on the one hand) stances that, to some minds, would seem to encourage scientific speculation and (on the other) actual accomplishments in the sciences remains tenuous at best. Crescas's call for disengagement was not motivated by restrictions placed on inquiry by theology. Rather, it was his perception that Aristotelian philosophy had weakened the commitment of some Jews to their faith, intensified no doubt by the violent anti-Jewish outbreaks of 1391 in Iberia as well as Christian missionary pressures, which spurred him on to his work of criticism.

Aristotle had one serious competitor in the Islamic orbit: the concatenation of doctrines known as *kalām* (scholastic theology, with a substantial admixture of metaphysics and natural philosophy), discussed briefly earlier. Many Jews were attracted to this body of thought, especially in the Near East.[35] Sa'adya Gaon (892–942), for example, accepted *kalām*'s rejection of the Aristotelian fifth element and the concomitant notion that the heavens do not undergo generation or corruption.[36] However, atomism, the distinctive physical theory of *kalām*, did not make inroads among rabbinite Jews. However, it was adopted by some Karaites, most notably the eleventh-century thinker Yūsuf al-Bāṣīr and his pupil Abū al-Faraj Furqān.[37] However, it should be added that some later Karaites rejected atomism, and Maimonides' refutation of atomism seems in any event to have been decisive.

There appears to have been considerably more interest in alchemy within Jewish communities located in Islamic regions than in those situated in a Christian environment; overall, the level of interest in both zones was rather low. Two prominent Hispano-Jewish thinkers, Yehudah ha-Levy and Baḥyā ibn Paqūda (eleventh century), both of whom wrote in Arabic, offer comparisons between alchemy and Judaism – each revealing along the way his own particular conception of his religion. In his *Cuzari*, ha-Levy argues that the precise numerical details inherent in Jewish rituals conform to proportionalities in the world of nature. These cannot be reproduced by the alchemist in his laboratory; hence rituals are more efficacious than alchemy in meeting this-worldly needs. As we shall see in the next section, utilitarianism is an important feature of ha-Levy's understanding of Judaism. Baḥyā devotes

35 Harry A. Wolfson, *Repercussions of the Kalam in Jewish Philosophy* (Cambridge, Mass.: Harvard University Press, 1979).

36 Other influences are discussed in Gad Freudenthal, "Stoic Physics in the Writings of R. Saadia Gaon al-Fayyumi and Its Aftermath in Medieval Mysticism," *Arabic Sciences and Philosophy*, 6 (1996), 113–36.

37 Haggai Ben-Shammai, "Kalam in Medieval Jewish Philosophy," in *History of Jewish Philosophy*, ed. Daniel H. Frank and Oliver Leaman (London: Routledge, 1997), pp. 115–48. Much new material relating to the doctrines of the kalām has recently come to light, and as it is studied the impact of these doctrines upon Jewish thought will have to be reappraised.

several pages of his ethical treatise *The Duties of the Heart* to a comparison between the very tenuous rewards of alchemy on the one hand and the security of putting one's trust in God (*tawakkul*) on the other. The alchemist requires hard-to-get raw materials, must perform dangerous operations, and lives in constant fear of discovery, robbery, and the potentates who support him. And even if he is successful, all that he gains is wealth, and that in itself cannot vouchsafe one's security.[38]

Dozens of fragments of alchemical treatises, mostly written in Judeo-Arabic, are extant, most of them coming from the Cairo *genizah*.[39] The total number of alchemical manuscripts in both Hebrew and Arabic is not large and comprises, in addition to the *genizah* fragments, excerpts and recipes. One original treatise is the *Epistle of Secrets*, falsely ascribed to Maimonides, which survives in a number of copies. Moreover, the allegorical interpretation of alchemical transmutation, which was an important theme in literature of other religions, is for all intents and purposes absent from Jewish writings – despite the strong appeal that allegory held throughout the diaspora. Kabbalists may have displayed a somewhat greater affinity for alchemy than did other segments of the Jewish public, particularly in later periods, but there is as yet no evidence for any major interaction between the kabbalah and alchemy during the medieval period.

THE IMPACT OF SCIENCE ON JEWISH THOUGHT

The medieval period saw a most meaningful engagement of Jewish thought with the sciences. Two considerations underlie the special significance of this encounter. First is the fact that during the medieval period – but not so in later times – the people who formulated Jewish responses to science were, for the most part, proficient in both the sciences (having mastered most or all of the scientific disciplines) and the traditional texts, especially the legal ones, upon which Judaism rests. Second, the encounter was meaningful in that the stances that were elaborated by the thinkers of this period remained, for the most part, valid options throughout subsequent Jewish history. These two factors reinforce each other; the perception that figures such as Moses Maimonides and Abraham ibn Ezra were first-rank scientists added considerable weight to the opinions that were voiced by or ascribed to them.

38 The surprising lack of interest in alchemy was noted in Freudenthal, "Science in the Medieval Jewish Culture of Southern France," p. 39. Much of the relevant material is assembled in Rafael Patai, *The Jewish Alchemists* (Princeton, N.J.: Princeton University Press, 1994), which, however, must be read with caution.

39 The Cairo *genizah* is an enormous horde of manuscripts and fragments uncovered in the nineteenth century in some Cairo synagogues. Several dozen items relating to medicine are listed in Haskell D. Isaacs and Colin F. Baker, *Medical and Para-Medical Manuscripts in the Cambridge Genizah Collections* (Cambridge: Cambridge University Press, 1994).

Some scientific teachings caused a real crisis in Jewish thought; for example, the persuasive arguments against creation ex nihilo.[40] The notion that human reasoning could on its own arrive at a coherent and, in the eyes of many, convincing understanding of the universe without the aid of revelation engendered some major reassessments of Jewish belief. Many Jews had believed that recognition of the deity and its design of and upon the cosmos were the exclusive privilege of their tradition, having been delivered directly from God through the Jewish prophets. The assertion of Maimonides and others that the strongest indications for God's existence come from science – and Hellenistic science at that – was shocking, and the reappraisal of notions such as divine providence and intervention, which was necessitated by the Maimonidean philosophy, seemed a difficult if not impossible task. However, the serious public controversies of the period, especially those associated with the reception of the writings of Maimonides, were limited to philosophy narrowly speaking; they do not seem to have impinged upon logic, astronomy, or medicine.

We shall now deal separately with the six most important Jewish thinkers of the period. Their significance for history lies both in their impact on succeeding generations and in the range of attitudes and achievements that they represent.

Saʿadya Gaon (882–942) was born in Egypt and studied at Tiberias and Aleppo before settling in Iraq. He was eventually appointed head of the ancient Talmudic academy at Sura. A wide-ranging scholar and fierce polemicist, Saʿadya took a very active role in communal affairs; he was extremely prolific in his literary activities, translating the Hebrew bible into Arabic and contributing to – indeed, in some sense founding – many branches of medieval Jewish literature. Saʿadya spent some time in Baghdad, where he tasted the intellectual life of that vibrant metropolis, most notably the *majālis-soirées* in which representatives of different religions and schools of thought debated intellectual issues. Saʿadya's *Book of the Choicest Beliefs and Opinions* is a handbook aimed at equipping Jews for participating in discussions of this sort. The chief scientific issue, and indeed the first topic to be discussed, is creation; Saʿadya supplies many proofs for creation and numerous refutations of a wide variety of cosmological theories. Moreover, he takes up some of the underlying epistemological questions, denying that there can be any essential contradiction between reason and revelation and calling for the allegorical interpretation of problematic revealed texts.

Saʿadya's approach has much in common with that of the Muʾtazilites, a rationalist school of *kalām*. Indeed, Hellenistic science presented Jewish, Muslim, and Christian believers with similar problems, and thinkers from all three traditions borrowed each others' ideas or independently arrived at

40 Herbert A. Davidson, *Proofs for Eternity, Creation, and the Existence of God in Medieval Islamic and Jewish Philosophy* (New York: Oxford University Press, 1987).

similar responses. Although they differed strongly in their stances on many specific issues, Saʿadya and the Jewish Aristotelians share the feature that they do not recoil at the thought of sharing the same natural philosophy – including its epistemological underpinnings – with adherents of other religions. Yehudah ha-Levy (to be discussed later) and others, by contrast, maintained that this posture leads inevitably to a blurring of the essential distinction between Judaism, with its unique grasp of the true path to God, and other religions, which, at best, are poor human imitations. Although Saʿadya's name is not often cited – his works are among the first to bear an author's imprint, but his thought was still amalgamated with the early, anonymous rabbinic tradition – his views resonate loudly in many later texts.

Another writing of Saʿadya's, his commentary on the early, anonymous *Sefer Yeṣira*, is instructive in a number of ways. Science was highly valued in Islamic civilization, and one of the main means for its propagation was the study of ancient texts written by or ascribed to figures from the distant past. Sometime shortly before Saʿadya's period of activity, Jewish thinkers began to exploit *Sefer Yeṣira* as an ancient nonsacral writing whose existence indicates the antiquity of Jewish mastery of the sciences and whose cryptic text invited elaboration and explication. Saʿadya's commentary was the first of a series that were written in Judeo-Arabic, all of them serving as springboards for the exposition of theories of cosmogony, cosmology, linguistics, biology, and other sciences. Particularly noteworthy is the nonreverential attitude displayed by Saʿadya; though he finds the theories of *Sefer Yeṣira* to be closer to the truth than those known to him from other sources, it, too, falls short.

Science also figures prominently in Saʿadya's lengthy commentary on the Pentateuch. His exposition of Genesis in particular is interrupted by long, digressive essays on astronomical, biological, and other scientific themes. The connection between these essays and the biblical text is associative; the essays are not necessary for the explication of a word or turn of phrase but rather are adduced to demonstrate the wisdom manifest in creation. This is an early instance of the use of a biblical commentary as an opportunity for discussing a scientific topic. As we shall see, later writers demanded a closer connection between the verse and the scientific theme that it evoked.[41]

The astrological treatises of Abraham ibn Ezra (1092–1167) played an important role in the dissemination of that art; however, his impact upon Jewish thought was exerted primarily through his biblical commentaries. He began his commentary on the Pentateuch with a critique of earlier

41 Alexander Altman, *Saadya Gaon, The Book of Doctrines and Beliefs*, abridged ed., trans. from the Arabic (Oxford: Phadion, 1946); Solomon Gandz, "Saadia Gaon as Mathematician," in *Saadia Anniversary Volume*, ed. Boaz Cohen (New York: American Academy for Jewish Research, 1943), reprinted in Gandz, *Studies in Hebrew Astronomy and Mathematics*, pp. 477–529; and H. Ben-Shammai, "Saadya's Goal in His Commentary on Sefer Yezira," in *A Straight Path*, ed. Ruth Link-Salinger (Essays in Honor of Arthur Hyman) (Washington, D.C.: Catholic University of America Press, 1988), pp. 1–9.

exegeses – specifically, the presence in Saʿadya's Pentateuch commentary of essays that ought to have been published separately as scientific works. Yet Ibn Ezra included in his own commentary more than one lengthy digression on astrology, numerology, or natural science. The difference between the two exegetes was that Saʿadya saw in biblical commentary an excellent instrument for educating a wide audience in the sciences – demonstrating, among other things, the ability of the sciences to reveal the wisdom of the Creator – whereas, in Ibn Ezra's view, certain verses encode scientific truths, and the task of the exegete is to indicate their true meaning.

The image of Ibn Ezra as a wise old man, expert in all that concerns the stars, has impressed itself deeply upon Jewish memory. His commentary spawned a whole industry of commentators on his writings, who struggled to understand the "secrets" that he had found in the Hebrew Bible. However, his outlook was grounded in a universalist, Neoplatonic spirituality within which Judaism had no essential superiority, a point that was not lost on his close friend and fellow poet Yehudah ha-Levy.[42]

Yehudah ha-Levy (d. 1141), considered by many the greatest of all Hebrew poets, authored one work of prose, the *Cuzari*. This book purports to recount discussions that led the king of the Khazars to adopt Judaism, rather than Islam, Christianity, or a nonreligious philosophical regimen. The last is the most serious competitor to Judaism, and the success of natural science, which was considered to be an adjunct of philosophy, was a key component of the philosopher's case. In order to compete with philosophy, ha-Levy must show that the Jews have a science of their own. Ha-Levy points to *Sefer Yeṣira* as the text that comprises the true ancient Jewish science. He also offers a utilitarian argument: Jewish ritual – at least under ideal circumstances, with the Temple standing and the Jews inhabiting the Holy Land – is more efficacious than the natural sciences in ensuring material prosperity. Ha-Levy is clearly reacting to the universalist trends that had made inroads among the elite of Hispano-Jewish society, many members of which saw in philosophy (especially Neoplatonism) a system that could supply their spiritual needs without the ritual demands and social disadvantages of Judaism. Abraham ibn Ezra, who shared this universalism, records a conversation that he had with ha-Levy concerning the first commandment of the decalogue, in which their opposite approaches were displayed quite clearly. Ha-Levy asks: If the best way to know God is through the study of his creation, as the philosophers assert, why didn't the deity reveal himself at Sinai as the God who created heaven and earth? Why instead did he identify himself as "the Lord, your God, who took you out of the land of Egypt"–a historical event of significance for Israel alone? Ha-Levy's forceful arguments for Jewish particularism dominated Jewish thought for many generations; however, his high valuation of the scientific enterprise within Judaism has generally gone

[42] Twersky and Harris, *Rabbi Abraham ibn Ezra.*

unnoticed.[43] Despite the importance of the three preceding thinkers, Moses Maimonides (1135–1204) dwarfs them and all other medieval Jews in his impact on Jewish thought. Born in Córdoba and educated in Andalusia and Morocco, Maimonides acquired thorough proficiency in all the branches of learning and contributed to the Arabic literature on mathematics, astronomy, logic, and especially medicine. For most of his adult life, he resided in Cairo, where he was nominal head of Egyptian Jewry and spiritual leader of Jews everywhere. He elevated the study of the natural sciences to a primary religious obligation, maintaining that only someone who has rigorously mastered them can achieve whatever knowledge of God is humanly attainable; in particular, only the person trained in the demonstrative sciences can distinguish between true insight and vain fantasies. He expounded these ideas in his philosophical masterpiece, *The Guide for the Perplexed*, which addressed many scientific questions, mainly those pertaining to astronomy. Although these questions engendered discussion, it was Maimonides' positive estimation of the sciences, rather than any particular set of scientific beliefs, that exercised the greatest influence. However, much more than the *Guide*, it was Maimonides' legal masterpiece, *Mishneh Torah*, that secured for its author his extraordinary stature within the Jewish communities. This is an encyclopedia of unparalleled breadth, clarity, and authority; the man who produced it had to be reckoned with, even when his philosophical views were hard to digest. By including in *Mishneh Torah* a concise synopsis of physics and astronomy, and also expounding a number of points of psychology and philosophy, Maimonides made it difficult for his readers to avoid exposure to his natural philosophy. These readers tended to treat his opinions as stimuli for debate rather than dogma. In later periods, just about every attempt to develop a rational Jewish perspective that took into account scientific opinion was considered to be "Maimonidean."[44] Maimonides' teachings, especially his strong endorsement of philosophy, caused much controversy. By contrast, Naḥmanides (Moses ben Naman) (1194–1270), who spent most of his career in his native Gerona, Spain, was a voice of moderation, recognizing a connection between some secularizing trends and the study of philosophy but opposing the complete ban under which some in the Jewish community had placed Maimonides' writings. The deteriorating political situation of Spanish Jewry, Christian missionary pressures, and the spread of the kabbalah, as well as his own education, of course, must be taken into account in understanding Naḥmanides' stance, which can be characterized as a devaluation of the natural sciences. Naḥmanides acknowledged their ability to provide a descriptive account of the phenomena. However, the truly significant connections are those between the phenomena and human behavior – specifically, the ethical

[43] Y. Tzvi Langermann, "Science and the Kuzari," *Science in Context*, 10 (1997), 495–522.

[44] See the very extensive lists compiled in Jacob Dienstag, "Art, Science and Technology in Maimonidean Thought: A Preliminary Classified Bibliography," *The Torah U-Madda Journal*, 5 (1994), 1–100; 6 (1995–1996), 138–204.

and ritual performance of the Jews. The natural sciences can tell us nothing about them, but Scripture and tradition explain how these events are manifestations of divine reward and punishment. The synthesis within which Naḥmanides develops these views is unique, but his ambivalence toward natural science – the simultaneous acceptance and devaluation of the scientific enterprise – is characteristic of many other thinkers. Indeed, it is an apt description of the rabbinic mainstream.[45] We have already called attention several times to the singular contributions to the scientific enterprise of Levi ben Gerson (1288–1344). It is important to bear in mind that most of Levi's scientific achievements are elaborated in a single theological work, *The Wars of the Lord*. This is a very tightly argued book, in which a wealth of detail, much of it Levi's own discoveries, is woven together. Thus Levi's astronomical research, for example, is included in its entirety in Book V (though most manuscripts and all printings skip over them). He argues that astronomy serves astrology, the fruit of the sciences, since only astrology can provide a teleological explanation of the variety of stellar motions; moreover, astrology serves theology by providing a mechanism for the exercise of divine providence. It may well be true that Levi's realistic epistemology motivated both the range and exactitude of his investigations: Levi taught that each person's eternal soul is the unique cluster of true facts that he has acquired during his earthly soujourn.[46]

Levi's impact on Jewish thought was minimal – owing, perhaps, to the difficult, advanced nature of his work, especially in astronomy. The accusation leveled by thinkers over the next two hundred years – that he conceded too much to philosophy – certainly did not help; one medieval Jewish writer claimed that his book ought to have been called *The Wars against the Lord*. However, there may be an additional explanation for his lack of influence. Unlike Maimonides, who dangled many tantalizing loose threads in his philosophy, and whose *Guide* was written in a deliberately esoteric style, Levi offered a complete and compact system in which every detail found its place. Maimonides left much room for interpretation, allowing a whole range of thinkers to consider themselves "Maimonidean," whereas Levi's system may have been so closely worked out that it could work only for him.

45 David Berger, "Miracles and the Natural Order in Na'manides," in *Rabbi Moses Na'hmanides (Ramban): Explorations in His Religious and Literary Virtuosity*, ed. Isadore Twersky (Cambridge, Mass.: Harvard University Press, 1983), pp. 107–28; and Y. Tzvi Langermann, "Acceptance and Devaluation: Na'manides' Attitude Towards Science," *Jewish Thought and Philosophy*, 1 (1992), 223–45.

46 Gad Freudenthal, "Sauver son âme ou sauver les phénomènes: sotériologie, épistémologie et astronomie chez Gersonide," in Freudenthal, *Studies on Gersonides*, pp. 317–52. All of the studies in this excellent volume, which includes an annotated bibliography of writings by and about Gersonides, are recommended.

7

SCIENCE IN THE BYZANTINE EMPIRE

Anne Tihon

Byzantine science was a product of two sources. First were the numerous scientific treatises from antiquity, which never ceased to be recopied through all the centuries of Byzantine history, often accompanied by commentaries emanating from the schools of Alexandria, Athens, or Syria at the end of antiquity. This was the constant sustenance for the scientific spirit of a Byzantine scholar. The second source was foreign material, mainly of Islamic origin but also Latin and Hebrew, which would be imported at various times. Entirely original Byzantine creations were rare, and the Byzantine adapters of foreign material were amateurs who never fully mastered the material they explicated.

To comprehend this state of affairs, we must understand it in the cultural context proper to Byzantium. In every period, Byzantine education was essentially based on the study of the trivium and quadrivium, a program inherited from late antiquity. The study of the *tetraktys tôn mathêmatôn* (the Greek equivalent of the mathematical quadrivium: arithmetic, geometry, astronomy, and music) might be treated only at a very rudimentary level, or it might attain a very high level, for the study of the sciences could be pursued without any time limit. Some fourteenth-century scholars devoted considerable efforts to it. But the study of the sciences as such was not seriously supported by imperial power. Generally speaking, the emperors encouraged erudition rather than genuine scientific research. Byzantine sovereigns never created institutions for purely scientific purposes, as was the case in Alexandria or in the Islamic world, except for hospitals, which doubled as medical schools. The Byzantine scholar was not a specialist but a learned man with encyclopedic knowledge, capable of making astronomical calculations or following Euclid's demonstrations but also able to comment on the Scriptures or ancient authors, dispute philosophy, compose poems or funeral orations for the occasion, or write pamphlets or polemical religious treatises – and all this in an artificial language that imitated ancient Greek.

Our sources consist mainly of texts since surviving scientific instruments are almost nonexistent. A number of Byzantine scientific treatises preserved in manuscripts still await critical editions, with translation and commentary, accessible to the modern reader. The manuscript tradition of ancient scientific treatises must therefore be carefully examined. Progress in Greek paleography and codicology often permits one to date precisely and determine the place of origin of a Byzantine copy. Such investigations offer precious evidence concerning the milieu in which a scientific matter was especially studied.

More than eleven centuries separate the foundation of the capital city of the Byzantine Empire, Constantinople, in 330, from its capture by the Turks in 1453. During the latest period of antiquity (fourth and fifth centuries), Alexandria was the main center of scientific study. The major ancient scientific works (Euclid, Archimedes, Apollonius, Ptolemy, and others) were edited and explained in extended commentaries: those of Pappus (ca. 323), who wrote a *Mathematical Collection* and a *Commentary on the Almagest*; Theon of Alexandria (ca. 364), who left voluminous commentaries on Ptolemy's works and edited Euclid; Eutocius of Ascalon (fifth to sixth centuries), who edited works of Apollonius and Archimedes; Proclus (d. 484), who left a commentary on Euclid's Book I and a survey of astronomy under the title *Hypotyposis of Astronomical Hypotheses*; and many anonymous teachers whose lectures are preserved in the form of scholia in the margins of the text being commented upon in their class.

The century of Justinian (sixth century), although illustrious in the arts, architecture, and letters, did not leave a great legacy of scientific works. The seventh and eighth centuries, disturbed by the Arabic invasions of the seventh century and the iconoclastic crisis of the eighth, were dismal periods for intellectual life, and scientific output was very slight.

The ninth century was the scene of an intellectual renewal that revealed itself especially in splendid manuscript copies, including many manuscripts containing ancient scientific works. Euclid, Ptolemy, Theon of Alexandria, and many others were recopied into luxurious manuscripts. Although advanced teaching was restored, few new works were composed, and the heritage coming to us from the ninth century consists above all in beautiful manuscripts. The tenth century was a period of encyclopedias, of which only a few survive. The eleventh and twelfth centuries, under the emperors Alexis, John, and Manuel Comnenus, formed a brilliant and cultivated period that witnessed important scientific achievements.

Unfortunately, the capture of Constantinople in 1204 by the Fourth Crusade of Latin Christendom caused a serious rupture in the evolution of the sciences in Byzantium. Important scientific efforts were cut short, and many works disappeared. During the period of Latin rule (1204–61), an intellectual restoration began at Nicaea, to which the court and the intellectuals had fled to escape the Latin yoke. After the restoration in 1261, scholars attempted to repair the ravages caused by the Latin occupation. Scientific manuscripts were recopied and restored, and the diffusion of scientific works was made easier by the massive introduction of paper, which was far less expensive than parchment. The great majority of Byzantine scientific works were written during this period, which runs from the end of the thirteenth century to the fall of Constantinople in 1453. Particularly brilliant and prolific, this Byzantine Renaissance under the *Palaeologoi* prepared the way for the Italian Renaissance.

MATHEMATICS

Certain branches of mathematics clearly exemplify the preceding pattern.[1] Interest in arithmetic was at a low ebb until the end of the thirteenth century. Very few treatises on arithmetic survive from before that time, whereas from the end of the thirteenth century to the collapse of the Byzantine Empire in 1453, many arithmetical treatises were composed. Children learned calculation by means of finger-reckoning, which had spread throughout the Mediterranean basin since antiquity. A striking illustration of the method of instruction was given by Nicolas Mesarites in his description of the Church of the Holy Apostles in Constantinople and its attached school: The fingers of the children "fly like birds when they learn to count, but the blows of the rod or of the whip rain down at the least error; for, it is said, the masters of reckoning are brutal and violent people." Finger-reckoning, which had spread throughout the Mediterranean since antiquity, was described by Nicolas Rhabdas (ca. 1340) in a letter on arithmetic dedicated to George Khatzykes.

Finger-reckoning has severe limits, and practical life and commerce demanded more elaborate procedures. These are expounded in detail in the two arithmetical letters of Nicolas Rhabdas.[2] The first contains, apart from an account of finger-reckoning, the so-called Palamede's Tables, which simplified the operations. The second letter is more advanced, explaining the operations with fractions with a sequence of unit fractions (of the type $3 + 1/3 + 1/14 + 1/42 \cdots$) and the extraction of the square root of a non-square number. This is followed by a procedure for calculating the date of Easter and a series of practical problems involving the money, weights, and measures used at that time.

These problems could equally well be solved by Hindu-Arabic arithmetic. These Hindu-Arabic procedures, known in the Latin West through

1 A survey of Byzantine mathematics appears in Anne Tihon, "La matematica bizantina," in her *Storia della Scienzia*, vol. 4: *Medioevo e Rinascimento* (Rome: Istituto della Enciclopedia Italiana, 2001), pp. 329–34; See also Anne Tihon, "Les sciences exactes à Byzance," *Byzantion*, 79 (2009), 380–434.

2 Paul Tannery, "Notice sur les deux lettres arithmétiques de Nicolas Rhabdas," in his *Mémoires scientifiques*, vol. 4: *Sciences exactes chez les Byzantins* (Toulouse: E. Privat, 1920), pp. 61–187.

translations of al-Khwārizmī (ca. 825) and explained by Leonardo of Pisa (Fibonacci) (ca. 1170–ca. 1240), were the object of an anonymous Byzantine treatise written in 1252 entitled *Calculation according to the Indians, called the Great Calculation.*[3] Composed during the Latin occupation of Constantinople, this treatise made use of Indian numerals in their Western form. The subject was taken up by Maximos Planudes in about 1293 in his treatise entitled *Indian calculation, called the Great Calculation,*[4] but this time with the Eastern form of the Hindu-Arabic numerals. It is unclear whether these numerals and related procedures were truly adopted in Byzantine practice; one finds only a few notes on the treatise of Planudes by Nicolas Rhabdas or Manuel Moschopoulos, or some anonymous scholia.[5] Manuel Moschopoulos (ca. 1300) also left a tract on so-called magic squares.[6] As for the scholars, they clearly remained attached to the traditional Greek notation (numbers denoted by letters), which precluded positional calculation. Several anonymous manuals intended for more practical use have been preserved.[7]

Learned arithmetic was conceived as an introduction to astronomy, but the explanations given by the authors go far beyond the needs of simple sexagesimal calculation. The methods were taught according to the ancient authors quoted previously, and especially Theon of Alexandria in his *Commentary on the Almagest.* Another source is an anonymous text entitled *Introduction to the Almagest,* probably authored by Eutocius of Ascalon (fifth–sixth century), which describes in detail the principal arithmetic operations in the sexagesimal system: addition, subtraction, multiplication, division, extraction of the square root, multiplication and division of fractions, and proportional interpolation.[8]

In the eleventh century, one finds an anonymous quadrivium written around 1007–8, attributed in some manuscripts to Romanos, judge of Seleucia, in which the arithmetical part is based on Euclid and Nikomachus.[9]

3 André Allard, "Le premier traité byzantin de calcul indien: classement des manuscrits et édition critique du texte," *Revue d'histoire des textes*, 7 (1977), 57–107.

4 André Allard, *Maxime Planude: Le grand calcul selon les Indiens* (Travaux de la Faculté de Philosophie et Lettres, 27) (Louvain-la-Neuve: Faculté de Philosophie et Lettres de l'Université Catholique de Louvain, 1981).

5 Paul Tannery, "Le scholie du moine Néophytos sur les chiffres Hindous," in Tannery, *Mémoires scientifiques*, vol. 4, pp. 20–6.

6 Paul Tannery, "Le Traité de Manuel Moschopoulos sur les carrés magiques," in Tannery, *Mémoires scientifiques*, vol. 4, pp. 27–60; and Jacques Sesiano, "Les carrés magiques de Manuel Moschopoulos," *Archive for History of Exact Sciences*, 53 (1998), 377–97.

7 Kurt Vogel, *Ein Byzantinisches Rechenbuch des frühen 14. Jahrhunderts* (Wiener Byzantinische Studien, 4) (Vienna: Österreichische Akademie der Wissenschaften, Kommission für Byzantinistik, Hermann Böhlaus Nachf., 1968); and Hermann Hunger and Kurt Vogel, *Ein byzantinisches Rechenbuch des 15. Jahrhunderts* (phil.- hist. Klasse, 78, no. 2) (Vienna: Österreichische Akademie der Wissenschaften, 1963).

8 Unedited. See Joseph Mogenet, *L'Introduction à l'Almageste* (Mémoires de l'Académie Royale de Belgique, Cl. Lettres, 51, fasc. 2) (Brussels: l'Académie Royale de Belgique, 1956).

9 J. L. Heiberg, *Anonymi logica et quadrivium cum scholiis antiquis, Kgl Danske Videnskabernes Selskab.* (Historisk-filologiske Meddelelser, 15, no. 1) (Copenhagen, 1929).

About 1300, George Pachymeres wrote a quadrivium that seems to have been very influential.[10] The first book is devoted to general arithmetic, whereas Book IV, the astronomical part, gives long explanations on sexagesimal calculation. The *Stoicheiôsis* of Theodorus Metochites (ca. 1300)[11] contains important arithmetical chapters, and Book I of the *Astronomical Tribiblos* of Theodorus Meliteniotes (ca. 1352) details arithmetical operations.[12] Anonymous compilations drawing from Metochites, Pachymeres, or Meliteniotes are found in the manuscripts. The extraction of the square root especially attracted the attention of Isaac Argyrus (ca. 1368), who devoted a short tract to it and tried to improve Hero's method.[13] Theoretical arithmetic, inspired by Books V and VII of the *Elements* of Euclid, was the object of the *Logistic* of Barlaam of Seminara (ca. 1330–40).[14]

The *Elements* of Euclid remained the fundamental source in geometry, supplemented by Hero of Alexandria, Pappus, and Proclus. Since the Byzantines had access to the text of Euclid, which, by contrast with Ptolemy, could be studied directly without recourse to numerous commentaries, they did not produce a great deal of geometrical literature. However, we do encounter two extraordinary practicing geometers during the reign of the Eastern Roman emperor Justinian (d. 565). Anthemius of Tralles and Isidore of Miletus displayed their geometrical talents as architects of the great Constantinople church Hagia Sophia. In addition, Anthemius wrote a book on paraboloidal burning mirrors,[15] and Isidore, along with Eutocius of Ascalon, edited, commented upon, and taught treatises by Archimedes, thereby playing a critical role in their survival.

Alongside the Euclidean tradition were many small treatises on geodesy – procedures for measuring regions of the earth. Representative of this genre were small tracts in verse by Michael Psellus (eleventh century), John Pediasmos (thirteenth century), and Isaac Argyrus (fourteenth century). We also find many anonymous practical treatises aimed at fiscal officials.[16]

10 Paul Tannery, *Quadrivium de Georges Pachymère*, texte revisé et établi par E. Stephanou (Studi e Testi, 94) (Vatican City: Biblioteca Apostolica Vaticana, 1960), pp. 44ff. On the author, see Stylianos Lampakis, *Georgios Pachymeris, Protekdikos and Dikaiophylax: An Introductory Essay* (National Hellenic Research Foundation, Institute for Byzantine Research, Monographs, 5) (Athens: National Hellenic Research Foundation, Institute for Byzantine Research, 2004).

11 Unedited. On the author, see B. Bydén, *Theodore Metochites' Stoicheiosis Astronomike and the Study of Natural Philosophy and Mathematics in Early Palaiologan Byzantium* (Studia graeca et latina Gothoburgensia, 56) (Göteborg, 2003).

12 Régine Leurquin, *Théodore Méliténiote: Tribiblos Astronomique*, Book 1 (Corpus des astronomes byzantins, 4) (Amsterdam: Hakkert, 1990).

13 André Allard, "Le petit traité d'Isaac Argyre sur la racine carrée," *Centaurus*, 22 (1978), 1–43.

14 Pantelis Carelos, *Barlaam von Seminara, Logistikè* (Corpus Philosophorum Medii Aevi, Philosophi Byzantini, 8) (Athens: The Academy of Athens; Paris: Vrin; Brussels: Editions Ousia, 1996).

15 George L. Huxley, *Anthemius of Tralles: A Study in Later Greek Geometry* (Cambridge, Mass.: Harvard University Press, 1959).

16 Jacques Lefort et al., collaborator J.-M. Martin, *Géométries du fisc byzantin* (Réalités Byzantines) (Paris: Editions P. Lethielleux, 1991), pp. 184–201.

ASTRONOMY

Byzantine astronomy was founded on the works of Ptolemy (fl. 130–150), especially the *Almagest* and *Handy Tables*, and also on commentaries on these works, especially that of Theon of Alexandria (ca. 364), and many other anonymous commentaries.[17] Ptolemy was studied and put into practice until the end of the Byzantine Empire. However, Byzantine astronomy was enriched over the centuries by a number of foreign imports, first Arabic and Persian, later Latin and Hebrew. During the early years of the Byzantine Empire, the sciences, and especially astronomy, were studied in Alexandria, the main center for mathematical science in late antiquity. Many anonymous documents, especially anonymous commentaries and scholia on Ptolemy's *Almagest* or Theon's commentaries, are preserved from the fifth and sixth centuries, probably originating in the teaching of the Neoplatonic schools.

The first truly Byzantine astronomical treatise is the manual of Stephen of Alexandria (ca. 610).[18] Apparently called to Constantinople by Emperor Heraclius shortly before the Arab conquest, Stephen, one of the last representatives of the school of Alexandria, wrote a manual of astronomy on the model of Theon's *Short Commentary* on the *Handy Tables* of Ptolemy – a set of instructions regarding the tables in question, illustrated by examples. For the first time, he introduced specifically Byzantine elements, such as tables for the latitude of Byzantium. At the end of the treatise, some chapters on chronology and a method for calculating the date of Easter were added, apparently by Emperor Heraclius himself.[19]

We know of little astronomy in the ninth and tenth centuries. In the eleventh century, the situation was altered by an influx of Islamic astronomical literature.[20] Anonymous scholia on the *Almagest*, written around 1032, quote Arabic observations made under the caliph al-Maʾmūn (830) and refer to the tables of Ibn al-Aʿlam (d. 985).[21] About 1072, an Arabo-Byzantine manual of astronomy used the commentary by Ibn al-Muthannā on al-Khwārizmī, borrowing its methods for calculating a solar eclipse, and

[17] A survey of Byzantine astronomy is given in Anne Tihon, "L'astronomia matematica a Bizanzio," in Tihon, *Storia della Scienza*, vol. 4, pp. 346–52.

[18] Unedited except for some chapters in Hermann Usener, "De Stephano Alexandrino," in *Kleine Schriften*, vol. 3 (Leipzig: Teubner, 1914), pp. 247–322.

[19] Anne Tihon, "Le calcul de la date de Pâques de Stéphanos-Héraclius," in *Philomathestatos, Studies in Greek Patristic and Byzantine Texts Presented to Jacques Noret*, ed. B. Janssens, B. Roosen, and P. van Deun (Leuven: Peeters, 2004), pp. 625–46.

[20] Anne Tihon, "Tables islamiques à Byzance," *Byzantion*, 60 (1987), 401–25.

[21] Joseph Mogenet, "Une scolie inédite sur les rapports entre l'astronomie arabe et Byzance," *Osiris* 14 (1962), 198–221; Joseph Mogenet, "Sur quelques scolies de l'Almageste," in his *Le Monde grec: Hommage à Claire Préaux* (Brussels: Editions de l'Université de Bruxelles, 1975), pp. 302–11; Anne Tihon, "Sur l'identité de l'astronome Alim," *Archives Internationales d'Histoire des Sciences*, 39 (1989), 3–21; and Raymond Mercier, "The Parameters of the Zīj of ibn al-Aʿlam," *Archives Internationales d'Histoire des Sciences*, 122, no. 39 (1989), 22–50.

the tables of Ḥabash al-Ḥāsib, from which trigonometric functions of sine and versine were borrowed.[22] The only extant Byzantine astrolabe comes from the eleventh century (dated 1062), made by a certain Sergius, of Persian origin and apparently revealing Oriental influence.[23]

The capture of Constantinople in 1204 by the Fourth Crusade cut short this activity. Throughout the Latin occupation (1204–61), the exiled court at Nicaea made serious attempts to recover the scientific heritage from the manuscripts destroyed by the pillage. However, the most brilliant period of Byzantine astronomy had to await the restoration under Michael VIII Palaeologus (r. 1261–82).

The astronomical part of the quadrivium of George Pachymeres (ca. 1300)[24] contains explanations of general astronomy and a description of the constellations, mainly their simultaneous risings and settings. About 1300, Theodorus Metochites wrote an enormous astronomical work, the *Stoicheiôsis* (*Elements*), in which his attempt to explain the *Almagest*, although verbose, had the merit of restoring the study of Ptolemy's astronomy. About 1330, Nicephorus Gregoras, pupil of Theodorus Metochites, went one step further, predicting several eclipses with the aid of Ptolemy's tables.[25] The prediction and calculation of eclipses with the aid of these tables became a fashionable activity, practiced with equal success by Nicephorus Gregoras and his rival, Barlaam of Seminara, who left two tracts devoted to the calculation of the solar eclipses of 1333 and 1337 with the aid of the *Almagest*.[26] Toward the middle of the fourteenth century, Nicholas Cabasilas wrote a Book III for the *Commentary on the Almagest* of Theon of Alexandria in order to replace the original, missing in the manuscripts.[27] The manuscript traditions of the *Almagest* and Theon's *Commentaries* reveal that those works were extensively studied, revised, and annotated by many Byzantine scholars throughout the fourteenth and fifteenth centuries.[28]

While Byzantine scholars energetically pursued the rediscovery and application of Ptolemy's astronomy, others turned eastward. At the end of the thirteenth century, following the Mongol invasion of Iran and Iraq, Hulagu Khan established the Maragha observatory, the reputation of which soon

22 Otto Neugebauer, "Commentary on the Astronomical Treatise Par. gr. 2425," *Mémoires de l'Académie Royale de Belgique*, 59, fasc. 4 (1969); and Alexander Jones, "An Eleventh-Century Manual of Arabo-Byzantine Astronomy," in *Corpus des astronomes byzantins*, vol. 3 (Amsterdam: J. C. Gieben, 1987).

23 O. M. Dalton, "The Byzantine Astrolabe at Brescia," *Proceedings of the British Academy, London*, (1926), 133–46.

24 Tannery, *Quadrivium de Georges Pachymère*, pp. 329–454.

25 Joseph Mogenet et al., *Nicéphore Grégoras: Calcul de l'éclipse de Soleil du 16 juillet 1330* (Corpus des astronomes byzantins, 1) (Amsterdam: J. C. Gieben, 1983).

26 Joseph Mogenet, Anne Tihon, and Daniel Donnet, *Barlaam de Seminara: Traités sur les éclipses de soleil de 1333 et 1337* (Louvain: Peeters, 1977).

27 Available only in the ancient edition of Joachim Camerarius, *Claudii Ptolemaei Magnae Constructionis, id est Perfectae caelestium motuum pertractationibus lib. XIII. Theonis Alexandrini in eosdem Commentariorum lib. XI* (Basel: apud Ioannem Walderum, 1538).

28 Anne Tihon, "Nicolas Eudaimonoioannes, réviseur de l'Almageste?" *Byzantion*, 73 (2003), 151–61.

reached Constantinople. In the preface of his *Persian Syntaxis*, written about 1347, George Chrysococces relates that a certain Gregory Chioniades had decided to travel to Persia via Trebizond in order to learn astronomy. Having done so, he returned to Trebizond with Persian works that he translated into Greek. These translations came into the possession of a priest of Trebizond named Manuel, who was the teacher of Chrysococces, author of this account. An extant collection of manuscripts from the end of the thirteenth century or later contains a corpus of Persian astronomy that seems to consist of translations by Chioniades and other scholars or collaborators.[29]

Around 1347, George Chrysococces wrote a treatise entitled *Persian Syntaxis* based on the *Zīj-i Īlkhānī* of Naṣīr al-Dīn aṭ-Ṭūsī. The *Persian Syntaxis* had a large diffusion, and its circulation was established in the second half of the fourteenth century and the fifteenth century. About 1352, Theodorus Meliteniotes composed a very substantial work, the *Astronomical Tribiblos*,[30] containing a justification of astronomy and an unconditional condemnation of astrology. Book I is devoted to arithmetical operations and the construction and use of the astrolabe, Book II to calculations using Ptolemy's *Almagest* and *Handy Tables*, and Book III to Persian astronomy. Meliteniotes, director of the Patriarchal School in Constantinople, apparently introduced mathematical astronomy into the training of the clergy of the Orthodox Church. In spite of the success of Persian astronomy, Ptolemy's tables were still used and adapted in 1368 by Isaac Argyrus in his *New Tables*. At the beginning of the fifteenth century, the striking fact is the presence of Jewish influence on Byzantine astronomy. Three Jewish astronomical works were the object of Byzantine adaptation: the *Six Wings* (*Shesh Kenaphayim*) of Emmanuel Bonfils of Tarascon (ca. 1365); the *Cycles* of Bonjorn (Jacob ben David Yom-Tob, Perpignan, ca. 1361); and the *Plane Way* (*Orah Selulah*) of Isaac ben Salomon ben Zaddiq Alhadib (ca. 1370–1426).[31] These treatises, which were the object of anonymous commentaries, continued to be used during the Ottoman period. Western astronomy was introduced in the Byzantine milieux of Cyprus around 1340 (treatise on the astrolabe, Toledan Tables), and in 1380 Demetrius Chrysoloras composed a Greek adaptation of the Alfonsine Tables. In the fifteenth century, one also finds the only truly original Byzantine astronomical treatise: the manual of

[29] David Pingree, "Gregory Chioniades and Palaeologan Astronomy," *Dumbarton Oaks Papers*, 18 (1964), 135–60; Otto Neugebauer, "Studies in Byzantine Astronomical Terminology," *Transactions of the American Philosophical Society*, n.s. 50, no. 2 (1960), 3–45; and Anne Tihon, "Les tables astronomiques persanes à Constantinople dans la première moitié du XIVe siècle," *Byzantion*, 57 (1987), 471–87. An edited version appears in David Pingree, *The Astronomical Work of Gregory Chioniades*, vol. 1: *The Zīj al-Alā'ī*, 2 vols. (Corpus des astronomes byzantins, 2) (Amsterdam: J. C. Gieben, 1985–1986).

[30] Book I, see n. 12; for Book II, see Régine Leurquin, *Théodore Méliténiote: Tribiblos Astronomique*, Book II (Corpus des astronomes byzantins, 5–6) (Amsterdam: J. C. Gieben, 1993).

[31] Anne Tihon, "L'astronomie byzantine à l'aube de la Renaissance (de 1352 à la fin du XVe siècle," *Byzantion*, 66 (1996), 244–80; P. Solon, "The Six Wings of Immanuel Bonfils and Michael Chrysococces," *Centaurus*, 15 (1970), 1–20.

George Gemistus Pletho.[32] The astronomy of Pletho is based on his ideas for the restoration of pagan antiquity, including a new calendar imitating that of the ancient Greeks and Romans. The months were strictly lunar; the year was luni-solar and began at the first new moon following the winter solstice. The starting points (new moon, winter solstice) were chosen because they mark the return of the light. The tables give the mean motions of the Sun, Moon, and planets. The tables of Pletho were certainly the most original astronomical works composed in the Byzantine world, the only ones that were not simple adaptations of foreign tables.

Except for the plane astrolabe, astronomical instruments were rarely made the subject of Byzantine treatises.[33] This instrument was described for the first time in a tract by John Philoponus (ca. 520–550). A poem by John Kamaterus dedicated to Emperor Manuel Comnenus (1143–1180) gives a description of an astrolabe. As already noted, the astrolabe of Brescia, the only extant Byzantine astrolabe, was constructed in 1062 for a certain Sergius, a man of Persian origin, and anonymous chapters "taken from a Saracen book" are found in the twelfth century. In the fourteenth century, Nicephorus Gregoras, Theodorus Meliteniotes, and Isaac Argyrus wrote important treatises on the construction and use of the plane astrolabe. Other instruments are not often described in Byzantine treatises. At the end of the fourteenth century, an anonymous treatise, probably produced in the astrological milieu centered on John Abramius, Demetrius Chlorus, or Eleutherius Elius, described an armillary sphere of Persian origin different from Ptolemy's. We also have an undated Greek translation of a treatise on the quadrant by John of Montpellier (or Robert the Englishman?).[34]

A source of abundant Byzantine astronomical literature was the calculation of the date of Easter (*computus*, or the Paschal calculation). Before the fourteenth century, texts on the subject are simple *computus* tracts, without scientific discussion, but in the fourteenth century, treatises on Easter included questioning about the exactness of the nineteen-year cycle, the length of the solar year, or the date of the spring equinox. The great scholars of the time, Nicephorus Gregoras, Barlaam,[35] Nicolas Rhabdas,[36] and Isaac Argyrus, devoted studies to the problem. The Metropolitan of Thessalonika, Isidorus Glabas, left us the calculation of a complete cycle of nineteen years (1390–1409), with mention of the eclipses. The importance of this question

32 Anne Tihon and Raymond Mercier, *Georges Gémiste Pléthon: Traité astronomique* (Corpus des astronomes byzantins, 9) (Louvain-la-Neuve: Academia Bruylant, 1998).

33 Anne Tihon, "Traités byzantins sur l'astrolabe," *Physis*, n.s. 32, fasc. 2–3 (1995), 323–57.

34 Paul Tannery, "Magister Robertus Anglicus in Montepessulano," *Mémoires Scientifiques*, 5 (1922), 112–203.

35 Anne Tihon, "Il Trattato sulla data della Pasqua di Barlaam comparato con quello di Niceforo Gregoras," in *Barlaam Calabro, L'Uomo, l'opera, il pensiero*, ed. Antonis Fyrigos (Rome: Gangemi editore, 1999), pp. 109–18. Text edited in Anne Tihon, "Barlaam de Seminara, Traité sur la date de Pâques," *Byzantion*, 81 (2011), 357–407.

36 Otmar Schissel, "Die Osterrechnung des Nikolaos Artabasdos Rhabdas," *Byzantinisch-Neugriechische Jahrbücher*, 14 (1937–1938), 43–59.

for the Orthodox Church was certainly a factor in promoting astronomical studies among its members.

ASTROLOGY

Educated Byzantines clearly distinguished between astronomy and astrology – the former being concerned with the theoretical study of celestial events, whereas the latter functioned as a practical art. Astronomy dealt with matters contained in Ptolemy's *Almagest*, astrology with matters studied in Ptolemy's *Tetrabiblos*. Despite condemnation by the fathers of the church (Gregory of Nyssa and Basil of Caesarea, for example),[37] astrology flourished in Byzantium. The attitudes of emperors and church authorities toward astrology varied, firm condemnations alternating with tolerance. This is evident from the many manuscripts containing astrological compilations.[38]

The principal sources of Byzantine astrology were ancient: Ptolemy, Valens (first to second centuries), Dorotheus of Sidon (first to second centuries), and Paul of Alexandria (fourth century), among others. In the seventh century, Theophilus of Edessa used Greek sources but also Hindu-Arabic material. Under the name "Stephanus," a horoscope of Islam and predictions for the Islamic world have come down to us, probably drawn up about 775, near the time when John of Damascus strongly condemned astrology. About the year 1000, Greek translations of Arabic astrologers such as Abū Maʿshar and Ahmad ibn Yūsuf appeared in works that were preserved in compilations of the eleventh and twelfth centuries. The popularity of astrology in those centuries is confirmed by historians such as Michael Psellus, Anna Comnena, and Nicetas Choniates. One text mentions astrological predictions about races in the hippodrome in 1132. A number of scholars or men of letters at that time dabbled in astrology. Some professed incredulity (Michael Psellus, Anna Comnena, Michael Italicus), whereas others, such as Symeon Seth, practiced it professionally. Emperor Alexis Comnenus prudently attempted to stem the fashion, but his grandson Manuel was enamored with astrology, which he defended in a letter that figures in a good many astrological manuscripts.

At the beginning of the fourteenth century, discussions of astrology flourished. Theodore Metochites accepted a "rational" astrology limited to natural events while rejecting the trivial influence of stars on the daily life of humans. It seems that at the beginning of the fourteenth century many astrological predictions of Oriental origin circulated in Constantinople. An anonymous

[37] Marie-Hélène Congourdeau et al., eds., *Les Pères de l'Église et l'astrologie* (Les Pères dans la Foi, 85) (Paris: J.-P. Migne, 2003).

[38] Paul Magdalino, *L'orthodoxie des astrologues, La science entre le dogme et la divination à Byzance (VIIe–XIVe siècles)* (Réalités byzantines, 12) (Paris: Lethielleux, 2006); and Paul Magdalino and Maria Mavroudi, eds., *The Occult Sciences in Byzantium* (Geneva: Editions de la Pomme d'Or, 2007).

astrological document, unique in its way, is an almanac for Trebizond in the year 1336, offering picturesque, marginal astrological predictions for the region of Trebizond.[39] This almanac was perhaps prepared by the priest Manuel of Trebizond, teacher of George Chrysococces, for the latter ended his *Persian Syntaxis* with several astrological chapters and a profession of faith according to which the stars are not the causes of events but received, by the Divine will, the power to foretell events. By way of contrast, the *Astronomical Tribiblos* of Theodorus Meliteniotes strongly condemned astrology.[40]

MUSIC THEORY

The final quadrivial art (along with arithmetic, geometry, and astronomy) was music theory, which Aristoxenus, Euclid, Ptolemy, and many other authors had developed in antiquity. Ancient music was based on the lyre, the tension of whose chords could be regulated so as to modify the pitch of the note. The starting point was the tetrachord, with two fixed notes (chords 1 and 4 on the outside) and two moving notes (chords 2 and 3 on the inside). Further tetrachords were added to create a more extended musical scale.

Ancient musical theory was, for the most part, a study of the mathematical ratios that represented musical intervals. These were measured using the famous Pythagorean "canon" – a ruler on which was stretched a cord that rested on a bridge (as on a violin). On comparing the pitch of the notes produced with the length of the cord, one could express the musical intervals by numerical ratios: 2:1 for the octave, 4:3 for the fourth, 3:1 for the fifth, and so on. Likewise, one had to define the intervals that are harmonious or discordant and explain in detail the naming of the notes, the systems, the genera, the modes, the tones, musical notation, and so forth. The study of harmonic ratios was also extended to cosmology, for particular notes and musical intervals could be associated with the different spheres of the planets. The Platonic idea that musical theory must reflect the universal harmony of the world, and that the latter is expressed by mathematical ratios, remained firmly anchored in mentalities well beyond antiquity and the Byzantine world.

Byzantine scholars continued to treat music as a branch of mathematics. The anonymous quadrivium of the eleventh century[41] presents a rather short musical section, in which it recalls, succinctly, the notes, intervals, tone, systems, genera, and modes. About the year 1300, George Pachymeres

39 Raymond Mercier, *An Almanac for Trebizond for the year 1336* (Corpus des astronomes byzantins, 7) (Louvain-la-Neuve: Academia, 1994).

40 On astrology in the fourteenth century, see Anne Tihon, "Astrological Promenade in the Early Palaeologan Period," in Magdalino and Mavroudi, *Occult Sciences in Byzantium*, pp. 265–90.

41 Heiberg, *Anonymi logica et quadrivium cum scholiis antiquis.*

devoted the second part of his treatise on the quadrivium to music.[42] His contemporary Manuel Bryennius left a voluminous treatise, *Harmonics*, in three books,[43] in which he also reviewed the standard rubrics of music theory in the tradition of the ancient theoreticians. Around 1330–1340, Nicephorus Gregoras wrote a complement to the *Harmonics* of Ptolemy to which Barlaam opposed a refutation. Byzantine treatises were extremely influential in the musical theory of the Renaissance.

GEOGRAPHY

The sources of Byzantine geography were also of ancient vintage – authors such as Strabo, Pausanias, Ptolemy, and the so-called minor geographers collected in manuscripts of the ninth and tenth centuries.[44] Strabo was often studied; extracts are found under the pens of various scholars, such as Michael Psellus (1018–1081), Eustathius (ca. 1170), Maximus Planudes (ca. 1300), or the copyist John Catrarius (ca. 1322).[45] At the very end of the Byzantine Empire, autograph notes of George Gemistus Pletho correct Strabo.[46] Pletho had new information on Scandinavia and northern Russia. For Scandinavia, he made use of a map by the Dane Claudius Clavus, shown to him by the Florentine Paolo Toscanelli. But he knew nothing of the travels of Marco Polo in the Far East, and his knowledge of that part of the world was still limited to Ptolemy.

Scientific geography was intimately linked with astronomy since the astronomical tables, especially the *Handy Tables*, included long lists of famous cities with their coordinates in longitude and latitude.[47] The manuscripts often included these lists, as well as notes on the climates, the continents, and so forth. In general, Ptolemy's coordinates are repeated without change. However, in the eleventh century, several documents, as well as the Brescia astrolabe (1062), used the latitude 41° for Constantinople in place of Ptolemy's 43°. In the fourteenth century, when the *Persian Tables* penetrated the Byzantine world, they were similarly accompanied by geographical lists. As for cartography properly speaking, one must await Maximus Planudes at the end of the thirteenth century to discover a Byzantine scholar interested in Ptolemy's *Geography*.[48]

42 Tannery, *Quadrivium de Georges Pachymère*.

43 G. H. Jonker, ed. and trans., *The Harmonics of Manuel Bryennius* (Groningen: Wolters-Noordhoff, 1970).

44 C. Müller, ed., *Geographi graeci minores*, 2 vols. (Paris: F. Didot, 1861–1885).

45 François Lasserre, "Étude sur les extraits médiévaux de Strabon suivie d'un traité inédit de Michel Psellus," *L'Antiquité Classique*, 28 (1959), 32–79.

46 Aubrey Diller, "A Geographical Treatise by Georgius Gemistus Pletho," *Isis*, 27 (1937), 371–81, 441–51.

47 Ernst Honigmann, *Die sieben Klimata und die πόλεις ἐπίσημοι* (Heidelberg: C. Winter, 1929).

48 J. L. Berggren and Alexander Jones, *Ptolemy's Geography: An Annotated Translation of the Theoretical Chapters* (Princeton, N.J.: Princeton University Press, 2000). For Ptolemy's maps, see the new

The conception of the inhabited world (*oikoumene*) for the Byzantines did not differ from that of the ancients, with the exception of a single unique work, the *Christian Topography* of Cosmas Indicopleustes.[49] Cosmas was a merchant from Alexandria whose field of activities covered the Mediterranean, the Red Sea basin, and the Persian Gulf. (In spite of his name, it is uncertain whether he traveled as far as India.) His book, *Christian Topography* (ca. 547–9), falls simultaneously under theology, cosmology, and geography. Cosmas defended the idea of a universe in the form of a chest with a vaulted cover, like the tabernacle of Moses. The Earth was conceived as a flat rectangle, surrounded by the ocean, with Paradise to the east. Days and nights were explained by the presence of mountains to the north. This work, which reflected Nestorian teachings inherited from Theodorus of Mopsuestia, contains numerous theological justifications and attacks on the partisans of a spherical cosmos. It also contains much information, taken from personal experiences of the author, regarding commerce with the Orient, exotic flora and fauna, and a description of the island of Ceylon. The cosmology of Cosmas was to have a long life in Slavic countries, and is still found in a few Byzantine texts of the twelfth century.

OPTICS AND MECHANICS

Byzantine scholars made a few contributions to optics and mechanics, both regarded as mathematical sciences. In the sixth century, Anthemius of Tralles, engineer and architect of the great church Hagia Sophia, wrote a treatise on burning mirrors.[50] In the ninth century, Emperor Theophilus (r. 829–42) owned a collection of automata. At the same time, Leo the Mathematician is credited with the construction of an optical telegraph whose signals were regularly transmitted to Constantinople from the eastern frontier of the empire (north of Tarsus).[51] Similarly, the imperial throne of Constantinople contained a mechanism that permitted it to rise to the ceiling while the bronze lions roared and shook their tails, griffins rose, and birds began to sing in golden trees.[52] These automata follow the tradition of Philo (second

edition of the *Geography* in Alfred Stückelberger and Gerd Grasshoff, eds., *Ptolemaios Handbuch der Geographie*, 2 vols. (Basel: Schwabe Verlag, 2006).

49 Wanda Wolska-Conus, *Topographie Chrétienne*, 3 vols. (Sources Chrétiennes 141, 159, 197) (Paris: Editions du Cerf, 1968–1973); and Eric O. Winstedt, *The Christian Topography of Cosmas Indicopleustes* (Cambridge: Cambridge University Press, 1909). See also Wanda Wolska, *La topographie chrétienne de Cosmas Indicopleustès: Théologie et science au VIe siècle* (Paris: Presses Universitaires de France, 1962).

50 Huxley, *Anthemius of Tralles*.

51 Paul Lemerle, *Le premier humanisme byzantin* (Paris: Presses Universitaires de France, 1971), pp. 154–5.

52 G. Brett, "The Automata in the Byzantine Throne of Salomon," *Speculum*, 29 (1954), 477–87.

century B.C.) or Hero of Alexandria (first century), whose success is attested by an abundant manuscript tradition (especially a superb Constantinopolitan manuscript of the tenth century). At the end of the fourteenth century, one finds a drawing of a Persian clock that marked the hours by means of a candle that caused drops of lead to fall.[53]

ALCHEMY AND CHEMISTRY

Byzantine alchemy is represented by three main tendencies:[54] the desire to preserve the heritage of Alexandrine alchemy and its commentators; a concern for combining this traditional culture with contemporary elements; and the integration of alchemical chapters into far greater compilations, including astronomy, astrology, magic, and the like. The identification and dating of ancient or Byzantine alchemical authors is very difficult owing to many pseudonyms, falsifications, and interpolations or glosses inserted into the texts. The most ancient compilation of alchemical works is the *Marcianus gr.* 299 (tenth century or the beginning of the eleventh), illustrated by precise drawings of alchemical instruments. Among the more recent texts, one may quote the works of Michael Psellus (ca. 1045–1046), Nicephorus Blemmydes (1197–1272), and the "Zuretti anonymous" (ca. 1300) based on Latin and Arabo-Latin sources.[55]

Byzantine chemistry made its greatest accomplishment in the realm of military weapons: the famous "Greek fire," a liquid of uncertain composition (perhaps a mixture of resin and sulfur, or oil from the Sea of Azov) that was propelled by a pump and a bronze tube onto enemy boats. The invention was improved by a Syrian called Callinicos, who used it to save Constantinople from the Arabs in 678. This invention ensured Byzantine naval superiority for centuries.

[53] Anne Tihon, "Un texte byzantin inédit sur une horloge persane," in *Sic itur ad Astra: Festschrift für den Arabisten Paul Kunitzsch zum 70. Geburtstag*, ed. Menso Folkerts and Richard Lorch (Wiesbaden: Franz Steiner Verlag, 2000), pp. 523–35.

[54] I owe most of this material on alchemy to Mrs. Andrée Colinet. Ancient alchemist texts are in Marcellin Berthelot and Charles-Émile Ruelle, eds., *Collection des anciens alchimistes grecs*, 3 vols. (Paris: Steinheil, 1888–1889); and Marcellin Berthelot and Charles-Émile Ruelle, eds., *Description of Manuscripts in Catalogue des Manuscrits alchimiques grecs*, 8 vols. (Brussels: Lamertin, 1924–1932). For Greek and Byzantine alchemy, see H. D. Saffrey, "Historique et description du manuscrit alchimique de Venise, Marcianus Graecus 299," in *Alchimie: art, histoire et mythes, Actes du 1er colloque international de la Société d'Étude de l'Histoire de l'alchimie*, ed. Didier Kahn and Sylvain Matton (Paris: S.E.H.A-Archè, 1995), pp. 1–10; J. Letrouit, "Chronologie des alchimistes grecs," in Kahn and Matton, *Alchimie*, pp. 11–93; and Christina Viano, dir., *L'Alchimie et ses racines philosophiques, La tradition grecque et la tradition arabe* (Histoire des doctrines de l'Antiquité Classique, 32) (Paris: Vrin, 2005).

[55] Edited with French translation in Andrée Colinet, *Les Alchimistes grecs: L'anonyme de Zuretti* (Paris: Les Belles Lettres, 2000); eadem, Les Alchimistes grecs: Recettes alchimiques (Par. gr. 2419; Holkamicus 109).

BOTANY

Botany and zoology were less systematically cultivated in Byzantium than the exact sciences. One explanation for this difference is that botany and zoology were not part of the traditional liberal arts curriculum, inherited from antiquity (as were arithmetic, geometry, astronomy, and music). As in antiquity, plants attracted interest primarily for their medicinal properties and their alleged magical power. The ancient sources of Byzantine botany are especially the learned poems of Nicander of Colophon (second century B.C.), which were transmitted in illuminated manuscripts of the mid-Byzantine period: the *Theriaka*, which described remedies against animal poisons; and the *Alexipharmaka*, which gave antidotes for food poisoning. But it was above all the *Materia Medica* of Dioscorides (first century) that constituted the foundation of Byzantine botany. This work was transmitted to Byzantium in a famous manuscript preserved in Vienna, copied in 512 by order of Julia Anicia and superbly illustrated.[56] The magnificent drawings of plants in this Vienna Dioscorides are the source of many illustrated Byzantine botanical manuscripts and offer an excellent impression of knowledge at the beginning of the sixth century in the domain of plants.

Byzantine medical use of plants is apparent in hospitals, of which Byzantium had several. Especially at the time of the Comneni (eleventh to twelfth centuries), we see, for example, the Hospital of Pantocrator, which included a pharmacy, where several herbalists worked. The monk Neophytos Prodromenos (ca. 1350) compiled a lexicon probably for the needs of the hospital founded by the Serbian king Uros II Milutin in the monastery of Petra in Constantinople. It gives a long list of ingredients for therapeutic use, mainly plants but also of mineral or animal origin, all in alphabetical order. No description is given of the plant itself or its medical use, for the main interest of such lists was to give the various names of each plant. For their medical use, one had to refer to collections of specifics, or medical manuals. These botanical lexicons were the usual means by which botanical knowledge circulated in Byzantium. Most of the lexicons are anonymous, except the lexicon of Neophytos Prodromenos (fourteenth century).[57] Certain late glossaries give, alongside the Greek names of plants, their Latin, Arabic, Italian, or Turkish names. These lexicons represent more than half of the preserved

56 Hans Gerstinger, *Dioscurides: Codex Vindobonensis med. gr. 1 der Oesterreichischeischen Nationalbibliothek. Kommentarband zu der Faksimileausgabe* (Codices Selecti Phototypice Impressi, 13) (Graz: Akademie Verlag, 1970); and Alain Touwaide, "Le traité de matière médicale de Dioscoride en Italie," in *From Epidaurus to Salerno*, ed. Antje Krug (Rixensart: PACT Belgium, 1992), pp. 275–305.

57 Armand Delatte, *Anecdota Atheniensia*, Bibliothèque de la faculté de philosophie et lettres de l'Université de Liège, fasc. 36 and 88, 2 vols. (Liège: Champion, 1927–1939), vol. 2, pp. 277–302; and Brigitte Mondrain, "Un lexique botanico-médical bilingue dans le Parisinus Gr. 2510," in *Lexiques bilingues dans les domaines philosophiques et scientifiques*, ed. Jacqueline Hamesse and Danielle Jacquart (Turnhout: Brepols, 2002), pp. 123–60.

texts of Byzantine botany.[58] Plants, herbs, and vegetables intervene also in writings on dietary regime.[59]

ZOOLOGY

Zoology seems to have excited the greatest interest among Byzantines in the tenth to twelfth centuries. Emperor Constantine IX Monomachus (r. 1042–55) established a zoo in Constantinople, and Michael Attaliates (eleventh century) described an elephant and a giraffe displayed there. In the learned circle of the Princess Anna Comnena (twelfth century), Michael of Ephesus commented on the zoological writings of Aristotle.[60]

Exotic animals had excited the interest of Timothy of Gaza (fl. ca. 491–518), author of a poem in four books on the subject – a mixture of zoology and legend based on Aristotle, Plutarch, Oppian, Elian, and others. This work survives only in a very popular prose summary from the eleventh century. In Book IX of *Christian Topography,* Cosmas Indicopleustes described the animals of Ethiopia, India, and Ceylon. The *Geoponica,* the encyclopedic compilation of rural life cited earlier, also describes various animals.[61] Around 1320, the poet Manuel Philes wrote a poem on the characteristics of animals, based on Elian. Later, Demetrius Pepagomenus, in the fifteenth century, wrote treatises on hunting dogs and falconry.[62] There were also anonymous treatises on birds (especially falcons). The ancient Greek veterinary treatises devoted to the horse were compiled in four different Byzantine recensions – the most ancient dating from the ninth or tenth century. Two of these manuscripts are illustrated.[63]

Finally, the *Physiologos,* a Christian bestiary composed in the second or fourth century of our era and of little or no scientific value, describes various

58 Armand Delatte, "Le lexique de botanique du Parisinus Graecus 2419," in *Serta Leodiensia* (Bibliothèque de la Faculté de Philosophie et Lettres de l'Université de Liège, fasc. 44 (Liège: Bibliothèque de la Faculté de Philosophie et Lettres de l'Université de Liège, 1930), 59–101; and Delatte, *Anecdota Atheniensia,* vol. 2, pp. 273–450.

59 On these texts, see Andrew Dalby, *Flavours of Byzantium* (Totnes: Prospect Books, 2003).

60 Aristotle and Michael of Ephesus, *On the Movement of Animals,* trans. Anthony Preus (New York: G. Olms, 1981).

61 Stella Georgoudi, *Des chevaux et des boeufs dans le monde grec. Réalités et représentations animalières à partir des livres XVI et XVII des Géoponiques* (Paris, 1990).

62 Claudii Aeliani, *Varia Historia, Epistulae, Fragmenta,* ed. Rudolf Hercher (Leipzig: Teubner, 1866; reprinted Graz: Akademische Druck- und Verlagsanstalt, 1971), pp. 335–516, 587–99.

63 Anne-Marie Doyen-Higuet, "The Hippiatrica and Byzantine Veterinary Medecine," *Dumbarton Oaks Papers,* 38 (1984), 111–20; Anne-Marie Doyen-Higuet, *L'Epitomé de la Collection d'hippiatrie grecque,* I (Publications de l'Institut Orientaliste de Louvain, 54) (Louvain-la-Neuve: Institut Orientaliste, 2006); Stavros Lazaris, "L'illustration scientifique à Byzance: le Parisinus graecus 2244, fol. 1–74," *Études balkaniques Cahiers Pierre Belon,* 2 (1995), 161–94; Stavros Lazaris, "L'illustration des traités hippiatriques byzantins: Le *De curandis equorum morbis* d'Hiéroclès et l'Epitomé," *Medicina nei secoli,* 11 (1999), 521–46; and Anne McCabe, *A Byzantine Encyclopaedia of Horse Medecine* (Oxford Studies in Byzantium) (Oxford: Oxford University Press, 2007). I owe much information on hippiatry in Byzantium to Mrs. Anne-Marie Doyen-Higuet.

real or fantastic animals from a Christian or symbolic perspective and is clearly intended not as a zoological manual but as a collection of allegories, employing animals and their customary behaviors for entertainment and instruction.[64]

CONCLUSION

In the field of Byzantine science, so many texts remain unedited or simply ignored that one cannot claim to give a complete account of Byzantine scientific achievements. Nevertheless, we can say that the scientific efforts of Byzantium have often been underestimated by modern historians of science. Although Byzantine scholars were deeply concerned with the preservation of the priceless scientific inheritance from antiquity, they were also receptive to the progress made by their nearest neighbors, especially Arabic, Persian, or Hebrew scientists. The European Renaissance owes to the efforts of the Byzantine scholars the preservation of major scientific texts from antiquity. But they did much more than just copying the ancient inheritance in many manuscripts. They kept it alive, attempting to understand the texts exactly, making new editions, training themselves in mathematical procedures or geometrical demonstrations, and commenting on and explaining endlessly mathematical treatises, astronomical tables, and musical theories. This is especially true in astronomy. Thanks to the persistent studies of Byzantine scholars and the efforts of émigrés like Cardinal Bessarion, important Greek astronomical works, including Ptolemy's *Almagest* and Theon's *Commentaries*, were transmitted in the fifteenth century to Latin astronomers, including Peurbach and Regiomontanus, the two towering figures of fifteenth-century mathematical astronomy.

64 Francesco Sbordone, *Ricerche sulle fonti e sulla composizione del Physiologus greco* (Naples: Torella, 1936); and Dimitris Kaimakis, ed., *Der Physiologus nach der ersten Redaktion* (Maisenheim am Glan: Anton Hain, 1974).

8

SCHOOLS AND UNIVERSITIES IN MEDIEVAL LATIN SCIENCE

Michael H. Shank

During the Latin Middle Ages, educational institutions underwent extraordinary transformations that produced long-lasting consequences for the history of learning. In this story, the scientific enterprise was both a beneficiary and a driving force. Between the disappearance of the last Roman urban schools in the sixth century and the enrollment boom of the fifteenth-century universities, the transformations were fundamental.

Overall, the numbers of masters and students experienced exceptional quantitative growth, new interests and specialties proliferated, and the Latin world witnessed a radical shift in the control of schools from bishops to guilds of masters organized into universities with tightly structured stepwise curricula. Each of these transformations proved to be a boon to the scientific enterprise, culminating in the proliferation and increasing specialization of the universities, which firmly anchored the inquiry into nature in a permanent institution.

During the medieval millennium, three overlapping phases and a coda characterize the schools of the Latin world. The three phases are monastic schools, urban schools, and universities, the last of which will be the focus of this chapter. The coda pertains to the growth of urban vernacular schools in the fourteenth and fifteenth centuries. In the first phase, during the early Middle Ages (sixth to ninth centuries),[1] the learned were often either monks or the products of decentralized rural monastic schools, which incorporated into their teaching useful elements of secular Roman education, from grammar to the elementary mathematical sciences. By the late eighth and early ninth centuries, a shift was under way. Charlemagne (d. 814) was promoting literacy more systematically throughout his empire and stimulating the development of monastic and cathedral schools. In the

[1] For the crucial social and economic background, see Chris Wickham, *Framing the Middle Ages: Europe and the Mediterranean, 400–800* (Oxford: Oxford University Press, 2005); and Chris Wickham, *The Inheritance of Rome: A History of Europe from 400 to 1000* (New York: Viking, 2009).

second phase, from 1000 onward, cathedral and other urban schools began their ascendancy over monastic schools. Latin masters were becoming keenly aware of Arabic excellence in astronomy, mathematics, and medicine, among other subjects. Their eagerness for this material was beginning to change the teaching in urban schools, the apex of which occurred in the twelfth century. Growing numbers of students and masters eagerly embraced the many new Latin translations of Greek and Arabic scientific and philosophical texts. The interests and independence of these urban masters provided the ferment out of which the universities emerged. During the thirteenth century, the scientific enterprise was offered a permanent home in a new kind of institution run by its members rather than the church hierarchy. As this "European institution *par excellence*"[2] diffused widely in subsequent centuries, it would both shape the inquiry into nature and firmly anchor the belief in a regular physical realm in the culture of Latin Europe. The narrative that follows focuses on these institutional developments, often illustrated by mathematical or astronomical examples, with some attention to the organization of the university and its relations with other sectors of society. To this day, the university remains the preeminent locus of excellence in scientific teaching and research and the leading model for the organization of higher learning, now emulated on a global scale.

To appreciate the significance of this development, one must understand its antecedents. The emblematic picture of cultural decline and fall that the narrative of the Roman Empire conjures up is more apposite for the Italian peninsula than for many other locales in this vast territory. Cultural domains that Roman culture had developed highly – Latin poetry, grammar, rhetoric, and law – arguably did see an early-medieval decline, but the case of science is not so clear. Although Roman civilization tolerated and imbibed some of the natural philosophy, mathematical sciences, and medicine of Greece, it neither appropriated nor extended nor enthused about them on any scale approaching that of the medieval civilizations of Islam and Latin Christendom.[3] Cicero translated Plato's *Timaeus* and Aratus's *Phaenomena* into Latin, but he had few peers. The Greek scientific heritage was largely unavailable in Latin, in part because interest in it was modest and in part because the few Romans who cared could read Greek. Latin encyclopedias and handbooks offered digests of Greek knowledge of nature (e.g., Pliny's massive first-century *Natural History*). The impressive second-century contributions of Ptolemy to the mathematical sciences and of Galen to medical theory were written, and remained, in Greek.[4]

[2] Walter Rüegg, "Forward" and "Themes," in *A History of the University in Europe*, vol. 1: *Universities in the Middle Ages*, ed. Hilde De Ridder-Symoens (Cambridge: Cambridge University Press, 1992), pp. xix–xxvii and 3–34, respectively, especially pp. xix, 8.

[3] Virgil, *Aeneid*, bk. VI, lines 848–53; and Edward Grant, *A History of Natural Philosophy from Antiquity to the Nineteenth Century* (Cambridge: Cambridge University Press, 2007), pp. 97–8.

[4] Roger French, *Natural History in Antiquity* (London: Routledge, 1994); William Stahl, *Roman Science: Origins, Development, and Influence to the Later Middle Ages* (Madison: University of Wisconsin Press, 1962), pp. 101ff; and Anthony Corbeill, "Education in the Roman Republic: Creating Traditions,"

Between the fourth and sixth centuries, the works of three Latin writers from North Africa left a deep imprint on the pedagogy of the liberal arts in subsequent centuries. Augustine's *De ordine* (*On Order*), Martianus Capella's *Marriage of Philology and Mercury*, and Macrobius's *Commentary on the Dream of Scipio* all valued the "trivium" of verbal disciplines (grammar, rhetoric, and logic). Like most scholars in the Platonic tradition, however, they gave pride of place to the four "nontrivial" mathematical sciences (arithmetic, geometry, musical proportions, and astronomy), later named the "quadrivium."[5]

Under Ostrogothic rule and patronage in late-fifth- and early-sixth-century Ravenna, the Roman statesman Boethius attempted to translate into Latin key Greek scientific and philosophical works. Under Rome, these domains had remained largely Greek in content and language. Since southern Italy and Sicily were former Greek colonies, the Greek language and contacts with Byzantium did not disappear completely in the Western Empire. A sixth-century document from Ravenna refers to a "physician from the Greek school," probably the source of some translations of Greek medical works into a Hellenized Latin. Nevertheless, only a small fraction of Greek medicine, mathematical science, and natural philosophy was available in Latin, and few people read the remainder.[6] It is easy to generate, as an older historiography has done, a precipitous early-medieval scientific decline by using Greek works from Hellenistic Alexandria or Athens as the standard. The situation looks very different if the baseline consists of the modest number of Latin works available in Rome at its height.

FROM BENEDICTINE EXPANSION TO THE URBAN SCHOOLS

After the Constantinian era, much of the western half of the Roman Empire was effectively in a postcolonial situation (see the Introduction). Its peoples and leaders began to chart their own courses piecemeal, melding their diverse traditions, their ingenuity, and whatever Roman cultural remnants

in *Education in Greek and Roman Antiquity*, ed. Yun Lee Too (Leiden: Brill, 2001), pp. 261–87, especially pp. 261, 267, 269. Some Greek medicine was available in Latin; see Monica H. Green, ed. and trans., *The Trotula: An English Translation of the Medieval Compendium of Women's Medicine* (Philadelphia: University of Pennsylvania Press, 2002), pp. 14–17.

5 Mark Joyal, Iain McDougall, and J. C. Yardley, eds., *Greek and Roman Education: A Sourcebook* (London: Routledge, 2009), pp. 265–6.

6 Pierre Courcelle, *Les Lettres grecques en Occident de Macrobe à Cassiodore* (Paris: Broccard, 1948), pp. 74–8; Innocenzo Mazzini and Nicoletta Palmieri, "L'École médicale de Ravenne," in *Les Écoles médicales à Rome*, ed. Philippe Mudry and Jackie Pigeaud (Geneva: Droz, 1991), pp. 286–93, especially p. 286; Danielle Jacquart, "Les traducteurs du XIe siècle et le latin médical antique," in *Le Latin médical, La constitution d'un langage scientifique: Réalités et langage de la médecine dans le monde romain*, ed. Guy Sabbah (Saint-Étienne: Université de Saint-Étienne, 1991), pp. 417–24; and Pierre Riché, *Écoles et enseignement dans le Haut Moyen Âge: Fin du Ve siècle – milieu du XIe siècle* (Paris: Picard, 1989), pp. 13–16.

they found useful, including the Latin literacy cultivated by the Church of Rome.

By the sixth century, the urban schools of the Roman model, which taught primarily grammar, poetry, and rhetoric, were disappearing. Between the fifth and tenth centuries, the new communal Benedictine monasticism proliferated rapidly beyond the mother abbey of Monte Cassino to Iberia, Ireland, the valleys of the Rhine and Danube, and beyond. The Benedictine Rule expected monks to be literate, for it prescribed daily reading. Monastic schools thus became the most important loci of early-medieval learning. The pupils, novices and some laymen, were often taught by single masters who at first were themselves monks. Boethius, a Christian layman, had been an advocate of the seven liberal arts, which the Benedictine schools would eventually embrace as aids to their religious life. This infrastructure of learning did more than preserve and expand Latin literacy in the early Middle Ages. It also introduced students to the mathematical sciences of the quadrivium, justified by their utility (for chanting the Psalms, accounting, or computing the date of Easter), and it helped to spawn cathedral schools. This instruction also kept alive the notion of an autonomous and regular physical realm most clearly expressed in the treatises *On the Nature of Things* by Isidore of Seville (sixth century) and Bede (early eighth century). Although modest in size and isolated, the early monastic schools were not completely on their own. Thanks to a network of visitation, traveling monks borrowed books, exchanged medical recipes, and carried medicinal seeds to far-flung locations. These contacts stimulated the copying of manuscripts in scriptoria, which expanded monastic libraries with secular as well as religious works and medicinal gardens with both local and exotic plants.[7]

Charlemagne (768–814) would build on and reinforce this educational foundation. From the late eighth century, he assembled a vast empire extending from Carinthia and Bohemia to northern Italy, the Pyrenees, and the North Atlantic. Throughout it, he instituted sweeping administrative and cultural reforms. Emulating Emperor Augustus's responsibility for Roman religion, Charlemagne intertwined his political and religious duties. His obsession with enforcing a uniform Latin liturgy among the many – sometimes forced – converts to Christianity in his realm largely motivated his educational policies. To implement it, he tapped the master of the York Minster school. At the court in Aachen/Aix-la-Chapelle, Alcuin (ca. 730–804) created a palace school that included boys and girls. Charlemagne's program, which later Carolingians also promoted, was to establish in every

[7] John Contreni, "Learning in the Early Middle Ages," in his *Carolingian Learning, Masters, and Manuscripts* (Hampshire: Variorum, 1992), pp. 1–21, especially pp. 10–11; M. M. Hildebrandt, *The External School in Carolingian Society* (Leiden: Brill, 1992), pp. 108–29; James J. O'Donnell, *Cassiodorus* (Berkeley: University of California, 1979), chap. 6; Riché, *Écoles et enseignement dans le Haut Moyen Âge*, pp. 25–41; and Barbara Obrist, *La cosmologie médiévale: Textes et images, I: Les fondements* (Florence: Edizioni del Galluzzo, 2004), pp. 38–40.

monastery and diocese a school with instruction in Latin grammar, writing, the Psalms, chant, musical notation, and *computus* (methods of computing the date of Easter from solar and lunar cycles – see the following chapters in this volume: McCluskey, Chapter 11; Eastwood, Chapter 12; and North, Chapter 19). Literate priests were in turn to teach the children in their villages. Although religious goals shaped their priorities, the Carolingian monastic and cathedral schools also taught classical secular works valued for their Latinity or learning. The association of these texts with the culture of imperial Rome was part of their appeal to Charlemagne, for they offered much needed political legitimation to the "Carolingian rulers, oath-breakers and usurpers."[8]

Recent scholarship illustrates new Carolingian trends in the study of nature. Although the earliest computists had emphasized practical numerical rules to compute feast days, their Carolingian successors increasingly transcended utilitarianism. Alcuin himself wrote the first Latin treatise of mathematical problems. Charlemagne's personal encouragement of the liberal arts led some computists to pursue greater precision in the measurement of time and interests irrelevant to the calendar, such as cosmology and planetary theory. Significantly, the court-sponsored *computus* collections of the early ninth century show a better understanding of the rudiments of Greek mathematical astronomy than did their Roman sources (Pliny, Martianus Capella, and Macrobius) and even kept alive a cosmology in which Mercury and Venus revolved around the Sun.[9] In addition, Carolingian records reveal ongoing exchanges of Mediterranean and Eastern plants, herbs, and spices.[10] Both directly and indirectly, Charlemagne's institutional initiatives raised the level of scientific discourse, expanded libraries, and promoted access to Mediterranean botanical specimens north of the Alps. His combination of personal patronage, institutional standardization, and ecclesiastical interference came to fruition in the extensive cultural network among the great

[8] Giles Brown, "Introduction: The Carolingian Renaissance," in *Carolingian Culture: Emulation and Innovation*, ed. Rosamond McKitterick (Cambridge: Cambridge University Press, 1994), pp. 1–51, especially pp. 1, 19–25, 28–45 (quotation at p. 45); John Marenbon, "Carolingian Thought," in McKitterick, *Carolingian Culture*, pp. 171–92; Riché, *Écoles et enseignement dans le Haut Moyen Âge*, pp. 349, 353; and Pierre Riché and Jacques Verger, *Des nains sur des épaules de géants: Maîtres et élèves au Moyen Âge* (Paris: Tallandier, 2006), p. 33.

[9] Bruce Eastwood, *Ordering the Heavens: Roman Astronomy and Cosmology in the Carolingian Renaissance* (Leiden: Brill, 2007), pp. 10–13, passim; David Singmaster, "The History of Some of Alcuin's Propositions"; and Harald Gropp, "*Propositio de lupo et capra et fasciculo cauli* – On the History of River-Crossing Problems," in *Charlemagne and His Heritage: 1200 Years of Civilization and Science in Europe*, ed. P. L. Butzer, H. Th. Jongenand, and W. Oberschelp (Turnhout: Brepols, 1998), pp. 11–29, 31–41 at p. 31; Bruce Eastwood and Gerd Grasshoff, "Planetary Diagrams for Roman Astronomy in Medieval Europe, ca. 800–1500," *Transactions of the American Philosophical Society*, 94, pt. 3 (2004), 1–158 at pp. 1–2; and Bruce Eastwood, "The Astronomies of Pliny, Martianus Capella, and Isidore of Seville in the Carolingian World," in *Science in Western and Eastern Civilization in Carolingian Times*, ed. P. L. Butzer and D. Lohrmann (Basel: Birkhäuser, 1993), pp. 161–80.

[10] Linda Voigts, "Anglo-Saxon Plant Remedies and the Anglo-Saxons," *Isis*, 70 (1979), 250–68, especially p. 260.

monastic schools of the ninth and tenth centuries, such as Bobbio (northern Italy), Fulda (western Germany), and St. Gall (northern Switzerland).[11]

Most monastic schools had a single master, but the larger ones had several, organized into a council (Fulda) or supervised by a headmaster. Once canon law restricted them to monks, however, the monastic schools lost their primacy. Monasteries nevertheless remained ongoing sites of scientific, medical, and technological activity into the late Middle Ages. The convent teaching of *magistra* Hildegard of Bingen surely exposed nuns to natural history and medicine in the twelfth century. Despite several conciliar decrees to the contrary, monk-physicians surfaced throughout the period. Finally, some monasteries remained active at the interface between technology and literacy. Abbot Richard of Wallingford made sure that St. Albans supported his fourteenth-century astronomical clockwork and writings. A century later, the Benedictine abbey of Subiaco hosted the first printers in Italy.[12]

For nonmonks around the millennium, the institutional locus of Latin education, including the study of nature, shifted to urban schools. In this genus, the cathedral school was at first the most prominent species, to the development of which individual Benedictine masters nevertheless contributed significantly. Here, too, single masters were at first the norm, but several locales became homes to multiple masters, drawing larger numbers of students.

By the tenth century, the Benedictine Gerbert of Aurillac's exposure to Arabic mathematics and astronomy in Catalonia stimulated his pedagogical innovations as master of the Rheims cathedral school. Imported from Muslim Spain, devices like the armillary sphere, the astrolabe, and the abacus supplemented the texts of the quadrivium with tactile teaching. Gerbert already embodied the later type of the famous master who stretches the cathedral-school curriculum beyond its original purpose and attracts students far outside the local diocese. When such individuals traveled, so did ideas, techniques, books, artifacts, and plants.[13] Already emphasized

[11] Rosamond McKitterick, "Script and Book Production," in McKitterick, *Carolingian Culture*, pp. 221–47, especially pp. 240–4; and Rosamond McKitterick, *The Carolingians and the Written Word* (Cambridge: Cambridge University Press, 1989), pp. 220–7.

[12] Hildebrandt, *External School in Carolingian Society*, pp. 72–3, 139–41; George B. Fowler, *Intellectual Interests of Engelbert of Admont* (New York: Columbia University Press, 1947), chap. 4; John D. North, *Richard of Wallingford: An Edition of His Writings, with Introduction, English Translation, and Commentary*, 3 vols. (Oxford: Oxford University Press, 1976); R. Graham, "The Intellectual Influence of English Monasticism between the Tenth and the Twelfth Centuries," *Transactions of the Royal Historical Society*, n.s. 17 (1903), 23–65, especially p. 54; and Nancy Siraisi, *Medieval and Early Renaissance Medicine: An Introduction to Knowledge and Practice* (Chicago: University of Chicago Press, 1990), pp. 43–4, 50.

[13] Uta Lindgren, *Gerbert von Aurillac und das Quadrivium* (Wiesbaden: Steiner, 1976); Marco Zuccato, "Gerbert of Aurillac and a Tenth-Century Jewish Channel on the Transmission of Arabic Science to the West," *Speculum*, 20 (2005), 742–63; Guy Beaujouan, "The Transformation of the Quadrivium," in *Renaissance and Renewal in the Twelfth Century*, ed. Robert Benson and Giles Constable (Cambridge, Mass.: Harvard University Press, 1982), pp. 463–87, especially p. 464; and Voigts, "Anglo-Saxon Plant Remedies and the Anglo-Saxons," p. 266.

in Charlemagne's prescriptions, the ethical component of this education devoted increasing attention to nature, from which it often drew moral lessons, relying on treatises from late antiquity on music and the cosmos (e.g., Plato's *Timaeus*).[14]

Gerbert's student Richer of Rheims (d. ca. 1010) testifies to additional ferment. A Benedictine who had read Hippocrates at the cathedral school of Chartres, Richer recounts a showdown at the French court between two physicians trained respectively in France and Salerno (south of Naples). The French doctor bests his Salernitan rival in both the disputation (theory) and the dramatic poison-antidote test (practice).[15]

The partisanship of this tall tale highlights the growing competition among masters and the early reputation of distant Salerno for medical learning. In southern Italy, the Greek language and contacts with Byzantium had not disappeared. It was there that Constantine the African (d. before 1098–1099), a multilingual Carthaginian educated in Cairo, would translate Greek, Arabic, and Persian medical literature from Arabic into Latin and become a Benedictine at Monte Cassino. Along the Salerno–Monte Cassino axis, Latin medicine complemented its status as a practical art with that of a theoretical discipline interacting with natural philosophy, to the lasting stimulation of the latter. The town of Salerno seems to have hosted a multiplicity of single-master schools that centered on medicine and related subjects. However mysterious the early structure of its teaching, Salerno offered an unusual intellectual and social setting. At its twelfth-century apex, the teaching masters in the city included learned women who specialized in, wrote about, and practiced medicine. Beyond developing successful pedagogical compendia of medicine (e.g., the *Articella*), the wide-ranging interests of the Salernitan masters in the twelfth century engaged with Aristotle's recently available natural philosophy and surfaced in the widely diffused thirteenth-century *Salernitan Questions*. Even though Salerno peaked in the twelfth century and did not gel into a university, Salernitan texts remained popular in university curricula into the sixteenth century.[16]

14 Léon Maître, *Les Écoles épiscopales et monastiques de l'Occident depuis Charlemagne jusqu' à Philippe-Auguste (768–1180)* (Paris: Dumoulin, 1866), pp. 186–7; C. Stephen Jaeger, *The Envy of Angels: Cathedrals Schools and Social Ideals in Medieval Europe, 950–1200* (Philadelphia: University of Pennsylvania Press, 1995), pp. 25, 164–79; and Nicholas Orme, *Medieval Schools from Roman Britain to Renaissance England* (New Haven, Conn.: Yale University Press, 2006), chaps. 1–5, especially p. 167.

15 Paul Oskar Kristeller, *Studi sulla scuola medica salernitana* (Naples: Istituto Italiano per gli Studi Filosofici, 1986), pp. 19–21; and Jason Glenn, *Politics and History in the Tenth Century: The Work and World of Richer of Reims* (Cambridge: Cambridge University Press, 2004), pp. 60–9.

16 Green, *Trotula*, pp. 6–13; Claudio Leonardi, "Intellectual Life," in *The New Cambridge Medieval History*, vol. 3: *c. 900–c. 1024*, ed. Timothy Reuter (Cambridge: Cambridge University Press, 1999), pp. 186–211, especially pp. 186–92; Charles Homer Haskins, *Studies in the History of Mediaeval Science* (Cambridge, Mass.: Harvard University Press, 1927), pp. 141ff; Mark Jordan, "The Construction of a Philosophical Medicine: Exegesis and Argument in Salernitan Teaching on the Soul," *Osiris*, 2nd ser., 6 (1990), 42–61, especially p. 61; Francis Newton, "Constantine the African and Monte Cassino: New Elements of the Text of the *Isagoge*," in *Constantine the African and ʿAlī ibn ʿAbbās al-Maǧūsī: The Pantegni and Related Texts*, ed. Charles Burnett and Danielle Jacquart (Leiden: Brill,

THE RISE OF GUILDS OF MASTERS AND STUDENTS

The era between Gerbert and 1150 witnessed powerful social and economic transformations intertwined with political ones: population growth, the rebirth and expansion of older towns and the emergence of many new ones, economic growth, and new trading practices. For the institutional aspects of science, however, the most fundamental changes were surely those in law. The days when Charlemagne could tell "his" bishops what to do were now past. The papacy and the empire fought pitched legal battles that rippled through politics and education. At all levels of society, law became a central concern; canon, Roman, Germanic, and many forms of customary law competed with and borrowed from one another. Lawyer-popes increasingly asserted their power, with notable repercussions for both the growth of bureaucracies and the systematization of canon law. These developments profoundly shaped institutions, systematic thought, and logic. Canon lawyers pioneered the organization of law into a deductive system, arranging disparate statutes under general principles, from which both these statutes and new laws could be deduced. This trend toward systematization undergirded the papacy's claims for the centralization of its power, which secular rulers countered with legal systems of their own.[17]

This competition on the terrain of legal principle left a deep mark on politics and culture. In the growing cities, corporate initiatives such as "communes," corporations, or guilds (Latin *communio*, *societas*, and *universitas*, respectively) extended from city politics to the formation of urban trades and professions. From the eleventh century on, the citizens of towns such as Bologna organized themselves into "communes" and took oaths to fight for local political privileges and their own laws, usually against such nearby embodiments of power as the local nobility and the bishop. The masters of various trades likewise formed guilds, or *universitates*, that, within their domain, were literally autonomous (i.e., governed by their own laws). The control that they exercised included setting standards of apprenticeship, competence, remuneration, and sound practice.

These corporate developments were the most salient social and political factors behind the expansion of the scientific enterprise in medieval Europe. The guild movement spawned a new class of autonomous educational corporations. Adapting this model for their own purposes, masters and students

1994), pp. 16–47, especially pp. 17–26; Brian Lawn, *The Salernitan Questions: An Introduction to the History of Medieval and Renaissance Problem Literature* (Oxford: Oxford University Press, 1963), pp. xii–xiii; Danielle Jacquart, "Aristotelian Thought in Salerno," in *A History of Twelfth-Century Western Philosophy*, ed. Peter Dronke (Cambridge: Cambridge University Press, 1988), pp. 407–28; and Siraisi, *Medieval and Early Renaissance Medicine*, pp. 13, 57–8.

17 Brian Tierney, *The Crisis of Church and State, 1050–1300* (Englewood Cliffs, N.J.: Prentice–Hall, 1964), pp. 97–8; and Harold Berman, *Law and Revolution: The Formation of the Western Legal Tradition* (Cambridge, Mass.: Harvard University Press, 1983), chap. 3.

formed "universities" (guilds) that governed new comprehensive schools.[18] These corporations would become the primary homes of natural philosophy, the mathematical sciences, and medicine in the late Latin Middle Ages.

The earliest universities emerged from modest learning communities that became increasingly more organized for idiosyncratic local reasons. Bologna was a center of legal teaching with many foreign students, Paris a city with many schools renowned for their teaching of the liberal arts and theology, Oxford a provincial town distinguished by the presence of royal and ecclesiastical courts.[19] These early models nicely illustrate how, from disparate circumstances, students and masters, respectively, shaped and gained control of curricula, examinations, and degrees, with lasting consequences for the scientific enterprise.

By the eleventh century, Bologna had become the most reputed center for the teaching of Roman law. Its masters attracted many foreign students who had no legal protection in the city. In the twelfth century, the Bolognese law students first secured imperial guarantees of safe travel and eventually organized themselves into several *universitates*, each with its own rector. These guilds protected the rights of students locally and regulated most aspects of teaching, including the duties and compensation of the masters. The latter had their own, obviously weak, *universitas*, which controlled only the awarding of degrees.[20] In the early thirteenth century, all of these guilds secured official recognition and jointly governed the University of Bologna independently of ecclesiastical authority. By the late thirteenth century, the payoff for the study of nature was clear. "Arts" and medicine merged into one faculty and became a major component of Bolognese teaching.

In contrast to Italy, the early-medieval church north of the Alps had effectively held a monopoly on education. Yet masters at Paris and Oxford eventually also achieved an autonomy analogous to that of the Italian universities, but with different trajectories and emphases. Although controlled by local bishops to educate their clergy, some early cathedral schools in northern France (Rheims, Chartres, and Paris, among others) had developed a constituency far wider than the priesthood, as the examples of Gerbert and Richer show. By the twelfth century, they had evolved from small, loosely institutionalized diocese-oriented communities into centers of learning

[18] John Baldwin, *The Scholastic Culture of the Middle Ages, 1000–1300* (Lexington, Mass.: D. C. Heath, 1971), pp. 6, 22–3, pace Makdisi's formal case for the origins of the university in the Islamic *madrasa* in George Makdisi, *The Rise of Humanism in Classical Islam and the Christian West, with Special Reference to Scholasticism* (Edinburgh: Edinburgh University Press, 1990).

[19] Rüegg, "Themes," in De Ridder-Symoens, *History of the University in Europe*, vol. 1, pp. 3–34, especially pp. 12–13; Frans van Liere, "The Study of Canon Law and the Eclipse of the Lincoln Schools, 1175–1225," *History of Universities*, 18 (2003), 1–13, especially pp. 9–10; and Vicente Cantarino, "Medieval Spanish Institutions of Learning: A Reappraisal," *Revue des Études Islamiques*, 44 (1976), 217–28, especially p. 218.

[20] Hastings Rashdall, *The Universities of Europe in the Middle Ages*, 2nd ed., 3 vols. (Oxford: Oxford University Press, 1936), vol. 1, pp. 143, 146, 154, 234.

overflowing with lay scholars from all over Europe and the British Isles. The foreign students who flocked to hear masters famous in the liberal arts and/or theology were committed to neither the local diocese nor the priesthood. Many were "adventurers seeking rare and difficult knowledge which would lead to personal advancement or the perfecting of a personal gift."[21] Their educational "grand tours" took them to a sequence of towns, as Peter Abelard, John of Salisbury, and many others had done.

By the time of Adelard of Bath, leading cathedral schoolmasters were advocates for *physica* (natural philosophy), the inquiry into the causes of natural phenomena. Masters like William of Conches argued that Genesis offered a woefully incomplete account of the origins of the universe. His colleagues at Chartres and other cathedral schools embraced Plato's *Timaeus*, with its picture of the cosmos as an organic concatenation of causes extending from the highest heavens to the earth. Bernard of Chartres captured the excitement of his contemporaries in his famous quip about their being "dwarves on the shoulders of giants. We thus see more and farther than they did."[22] Seeing more and farther, especially in their search for natural causes, were the new goals of learning, effectively ends in themselves. The masters' interest in understanding nature proved to be long-lived. It not only motivated the scientific translation movement from Arabic into Latin and ensured its success but also shaped the curricula of the emerging universities.

Alongside the established cathedral school of Notre Dame, various "free schools" emerged in early-twelfth-century Paris. They often consisted of little more than eager students, a meeting place, and a charismatic master who excelled in disputation. When Peter Abelard became known as a fearsome logician by besting his own masters in philosophical combat, he drew many students away from Notre Dame to the left bank of the Seine, where he began to teach, ostensibly outside the bishop's educational jurisdiction. The coexistence of established institutions with such upstarts made Paris a particularly lively environment, which attracted ever more students and masters. Around mid-century, Master Gilbert of Poitiers reportedly held forth in Paris before almost three hundred people. The bishop of Paris tried, through his chancellor, to control these centrifugal activities, most notably by granting to (or withholding from) masters the traditional license to teach.[23]

[21] Richard Southern, "The Schools of Paris and the School of Chartres," in *Renaissance and Renewal in the Twelfth Century*, ed. Robert Benson and Giles Constable (Cambridge, Mass.: Harvard University Press, 1982), pp. 113–37, especially pp. 115, 124.

[22] Tullio Gregory, "La nouvelle idée de nature et de savoir scientifique au XIIe siècle," in *The Cultural Context of Medieval Learning*, ed. John E. Murdoch and Edith D. Sylla (Boston Studies in the Philosophy of Science, 26) (Boston: Reidel, 1975), pp. 192–218, especially pp. 198–9; and Jacques Verger, "À propos de la naissance de l'université de Paris: contexte social, enjeu politique, portée intellectuelle," in his *Les Universités françaises au moyen âge* (Education and Society in the Middle Ages and Renaissance, 7) (Leiden: Brill, 1995), pp. 17–18.

[23] John Baldwin, "Masters at Paris from 1179 to 1215: A Social Perspective," in Benson and Constable, *Renaissance and Renewal in the Twelfth Century*, pp. 138–72, especially pp. 138–9; and Verger, "À propos de la naissance de l'université de Paris," p. 18.

THE UNIVERSITY AS GUILD

Within a few generations, the bishop had lost the power of licensing in all but name. In the intervening years, the masters of the various Parisian schools had become keenly aware of their common concerns. By the 1170s, they were using the law to secure their collective privileges (*privi/legium* = a body of private law) and to unite in a guild. As their late-twelfth-century initiation ceremonies show, their guild was already functioning two generations before officialdom recognized its legal status.[24] By the mid-thirteenth century, the guild (*universitas*) of masters of arts controlled its own teaching and degrees. Theology, medicine, and law also became highly structured and quasi-professionalized in that master-practitioners certified competence in each field.

Bologna and Paris were the two oldest centers of learning, but Oxford came into its own in the late twelfth and early thirteenth centuries, stimulated by the presence of ecclesiastical and royal courts and the teaching of law by a Bologna-trained master. A crisis in 1209 precipitated a student strike that eventually led to both a new university in Cambridge and new academic privileges in Oxford. By the late thirteenth century, the guild of Oxford masters elected its own chancellor, whom the bishop of Lincoln now appointed pro forma.[25]

As Bologna, Paris, and Oxford illustrate, guilds of masters and/or students in very different locations achieved an autonomy that resulted in a new kind of educational institution: a school with universally valid privileges (*studium generale*) and usually incorporating multiple disciplines or faculties.[26] Although we now restrict the term "university" to an educational institution, it originally designated its corporate legal status, which gave control of its teaching and certification to its members. Neither the cathedral schools nor Salerno disappeared immediately; as centers of education, however, they clearly lost out to the new universities. The cathedral school of Lincoln bowed to the new university in Oxford, that of Notre Dame to the *studium generale* of Paris. Even Salerno ceded its primacy in medicine to the universities of Bologna, Montpellier, and Padua.[27]

However modestly and chaotically this institutional shift began, it transformed the history of education and left an indelible mark on world history.

[24] Nancy Spatz, "Evidence of Inception Ceremonies in the Twelfth-Century Schools of Paris," *History of Universities*, 13 (1994), 3–19, especially pp. 7–10; and Alexander Murray, *Reason and Society in the Middle Ages* (Oxford: Clarendon Press, 1985), pp. 283–7.

[25] Richard Southern, "From Schools to University," in *A History of Oxford University*, ed. Jeremy Catto, 8 vols. (Oxford: Oxford University Press, 1984–1994), vol. 1, pp. 1–36 at pp. 25–7, 34–5.

[26] Jacques Verger, "Patterns," in De Ridder-Symoens, *History of the University in Europe*, vol. 1, pp. 35–68, especially pp. 35–7.

[27] Frans van Liere, "The Study of Canon Law and the Eclipse of the Lincoln Schools, 1175–1225," *History of Universities*, 18 (2003), 1–13, especially pp. 5–8; and Baldwin, "Masters at Paris from 1179 to 1215," pp. 138–72.

The fundamental fact about the early universities is their origin in struggles for legal autonomy, which led to corporations of masters or students acting under their own jurisdiction. Since they now controlled curricula, set up exams, and established requirements for degrees, their emphases were not external impositions.[28] Oxford witnessed a smooth transition of its teaching to the newly translated natural philosophy of Aristotle. Matters went otherwise in Paris. Since "conflicts are inherent in social roles," it is not surprising that changing roles sharpened tensions.[29] Paris witnessed the testiest clashes of will between the masters' guild and the bishop, who eventually lost control of most local teaching.

The victory of the masters of arts set a precedent for later universities, with profound consequences for the scientific enterprise. Newly empowered, the masters enacted a curriculum that reflected their interests, which since the early twelfth century had become increasingly oriented toward nature. Well before the universities, scholars were cultivating theoretical and practical medicine at Salerno, studying logic, and embracing the natural philosophy of Plato's *Timaeus* in the cathedral schools of Northern Europe. Thus whetted, their appetite was ready for the cornucopia of scientific materials freshly translated from the Arabic, from Euclid through the vast corpus of Aristotelian natural philosophy and logic to mathematical astronomy, astrology, optics, and medicine.

Without the masters' craving for them, these texts would have been ignored. As it happened, most of the new Greek and Arabic scientific material had become available in Latin as the masters were beginning to flex their institutional muscles and to protect their interests with legal privileges. Once the first universities formally came into being around 1200, the masters had the wherewithal to protect their intellectual interests as well. Accordingly, the newly translated natural philosophy and logic of Aristotle and his Islamic commentators eventually not only dominated the curricula of arts faculties but also profoundly affected the higher faculties, especially medicine and theology.

Thanks to their university privileges, the masters' appropriation of this material gave to natural philosophy, medicine, and the mathematical sciences a secure and enduring institutional home. Alone, the earliest universities would have made a modest impact on European culture, but they did not remain alone. Some sixty were in operation by 1500, making their influence pervasive and effectively permanent.[30] Despite downturns and adjustments, the marriage of the university to the scientific enterprise has endured to the present day. Later changes in the content and practices of the older

[28] Antony Black, *Guilds and Civil Society in European Political Thought from the Twelfth Century to the Present* (London: Methuen, 1984), p. 50.

[29] Rüegg, "Foreword," p. xxvi.

[30] Verger, "Patterns," pp. 62–5.

sciences and the emergence of new ones should not obscure the fundamental significance of this long institutional association.

THE STUDY OF NATURE AND THE FACULTY STRUCTURE

It is remarkable that the universities eventually joined in one institution natural philosophy, medicine, law, and theology. None of these endeavors belonged to the classical seven liberal arts. The universities nevertheless made natural philosophy, however ill-fitting, a central feature of their "arts" curriculum, and they turned into higher faculties the three disciplines of law, theology, and medicine, each of which had recently undergone a major transformation. Each, including theology, was now controlled by its practitioners, who shared a fundamental methodological commitment to reasoned argumentation, even if they sometimes disagreed vigorously about the premises and limitations of their colleagues' arguments.[31]

Each faculty organized the various aspects of its teaching in a stepwise curriculum. At its apex was the certification of competence by awarding degrees (*gradus* = step, degree) to students who had satisfactorily heard required lectures and engaged in prescribed disputations. This progression proved highly significant, for it diffused among the learned elite an outlook that stressed explanations of natural phenomena by using ordinary causes (philosophical naturalism). The most foundational faculty ("arts and medicine" in Italy, "arts" in Paris and elsewhere) was usually the largest, comprising 80–90 percent of the academic population north of the Alps.[32] A large proportion of its basic curriculum dealt with knowledge of the natural world, typically including a wide spectrum of natural philosophy, the logical tools necessary to pursue it, and a smattering of the mathematical sciences.[33] In the many universities modeled on Paris, a command of these subjects certified by a master of arts degree was usually required for study in law, medicine, or theology.

When they moved to the higher faculty, masters did not abandon their interests or the naturalistic assumptions of their earlier training. Indeed, some could not afford to do so. To finance their advanced studies, the new masters

[31] Edward Grant, "Science and the Medieval University," in *Rebirth, Reform, and Resilience: Universities in Transition, 1300–1700*, ed. James Kittelson and Pamela Transue (Columbus: Ohio State University Press, 1984), pp. 68–102 at p. 84; and J. M. M. H. Thijssen, *Censure and Heresy at the University of Paris, 1200–1400* (Philadelphia: University of Pennsylvania Press, 1998), p. 113.

[32] Katharine Park, *Doctors and Medicine in Early Renaissance Florence* (Princeton, N.J.: Princeton University Press, 1985), pp. 51–3; Rainer Schwinges, *Deutsche Universitätsbesucher im 14. und 15. Jahrhundert: Studien zur Sozialgeschichte des alten Reiches* (Stuttgart: Steiner Verlag, 1986), pp. 467–8; and Iolanda Ventura, "Quaestiones and Encyclopedias: Some Aspects of the Late-Medieval Reception of the Pseudo-Aristotelian Problemata," in *Schooling and Society: The Ordering and Reordering of Knowledge in the Western Middle Ages*, ed. Alasdair MacDonald and Michael W. Twomey (Leuven: Peeters, 2004), pp. 23–42, especially pp. 24, 39–42.

[33] Grant, "Science and the Medieval University," pp. 68–71.

of arts often continued to teach in the arts faculty, sometimes for a decade or more. Far from atrophying, their skills in logic and natural philosophy remained sharp or improved, shaping in turn their disputations in medicine or theology. This institutional arrangement helps to explain the vigorous interaction between the curriculum of the lower faculties and the content of the higher ones.[34] Despite the complaints of traditionalists, budding theologians also brought much natural philosophy, logic, and mathematics into their work, most obviously when discussing creation but also when using the specialized logical tools developed in natural philosophical contexts.[35] Even friars, who could study theology without an arts degree, encountered the new scientific material in the schools of their orders.[36]

Despite their obsolete name, the faculties of arts in Northern Europe and of arts and medicine in Italy were no longer tied to the classical schema. As Aquinas noted, "the seven liberal arts do not adequately divide theoretical philosophy." The old terminology of the liberal arts nevertheless lent the new curriculum its traditional justification for studying the writings of pagans and infidels: their utility for understanding the Bible (later theology) and for defending the faith.[37] The new faculties of arts made a similar point by choosing as their patron St. Katherine of Alexandria, who reputedly converted fifty pagan sages to Christianity by rational argumentation, thanks – presumably – to her mastery of their views.

Nevertheless, the extensive new library revealed by the translation movement dramatically changed the content and quantity of secular learning, and the university structure strained the old justifications. The masters now grappled not only with mathematical and logical writings vastly more sophisticated than those of the Roman liberal arts but also with Aristotle's impressive philosophy (natural, moral, and "first" philosophy or metaphysics) and with the commentaries of Averroes, the great twelfth-century Spanish Muslim philosopher. Theologians could continue to claim that the verbal disciplines

34 William J. Courtenay, "Parisian Theology, 1362–1377," in *Philosophie und Theologie des ausgehenden Mittelalters: Marsilius von Inghen und das Denken seiner Zeit*, ed. Maarten J. F. M. Hoenen and Paul J. J. M. Bakker (Leiden: Brill, 2000), pp. 3–19, especially p. 3.

35 John E. Murdoch, "From Social into Intellectual Factors: The Unitary Character of Medieval Learning," in Murdoch and Sylla, *Cultural Context of Medieval Learning*, pp. 271–348, especially pp. 276–7.

36 John North, "The Quadrivium," in De Ridder-Symoens, *History of the University in Europe*, vol. 1, pp. 337–59, especially p. 354; Alfonso Maierù, *University Training in Medieval Europe*, trans. D. N. Pryds (Leiden: Brill, 1994), pp. 12–14; Grant, "Science and the Medieval University," pp. 84–5; and James A. Weisheipl, "Ockham and the Mertonians," in Catto, *History of Oxford University*, vol. 1, pp. 607–58, especially pp. 614–15.

37 James Weisheipl, "The Place of the Liberal Arts in the University Curriculum during the XIIIth and XIVth Centuries," in *Arts libéraux et philosophie au moyen âge* (Actes du Quatrième Congrès International de Philosophie Médiévale) (Montreal: Institut d'Études Médiévales; Paris: Vrin, 1969), p. 209; Armand Maurer, *St. Thomas Aquinas: Division and Methods of the Sciences* (Toronto: Pontifical Institute of Mediaeval Studies, 1963), p. 11; Grant, "Science and the Medieval University," pp. 71–5; Murray, *Reason and Society in the Middle Ages*, pp. 287–9; and Pearl Kibre and Nancy Siraisi, "The Institutional Setting: The Universities," in *Science in the Middle Ages*, ed. David C. Lindberg (Chicago: University of Chicago Press, 1978), pp. 120–44, especially p. 139.

and elementary mathematical sciences were important for theology; this was not so obvious for the recently translated natural philosophy and the more advanced mathematical sciences. In any case, most masters of arts did not go on in theology, a minority field despite its prestige. From the thirteenth century on, the masters (following Aristotle) recast their domain as "theoretical (or speculative) philosophy" and thought of themselves as philosophers, with the duty of teaching the unvarnished Aristotle, whatever the theologians might think.[38] For those in the arts faculties who yearned to know the causes operating in nature, mastering the extensive natural philosophy of Aristotle – its treatment of space, time, change, local motion, generation, animals, the soul, and the structure of the cosmos – was a goal in itself. Initially at least, these subjects were not obviously preparatory for a higher discipline. The tensions between study in "arts" for its own sake (wisdom) and for a professional goal – whether theology, medicine, or law – shaped the conflicts between the faculties.

These deep changes in the intellectual landscape and the new institutional framework of the university together embedded the scientific enterprise in late-medieval culture. Methodologically, the materials studied in faculties of arts and medicine and the masters' approach to them were self-consciously based on the use of tools accessible in principle to everyone (i.e., reason, experience).[39] On this terrain, despite occasional carping, the writings of pagans, Muslims, Jews, and Christians largely competed on an equal footing. The net result was that methodological naturalism – no appeals to the supernatural – pervaded the largest faculty of the universities. The specialization inherent in the faculty structure reinforced this naturalism, which even the theologians widely recognized as proper to the lower faculty. Mere masters of arts were to avoid theological issues altogether and to defer to theologians when they could not (e.g., when discussing the eternity of the world or the nature of the soul).

Conversely, the structure of the university channeled a largely unimpeded flow of natural philosophy, logic, and mathematics into the higher faculties. Since masters of arts had studied logic and natural philosophy rigorously, those who advanced to the faculties of theology or medicine often continued to teach these subjects for many years in the arts faculty. By the fourteenth century, the sophisticated logic and natural philosophy of the arts faculties furnished the analytical techniques and the conceptual tools of the higher faculties. Some of the most intriguing natural philosophy and logic of the fourteenth century occurs in commentaries on the *Sentences* of Peter Lombard. In effect, these medieval "doctoral dissertations" in theology culminated

[38] Mariateresa Fumagalli Beonio Brocchieri, "The Intellectual," in *Medieval Callings*, ed. Jacques Le Goff, trans. Lydia Cochrane (Chicago: University of Chicago Press, 1990), pp. 180–209, especially pp. 194–6.

[39] Edward Grant, *God and Reason in the Middle Ages* (Cambridge: Cambridge University Press, 2001), pp. 160–4.

at once theological studies and years of teaching and disputing about natural philosophy and mathematics in the arts faculty.[40] Not even lectures on the Bible were immune from the latter. In the 1390s, Henry of Langenstein's lectures on Genesis gave his Viennese theology students a detailed summary of the full range of late-medieval science.[41]

A similar pattern surfaced in medicine, most clearly in Italy, where the universities had developed around strong traditions of law and medicine. At Bologna, students who completed an initial degree usually did so jointly in "arts and medicine," a curriculum that integrated the teaching of mathematics, astronomy, and medical astrology into more traditional medical courses.[42] In fourteenth-century Padua, Pietro d'Abano wrote his *Conciliator* to reconcile the differences between natural philosophers and physicians; significantly, his will identified him ecumenically as "professor of the art of medicine, of philosophy, and of astrology."[43] The impact of natural philosophy extended even to alchemy, a nonuniversity subject that its practitioners tried to bring into the Aristotelian framework in the twelfth and thirteenth centuries.[44]

By the late Middle Ages, then, the university's hierarchical degree structure had turned the ill-fitting rhetoric of the preparatory arts into a self-fulfilling prophecy. It was as difficult then as it is now to understand the language of fourteenth- and fifteenth-century medicine and theology without a thorough grounding in the logic and natural philosophy of the arts faculty.

TEACHING AND LEARNING: LECTURES, COMMENTARIES, AND DISPUTATIONS

Although a master of arts could in principle lecture on anything in the curriculum of his faculty, teaching in fact followed a hierarchy. At the top were the regent masters who, by seniority, selected the required book on which they wanted to lecture. Held in the morning, these "ordinary" lectures offered

40 Murdoch, "Unitary Character of Medieval Learning," especially pp. 272–83.

41 Nicholas Steneck, *Science and Creation in the Middle Ages: Henry of Langenstein (d. 1397) on Genesis* (Notre Dame, Ind.: University of Notre Dame Press, 1976).

42 Charles B. Schmitt, "Science in the Italian Universities in the Sixteenth and Seventeenth Centuries," in *The Emergence of Science in Western Europe*, ed. Maurice Crosland (New York: Science History Publications, 1976), pp. 35–56, especially p. 37; and Jole Agrimi and Chiara Crisciani, *Edocere medicos: Medicina nei secoli XIII–XV* (Naples: Guerini, 1988).

43 Antonio Favaro, "I lettori di matematiche nella Università di Padova dal principio del secolo XIV alla fine del XVI," in *Memorie e documenti per la storia della Università di Padova*, 2 vols. (Padua: La Garangola, 1922), vol. 1, pp. 1–70 at pp. 7–8, 10–11; and Nancy Siraisi, "Two Models of Medical Culture: Pietro d'Abano and Taddeo Alderotti," in her *Medicine and the Italian Universities* (Leiden: Brill, 2003), pp. 79–99.

44 William R. Newman, *Atoms and Alchemy: Chymistry and the Experimental Origins of the Scientific Revolution* (Chicago: University of Chicago Press, 2006), pp. 25–6.

expositions of the standard texts to scores of students. In natural philosophy (devoted to "moveable/changeable body"), the core texts included Aristotle's *Physics, On the Heavens, On Generation and Corruption,* and *On the Soul,* on which many regent masters eventually wrote commentaries.[45] In the afternoon, advanced bachelors gave "cursory" lectures that reviewed the morning material, and junior masters discussed shorter or optional books of natural philosophy and the mathematical sciences (especially on feast days).

Also held in the afternoon, the disputations served at once to teach, to hone logical skills, and to advance knowledge. With origins in Roman law, an endorsement by Augustine, and clear use in the ninth century, the "disputed question" came into its own in the eleventh century. In natural philosophy and medicine, it had its heyday between the twelfth and seventeenth centuries. Since the disputations involved the careful analysis of arguments, they powerfully stimulated logic, which developed exceptional sophistication in the late Middle Ages, often in close interaction with natural philosophy. Attempts to answer a given disputed question began by examining arguments for the two possible contradictory answers to it and ended with a resolution, often a third alternative.[46]

Disputations were required of both students (for obtaining degrees) and masters (for pedagogical purposes). Masters assigned the topics of student disputations, which in some Parisian colleges took place almost daily before an audience expected to participate (poor performers and the taciturn were fined). The masters' disputations were usually weekly events that younger students attended for two years before taking part in their own.[47] Students learned by hearing the material discussed and the arguments dissected repeatedly and from several angles. They witnessed the airing of problems with objections and responses, including genuine disagreements among their masters, who sometimes changed their minds. In fifteenth-century Bologna, masters normally could dispute only in areas that they had taught.[48] Not

[45] Grant, *History of Natural Philosophy from Antiquity to the Nineteenth Century*, chap. 5, especially p. 179.

[46] John Contreni, "Education and Learning in the Early Middle Ages: New Perspectives and Old Problems," *International Journal of Social Education*, 4 (1989), 9–25, especially pp. 16–17; Brian Lawn, *The Rise and Decline of the Scholastic 'Quaestio Disputata' with Special Emphasis on Its Use in the Teaching of Medicine and Science* (Leiden: Brill, 1993), pp. 3–12; and Edith Sylla, "The Oxford Calculators," in *The Cambridge History of Later Medieval Philosophy*, ed. Norman Kretzmann, Anthony Kenny, and Jan Pinborg (Cambridge: Cambridge University Press, 1982), pp. 542–8.

[47] Gordon Leff, "The *trivium* and the Three Philosophies," in De Ridder-Symoens, *History of the University in Europe*, vol. 1, pp. 307–36, especially p. 326.

[48] Serge Lusignan, "L'enseignement des arts dans les collèges parisiens au moyen âge," in *L'enseignement des disciplines à la faculté des arts (Paris et Oxford, XIIIe–XVe siècles)*, ed. Olga Weijers and Louis Holtz (Studia artistarum, 4) (Turnhout: Brepols, 1997), pp. 43–9, especially p. 47; Olga Weijers, "La 'disputatio,'" in Weijers and Holtz, *L'enseignement des disciplines à la faculté des arts*, pp. 393–404, especially pp. 393–4, 401–2; and Carlo Malagola, ed., *Statuti delle università e dei collegi dello studio bolognese* (Bologna: Zanichelli, 1888), pp. 261–4.

least, in universities like that of Pavia, the "circle" (*circulus*) of the master met in the evening as a kind of advanced seminar.[49]

The university thus institutionalized a culture of criticism and argumentation. Unlike the humanists who later scorned it as pointless haggling, the masters valued the disputation highly as a remarkable road to the truth, to which all disputants were to be committed "for its own sake." Jewish observers of the university, such as the astronomer Levi ben Gerson, also commented on the power of the disputation, praised its approach as Talmud-like, and used it in their writings. The oral disputation also shaped the form and content of many university writings, which adopted the *quaestio* format (poems and literary dialogues had been leading genres in twelfth-century discussions of nature).[50] Natural philosophers often defended their views in written disputed questions that engaged hypothetical and actual opponents, both living and dead. Masters working in the mathematical sciences sometimes used *quaestiones*; more typically, however, their treatises emulated the axiomatic approach of geometry (notably in optics and statics).

CLERICAL STATUS AND SOCIAL PARAMETERS

Learning in the Middle Ages is often called clerical – a misleading generalization if one overlooks the changing meanings and ambiguities of the term. In leading Italian cities, the remnants of classical schools always had been free from ecclesiastical control, a tradition on which Italian universities built. In contrast, the monastic and cathedral schools were church institutions controlled respectively by abbots to educate monks and by bishops to train literate parish priests. The new universities were clearly not clerical in this sense, for self-governing guilds or corporations of masters or students administered them independently of local ecclesiastical officials.[51] Their members'

49 Agostino Sottili, "The University at the End of the Middle Ages," in his *Humanismus und Universitätsbesuch / Renaissance Humanism and University Studies* (Education and Society in the Middle Ages and Renaissance, vol. 26) (Leiden: Brill, 2006), pp. 1–14, especially p. 4.

50 Mikko Yrjönsuuri, *Obligationes: Fourteenth Century Logic of Disputational Duties* (Acta Philosophica Fennica, 55) (Helsinki: The Philosophical Society of Finland, 1994), pp. 22–4; Mary Martin McLaughlin, *Intellectual Freedom and Its Limitations in the Universities of Paris in the Thirteenth and Fourteenth Centuries* (New York: Arno Press, 1977), pp. 8–9; Michael Shank, *"Unless You Believe, You Shall Not Understand": Logic, University, and Society in Late Medieval Vienna* (Princeton, N.J.: Princeton University Press, 1988), pp. 151–2; Edith Sylla, "The Oxford Calculators in Context," *Science in Context*, 1 (1987), 257–79, especially pp. 272–3; Colette Sirat, "L'enseignement des disciplines dans le monde hébreu," in Weijers and Holtz, *L'enseignement des disciplines à la faculté des arts*, pp. 495–509, especially pp. 504–8; Edith Sylla, "Science for Undergraduates in Medieval Universities," in *Science and Technology in Medieval Society*, ed. Pamela O. Long (Annals of the New York Academy of Sciences, 441) (New York: New York Academy of Sciences, 1985), pp. 171–86, especially p. 171; and Anthony Grafton and Lisa Jardine, *From Humanism to the Humanities: Education and the Liberal Arts in Fifteenth- and Sixteenth-Century Europe* (Cambridge, Mass.: Harvard University Press, 1986), pp. xii–xiii.

51 Grant, "Science and the Medieval University," p. 86.

oaths of loyalty to the university (or rector) were powerful obligations that usually excluded oaths to other authorities, including the bishop, the city, or the realm.[52]

With papal power at its height in the thirteenth century, it was the pope, as a universal authority, who claimed the right to turn a university into a *studium generale* by confirming the universality of its privileges (the emperor could do so, too).[53] Yet the papacy exercised remarkably little control over the teaching, reading, and production of medieval science, to say nothing of its details (the mortality of the soul was a notorious exception). On the contrary, important masters treated the pope as a patron to whom they dedicated major works. When the bishop of Paris contested the masters' privileges, the papacy usually sided with the university. Many new universities of the late Middle Ages were deliberately founded in cities that were not episcopal sees. In the early fifteenth century, the statutes of the University of Louvain/Leuven ensured that the bishop of Liège had no power over the university. The rector's judicial control over students and masters was total until the university eventually ceded to the duke of Brabant jurisdiction over the most serious crimes of its members.[54]

As for members of the universities, students in Italy generally did not have clerical status by virtue of their matriculation. North of the Alps, where they did, students took no vows (their oaths were to the university) and performed no specific religious duties (most were not even tonsured).[55] Their clerical status was neither a social role nor a religious vocation; it was primarily a legal privilege with valuable judicial and economic benefits. Some university students were as young as twelve years old; many were in their teens. When they got into drunken brawls or committed murder, their clerical status kept these juvenile delinquents out of the secular courts. University authorities themselves or lenient ecclesiastical courts tried the offenders, to the perpetual disgust of the townsfolk.

For some masters, their clerical status also meant financial support. On behalf of their masters, universities collectively petitioned newly elected popes for benefices, which gave the recipients income for life. Although these posts usually carried religious and administrative obligations, the masters often held them in absentia, delegating their daily duties to vicars. From the twelfth century onward, however, the number of masters quickly outpaced available benefices. The tradition of dispensing wisdom freely gave way to

52 Pearl Kibre, "Academic Oaths at the University of Paris in the Middle Ages," in *Essays in Medieval Life and Thought Presented in Honor of Austin Patterson Evans*, ed. John Mundy, Richard Emery, and Benjamin Nelson (New York: Columbia University Press, 1955), pp. 123–37, especially pp. 123–5, 136–7; and Rashdall, *Universities of Europe in the Middle Ages*, vol. 1, pp. 152–3.

53 Verger, "Patterns," pp. 35–6.

54 E. J. M. van Eijl, "The Foundation of the University of Louvain," in *The Universities in the Late Middle Ages*, ed. Jozef IJsewijn and Jacques Paquet (Leuven: Leuven University Press, 1978), pp. 37–8. Exceptions included Uppsala and St. Andrews.

55 Murray, *Reason and Society in the Middle Ages*, pp. 263–4.

the urban ethos of the guild, as nonbeneficed masters taught the growing number of students in exchange for fees.[56]

A university degree did not necessarily lead to a traditional church career. A new university-educated class played a dominant role in running the growing bureaucracies, administrations, and courts of law of princes, ecclesiastics, and towns. These "intellectuals" (to use Jacques Le Goff's felicitous anachronism) not only had a *métier* – a profession and a vocation – but also were in such demand that some universities required new masters of arts to teach for two years before they could leave.[57]

By the late Middle Ages, *clericus* often denoted a learned man generically, with no more religious connotation than the modern English "clerk."[58] This usage was also widespread in the vernaculars, which designated the learning of such pagans as Plato and Virgil as "clergy" or "clergie" ("Virgile... A Mirour made of his clergie").[59] Like Plato and Virgil, the vast majority of medieval university students were learned without being priests, monks, friars, or theologians. The faculties of arts and medicine did not, indeed could not, touch theology or the Bible; they focused overwhelmingly on the writings of pagans and infidels such as Aristotle, Averroes, and Avicenna.

Like those of the clergy, the social origins of medieval students of nature ranged from the highest nobility to the children of burghers, craftsmen, peasants, and paupers. A financial snapshot of Paris around 1330 reveals a majority of middle-income students, a small minority of poor, with the affluent and very affluent constituting a quarter of the population.[60] In late-medieval New College, Oxford, more than 60 percent of the students were the sons of small landowners. In fifteenth-century Cracow, however, the sons of burghers constituted 50 percent of the student body and the peasantry

[56] John W. Baldwin, "Introduction," in *Universities in Politics: Case Studies from the Late Middle Ages and Early Modern Period*, ed. John W. Baldwin and Richard A. Goldthwaite (Baltimore: The Johns Hopkins Press, 1972), pp. 9–10.

[57] John Baldwin, "The Penetration of University Personnel into French and English Administration at the Turn of the Twelfth and Thirteenth Centuries," *Revue des Études Islamiques*, 44 (1976), 199–215; Julio Valderon Baruque, "Universidad y sociedad en la Europa del los siglos XIV y XV," in *Universidad, Cultura y Sociedad en la Edad Media*, ed. Santiago Aguadé Nieto (Alcalá de Henares: Universidad de Alcalá de Henares, 1994), pp. 15–24, especially p. 20; and Alain Boureau, "Intellectuals in the Middle Ages, 1957–1995," in *The Work of Jacques LeGoff and the Challenges of Medieval History*, ed. Miri Rubin (Woodbridge: Boydell and Brewer, 1997), pp. 145–56, especially pp. 145–7.

[58] Jacques Verger, *Men of Learning in Europe at the End of the Middle Ages*, trans. Lisa Neal and Steven Rendell (Notre Dame, Ind.: University of Notre Dame Press, 2000), p. 187, n. 3.

[59] John Gower, *Confessio amantis*, in *The English Works of John Gower*, ed. George Campbell Macauley, 5 vols. (London: Oxford University Press for the Early English Text Society, 1900), vol. 1, bk. V, lines 2032–3; Serge Lusignan, *Parler vulgairement: Les intellectuels et la langue française aux XIIIe et XIVe siècles*, 2nd ed. (Paris: Vrin, 1986), pp. 161–4; and Christine Silvi, *Science médiévale et vérité: Étude linguistique de l'expression du vrai dans le discours scientifique en langue vulgaire* (Paris: Honoré Champion, 2003), pp. 64–5.

[60] William J. Courtenay, *Parisian Scholars in the Early Fourteenth Century: A Social Portrait* (Cambridge: Cambridge University Press, 1999), pp. 95–8.

only 10 percent.[61] University attendance nevertheless resonated strongly with the fate of the latter. From the late fourteenth to the early sixteenth century in the Holy Roman Empire, matriculations fluctuated according to the yearly and longer-term rhythms of the agricultural economic cycle.[62]

Being socially the most diverse of guilds, the university gently tampered with the prevailing social hierarchy. Its valuation of intellectual and administrative competence permitted significant social inversions with real-world consequences. The son of a pauper could not only dispute with the son of a count but also, if elected rector, have judicial authority over him. By the fifteenth century, however, a university education had become a desideratum for the well-to-do and the nobility, to which some universities catered (e.g., Pavia). Even universities that did not (e.g., Vienna) witnessed increasing enrollments from this population. Padua undercut the influence-peddling of the nobility by requiring noble professors and students to leave the city during the traditional balloting for professorial chairs.[63]

Whatever their social class, students and masters were almost exclusively male and Christian, and overwhelmingly remained such until recent times. A few remarkable incidents nevertheless testify to subtle pressure on these categories. In the fourteenth century, two Jewish physicians unsuccessfully petitioned the university in Perugia for a doctorate in medicine, but later petitioners succeeded when "eminent science" justified a relaxation of the exclusionary law. Padua licensed a Jewish physician "in the vernacular." In early-fifteenth-century Cracow, one woman studied successfully for two years, disguised as a man.[64]

When conflicts with the ecclesiastical authorities arose, it was usually not "the church" that issued condemnations but local bishops, who did so for local reasons. In Paris, Aristotle's arguments for the eternity of the world and his views on the human soul faced suspicion. The bishop banned the reading of Aristotelian natural philosophy on pain of excommunication (1210/1215).

61 Guy Fitch Lytle, "The Social Origins of Oxford Students in the Late Middle Ages: New College, c. 1380–c. 1510," in IJsewijn and Paquet, *Universities in the Late Middle Ages*, pp. 426–54, especially p. 432; and Mieczsław Markowski, "Philosophie und Wissenschaft an der Krakauer Universität im 15. Jahrhundert," *Studia Mediewistyczne*, 22 (1983), 79–103, especially pp. 86–7.

62 Rainer Schwinges, *Deutsche Universitätsbesucher im 14. und 15. Jahrhundert: Studien zur Sozialgeschichte des alten Reiches* (Stuttgart: Steiner Verlag, 1986), pp. 207–20.

63 Agostino Sottili, ed., *Lauree pavesi nella seconda metà del '400*, vol. 1: *1450–1475* (Fonti e studi per la storia dell' Università di Pavia, 25) (Bologna: Cesalpino, 1995), p. 9; Pearl Kibre, *Scholarly Privileges in the Middle Ages: The Rights, Privileges, and Immunities of Scholars and Universities at Bologna, Padua, Paris, and Oxford* (Cambridge, Mass.: Mediaeval Academy of America, 1962), pp. 76, 78–9; and Hilde De Ridder-Symoens, "Rich Men, Poor Men: Social Stratification and Social Representation at the University (13th–16th Centuries)," in *Showing Status: Representations of Social Position in the Late Middle Ages*, ed. Wim Blockmans and Antheun Janse (Medieval Texts and Cultures of Northern Europe, 2) (Turnhout: Brepols, 1999), pp. 159–75.

64 Diego Quaglioni, "*Orta est disputatio super matheria promotionis inter doctores*: L'ammissione degli ebrei al dottorato," in *Gli Ebrei e le scienze* (Micrologus, 9) (Florence: SISMEL-Edizioni del Galluzzo, 2001), pp. 249–67, especially pp. 256–61; Park, *Doctors and Medicine in Early Renaissance Florence*, p. 65; and Michael H. Shank, "A Female University Student in Fifteenth-Century Kraków," *Signs*, 12 (1987), 373–80.

It is crucial to emphasize that these early Parisian condemnations did not affect other universities. Although founded by the pope, the new university in Toulouse (1229) courted Parisian students by advertising its unrestricted study of Aristotle.[65] The significance of the early condemnations is not that they happened but that they led to the university's autonomy from the bishop, a model that would multiply north of the Alps. By 1255, the statutes required candidates in arts to read the very books of Aristotle's natural philosophy that the bishop had prohibited a generation earlier.[66]

In the history of medieval science, the most famous condemnations occurred in 1277, when the bishop of Paris censured 219 motley propositions that had surfaced in discussions in the faculty of arts.[67] Again, this was a local struggle between factions inside the Parisian faculty of theology and between the latter and the faculty of arts. Allied with conservative theologians, the bishop took on certain themes (astrological determinism, natural necessity, the limits of divine power) and individuals (notably Thomas Aquinas), as well as the cockiness of some masters of arts (who allegedly claimed that "the only wise men are philosophers" or that "nothing is known better by knowing theology"). As "philosophers," the masters of arts considered it their prerogative to raise controversial theses and to discuss even errors.[68] Although complicated by the intervention of both bishop and pope, the controversy was effectively a turf battle about faculty boundaries. Does a mere philosopher have the right to determine a conclusion with theological implications? Can a theologian set the limits of inquiry in natural philosophy? Who judges the outcome?

In the late Middle Ages, other universities sporadically witnessed condemnations as well. Institutionally, however, these local disputes are far less important than the larger story of the universities' growth in autonomy, in numbers, and in prestige, multiplying all the while their students' exposure to natural philosophy, the mathematical sciences, and medicine.

THE EXPANSION OF THE UNIVERSITY

The late-medieval universities were remarkable for both their diffusion and their internal growth. The mobility of students and masters greatly

65 Lynn Thorndike, *University Records and Life in the Middle Ages* (New York: Columbia University Press, 1944), pp. 32–5.

66 Gordon Leff, *Paris and Oxford Universities in the Thirteenth and Fourteenth Centuries: An Intellectual and Institutional History* (New York: John Wiley and Sons, 1968), pp. 140–1.

67 J. M. M. H. Thijssen, *Censure and Heresy at the University of Paris, 1200–1400* (Philadelphia: University of Pennsylvania Press, 1998); and John E. Murdoch, "1277 and Late Medieval Natural Philosophy," in *Was Ist Philosophie im Mittelalter?* ed. Jan Aertsen and Andreas Speer (Miscellanea Mediaevalia, 26) (Berlin: De Gruyter, 1998), pp. 111–21, especially p. 121.

68 McLaughlin, *Intellectual Freedom and Its Limitations in the Universities of Paris in the Thirteenth and Fourteenth Centuries*, pp. 85ff, 306.

facilitated the proliferation of the institution during the fourteenth and fifteenth centuries and helps to explain the diffusion of customs, texts, and legal privileges to far-flung universities. The masters' "right of teaching anywhere" (*ius ubique docendi*) was sometimes empty.[69] The utility of their common language was not. Far from being the dead language of a bygone empire, Latin was the lingua franca of working intellectuals, peppered with neologisms and imports from Arabic, Greek, and the vernaculars.

The sheer number of its specimens soon made the university as a species impossible to extinguish. Unlike their spontaneous predecessors, dozens of later universities began at the initiative of a city or a ruler. The kings of Castile sponsored universities at Palencia (1214–16) and Salamanca (before 1230); Emperor Charles IV of Luxemburg founded not only a university in Prague (1348) but also seven others, primarily in northern Italy (1353–69); the Habsburg dukes promoted the university in Vienna (1365, 1389); and the city of Erfurt successfully worked to establish its own university (1392).[70] These efforts say much less about political control than they do about the growing prestige and perceived utility of the universities. The successful ones wisely imitated the model of magisterial autonomy. Emperor Frederick II's foundation of the "prototype of a state university" under his own control in Naples (1224) failed in his day, as did his attempts to eliminate its most formidable competitor, the university in Bologna.[71]

Less drastic efforts by secular or ecclesiastical powers to tamper with their local university sometimes aided the movement as a whole, as objectors dispersed and spread the institutional model to new cities. Troubles at Oxford and Paris stimulated new universities at, respectively, Cambridge (1209) and Toulouse (1229). During the papal Great Schism of the late fourteenth century, many German masters who left Paris after the French crown's heavy-handed treatment of the university helped to establish such new *studia generalia* as Heidelberg, Cologne, and Erfurt and to reinvigorate Vienna, aided by the authorities' motivations of prestige and local student spending.[72] Although papal bulls now made the new foundations official, university autonomy was by this time well established.

By the mid-fifteenth century, enrollments in many of the new German universities were so large that they would not be surpassed until the late nineteenth or early twentieth centuries. In little more than a century, they

[69] Alan B. Cobban, *The Medieval Universities: Their Development and Organization* (London: Taylor and Francis, 1975), pp. 27–33.

[70] Verger, "Patterns," pp. 57–8; Cantarino, "Medieval Spanish Institutions of Learning," p. 220; and Erich Kleineidam, *Universitas studii erffordensis: Überblick über die Geschichte der Universität Erfurt im Mittelalter, 1392–1521*, 2 vols. (Leipzig: St-Benno Verlag, 1964–1969), vol. 1, pp. 4–9.

[71] Rüegg, "Themes," p. 18; Rashdall, *Universities of Europe in the Middle Ages*, vol. 2, pp. 22–6; and Paolo Nardi, "Relations with Authority," in De Ridder-Symoens, *History of the University in Europe*, vol. 1, pp. 77–107, especially pp. 86–91.

[72] Rüegg, "Themes," p. 18; Kleineidam, *Universitas studii erffordensis*, vol. 1, pp. 4–12, 216–20.

had matriculated more than a quarter of a million students.[73] The universities' remarkable proliferation and internal growth created a literate elite of unprecedented size, which had been exposed to the rigors of logic, the fundamental concepts of Greek and Arabic natural philosophy, the elementary mathematical sciences, and the naturalist philosophical assumptions that pervaded these disciplines and medicine.

CURRICULAR TRADITION, INNOVATION, AND SPECIALIZATION

When the universities emerged in the thirteenth century, their curricula were not ancient and medieval (as we think of them) but innovative. Most of their early lectures, concepts, and texts – from Aristotle's natural philosophy through Ibn al-Haytham's optics to Avicenna's medicine – were recent imports into Latin culture from the Arabic. In addition, the university stimulated and adopted new works written for teaching purposes, such as Sacrobosco's *Sphere* for elementary astronomy and the Salerno-trained Giles of Corbeil's *De pulsibus* (On Pulses) and *De urinis* (On Urines) for medicine.[74]

That some late-medieval universities continued to teach much of the new thirteenth-century curriculum seems to confirm the stereotype of late-medieval university stagnation or decline. Yet techniques, textbooks, courses, and chairs that did not exist in the early thirteenth century spread and multiplied in the new foundations of the fourteenth and fifteenth centuries. Mondino de' Liuzzi's (d. 1326) *Anatomia* diffused in universities, in which anatomy and surgery became academic subjects sometimes associated with new chairs.[75] In logic, a subject of central importance to natural philosophy, demanding new work pressed far beyond Aristotle's contributions.[76] The insight that qualities could be quantified brought mathematical techniques and geometrical representations to bear on the problem of change. In the new universities of the Holy Roman Empire, this approach also generated new textbooks and courses (the "latitude of forms" – see the following chapters in this volume: Laird, Chapter 17; and Molland, Chapter 21).

Many of these innovations derived from specialization – and sometimes from partisanship. Some fifteenth-century German universities in particular became embroiled in philosophical controversies between supporters of the *via antiqua* and the *via moderna*, associated respectively with philosophical

73 Schwinges, *Deutsche Universitätsbesucher im 14. und 15. Jahrhundert*, pp. 487–8.

74 Lynn Thorndike, *The Sphere of Sacrobosco and Its Commentators* (Chicago: University of Chicago Press, 1949); and Mireille Ausécache, "Gilles de Corbeil ou le médecin pédagogue au tournant des XIIe et XIIIe siècles," *Early Science and Medicine*, 3 (1998), 187–215.

75 Park, *Doctors and Medicine in Early Renaissance Florence*, pp. 62–6.

76 Henrik Lagerlund, *Modal Syllogistics in the Middle Ages* (Leiden: Brill, 2000), chap. 5.

realism and nominalism. Although primarily logical and metaphysical, these disputes also touched natural philosophy. Were time, space, and motion real, independently existing "things," as the *antiqui* believed? Or were they convenient terms designating the relations of individual objects, as the *moderni* thought? Freiburg im Breisgau effectively created a two-track arts faculty. Each student chose the *via* ("way") in which he would be taught and examined. Other universities offered only one *via*. After 1427, Louvain admitted as regent masters only those who "uphold the doctrine of Aristotle except in cases that went against the faith" and "swear never to teach [Jean] Buridan, Marsilius of Inghen, [William of] Ockham, and their followers."[77] Although undesirable at Louvain because they undermined Aristotle, these *moderni* were lionized as leading contributors to logic and natural philosophy in Cracow, Vienna, and other new universities.[78]

Less ideological symptoms of specialization also appear in teaching concentrations, some of them mandated, others freely chosen. In mid-fourteenth-century Paris, Jean Buridan did his pioneering work in logic and natural philosophy during an exceptional lifelong career in the arts faculty. At Vienna between 1390 and 1460, fifty-nine regent masters taught for more than ten years in the arts faculty. Of the twenty-five (42 percent) who specialized, more than half taught science-oriented courses: natural philosophy (five), psychology (three), and sometimes the mathematical sciences (six) – to say nothing of logic, which included substantial natural-scientific and mathematical material.[79]

Both colleges and the institutional practice of small-group supervision also favored specialization. In fourteenth-century Paris, individual masters worked closely with groups averaging two to six students, whose academic progress they supervised and whom they sometimes housed. Medical masters in particular made use of nuptial and familial metaphors when discussing teacher–student relations. The new master, now wedded to the medical science he possessed, could generate pupils that would resemble him as he resembled his own master.[80] Although the content of such intimate teaching

77 Rashdall, *Universities of Europe in the Middle Ages*, vol. 2, pp. 273–4. The quotations are from two separate statutes cited by Anton G. Weiler, "Les relations entre l'université de Cologne et l'université de Louvain au XVe siècle," in IJsewijn and Paquet, *Universities in the Late Middle Ages*, pp. 49–81, especially p. 60 n.; Martin J. M F. Hoenen, *Marsilius of Inghen: Divine Knowledge in Late Medieval Thought* (Leiden: Brill, 1993), pp. 11–15; and Robert F. Seybolt, trans., *Manuale scolarium: An Original Account of Life in the Mediaeval University* (Cambridge: Cambridge University Press, 1921), pp. 40–5.

78 Luca Bianchi, *Studi sull' aristotelismo del Rinascimento* (Padua: Il Poligrafo, 2003), pp. 114–18.

79 Claudia Kren, "Patterns in Arts Teaching at the Medieval University of Vienna," *Viator*, 18 (1987), 321–36.

80 Figures for Paris; see William J. Courtenay, "The Arts Faculty at Paris in 1329," in Weijers and Holtz, *L'enseignement des disciplines à la faculté des arts*, pp. 55–69 at pp. 59–60, 62; and Chiara Crisciani, "Teachers and Learners in Scholastic Medicine," *History of Universities*, 15 (1997–1999), 75–101, especially pp. 80–4.

rarely leaves official traces, both the practice and the outlook explain enduring clusters of specialized instruction around some masters.

Also contributing to this trend were the colleges, which benefactors from the 1180s onward endowed to provide room and board for poor students. A "site of academic sociability par excellence,"[81] the college sometimes fostered striking intellectual results by developing specific emphases. In Paris, Queen Joanna's Collège de Navarre (1305) supported several leading natural philosophers (Nicole Oresme spent eighteen years there). Gervais Chrétien, a master of medicine and King Charles V's physician, founded the Collège of Maître Gervais (1371), to which the king, with papal approval, added two fellowships in astrology.[82]

In the fourteenth century, Merton College (Oxford) famously stood for a particular way of bringing mathematics to bear on natural philosophy. Several Merton masters pioneered new mathematical tools in both logic and the study of change, including Thomas Bradwardine's *On the Proportions of Velocities in Motions* (1328) and the "Merton mean speed theorem" (see in this volume Laird, Chapter 17; Molland, Chapter 21; and Ashworth, Chapter 22). As these masters prepared first- and second-year Oxford students for the disputations on "sophismata" (counterintuitive propositions requiring proof or disproof), many of the propositions they analyzed touched on change, the heart of natural philosophy (maxima and minima; the limits of forces, resistances, and motions; etc.).[83] No higher faculty required the highly specialized works that were sometimes taught as exercises for young students (e.g., the very difficult blend of logic, natural philosophy, and mathematics in William Heytesbury's *Rules for Solving Sophismata*). They were ends in themselves for brilliant masters who developed their ideas in shared living quarters and used them in small-group instruction. For them, "the process of teaching [was] not sharply distinguishable from the process of doing science."[84] The original mathematical techniques that the Merton

81 Verger, *Men of Learning in Europe at the End of the Middle Ages*, p. 58.

82 Richard Lemay, "Teaching of Astronomy in Medieval Universities Principally at Paris in the Fourteenth Century," *Manuscripta*, 20 (1976), 197–217; Grant, "Science in the Medieval University," p. 75; Nathalie Gorochov, *Le collège de Navarre de sa fondation (1305) au début du XVe siècle* (Paris: Honoré Champion, 1997), pp. 339–42, 680–1; and Joan Cadden, "Charles V, Nicole Oresme, and Christine de Pizan: Unities and Uses of Knowledge in Fourteenth-Century France," in *Texts and Contexts in Ancient and Medieval Science: Studies on the Occasion of John E. Murdoch's Seventieth Birthday*, ed. Edith Sylla and Michael McVaugh (Leiden: Brill, 1997), pp. 208–44, especially pp. 237–9.

83 Sylla, "Oxford Calculators in Context," pp. 259–63; Weisheipl, "Ockham and the Mertonians," pp. 631–42; Norman Kretzmann, "Syncategoremata, Exponibilia, Sophismata," in Kretzmann, Kenny, and Pinborg, *Cambridge History of Later Medieval Philosophy*, pp. 211–53; John E. Murdoch, "The Involvement of Logic in Late Medieval Natural Philosophy," in *Studies in Late Medieval Natural Philosophy*, ed. Stefano Caroti (Florence: Olschki, 1989), pp. 3–28 at p. 25; Marshall Clagett, *The Science of Mechanics in the Middle Ages* (Madison: University of Wisconsin Press, 1959), chap. 11; and Grant, *God and Reason in the Middle Ages*, pp. 121–40.

84 Guy Beaujouan, "Le *quadrivium* et la faculté des arts," in Weijers and Holtz, *L'enseignement des disciplines à la faculté des arts*, pp. 185–94, especially pp. 192–4; and Sylla, "Science for Undergraduates in Medieval Universities," pp. 179, 182.

"calculators" pioneered in late medieval natural philosophy radiated to Paris, where at mid-century Jean Buridan, Nicole Oresme, Albert of Saxony, and others amplified them. Thanks to the institutional boom of the late Middle Ages, they diffused widely – to universities in Spain, Italy, and especially the Holy Roman Empire, where Paris-trained masters put them on the curriculum.[85]

In the late Middle Ages, most universities kept their lectures on advanced mathematical subjects "extraordinary," but others made them "ordinary" by establishing new chairs in them during the fifteenth century. These developments paralleled and fed the burgeoning interest in astrology. Courts, towns, and individuals all found ready use for university masters competent in this material, whatever their means of specialization.[86]

Whereas Paris had perfunctorily allowed Sacrobosco's thin *Sphere* to satisfy its hundred-lecture requirement in "mathematica," fifteenth-century Vienna covered that book in twenty lectures, supplemented by Euclid, optics (*perspectiva*), and elementary planetary theory.[87] Curricular prescriptions tell only part of the story, however, for astronomical activity extended beyond lectures. In late-thirteenth- and early-fourteenth-century Paris, for example, several specialized astronomers engaged in extracurricular practices (e.g., they measured the precession of the equinoxes, built a giant quadrant, and observed eclipses using a pinhole camera). Only after reaching Paris did the thirteenth-century Castilian Alfonsine Tables diffuse widely, thanks to the extracurricular adaptations and explanatory canons of early-fourteenth-century masters of arts. At Padua, the medical master Giovanni Dondi (d. 1359) followed no statute when he built one of the two most complex machines of the era: a clockwork-driven *astrarium* that gave the positions of all the planets in real time.[88] As these extracurricular activities suggest,

85 John E. Murdoch, "*Mathesis in philosophiam scolasticam naturalem introducta*: The Rise and Development of the Application of Mathematics in Fourteenth Century Philosophy and Theology," in *Arts libéraux et philosophie au moyen âge*, pp. 238–46; and Joel Kaye, *Economy and Nature in the Fourteenth Century: Money, Market Exchange and the Emergence of Scientific Thought* (Cambridge: Cambridge University Press, 1998).

86 Hilary Mary Carey, *Courting Disaster: Astrology at the English Court and University in the Later Middle Ages* (New York: St. Martin's Press, 1992); and Michael H. Shank, "Academic Consulting in Late Medieval Vienna: The Case of Astrology," in Sylla and McVaugh, *Texts and Contexts in Ancient and Medieval Science*, pp. 245–70, especially pp. 248–51.

87 Pearl Kibre, "Arts and Medicine in the Universities of the Later Middle Ages," in IJsewijn and Paquet, *Universities in the Late Middle Ages*, pp. 213–27, especially p. 222; and Michael H. Shank, "The Classical Scientific Tradition in Fifteenth-Century Vienna," in *Tradition, Transmission, Transformation: Proceedings of Two Conferences Held at the University of Oklahoma*, ed. Jamil Ragep and Sally Ragep, with Steven Livesey (Leiden: Brill, 1996), pp. 115–36, especially p. 120.

88 Pierre Duhem, *Le Système du monde: Histoire des doctrines cosmologiques de Platon à Copernic*, 10 vols. (Paris: Hermann, 1913–1959), vol. 4, pp. 16–17; Max Lejbowicz, "Les disciplines du *quadrivium*: l'astronomie," in Weijers and Holtz, *L'enseignement des disciplines à la Faculté des arts*, pp. 195–211, especially pp. 209–11; Guy Beaujouan, "Le *quadrivium* et la Faculté des arts," in Weijers and Holtz, *L'enseignement des disciplines à la Faculté des arts*, pp. 185–94 at p. 194; John D. North, *Cosmos* (Chicago: University of Chicago Press, 2008), pp. 251–5; Giovanni Dondi dall' Orologio, *Tractatus astrarii*, ed. Emmanuel Poulle (Geneva: Droz, 2003); Michael H. Shank, "Mechanical

universities also played a crucial role in crystallizing and growing communities of specialization.

Building on this foundation, several fifteenth-century universities turned the mathematical sciences into official areas of teaching specialization. Oxford reportedly offered "four years of medicinals and astronomy," while Erfurt also gave significant attention to the mathematical sciences. Bologna and Cracow each eventually assigned two chairs to the mathematical sciences; Salamanca created one as well.[89] Vienna did not, but John of Gmunden's (d. 1442) twenty-six-year career of teaching primarily the mathematical sciences constituted de facto specialization. Prague, Cracow, and Vienna all offered new or improved courses on the astrolabe.[90] The pioneering Italian *Algebra* (ca. 1440) of Giovanni Bianchini made its way into Latin and the university circles of Cracow and northern Italy later in the century. Georg Peuerbach had first taught his *Theoricae novae planetarum* (New Theories of the Planets) (1454) at the "citizens' school" in Vienna. Within a generation, masters commented on it at Cracow and Padua, and it eventually replaced its older competitors. Ptolemy's highly specialized *Almagest* largely remained off the official curriculum for reasons of competence, demand, and remuneration. The chair of astrology at Bologna lectured only once every four years on the third of the *Almagest*'s thirteen books. Lecturing on feast days, Jan Schindel (d. 1456) at Prague did complete the *Almagest* – in six years (1412–18).[91] Without his day job as royal physician, he would have starved. At Vienna, Georg Peuerbach and Johannes Regiomontanus mastered the *Almagest* around 1450–60 without leaving an official trace.[92] Some highly specialized learning clearly took place informally.

Thinking in European Astronomy (13th–15th Centuries)," in *Mechanics and Cosmology in the Medieval and Early Modern Period*, ed. Massimo Bucciantini, Michele Camerota, and Sophie Roux (Biblioteca di Nuncius, 64) (Florence: Leo Olschki, 2007), pp. 3–27; and Paul F. Grendler, *The Universities of the Italian Renaissance* (Baltimore: Johns Hopkins University Press, 2002), p. 416.

89 Kleineidam, *Universitas studii erffordensis*, vol. 2, pp. 64–77; Mieczsław Markowski, "Astronomie an der Krakauer Universität," in IJsewijn and Paquet, *Universities in the Late Middle Ages*, pp. 256–75, especially pp. 257–60; and José Chabás, "Astronomy in Salamanca in the Mid-Fifteenth Century: The *Tabulae resolutae*," *Journal for the History of Astronomy*, 29 (1998), 167–75, especially pp. 167, 172–3.

90 George Molland, "The *Quadrivium* in the Universities: Four Questions," in *Scientia und Ars im Hoch- und Spätmittelalter*, ed. Ingrid Craemer-Ruegenberg and Andreas Speer, 2 vols. (Miscellanea Mediaevalia, 22) (Berlin: De Gruyter, 1994), vol. 1, pp. 66–78, especially p. 77; Claudia Kren, "Astronomical Teaching at the Late Medieval University of Vienna," *History of Universities*, 3 (1983), 15–30; and Petr Hadrava and Alena Hadravová, "Das Albion des Richard von Wallingford und seine Spuren bei Johannes von Gmunden und Johannes Schindel," in *Johannes von Gmunden (ca. 1384–1442), Astronom und Mathematiker*, ed. Rudolf Simek and Kathrin Chlench (Vienna: Fassbinder, 2006), pp. 161–7.

91 Grażyna Rosińska, "The 'Italian Algebra' in Latin and How It Spread to Central Europe: Giovanni Bianchini's *De Algebra* (ca. 1440)," *Organon*, 26–27 (1997–1998), 131–45, especially pp. 134–5, 139; Malagola, *Statuti delle università e dei collegi dello studio bolognese*, p. 76; and Grażyna Rosińska, *Scientific Writings and Astronomical Tables in Cracow (XIVth–XVIth Centuries)* (Studia Copernicana, 22) (Wrocław: Ossolineum, 1984), p. 377.

92 Kren, "Astronomical Teaching at the Late Medieval University of Vienna," pp. 19ff; and Shank, "Classical Scientific Tradition in Fifteenth-Century Vienna," pp. 121–4, 131–2.

THE CIRCULATION OF KNOWLEDGE ABOUT NATURE

As Otto Neugebauer once noted, mathematical-astronomical material often offers the clearest evidence of cultural transmission.[93] It is remarkable to reflect on the fact that the Alfonsine Tables, calculated by thirteenth-century Spanish Jews working in Toledo for King Alfonso X, diffused throughout Europe in a new fourteenth-century Parisian format and thanks to canons (instructions) by a German master from the University of Paris, where they were not yet on the curriculum. By the fifteenth century, the Tables and accompanying canons were on the curricula of universities like Prague, Cracow, and Vienna, suitably adapted to new coordinates. Pedagogy and ease of use drove innovations in tabulation and computation across Europe, from England to Italy. When a Polish master in the mid-fifteenth century brought the Alfonsine-based *Tabulae resolutae* with him from Central Europe to Salamanca, they reinvigorated the diffusion of the Alfonsine Tables in Spain two centuries after they had been commissioned. These wanderings illustrate vividly the interplay between court and university, between intellectuals of diverse religions, between traveling scholars and distant locales, and between aspirations to universal knowledge and local needs. In short, this complex give-and-take between institutional, geographical, and social factors demonstrates clearly the power of the late-medieval universities to adopt, to adapt, to use, and to diffuse what began as a local, nonacademic innovation.[94]

Although universities did not control the travels of masters, they often regulated the local diffusion of manuscripts, setting their prices and the cost of making copies, pioneering new copying techniques, and adopting the new medium of paper. High-demand university manuscripts were quickly copied according to the *pecia* system, well attested in thirteenth-century universities. "Stationers" rented error-free exemplar manuscripts one *pecia* ("piece") of two to four folios at a time, allowing many scribes to copy the same work on a staggered schedule. Paper making, machine-assisted by the late thirteenth century, compounded the effects of this innovation, reducing book prices by factors of five to thirteen.[95]

93 Otto Neugebauer, *The Exact Sciences in Antiquity* (Princeton, N.J.: Princeton University Press, 1952), p. 1.

94 Emmanuel Poulle, *Les Tables alphonsines avec les canons de Jean de Saxe: édition, traduction et commentaire* (Paris: Éditions du Centre National de la Recherche Scientifique, 1984); José Chabás and Bernard Goldstein, *The Alfonsine Tables of Toledo* (Archimedes, 8) (Dordrecht: Kluwer Academic Publishers, 2003); José Chabás, "The University of Salamanca and the Renaissance of Astronomy during the Second Half of the 15th Century," in *Universities and Science in the Early Modern Period*, ed. Mordechai Feingold and Victor Navarro-Brotons (Dordrecht: Springer, 2006), pp. 29–36; and Chabás, "Astronomy in Salamanca in the Mid-Fifteenth Century," pp. 167–75.

95 Rashdall, *Universities of Europe in the Middle Ages*, vol. 1, p. 179; Jean Destrez, *La pecia dans les manuscripts universitaires du XIIIe et du XIVe siècle* (Paris: Editions Jacques Vautrain, 1935); Graham Pollard, "The *pecia* System in the Medieval Universities," in *Medieval Scribes, Manuscripts, and Libraries: Essays Presented to N. R. Ker*, ed. M. B. Parkes and Andrew G. Watson (London: Scolar Press, 1978), pp. 145–61, especially pp. 146–51; Thorndike, *University Records and Life in the Middle*

Universities both benefited from and drove these trends. In the Holy Roman Empire, manuscripts multiplied exponentially. The thirteenth century produced twice as many as the twelfth, the fifteenth six times as many as the thirteenth. Some scientific best-sellers still survive in hundreds of manuscripts. Other texts repackaged academic science in Latin and vernacular encyclopedias that often retained a *quaestio* format while focusing on the "useful."[96]

The protests of scribes did not stop some university-trained masters from taking up printing soon after the invention of movable type. In Nuremberg, the astronomer Johannes Regiomontanus, a Vienna master of arts, started the first press specializing in the mathematical sciences (1471–5). His premature death left his ambitious plans incomplete, but his initial efforts underlay the upgrading of elementary planetary theory in sixteenth-century universities. His successors followed through with editions of – among others – Euclid and compendia of elementary astronomy, texts with an obvious university clientele.[97]

BEYOND THE HALLS OF THE UNIVERSITY

From the thirteenth century on, the universities hosted the bulk of the scientific enterprise in Europe, but they were limited to Christian men literate in Latin and to a specific set of subjects. Other venues could and did ignore such constraints. Although courts and vernacular schools drew on university expertise, they also reached beyond academic personnel, subject matter, and reading material. Many non-Latin works – from translations of university textbooks to encyclopedias through handbooks of medicine or hunting – reflected and fed the widespread interest in the operations of nature among the growing nonacademic audiences.[98]

Ages, pp. 112–17; Richard H. Rouse and Mary A. Rouse, "The Book Trade at the University of Paris, 1250–1350," in *La Production du livre universitaire au moyen âge: Exemplar et pecia*, ed. Louis Jacques Bataillon, Bertrand Guyot, and Richard Rouse (Paris: Editions du Centre National de la Recherche Scientifique, 1988), pp. 13–15; Louis Jacques Bataillon, "Le fonds Jean Destrez-Guy Fink Errera à la Bibliothèque du Saulchoir," in Bataillon, Guyot, and Rouse, *La Production du livre universitaire au moyen âge*, pp. 41–85, especially p. 67; and Verger, *Men of Learning in Europe at the End of the Middle Ages*, p. 66.

96 Thorndike, *University Records and Life in the Middle Ages*, pp. 120–3; Colin Clair, *A History of European Printing* (London: Academic Press, 1976), p. 59; Edwin Hunt and James Murray, *A History of Business in Medieval Europe* (Cambridge: Cambridge University Press, 1999), pp. 198–203; Kibre, *Scholarly Privileges in the Middle Ages*, pp. 220–1, 255, 257; Uwe Neddermeyer, *Von der Handschrift zum gedrucktem Buch. Schriftlichkeit und Leseinteresse im Mittelalter und in der frühen Neuzeit: Quantitative und qualitative Aspekte*, 2 vols. (Wiesbaden: Harrassowitz Verlag, 1998), vol. 2, pp. 711–12; and Ventura, "Questiones and Encyclopedias," pp. 27–8, 39–42.

97 Henry Lowood and Robin Rider, "Literary Technology and Typographic Culture: The Instrument of Print in Early Modern Science," *Perspectives on Science*, 2 (1994), 1–37, especially pp. 4–8; and Olaf Pedersen, "The *Theorica planetarum* Literature of the Middle Ages," *Classica et Mediaevalia*, 23 (1962), 225–32.

98 William Crossgrove, "The Vernacularization of Science, Medicine, and Technology in Late Medieval Europe: Broadening Our Perspectives," *Early Science and Medicine*, 5 (2000), 46–63; and Faye M.

Courts promoted some subjects that universities neglected or did not touch. Emperor Frederick II (d. 1250) in Sicily actively carried out his own inquiries into nature. His *On the Art of Hunting with Birds* showcased his critical skills and detailed field observations of raptors and their prey – topics of scientific inquiry in their own right (see in this volume Grant, Chapter 18; and Reeds and Kinukawa, Chapter 24). His court included, among others, Michael Scot (d. ca. 1236), astrologer and translator from the Arabic in Toledo, who in turn had ties with Leonardo Fibonacci, a Pisan merchant versed in Arabic mathematics and the most impressive mathematician of the era. Although practiced unofficially by university-trained men on the fringes of the university, endeavors such as alchemy, natural magic, and physiognomy openly drew the interest and patronage of rulers.[99]

Courts also enabled the direct interaction of Latin, Arabic, Hebrew, and vernacular cultures, often with repercussions on the universities. In multicultural Toledo, King Alfonso X ("the Wise") sponsored the original Castilian Alfonsine Tables, the work of Jewish astronomers who had achieved high competence outside the universities. In 1342, Pope Clement VI commissioned the translation from Hebrew of the *Astronomy* of Levi ben Gerson, the most creative astronomer of fourteenth-century Europe. Levi would dedicate it to the pope, who was also the most prominent client of his brother, a physician.[100] These works soon made their way into university circles.

In addition, courts sponsored not only original vernacular works but also vernacular translations of academic works. A century after Alfonso the Wise's sponsorship of Castilian works, the court of the French King Charles V (also "the Wise") was the context for Nicole Oresme's (d. 1382) *Livre de divinacions*, which attacked excessive reliance on astrology even as Christine de Pizan, daughter of an astrologer to the king, defended the subject in her French writings.[101] The king also commissioned vernacular translations of Aristotle's works. Oresme's *Livre du ciel et du monde* (Book of the Heavens and Earth) introduced the court to university critiques of

Getz, *Healing and Society in Medieval England: The Middle English Translation of the Pharmaceutical Writings of Gilbertus Anglicus* (Madison: University of Wisconsin Press, 1991).

99 See the valuable volume *I sapere nelle corti/Knowledge at the Courts* (Micrologus, 16) (Florence: SISMEL-Edizioni del Galluzzo, 2008); the essays by Robert Halleux ("L'alchimia"), Danielle Jacquart ("La fisiognomica: il trattato da Michele Scoto"), and David Pingree ("La magia dotta") in *Federico II e le scienze*, ed. Pierre Toubert and Agostino Paravicini Bagliani (Palermo: Selerio, 1994), pp. 152–61, 338–53, 354–70, respectively; Cadden, "Charles V, Nicole Oresme, and Christine de Pizan," pp. 236–42; and Benedek Láng, *Unlocked Books: Manuscripts of Learned Magic in the Medieval Libraries of Central Europe* (University Park: Pennsylvania State University Press, 2008), chap. 8.

100 Bernard R. Goldstein, *The Astronomy of Levi ben Gerson (1288–1344)* (New York: Springer, 1985), pp. 1–2; Bernard R. Goldstein, "Levi ben Gerson: On Astronomy and Physical Experiments," in *Physics, Cosmology, and Experiment, 1300–1700: Tension and Accommodation*, ed. Sabetai Unguru (Boston Studies in the Philosophy of Science, 126) (Dordrecht: Kluwer Academic Publishers, 1991), pp. 75–82, especially pp. 76–7.

101 Nicole Oresme, *Le livre du ciel et du monde*, ed. Albert Menut and Alexander Denomy (Madison: University of Wisconsin Press, 1968); G. W. Coopland, *Nicole Oresme and the Astrologers: A Study of his Livre de divinacions* (Liverpool: The University Press, 1952); and Cadden, "Charles V, Nicole Oresme, and Christine de Pizan," pp. 211–12, 216–31.

Aristotelian cosmology, including new arguments for the Earth's possible rotation. The king's interests were not a fleeting fancy: 450 manuscripts in his 1100-book library dealt with science and medicine.

Quantitatively, the greatest demand for academic works in the vernacular came not from courts but from city schools, increasingly staffed by university masters who diffused elementary science beyond both universities and courts. Sacrobosco's *Sphere* appeared in Italian in 1314 and in a German translation by Conrad of Megenberg (d. 1376). This Paris master of arts also wrote *Das Buch der Natur* (The Book of Nature), a derivative summa of natural science probably intended for his students in the Viennese "citizens' school." In many locales, city schools had replaced monastic and cathedral schools; in Italy, the latter had disappeared by 1300. By the fifteenth century, vernacular schools from Italy to England, the Low Countries, Denmark, and Germany were teaching elementary mathematics to the sons and a few daughters of townsfolk. Whether or not Piero della Francesca attended one, his vernacular treatises on the abacus, the five regular solids, and perspective in painting point to both his seriousness with mathematics and an audience of colleagues who were adapting university *perspectiva* (optics) to what we now call "perspective."[102]

CONCLUSION

It was during the Middle Ages, in both Islamic civilization and Latin Christendom, that the sciences of nature as distinct bodies of learning became widely diffused. In Europe, the boost came from the firm institutional anchorage of natural philosophy, the elementary mathematical sciences, and medicine in the curricula of the universities, and of the most specialized mathematical sciences in their aura. This learning was clearly in demand, for it spread with the proliferating universities, developing internally and diffusing to readers of the vernaculars.

Noteworthy on intellectual grounds alone, these developments were magnified by their quantitative aspects. Of the ninety universities founded since 1200, some sixty were still active by 1500. They had created a vast learned class that had grown sufficiently to make a substantial impact on European

102 Konrad von Megenberg, *Die deutsche Sphaera*, ed. Francis Brévart (Tübingen: Max Niemeyer Verlag, 1980); Sacro Bosco, *Il Trattato de la spera, volgarizzato da Zucchero Bencivenni*, ed. Gabriella Ronchi (Quaderni degli Studi di Filologia Italiana, 15) (Florence: Accademia della Crusca, 1999); Hilde De Ridder-Symoens, "The Changing Face of Centres of Learning, 1400–1700," in MacDonald and Twomey, *Schooling and Society*, pp. 115–38, especially pp. 117–19; Paul F. Grendler, *Schooling in Renaissance Italy: Literacy and Learning, 1300–1600* (Baltimore: Johns Hopkins University Press, 1989), pp. 5–7, 306–29; James Banker, *The Culture of San Sepolcro during the Youth of Piero della Francesca* (Ann Arbor: University of Michigan Press, 2003), pp. 85–9; and J. V. Field, "Piero della Francesca's Mathematics," in *The Cambridge Companion to Piero della Francesca*, ed. Jeryldene M. Wood (Cambridge: Cambridge University Press, 2002), pp. 152–70.

society.[103] Even though the population of Europe was decreasing during the fifteenth century, the enrollments of several universities peaked at mid-century, in some cases reaching levels unmatched until the late nineteenth and early twentieth centuries. Since universities devoted a third or more of their required curriculum to scientific subjects broadly understood, they made knowledge of nature and the methodological naturalism of their faculties of arts and medicine a central part of what it meant to be educated. Not least, they did so on a grand scale. Hundreds of thousands of students were directly exposed to their teachings, and many more came into contact with these students and their views. The interest in scientific concerns that the medieval university nurtured and made its own not only diffused widely but also endured. This alliance opened a new and still expanding chapter in the history of the scientific enterprise. In this respect, the Middle Ages have not yet ended.

103 Verger, *Men of Learning in Europe at the End of the Middle Ages*, p. 3.

9

THE ORGANIZATION OF KNOWLEDGE

Disciplines and Practices

Joan Cadden

Carved on the west facade of Chartres cathedral is the most familiar and durable representation of the learned disciplines in the Middle Ages: the seven liberal arts.[1] Along with the allegorical figure of Grammar (who deploys a switch against two sleepy little boys), the six other branches of systematic knowledge appear, accompanied by their founders or main authorities – Geometry with Euclid, for example. Sculpted in the mid-twelfth century, these figures express at once the broad cultural acceptance of this particular picture of how learning was organized and also some of the problems associated with taking such cultural consensus at face value. On the one hand, the cathedral's school, famous for its academic excellence since the early twelfth century, continued to associate the seven arts with the curriculum for beginning students – first the three verbal disciplines (grammar, rhetoric, and logic) and then the four mathematical disciplines (arithmetic, geometry, astronomy, and music) (Figure 9.1).[2] On the other hand, this template had never entirely fit the shape of scientific enterprises in the early Middle Ages, and, by the time the portal was carved, changes within and beyond the school were making the taxonomy obsolete. The Chartres portal to the contrary notwithstanding, medieval disciplines were not written in stone. Both the fluidity of disciplinary divisions over time and their flexibility at any given moment pose problems for constructing an overview of the borders and compartments of medieval science. By their very nature, however, these uncertainties and variabilities do provide opportunities for understanding

[1] Philippe Verdier, "L'iconographie des arts libéraux dans l'art du moyen âge jusqu'à la fin du quinzième siècle," in *Arts libéraux et philosophie au moyen âge: Actes du quatrième Congrès International de Philosophie Médiévale, Université de Montréal, Montréal, Canada, 27 août–2 septembre 1967* (Montreal: Institut d'Études Médiévales; Paris: Librairie Philosophique J. Vrin, 1968), pp. 306–55.

[2] On this figure, see John E. Murdoch, *Album of Science: Antiquity and the Middle Ages* (New York: Charles Scribner's Sons, 1984), fig. 172, p. 191.

The author is grateful to the Max-Planck-Institut für Wissenschaftsgeschichte, Berlin, to the members of the 1997–8 Abteilung II research group, and to its Director, Lorraine Daston, for support and advice.

Figure 9.1. Allegorical representations of quadrivial arts with attributes. The quadrivium, or four mathematical arts, appears in a ninth-century copy of a tract on arithmetic, accompanied by identifying objects. From left to right, Music holds a stringed instrument; Arithmetic has a number cord in her right hand and displays a technique of finger-reckoning with her left; Geometry holds a measuring rod, or *radius*, and looks down at a tablet inscribed with geometrical figures; and above Astronomy's head are the stars, Moon, and Sun. Because of differences among authorities, as well as different readings or misreadings of texts or earlier images, such depictions did not follow set formulas. The column in this illustration may reflect Martianus Capella's description of geometry or may be an allusion to the role of geometry in architecture; the torches held by Astronomy have not been explained. By the end of the Middle Ages, new symbols were available: Arithmetic sometimes carries an abacus board, and Astronomy sometimes has an astrolabe or an armillary sphere. By permission of Staatsbibliothek Bamberg, MS HJ.IV.12, fol. 9v.

how the various sciences were both delimited and related, and the extent to which natural science constituted a coherent endeavor in the Middle Ages.

The purpose of this chapter is to investigate the boundaries and relations among medieval disciplines dealing with the natural world. Since medieval intellectuals themselves sought to organize the knowledge they inherited or produced about the natural world, their own views serve as a point of departure. The systems of classification they articulated reflected at once a respect for the programs of their ancient sources, an attentiveness to the problems of coordinating various traditions, and a concern for the ways in which learning could be used. In their prefaces or in the arrangement of their works, medieval authors named, defined, and diagramed the relationships among the disciplines that embodied what we have come to regard as "science." Yet, alongside the formal and explicit mapping of knowledge, other lines of organization, often informal and unspoken, emerged. Understanding the taxonomy of the sciences therefore requires placing them in the context of medieval scientific practices; that is, in terms of the ways medieval people acquired, transmitted, and applied ideas about nature.

Given the changes over time and the slippage between theory and practice, the result is not a clear and fixed map of the sciences but rather a set of perspectives from which to approach the question, "What was medieval science?" The first section of the chapter surveys general notions about disciplines and their relations to one another as they were laid out before the twelfth century. For scholars of that period, retaining and transmitting the outlines of received wisdom was often a difficult task. In such an environment, however, scholars were free to try out various strategies and new uses for old knowledge. The second section sketches some of the changes that rendered the older formulations obsolete. Starting in the late eleventh century, new social conditions for learning and the translation of Greek and Arabic texts introduced not only new subject matter but also new methods and even new goals for the sciences. Finally, the third section deals with the ways in which these changes shaped and were shaped by new conditions, especially the organization of learning within the university, between the thirteenth and fifteenth centuries.

THE ERA OF THE LIBERAL ARTS: FIFTH TO TWELFTH CENTURIES

The Latin terms *ars*, *disciplina*, and *scientia* all signified elements of *philosophia* and, as such, were manifestations of ordered thought. They were frequently associated with specific definitive texts and with characteristic rules by which they operated – that is, both with what was to be known and what was to be done. When arts, disciplines, and sciences were distinguished from each other, they usually formed a hierarchy of abstraction or of certainty. For example, the encyclopedist Bishop Isidore of Seville (ca. 560–636) assigned the terms *scientia* and *disciplina* to what was known with certainty. They are about things that cannot be other than they are. Arts, in contrast – including tenets of natural philosophy, such as the belief that the Sun is larger than the Earth – were the domain of mere opinion. Such distinctions were not, however, either fixed or enforced. An author indebted to Isidore reported a variant: Disciplines deal with what can be produced by thought alone, whereas arts, such as architecture, are expressed in material media.[3] And Isidore himself went on to conflate disciplines and arts, saying, "There are seven disciplines of the liberal arts."[4]

Medieval authors often employed one of these three terms, which will be used interchangeably here, to designate the principal divisions of "philosophy," as the recognized body of systematic learning was persistently called. But just as the meanings and relations of "sciences" and "arts" varied, so did their membership and order – and indeed the principles upon which they were arranged. Medicine, for example, might be located according to its subject matter (e.g., the maintenance of health), according to the type of study it represented (e.g., a practical art), or according to texts in which its substance was contained (e.g., Galen's *Art of Medicine*). Furthermore, both architecture and medicine were classified sometimes as mechanical or practical and sometimes as liberal or theoretical arts. Even the familiar names of individual disciplines could be problematic: "*astronomia*" and "*astrologia*" were sometimes synonymous and sometimes quite distinct.

This tangle of terms suffices to illustrate some of the issues involved in concepts about the constellation of knowledge. The structures were not simple; the articulations of them were not formulaic. The utterances of an authority like Isidore or the representations of a source like the cathedral at Chartres were only a part of what was involved, but they convey some of the difficulties of drawing a map of natural knowledge in the early Middle Ages.

THE LIBERAL ARTS AND THEIR SISTERS

Medieval authors did draw such maps, however. Divisions of the sciences have a long and intricate history, borrowing from a variety of traditions and reflecting the dynamics of the intellectual scene.[5] Even before the influence

[3] Hugh of Saint Victor, *The Didascalicon of Hugh of Saint Victor*, trans. Jerome Taylor (New York: Columbia University Press, 1961), bk. II, chap. 1, p. 62.

[4] Isidore of Seville, *Etymologiarum sive originum libri XX*, ed. W. M. Lindsay, 2 vols. (Scriptorum Classicorum Bibliotheca Oxoniensis) (Oxford: Oxford University Press, 1911),vol. 1, bk. II, chap. 2, sec. 2; bk. I, chaps. 1 and 2; see also bk. III, chap. 3, p. 86. All translations are mine unless otherwise noted. See also Boethius, *De trinitate*, in *The Theological Tracts, The Consolation of Philosophy*, ed. E. K. Rand (Loeb Classical Library, 74) (Cambridge, Mass.: Harvard University Press, 1973), pp. 2–31 at Prologue, p. 4.

[5] The basic treatments of the medieval disciplines and their classification are: Richard William Hunt, "The Introduction to the 'Artes' in the Twelfth Century," in *Studia mediaevalia in honorem admodum*

of Arabic science and the wholesale introduction of Aristotle's natural works, some basic elements of what was to be a continuing medieval conversation about disciplines were already present. The most important of these were (1) the seven liberal arts and, sometimes, their stepsisters, the mechanical arts; (2) the distinction between theoretical and practical sciences, with its subdivision of the theoretical into divine, mathematical, and natural sciences; and (3) the schema of physical, logical, and ethical knowledge.

A highly influential work by Martianus Capella (fl. ca. 365–440) enumerated seven liberal arts and offered an introduction to (as well as a personification of) each, including the four "mathematical" arts (later named the "quadrivium"): arithmetic, geometry, astronomy, and music. For early medieval authors, these illuminated nature in various ways. Mathematical relations represented the essence of the created world, the subject of mathematical sciences was quantity separated in thought from the (natural) matter in which it actually inhered, and the quadrivium had functions and uses related to material objects. The particulars of arithmetic, geometry, and astronomy are treated in other chapters of this volume (see North, Chapter 19; Molland, Chapter 21). Against its persistent inclusion by medieval authors, historians of science have generally declined to treat music seriously in this context. Its claim to a place among the mathematical sciences rests on its central concern with intervals and thus with ratios. In addition, through such notions as harmony or proportion, which applied not only to sounds but also to the macrocosm of the heavens and the microcosm of the human body, the discipline of music sometimes incorporated significant natural-philosophical as well as mathematical material.

The other group of arts, the "trivium" – grammar, rhetoric, and logic – bore virtually no formal relation to the pursuit of natural knowledge in its medieval or modern senses. In practice, however, the verbal sciences were relevant in three ways. First, medieval authors used literary skills, represented by grammar and rhetoric, to analyze the natural questions contained in authoritative texts, from the book of Genesis (in which the six days of Creation became a traditional site for discussions of the natural world) to the *Timaeus* of Plato. In addition, literary sources contained valuable wisdom: Virgil's *Georgics* contained agricultural information, and one twelfth-century author referred to Hesiod as a "teacher of natural science."[6] Finally, although it took on its full prominence only later, the discipline of logic became

reverendi patris Raymundi Josephi Martin, Ordinis Praedicatorum, S. Theologiae Magistri, LXXum natalem diem agentis (Bruges: De Tempel, [1948]), pp. 85–112; James A. Weisheipl, "Classification of the Sciences in Medieval Thought," *Mediaeval Studies*, 27 (1965), 54–90; and James A. Weisheipl, "The Nature, Scope, and Classification of the Sciences," in *Science in the Middle Ages*, ed. David C. Lindberg (Chicago History of Science and Medicine) (Chicago: University of Chicago Press, 1978), pp. 461–82. See also A. J. Minnis, *The Medieval Theory of Authorship*, 2nd ed. (Philadelphia: University of Pennsylvania Press, 1988), especially chap. 1.

6 Brian Stock, *Myth and Science in the Twelfth Century* (Princeton, N.J.: Princeton University Press, 1972), pp. 43, 77.

relevant as a subject and as a method, bearing on such questions as how much certainty a science could attain.

Close to but always in the shadow of the liberal arts stood what came to be called the mechanical arts, also often numbered seven, though their exact membership varied.[7] The usual candidates included textiles, arms, commerce, agriculture, hunting, medicine, theater, architecture, and sports. (Later enumerations included navigation, alchemy, and various forms of divination.) They played two roles in the conceptualization of medieval disciplines. The first was negative. In contrast to the liberal arts, the mechanical or "adulterate" arts engaged the body as well as the mind, and their subject was "merely human works."[8] The superiority of the former was reinforced by the social distinctions between those who work with their hands and those who do not – "the populace and sons of men not free," in contrast to "free and noble men."[9] More positively construed, the mechanical arts supplemented the liberal arts, particularly with respect to their engagement with the natural world. This very involvement with objects, which placed them outside the domains of philosophy (they were regularly denied the status of "discipline"), made them potentially useful for expanding the systematic understanding of nature. Some of the links between the mechanical and liberal arts manifested themselves in practices and instrumentation. Thus, a pair of compasses not only regularly accompanies the allegorical figure of Geometry but also appears as an emblem of stonemasons.

TRADITIONS OF CLASSIFICATION

Although the notion of the seven liberal arts was the most widely known basis for classifying knowledge, including that concerned with the natural world, it coexisted with other persistent schemata. The existence of alternatives invited scholars to choose, combine, or modify their elements in ways that suited them. The second major framework distinguished between theoretical (or speculative) sciences and practical (or active) sciences. This division was most influentially articulated in the Latin works of Boethius (ca. 480–525), who depicted the Lady Philosophy with the Greek letters theta (for "theory") and pi (for "practice") on her garment.[10] The so-called practical sciences, however, concerned not the efforts of artisans but rather the responsibilities of the aristocracy – ethics, household management ("economics"), and politics. Under the influence of Platonism and Christianity, however, the contemplative enjoyed a higher value than the active.

7 Elspeth Whitney, "Paradise Restored: The Mechanical Arts from Antiquity through the Thirteenth Century," *Transactions of the American Philosophical Society*, 80, no. 1 (1990), 1–170 at chap. 4.

8 Hugh of Saint Victor, *Didascalicon*, bk. I, chap. 8, p. 55.

9 Ibid., bk. II, chap. 20, p. 75.

10 Boethius, *De consolatione philosophiae*, in Rand, *The Theological Tracts, The Consolation of Philosophy*, pp. 130–435 at bk. I, prosa 1, ll. 18–19, p. 132.

The three constituents of "theory" likewise formed a value hierarchy: Theology was concerned with a subject that existed independently of matter; mathematics with the formal relations abstracted from their material subjects (e.g., dimensions abstracted from the land they measured); and *physica* (that is, natural philosophy) with the properties of material objects. Just as association with manual labor devalued the dignity of the mechanical arts vis-à-vis the liberal arts, so association with matter placed mathematical and natural sciences in descending order below theology. Such a ranking, which, according to Boethius, corresponded to different ways of knowing,[11] suited a Christian sensibility that emphasized the triumph of immaterial spirit over material flesh. It was not, however, static. In the intellectual as in the spiritual realm, the mundane could be a stepping stone to higher levels, thus lending dignity to natural and mathematical sciences.

The third and less influential arrangement of the disciplines derived from an ancient Stoic tradition and was passed on by Isidore of Seville. It distinguished ethics (that is, the active sciences of the second scheme), *physica* (including the quadrivium), and logic (including the trivium).[12] Whereas Boethius had separated the mathematical disciplines from natural philosophy, here mathematics is part of it. Indeed, this arrangement sometimes also included in the category of *physica* the more practical arts of astrology, mechanics (meaning certain kinds of craft production), and medicine.[13] Although it, too, found expression within monastic schools, this taxonomy was less hierarchical than the previous one and less closely associated with a program of spiritual ascent. In these respects, it placed a higher and more independent value on at least some natural and verbal sciences.

Throughout the early Middle Ages and beyond, tensions among the various schemata, along with the variety of traditions that nourished them, gave rise to a fluid and eclectic outlook on the divisions and relations of scientific disciplines. The work of individual scholars often represented compromises among the various options. For example, the abbess Herrad of Landsberg (ca. 1130–1195) represented Philosophy as a queen encircled by figures of the seven liberal arts, wearing a crown with figures of ethics, logic, and physics.[14]

Such reworkings have contributed to the perception that "nobody knew what to make of 'philosophy' or 'science,'" and scholars have suggested

11 Boethius, *De trinitate*, chap. 2, p. 8.

12 Weisheipl, "Classification of the Sciences in Medieval Thought," pp. 63–4.

13 Manuel C. Díaz y Díaz, "Les arts libéraux d'après les écrivains espagnols et insulaires aux VIIe et VIIIe siècles," in *Arts libéraux et philosophie au moyen âge*, pp. 37–46; Murdoch, *Album of Science*, fig. 29, p. 40; Hrabanus Maurus, *De universo, in Opera Omnia*, ed. Martin Mabillon, (Patrologia Latina, 111) (Paris: J.-P. Migne, 1864), vol. 5, cols. 9–614 at bk. II, chap. 1, col. 113.

14 Herrad of Hohenbourg, *Hortus deliciarum*, reconstructed with commentary by Rosalie Green, Michael Evans, Christine Bischoff, and Michael Curschmann with T. Julian Brown and Kenneth (Levy Studies of the Warburg Institute, 36), 2 vols. (London: The Warburg Institute; Leiden: Brill, 1979), vol. 2, pl. 18. The manuscript was destroyed in the Franco-Prussian War of 1870. On this figure, see Murdoch, *Album of Science*, fig. 31, p. 41.

that *physica* remained a virtually empty category until the assimilation of Aristotle's natural works starting in the twelfth century.[15] In one sense, this is true, in that unlike the liberal arts, each of which was regularly linked to a basic text (Porphyry on logic, Boethius on music, and so forth), natural philosophy had no standard introductory authority. But this perspective ignores not only the extent to which other kinds of texts – Genesis and Plato's *Timaeus* – provided textual grist for the natural-philosophical mill but also the extent to which subject matter was imported from a variety of other categories. Latin authors not only arranged and rearranged but also added to the list of disciplines, a process that further illustrates the malleable and living nature of medieval classifications. In the ninth century, an encyclopedic work by Hrabanus Maurus (ca. 780–856) full of information about natural philosophy included under the heading of *physica* not only arithmetic, geometry, astronomy, and music but also such "practical" or "mechanical" arts as astrology, medicine, and mechanics.[16] This realignment reflects a process by which information and ideas migrated across putative boundaries.

Especially in an environment in which authoritative texts were scarce, scholars often appealed to learning in one domain to illuminate another. Medicine in particular was a resource for those seeking to explore the principles of nature. Isidore of Seville had likened medicine to philosophy itself because it drew upon all of the liberal arts.[17] In the course of the early Middle Ages, standard medical texts mentioned the constituents of both the body and the environment; materia medica spoke of plants, animals, and stones; and tracts on obstetrics touched on principles of reproduction as well as practical advice. The intellectual cross-fertilization suggested by the permutations of classification is confirmed by material evidence. For example, book owners bound medical, mathematical, and natural-philosophical texts together in the same manuscript books.

The absence of specialization enhanced these processes. The Venerable Bede (672–735) wrote on geographical subjects in addition to mathematical disciplines. Practitioners of medicine might be socially distinguishable, but its content was accessible to others. After giving a stranger some advice on his health, for example, Gerbert of Aurillac (945–1003) offered this disclaimer: "Do not ask me to discuss what is the province of physicians, especially because I have always avoided the practice of medicine even though I have

[15] Weisheipl, "Nature, Scope, and Classification of the Sciences," p. 472. See also Weisheipl, "Classification of the Sciences in Medieval Thought," p. 62; and Whitney, *Paradise Restored*, p. 59 and n. 8.

[16] Hrabanus Maurus, *De universo*, bk. V, chap. 1, col. 413. See Díaz y Díaz, "Les arts libéraux d'après les écrivains espagnols et insulaires," p. 41, nn. 17 and 43; Murdoch, *Album of Science*, fig. 29, p. 40; Whitney, *Paradise Restored*.

[17] Isidore of Seville, *Etymologie*, bk. IV, chap. 13; cf. Bruce S. Eastwood, "The Place of Medicine in a Hierarchy of Natural Knowledge: The Illustration in Lyon Palais des Arts, ms. 22, f. 1r, from the Eleventh Century," *Sudhoffs Archiv*, 66, no. 1 (1982), 20–37.

striven for a knowledge of it."[18] Taken together, this diverse body of evidence bears witness to the gradual formation of a loosely associated body of knowledge about the constituents, causes, and arrangements of the natural world rather than the scholarly void that has been suggested.

CULTURAL FUNCTIONS OF DISCIPLINARY IDEALS

Both the attempts to define and arrange specific disciplines and the conditions that moved or eroded boundaries manifested themselves in the various uses to which medieval authors put the sciences. During the early Middle Ages, even the reiteration of fixed names and definitions could serve a variety of cultural, religious, and political functions. For example, in her drama about the conversion of a prostitute, the abbess Hrotswitha of Gandersheim (ca. 935–1000) has the saintly Paphnutius name the quadrivial arts and define the discipline of music as he explains to his students the harmony of the elements in the human body.[19] Both for Hrotswitha, whose own familiarity with the liberal arts was extensive, and for her protagonist, the preservation of learned traditions was a significant project in itself. Similarly, something as simple as a shared terminology facilitated more complex scholarly exchanges, as when Gerbert, the future Pope Sylvester II, sought help from correspondents across Europe in acquiring old and new works on *astrologia*.[20]

Such cultural reproduction played a role in social and political developments, such as the construction of the Carolingian Empire and the evolution of clerical power. The prominence of the seven arts in the early Middle Ages is as much a product of a political agenda as it is a reflection of the intellectual projects and practices of the time. Charlemagne's biographer Einhard emphasized that he had his sons and daughters educated in the liberal arts (of which the ruler's own favorite was astronomy).[21] A classicizing curriculum, like a classicizing biography, suited Carolingian claims to be successors to the Roman Caesars and protectors of the Roman Church. Alcuin of York (ca. 735–804), master of Charlemagne's palace school and Bede's intellectual heir, was thus advancing a broad cultural and political program, as well as following his own scholarly trajectory, when he gave the subjects of the quadrivium a respectable (though not a prominent) place in the curriculum.

18 Gerbert of Aurillac, *The Letters of Gerbert with His Papal Privileges as Sylvester II*, trans. Harriet Pratt Lattin (Records of Civilization, Sources and Studies) (New York: Columbia University Press, 1961), no. 159, p. 187; cf. no. 178, pp. 207–8.

19 Hrotswitha von Gandersheim, "Conversio Thaidis meretricis" [or "Pafnutius"] 2.5 in *Opera*, ed. H. Homeyer (Paderborn: Ferdinand Schöningh, 1970), pp. 328–49, scene 1, lines 4–22.

20 Gerbert of Aurillac, *Letters of Gerbert*, no. 15, pp. 54–5; cf. no. 138, pp. 168–9.

21 Einhard, *Vita Karoli Magni: The Life of Charlemagne*, ed. and trans. Evelyn Scherabon Firchow and Edwin H. Zeydel (Miami: University of Miami Press, 1972), chap. 19, pp. 78–9 and chap. 25, pp. 92–3.

Hrotswitha of Gandersheim illustrates how command of the terminology and substance of scientific disciplines conveyed and even constituted clerical superiority over the laity. In her allegorical Latin drama on the martyrdom of virgins named Faith, Hope, and Charity, one of their persecutors inquires about the girls' ages. Their mother, Wisdom, asks, "Does it please you, my daughters, that I should exhaust this fool with an arithmetic disputation?" and she proceeds to overwhelm him with a long and learned exposition on numbers, derived from Boethius.[22] Although the classical disciplines are not as powerful as the Christian virtues, the allegorical figure Wisdom and the abbess Hrotswitha wield the two sets of weapons in close coordination, appropriating and thereby according dignity to the arts. As the cases of Hrotswitha and Alcuin (a Benedictine monk) suggest, the naming and arrangement of the disciplines belonged first and foremost in the early Middle Ages to monastic environments that played a central role in the transmission and validation of the disciplines.

BEYOND DISCIPLINARY IDEALS

The political and cultural uses of the scientific disciplines depended in part on their clarity, stability, and links with recognized authority. To that extent, they were conservative – in tension with the dynamics by which the definitions and materials of individual disciplines and natural knowledge more generally were shifting and expanding. Gerbert expressed an awareness of precisely this problem as he sought to enhance the texts and practices available to a student of arithmetic by laying out rules for the use of an abacus – a tool of practical calculation: "Do not let any half-educated philosopher think that [these rules] are contrary to any of the arts or to philosophy."[23] The "half-educated" purists did not prevail. An eleventh-century tract on geometry incorporated not only passages from Euclid but also discussions of the abacus, land measurement, map making, and land tenure.[24]

The practices discussed so far, even those relating to calculation and cartography, were textual in nature. They involved the transmission and elaboration of written traditions, whether associated with an ideal curriculum or with more immediate and mundane matters. As Gerbert's apparently contested interest in the abacus suggests, we have evidence of nontextual practices inscribed in sources ranging from the geometrical artifacts of stonemasons to records of the heuristic methods used in schools. A monastic teacher of the twelfth century took his pupils out in front of the church in the middle of the night, extending his arm and using his fingers to show them how to

[22] Hrotsvitha von Gandersheim, "Passio sanctarum virginum Fidei Spei et Karitatis" [or] "Sapientia," 2.6 in *Opera*, pp. 181–99, scene 3, secs. 8–22 on pp. 184–7.

[23] Gerbert of Aurillac, *Letters of Gerbert*, no. 7, p. 45; see also pp. 46–7, n. 5.

[24] John E. Murdoch, "Euclid," in *Dictionary of Scientific Biography*, IV, cols. 437–459 at cols. 443a–b.

observe the course of the stars.[25] Gerbert himself illustrates the comfortable coexistence of textual and manual practices. He was famous for the swiftness with which he could do calculations using an abacus, and he described to a student how to construct and use a tube for astronomical observations.[26]

Determination of the date of Easter generated both active and contemplative science.[27] It was a source of perennial concern (as well as sectarian discord) and required the use of astronomical data and mathematical calculations. The problem being solved was essentially liturgical in that its purpose was to answer not a question about nature but rather one about ritual observance seen through the lens of natural phenomena. Thus the fixing of Easter bore a limited relationship to quadrivial and natural-philosophical disciplines as formally defined. Nevertheless, just as artists depicted both stonemasons and the allegorical figure of Geometry with a compass, so copyists and librarians perceived some link when they copied and bound these texts on calendrical calculation along with a variety of materials treating quadrivial and natural-philosophical subjects.[28] The abacus and the astrolabe may often have been instruments more of intellectual fascination than of practical application, but they were understood and even used to illustrate principles and perform specific operations.

CULTURAL CONFLUENCES AND TRANSFORMATIONS OF THE ARTS: TWELFTH CENTURY

The existence of a variety of tools adds complexity to our picture of early-medieval practices, suggesting not only a manual but probably also an oral dimension to the pursuit and transmission of natural knowledge – from eclipse prediction to surgery, from numerology to divination. Yet the most powerful scientific instrument in the Middle Ages remained the book. And within the book, though illustrations and diagrams played a variety of important roles, written words did the lion's share of the work. The period from the late eleventh to the early thirteenth centuries witnessed a proliferation of text-based analytical and argumentative techniques. These accompanied the formation of the universities, which dominated the intellectual scene in the thirteenth and fourteenth centuries. The enrichment of the substance, methods, and taxonomies of the sciences during this transition in the Latin West depended on two closely related processes: the selection, translation, adaptation, and incorporation of Greek and Arabic learning; and the expansion

25 Philippe Delhaye, "L'organisation scolaire au XIIe siècle," *Traditio*, 5 (1947), 211–68 at p. 252, n. 82.

26 Gerbert of Aurillac, *Letters of Gerbert*, letter of Richier quoted at no. 7, p. 46, n. 1; no. 2, pp. 36–9.

27 Stephen C. McCluskey, *Astronomies and Cultures in Early Medieval Europe* (Cambridge: Cambridge University Press, 1997), especially chap. 5.

28 Delhaye, "L'organisation scolaire au XIIe siècle."

of literary and philosophical activity associated with the "Renaissance of the Twelfth Century."

CONVERGING TRADITIONS

European access to Greek, Arabic, and Hebrew learning was concentrated in southern Italy and Spain – two multicultural crossroads of Mediterranean societies. Constantine the African (fl. 1065–85), for example, a converted Muslim who became a monk at Monte Cassino, both carried Arabic medical books from North Africa to Italy and rendered them intelligible to a Latin audience. Consolidating a huge body of learning from the Aegean, West Asia, and North Africa, Arabic works by philosophers and physicians especially fostered the adoption of Aristotle's ideas and methods and provided interpretations of the Aristotelian natural world. With respect to practices, Latins learned about specific instruments, such as the astrolabe and the zero, and an array of ways to treat and order texts – from structures for medical formularies to modes of philosophical commentary. With respect to the organization of knowledge, Europeans confronted a number of serious challenges that opened new areas of inquiry and revivified old ones. Arabic authors not only proposed their own versions of how disciplines were constituted and arranged but also made massive, highly developed substantive contributions to subjects that had commanded little or no place in older Latin systems. Areas such as optics or alchemy, hardly discussed in early Latin schemata for dividing the sciences, became impossible to ignore. Natural science had become more important, while the substance of its diverse parts had become richer and even harder to map.

In ways that varied with local conditions, many twelfth-century scholars not only welcomed but actively pursued the lush profusion of possibilities contained in newly available texts. In some areas of Southern Europe, for example, medicine was the intellectual seed around which natural questions crystallized. At Salerno and Monte Cassino in the late eleventh century, and later in northern Italy and southern France, it reshaped Latin inquiries into the natural world. First, medicine as received from Greek and Arabic sources offered explicit models for the relation between theory and practice in the arts. Second, as medical writers sought to elaborate and strengthen the theoretical foundations of their knowledge, they directly addressed the form and content of natural philosophy. In doing so, authors like Constantine the African not only shaped medicine but also conveyed to natural philosophy a flood of material that was to be put to many uses.[29] The theory of the four elements, for example, which finds no specific place in the older taxonomies of natural knowledge, occupied a pivotal position between the

[29] Paul Oskar Kristeller, "The School of Salerno: Its Development and Its Contribution to the History of Learning," *Bulletin of the History of Medicine*, 17 (1945), 138–94.

constitution of the world in general and the physiological principles of medical science. Along with the body of knowledge, textual and pedagogical practices developed. Prominent among them were loosely organized series of short queries that came to be known as "Salernitan Questions."[30] These bear the marks of a method in which rote learning was coupled with medical apprenticeship. As they were disseminated, the questions acquired written answers, which in turn became more elaborate, not only incorporating more natural-philosophical material but also making room for the seeds of debate. For example, one such text summarizes Hrabanus Maurus's explanation for the deadly look of the basilisk, then states that it is *not* the creature's look but rather its ability to poison the air that makes it dangerous.[31]

By the late twelfth century, such Salernitan Questions flourished in the very different cultural climate of the Île-de-France, Normandy, and England, where natural philosophy had previously drawn much of its content and its methods from literary studies. Indeed, northern learning about the natural world owed more to the practices associated with the trivium, especially grammar and rhetoric, than to those associated with the quadrivium. In particular, the glosses produced at Chartres and elsewhere, not only on Plato's cosmogonical myth, the *Timaeus*, but also on the works of Macrobius (fl. early fifth century) and Martianus Capella, brought to the intellectual stage such powerful concepts as prime matter and the four elements. Scholars in this environment applied a variety of textual techniques to topics such as the emergence and differentiation of the cosmos. William of Conches (ca. 1100–1154), for example, after writing a formally conventional gloss on the *Timaeus*, produced a work that combined aspects of Plato's account of nature with Salernitan material. Whereas some of William's contemporaries mustered the quadrivial and natural-philosophical material to serve literary purposes, he struggled to define, give shape to, and legitimize the discipline of natural philosophy (*physica*) as "the true understanding of what exists and is seen and of what exists and is not seen."[32] And whereas William drew upon the Northern academic culture of grammar and rhetoric, others, most notably Peter Abelard (1079–1142), advanced the third member of the trivium, logic.

The variety of academic practices (question-and-answer and textual explication, mythological poetry and practical prose) and the diversity of interests (medical and cosmological, natural-philosophical and metaphysical) formed

30 Brian Lawn, *The Salernitan Questions: An Introduction to the History of Medieval and Renaissance Problem Literature* (Oxford: Clarendon Press, 1963).

31 Brian Lawn, *The Prose Salernitan Questions Edited from a Bodleian Manuscript (Auct. F.3.10): An Anonymous Collection Dealing with Science and Medicine Written by an Englishman c. 1200 with an Appendix of Ten Related Collections*, ed. Brian Lawn (Auctores Britannici Medii Aevi, 5) (London: for the British Academy by Oxford University Press, 1979), B.105, p. 49 and P.7, p. 209.

32 William of Conches, *Philosophia*, ed. Gregor Maurach (Studia, 16) (Pretoria: University of South Africa, 1980), bk. I, chap. 1, sec. 4, p. 18. See Stock, *Myth and Science in the Twelfth Century*, especially pp. 173–83; and Helen Rodnite Lemay, "Guillaume de Conches' Division of Philosophy in the *Accessus ad Macrobium*," *Medievalia*, 1 (1977), 115–29.

one axis of the twelfth-century legacy; the wholesale importation of texts formed the other. These changes in turn produced new challenges and opportunities for European intellectuals seeking to order knowledge and organize education.

A NEW CANON OF THE ARTS

One of the earliest and most influential Latin treatises to reflect the changing intellectual climate was *On the Division of Philosophy* by Dominicus Gundissalinus. Active in Toledo (Spain) in the late twelfth century, he had participated in the translation efforts that brought so much previously unavailable scholarship into the West. His classification of the sciences reflects lasting reorientations in Western thinking about the scientific disciplines: (1) direct indebtedness to Arabic ideas about the arrangement of systematic knowledge; (2) adjustment to the introduction of massive new material and even new sciences; and (3) involvement of classification in fundamental questions about the order of nature and the path to secure knowledge.

Gundissalinus's work drew heavily upon a treatise of al-Fārābī (ca. 873–950), which he had translated into Latin, not only with respect to the enumeration of specific branches of learning but also with respect to the nature and order of the world that natural science sought to describe. Several of Gundissalinus's moves were far from revolutionary, as the improvisations of the early Middle Ages attest. He expanded mathematics beyond the traditional quadrivium (arithmetic, music, geometry, and astronomy) to include the science of weights (statics) and the science of engines (i.e., using natural bodies and mathematical principles to some end).[33] These last two were among the areas virtually unexplored by earlier Latin authors and amply developed within the Arabic tradition. In addition, though on a much more modest scale, he offered subdivisions of *physica* or, as he called it, "natural science": medicine, omens, necromancy, magical images, agriculture, navigation, optics ("mirrors"), and alchemy. The strong presence of sciences of divination and control is an indication of the powerful influence of new intellectual appetites and materials.

Beginnings of deeper structural changes also appear in Gundissalinus, among them the organization of the sciences around Aristotelian texts. His eight-part division of natural philosophy bypasses his own list of subdisciplines just mentioned and sets up a sequence of subjects ranging from the study of bodies in general through the more particular properties of minerals, plants, and animals. He names a text newly available in Latin as an element in his characterization of each subdivision of science.[34] Although

[33] Dominicus Gundissalinus, *De divisione philosophiae*, ed. Ludwig Baur, *Beiträge zur Geschichte der Philosophie des Mittelalters*, 4, fasc. 2–3 (Münster: Aschendorff, 1903), pp. 122–3.

[34] Ibid., pp. 19–27.

the specific texts did not all remain the same, this was the way in which natural philosophy came to be structured in European universities.

In some respects, the way Gundissalinus presents the relations among the sciences reflects ideas about pedagogical process – one must learn grammar before turning to more complicated subjects – but this sequence also mirrors his ideas about the ranking of objects of knowledge and the ways in which they are known. In Gundissalinus's day, long-familiar sources ranging from Plato to Augustine, and new works by Arabic authors like al-Fārābī and Ibn Sīnā (Avicenna; 930–1037), lent a strong Platonic color to Latin philosophy, one aspect of which was the conceptualization of a hierarchy of substances and thus of the academic subjects treating them. Physics studies the general principles of change without reference to any particular bodies; cosmology studies change of place in otherwise changeless bodies; generation and corruption studies the changes of bodies coming to be and passing away; and the lowest subjects are concerned with the specific properties and operations of particular bodies in the elemental world. This ladder of value resonates with some of the older classifications, such as Boethius's view that theology is more exalted than mathematics and that mathematics is higher than natural philosophy.[35] At the same time, it seems to undermine the status Gundissalinus lent to medicine, alchemy, and other arts concerned with the material and the particular rather than the formal and the general.[36]

Gundissalinus's attempts to sort out the subjects, relations, and values of knowledge about nature, like those of earlier medieval authors, bespeak the variety, flexibility, and mobility of the disciplines and reflect an active intellectual scene. Furthermore, older textual practices, from the examination of etymologies to the preparation of compilations, continued to play a role in scientific learning. By the early thirteenth century, however, much had changed in the substance, methods, and conditions of the sciences. The work done by Isidore of Seville or Hrotswitha of Gandersheim simply to name, define, and iterate the fundamentals of the disciplines was no longer called for in a world in which thousands of students traveled from one European city to another to hear masters lecture, call out questions at disputations, and purchase and annotate books.

THE ERA OF THE FACULTIES OF ARTS: THIRTEENTH TO FOURTEENTH CENTURIES

The works of Aristotle, whose titles became metonymic for many disciplines in the later Middle Ages, had displaced (though not entirely erased) the

35 Weisheipl, "Classification of the Sciences in Medieval Thought," pp. 68–81.

36 George Ovitt, Jr., *The Restoration of Perfection: Labor and Technology in Medieval Culture* (New Brunswick, N.J.: Rutgers University Press, 1986), pp. 107–36.

liberal arts as critical landmarks on the map of learning. More technical, specialized, and advanced, they never took on the iconographic status of their predecessors – arithmetic with her number cord or astronomy with her quadrant. Some general classificatory principles, however, persisted. Writers continued to distinguish in principle between "sciences" and "arts." According to Thomas Aquinas (ca. 1225–1274), the former (e.g., metaphysics and physics) involve "only knowledge," whereas the latter (e.g., logic, which constructs syllogisms, and astronomy, which calculates planetary positions) "involve not only knowledge but also a work that is directly a product of reason itself" or, in the case of nonliberal arts (e.g., medicine and alchemy), "involve some bodily activity."[37] As in the earlier period, however, these distinctions were not widely enforced in the language or institutions of the late Middle Ages; thus students in the "arts" faculties of universities attended lectures on both physics and logic.

The world in which knowledge about nature was shaped and transmitted had also changed considerably by the early thirteenth century. The growth of towns, for example, had created demand for higher levels of practical knowledge in such areas as calculation and medicine, and new forms of political administration had created demand for training not only in law but also in astrology. With support from civil or ecclesiastical authorities (or both), universities took shape. Through the formulation of curricula, the support of advanced investigation, and the position of natural sciences within the larger institutional structure, they provided both opportunities and constraints for defining and pursuing scientific disciplines.

ARTS AND METHODS

Questions about curriculum and pedagogy, challenges associated with the profusion of disciplines, and debates contained in the works of newly available authorities all contributed to a sense of urgency about the methods of the sciences. Did each have its own rules of investigation, forms of argumentation, and degree of certainty? Boethius's assertion that divine, mathematical, and natural sciences were known differently[38] no longer sufficed for thirteenth- and fourteenth-century scholars interested in the distinction between "natural philosophy" and "mathematics." The latter had once meant the quadrivium, but mathematical developments in the Islamic world not only revolutionized old categories, such as arithmetic, but also introduced new ones. In particular, the distinction between mathematical and natural

[37] Thomas Aquinas, *The Division and Methods of the Sciences: Questions V and VI of His Commentary on the De Trinitate of Boethius*, trans. Armand Maurer, 3rd ed. (Toronto: Pontifical Institute of Mediaeval Studies, 1963), q. 5, art. 1, p. 12.

[38] Boethius, *De trinitate*, chap. 2, p. 8.

knowledge, unsettled even in the early Middle Ages (see Figure 9.2), receded with the incorporation of what came to be called "middle sciences." Optics, the science of weights, and astronomy (the last of these once housed in the quadrivium) dealt with specific properties of natural objects but employed mathematical representations and demonstrations. Some of the issues raised by these changes were formal: Do the middle sciences actually constitute a subcategory of mathematics?[39] Others were epistemological: What degree of certitude can astrology or medicine attain?[40]

Theoretical debates on the relation of subject to method took a number of forms. From one perspective, the crux of the matter was what kind of demonstration each group of sciences could muster. The conviction that geometry (as represented by Euclid) could produce airtight proofs and hence incontrovertible explanations enjoyed wide acceptance, as did the complementary view that natural philosophy, insofar as it dealt with material objects and was thus burdened by the attendant contingencies, could not aspire to give a complete and certain account. Disagreement nevertheless abounded. For some scholars, such as Albertus Magnus (ca. 1200–1280), the physical world, in which form and matter were actually inseparable, posed questions to which mathematical methods could offer only partial solutions because they treated just a small number of properties abstracted from the actual natural body. For others, such as Roger Bacon (ca. 1219–ca. 1292), natural objects could not be properly understood without mathematics.[41]

Such disagreements illustrate the extent to which classification of the sciences had become implicated in debates about the nature of scientific thought itself. Yet when scholars were working on specific problems, the theoretical divisions often blurred. Albertus Magnus, for example, was committed in general to clarifying the independence of natural philosophy and mathematics. When discussing the generation of a surface by the motion of a line, however, he saw number not only as located in the mind of the mathematician but also as inhering materially in numbered things.[42] Conversely, Roger Bacon articulated a strong theoretical program for the subordination

39 Murdoch, *Album of Science*, fig. 32, p. 44.

40 Edward Grant, "Nicole Oresme on Certitude in Science and Pseudo-Science," in *Nicolas Oresme: Tradition et innovation chez un intellectuel du XIVe siècle*, ed. P. Souffrin and A. Ph. Segonds (Science et Humanisme) (Paris: Les Belles Lettres, 1988), pp. 31–43; and Michael R. McVaugh, "The Nature and Limits of Medical Certitude at Early Fourteenth-Century Montpellier," *Osiris*, 2nd ser., 6 (1990), 62–84.

41 Weisheipl, "Classification of the Sciences in Medieval Thought," pp. 72–89; David C. Lindberg, *Roger Bacon and the Origins of Perspectiva in the Middle Ages: A Critical Edition and English Translation of Bacon's "Perspectiva"* (Oxford: Clarendon Press, 1996), Introduction, pp. xxxvii–xliv; and David C. Lindberg, "On the Applicability of Mathematics to Nature: Roger Bacon and His Predecessors," *British Journal for the History of Science*, 15 (1982), 3–25.

42 A. G. Molland, "Mathematics in the Thought of Albertus Magnus," in *Albertus Magnus and the Sciences: Commemorative Essays 1980*, ed. James A. Weisheipl (Texts and Studies, 49) (Toronto: Pontifical Institute of Mediaeval Studies, 1980), pp. 462–78 at pp. 475–6.

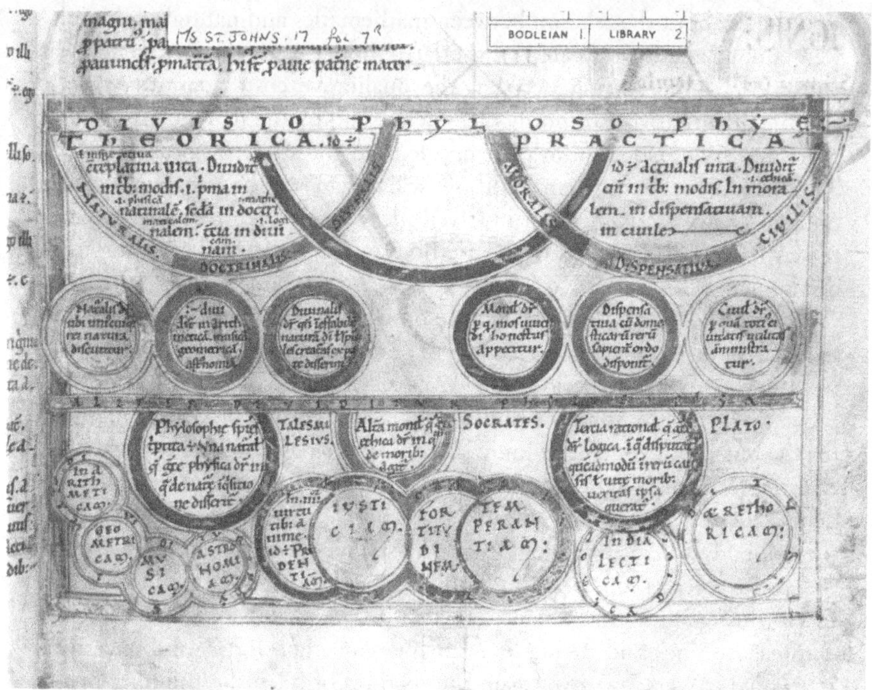

Figure 9.2. Combined divisions of philosophy. This twelfth-century diagram illustrates the mix-and-match character of medieval maps of scientific knowledge. The top half divides philosophy into theory and practice, with the former (on the left) constituted of (from left to right) natural, mathematical, and divine sciences. The circle of the mathematical sciences contains the quadrivium from the liberal arts: arithmetic, music, geometry, and astronomy. The bottom half starts with a three-part division (articulated by Augustine and attributed to Plato): natural, moral, and rational sciences (left to right). To the right of each of these almost-circles, a scribe has carried on the tradition of associating disciplines with their founders, inserting the names of Thales of Miletus, Socrates, and Plato, respectively. Little circles containing the members of the quadrivium are here clustered around natural science, or *physica*, rather than belonging to a separate mathematical division as in the top half. On the right, rational science is flanked by circles for dialectic (logic) and rhetoric. Six of the seven liberal arts are thus represented, with grammar, the most elementary, omitted. Reproduced with the permission of the President and Scholars of St. John's College, Oxford, MS 17, fol. 7r.

of natural philosophy to mathematics, but his accounts of specific phenomena sometimes contained elements that were not reducible to mathematics. Thus his treatment of refraction, while deeply mathematized, depended nevertheless upon his understanding of the physical properties of light and upon a metaphysical principle of uniformity.[43]

[43] Lindberg, *Roger Bacon and the Origins of Perspectiva in the Middle Ages*, Introduction, pp. l–lii.

Debates about the relation between mathematics and natural philosophy were among the most heated in the late Middle Ages. Their urgency was enhanced by the tension between the qualitative cosmology of Aristotle and the quantitative astronomy of Ptolemy, and by the mathematization of more and more fields, from pharmacology to the study of local motion.[44] Even within the Aristotelian tradition, which had traditionally bypassed the middle sciences, classification involved ideas about the order and value of the entities studied and about the methods proper to each or common to all disciplines.

Indeed, the Aristotelian perspective on what constituted appropriate and secure demonstration was at the heart of one of the most striking disciplinary rearrangements of the period: the elevation of logic as the most important preparation for the study of philosophy, as the source of critical methods for the pursuit of systematic knowledge, and even as a subject for advanced research in its own right. The privileged position of logic had earlier precedents, but it acquired new meaning and force through the availability of the full body of Aristotle's logical writings. The curricula of universities, as well as the declarations of natural philosophers and learned physicians, testify to this reconceptualization of the starting point for higher learning. As Thomas Aquinas said, citing first Aristotle and then Ibn Rushd (Averroes; 1126–1198), "We must investigate the method of scientific thinking before the sciences themselves. And . . . before all sciences a person should learn logic, which teaches the method of all the sciences; and the trivium concerns logic."[45] Logic precedes natural philosophy not because its subject matter is more exalted but because it offers tools necessary for the pursuit of the other sciences.

Collections of texts and university curricula embodied the methodological principle that logic comes before the sciences, but at the same time they subscribed to two other ways of ordering knowledge: the principle that higher beings have precedence (and power) over lower beings and the principle that one should move from the general to the particular. The placement of Aristotle's *On the Heavens* before his *Generation and Corruption* reflects the first of these, for the celestial subject matter is more exalted than the earthly. But the placement of *On Vegetables* before *On Animals* reflects the second, for plants are not superior to animals. Rather they embody the defining fundamentals of life – nutrition, growth, and reproduction.

The priority of logic and the high value placed on what was general did not preclude either a role for sense experience or attention to the particulars of nature. From the early Middle Ages onward, there is evidence of

[44] Murdoch, *Album of Science*, especially pts. 2, 3, and 6; Michael R. McVaugh, "The Development of Medieval Pharmaceutical Theory," in *Aphorismi de gradibus*, vol. 2 of *Arnaldi de Villanova opera medica omnia*, ed. Luis García-Ballester et al. (Seminarium Historiae Medicae Granatensis) (Granada: n.p., 1975), pp. 1–136.

[45] Aquinas, *Division and Methods of the Sciences*, q. 5, art. 1, p. 11.

the purposeful examination of natural phenomena. With a simple tube that sheltered the eye from ambient light, the curious could focus their attention on a star or a planet; with a complex astrolabe, keyed to the local latitude, the trained observer could make measurements and calculations relating to the same object. In Latin and Hebrew compilations of herbal remedies, terse expressions of approval follow some recipes – but not all – with phrases like, "This has been tested." Occasionally observations are singular – reports of specific events or conditions at a particular place and time. Some astronomical data, including those incorporated into "nativities," or horoscopes, are of this kind, as are autopsy reports, which proliferate in the late Middle Ages. However, first-person accounts were not necessarily based on singular experiences. In late-medieval Italy, compilations of clinical reports by prominent physicians became a genre of scientific literature, and it is likely that some of the cases recorded were encapsulations of medical theory or more general clinical experience. Nevertheless, the existence of such works is evidence that experience had a certain status in the profession, as was the fact that medical students at the University of Paris received bedside training as well as lectures. Particular disciplines, such as astronomy, were more oriented than others toward seeking and using data directly related to the questions addressed.

Most often, the observations invoked in natural philosophy are of a general character, even if they may have been built on personal and perhaps hands-on experience. Albertus Magnus, for example, in his explanation of the phenomena of growth, makes use of the fact that lower creatures are able to regenerate more of their bodily parts than higher creatures.[46] Furthermore, most works with significant empirical content blended material from ancient authorities, contemporary informants, and personal experience, as was the case with the book on hunting with birds compiled by the Holy Roman Emperor Frederick II (1195–1250). Occasionally, however, the context and wording of an appeal to experience strongly suggest a specific observation or series of observations deliberately and personally undertaken. Such is the case when Roger Bacon gives detailed instructions for constructing and using an apparatus to demonstrate the phenomenon of double vision.[47]

Observation served a number of functions. Reports of anomalous occurrences, especially those regarded as "marvels," excited wonder and gave rise to reflections about what could and what could not be brought within the fold of natural sciences.[48] On the other end of the spectrum, everyday

[46] Albertus Magnus, *De generatione et corruptione*, in *Opera omnia*, ed. August Borgnet (Paris: Louis Vivès, 1890), vol. 4, bk. I, tract 3, chap. 8, pp. 383–4.

[47] Roger Bacon, "Perspectiva," in Lindberg, *Roger Bacon and the Origins of Perspectiva in the Middle Ages*, pt. 2, distinction 2, chap. 2, pp. 182–7 and lxi–lxii.

[48] Bert Hansen, *Nicole Oresme and the Wonders of Nature: A Study of His "De mirabilium"* (Toronto: Pontifical Institute of Mediaeval Studies, 1985); and Lorraine Daston and Katharine Park, *Wonders and the Order of Nature, 1150–1750* (New York: Zone Books, 1998).

experiences – whether directly relevant or in the form of analogies – often served heuristic and persuasive purposes. Observations were frequently made and called upon to confirm or illustrate preexisting knowledge. This was the case with the practice of human dissection when it first became integrated into university curricula. Although conservatively framed, such practices sometimes slid from illustration to clarification to revision to critique, as occurred in the field of anatomy. In addition, experience often occupied a place within the structure of an argument. For example, the size of a human body increases either because material is added to it or because its original material gets rarified; but we see that a man's flesh is denser, not rarer, than a boy's; therefore growth occurs by the addition of material.[49] Although the practice was not common, more specialized observations could be similarly invoked to confirm or rule out a theory or to choose among competing premises. To establish that refracted rays of light are involved in vision, Bacon offers the evidence of a thin straw held close to the face against a distant background. The straw does not block our perception of the background, which it would if only direct rays were involved.[50] The invocations of experience are too varied to constitute a single scientific method, but the profusion of observations and attentiveness to the particulars of nature attest to the seriousness with which scholars approached the phenomena that their disciplines undertook to record and explain.

In spite of the diverse roles played by experience, the differences among these Aristotelian hierarchies, and the disagreements about the role of mathematics, late-medieval sciences achieved a certain coherence when it came to scholarly practices.[51] As in the earlier Middle Ages, these were, first and foremost, textual. Now, however, the new bodies of knowledge, the lessons from Greek, Arabic, and Hebrew scholarship, and especially the development of the universities contributed to the creation of far more varied and technically sophisticated ways of dealing with the corpus of authoritative texts and generating a corpus of modern texts. Some modes of university teaching and research, such as the explication of an authority's literal meaning, were indebted to older habits of exegesis. Others, such as the public debate of disputed questions (often in a raucous environment), were unique to the new conditions. Masters had to be able to take and defend positions on a variety of topics: philosophers on whether the Earth is always at rest in the middle

49 Albert of Saxony, *Questiones de generatione et corruptione*, in *Questiones et decisiones physicales* (Paris: Iodocus Badius and Conrardus Resch, 1518), bk. I, question 9, fol. 135ra.

50 Bacon, "Perspectiva," bk. III, distinction 2, chap. 1, pp. 293 and lxiv–lxv.

51 John E. Murdoch, "From Social to Intellectual Factors: An Aspect of the Unitary Character of Late Medieval Learning," in *The Cultural Context of Medieval Learning: Proceedings of the First International Colloquium on Philosophy, Science, and Theology in the Middle Ages, September 1973*, ed. John E. Murdoch and Edith D. Sylla (Boston Studies in the Philosophy of Science, 26; Synthèse Library, 76) (Dordrecht: Reidel, 1975), pp. 271–348; and Jole Agrimi and Chiara Crisciani, *Edocere Medicos: Medicina scolastica nei secoli XIII–XV* (Istituto Italiano per gli Studi Filosofici, Hippocratica Civitas, 2) (Naples: Guerini, 1988).

of the heavens or whether it can be moved, physicians on whether men or women experience greater pleasure in sexual intercourse, and so forth. In an intricate structure, a master preparing responses had to present arguments for and against each proposition, raise objections to the arguments, and provide responses to the objections, as well as muster the relevant evidence from authoritative texts. Many such practices were widespread, deployed not only in a variety of disciplinary areas, from mathematics to meteorology, but also in all parts of Europe at institutions that differed in other respects.

USES OF THE ARTS

University students encountered these patterns of scholarly inquiry in the faculty of arts, where all began their education with Aristotelian logic and natural philosophy. Although in the early Middle Ages scientific ideas and practices had fulfilled a number of social functions, from the calculation of Easter to the enhancement of cultural prestige, in the changing demographic, economic, and political scene of the late Middle Ages people with scientific knowledge became more common and more prominent. Some went on to advanced degrees in theology, medicine or law (civil or canon); others moved more quickly into opportunities available to this literate elite.

As the new class of university-trained men pursued a variety of newly developing careers, not only in the professions but also in the management of secular and ecclesiastical government, the old distinctions between theory and practice underwent radical revisions. Boethius had distinguished theory (theology, mathematics, and natural philosophy) from practice (ethics, economics, and politics); encyclopedists had valued the liberal arts above the mechanical arts because the latter had involved the use of the hands. Now texts converged with social conditions to produce a growing respect for action in the world, including the mechanical arts.[52] Under the influence of Arabic traditions, Westerners began to take seriously the idea that each art had a theoretical and a practical part.[53] More important, those same traditions had been the source of significant bodies of "practical" learning in such areas as mathematical calculation, observational astronomy, magic, and medicine. Signs of this shift appear in specific institutional changes. For example, what had originally distinguished the university-trained physician from other medical practitioners was his mastery of classic Latin texts. By the end of the Middle Ages, however, surgery – the most manual branch of medicine – had acquired a place within the university curriculum itself. Manuscripts of astronomical tables abounded in the libraries of princes as

52 Pamela O. Long, "Power, Patronage, and the Authorship of *Ars*: From Mechanical Know-How to Mechanical Knowledge in the Last Scribal Age," *Isis*, 88 (1997), 1–41; and Whitney, *Paradise Restored*, pp. 129–49.

53 Hunt, "Introduction to the 'Artes' in the Twelfth Century," pp. 98–105.

well as those of schools, and enough horoscopes survive to indicate that they were not there just for show. A passing reference to "alchemists' books" in a letter from Christine de Pisan (fl. 1399–1429) to a member of the French court suggests that people were ready to put these texts to use: "Some read and understand them one way and others completely differently. . . . And on this basis they open and prepare ovens and alembics and crucibles, and they blow hard for a little sublimation or congelation."[54]

The mushrooming of the middle sciences, the enrichment of the applied dimensions of theoretical sciences, the articulation of institutions dedicated to the development and transmission of learning, and the multiplication of social functions for scientific knowledge all contributed to a situation in which the most advanced study in many fields was highly technical. These changes, too, are reflected in late-medieval divisions of the sciences. According to a diagram in one fifteenth-century manuscript, for example, mathematics has eleven distinct parts, some of which are parts of parts of parts.[55] The intricacies of such divisions and subdivisions reflect a real situation in which not only the specificity but also the sophistication of advanced scientific work is inscribed. Few students or even masters in the faculties of arts or medicine actually read Ptolemy's *Almagest*. Likewise, in other fields, works of comparable complexity (if not always of comparable stature) were accessible to only the most advanced scholars. This situation gave rise to a degree of specialization and thus a hardening of disciplinary lines. The commentators on Ibn Sīnā's *Canon of Medicine* typically did not expound theories of the rainbow. Gerbert, who in the tenth century had access to a very modest collection of texts, had busied himself producing textbooks and instruments for teaching rhetoric, astronomy, and music, and enjoyed a reputation for his astonishing calculational abilities. By contrast, Albertus Magnus, the"Universal Doctor" of the thirteenth century, had available a vastly larger library but produced little to suggest proficiency in the mathematical sciences.[56] At the same time, as Arabic arithmetic techniques, growing academic interest in mathematics, and flourishing urban commerce all converged, new systems of calculation joined, if they did not entirely displace, Gerbert's counting method. By the thirteenth century, for example, scholars in Paris did what were recognized as Arabic "algorithms," dealing

54 Christine de Pisan, "A maistre Pierre Col, Secretaire du roy nostre sire," in *Christine de Pisan, Jean Gerson, Jean de Montreuil, Gontier Col, and Pierre Col, Le débat sur le Roman de la rose*, ed. Eric Hicks (Bibliothèque du XVe Siècle, 43) (Paris: Honoré Champion, 1977), no. 4, pp. 115–50 at p. 126.

55 On this figure, see Murdoch, *Album of Science*, fig. 32, p. 42.

56 M. J. E. Tummers, "The Commentary of Albert on Euclid's Elements of Geometry," in Weisheipl, *Albertus Magnus and the Sciences*, pp. 479–99; and Molland, "Mathematics in the Thought of Albertus Magnus." Compare David C. Lindberg, "Roger Bacon and the Origins of Perspectiva in the West," in *Mathematics and Its Applications to Science and Natural Philosophy in the Middle Ages: Essays in Honor of Marshall Clagett*, ed. Edward Grant and John E. Murdoch (Cambridge: Cambridge University Press, 1987), pp. 249–68 at pp. 256–7.

with remainders and carrying by writing and erasing digits in sand on a table, whereas their predecessors had used tokens marked with numerals on a board laid out as an abacus.[57]

Although much of what medieval scholars *did* when they applied or enacted their knowledge is inaccessible to us, some evidence points to lively economic, social, and even mechanical activities. The construction of clocks called upon both mathematical knowledge and mechanical know-how.[58] The horoscopes and other forms of astrological counsel offered for a fee by university mathematicians of fifteenth-century Vienna represented at once expert calculations and useful products.[59] Similarly, the *consilia*, or case histories, written down by physicians constituted not only texts for instruction but also representations (if not always transparent) of their careers as medical practitioners.[60]

THE ARTS AND THE BODY OF MEDIEVAL SCIENCE

Although late-medieval classification schemes were mainly concerned with the internal structure of systematic learning – with the functions and relations of its parts – they also served to delineate what constituted the body of legitimate knowledge as a whole. Whether explicitly or implicitly, the taxonomies marked off what might (or had to) be excluded from consideration. The same diagram that so intricately parsed mathematics also divided astronomy into two parts: the study of heavenly motions and the study of their effects. The second of these is bluntly divided into "prohibited" (with no further elaboration as to the subjects and texts implicated) and "not prohibited" (see Figure 9.3).

Medieval authors did not always agree about which inquiries were licit, but wherever the line was drawn, some ways of knowing and dealing with nature were left outside of a boundary that thus defined the proper domains of natural science in general. Distinctions between permitted and prohibited, or proper and improper, were not limited to astrology. Medical treatises, for example, reflect controversies about what aspects of sexual experience a physician ought properly to consider.[61] Much of the excluded material

[57] Guy Beaujouan, "L'enseignement de l'arithmétique élémentaire à l'université de Paris aux XIIIe et XIVe siècles: De l'abaque à l'algorisme," in *Homenaje a Millás-Vallicrosa*, 2 vols. (Barcelona: Consejo Superior de Investigaciones Cientificas, 1954), vol. 1, pp. 93–124 at pp. 94–6.

[58] Richard of Wallingford, *Tractatus horlogii astronomici*, in *Richard of Wallingford*, ed. and trans. John D. North, 3 vols. (Oxford: Clarendon Press, 1976), vol. 1, pp. 441–523; vol. 2, pp. 309–60; vol. 3, pp. 63–74.

[59] Michael H. Shank, "Academic Consulting in Fifteenth-Century Vienna: The Case of Astrology," in *Texts and Contexts in Ancient and Medieval Science: Studies on the Occasion of John E. Murdoch's Seventieth Birthday*, ed. Edith Sylla and Michael McVaugh (Leiden: Brill, 1997), pp. 245–70.

[60] Jole Agrimi and Chiara Crisciani, *Les consilia médicaux*, trans. Caroline Viola (Typologie des Sources du Moyen Âge Occidental, 69) (Turnhout: Brepols, 1994).

[61] Joan Cadden, "Medieval Scientific and Medical Views of Sexuality: Questions of Propriety," *Medievalia et Humanistica*, n.s. 14 (1986), 157–71.

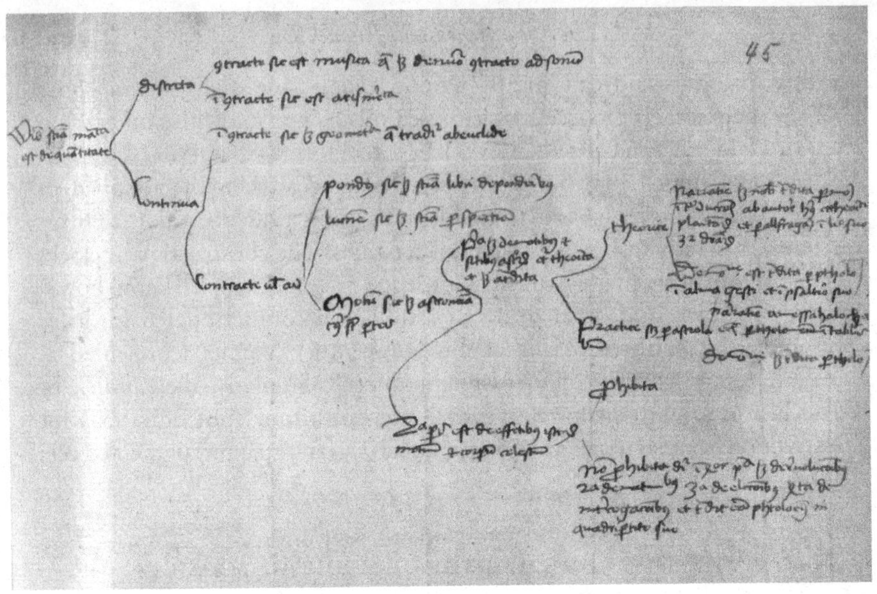

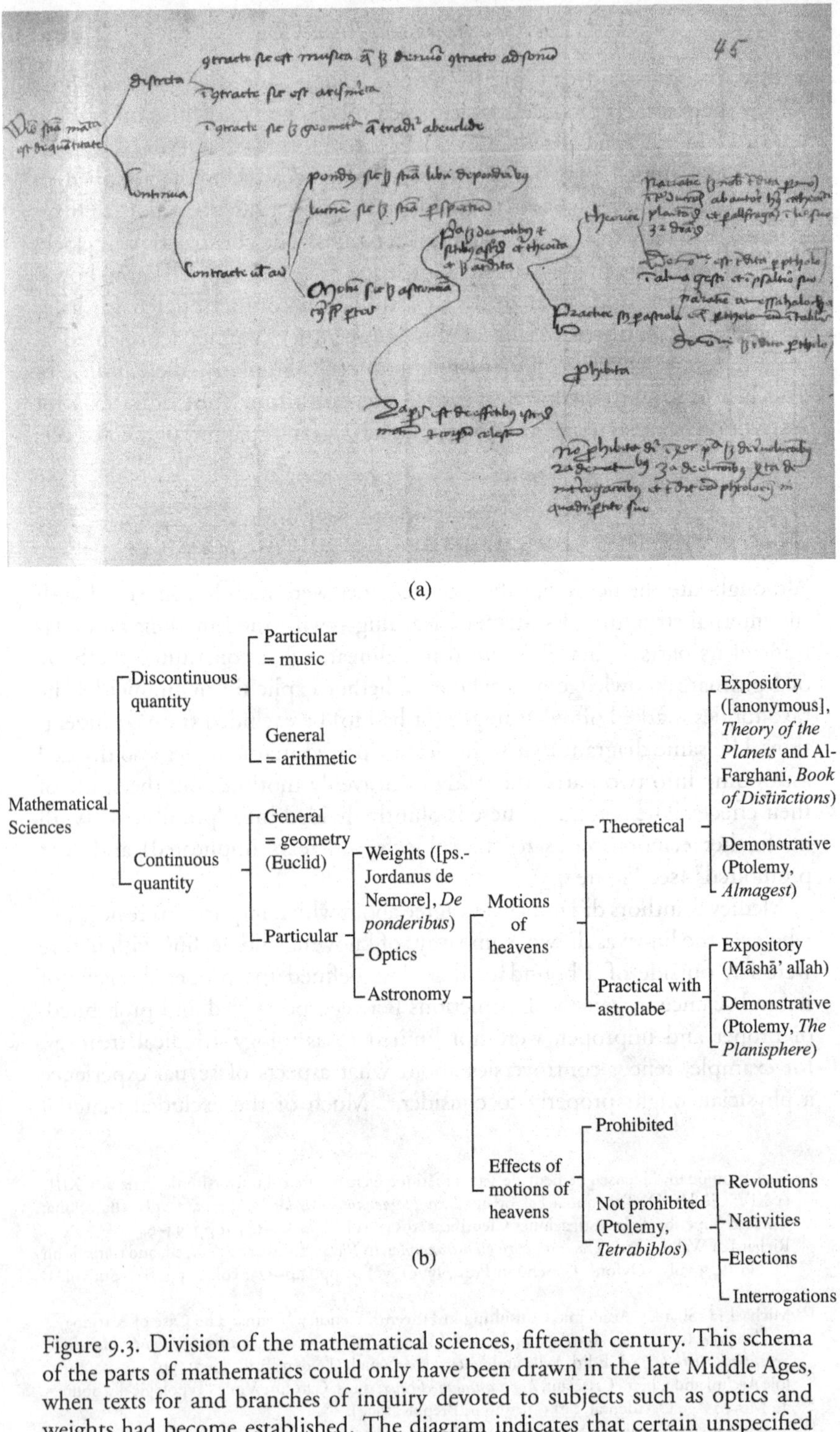

(a)

- Mathematical Sciences
  - Discontinuous quantity
    - Particular = music
    - General = arithmetic
  - Continuous quantity
    - General = geometry (Euclid)
    - Particular
      - Weights ([ps.-Jordanus de Nemore], *De ponderibus*)
      - Optics
      - Astronomy
        - Motions of heavens
          - Theoretical
            - Expository ([anonymous], *Theory of the Planets* and Al-Farghani, *Book of Distinctions*)
            - Demonstrative (Ptolemy, *Almagest*)
          - Practical with astrolabe
            - Expository (Māshā' allah)
            - Demonstrative (Ptolemy, *The Planisphere*)
        - Effects of motions of heavens
          - Prohibited
          - Not prohibited (Ptolemy, *Tetrabiblos*)
            - Revolutions
            - Nativities
            - Elections
            - Interrogations

(b)

Figure 9.3. Division of the mathematical sciences, fifteenth century. This schema of the parts of mathematics could only have been drawn in the late Middle Ages, when texts for and branches of inquiry devoted to subjects such as optics and weights had become established. The diagram indicates that certain unspecified areas of astrology are prohibited; the licit portion includes horoscopes ("nativities"). By permission of Basel, Öffentliche Bibliothek der Universität, MS F.II.8, fol. 45r.

was what opponents labeled "divination" or "sorcery," from the casting of lots to the manipulation of images to achieve specific results. Many of the works associated with these arts were (or were purported to be) from Arabic, Hebrew, "Chaldean," or other exotic traditions, making them both more interesting and more suspect. Curricular statutes and learned arguments, as well as rhetorical attacks, acted to contain the pursuit of such sciences, but they were by no means successfully suppressed or even marginalized. Their survival was due not only to the wealth of texts but also to their perceived utility. In the early thirteenth century, the Holy Roman Emperor received a commentary on a work supposedly written by Aristotle for Alexander the Great. It included material on judging a person's character from physical traits – physiognomy, "the science of which should really be kept secret, because of its great effectiveness. It contains secrets of the art of nature that meet the need of every astrologer. . . . [A]mong other things of which you should be mindful is the science of good and evil."[62] An array of evidence attests to diverse, flourishing, learned, and occasionally highly technical activity in precisely the domains targeted, such as geomancy and chiromancy, suggesting the futility of medieval (and modern) attempts to exclude these subjects from the canon of medieval natural knowledge.[63]

In addition to such hotly contested lines of demarcation, other signs point to the ambiguous relationships of individual sciences to the central body of scientific knowledge. This situation was intensified by the newness of some subjects and texts for scholars in the Latin West. For example, works on physiognomy, of which there was hardly a trace in the early Middle Ages, were sometimes enshrined with the Aristotelian natural corpus and adorned with learned commentaries, sometimes copied into manuscripts containing medical or magical texts, and sometimes reproduced in the company of religious and moral writings. The boundaries of exclusion and inclusion, whether indefinite (as in the case of physiognomy) or contested (as in the case of certain branches of astrology), thus manifested the same sorts of flexibility and fluidity as the internal lines dividing the constituent parts of natural knowledge from each other.

CONCLUSION

Divisions and classifications (whether explicit or implicit) reflected, embodied, or activated, but did not determine, the ways in which knowledge about nature was received, created, shaped, and transmitted. Even in the earlier part of the period, alternative models and cheerful syncretism left authors

[62] Michael Scot, *De secretis secretorum*, cited in Charles H. Lohr, "Medieval Latin Aristotle Commentaries, Authors: Johannes de Kanthi-Myngodus," *Traditio*, 27 (1871), 251–351 at p. 349, no. 1.

[63] The greatest mass of evidence is contained in Lynn Thorndike, *A History of Magic and Experimental Science*, 8 vols. (New York: Columbia University Press, 1923–1958).

much freedom to rearrange the components of the intellectual map to suit their purposes. The dyad of theory and practice and the triad of theology, mathematics, and natural philosophy both intersected with the seven liberal arts. Later, pluralism and outright conflict prevented the hegemony of any particular system. For example, scholars categorized questions about the motions of the heavenly bodies differently, depending upon whether they arose from Aristotle's *On the Heavens* or texts on mathematical astronomy. More important, at no time did the most favored taxonomies encompass all of the activities that medieval scholars themselves called "sciences" and that they associated with the objects and operations of the created world. In the early Middle Ages, the theory of the four elements did not occupy a secure position; in the later Middle Ages, the proper place of physiognomy was unclear. For these reasons, not only the internal organization but also the external boundaries of natural knowledge were flexible and fluid, contested and contextual, in the Middle Ages.

Changing material, institutional, and intellectual conditions, from urbanization to the accessibility of Arabic science, added a chronological dimension to this variability. After the twelfth century, the number of areas of investigation that were candidates for the denomination"science" had multiplied dramatically, as had the kinds of issues that denomination raised. Not only did new subjects, such as alchemy, challenge the boundaries of the natural and mathematical sciences and new texts, such as the *Optics* of Ibn al-Haytham, test the capacities of individuals and even curricula to reach the most advanced levels in all fields, but new questions concerning the foundations of knowledge about the world, such as the role of mathematics, demanded increased attention to how sciences were conducted. In this intellectual and social environment, the stabilizing and conservative functions of dividing and classifying the sciences characteristic of the early Middle Ages gave way to more dynamic functions, such as creating institutional space for competing bodies of knowledge and providing a medium for debates about substances and methods.

No matter how differently scholars construed and used the arrangements of the various fields of natural knowledge before and after the changes centered on the twelfth century, notions of disciplinary distinctions and order played certain continuing roles throughout the Middle Ages. First, they provided a vocabulary with which to express successive attempts to organize not just concepts but also books, curricula, and activities. Second, they highlighted, even as they circumscribed, certain persistent distinctions that precluded a simple, static, and unified science of nature – divisions between mathematics and natural philosophy, for example, or between theory and practice. Systems of classification thus brought order to a diverse set of activities and helped to create a foundation and a map for a wider range of knowledge and practices. At the same time, because of the many purposes they served, because of the variety of traditions and outlooks they encompassed, and because they were neither complete nor consistent with each other, their lacunas, tensions, and fissures constituted an aspect of the productive, open-ended environment in which medieval science thrived.

10

SCIENCE AND THE MEDIEVAL CHURCH

David C. Lindberg

Two extreme positions have dominated popular conceptions of the relationship between science[1] and the medieval church. At one extreme is an opinion that emerged from powerful antireligious currents within the French Enlightenment. It equates the rise of scientific rationality with the decline of organized religion and believes that the medieval church engaged in all-out war against free scientific inquiry, erecting insurmountable obstacles to the development of genuine science for a thousand years.[2] This is commonly referred to as "the warfare thesis." Situated at the opposite pole, and equally the product of a religious agenda, is the antithetical claim that the Christian church was not the adversary of science but its partner and patron, furnishing theological assumptions without which genuine science would have been impossible.[3]

Few scholars now take either of these positions seriously, but both have wide circulation among the educated public; moreover, scholarly attempts to explore the great middle ground between them and articulate a convincing alternative are still in the preliminary stages. The purpose of this chapter is to summarize these recent efforts and to point the way toward what I believe to be a more dispassionate, balanced, and nuanced understanding

[1] I am well aware that the medieval Latin term *scientia* had a meaning broader than that of the modern English term "science" and that nothing corresponding precisely to modern science existed during the Middle Ages. Nonetheless, I employ the term "science" in this chapter as a convenience to designate the medieval ancestors of modern science (principally mathematics, natural philosophy, natural history, and medicine), which bear a family resemblance to their modern offspring without being identical to them (see Shank and Lindberg, Introduction, and Cadden, Chapter 9, this volume).

[2] See especially the nineteenth-century codifiers and popularizers of this view: John William Draper, *History of the Conflict between Religion and Science* (New York: Appleton, 1874); and Andrew Dickson White, *A History of the Warfare of Science with Theology in Christendom*, 2 vols. (New York: Appleton, 1896). For a recent example, see Charles Freeman, *The Closing of the Western Mind: The Rise of Faith and the Fall of Reason* (New York: Alfred A. Knopf, 2003).

[3] Stanley L. Jaki, *The Road of Science and the Ways to God* (Chicago: University of Chicago Press, 1978); and Rodney Stark, *For the Glory of God: How Monotheism Led to Reformations, Science, Witch-Hunts, and the End of Slavery* (Princeton, N.J.: Princeton University Press, 2003).

of the relationship between religion and the natural sciences during the Middle Ages than we have yet achieved.[4] Several methodological precepts are required for the success of this venture, and with these I begin.[5]

METHODOLOGICAL PRECEPTS

First, we must continually remind ourselves that "science," "Christianity," "theology," and "the church" are abstractions rather than really existing things, and it is a serious mistake to reify them. What existed during the Middle Ages were highly educated scholars who held beliefs about both scientific and theological matters. Science and theology cannot interact, but scientists and theologians can. Therefore, when the words "science," "theology," and "the church" appear in the remainder of this chapter, they must be understood as shorthand references to the beliefs and practices of scientists, theologians, and the people who populated the institutions of organized Christianity.

Second, scholars who made scientific beliefs their main business and scholars who made religious or theological beliefs their main business were not always rigidly separated from each other by disciplinary boundaries. It is true that the nontheologian who entered theological territory ran certain risks (the severity of which varied according to time and place), but all medieval scholars were both theologically and scientifically informed – and entitled to opinions in both areas. It was customary to acknowledge that theological beliefs entailed scientific consequences and conversely. Indeed, the scientist and the theologian were occasionally the very same person, educated in the full range of medieval disciplines – capable of dealing with both scientific and theological matters, and generally eager to find ways of integrating scientific and theological beliefs.

[4] For other recent attempts, see especially Edward Grant, "Science and Theology in the Middle Ages," in *God and Nature: Historical Essays on the Encounter between Christianity and Science*, ed. David C. Lindberg and Ronald L. Numbers (Berkeley: University of California Press, 1986), pp. 49–75; Darrel W. Amundsen, *Medicine, Society, and Faith in the Ancient and Medieval Worlds* (Baltimore: Johns Hopkins University Press, 1996); Edward Grant, *God and Reason in the Middle Ages* (Cambridge: Cambridge University Press, 2001); David C. Lindberg, "Religion and Science: Medieval," in *Reader's Guide to the History of Science*, ed. Arne Hessenbruch (London: Fitzroy Dearborn, 2000), pp. 643–5; and David C. Lindberg, "Early Christian Attitudes toward Nature" and "Medieval Science and Religion," in *The History of Science and Religion in the Western Tradition: An Encyclopedia*, ed. Gary B. Ferngren (New York: Garland, 2000), pp. 243–7 and 259–67, respectively.

[5] Although the bulk of this chapter is new, certain portions, including the first, second, and fourth of these precepts, are borrowed (usually with significant modification but occasionally in identical words) from two other publications of mine: "Medieval Science and Its Religious Context," *Osiris*, 2nd ser., 10 (1995), pp. 61–79; and "The Fate of Science in Patristic and Medieval Christendom," in *The Cambridge Companion to Science and Religion*, ed. Peter Harrison (Cambridge: Cambridge University Press, 2010), pp. 21–38.

Third, at no point during its history has science been immune to influence from the culture in which it was situated. Any product of the human mind is subject to influence from any other relevant belief in the mind that produced it; this was true during the Middle Ages, and it is true today. I stress the applicability of this claim to the present because of the need to counter the widespread supposition that medieval natural philosophy suffers in this respect compared with modern science. The quantity of relevant belief in the supporting culture capable of impinging on science may have been greater during the Middle Ages than it is today (though I am far from certain about this), but the truth of the claim that relevant belief routinely impinges is timeless.

Fourth, medieval science as a collection of theories was not uniform or monolithic, and, as an activity, its pursuit was as varied as the scholars who pursued it. Similar claims can be made on behalf of medieval theology and religion. The relations between medieval natural philosophy and medieval Christianity therefore varied radically over time, from place to place, from one scholar to another, and with regard to different issues. We must, of course, generalize, but generalizations reducible to a phrase or a slogan will never be sufficient. A useful historical account must take the variations seriously, make distinctions, and reveal nuance and change. In short, the interaction between science and religion in the Middle Ages was not an abstract encounter between opposing epistemologies or fixed metaphysical or cosmological theories but part of the human quest for understanding. It was characterized by the same vicissitudes and the same rich variety as all other human endeavor.

Fifth, it follows that the interactions (plural, for the interactions were numerous and frequently distinct) between Christianity and science were local, historically contingent phenomena. We must resist the temptation to limit our search for causes to universal, ideological factors. We must pay equal attention to the local circumstances impinging on individual historical actors – fears, rivalries, ambitions, personalities, power struggles, political context, socioeconomic circumstances, and so forth.

Sixth and last, there was nothing in the worldview or the social fabric of the Middle Ages that would have led medieval scholars, whatever their ideological leanings, to deny the church and its theology an epistemological role or a substantial degree of epistemological authority with regard to questions about the visible world.[6] It follows, if we are to judge medieval behavior by medieval values, that religious influence on medieval science was not an inappropriate intrusion, to be regretted by the historian.[7] Religion was not

6 Arguably, even as allegedly modern a thinker as Galileo disputed not so much the church's epistemological authority in natural philosophy as its exegetical principles.

7 If we hope to learn something about the Middle Ages rather than merely condemn it for not being modern, then we are obliged to judge medieval behavior by medieval behavioral norms.

a bothersome meddler but an inevitable and (in medieval terms) legitimate player in the game.

AUGUSTINE AND THE HANDMAIDEN FORMULA

Attitudes of the early church fathers toward pagan learning were ambivalent. There was, on the one hand, much to be feared about the classical philosophical tradition.[8] Tertullian (d. ca. 230), one of those who feared it most, saw it as the incubator of heresy:

> These are the teachings of men and of demons, supplied for ears that itch for the cleverness of worldly wisdom. The Lord, calling this "foolishness," chose the foolish things of the world to confound even philosophy. For philosophy is the substance of worldly wisdom, the reckless interpreter of the divine nature and divine ordination. Indeed, heresies themselves receive their instruction from philosophy.

Tertullian proceeded to detail some of the heretical beliefs spawned by Platonists, Stoics, Epicureans, and others. He lashed out at "wretched Aristotle," who "invented dialectic for these men – the art of constructing and destroying, elusive in its claims, contrived in its conjectures, harsh in argumentation, prolific in contentions, a nuisance even to itself, retracting everything, and resolving absolutely nothing." And he concluded with the admonition: "No curiosity is required of us after Christ Jesus, no investigation after the gospel. For once we believe [the gospel], we should have no desire for further belief."[9]

Tertullian was not alone in expressing hostility toward Greek learning. Tatian, a second-century Syrian, ridiculed the philosophers for their arrogance and cautioned against taking them seriously. Around the beginning of the third century, Hippolytus surveyed the history of philosophy (both Greek and non-Greek) in order to demonstrate that heresy and atheism take their origins "from the wisdom of the Greeks, from the conclusions of those who have formed systems of philosophy, and from would-be mysteries and the vagaries of astrologers."[10] Echoes of such attitudes reverberated through the Middle Ages and beyond.

[8] I employ the terms "philosophy" and "philosophical" in this chapter in a sense that includes natural philosophy or science.

[9] Tertullian, *De praescriptione haereticorum*, bk. VII, in *Opera*, ed. Nic. Rigaltius (Paris: Societas Typographica, 1664), pp. 204, 205.

[10] Tatian, *Address to the Greeks*, bk. III, trans. J. E. Ryland, in *The Ante-Nicene Fathers: Translations of the Writings of the Fathers down to A.D. 325*, ed. Alexander Roberts and James Donaldson, 10 vols. (Grand Rapids: Eerdmans, 1985–1987), vol. 2, p. 66; and Hippolytus, *The Refutation of All Heresies*, bk. I, proemium, trans. J. H. MacMahon, in Roberts and Donaldson, *Ante-Nicene Fathers*, vol. 5, p. 10 (with one change of punctuation).

But it does not follow that early Christians totally rejected the Greek philosophical and scientific tradition. If pagan learning appeared dangerous, it also proved indispensable, and the level of hostility expressed by Tertullian and those who shared his view was the exception rather than the rule. There were various reasons why Greek learning could not (as a practical matter) be repudiated. In the first place, many of the early church fathers had acquired a solid philosophical education before their conversion to Christianity (which had philosophical roots of its own in Pauline teaching). It would have been no easy matter to lay aside deeply ingrained philosophical commitments and habits of mind. When such men participated in the formulation of Christian doctrine, it was natural (and probably inevitable) that they should build on a Greek foundation.

Second, further exploration of the implications of Christian theology and the development and clarification of its doctrines simply could not be carried out without utilization of the rules of rational thought and discourse – that is, without the logical tools of Greek philosophy. Third, from early in the second century, Christian doctrine came under increasing attack from pagan critics on intellectual grounds. In response to these critics, Christians mounted a major apologetic campaign, which by its very nature was philosophical and, consequently, could not be carried out successfully without a substantial investment in philosophical analysis and argumentation.

Fourth, there was the practical matter of Greek and Roman learning in subject areas where no conflict with Christian doctrine was likely or possible. It was unthinkable that educated Christians should repudiate the intellectual riches of the classical tradition in geometry, astronomy, physics, meteorology, mineralogy, metallurgy, natural history, agriculture, medicine, pharmacology, grammar, rhetoric, literary criticism, philology, history, geography, and other subjects, thereby returning to a state of barbaric ignorance. From the fact that Christians declined to participate in the traditional religions of the Greco-Roman world, it certainly did not follow that they were prepared to opt out of all other aspects of Greco-Roman culture.

The person who most influentially defined the proper attitude of Christians toward pagan learning was Augustine (354–430), whose views we will examine at some length in his many influential writings. Augustine made no attempt to conceal his occasional worries about the Greek philosophical tradition, warning against the dangers of vain curiosity, expressing regret for the effort he had devoted to mastering the liberal arts, and firmly elevating divine wisdom, as revealed in Scripture, over the results of human rational activity.[11] As for nature, Augustine noted in his *Literal Meaning of Genesis* that scholars frequently present long discussions on the form and shape of the heavens, matters that "the sacred writers," in their profound wisdom, "have omitted." "Such subjects," he continued, "are of no profit for those

[11] See Augustine, *Confessions*, IV.16 and X.35, trans. J. G. Pilkington, in *The Basic Writings of Saint Augustine*, ed. Whitney J. Oates, 2 vols. (New York: Random House, 1948), vol. 1, pp. 56, 174.

who seek beatitude, and, what is worse, they take up very precious time that ought to be given to what is spiritually beneficial."[12] Ignorance of the workings of nature on the part of Christians is not, in Augustine's opinion, a cause for concern; it is sufficient for them to understand that all things issue from the Creator.[13]

But this is only part of the story. Elsewhere in his *Literal Meaning of Genesis*, Augustine noted that scriptural knowledge of created things is to our sensory knowledge of those same things as day is to night. However, sensory knowledge of created things, in turn, is "so different from the error or ignorance of those who know not even the creature, that in comparison with this darkness it deserves to be called day."[14] Moreover, Augustine expressed concern that Christians, naively interpreting scripture, might utter absurd opinions on cosmological issues, thus provoking ridicule among better-informed pagans and bringing the Christian faith into disrepute. "Even non-Christians," he argued, know

> something about the earth, the heavens, and the other elements of this world, about the motion and orbit of the stars and even their size and relative positions, about the predictable eclipses of the sun and moon, the cycles of the years and the seasons, about the kinds of animals, shrubs, stones, and so forth. . . . Now, it is a disgraceful and dangerous thing for an infidel to hear a Christian, presumably giving the meaning of Holy Scripture, talking nonsense on these topics; and we should take all means to prevent such an embarrassing situation, in which people show up vast ignorance in a Christian and laugh it to scorn.[15]

It is clear, then, that there were contexts in which Augustine's attitude toward pagan scientific learning was relatively favorable. In *On Christian Doctrine*, he admonished: "If those who are called philosophers, especially the Platonists, have said things which are indeed true and are well accommodated to our faith, they should not be feared; rather, what they have said should be taken from them as from unjust possessors and converted to our use." Pagan learning contains not only "superstitious imaginings" but also "liberal disciplines more suited to the uses of truth." A good Christian, he concluded, "should understand that wherever he may find truth, it is his Lord's."[16]

Augustine valued mathematics and the natural sciences primarily for their ability to assist in the interpretation of Scripture. Mathematics was important because "numbers and patterns of numbers are placed by way of similitudes

[12] Augustine, *The Literal Meaning of Genesis*, II.9, trans. John Hammond Taylor, S.J., ed. Johannes Quasten, Walter J. Burghardt, and Thomas C. Lawler (Ancient Christian Writers, 41–42) (New York: Newman Press, 1982), vol. 1, p. 59.

[13] See Augustine, *Enchiridion*, bk. X, trans. J. F. Shaw, in Oates, *Basic Writings of Saint Augustine*, vol. 1, pp. 661–2.

[14] Augustine, *Literal Meaning of Genesis*, IV.23, vol. 1, p. 131.

[15] Augustine, *Literal Meaning of Genesis*, I.19, vol. 1, pp. 42–43. Compare ibid., II.1, vol. 1, pp. 47–8.

[16] Augustine, *On Christian Doctrine*, II.40 and II.18, trans. D. W. Robertson, Jr. (Indianapolis: Bobbs–Merrill, 1958), pp. 75 and 54, respectively.

in the sacred books as secrets which are often closed to readers because of [their] ignorance of numbers." Likewise, some knowledge of natural history was obligatory because Scripture often employs the characteristics of animals, plants, and stones as similitudes. The reader ignorant of the nature of serpents and doves, for example, will not comprehend the biblical injunction to "be wise as serpents and innocent as doves."[17]

Augustine most copiously illustrated the exegetical utility of the natural sciences, while revealing his own command of pagan scientific literature, in his *Literal Commentary on Genesis*, where he brought it to bear on the interpretation of the biblical Creation story. Here we encounter Greco-Roman ideas about lightning, thunder, clouds, wind, rain, dew, snow, frost, storms, tides, plants and animals, matter and form, the four elements, the doctrine of natural place, seasons, time, the calendar, planetary motion, sensation, light and shade, and number theory. For all of his worry about overvaluing the natural philosophy of the classical tradition, Augustine applied it with a vengeance to the exegesis of the Creation story.

In Augustine's view, then, natural philosophy was not to be loved but to be used. "We should use this world and not enjoy it," he wrote in *On Christian Doctrine*, "so that . . . by means of corporeal and temporal things we may comprehend the eternal and spiritual."[18] The study of natural things was legitimized by the service it could perform for the faith and in particular for the assistance it could lend to the process of scriptural exegesis. The natural sciences must serve as the handmaidens of theology and religion.

EARLY-MEDIEVAL SCIENCE AND THE RECOVERY OF THE CLASSICAL TRADITION

It was largely this attitude that motivated pursuit of the natural sciences through the early Middle Ages. During this period of political disintegration, social turmoil, and intellectual decline, natural philosophy was an item of low priority, but when it was cultivated, as it sometimes was, it was primarily by people in positions of religious authority or with a religious agenda, motivated by its perceived religious or theological utility. Consider, for example, the following five scholars from the sixth through tenth centuries (one to a century): Cassiodorus, Isidore of Seville, the Venerable Bede, John Scotus Eriugena, and Gerbert of Aurillac (each of them, except Cassiodorus, arguably the outstanding scholar of his century). All five had religious vocations. Moreover, all had monastic educations, except Cassiodorus, who became a monastic educator, and it was out of the educational experience in the monastery, rather than in repudiation of it, that their interest

[17] Ibid., pp. 52 and 50 (quoting Matt. 10:16).
[18] Augustine, *On Christian Doctrine*, p. 10.

in the natural sciences grew. All wrote treatises that revealed their interest in natural philosophy and helped to enlarge its role in European culture. And finally, it can be plausibly argued in each case that the motivation for writing about the natural world was supplied by the handmaiden formula. Natural philosophy is worth pursuing because ultimately it is a religious necessity.[19]

Europe experienced dramatic political, social, and economic recovery in the eleventh and twelfth centuries. The causes were complex, and we cannot go into them here, but they led (among other things) to rapid urbanization, the multiplication and enlargement of schools, and the growth of intellectual culture. Education shifted from the countryside to the cities, as cathedral and municipal schools replaced the monasteries as the principal educational institutions. Although cathedral schools shared the monastic commitment to education that was exclusively religious, the curriculum of the cathedral schools reflected a far broader conception of the range of religiously beneficial studies that might be legitimately taught and learned. Finally, from the middle of the twelfth century and continuing into the thirteenth and fourteenth, universities began to take shape, offering education at an advanced level.

Within this new intellectual world, three developments require our attention. First, the eleventh- and twelfth-century schools rediscovered Plato's *Timaeus* and presided over an attempt to master its natural philosophy and harmonize its cosmogony with the biblical account of Creation. Second, philosophical method began to percolate into all realms of human activity, including theology. Representative of this development are the attempt by Anselm of Canterbury (1033–1109) to prove the existence of God by rational means alone, without reference to revelation, and also the obvious rationalism and heavy investment in the techniques of Aristotelian logic evident in the teaching of Peter Abelard (ca. 1079–ca. 1142). Third, the educational revival created a thirst for knowledge that was ultimately satisfied by the translation, from Greek and Arabic into Latin, of much of the classical tradition (only fragments of which had been available during the early Middle Ages). The most important of the translated works, for our purposes, were those that made up the Aristotelian corpus.

This newly translated literature provoked a crisis in European intellectual life. At the heart of the crisis were the works of Aristotle, which ran counter to the prevailing synthesis of Platonic philosophy and Christian theology in both intellectual tone and doctrinal content. Central to Aristotelian philosophy was a theory of knowledge based solely on natural human capacities – a theory, moreover, that aimed at demonstrative certainty, obtainable only by applying the technical apparatus of Aristotelian logic. Among the claims of Aristotelian philosophy were many that were in tension, if not outright

[19] These five cases are dealt with more fully in my forthcoming "Medieval Science and Christianity."

conflict, with Christian doctrine. These include Aristotelian naturalism (the only causes that exist are the natural powers of things), the repudiation of divine providence, arguments for the eternity of the world (tantamount to a contradiction of the Creation story), and Aristotle's theory of the soul, with its denial of personal immortality. If Aristotelian philosophy were to be made acceptable to the educated elite of medieval Christendom, major negotiations and, possibly, substantial accommodation would be required.

Given the seriousness of the issues and the background of suspicion emanating from the church fathers, I do not believe that one could have predicted in advance that the negotiations would be successful and the necessary accommodations made. But, in fact, they were. Despite profound differences of outlook and teaching, peace was ultimately made, Aristotelian philosophy and Christian theology became allies (even allies have their differences), and in the long run Aristotle became the official philosopher of the Catholic Church. And, of course, along with Aristotle came a large piece of the Greek tradition in the natural sciences. How did this extraordinary turn of events come about?

ACCOMMODATION IN THE THIRTEENTH CENTURY

What was there in the classical tradition (Aristotelian philosophy in particular), or in the circumstances of its reception, that led to its ultimate appropriation by a culture that ought to have been profoundly opposed, for religious reasons, to many of its central tendencies and certain portions of its content? We must understand, to begin, the intellectual richness of this body of learning, brought to Western Europe through its translations. In the mathematical sciences, natural history, medicine, and much of natural philosophy, the newly translated learning was superior to anything hitherto available in Latin Christendom; it contained no conclusions that were problematic from a theological standpoint; it filled an intellectual void; and it was eagerly adopted by an intellectual community hungry for additional meat on which to chew. The only danger such materials ever faced was the risk of guilt by association with other, more dangerous, portions of the classical tradition, and that risk seems, in retrospect, not to have been serious.[20]

Even the Aristotelian corpus, despite its manifest dangers, provided a satisfying account of so many aspects of the created world, as well as tools of unprecedented power for the further exploration and analysis of that world, that scholars, including those of undoubted orthodoxy, were loathe to relinquish it. In short, we may, I believe, safely infer that the classical tradition was valued for its broad contributions (actual and potential) to

20 On the Western reception of Aristotle's philosophy, see Fernand Van Steenberghen, *Aristotle in the West: The Origins of Latin Aristotelianism* (Louvain: Nauwelaerts, 1955).

European intellectual life. It was simply too attractive and useful to be tossed aside lightly. The goal was never to eradicate it but rather to appropriate it and domesticate it within a Christian culture.

Another factor that undoubtedly contributed to the favorable reception of Aristotelian philosophy (especially its methodology) was the drive for professional status among scholars within the medieval universities. Scholars eager to differentiate themselves from people of lesser intellectual accomplishment had every reason to be attracted to a rational methodology that required years of intensive education in a university setting. This was true even of theologians, the university scholars who risked the most by deciding to apply the techniques of Aristotelian demonstration to the subject matter over which they claimed proprietorship. We thus see, beginning in the twelfth century and continuing through the Middle Ages and beyond, the growth of a speculative theology that invested heavily in the rational methodology of Aristotelian logic, viewed as the key to any serious theoretical inquiry into theological matters.[21] By the end of the thirteenth century, Aristotelian philosophy, with its emphasis on logical technique, had come to dominate the curriculum of the medieval universities – not only in the faculties of arts, law, and medicine but also in the faculty of theology.

But there was also an explicitly religious motivation for adopting and domesticating the works of Aristotle (and other portions of the classical tradition). This motivation was most likely to be articulated when the pursuit of philosophy and the natural sciences was questioned or challenged. I refer to arguments defending the classical tradition as the valued handmaiden of religion and theology. Such justifications became particularly common in the thirteenth century, when medieval scholars of all ideological stripes were induced to declare their loyalties, either for or against the legitimacy of the new learning. For example, Roger Bacon (ca. 1220–ca. 1292), a Franciscan friar notorious for his attack on authority, wrote a very long book, the main purpose of which was to defend the new science as the faithful handmaiden of theology and the church. "One science," he argued, "is mistress of the others – namely, theology, for which the others are integral necessities and which cannot achieve its ends without them. And it lays claim to their virtues and subordinates them to its nod and command." Among the disciplines that Bacon considered capable of serving theology and the church were mathematics, astronomy, astrology, optics, medicine, and "experimental science." The latter, for example, produces technological wonders that will rescue the

[21] On theological method, see G. R. Evans, *Old Arts and New Theology: The Beginnings of Theology as an Academic Discipline* (Oxford: Clarendon Press, 1980), especially chap. 3; Steven P. Marrone, "Speculative Theology in the Late Thirteenth Century and the Way to Beatitude," in *Les philosophies morales et politiques au Moyen Âge (Actes du IXe Congrès International de Philosophie Médiévale, Ottawa, 1995)*, ed. B. Carlos Bazán, Eduardo Andújar, and Léonard G. Sbrocchi, 3 vols. (New York: LEGAS, 1995), vol. 2, pp. 1067–80; and William J. Courtenay, *Schools and Scholars in Fourteenth-Century England* (Princeton, N.J.: Princeton University Press, 1987), pp. 262–4, 276–82.

church from the Antichrist. And mathematics is of critical importance for biblical exegesis. According to Bacon, a grasp of the literal sense of Scripture demands an understanding of the natures and properties of things. Therefore, "the theologian requires an excellent knowledge of created things. But it has been shown that without mathematics created things cannot be known; it follows that mathematics is altogether necessary for sacred knowledge."[22]

Bacon is but one representative of an outpouring of handmaiden rhetoric in the thirteenth century. Among the many others who expressed similar views (though rarely as copiously or as enthusiastically as Bacon) were William of Auvergne, bishop of Paris from 1228 to 1249; Bonaventura, theologian and minister general of the Franciscan Order from 1257 to 1274; Thomas Aquinas (ca. 1224–1274), famed Dominican theologian; and John Pecham, archbishop of Canterbury from 1279 to 1292 and author of the standard medieval textbook on mathematical optics.

THE COURSE OF EVENTS

But we need to situate these ideas and the men who articulated them in the real world – that is, in time, space, and institutional context. All five of these defenders of the handmaiden status of philosophy and the natural sciences had significant connections with the University of Paris (one of Europe's oldest and most distinguished universities, the only one on the continent at the time that had a faculty of theology). And Paris is where the controversy that we are about to investigate was centered during the first three-quarters of the thirteenth century.

Most of Aristotle's works (accompanied by various Islamic commentaries) were available in Paris by the beginning of the century. By 1210, the council of bishops responsible for Paris considered the dangers of Aristotelianism serious enough to justify a ban (renewed in 1215) on the teaching of Aristotle's natural philosophy in the faculty of arts. A papal bull (and subsequent letter) issued by Pope Gregory IX in 1231 acknowledged both the dangers and the value of Aristotelian natural philosophy, mandating that the books banned in 1210 and 1215 should not be taught at Paris "until these shall have been examined and purged from all suspicion of errors," so that, suspect material having been removed, "the rest may be studied without delay and without offense."[23] Within a decade, however, these early bans ceased to be enforced. By the late 1230s or early 1240s, lectures on Aristotle's natural philosophy

22 Roger Bacon, *Opus maius*, pt. 4, in *The Opus Majus of Roger Bacon*, ed. John H. Bridges, 3 vols. (London: Williams and Norgate, 1900), vol. 1, p. 175. See also David C. Lindberg, "Science as Handmaiden: Roger Bacon and the Patristic Tradition," *Isis*, 78 (1987), 518–36.

23 Lynn Thorndike, ed., *University Records and Life in the Middle Ages* (New York: Columbia University Press, 1944), pp. 38, 40. There is no reason to believe that the commission appointed by the pope ever met, and no purged version of Aristotle's works has ever been discovered.

made their appearance in the faculty of arts, and by 1255 lectures on all of Aristotle's *libri naturales* were mandated for the MA degree. In a remarkable turning of the tables, Aristotle's natural philosophy, which had formerly been excluded, had now, thanks to pressures within the faculty of arts, become the centerpiece of the arts curriculum.

With the rehabilitation of Aristotle came the obligation to confront the serious problems that had led to the exclusion of the Aristotelian corpus from the curriculum in the first place. Problematic Aristotelian claims, such as the eternity of the world, a deterministic universe with its exclusion of divine providence, and apparent denial of personal immortality (all wrapped up in a naturalistic methodology), had to be harmonized with theological doctrine. This task was undertaken by a number of scholars in the thirteenth century, most notably (but by no means single-handedly) two theologians, both with Parisian connections: Albert the Great (ca. 1200–1280) and Thomas Aquinas (ca. 1225–1274). What Albert and Thomas and like-minded scholars did was work their way through the Aristotelian corpus, systematically confronting troublesome doctrines – interpreting, supplementing, refuting, and correcting them as required. This was a massive undertaking: Albert the Great devoted about 8,000 pages (in the modern edition of his works) to the task; Thomas Aquinas devoted 3,000 pages to the effort in his *Summa theologica* alone.

While moderates like Albert and Thomas were taking advantage of philosophical freedom at Paris to harmonize Christian theology and Aristotelian philosophy, certain radical members of the Parisian arts faculty went further, testing the limits of that freedom. This faction, led by Siger of Brabant (ca. 1240–1284), began to teach such controversial doctrines as the eternity of the world and Aristotle's theory of the soul, with little or no regard for potential theological implications. They were professional philosophers, determined to extend the reach of their discipline as far as it would go. They were not, as far as one can determine from this distance, denying any central Christian doctrines but were simply ignoring them. Their purpose was to "argue philosophically," employing Aristotle's rationalistic and naturalistic agenda. It was somebody else's problem to bring about reconciliation between philosophical and theological conclusions.[24]

Conservative theologians in the faculty of theology and local ecclesiastical authorities were not amused. Decisive action seemed appropriate, and this came in 1270 and 1277, when the bishop of Paris, Etienne Tempier, condemned two collections of propositions that he considered dangerous (numbering 13 and 219, respectively) and that were allegedly debated in the faculty of arts. Included were propositions representing Aristotelian

[24] Fernand Van Steenberghen, *Aquinas and Radical Aristotelianism*, trans. John F. Wippel, Dominic O'Meara, and Stephen Brown (Washington, D.C.: Catholic University of America Press, 1980), chap. 3.

rationalism and naturalism, the eternity of the world, denial of personal immortality, and so forth. Any member of the university who had taught any of the condemned doctrines (or even listened to such teaching) was ordered to confess within seven days or risk excommunication.[25]

Viewed superficially, this story of thirteenth-century struggle, culminating in the condemnations of 1270 and 1277, may appear to reinforce the old stereotype of medieval theological repression snuffing out the flame of philosophical and scientific achievement. It is unquestionably true that in 1210, 1215, 1231, 1270, and 1277 theological restrictions were imposed on philosophical and scientific thought. But that is neither the end nor the whole of the story. A more careful examination yields several further conclusions. First, the various bans were only temporarily or partially effective in slowing the march of Aristotelian philosophy. Portions of the condemnation of 1277 were rescinded in 1325, and the remainder, though periodically appealed to, lacked the power to extinguish the Aristotelian tradition. Second, the condemnation of 1277 banned only a short list of Aristotelian doctrines. While aiming to nip philosophical arrogance in the bud, it never had as a serious aim the end of philosophical and scientific effort. This was not a battle to the death between a pair of intractable foes but a serious struggle in which the boundaries and power relations between Christian theology and the Aristotelian philosophical tradition were being negotiated. Third, this theological attack on certain Aristotelian philosophical doctrines was not wholly negative in its outcome, for it had the unforeseen benefit of encouraging the investigation of non-Aristotelian philosophical alternatives, some of which contributed importantly to future scientific developments. Finally, despite the condemnations, the church did not in the end repudiate the Aristotelian tradition but rather Christianized it. Purged of a number of troublesome doctrines, Aristotelianism became the official philosophy of the Catholic Church, and the church became the most powerful and generous patron of serious philosophical and scientific effort the West had ever seen.[26]

LATE-MEDIEVAL DEVELOPMENTS

The events of 1277 did not lead to an abrupt change in the basic contours of the relationship between Christianity and science. The fundamental power relations, political and epistemological, established by virtue of the struggles

[25] On the condemnations, see John F. Wippel, "The Condemnations of 1270 and 1277 at Paris," *Journal of Medieval and Renaissance Studies*, 7 (1977), 169–201; Edward Grant, "The Condemnation of 1277, God's Absolute Power, and Physical Thought in the Late Middle Ages," *Viator*, 10 (1979), 211–44; and Roland Hissette, *Enquête sur les 219 articles condamnés à Paris le 7 mars 1277* (Philosophes médiévaux, 22) (Louvain: Publications Universitaires, 1977).

[26] David C. Lindberg, *The Beginnings of Western Science: The European Scientific Tradition in Philosophical, Religious, and Institutional Context, 600 B.C. to A.D. 1450* (Chicago: University of Chicago Press, 1992), pp. 234–41.

of the thirteenth century, changed little in the fourteenth. Theology was still queen of the sciences, and the natural sciences were promoted and practiced as diligent and productive handmaidens.

Although the broad contours remained unchanged, much changed within them. Many of these changes resulted from the maturation of philosophy and theology after a century of access to the whole of Aristotelian philosophy and other portions of the classical tradition. Scholars continued to attack old problems, but now using more sophisticated philosophical and theological weaponry. Ontological and epistemological questions, for example, became the objects of penetrating analyses by such towering scholars as William of Ockham (ca. 1285–1347). To take a more specific case, the question of the eternity of the world was not laid to rest by the condemnation of 1277 but continued to attract attention into the early decades of the fourteenth century. Also significant in fourteenth-century discussions (especially at Oxford) was the tendency to transform the problem of eternity into an analysis of the nature of the infinite.[27]

The potential for conflict between "faith and reason" did not evaporate in the fourteenth century, and fourteenth-century thinkers continued to search for ways of resolving or at least diminishing the tensions. One influential solution, emanating from John Duns Scotus (ca. 1266–1308) and Ockham, was to cast doubt on the competence of philosophy to address theological claims. This identification of epistemological differences between philosophy and theology had the effect of defining separate spheres of influence – thus limiting the scope of philosophy and protecting theology from philosophical encroachment.

The disengagement of philosophy and theology at one level proved no obstacle to heavy borrowing between the two disciplines at other levels – a continuation of the methodological unification of philosophy and theology begun in the twelfth and thirteenth centuries. For example, stress within theology on the power of an omnipotent God to have created any sort of world he wished (short of self-contradiction) led to a dramatic expansion of the horizons of fourteenth-century natural philosophers, who took up the challenge of exploring the nature and implications of imaginary worlds that God had not, but (in view of his absolute power) could have, created.[28]

Methodological principles were also transferred from natural philosophy to theology. A group of "analytical languages" that emerged within fourteenth-century natural philosophy and logic came to be applied with a vengeance to theological problems. These included the languages of intensification and remission of forms, of first and last instants of being and non-being, and of *suppositio*, which were "plied as analytical tools in all corners

[27] Richard C. Dales, *Medieval Discussions of the Eternity of the World* (Leiden: Brill, 1990), pp. 178–253.

[28] John E. Murdoch, "The Development of a Critical Temper: New Approaches and Modes of Analysis in Fourteenth-Century Philosophy, Science, and Theology," *Medieval and Renaissance Studies*, 7 (1978), 51–79, especially p. 53.

of philosophy, medicine, and theology."[29] The quantification of theology, resulting from application of the language of intensification and remission of forms, as well as the problems of infinity that emerged in the consideration of first and last instants, is a remarkable example.

Finally, running through these developments was a current of secularization. I am not referring to full-fledged secularization of the sort that might deny the existence of a Creator or the epistemological authority of revelation in matters bearing on metaphysics and cosmology. The secularization of the late Middle Ages was of a much milder form. The practitioners of philosophy and the various scientific disciplines did not abandon the fundamentals of Christian belief, but they did spend less time defining their enterprise in theological terms and were far less likely to allow God a causal or explanatory role in their theories. Moreover, university scholars found themselves increasingly free from close theological or religious supervision, and able to engage in scholarship without continually articulating its religious justification. This was not yet the secularized Aristotelianism of the Italian Renaissance, but steps in that direction were clearly being taken.

CONCLUDING GENERALIZATIONS

Generalizing is always dangerous. But it is also necessary if we hope to lay claim to genuine understanding. The broad generalizations that follow should be taken as a first attempt to summarize what we know (or think we know) about the relationship between science and the medieval church.

Most fundamentally, medieval science was an import – a continuation of the classical tradition, Greek in origin but with important Muslim additions. This tradition provided medieval scholars with written texts that offered both a sophisticated account of the cosmos and a methodology for its further investigation. It does not follow from the claim that medieval science belonged to a textual tradition that medieval scholars were incapable of investigating nature for themselves and never did so. My point is simply that medieval scholars learned how to think about nature by imitating the most advanced scientific culture to which they had access. It is also important to note that the imported classical sciences were received not as finished products but as works-in-progress, to be assimilated, clarified, criticized, corrected, and extended. I am thus stating my vehement opposition to the opinion, held in some circles, that Christian religion or theology was the nursery out of which Western science emerged.

[29] Ibid., pp. 54–5. See also John E. Murdoch, "From Social into Intellectual Factors: An Aspect of the Unitary Character of Late Medieval Learning," in *The Cultural Context of Medieval Learning*, ed. John E. Murdoch and Edith D. Sylla (Boston Studies in the Philosophy of Science, 26) (Dordrecht: Reidel, 1975), pp. 289–303; Laird, Chapter 17; and Ashworth, Chapter 22, this volume.

But Christian influence was not absent. If the defining feature of medieval scientific knowledge was its classical origin, surely a powerful shaping force was the Christian context within which scientific investigations rooted in the classical tradition took place: sometimes nurtured, sometimes neglected or opposed, but in all cases noticeably shaped by the institutions, values, and beliefs of that culture.

An important example is the handmaiden formula, which provided the justification for scientific activity during the early Middle Ages and continued to be asserted thereafter, whenever the value or legitimacy of scientific knowledge was called into question. The overwhelming majority of medieval scientific achievements, including those that we most admire, were produced by scholars who subscribed to the Augustinian formula of science as the handmaiden of theology and the church.

It does not follow that every piece of philosophical, mathematical, or observational activity during the Middle Ages was carried out by somebody conscious at that moment of its religious utility. The motivation for most human activities is mixed, rather than simple, and not easily discerned even in cases where we have a great deal more data than we have for the medieval period. Moreover, once the classical tradition was firmly institutionalized in the schools and universities, the question of motivation for its pursuit lost its saliency. Scholars could then pursue natural science simply because it had acquired a satisfactory level of social sanction, offered various satisfactions (including career opportunities), and because it interested them or they were good at it. Under such circumstances, knowledge was being sought neither as handmaiden of theology and religion nor for its own sake, but for the sake of personal satisfaction and reward. In short, the handmaiden formula was being deployed not at the personal level, as motive, but at the cultural level, as justification – the centerpiece of a campaign directed at the achievement of social sanction for the natural sciences.

The success of the handmaiden formula neither eliminated worries about problematic aspects of the classical tradition nor eradicated all tensions between scientific theory and Christian doctrine. Discord burst into public view on a number of occasions – most notably in the condemnation of 1277 – and it was probably never far below the surface. Unanimity of opinion on the legitimacy and proper role of the natural sciences and their relationship to theological tradition was no closer to being achieved during the Middle Ages than it was at the beginning of the twenty-first century.

But we must not allow the existence of ongoing tensions to obscure the larger truth. If the natural sciences agreed to serve as the handmaidens of theology and the church, the church's part of the bargain was to support and protect its handmaidens. How exactly did the church do this? The church became the patron and protector of the sciences in several ways. In the first place, it presided over a synthesis of Christian theology and Aristotelian natural philosophy. I have in mind the achievements of Albert the

Great, Thomas Aquinas, and like-minded scholars from the twelfth century onward, who worked their way laboriously through the writings of Aristotle and others, identifying errors, reinterpreting texts, crafting compromises, and resolving tensions.[30] The church did *not* reject the classical tradition but appropriated it, creating what we must regard as a new worldview and setting Christendom on a course that had profound implications for the subsequent history of Western science.

The church also became the patron of the sciences through its support of schools and universities, many of which (in Northern Europe) were under its authority and protection. In the universities, natural science found a secure institutional home for the first time in history and became the common intellectual property of the educated classes (see Shank, Chapter 8, this volume).

Finally, the church patronized the classical scientific tradition in its role as an agency for the redistribution of excess wealth – some of which was channeled to promising scholars. The mechanism of this remarkable arrangement was the system of benefices – revenue-producing properties or offices, controlled by the church, that provided support (typically lifelong) for their recipients. The scholarly holder of a benefice was frequently allowed to draw this support in absentia while living, for example, at court or in a university setting. Because of the enormous scale of the system of benefices and the resulting impossibility of close supervision, such scholars had considerable latitude in their choice of scholarly pursuits.[31]

Thus, in the long relationship between Christianity and the natural sciences, the medieval chapter is one in which (contrary to the old stereotype of bloody suppression) Christianity and the classical tradition made peace. The arguments of William of Auvergne, Roger Bacon, Thomas Aquinas, and those who shared their opinions prevailed, and in the end this body of learning, pagan in origin, became the centerpiece of university education in medieval Christendom. There were, of course, many specific points of controversy that required adjudication. But in the end the negotiations were mostly successful, and Christendom appropriated virtually the whole of classical science and natural philosophy, including the natural philosophy of Aristotle. An amicable peace was achieved – perhaps more so than in any other period of European history.

But was this peace achieved at an unacceptable price? Is handmaiden status compatible with the existence of genuine science? The answer to the latter question is an emphatic "of course!" Patronage without "strings" of any kind is a figment of the imagination. In the real world, to acquire support is (in all but the most exceptional circumstances) to give up some measure of

30 Albert the Great devoted 8,000 pages (in the standard nineteenth-century edition of his writings) to this project.

31 I thank Robert G. Frank, Jr., for inducing me to think about benefices.

autonomy. Surely, science in its normal state, from antiquity to the present, has been justified by the service it offers to some ideology, social program, political goal, practical end, or profit-making venture.

But how tight were the strings? During the Middle Ages, the mistress, theology, was surprisingly liberal with her handmaidens, the natural sciences, offering them sufficient freedom for the serious and productive pursuit of scientific knowledge. The medieval scholar who specialized in the natural sciences certainly did not regard himself as suffering from overly close supervision by theologians or any other party within the church. To dispel an old myth, during the Middle Ages scientific conclusions were *not* dictated by a crude, literal reading of the Bible. For example, the often-repeated claim that medieval scholars were induced by biblical cosmology to regard the Earth as flat is pure legend.[32] With few exceptions, the medieval scholar could pursue any question he liked and defend whatever conclusions emerged from the application of reason and experience, without fear of coercion or reprisal.[33] His freedom of thought and expression was roughly comparable to that of his successors in the seventeenth century. If his achievement in the natural sciences fell short of theirs, this was the result not of theological oppression or religious intolerance but of different stages in the development of a tradition, and of a complex concatenation of intellectual, religious, institutional, and socioeconomic circumstances, which historians are still endeavoring to unravel.

[32] The principal creator of the legend was the American essayist, satirist, and humorist Washington Irving (1783–1859) in his four-volume *History of the Life and Voyages of Christopher Columbus* (London: Murray, 1828). See Jeffrey B. Russell, *Inventing the Flat Earth: Columbus and Modern Historians* (New York: Praeger, 1991).

[33] This was true especially of the mathematical sciences and other technical disciplines that had few worldview implications. For example, although Roger Bacon (ca. 1214/1220–ca. 1292) defended his investigation of light and vision based on its theological utility, nobody has discovered any theoretical claim in his extensive writings on *perspectiva* (totaling about 90,000 words) that reflects theological influence. See David C. Lindberg, *Roger Bacon and the Origins of Perspectiva in the Middle Ages* (Oxford: Clarendon Press, 1996), pp. 321–35.

11

NATURAL KNOWLEDGE IN THE EARLY MIDDLE AGES

Stephen C. McCluskey

We sometimes ask whether there was anything distinctive about the science of the early Middle Ages, whether medieval scholars developed their own scientific view of the world. If we restrict ourselves to natural philosophy or those exact sciences studied within the classical framework of the seven liberal arts (grammar, logic, rhetoric, geometry, arithmetic, astronomy, and music), early-medieval science often appears as little more than an attenuated version of ancient science. From this perspective, the history of science from the fifth to the eleventh centuries is reduced to "the repeated recovery of the legacy of antiquity."[1]

If we redirect our attention from the narrow concept of natural philosophy to the broader concept of natural knowledge, we find an understanding of nature that reflected the intellectual concerns and practical needs of early-medieval society. In adopting this broader focus, we should note that in the early Middle Ages there were no scientists in the modern sense of the word; neither were there natural philosophers. Knowledge of the natural world was an integral part of a broader kind of learning. Thus, medieval natural knowledge juxtaposed practical knowledge with the theoretical findings of classical antiquity, the latter embodied in the classical pedagogical framework of the seven liberal arts. (See also the following chapters in this volume: Cadden, Chapter 9; Eastwood, Chapter 12.)

ANTIQUE LEARNING IN OSTROGOTHIC ITALY

In the late fifth century, as the Roman Empire was slowly disintegrating, the last emperor in the West gave up his nominal power, and the Ostrogothic leader Theodoric established an independent kingdom in Italy. Roman government was effectively at an end; the attenuated murmurs of Roman

[1] N. L. Swerdlow, "Review of Olaf Pedersen, *Early Physics and Astronomy*," *Speculum*, 72 (1997), 875–7.

learning and culture, which continued to echo from generation to generation, became fainter and less distinct with each repetition. Yet Theodoric, who had received some education in Constantinople, valued the Roman ideal of learning. Cassiodorus described Theodoric's grandson, Athalric, as a virtual philosopher in regal purple.[2] Even if the reality did not match this courtly flattery, the ideal of the learned ruler was current in Ostrogothic Italy.

The same ideal appears in the influential *Consolation of Philosophy* of Anicius Manlius Severinus Boethius (ca. 480–524). The *Consolation* is set in prison, where Boethius awaited his fate on the charge of conspiracy against Theodoric. Lady Philosophy appears, reminding him that just as the astronomers have shown the Earth to be an insignificant point, earthly fame and suffering are likewise of no real significance. For Boethius, the model of the cosmos illuminated the intrinsic value of public service. He recalled the words of Plato, "that commonwealths would be good if wise men were appointed rulers, or if those appointed to rule would study wisdom." But there were Christian as well as Platonic overtones when Boethius praised the "Creator of the star-filled universe [who gives] the stars their laws." Both in the original Latin and in Anglo-Saxon and German translations, Boethius's moral lessons impressed the image of the tiny Earth, resting like a point at the center of an intelligible and orderly spherical cosmos, upon the minds of medieval thinkers.[3]

Boethius's cosmological image was complemented by the *Introduction to the Divine and Human Readings* of his younger contemporary Cassiodorus Senator (ca. 490–ca. 583). Sometime after the year 551, Cassiodorus, who had served as consul and master of the offices under Theodoric, retired to a monastery on his estate in southern Italy and there wrote the *Introduction* as a guide for monks. Cassiodorus's treatise presents little in the way of scientific content. Yet by combining brief summaries of each topic with lists of the books a monk ought to study, Cassiodorus preserved the model of the liberal arts.

NATURAL KNOWLEDGE IN THE VISIGOTHIC COURT

The most widely read compiler of the early Middle Ages was the Spanish bishop Isidore of Seville (ca. 560–636). With Isidore we see the most fundamental transformation that distinguishes the scientific writing of the early Middle Ages: that he and all of his successors were churchmen. We are no

[2] Cassiodorus Senator, *Variarum Libri XII*, IX.24.8, ed. T. Mommsen (Monumenta Germaniae Historica, Auctores Antiquissimi, 12) (Berlin: Weidmann, 1894), p. 290.

[3] Margaret Gibson, ed., *Boethius: His Life, Thought, and Influence* (Oxford: Basil Blackwell, 1981); Boethius, *De consolatione philosophiae*, 1, pros. 4.18–25; 1, vers. 5; 2, pros. 7.10–24; ed. H. F. Stewart (Loeb Classical Library, 74) (Cambridge, Mass.: Harvard University Press, 1936), pp. 142–5, 154–9, 212–15.

longer in the world of educated Roman laymen; the teachers of the early Middle Ages were clerics. They discussed nature within the framework of a Christian order and applied this knowledge to practical problems facing a Christian community.

Yet we also see the cultural interplay in which the heirs of the Roman tradition provided philosophical support for the dominion of Germanic rulers. Isidore came from an old Hispano-Roman family, succeeding his brother as bishop of Seville in 599. He presented copies of his two principal works dealing with natural phenomena, *On the Nature of Things* and the *Etymologies*, to the Visigothic king Sisebut.[4]

The first of these was stimulated by a query from the king asking the significance of a recent series of solar and lunar eclipses visible in Spain. Isidore responded with an elementary discussion of the structure of the universe, noting that these eclipses followed a divinely established and intelligible natural order. The bishop urged King Sisebut to avoid the superstitious "learning" of the pagans and follow the Old Testament ideal of Solomon, to whom God had given "true knowledge of the order of the heavens and the powers of the elements, the divisions of time, the course of the year and the configurations of the Stars."[5]

Isidore's *Etymologies* is a larger work in twenty books. It begins with three books discussing the seven liberal arts, but the remaining text deals with an encyclopedic range of practical knowledge, including medicine, architecture, agriculture, household implements, clothing, and games. But even a single subject could be dispersed throughout the *Etymologies* under several topical headings. Parts of astronomy remained among the four arts of the quadrivium; practical matters concerning units of time, the calendar, and the solstices and equinoxes appear in a section on chronology following his discussion of law. The astronomy underlying the date of Easter is summarized in a section on the offices of the church, and astrologers and astrology are mentioned in a section devoted to various religious and philosophical sects.[6]

At no point, however, did Isidore go into great detail; he presented only the merest sketch of ancient learning. For further details, his readers would have needed such sources as those mentioned by Cassiodorus. But in the early Middle Ages many of those texts in the liberal arts had been lost; only the general outlines of classical learning remained. The areas in which learning would endure were not limited to the three books of the *Etymologies* dealing with the liberal arts; equally important was the more practical knowledge outlined in the remaining seventeen books.

4 Pierre Riché, *Education and Culture in the Barbarian West from the Sixth through the Eighth Century*, trans. John Contreni (Columbia: University of South Carolina Press, 1978), pp. 258–9.

5 Isidore of Seville, *De Natura Rerum*, praef. 1–2, ed. Gustavus Becker (Berlin: Weidmann, 1857; reprint Amsterdam: Adolf M. Hackert, 1967), pp. 1–2; cf. Wisdom 7:17–21.

6 Isidore of Seville, *Etymologiae*, III.34–61, V.28–36, VI.17, VIII, IX.22–27, ed. W. M. Lindsay, 2 vols. (Oxford: Clarendon Press, 1911), vol. 1, pp. 147–64, 199–207, 236–43, 325–6.

MIRACLES AND THE NATURAL ORDER

When Isidore told Sisebut how the regular course of the Sun and Moon governed the occurrence of eclipses, he demonstrated one side of the tension between a stable natural order and miraculous suspensions of that order. This tension between miracles, as occasional suspensions of the order of nature for divine purposes, and the trustworthy course of nature, established by divine command and sometimes seen as wondrous in itself, was an enduring theme in Christian understandings of nature.

Valuable insights into early-medieval understandings of nature can be drawn from scriptural commentaries and sermons. These expressed scholarly interpretations of those biblical texts that had supplanted the works of antique authors as the core of the education for both medieval clerics and the educated laity.

Medieval scholars saw the order of nature as an expression of the Creator's activity. To the psalmist, the order of nature reflected the wisdom of God's design (Ps. 103:5–24). The prophet Jeremiah saw testimony to the constancy of God's covenant with Israel in the law that governs the motions of the heavens and the seas: "Were this established order ever to pass away . . . only then would the race of Israel also cease to be a nation in my presence forever" (Jer. 31:36). Saint Jerome (ca. 327–420) saw this passage as evidence that the natural order cannot change. The unchangeable laws to which the prophet referred are not the laws given to Moses but the very constitution and order of nature.[7]

One text takes pride of place in medieval understandings of the order of nature because of its central role in the daily monastic liturgy. Every morning members of monastic communities rose before dawn to praise God for his creation, chanting Psalms 148 to 150, including the significant lines:

> Praise him, Sun and Moon;
> praise him, every star that shines.
> Praise him, you highest heavens,
> you waters beyond the heavens.
> Let all these praise the Lord;
> it was his command that created them.
> He has set them there unageing for ever;
> given them a law that cannot be altered.
> Give praise to the Lord on earth,
> monsters of the sea and all its depths;
> fire and hail, snow and mist,
> and the storm-wind that executes his decree.
> (Ps. 148:3–8)

[7] Jerome, *In Hieremiam*, VI.27.3, ed. S. Reiter (Corpus Christianorum Series Latina, 74) (Turnhout: Brepols, 1960), p. 321.

For centuries, the daily repetition of this passage reinforced the belief that the heavens and the elements are governed by unchanging law, firmly established by God.

Even St. Augustine of Hippo (354–430), who usually restricted the value of natural knowledge to such practical uses as guiding ships or understanding Scripture, saw in Psalm 148 a sign that intelligent creatures render the most appropriate praise to God by directing their intellect toward the works of the Creator. Here Augustine saw that the study of nature could transcend vain curiosity to become a form of worship.[8]

Theodore of Mopsuestia (d. 428) saw knowledge of nature as a special way to come to a knowledge of God. Since "The heavens proclaim the glory of God" (Ps. 18:1–2), it follows that all peoples, whatever languages they speak, can confirm His providence and works through their natural intellect. An Old Irish gloss carried Theodore's assertion a step further, making this natural knowledge a divine mandate: "[I]t is a law that God should be known through them."[9]

The relation between God's dominion and the natural order was elaborated in an eighth-century gloss on the Psalms by an unknown Northumbrian or Irish author. This commentator made explicit the psalmist's theme that the heavens praise God when they minister to Him in their usual order, as well as when they occasionally obey Him contrary to their nature.[10] This allusion to actions contrary to the nature of a thing addresses the role of miracles in early-medieval understandings of nature.

Throughout the Middle Ages, Christian writers would point to cases in which God had miraculously intervened to alter the established order of nature. The classic examples were the Star of Bethlehem (Matt. 2:1–2, 9–10), the darkening of the Sun at Jesus' crucifixion (Matt. 27:45; Mark 15:33; Luke 23:44–45), the pause of the Sun and Moon at the Battle of Gibeon (Joshua 10:12–14), and the turning back of the shadow on a sundial at the prayer of Isaiah (2 Kings 20:1–11).

The paradox that God would suspend the regular course of the Sun, which He Himself had established, was usually resolved by maintaining that although the created order was unchanging, human minds might not be able to comprehend it fully.[11] Although Augustine acknowledged the

[8] Augustine, *Enarrationes in Psalmos*, 103, Sermo 4.2; 148, Sermo 3, ed. E. Dekkers and J. Fraipont (Corpus Christianorum Series Latina, 40) (Turnhout: Brepols, 1956), pp. 1522, 2167.

[9] Theodore of Mopsuestia, *Expositiones in Psalmos*, 18.5a–5b, ed. L. De Connick (Corpus Christianorum Series Latina, 88A) (Turnhout: Brepols, 1987), pp. 99–100; and "The Milan Glosses on the Psalms," fol. 42d, 2, in *Thesaurus Palaeohibernicus*, vol. 1, ed. Whitley Stokes and John Strachan (Cambridge: Cambridge University Press, 1901), p. 118.

[10] *Glossa in Psalmos: The Hiberno-Latin Gloss on the Psalms of Codex Palatinus Latinus 68*, CXLVIII.1–7, CXLVI.4, ed. Martin McNamara (Studi e Testi, 310) (Vatican City: Biblioteca Apostolica Vaticana, 1986), pp. 72–3.

[11] Augustine, *De Genesi ad litteram* VI.16–17, ed. Joseph Zycha (Corpus Scriptorum Ecclesiasticorum Latinorum, 28) (Vienna: F. Tempsky, 1894), pp. 190–2.

merit of studying nature, in his view the created order was not transparent to the human intellect. In discussing the miracle by which God turned back the sundial, Augustine pointed out that God knew from the beginning of time that Isaiah would pray, and that He would answer those prayers in that manner. Hence, this miraculous action did not disturb the created order. The human intellect can know the hidden natures of things that govern their regular development, but only God, through His unchanging foreknowledge, knows the hidden pattern of specific events that humans perceive as contingent.[12]

An extreme example of naturalistic exegesis is found in the mid-seventh-century treatise *De mirabilibus sacrae scripturae*, whose author is known as the Irish Augustine. The author uses numerous biblical examples to illustrate Augustine's position that miracles do not alter the order of nature that God had established at the Creation. These miracles, he maintains, were performed within the constraints of that created natural order.

A central premise of the Irish Augustine's approach was the Aristotelian claim that everything has a fundamental nature that defines its properties and governs its behavior. To act in a way inconsistent with this nature would be contrary to nature or, in a word, unnatural. The author's concern was to determine whether miracles were, in this sense, unnatural or whether they could be explained as simple, if uncommon, extensions of the natures of things. His conclusion was that all of the biblical miracles could be explained as examples of God's everyday administration of the ordinary course of nature rather than the result of any extraordinary exercise of divine power.

Thus he maintained that in all of these miracles the unchanging natures of things were modified naturally without the production of something new. But how could such a change as the transformation of Lot's wife, a sentient, rational, human being, into a pillar of salt, an insentient, nonrational, mineral object, come about without the production of something new? His answer was simple. The human body contains a substantial amount of salt; tears, phlegm, and sputum are all salty. Hence all that was needed in this case was for God to allow the nature of salt, latent in parts of the body, to grow naturally until it dominated the whole.[13]

The Irish Augustine explained even the virgin birth of Jesus in naturalistic terms. He believed that bees, many species of birds and fish, and the worms that arise in rotting flesh are all generated without copulation. Although it might be outside the usual course of nature for a human being to be conceived without intercourse, Jesus' human body arose from the flesh provided by his

[12] Augustine, *De civitate Dei*, XXI.8, ed. B. Dombart and A. Kalb (Corpus Christianorum Series Latina, 48) (Turnhout: Brepols, 1955), pp. 771–3.

[13] Pseudo-Augustine, *De mirabilibus sacrae scripturae*, I.11, ed. J.-P. Migne (Patrologia Latina, 35) (Paris: J.-P. Migne, 1845), cols. 2161–2; and Carol Anderson, "Divine Governance, Miracles and Laws of Nature in the Early Middle Ages: the *De mirabilibus Sacrae Scripturae*" (PhD diss., University of California, Los Angeles, 1982, p. 166).

mother. He "naturally acquired the nature [of a human being] from the nature" of his mother.[14]

For most medieval people, however, miracles were not confined to the biblical past. Gregory of Tours (ca. 539–594) collected many volumes of miracle stories associated with the tombs and relics of various saints, especially of the great patron of his city, St. Martin of Tours. Gregory was not concerned with explaining miracles. He valued them as signs of God's enduring power, of the special role of the saints as intermediaries between God and His people, and consequently, of the power of their tombs and relics.[15]

Some of the miraculous cures Gregory describes display his ambivalence toward conventional and miraculous healing. In one case, a cleric named Leunast, who was suffering from cataracts, fasted and prayed at the church of St. Martin and began to recover his sight. He then went off to a Jewish physician for further treatment and became as blind as he had been before. He then returned to the care of St. Martin, but in vain. Gregory saw this as a sign of two failings: first, that Leunast sought healing from an unbeliever; and, second, that he sought earthly remedies after having already received healing from heaven.[16]

But Gregory did not totally dismiss natural remedies; he praised the nun Monegundis, who during her lifetime had healed the sick using prayer and herbs. After her death, miraculous cures continued at her tomb.[17] Gregory accepted natural medicines in the hands of a prayerful healer; they are less dependable in the hands of an unbeliever. Ultimately, the gift of healing flows from God.

Early Christian writers did not develop a common position on the relationships between miracles and the natural order. Yet a steady theme in their writings was a recognition that although God may, at times, miraculously intervene to alter that order, a natural order does exist that can be understood and employed for practical purposes.

CHRISTIAN FEASTS AND THE SOLAR CALENDAR

One widespread practical use of the concept of natural order in the early Middle Ages was to provide a guide for religious rituals. Thus the solstices and equinoxes provided a framework for a cycle of Christian feasts marking

14 Pseudo-Augustine, *De mirabilibus sacrae scripturae*, III.2, col. 2193; and Anderson, "Divine Governance, Miracles and Laws of Nature in the Early Middle Ages," pp. 132–7.

15 Peter Brown, *The Cult of the Saints: Its Rise and Function in Latin Christianity* (Chicago: University of Chicago Press, 1981), especially pp. 76–9, 106–27.

16 Gregory of Tours, *Historia Francorum*, V.6, trans. O. M. Dalton (Oxford: Clarendon Press, 1927), vol. 2, pp. 175–6.

17 Gregory of Tours, *Glory of the Confessors*, 24, trans. R. van Dam (Liverpool: Liverpool University Press, 1988), pp. 39–40; Gregory of Tours, *Life of the Fathers*, 19, trans. Edward James (Liverpool: Liverpool University Press, 1988), pp. 118–25.

the turnings of the year. As early as the fourth century, a Christian author set out to demonstrate that Jesus's birth should be celebrated at the winter solstice. He asserted that John the Baptist was conceived at the autumnal equinox when the days were growing shorter and that Jesus was conceived six months later at the vernal equinox when days were growing longer, reflecting the Baptist's statement that Jesus "must grow, while I must diminish" (John 3:30). Consequently, John, the "lamp lit to show the way" (John 5:35), was born at the summer solstice and Jesus at the winter solstice. The discussion concluded by contrasting the pagan Roman celebration of the birth of *Sol Invictus* at the winter solstice with the birth of Christ, the truly unconquered Sun of Justice.[18]

Throughout the Middle Ages, popular feasts on St. John's Day and Christmas marked the turnings of the year, and writers from Augustine to Bede to Sacrobosco accepted these dates as having marked the solstices at the time of Jesus's birth.[19] Diagrams in astronomical texts connected these feasts to observable phenomena by using "the birth of the Lord" and "the birth of St. John" to mark the places of sunrise and sunset at the solstices.

Besides these solstitial and equinoctial feasts of the universal church, medieval communities developed feasts of local saints to supplant pagan festivals of an earlier observational solar calendar. Most striking are the feasts of St. Justus of Lyons, St. Oswald of Northumbria, and St. Brigit of Kildare, which, by marking days midway between the solstices and the equinoxes, display the continuity of pre-Christian solar festivals into the early Middle Ages.

The cults of these saints display certain common traits reflecting their pagan antecedents. Their biographers connected them with attributes of Celtic deities, and their sanctity included distinct solar elements. Their feasts were tied to the communities in which the saints had been powerful leaders, and were sponsored by local elites. They represented local centers of temporal and spiritual power, just as the pre-Christian festivals that they replaced had been regulated by local solar observations.[20]

Although these feasts were now fixed in the Julian calendar rather than regulated by solar observations, familiarity with the changing positions of sunrise and sunset continued. The Northumbrian monk Bede of Jarrow (672/673–735) noted that one could demonstrate the necessity of a leap year in the church calendar by carefully noting the place of sunrise on 21 March when, he claimed, the Sun would rise exactly in the east. Each successive year

[18] Bernard Botte, *Les Origines de la Noël et de l'Épiphanie* (Louvain: Abbaye du Mont César, 1932), pp. 93–105.

[19] Augustine, *Sermones*, 194.2, 287.3, 288.5, ed. J.-P. Migne (Patrologia Latina, 38) (Paris: J.-P. Migne, 1845), cols. 1016, 1302, 1308; Bede, *De temporum ratione*, 30.47–55, ed. Charles W. Jones (Corpus Christianorum Series Latina, 123C) (Turnhout: Brepols, 1980), p. 374; and Olaf Pedersen, "In Quest of Sacrobosco," *Journal for the History of Astronomy*, 16 (1985), 175–221 at p. 210.

[20] Stephen C. McCluskey, *Astronomies and Cultures in Early Medieval Europe* (Cambridge: Cambridge University Press, 1998), pp. 60–76.

the Sun would rise somewhat more to the southeast until after four years one must insert an extra day to restore the rising Sun to its former place.[21]

Near the end of the ninth century, Helperic of Auxerre proposed a similar empirical method to demonstrate the dates of the equinoxes. He went beyond Bede's simple horizon observations to recommend that his reader should note each day at sunrise where sunbeams passing through an aperture fell on the western wall of a room. Helperic claimed that such observations would demonstrate that the Sun first turns back at the winter solstice on 22 December and again at the summer solstice 182 days later, on 21 June.[22]

Neither Bede's nor Helperic's observations can be described as exact. When they wrote, the equinoxes and solstices fell closer to the seventeenth than the twenty-first. But they were not trying to make precise measurements; they wanted to use traditional observations of the rising Sun to illustrate the divisions of the year. Given the inherent lack of precision of these observations, the results could be read as confirming their expectations.

Bede and Helperic present the kind of simple observations of the positions of the Sun and Moon that have been documented in a wide range of ethnographic and archaeological contexts.[23] That they raised these practical techniques in the context of the church calendar reflects the applications of observations of nature to religious ritual that continued through the early Middle Ages.

COMPUTUS AND THE DATE OF EASTER

Computation, rather than observation, governed the problem of finding the date of Easter, a problem that in the eighth century became central to medieval teaching about the natural world. As early as the fourth century, several different methods for calculating the date of Easter in advance had been developed in Rome, Alexandria, and elsewhere. Thus, although later medieval scholars did not have to develop new techniques, they did need to resolve conflicts among competing computistical methods and to provide texts that would demonstrate the historical, calendric, astronomical, and arithmetical principles underlying these computations to generations of students. The disciplines drawn together in these texts defined *computus*, a distinctly medieval science. *Computus* became an essential part of the

[21] Bede, *De temporibus liber*, 10.14–21, ed. Charles W. Jones (Corpus Christianorum Series Latina, 123C) (Turnhout: Brepols, 1980), p. 593; and Bede, *De temporum ratione*, 38.47–58, ed. Charles W. Jones (Corpus Christianorum Series Latina, 123B) (Turnhout: Brepols, 1980), p. 401.

[22] Helperic, *Liber de computo*, 31, ed. J.-P. Migne (Patrologia Latina, 137) (Paris: J.-P. Migne, 1853), cols. 40–3.

[23] Clive L. N. Ruggles, *Astronomy in Prehistoric Britain and Ireland* (New Haven, Conn.: Yale University Press, 1999); and Michael Zeilik, "The Ethnoastronomy of the Historic Pueblos, I: Calendrical Sun Watching," *Archaeoastronomy*, no. 8, supplement to *Journal for the History of Astronomy*, 16 (1985), S1–S24.

education of clerics and guided much of the preservation, transmission, and development of natural knowledge from the time of Bede to the rise of the universities.

Although the methods of determining the date of Easter are arithmetical and astronomical, underlying the development of *computus* was the ideal of ritual uniformity. This was usually expressed in terms of the spiritual imperative that all Christians be united in prayer. Unity of ritual has political as well as spiritual implications; the determination of the single uniform day requires the acknowledgment of some competent authority. From Constantine to Charlemagne, rulers would employ this concern for a uniform date of Easter as an element of their religious, educational, and political agendas.[24]

The historical basis for the date of Easter is found in the Gospels' accounts that Jesus was crucified on the Friday following a Passover meal and rose from the dead on the following Sunday. A practice soon emerged within the early church of celebrating Easter on the fourteenth day of the Hebrew month of Nisan, the first month of spring. Thus Easter, as the commemoration of the Resurrection at which humankind was spiritually renewed, was seen, like the equinox of spring, as a triumph of light over darkness, in which Christ, the true Sun of Justice, had risen and shone forth.[25]

This symbolism required that Easter be celebrated after the vernal equinox, when the length of day was greater than the length of night, and after the Paschal full moon, when the Moon shed its fullness of light. Sometime before 590, one author noted that: "It surely is impossible that at Easter any part of darkness should rule over the light, since the feast of the Lord's Resurrection is light, and there is no communion of light with darkness."[26]

It may be tempting to dismiss these doctrinal details as irrelevant to the astronomy and mathematics of *computus*. But the central issue for medieval computists was not to find *a time* for a celebration but to return to *the time* of salvation in which mankind is renewed. The astronomy and mathematics served a sacred purpose, and the central place that early-medieval scholars granted to the study of *computus* reflects the importance of that purpose.[27]

The quest for uniformity in ritual led to several competing mathematical techniques for establishing in advance the dates of the vernal equinox, the subsequent Paschal full moon, and Easter Sunday. The problem required

[24] Eusebius, *Vita Constantini*, 3.18, ed. J.-P. Migne (Patrologia Graeca, 20) (Paris: J.-P. Migne, 1857), cols. 1073–6; and Charlemagne, *Admonitio generalis*, cap. 72, ed. A. Boretius (Monumenta Germaniae Historica, Leges, Capitularia regum Francorum, I) (Hannover: Hahn, 1883), p. 60.

[25] Charles W. Jones, *Bedae Opera de Temporibus* (Cambridge, Mass.: Medieval Academy of America, 1943), pp. 9–10; and Jerome, *In die Dominica Paschae*, ed. G. Morin (Anecdota Maredsolana, 3, pars. 2) (Maredsous/Oxford: J. Parker, 1897), p. 418.

[26] Pseudo-Anatolius, *De ratione paschali*, 4, in B. Krusch, *Studien zur Christlich-mittelalterlichen Chronologie: Der 84-jährige Ostercyclus und seine Quellen* (Leipzig: von Zeit, 1880), p. 320.

[27] Bede, *The Reckoning of Time*, ed. and trans. Faith Wallis (Liverpool: Liverpool University Press, 2004), pp. xviii–xxxiv.

consideration of the astronomical fact that full moons recur on the same date every 19 years, or less accurately every 8 or 11 years, and the calendric fact that Sundays recur on the same date every 28 years. Whereas in the East the practice of Alexandria, based upon a 19-year lunar cycle, became dominant, the church at Rome first settled on a more convenient 84-year cycle as the basis for its Easter reckoning.

The more accurate 19-year lunar cycle, which ultimately came to dominate medieval *computus*, appeared in a number of guises. Isidore of Seville and other early writers combined five such 19-year cycles to form a 95-year cycle, which Isidore erroneously believed to be a perpetual Easter cycle. In 525, the definitive medieval Easter cycle, which combined the 19-year lunar cycle with the 28-year cycle of the days of the week to form a 532-year cycle, was proposed by Dionysius Exiguus on the basis of Alexandrian practice.[28]

These discrepant techniques for calculating the date of Easter led to recurring controversies, especially in Ireland and Britain, which, with the rest of Western Christendom, had first adopted an 84-year Easter cycle. As Irish Easters came to disagree with those computed on the continent, the Irish held a synod at Mag Léne (ca. 630), which recommended the universally accepted 532-year cycle employed at Rome. When the same issue arose in England, King Oswiu of Northumbria called a synod at Whitby (in 664), where the Easter question was resolved in favor of the 532-year cycle. Like the Irish, the British were motivated more by the unity of Christendom and the power of St. Peter and his successors as keepers of the keys of heaven than by any evaluation of astronomical precision.[29]

As Irish and British scholars addressed the Easter question, they drew into the discussion topics heretofore taken for granted. Whereas the church fathers had addressed Roman audiences, for whom the Julian calendar was part of their everyday experience and the rudiments of astronomy part of their background in the liberal arts, Irish and British scholars had to present the elements of arithmetic, the divisions of time, and the basics of astronomy to students who found them all new and foreign. Bede of Jarrow provided the standard treatment of Easter *computus*, drawing on both secular and religious authorities in his discussions of the zodiac, the solstices and equinoxes, the causes of lunar and solar eclipses, the motions of the planets, the sphericity of the Earth, and the changes of the seasons. These new materials are not really necessary to calculate the date of Easter, yet they expanded *computus* so that understanding the principles underlying these computations became the guiding rationale for the further study of astronomy and cosmology.

28 Isidore of Seville, *Etymologiae*, 6.17.9, p. 239; and Jones, *Bedae Opera de temporibus*, pp. 61–75.

29 Maura Walsh and Dáibhí Ó Cróinín, eds., *Cummian's letter De controversia paschali . . . Together With a Related Irish Computistical Tract, De ratione conputandi* (Studies and Texts, 86) (Toronto: Pontifical Institute of Mediaeval Studies, 1988), pp. 7–15; and Bede, *Historia ecclesiastica gentis Anglorum*, 3.25, ed. Bertram Colgrave and R. A. B. Mynors (Oxford Medieval Texts) (Oxford: Clarendon Press, 1969), pp. 306–7.

Computus soon became a major part of the educational reform sponsored by Charlemagne and his successors, and was elaborated in a series of computistical anthologies and textbooks. The Carolingian anthologies incorporated material from Macrobius and Martianus Capella to show how measurements, or more properly purported measurements, could yield the dimensions of the Earth, the Sun, and the Moon.[30]

These anthologies were accompanied by new textbooks composed to teach the astronomy of *computus*. In the summer of 820, Rabanus Maurus, master of the monastic school at Fulda, wrote a dialogue between master and pupil combining traditional computistical material with such topics as Aratus's newly recovered descriptions of the classical constellations. Like Bede before him, Rabanus demonstrated that medieval natural knowledge was not mere book learning. In his *computus*, he recorded the positions of the planets in July 820. Rabanus calculated the positions of the Sun and Moon, observed the positions of Mars, Jupiter, and Saturn, and noted that Venus and Mercury were too close to the Sun for him to determine their positions.[31]

Along with such pedagogical observations, there was also a long tradition of inserting descriptions of noteworthy astronomical phenomena into monastic chronicles and calendars. Solar and lunar eclipses, conjunctions of the planets, occurrences of retrograde motion, and even an occultation of Jupiter by the Moon and a purported transit of Mercury across the face of the Sun are mentioned. Yet these are all unusual events, and they are not recorded with any great precision; these are not the kinds of systematic observations used to develop astronomical theories. They do, however, reflect a familiarity with the heavens and an interest in what happens there.[32]

Abbo of Fleury (ca. 945–1004), attempted to extend the methods of *computus* to calculations of the positions of the five known planets.[33] Abbo's results could never have survived comparison with even the crudest observations of the planets. The planetary periods he used were based on numerological principles, while the computistical assumption of uniform motion made no attempt to account for the planets' retrograde motions, of which he

[30] Vernon H. King, "An Investigation of Some Astronomical Excerpts from Pliny's *Natural History* Found in Manuscripts of the Earlier Middle Ages" (B. Litt. thesis, Oxford University, 1969, pp. 3–45, 57–78); and Arno Borst, "Alkuin und die Enzyklopädie von 809," in *Science in Western and Eastern Civilization in Carolingian Times*, ed. Paul L. Butzer and Dietrich Lohrmann (Basel: Birkhäuser Verlag, 1993), pp. 53–78.

[31] Rabanus Maurus, *De computo*, 48.14–19, ed. J. McCulloh and W. Stevens (Corpus Christianorum Continuatio Mediaevalis, 44) (Turnhout: Brepols, 1989), p. 259.

[32] Dietrich Lohrmann, "Alcuins Korrespondenz mit Karl dem Grossen über Kalendar und Astronomie," in Bützer and Lohrmann, *Science in Western and Eastern Civilization in Carolingian Times*, pp. 79–114 at pp. 113–14.

[33] Abbo of Fleury, *De ratione spere*, 1–67, 170–185, ed. R. B. Thomson, in *The Light of Nature: Essays in the History and Philosophy of Science Presented to A. C. Crombie*, ed. John D. North and J. J. Roche (Dordrecht: Martinus Nijhoff, 1985), pp. 120–3, 132–3; and David Juste, "Neither Observation nor Astronomical Tables: An Alternative Way of Computing the Planetary Longitudes in the Early Western Middle Ages," in *Studies in the History of the Exact Sciences in Honour of David Pingree*, ed. Charles Burnett et al. (Leiden: Brill, 2004), pp. 181–222.

was well aware. But Abbo's concern with the problem of the planets reflects an active curiosity about astronomical matters, though his simple arithmetical approach was not able to deal with even the most obvious elements of planetary motion.

Early in the eleventh century, at the English monastery of Ramsey, the monk Byrhtferth wrote his *Manual*, with portions in Anglo-Saxon for the benefit of those clerks who might not know Latin. He placed *computus*, or *rimcræft*, at the heart of a broader study of the natural world. The basics of the *Manual* are taken from Bede and Rabanus, but Byrhtferth drew upon other areas of the quadrivium in his discussion of the natural and mystical significance of numbers in the universe, which God had established "in measure, in number, and in weight."[34]

Marking the culmination of this computistical tradition is a manuscript written near the end of the century only a few miles from Ramsey. It goes beyond the Carolingian computistical anthologies to include the work of Abbo, Byrhtferth, and Helperic, as well as such earlier authorities as Bede, Isidore, and Pliny. This visually striking manuscript is known for its many diagrams that illustrate a wide range of scientific topics.[35]

This manuscript's most complex illustration (see Figure 11.1) reveals how the number four binds together the four elements, the four sensible qualities, the four seasons, the four ages of man, the four directions (east = *Anatole*, west = *Disis*, north = *Arcton*, and south = *Mesembrios*), and the four letters of the name of the first-formed man, ADAM.[36] In the teachings of Abbo, Byrhtferth, and their colleagues in monastic schools, number and *computus* were at the core of the understanding of nature.

MONASTIC TIMEKEEPING

Monastic scholars played an active role in preserving and transmitting ancient learning through the early Middle Ages, and their communities often shaped this knowledge to meet their spiritual and secular needs. The regular assembly for communal prayer is a central element of monastic ritual, justified by the psalms. "Seven times a day I have praised you. . . . At midnight I arose to give you praise" (Ps. 118:164, 62). In his *Rule*, St. Benedict considered the changing length of daylight, noting that there should be fewer readings from the psalms at night during the summer because the nights are shorter.[37]

34 Byrhtferth of Ramsey, *Byrhtferth's Manual*, ed. S. J. Crawford (Early English Text Society, 177) (London: H. Milford, Oxford University Press, 1929), pp. 8.16–18.

35 Faith Wallis, "MS Oxford, St. John's College 17: A Medieval Manuscript in Its Context" (PhD diss., University of Toronto, 1985, pp. 216–19). For an extensive discussion of the manuscript with a digital facsimile, see Faith Wallis, *The Calendar and the Cloister: Oxford – St. John's College MS17*, http://digital.library.mcgill.ca/ms-17/index.htm.

36 Oxford, St. John's College, MS 17, fol. 7v.

37 Benedict of Nursia, *Regula*, 10, ed. and trans. T. Fry (Collegeville, Minn.: The Liturgical Press, 1980), pp. 204–5.

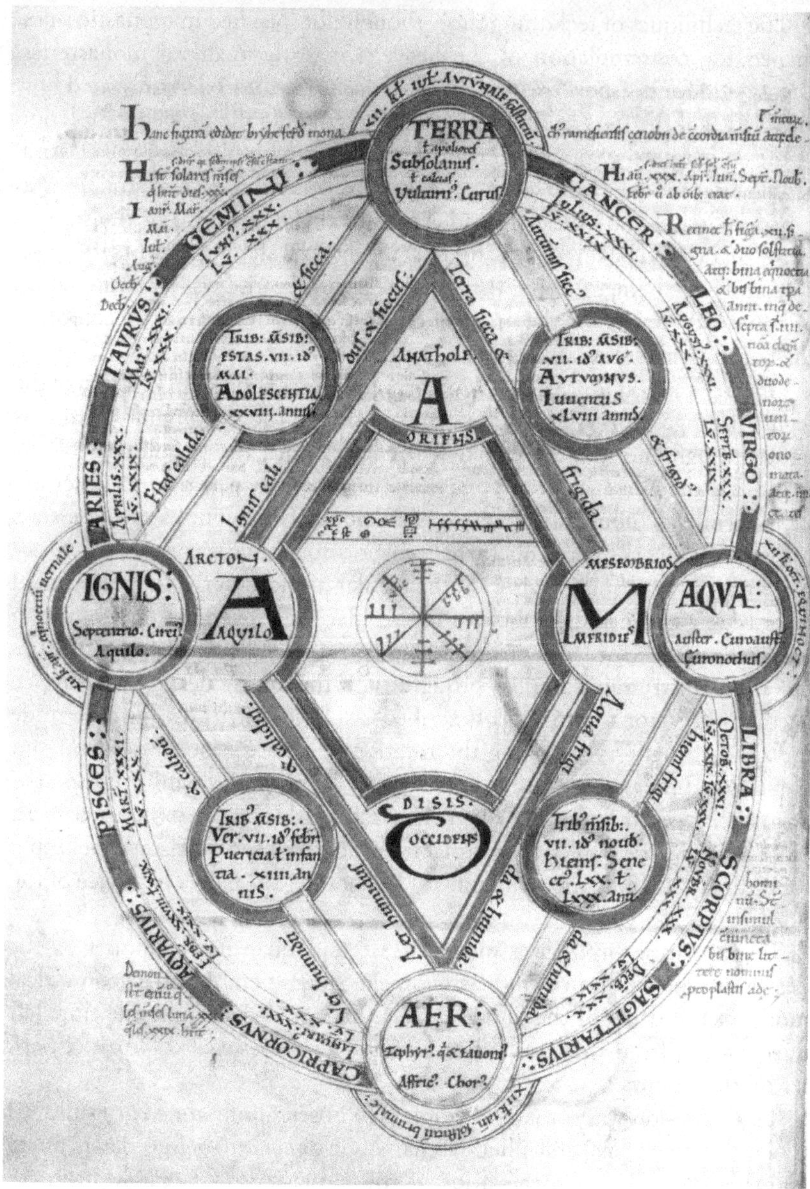

Figure 11.1. Cosmic symbolism of the number four. Oxford, St. John's College, ms 17, fol. 7v. Used by permission of the President and Scholars of St. John's College, Oxford.

This regular pattern of prayer presupposed a regular pattern of timekeeping, which, in turn, called for a rudimentary knowledge of the changing length of daylight and nighttime during the year and a familiarity with the regular motions of the Sun, stars, and Moon.

The techniques of reckoning time, though not specified in monastic rules, shaped the contemplation of the heavens in early-medieval monasteries. Gregory of Tours's short treatise *On the Course of the Stars* demonstrated how knowledge of the stars could serve God by regulating the proper completion of the daily round of prayer. Gregory revealed little influence of the classical geometrical model of the cosmos; he did not mention such abstractions as spheres, circles, and poles. Instead, he defined the positions of the stars in terms that would be directly useful to an observer watching them as they rise. The unifying theme running through the entire treatise was a consideration of how God established the order of creation and how it, in turn, can order the worship of the Creator.[38]

Gregory's astronomical discussion begins with a simple computation of the regular changes of the length of daylight through the seasons and of moonlight through the month, which will endure until the dissolution of the world. He then describes fourteen constellations and their rising and setting each month, demonstrating how observations of the courses of these stars can regulate the course of nocturnal prayer throughout the year.

Through the ensuing centuries, the regular motions of the stars continued to provide a guide and symbol for the regular order of monastic life, which itself reflected the order of creation. This practice remained open to the use of various instruments, which brought with them a model of the heavens. Pacificus of Verona (776–844) described a simple instrument to determine the time of night by watching the rotation of Polaris around the faint star that marked the pole about the year 800. At the turn of the millennium, the astrolabe, recently imported from the Arab world, began to be used in its place. One description of the astrolabe considered it too precise a timekeeper for ordinary use but "most suitable for celebrating the daily . . . service of the Lord."[39]

The astrolabe brought a more precise and quantitative model of the heavens to the tradition of monastic timekeeping. It embodied a geometrical model of the rising of the Sun and stars that could determine the time left until sunrise from the observation of a single star (see also North, Chapter 19, this volume).

This precision was sometimes applied to observations aimed at improved understanding of natural phenomena. About an hour before dawn on 18 October 1092, Walcher, the prior of the monastery of Great Malvern in England, saw the Moon begin to grow dark. For some time he had wanted

[38] Stephen C. McCluskey, "Gregory of Tours, Monastic Timekeeping, and Early Christian Attitudes to Astronomy," *Isis*, 81 (1990), 8–22; and Gregory of Tours, *De cursu stellarum*, 16, ed. W. Arndt and B. Krusch (Monumenta Germaniae Historica, Scriptores rerum Merovingicarum, 1, pt. 2) (Hannover: Hahn, 1885), p. 863.

[39] Joachim Wiesenbach, "Pacificus von Verona als Erfinder einer Sternenuhr," in Butzer and Lohrmann, *Science in Western and Eastern Civilization in Carolingian Times*; *De utilitatibus astrolabii*, 5.4, in *Gerberti postea Silvestri II papae Opera mathematica*, ed. N. Bubnov (Berlin: R. Friedländer & Sohn, 1899; repr. Hildesheim: Georg Olms, 1963), pp. 129–30.

to measure the time of a lunar eclipse as the basis for more precise lunar tables than those found in *computus* texts. Taking his astrolabe, he measured the height of the Moon at its darkest and used this single observation to determine the time of the lunar eclipse to within a quarter-hour. Walcher then computed a set of tables that he hoped would give the exact time of each new Moon from 1036 through 1111. The failure of his tables to achieve such precision led him to study the new astronomy of the Arabs taught by his master, Peter Alfonsi.[40]

But such scientific concerns remained peripheral to monastic life; the practical concerns of governing the time for prayer remained primary. In the last quarter of the thirteenth century, the records of great monasteries and cathedrals begin to mention major expenditures for the construction and maintenance of mechanical clocks. Richard of Wallingford left a detailed account of such a clock, including the underlying mathematical and astronomical principles, that was built while he served as abbot of St. Albans from 1327 to 1336. His clock was a veritable model of the heavens, for the seasonal hours were indicated by a rotating astrolabe dial, and other mechanisms depicted the phases of the Moon and its motions above and below the ecliptic.[41]

With mechanical clocks, astrolabes, and the investigation of newly translated texts, we emerge from the early Middle Ages and enter a different world of learning. But this new world of the "high" Middle Ages, with its universities, where nature was studied within the framework of scholastic natural philosophy, rested on foundations laid by the practical natural knowledge of the early Middle Ages.

[40] McCluskey, *Astronomies and Cultures in Early Medieval Europe*, pp. 180–4.

[41] C. F. C. Beeson, *English Church Clocks, 1280–1850: History and Classification* (London: Phillimore, 1971), pp. 13–22; and John D. North, *Richard of Wallingford: An Edition of His Writings with Introductions, English Translation and Commentary*, 3 vols. (Oxford: Clarendon Press, 1976), vol. 1, pp. 441–526; vol. 2, pp. 321–70.

12

EARLY-MEDIEVAL COSMOLOGY, ASTRONOMY, AND MATHEMATICS

Bruce S. Eastwood

Cosmology, astronomy, and mathematics entered the early Middle Ages through many channels. The Roman aristocrat Boethius (ca. 480–ca. 525) stressed their mathematical character and their crucial place in the foundation of a philosophical order associated with Plato; this identified them as elite studies, and therefore as having limited utility and importance in a culture where the traditional Roman aristocracy was in decline. Their use for timekeeping reoriented the meanings of these disciplines toward a focus on numbers and discrete events, especially in the life of the Christian church, which was their most important preserver and patron. And in new ways they offered an understanding of the establishment of God's order in the world of routine experience. Each discipline developed a many-sided history in the uses and problems to which it was applied across Western Europe and over the long span of six or seven hundred years before the twelfth century. Cosmology, astronomy, arithmetic, and geometry meant very different things in a fifth-century Roman villa, a seventh-century Irish monastery, a ninth-century royal library, and an eleventh-century ecclesiastical school.[1]

COSMOLOGY

Understanding the cosmos involved placing its important parts, or inhabitants, in some kind of order, both spatially and in terms of value. The spatial framework in medieval cosmology contained a central spherical Earth and an enclosing sphere with stars fixed to it. Between the Earth and this outer sphere, the seven planets (including the Moon and Sun) circled in orbits.

[1] Philip Merlan, *From Platonism to Neoplatonism* (The Hague: Martinus Nijhoff, 1960); Claudio Leonardi, Lorenzo Minio-Paluello, Pierre Courcelle, and Ubaldo Pizziani, "Boezio," *Dizionario biografico degli Italiani*, XI (1969), 142–65; Pierre Courcelle, *Les lettres grecques en Occident de Macrobe à Cassiodore* (Paris: Boccard, 1948); and Stephen Gersh, *Middle Platonism and Neoplatonism: The Latin Tradition*, vol. 2 (Notre Dame, Ind.: University of Notre Dame Press, 1986).

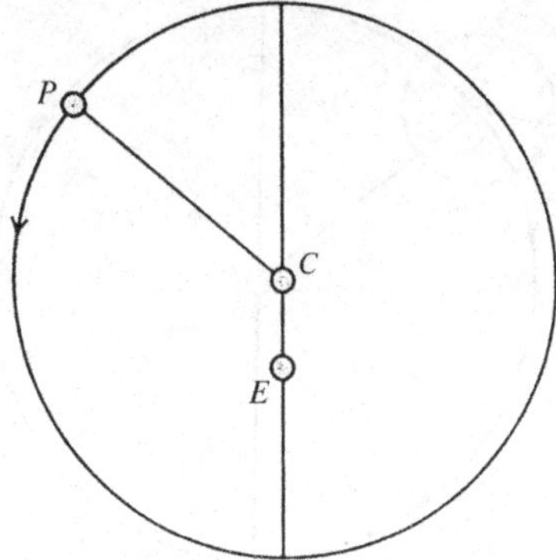

Figure 12.1. Eccentric orbit of planet. C = center, P = planet, E = eccentric earth.

The value system attached to this image gave highest status to the outermost sphere and least status to the central Earth. Accordingly, various pagan and Christian cosmologies populated the spatial frame with deities, spirits, and demons; the lowest of these, near the Earth and humans, were most limited in perfection and powers or – in Christian cosmologies – clearly evil.

Plato's *Timaeus* was a major foundation stone of late-ancient and early-medieval cosmology. The first two-thirds of this work was translated into Latin and commented upon in the fourth century by the North African philosopher Calcidius, who argued for its agreement with late-ancient astronomy. He set forth the centrality and sphericity of the Earth, the description of circles on the celestial sphere, and the stellar and planetary motions. Where Plato had used concentric circles for planetary motions, Calcidius used off-centered circles (eccentrics) (Figure 12.1) and combinations of circle-upon-circle (epicycles) (Figure 12.2) to explain these phenomena, showing how these models of the astronomers reduced the apparent irregularity of planetary motions to a truer regularity. In this way, he introduced Hellenistic elements, the eccentrics and epicycles, to explain the nonuniform apparent motions of planets. As part of the created world order, these models helped to establish the mathematical rationality of the Creator. The subjection of all celestial arrangements and movements to the World Soul produced a well-ordered physical cosmos and gave a thoroughly rational character to time. This reasoned order of the macrocosm was reflected in the microcosm, or individual man. Thanks to these associations, the commentary of Calcidius on Plato offered much food for Christian theological and philosophical thought.[2]

[2] The text of Calcidius, which has not yet been translated, is in *Timaeus a Calcidio translatus commentarioque instructus*, ed. J. H. Waszink (Plato Latinus, 4) (London: Warburg Institute, 1962). See Gersh, *Middle Platonism and Neoplatonism*, vol. 2, pp. 421–92, and the literature cited therein.

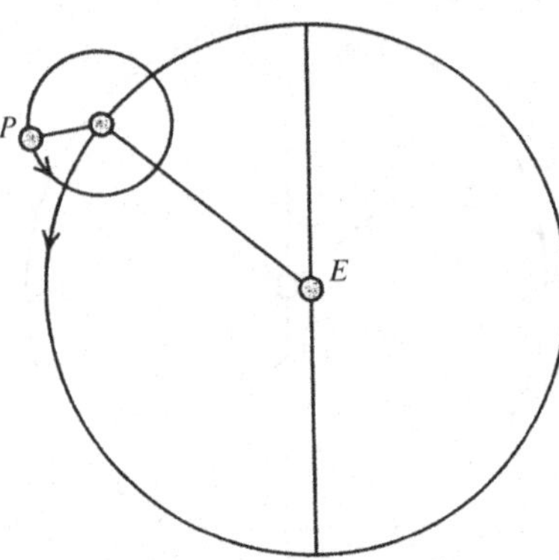

Figure 12.2. Epicyclic orbit of planet. E = central earth, P = planet on epicycle.

Early in the fifth century, a Roman aristocrat named Macrobius developed certain Platonic themes in his encyclopedic commentary on Cicero's *The Dream of Scipio*. Macrobius focused attention on the simplest questions of celestial order, spending a good bit of time on such questions as the different scholarly opinions about the relative positions of the planets Venus and Mercury with respect to the Sun. To solve this problem, he assumed, with Plato, a direct relationship between planetary distance from the Earth and planetary orbital time. Because these two planets moved with about the same period as the Sun, they apparently had similar radial distances as well. To determine the final answer to this problem, Macrobius invoked the greater wisdom of Plato and the Egyptians, making the planets Mercury and Venus, in that order, next beyond the Sun in radial distance. However, later in his commentary, Macrobius gave these two planets the same radial distance and the Sun a remarkably different distance, ignoring the contradiction between the separated arguments. With regard to the geometry of celestial movements, he was consistently imprecise, apparently accepting eccentric planetary spheres but excluding epicycles because of their lack of physical reality. He followed Pliny the Elder in some of his physical cosmology and in at least one place described the Sun as the body regulating all other planetary motions, including retrograde motion. Macrobius wished to preserve a spectrum of late-ancient Platonist views about the world, representing the knowledge of an aristocrat with a proper philosophical education. As a collection of many cosmological and cosmographical doctrines, rather than as an integrated philosophical system, his commentary on *The Dream of Scipio* found a place in late antiquity and again during the ninth and later centuries.[3]

3 Jacques Flamant, *Macrobe et le néoplatonisme latin à la fin du IVe siècle* (Leiden: Brill, 1977), especially chaps. 7–9; Gersh, *Middle Platonism and Neoplatonism*, vol. 2, pp. 493–595. On Macrobius and heliocentrism, see Bruce Eastwood, "Kepler as Historian of Science: Precursors of Copernican

In the Christian literary tradition, commentaries on the biblical book of Genesis used Greek cosmological doctrines extensively. Exemplifying this tradition, the *Literal Commentary on Genesis in Twelve Books* (completed in 416) by St. Augustine (354–430) became both an interpretive standard and a source of doctrines for the early Middle Ages. As in his even more influential *On Christian Teaching* (completed in 426), Augustine attempted to apply to the understanding of Holy Scripture whatever classical knowledge of the world seemed useful and correct. In his interpretation of the biblical Creation, he found Greek natural philosophy a useful guide. As he proceeded, Augustine recommended that a reader should never commit himself solely to a single scientific interpretation since various acceptable interpretations were always possible. Although committed to using only what was known "by most certain reasoning or experience," he proposed no method for finding unique truths about nature. He preferred to include rather than exclude anything useful in order to avoid discrediting the Faith. For Augustine as well as others living in a world of competing religious systems, natural philosophy seemed to offer an advantage in the competition.

How did God establish "the natures of things" (a favored medieval book title)? Augustine chose this question as his line of investigation rather than simply appealing to an arbitrary and incomprehensible divine omnipotence to explain the order of Creation. The numbered days of Creation signified the succession of kinds of things. God created according to measure, number, and weight, which were standards preceding creation and referring to God as the source of all creation. And even after the sixth day, when no more natures were created, God continued to exercise providential rule over His creation since it could not otherwise subsist. To explain the biblical passage that placed water above the firmament as well as below it (Gen. 1:7), Augustine appealed to Greek philosophy to show that a heavy body could become light through indefinite division, thereby allowing water in the form of moist clouds to be high above much of the air, although we think of air as lighter than water. More minutely divided particles of water would be able to rise even higher and pass above the celestial layer of stars. And so he proceeded to use late-ancient natural philosophy to describe a reasoned cosmos that, in its basic structure and processes, was largely Platonic except where Genesis prohibited it.

The presence of demons in the elemental layer of air, of which they were materially composed, was one of Augustine's many points of agreement with ancient cosmology. In Christian terms, demons were fallen angels, Lucifer and his companions, who had real, though limited, powers to mislead humans willing to listen to them. The genuine effects of magic and judicial astrology were demonic effects, created to trick men and women

Heliocentrism according to *De Revolutionibus*, I, 10," *Proceedings of the American Philosophical Society*, 126 (1982), 367–94, especially pp. 383–91.

into enslaving their wills to demons in the hope of knowing and influencing the future. These demonic inhabitants of both the pagan and the Christian cosmos throughout late antiquity and the Middle Ages were fundamental to the explanation of sin and evil.[4]

Isidore, bishop of Seville (ca. 570–636), composed *The Nature of Things* (ca. 613) as a brief cosmology, divided into sections that dealt successively with the elements of time and with each of the four elemental realms. Frequently, as he described the levels from outermost stars to the central Earth and its parts, he taught his reader much Christian allegory. Isidore believed that just as we can find moral and mystical meanings in the Bible, so can we "read" the visible book of nature, exposed through the accounts of classical writers.

Isidore's account of the macrocosm followed Platonic and Stoic doctrines at a simple and easily absorbed level, explaining the regions and divisions of the celestial sphere, the order and the rationality of stellar and planetary motions, and some details of solar and lunar eclipses. A variety of meteorological events, the seas and rivers, and certain geological and geographical explanations completed his picture, with Christian writings providing important information derived ultimately from pagan sources. Even a selection of weather prognostications appeared, recalling Augustine's approval of the practical knowledge of sailors and farmers. In fact, Isidore's *The Nature of Things*, geared to the less sophisticated audience of the distant seventh-century Roman province of Spain, corresponded in important ways to the earlier Genesis commentary of Augustine, the theme of which was the reasoned order of God's Creation. Isidore differed from Augustine in his direct quotation of pagan literary writings and in his frequent, though not constant, use of Christian allegory. But he followed the recommendations of Augustine's *On Christian Teaching*, legitimizing the study of pagan learning for Christian purposes.[5]

In the interval between Isidore and the English monk Bede (673–735), Irish scholars produced many anonymous works on *computus* (the study of the rules and the arithmetical calculations for establishing the date of Easter) and generally on the reading and understanding of the Scriptures. Bede himself was aware of and used some of them. Their foundations were various but clearly reveal the influence of Isidore. Irish writing presented a strong sense of alienation from the classical heritage, as if it were a strange

4 Robert A. Markus, "Marius Victorinus and Augustine," in *The Cambridge History of Later Greek and Early Medieval Philosophy*, ed. A. H. Armstrong (Cambridge: Cambridge University Press, 1970), pp. 331–419, especially pp. 395–405. See the critical notes in Augustine, *La Genèse au sens littéral en douze livres*, ed. P. Agaesse and A. Solignac, 2 vols. (Oeuvres de Saint Augustin, 48) (Paris: Desclée de Brouwer, 1972); Augustine, *The Literal Meaning of Genesis*, trans. John H. Taylor, 2 vols. (Ancient Christian Writers, 41–42) (New York: Newman Press, 1982); and Augustine, *On Christian Teaching*, trans. Roger P. H. Green (Oxford: Oxford University Press, 1997).

5 Isidore of Seville, *Traité de la nature*, ed. Jacques Fontaine (Bordeaux: Féret, 1960); and Jacques Fontaine, *Isidore de Séville et la culture classique dans l'Espagne wisigothique*, 3 vols. (Paris: Études Augustiniennes, 1959–1983).

body of texts dropped from the sky. In this the Irish differed greatly from Isidore, who still showed a sense of intellectual kinship with pagan classical writings.[6]

At the beginning of the eighth century, Bede wrote his own *The Nature of Things* (ca. 701) to offer his monastic students an intensive, though still largely descriptive, account of the cosmos. In its deeply Augustinian opening, he distinguished succinctly the eternal and uncreated divine Word from three different temporal frameworks, within which form, matter, elements, cause and effect, and delayed causation all came into being. Bede's list of created things was simple: heaven and earth, air and water, and angels – all created from nothing. He then set out the ordered layers of the cosmos from top to bottom, following closely the chapter headings of Isidore's *The Nature of Things*, written a century earlier.

Unlike Isidore, Bede avoided allegorical discussion and concentrated on physical cosmology, adding more topics in each of the realms of fire (the heavens), air (meteorology), water, and earth than had Isidore. The result was more nuanced and complete, with the planets, the Sun and Moon, and fundamentals needed for a subsequent work on time reckoning receiving notable attention. Bede introduced eccentric planetary circles (but said nothing of epicycles), used Pliny's locations of planetary apogees as well as his numerical values for planetary latitudes, and included a Plinian solar radial force as the cause of planetary retrogradations. Bede's work on the "natures" (the four elements and their subordinate parts) created by God represented a new stage in early-medieval Christian cosmology, informed by Augustinian categories and classical contents in extended detail. It became a model for a purely physical description of the results of divine creation, devoid of allegorical interpretation, and using the accumulated teachings of the past, both Christian and pagan.[7]

On the European continent, Isidore's *The Nature of Things* was more widely read during the eighth century, but Bede's work by the same title dramatically surpassed it in the ninth century and later. In the Carolingian Empire (eighth and ninth centuries), clerical education improved sufficiently to encourage a much wider consultation of the Latin classics and patristic writings along with the introduction of a few Greek Christian texts. In cosmology, not only Pliny's *Natural History* but also the later Latin works by Calcidius and Macrobius became current. These works were excerpted and used in a variety of contexts during the ninth century. The most original

[6] The popular book by Thomas Cahill, *How the Irish Saved Civilization* (New York: Talese/Doubleday, 1995), which deals largely with the sixth and seventh centuries, does not do justice to the historical record reconstructed by scholars over the past half century and should not be regarded as serious historical scholarship. The limits and outlook of Irish cosmological thought in the seventh century are carefully discussed by Marina Smyth, *Understanding the Universe in Seventh-Century Ireland* (Woodbridge: Boydell Press, 1996), pp. 104–75.

[7] Bede, *De natura rerum*, in *Opera pars VI: Opera didascalica*, ed. Charles W. Jones, 3 vols. (Corpus Christianorum, Series Latina, 123A–123C) (Turnhout: Brepols, 1975–1980), vol. 1, pp. 173–234.

thinker of the century produced a monumental work, *On Nature* (*Periphyseon*), which encompassed Christian cosmogony, cosmology, anthropology, and a theory of salvation. Its author, John Scotus Eriugena (ca. 810–ca. 877), stood out among a group of Irish scholars in France of the 850s and 860s.

Using the form of the Platonic dialogue as well as contemporary pedagogical dialogues, Eriugena's *On Nature* grew out of reflections upon the *Timaeus*, Augustine's literal commentary on Genesis, and Neoplatonic works available to Eriugena from Greek and Latin traditions. Beginning with the statement that nature comprises everything – that which is unknown as well as that which can be known by reason – he set out being and nonbeing as the primary objects of philosophical consideration. God, considered as such, is unknowable and beyond being, but can be partially known through creation. Creation is a procession into being, and we understand it dialectically as a series of divisions. This "division of nature" is our way of analyzing the divine process of creation, but it is also that process itself. The division proceeds through four stages: (1) from that which is uncreated and creates, to (2) that which is created and creates, to (3) that which is created and does not create, and finally to (4) that which is not created and does not create. The first and the last of these are God, understood as source and as end, or goal. The second refers to the primordial causes of all the things we know in this perceptible world, which is, in turn, the third stage in the process of division. The complete process is the outflowing from and return of all things to God. Each stage is a manifestation of God. Each stage is in process, not static.

Eriugena presents the perceptible cosmos, created nature, through a discussion of the six days of the book of Genesis. Rather than equate the nothing, from which God created, with absolute privation, he makes it the Divine Wisdom from which the created flows, analogous to the monad from which numbers proceed and to which they return. Eriugena finds both the intelligibility and the eternity of numbers in the monad; they are created and are contemplated in the many objects of the world that we perceive. In his opinion, both the primordial causes of things and the created effects have existed in Divine Wisdom from eternity.

Although each day of the Creation week is carefully treated, the Second and Fourth days are of special importance for the order of the created cosmos. Since each day in Eriugena's Genesis account witnesses both a primordial cause and its created effects, the Second Day's supercelestial waters are for Eriugena the spiritual creatures, or causes, for this day. The lower waters he calls the physical bodies, or effects, with the firmament between the waters being the four elements that unfold into the physical bodies created. The four elements are the medium between causes and effects. On the Fourth Day, the elements proceed through their qualities to materialize the forms of all bodies below the firmament down to the earth. Eriugena lays out the planets in their circles around the Earth, making the interplanetary distances harmonic intervals. He uses small whole numbers or their multiples (or powers) to define the size of the Earth, the size of the Moon, the distance of the Moon, and the size of the solar orbit. His only comment on the order of the planets is the following. The Sun is midway from Earth to the stars; Saturn is closer to the stars than to the Sun, but the four other planets (excluding the Moon), by virtue of proximity, partake more or less of the Sun's qualities as they circle around the Earth in orbits closer to the Sun's orbit than to either Earth or the stars. The physical order of the visible cosmos he believes to be thoroughly rational and describable by numbers.[8]

Eriugena's *On Nature* contained the fullest synthesis of philosophical themes in Christian cosmology before the twelfth century, but its major influence was only felt at that later time. Among early-medieval cosmological writers, Eriugena did not compete in importance with Bede. Contemporaries and successors through the eleventh century gave the latter's *The Nature of Things* greater attention, widely copying, excerpting, and imitating its thoroughly physical description of the cosmos.

ASTRONOMY

The meaning of "astronomy" in the Roman world of the fifth and sixth centuries varied widely. An education at a public school in Rome or Milan, in Bordeaux or Lyon, at the beginning of this period would have included more than one opportunity for study of the stars, the planets, and especially the Sun and Moon. The teacher known in such schools as *grammaticus* led students through a wide variety of literature, studying models of Latin poetry and prose. Encounters with Virgil's *Georgics* and one of the Latin translations of Aratus's *Phaenomena* were among the natural points of intersection for the study of both language and astronomy. An aristocrat's pride in the breadth of his Latinity, a landowner's concern with the weather and its effects on his crops, and a navigator's or traveler's need to understand the changes in positions and appearances of the stars as he changed latitude all required practical and technical explanations, which could be learned from a *grammaticus* who had made these topics part of his knowledge.

Whereas the study of grammar, in the largest sense, was the one universal attribute of late-Roman education, the presence of any advanced mathematical discipline, including astronomy, either at a municipal school or in the preparation of a private tutor, was quite unpredictable. A technically limited and largely qualitative astronomical awareness characterized landlords, navigators, and *grammatici* in the fifth century. And St. Augustine, who

[8] Willemien Otten, *The Anthropology of Johannes Scottus Eriugena* (Leiden: Brill, 1991). For background, see John J. Contreni, "The Carolingian Renaissance: Education and Literary Culture," in *The New Cambridge Medieval History*, vol. 2: *Ca. 700–ca. 900*, ed. Rosamond McKitterick (Cambridge: Cambridge University Press, 1995), pp. 709–57, 1013–24.

discouraged the study of technical astronomy in general as being of little importance for understanding the Bible, made exceptions only for knowing the following useful items: the annual rounds of the signs of the zodiac and the other constellations, eclipses of the Sun and the Moon, and reckoning the course of the Moon with sufficient precision to identify the proper Sunday for the Christian festival of Easter.[9]

The disappearance of public schools in the sixth century, both in Gaul and in Italy, coincided with other threats to the dissemination of astronomical knowledge. Boethius (ca. 480–ca. 525), who had only enough time (before his imprisonment and execution) to begin an ambitious program of translating into Latin the standard Greek works for each of the four mathematical sciences, which he canonized with the name of quadrivium, did not produce an astronomy. At least, we have no trace of such a work. No scholar in the West after Boethius even proposed such a goal until the twelfth century.

Later in the sixth century, the Roman aristocrat and former public official for the Ostrogothic kings in Italy, Cassiodorus (ca. 480–ca. 575), produced his *Introduction to Divine and Human Readings* (ca. 562), written for monks rather than for secular students. He portrayed the liberal arts and especially astronomy as pathways to the higher wisdom of Holy Scripture. To the mathematical disciplines he ascribed unusual significance when he reminded students that God disposed His works in accord with number, weight, and measure, whereas the iniquitous works of the devil could not be founded on weight or number or measure. In his brief survey of astronomy, Cassiodorus listed the basic categories of astronomical vocabulary and briefly distinguished the kinds of computation based on numerical tables that one should expect to encounter. He added varieties of geographical knowledge and the variations of sundials, both as functions of changes in latitude, along with an account of the weather and the seasons for navigation and agriculture. At the end of the chapter, Cassiodorus warned against astrology, adding the explicit statement that believers in astrology lost their freedom of will and also judged falsely because God might at any time interrupt the ordered course of events with a miracle.[10]

The likelihood of miraculous intervention in nature became a familiar theme in writings of the latter half of the sixth century, for example by Cassiodorus, Gregory of Tours, and Pope Gregory I. But with Isidore of Seville an emphasis upon lawfulness in the physical order, and certainly in astronomy, dominated. His *Etymologies*, the most widely read encyclopedia of the entire Middle Ages, began with three books on the seven liberal arts,

9 Stanley F. Bonner, *Education in Ancient Rome* (Berkeley: University of California Press, 1977), pp. 77–9, 180–8, and chaps. 16–18; Robert A. Kaster, "Notes on 'Primary' and 'Secondary' Schools in Late Antiquity," *Transactions of the American Philological Association*, 113 (1983), 322–46; and Henri I. Marrou, *Saint Augustin et la fin de la culture antique* (Paris: Boccard, 1938).

10 Cassiodorus Senator, *Institutiones*, ed. R. A. B. Mynors (Oxford: Clarendon Press, 1937), pp. 90, 111, 130–2, 153–63.

devoting twice as much space to astronomy as to any other mathematical discipline. Using the *Astronomy* of Hyginus (first century) as his major source, Isidore proceeded to define terms more as items in a literary vocabulary than as working scientific definitions. His literary view of astronomy obscured many distinctions, such as that between planet and star, and produced a body of information that could be of practical use only with regard to solar and lunar phenomena. The inclusion of discussions of chronology and the Easter cycle elsewhere in the *Etymologies* ignored the strictly astronomical aspects of these topics. Chronology became no more than the correct assignment of dates to events, and arithmetical calculation of the date of Easter (known as "*computus*" throughout the Middle Ages) was the only part of the luni-solar cycle defining this central Christian feast that Isidore discussed. Here, in the first half of the seventh century, we have reached the low point in the presentation of astronomy as one of the seven liberal arts.[11]

Early in the next century, the English monk Bede (673–735) developed the subject of *computus* with *The Reckoning of Times* (725). Lacking access to almost all ancient astronomical writings except Pliny's work, Bede depended heavily upon Isidore and the computistical tradition available to him. Bede first introduced numbers and arithmetic and the creation of time and its measures by God, then embarked upon an extensive presentation of luni-solar time calculation, and closed with a lengthy world chronicle, culminating in discussions of the end of time and the Last Judgment. Within this context, Bede focused upon calculating the number of days within various cycles of the Moon and the locations of new moon and full moon in the calendar. These were the bases for identifying the correct full moon (after the spring equinox) to determine the specific Sunday on which all Christians should observe Easter.

The tying of lunar intervals to positions along the zodiac allowed Bede to present the luni-solar temporal relationship more clearly and more accurately than previous writers, who had not done this. Nonetheless, it is remarkable that he rarely referred to specific zodiacal positions, and he subdivided the zodiac according to solar days rather than degrees. His absolutes were temporal rather than spatial since the locations of Creation and all subsequent events, both in their intervals since Creation and in their places in the cycle of the yearly calendar, depended only upon solar and lunar time intervals. At the same time, Bede included extensive astronomical matter that he thought relevant, such as a long extract from Pliny on the seven canonical climatic zones (*climata*) with their different maximum and minimum lengths of daylight, along with their shadow lengths cast by a gnomon at equinox. Bede

[11] On chronology and the Easter cycle, see Isidore, *Etymologiae sive Origines libri XX*, ed. W. M. Lindsay (Oxford: Clarendon Press, 1911), V.28–33 and VI.17, respectively; Jacques Fontaine, "Isidore de Séville et la mutation de l'encyclopédisme antique," *Cahiers d'histoire mondiale*, 9 (1966), 519–38; and Jacques Fontaine, *Isidore de Séville et la culture classique dans l'Espagne wisigothique*, 3 vols. (Paris: Etudes Augustinniennes, 1959–1983), vol. 2, pp. 453–589.

clearly found it difficult to identify the equinox precisely, and he did not attempt to include personally obtained data. In fact, although he noted various computistical preferences for the date of the equinox, he chose 21 March, thereby placing his trust with the church of Alexandria on this question.

As the crucial turning point in the early-medieval history of *computus*, Bede's *Reckoning* began the process of integrating more and more of what we would call astronomical material into calendrical discussions. However, Bede concentrated on time – God's times – in solar and lunar terms and with a certain amount of allegorical comment. In a modern sense, some of this work was astronomical, but Bede's *Reckoning* was not astronomy, nor did he call it that. The genre of "*computus*" was in flux before and after him, and one of his contributions was to ensure a place for a strictly astronomical component.[12] (On *computus*, see also McCluskey, Chapter 11, this volume.)

The political, ecclesiastical, and cultural revival that began in Western Europe even before Charlemagne (768–814) continued, at least on its cultural side, to broaden and deepen through the third quarter of the ninth century. With its roots in royal demands for more uniform and dependable ecclesiastical education in *computus*, the expansion of clerical training beyond *computus* and into traditional areas of astronomy began in earnest with the idea that a fuller understanding of the order of God's Creation would ensue from studying the seven liberal arts. This expansion was possible only because of the energetic search for texts and their accumulation and copying at royal and episcopal courts and at well-endowed monasteries.

With the availability of the relevant Roman works of Pliny, Macrobius, and Calcidius at the beginning of the ninth century, attention focused on various questions of planetary astronomy that had not arisen in the West since the sixth century. In addition to these three works, there reappeared during the second quarter of the ninth century the work of a fifth-century Carthaginian, Martianus Capella, composed of two allegorical books joined to seven books on the liberal arts, resulting in *The Marriage of Philology and Mercury*. In his book on astronomy, Capella used eccentric planetary orbits for all planets but the Moon (considered concentric with the Earth) and Venus and Mercury, which he said traveled on circles around the Sun, and he called these circles epicycles. He did not assign epicycles to any other planets. He used eccentrics to explain the variation in the Sun's speed and the consequent variations in the lengths of the seasons. Moreover, Capella said that the power of solar rays restricted the elongations of the planets Venus and Mercury from the Sun. This account of the limited distance of the circumsolar planets from the Sun made their epicyclic paths nothing more than the results of solar radial force. As such, epicycles were descriptive devices, used solely for these inner planets,

12 Wesley M. Stevens, *Cycles of Time and Scientific Learning in Medieval Europe* (Aldershot: Variorum, 1995), chaps. 2, 4, 5; and Bede, *De temporum ratione liber*, in Jones, *Opera pars VI: Opera didascalica*, vol. 2, pp. 238–544.

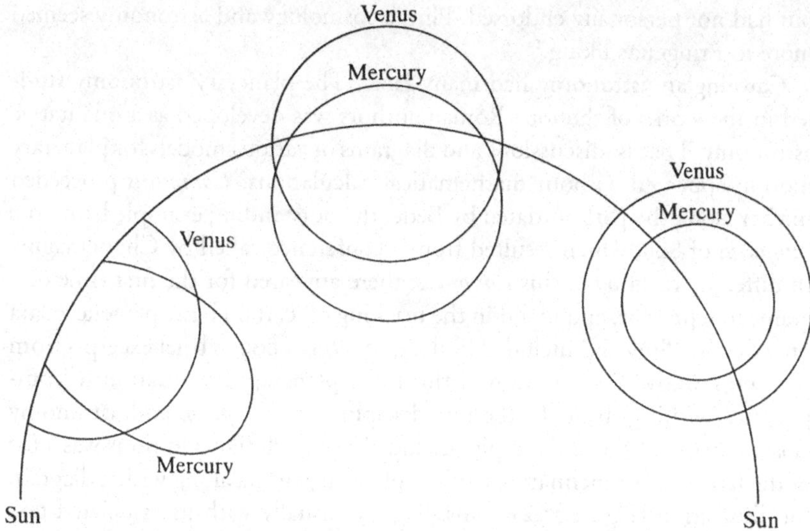

Figure 12.3. Three versions of circumsolar orbits for Mercury and Venus, designed by ninth-century scholars and attributed to Pliny, Martianus Capella, and Plato, from left to right.

and were excluded from application to the outer planets since the forces of solar rays produced different effects there. The text of Capella stimulated an investigation into other sources that treated the relationship of the orbits of Mercury and Venus to the Sun. Patterns derived from Calcidius and Pliny were compared with Capella's, and diagrams of alternative planetary models for the inner planets appeared in the margins of Capella's text by mid-century. These became canonical in a diagram (Figure 12.3) showing the three versions derived from the three authors, which many scholars of the middle and late ninth century copied into their own volumes of Capella.[13]

John Scotus Eriugena, one of the more controversial scholars of the ninth century, made his own study of both Plinian and Capellan astronomy and produced two commentaries on the latter. In *On Nature*, he used elements borrowed from both authors. Despite modern attempts to find unusual and ingenious planetary patterns in these writings of Eriugena, in fact his commentaries on *The Marriage of Philology and Mercury* preserved a straightforward Capellan system. In *On Nature*, Eriugena chose to support Pliny by adopting his pattern of eccentric orbits for all planets except the Moon and omitting any reference to circumsolar Mercury and Venus. Here he abandoned Capellan circumsolarity, which he had described in his commentaries

[13] Bruce S. Eastwood, "Astronomical Images and Planetary Theory in Carolingian Studies of Martianus Capella," *Journal for the History of Astronomy*, 31 (2000), 1–28; and Bruce S. Eastwood, *Ordering the Heavens: Roman Astronomy and Cosmology in the Carolingian Renaissance* (Leiden: Brill, 2007), pp. 238–59, 309–11.

but had not personally endorsed. Pliny's cosmology and astronomy seemed more to Eriugena's liking.[14]

Carolingian astronomy had many faces. The planetary astronomy studied in the works of the four Roman authors was developed as a qualitative astronomy. That is, discussions and diagrams of various models for planetary motion appeared without mathematical calculations. *Computus* proceeded further along the path initiated by Bede, the outstanding example being the *computus* of 809, which resulted from a conference called by Charlemagne. In different versions of this *computus*, there appeared for the first time diagrams to represent, and to aid in the teaching of, certain basic planetary data provided by Pliny and included in the *computus* of 809 as brief excerpts from *Natural History*. The addition of this fuller planetary information to computistical writings brought the two disciplines of *computus* and astronomy closer together. One of the topics included in the Plinian additions was a list of the latitudes, or inclinations, of the planetary orbits along with a diagram for illustration (Figure 12.4). This diagram, usually with its associated text from Pliny, continued as an element of *computus* as well as astronomical and cosmological textbooks for over 500 years beyond its invention in the 800s. This focus upon planetary latitude as a basic property of planetary motion reinforced the use of measurement by degrees rather than by days of solar motion, as done in traditional *computus*. The often-mentioned phenomenon of limited distance between either Mercury or Venus and the Sun also made the use of angular rather than temporal measurements more important in the study of the heavens.[15]

Reports of unusual celestial phenomena appeared in annals throughout the Carolingian era, as might be expected, since anything potentially ominous always received attention. Comets and eclipses, both lunar and solar, were often carefully recorded. Records of stellar and planetary events or positions do not appear nearly as frequently in such sources, although notable occurrences like planetary conjunctions in 770 and 807 are found in a calendar and a set of annals. In 798, a letter of Alcuin (737–804) described the course of the Moon in the zodiac. Alcuin's readiness to relate lunar motion

[14] Edouard Jeauneau, *Quatre thèmes érigéniens* (Montréal: Institut d'Etudes Médiévales Albert-le-Grand, 1978), p. 114; Iohannis Scottus, *Annotationes in Marcianum*, ed. Cora E. Lutz (Cambridge, Mass.: Mediaeval Academy of America, 1939), pp. 22–3, modification based on Paris Bibliothèque Nationale de France ms. lat. 12960, fol. 56r, line 18, where an erasure and corruption exist; Bruce Eastwood, "The Astronomies of Pliny, Martianus Capella and Isidore of Seville in the Carolingian World," in *Science in Western and Eastern Civilization in Carolingian Times*, ed. P. L. Butzer and D. Lohrmann (Basel: Birkhäuser, 1993), pp. 169–74; and Bruce S. Eastwood, "Johannes Scottus Eriugena, Sun-Centred Planets, and Carolingian Astronomy," *Journal for the History of Astronomy*, 32 (2001), 281–324.

[15] Arno Borst, "Alcuin und die Enzyklopädie von 809," in Butzer and Lohrmann, *Science in Western and Eastern Civilization in Carolingian Times*, pp. 53–78; Bruce Eastwood, "Plinian Astronomical Diagrams in the Early Middle Ages," in *Mathematics and Its Applications to Science and Natural Philosophy in the Middle Ages*, ed. Edward Grant and John E. Murdoch (Cambridge: Cambridge University Press, 1987), pp. 141–72; and Bruce S. Eastwood, *Astronomy and Optics from Pliny to Descartes* (London: Variorum, 1989), chaps. 3, 5, 6.

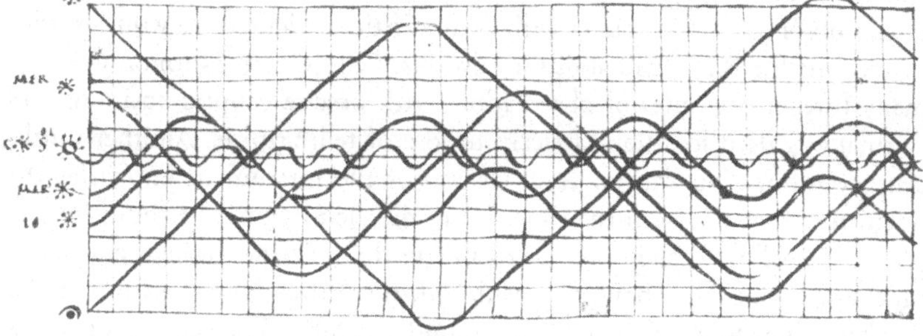

Figure 12.4. Planetary latitudes, drawn on a rectangular grid (not a graph) of thirteen horizontal lines for the 12° of width of the zodiacal band. From top to bottom, the planetary lines (names provided at left side) are: Venus, Mercury, Saturn, Sun, Mars, Jupiter, Moon. Bodl. Libr. MS Canon. class. lat. 279, fol. 34r, from the late ninth or early tenth century, by permission of the Bodleian Library, University of Oxford.

to zodiacal position for computistical purposes went a definite step beyond Bede's careful inclusion of zodiacal position only as an artificial framework. Ready willingness to use the signs reveals to us both a greater familiarity with classical astronomical assumptions and also a reduced fear of entanglement with astrology.[16]

Visual recognition of the zodiacal signs required repeated instruction of students by an experienced observer, which would give much greater practical meaning to the descriptions of these constellations found in works like the Latin version of Aratus's *Phaenomena* made by Germanicus (early first century). Rabanus Maurus described such visual instruction in his *computus* of 820. Three centuries later, the schoolmaster Odardus of Tournai still

[16] Dietrich Lohrmann, "Alcuins Korrespondenz mit Karl dem Grossen über Kalendar und Astronomie," in Butzer and Lohrmann, *Science in Western and Eastern Civilization in Carolingian Times*, pp. 79–114; and Isabelle Draelants, *Eclipses, comètes, autres phénomènes célestes et tremblements de terre au moyen âge* (Louvain-la-Neuve: Presses Universitaires de Louvain, 1995), pp. 114–20.

used the same sort of direct oral instruction of students, pointing out the constellations of the zodiac at night. In fact, we should assume this sort of instruction in the time of Bede since he probably meant his limited reference to the zodiacal signs to be more than a bookish embellishment upon the practical calculation of lunar days, which did not require the ability to identify the signs in the sky.[17]

In at least two cases, Carolingian texts made full use of the signs to signify exact dates. An image in an elaborately decorated manuscript portrayed the labors of the months and the mythological figures for planets and signs, and placed the seven known planets under the signs for the date of 18 March 816. Likewise, Rabanus Maurus gave the date 9 July 820 in his *computus* and then located the corresponding signs for the three outer planets and the exact degrees in the respective signs for the Sun and the Moon. Direct observation combined with computistical calculations would suffice to account for Rabanus's ability to do this. Carolingian astronomy reveals many further examples of observation, with precise ecliptic positions and calculations used for the practical purpose of telling time. Another ninth-century innovation was an early version of the instrument known as the nocturnal. Invented near the time and in the vicinity of Pacificus of Verona (d. 844), it allowed the observer to isolate the polestar through a sighting tube and to note the changing position through the night of a circumpolar star according to a graduated ring around the tube. The star's position on this ring gave the observer the number of hours into the night, adjusted for the time of year.[18]

For the late ninth century to the third quarter of the tenth, renewed invasions and political disturbances have left us with very limited evidence of intellectual activity. Across this period, a monastery like St. Gallen (located in present-day Switzerland) thrived until about 920, after which the Hungarians and fire brought very hard times and little scholarly production for many decades. But with Notker Labeo (ca. 950–1022), who translated Books I and II of Martianus Capella and Boethius's logical works into German, we see renewed activity in all the liberal arts and *computus* at St. Gallen. We know of more schools in the tenth century than in the ninth, and the names of more scholars in the later period as well. Certainly the late tenth century had available much of the fruit of Carolingian labor and could build its teaching and studies on this basis. Cologne, Bonn, Reims, Liège, and other cathedral schools in the late tenth century saw scholars who attained renown

17 Rabanus Maurus, *Martyrologium, De computo*, ed. John McCulloh and Wesley Stevens (Corpus Christianorum, Continuatio Mediaevalis, 44) (Turnhout: Brepols, 1979), p. 252; and Hermannus Abbas, "Narratio restaurationis abbatiae S. Martini Tornacensis," in *Eugenii III Romani Pontificis Epistolae et Privilegia*, ed. J.-P. Migne (Patrologia Latina, 180) (Paris: J.-P. Migne, 1855), col. 41.

18 Bruce Eastwood, "Latin Planetary Studies in the IXth and Xth Centuries," *Physis*, n.s. 32 (1995), 217–26; Rabanus, *Martyrologium, De computo*, p. 259; and Joachim Wiesenbach, "Pacificus von Verona als Erfinder einer Sternenuhr," in Butzer and Lohrmann, *Science in Western and Eastern Civilization in Carolingian Times*, pp. 229–50.

for specialized knowledge, often in astronomy. At Liège, for example, two bishops in a row, Euraclius (959–971) and Notker (971–1008), shared an interest in and support of astronomical study.[19]

Gerbert of Aurillac (ca. 945–1003) was the most famous scholar of the late tenth century. At Reims (from 972 to 991), he not only reorganized and improved the study of logic but also developed different physical models for astronomical instruction, which included direct observation of the major stars and constellations. Gerbert's name has often been attached to the introduction of the astrolabe into Northern Europe from Spain. However, his letters to Lupitus of Barcelona and others reveal only his extensive knowledge of the authors known to the Carolingians, along with other classical sources, but no new Arabo-Latin astronomical material. Although Lupitus's own work on the astrolabe reached Gerbert, there is no clear evidence that Gerbert learned enough to compose a work on the instrument himself.[20]

The appearance of the astrolabe (see North, Chapter 19, this volume) in Latin Europe by the turn of the millennium brought much greater precision to observations that were already being made. Reckoning the time of day was the most common early use of the instrument. Replacing traditional computistical methods for identifying times during the month and year gave the astrolabe immediately recognizable value. Hermann of Reichenau (1013–1054) offered the earliest reasonably clear and complete description of it in *The Construction of the Astrolabe* (ca. 1045). Just after mid-century, William of St. Emmeram (of Hirsau, after 1069) made unusually careful observations and revised the date of the vernal equinox to 16–17 March. Bernold of Constance (1074) followed and confirmed William in this dating, noting as well that others, whom he named simply "the moderns," were finding a date around 17–18 March. These moderns were largely the children of the astrolabe.[21]

For planetary astronomy and the models of planetary motion, the four Roman works used in the Carolingian era were the objects of extensive study. More than thirty manuscripts of Macrobius's commentary survive from the eleventh century. Although use of the work of Martianus Capella continued, the astronomy of Calcidius (in his *Timaeus* commentary) was much more

[19] L. M. de Rijk, "On the Curriculum of the Arts of the Trivium at St. Gall from c. 850–c. 1000," *Vivarium*, 1 (1963), 41–51; and Reinerus, "Euracli episcopi Leodiensis vita," in *Thesaurus anecdotorum novissimus*, ed. Bernhard Pez, 6 vols. (Vienna: Veith, 1721–1729), vol. 4, pt. iii, cols. 155–66.

[20] Richer, *Histoire de France 888–995*, ed. Robert Latouche (Paris: Les Belles Lettres, 1937), vol. 2, pp. 54–62; and *Gerberto scienza storia e mito, Atti del 'Gerberti Symposium' Bobbio 25–27 luglio 1983*, ed. Michele Tosi (Archivum Bobiense – Studia, 2) (Bobbio: Archivi Storici Bobiensi, 1985), pt. 2, passim.

[21] Werner Bergmann, *Innovationen im Quadrivium des 10. und 11. Jahrhunderts* (Sudhoffs Archiv, Beiheft 26) (Stuttgart: Steiner, 1985); Arno Borst, *Astrolab und Klosterreform an der Jahrtausendwende* (Sitzungsberichte der Heidelberger Akademie der Wissenschaften, Philosophisch-historische Klasse [1989/1]) (Heidelberg: Winter, 1989); and Joachim Wiesenbach, "Wilhelm von Hirsau, Astrolab und Astronomie im 11. Jahrhundert," in *Hirsau St Peter und Paul, 1091–1991*, vol. 2: *Geschichte, Lebens- und Verfassungsformen eines Reformklosters*, ed. Klaus Schreiner (Stuttgart: Theiss, 1991), pp. 109–54.

widely copied and excerpted during the same century. Not only were the planetary models of eccentrics and epicycles thus learned, but also certain scholars discovered errors in the diagrams found in the manuscript tradition of Calcidius's account of the bounded elongation of Venus from the Sun. They corrected these diagrams and understood clearly the two different models discussed by Calcidius – thus preceding modern rediscovery by over 900 years and revealing their mastery of the models described in the text.[22]

By the late eleventh century, the astrolabe had provided much help in measurement and had improved the precision of luni-solar calculations greatly. At the same time, the alertness to qualitative geometrical models and the use of zodiacal positions in degrees continued to develop in planetary astronomy. Models and applied mathematics together prepared Latin astronomers for the research of the twelfth century into the Greco-Arabic traditions of their discipline.[23]

ARITHMETIC AND GEOMETRY

Saint Augustine, in a discussion of the order found in each of the liberal arts, explained how measurement, or number, lay beneath the essential character of grammar, astronomy, and the other arts as well. Grammar, he explained, measures the lengths of sounds and syllables, establishes numerical patterns in the accents of words, and uses number to define the meter of poetry. When he wrote to advise clerics and others looking for the basic needs of Christian wisdom, in *On Christian Teaching*, Augustine limited his recommendations severely, but the study of number remained extremely important for interpreting scriptural references, for understanding the temporal order, and for learning the various kinds of measure in language, in music, in motion, and so on. In the late sixth century, Cassiodorus reminded clerical students that number and measure were fundamental to God's Creation and made explicit the inability of the devil to make anything according to number and measure. Introducing the discipline of arithmetic, he pointed to the foundations of the four mathematical arts in numbers. Combining these justifications for the study of number and arithmetic with the very practical counting and arithmetical techniques of Roman primary schooling, early-medieval mathematics had no need or desire to go further beyond the knowledge available in Boethius's *Arithmetic* (based on the Greek *Arithmetic* of Nicomachus,

22 Bruce Eastwood, "Heraclides and Heliocentrism: Texts, Diagrams, and Interpretations," *Journal for the History of Astronomy*, 23 (1992), 233–60; and Bruce Eastwood, "Calcidius's Commentary on Plato's *Timaeus* in Latin Astronomy of the Ninth to Eleventh Centuries," in *Between Demonstration and Imagination: Essays in the History of Science and Philosophy Presented to John D. North*, ed. Lodi Nauta and Arjo Vanderjagt (Leiden: Brill, 1999), pp. 171–209.

23 Bruce Eastwood, "Astronomy in Christian Latin Europe c. 500–c. 1150," *Journal for the History of Astronomy*, 28 (1997), 250–6.

ca. 100) than the practicalities of finger-reckoning and an ability to manipulate the abacus.[24]

Boethius's *Arithmetic* began by defining philosophy as a search that must ascend from objects perceivable by the senses to unchanging, incorporeal substances, the first and most basic of which is number. In his first book of arithmetic, he explained the regularities of number. He defined even and odd and primes. Perfect numbers, which are equal to the sums of their factors, he compared to virtues. He showed how different kinds of numerical sequences preserve the same numerical relationships, and he gave the superparticular (an inequality in which the larger number contains within itself the smaller number plus a single part, such as 3:2 or 4:3) and the superpartient (an inequality in which the larger number contains within itself the smaller number plus several parts, such as 5:3 or 7:4) as important examples of such relationships. Book II proceeded to discuss shapes based on number and then addressed numerical proportions; in other words, geometrical and harmonic theory. Boethius defined and explained the generation of basic plane and solid shapes and then turned to proportions; there he explained and illustrated arithmetic, geometric, and harmonic series of numbers. Those scholars who passed on to Boethius's *Music* also found access to the harmonic theories of Euclid and Ptolemy. Here Boethius made significant changes in the epistemological foundations of harmonics, excluding any basis in the senses by claiming that reason alone determined harmonic truth, a very different conception from Ptolemy's dual criteria of reason and experienced hearing.[25]

The practice, as distinguished from the theory, of arithmetic had a largely oral rather than a written history through the early Middle Ages. Finger-counting and finger-calculation in a strictly decimal system with place value made possible the addition and subtraction of Roman numerals; multiplication and division were serial additions and subtractions. Although the Romans used and casually referred to the finger-technique, it was only through Bede's concern to ensure the proper instruction of his students that we now have an early explanation of counting and tabulating procedures on the fingers. Bede made it clear that one used finger-reckoning not only for private arithmetical work but also for communicating numbers to others, hence an arrangement of units, tens, hundreds, and thousands from left to right across the two hands held up toward the viewer. The memorization of sums and multiples was evidently essential for success in the technique.

24 Augustine, *De doctrina christiana*, II.16.25, II.38.56, ed. Joseph Martin (Corpus Christianorum Series Latina, 32) (Turnhout: Brepols, 1962); Augustine, *St. Augustine On Education*, ed. George Howie (South Bend, Ind.: Gateway, 1969), pp. 245–61; and Cassiodorus, *Institutiones*, pp. 90, 132–3.

25 Boethius, *De institutione arithmetica, De institutione musica*, ed. Gottfried Friedlein (Leipzig: Teubner, 1867); and Alan C. Bowen and William R. Bowen, "The Translator as Interpreter: Euclid's *Sectio canonis* and Ptolemy's *Harmonica* in the Latin Tradition," in *Music Discourse from Classical to Early Modern Times*, ed. Maria R. Maniates (Toronto: University of Toronto Press, 1997), pp. 97–148.

Bede mentioned the common use of Greek letters as numbers. A famous example of this appeared in Martianus Capella's book on arithmetic, where he described the muse Arithmetic using finger-reckoning and flashing the number 717, which spelled in Greek the philosophical name of Jupiter, *arche*, as chief and origin of the gods.[26]

Ninth-century scholars read and immediately understood this example in Capella. Some practical, and sometimes entertaining, problems appeared in later texts. From the ninth century, we have various sets of classroom problem-texts, at least one of which was attributed to Alcuin (737–804), a scholar at Charlemagne's court and later an abbot of Tours. These suggest a nice admixture of ingenuity, humor, common sense, and calculation, both arithmetical and geometrical, even if at an elementary level. For example, when a leech invites a slug for lunch a league away, how long must the slug travel if his speed is one inch per day? Since the distance is 90,000 inches, the slug must travel for 246 years and 210 days.[27]

The abacus, rather like finger-reckoning, was in use in Roman times but was not described by medieval scholars before the tenth century. By the turn of the millennium, Hindu-Arabic numerals began to be used in Latin Europe and were even combined with the abacus.

The situation of geometry in the early Middle Ages was less precise than that of arithmetic. While the biblical Wisdom of Solomon set forth the canonical reference to the creation of the world according to weight, number, and measure, medieval scholars distinctly subordinated geometry to arithmetic, for they considered number fundamental to both. For example, number is required to define and name any polygon. The admonition of Augustine to learn the arts insofar as they were useful to understanding Scripture put far greater stock in the study of numbers and arithmetic than of geometry. Although Boethius made a translation of Euclid's geometry – we do not know how complete – which Cassiodorus still knew later in the sixth century, only Boethius's arithmetic text survived intact; his geometry was known by the ninth century only in fragmentary form. And we should recall that Boethius treated geometrical shapes, which he called "figured numbers," in the first half of Book II of his *Arithmetic*. Both Cassiodorus in the sixth century and Isidore in the seventh had only two major things to say about geometry, even though the former could still recommend reading Euclid in Latin. Both emphasized the origins of geometry in land measurement, and

[26] Burma P. Williams and Richard S. Williams, "Finger Numbers in the Greco-Roman World and the Early Middle Ages," *Isis*, 86 (1995), 587–608; and James Willis, ed., *Martianus Capella* (Leipzig: Teubner, 1983), p. 261, lines 10–15. This passage is explained in Remigius of Auxerre, *Commentum in Martianum Capellam*, 36.20, ed. Cora Lutz, 2 vols. (Leiden: Brill, 1965), vol. 2, pp. 178–9.

[27] Menso Folkerts, ed., *Die älteste mathematische Aufgabensammlung in lateinischer Sprache: Die Alkuin zugeschriebenen 'Propositiones ad acuendos iuvenes'* (Österreichische Akademie der Wissenschaften, mathematisch-naturwissenschaftliche Klasse, Denkschriften 116, Band 6, Abhandlung) (Vienna: Springer, 1978). Latin text in John Hadley and David Singmaster, trans., "Problems to Sharpen the Young," *The Mathematical Gazette*, 76 (1992), 102–26.

both spoke of figured numbers and of geometry's description of line, plane, and space. Bede did not write about geometry.

A fuller account of geometry than that suggested by Cassiodorus or Isidore appeared in the geometrical book (Book VI) of Martianus Capella's handbook of the liberal arts. It did not present Euclidean geometry in any detail, and only the final one-seventh dealt with geometry in the Euclidean sense, yet it introduced Euclidean definitions, the five postulates, the three clearly genuine axioms, the five steps in a theorem (incidentally mentioning the relationship between dialectic and geometrical demonstration), commensurability and incommensurability, and the thirteen irrational lines. By ending with the enunciation and beginning of Euclid's first proposition, Capella made it evident that all of this was preliminary to the actual practice of geometry.[28]

What else did Capellan "geometry" include? According to its author, the first six-sevenths of the book built upon a notion of *geometria* as "traversing and measuring out the earth." The resulting extended geographical description offered something remarkably like the contents of later books entitled *The Nature of Things* by Isidore, Bede, and Rabanus Maurus. Opening with a discussion of measurement, specifically the measurement of the cosmos using both astronomical and mercantile instruments, a personified Geometry referred to a manufactured celestial globe and a powdered board for making calculations. The shape and traditional divisions of the Earth followed a brief statement of the concentric elemental spheres and the surrounding planetary spheres. Geometry discussed the size of the earthly globe according to both Eratosthenes and Ptolemy. Lengths of daylight at the summer solstice in four geographical locations and varying positions of different constellations from different locations on the Earth's globe were marshaled to indicate the variety of effects of terrestrial sphericity. Finally, after more description of the measured zones of Earth, there ensued a very long description of very specific parts of the Earth. This section of Capella's book is the longest, proceeding around the known world of antiquity according to various written itineraries, primarily that of Solinus. After progressing through Africa, Asia, and Europe – with the familiar arrangement of three continents around a "T" formed by the Mediterranean, the Nile River, and the Don River, later to be found in many medieval world maps – Capella's lady Geometry ended her tale of travels and introduced the discipline of geometry proper.

In the ninth century, Capella's work and fragments of Boethius's Euclid served as the models for geometry. Actual land measurement was a widely possessed skill, and agrimensorial books were readily available. In Euclidean geometry proper, the knowledge of the day seems to have encompassed major

[28] Willis, *Martianus Capella*, pp. 201–58. See William H. Stahl and Richard Johnson, trans., *Martianus Capella and the Seven Liberal Arts*, 2 vols. (New York: Columbia University Press, 1977), vol. 2, pp. 215–72, with many useful notes.

parts of the first five books of the *Elements*, for which the propositions were simply enunciated with no demonstrations. This suggests that, somewhat like the oral instruction and practice in *computus* and astronomy mentioned by Bede and Rabanus, students may well have learned how to develop the propositions, perhaps using the same kind of powder-covered boards mentioned by Capella. In the late tenth and eleventh centuries, Gerbert and others compiled additional, though still limited, Euclidean and agrimensorial materials. Nonetheless, the fuller text of the early books as well as the later books of Euclid's *Elements* were not known in Europe until their twelfth-century translations.[29]

Arithmetic and geometry in the early Middle Ages described an orderly world based on simple numbers and shapes. Time intervals, land measurements, travel distances, business transactions, and the various measuring instruments of the day all assumed the numbers and shapes of the basic arithmetic and geometry that ordered their cosmology as well as their practical interests. The rationality of the created cosmos – its celestial and its terrestrial structures – and the workings of its parts depended upon numerical and geometrical order.

[29] Menso Folkerts, "The Importance of the Pseudo-Boethian *Geometria* during the Middle Ages," in *Boethius and the Liberal Arts*, ed. Michael Masi (Bern: P. Lang, 1981), pp. 187–209, and sources cited therein; and Wesley M. Stevens, "Euclidean Geometry in the Early Middle Ages: A Preliminary Reassessment," in *Villard's Legacy: Studies in Medieval Technology, Science and Art in Memory of Jean Gimpel*, ed. Marie-Thérèse Zenner (AVISTA Studies in the History of Science, Technology and Art, 2) (Aldershot: Ashgate, 2004), pp. 229–63.

13

EARLY-MEDIEVAL MEDICINE AND NATURAL SCIENCE

Vivian Nutton

CHRISTIANITY AND PAGAN MEDICINE

Dionysius the doctor had an eventful career in the early years of the fifth century. Ordained a deacon, he continued to practice medicine, "despising sordid gain and treating without fee all who came to him," a task made easier by his own substantial wealth. But when Rome was sacked by the Goths in 410, he was taken prisoner and his riches seized. He did not despair, and by his medical art he convinced his captors to entrust themselves to him. The size and eloquence of his tombstone implies that their trust was accompanied by some monetary appreciation. This brief summary of Dionysius's career is instructive in several ways. It attests to the increasing fragility of Roman society around 400 , the wealth that might be made through medicine, and, most important, the intertwining of medicine and Christianity. It is not just that Dionysius was a deacon, for whom "his art proclaimed his faith, and the glory of his faith enhanced his art." His Christianity also prevented him from falling into the trap of the "illicit arts" of magic, graphically described by the historian Ammianus Marcellinus only a few years before.[1]

The arrival of Christianity in 313 as an approved religion of the Roman state, and its increasing dominance, introduced new relationships into medicine and natural science, in both the Latin West and the Greek-speaking East (the later Byzantine Empire). For example, explanations in terms of miracles were given alongside, or instead of, those of natural causes. If Christianity had to come to terms with an ambient non-Christian medical tradition, non-Christians had to be wary of offending Christian sensitivities. Above all, the gradual redefinition of society in terms of a community of Christian belief marginalized those who did not share that belief or whose brand of Christianity was viewed as less than orthodox.

[1] Ernst Diehl, *Inscriptiones latinae Christianae veteres* (Berlin: Weidmann, 1925–1931), no. 1233; and Sir Ronald Syme, *Ammianus and the Historia Augusta* (Oxford: Clarendon Press, 1968), pp. 31–3.

Practitioners of medicine fell under various degrees of suspicion. Several leading doctors, from Oribasius in the late fourth century to Asclepiodotus of Aphrodisias and Jacobus Psychrestus in Constantinople at the end of the fifth, were notorious pagans, and the medical teachers and students at Alexandria were often suspected of adherence to the old ways. Sophronius, patriarch of Jerusalem, rejoiced at the discomfiture of Professor Gesius (fl. 500), forced by his own illness to abjure his paganism and to admit the superiority of the Christian healers, Saints Cyrus and John.[2] Accusations of magic and divination were directed at doctors across the Roman world. Books on astrology read by Peter, town doctor of Constantina, were the chief evidence in a sorcery trial in 449. A century later, Procopius of Gaza warned his flock against doctors who used chants and charms, or whose confident predictions of success left no place for hope in God. He was not tilting at windmills. His contemporary, the Christian Alexander of Tralles, included amulets, phylacteries, and religious charms among his therapeutic recommendations, although he dared not include all that he had found effective.[3]

But the extent of this pagan opposition should not be exaggerated. For every Eusebius of Emesa, a doctor who became an interpreter of a pagan oracle in Syria at the end of the fifth century, one can point to a Gerontius, an Italian exile, whose supporters defied the attempts of John Chrysostom to depose him as bishop of Nicomedia in 401 by citing his unstinting use of his medical skills among them; to an Elpidius, deacon and personal physician to Theodoric the Ostrogoth in 526; or, at a much humbler level, to John, deacon and doctor at Erythrae in the fifth or sixth century.[4] The healings of the saints replaced, sometimes in the same location, the healings of pagan gods like Asclepius or Nodens, a local British deity from the Forest of Dean. Chants, charms, phylacteries, and a whole range of folk medical practices can be found deprecated in the pages of Pliny or Galen just as much as in the sermons of St. Augustine or St. Caesarius of Arles four centuries later. Nor did theologians' ideas on the superiority of faith to human healing, their disdain for the physical body compared with the eternal life of the soul, and their exaltation of the nobility of suffering prevent other Christians from

[2] Vivian Nutton, "From Galen to Alexander, Aspects of Medicine and Medical Practice in Late Antiquity," *Dumbarton Oaks Papers*, 38 (1984), 1–14 at p. 6, reprinted with identical pagination as Chapter X in Vivian Nutton, *From Democedes to Harvey* (London: Variorum Reprints, 1988).

[3] Ernst Honigmann, "A Trial for Sorcery on August 22, A.D. 449," *Isis*, 35 (1944), 281–4; Procopius of Gaza, *Commentary on Kings, Book 3*, ed. J.-P. Migne (Patrologia Graeca, 87) (Paris: J.-P. Migne, 1865), col. 1165; and Alexander of Tralles, *Medical Writings*, ed. Theodor Puschmann, 2 vols. (Vienna, 1878–1879), vol. 1, p. 573; vol. 2, p. 375.

[4] Respectively, Photius, *Bibliothèque*, 348 B, ed. René Henry, 8 vols. (Paris: Société d'édition Les Belles Lettres, 1959–1977); Sozomen, *Kirchengeschichte*, VIII.6, ed. Joseph Bidez (Berlin: Akademie-Verlag, 1960), pp. 358–9; Procopius of Caesarea, *Opera omnia*, V.1.38, ed. Jacob Haury, 4 vols. (Leipzig: Teubner, 1905–1913); and *Die Inschriften von Erythrae*, ed. Heinz Engelmann and Reinhold Merkelbach (Bonn: R. Habelt, 1972), no. 142.

taking whatever measures they thought necessary when they fell ill, or from traveling miles to consult a famous doctor.[5]

In one major respect, however, the coming of Christianity introduced a new element into medicine. Building upon Jewish attitudes of charity toward fellow Jews, Christian leaders placed charity toward all as a central element in their message. Third-century bishops, like St. Cyprian of Carthage, took the needy (however defined) into their houses, while at the same time the church in Rome was famous for its extensive programs for charitable relief. In the middle years of the fourth century, these charitable initiatives were given permanence in the form of buildings. Hospitals, hostels, poor-houses, sick places – the variety of names indicates a multitude of frequently overlapping functions – all catered to the needs of those without immediate family support. Beginning in Asia Minor, and burgeoning in the Eastern half of the Roman Empire from the fifth century onward, the hospital became a common feature of Christian life. By 600, the hospital was ubiquitous in the Eastern Empire. In size it ranged from a mere room off a courtyard where a traveler could find rest and refreshment to large establishments with hundreds of beds. Some were served, and even owned, by medical men, others specifically excluded the sick or certain types of sufferers. In large metropolitan hospitals, as in Constantinople, Antioch, or Jerusalem, one can detect the beginnings of specialization with the division into male and female wards or into surgical and nonsurgical areas, but such grandeur was far from the minds of the legislators of the Syriac church when, from the fifth century onward, they frequently demanded that every community should have its own hospital, run by a devout and honest administrator.[6]

What impact these institutions had on medicine is hard to determine. The mere provision of food and shelter, of care and concern, is part of the healing process, but one should be wary of reading back into late antiquity developments first found in the Muslim world of the tenth century or in later Byzantium. The hospital spread much more slowly in the West, where its Greekish name, *xenodochium*, revealed its Eastern origins. The first hospitals were set up at the end of the fourth century in Rome and its port, Ostia, where there were many pilgrims, but traces of these institutions elsewhere are difficult to find. Although by 600 many small hostels dotted the main

[5] Ramsay MacMullen, *Christianity and Paganism in the Fourth to Eighth Centuries* (New Haven, Conn.: Yale University Press, 1997). Owsei Temkin, *Hippocrates in a World of Pagans and Christians* (Baltimore: Johns Hopkins University Press, 1991), stresses ethical continuities.

[6] Timothy S. Miller, *The Birth of the Hospital in the Byzantine Empire*, rev. ed. (Baltimore: Johns Hopkins University Press, 1996), collects the early Christian evidence, but not all have been convinced by his arguments. See the essay review in Vivian Nutton, *Medical History*, 30 (1986), 218–21; and Peregrine Horden, "The Byzantine Welfare State: Image and Reality," *Society of the Social History of Medicine Bulletin*, 37 (1985), 7–10. See also Constantina Mentzou-Meimari, "Religious Charitable Foundations, Up to the End of the Iconoclastic Controversy," *Byzantina*, 11 (1982), 255–308; and Nigel Allan, "Hospice to Hospital in the Near East," *Bulletin of the History of Medicine*, 64 (1990), 446–62.

pilgrimage routes, and small infirmaries began to appear within the walls of Benedictine monasteries, it remains unclear how far they served the needs of the surrounding communities.[7]

THE DECLINE OF MEDICINE?

Equally controversial is any attempt to estimate changes in the role, status, or numbers of doctors, surgeons, and the like between the years 300 and 600. A good case can be made, at least for the Greek world, for saying that little indeed changed. There are many records of rich physicians, and, on the Syrian frontier, their multilingualism enabled several of them to gain fame and fortune as ambassadors or translators. Inscriptions referring to medical men can be found in towns of all sizes, and the intellectual productions of Alexandrian medical professors show a direct continuity, in quality and occasionally in wording, with what had been taught there centuries before. Similar continuities can be found in the Latin world, where law codes of the sixth century repeat, as if still valid, centuries-old rules for appointing town physicians, securing redress for incompetent treatment, and granting tax immunity to physicians and surgical specialists (but not to exorcists, even when some patients claimed to have been cured by them). Bearers of the title *medicus* (physician) can be found in remoter parts of Wales and Spain long after direct rule from Rome had ceased. In North Africa in the fifth century, a doctor might earn enough to ransom a governor's nephew for a sum equivalent to ten years' pay for a soldier. Pulpit orators took for granted that their hearers would understand their medical imagery and allusions, and that they had seen, and even felt, the painful operations they described. In this interpretation of medicine, intimations of decline reflect, in particular, the absence of scientific authors of the quality of Galen, Euclid, or even Oppian and Pliny. The medical life of North Africa, apparently flourishing around 400, is lost to view in 500 because there is no longer a St. Augustine to record it.[8]

The alternative view proposes that the qualitative changes in the types of medical and scientific works produced correspond to wider realities. Books and authors become rarer and change character as compilation and preservation take over from intellectual inquiry. References to experiment or to continued anatomical dissection are either literary repetitions of

7 Thomas Sternberg, *Orientalium more secutus. Räume und Institutionen der Caritas des 5. bis 7. Jahrhunderts in Gallien* (Münster: Aschendorff, 1991); and Katharine Park, "Medicine and Society in Medieval Europe, 500–1500," in *Medicine in Society: Historical Essays*, ed. Andrew Wear (Cambridge: Cambridge University Press, 1992), pp. 70–2.

8 Vivian Nutton, "From Galen to Alexander," pp. 10–13. For North Africa, see Nacéra Benseddik, "La pratique médicale en Afrique au temps d'Augustin," in *L'Africa romana: Atti del VI convegno di studio, Sassari, 16–18 dicembre 1988*, ed. Attilio Massimo (Sassari: Edizioni Gallizzi, 1989), pp. 663–82.

observations made centuries before or striking because of their singularity. Likewise, the apparent continuity of the law codes is as anachronistic as the medical writers' prescription of gladiator's blood centuries after the gladiatorial games had ended. Alexandrian medical and mathematical teaching is the exception, an ideal almost never attained in Latin Europe, save, briefly, at sixth-century Ravenna. In its place we find self-help advice, sometimes openly declaring an absence of doctors and implying that proper treatment and competent practitioners are to be found only in a few urban centers. Previously widespread rational explanations for disease are doubted or superseded by others that invoke demons or sinfulness. The doctor Posidonius, who lived around 400, is represented by the historian Philostorgius as unusual for believing that madness was the result of humoral changes in the body, not demonic possession. Even so, Posidonius disputed only the effectiveness of the demons, not their existence. In this view, changes in medicine correspond to wider changes in society, as the Roman Empire first fragments and then collapses entirely in the sixth century. Although there are still Greek doctors in sixth-century Spain, and a bishop in England in 754 could still obtain, albeit with difficulty, some of the Mediterranean ingredients his recipe book advised, the economic, political, and military decline of the Roman Empire entailed in the Latin West the decline of ancient science.[9]

To decide between the two interpretations is not easy, and the next two sections will set out in detail some of the evidence that lies behind them.

THE TRIUMPH OF GALENISM IN THE EAST

East and West split apart, to follow different scientific routes. In the Greek East, the period after 300 is marked by the increasing dominance of the ideas and approaches of the great Greek physician and doctor to the Roman emperors Galen of Pergamum (129–ca. 216). They were based on the doctrine of the four humors – blood, phlegm, bile, and black bile – whose balance or imbalance determined the body's health or illness. Galen's claim that this theory went back to Hippocrates, around 400 B.C., gave his own arguments the authority of great antiquity. An increasing Galenism is most evident in the development of medical encyclopedias. Oribasius (ca. 325–400), the earliest such author, compiled one in forty books, another in nine, one in four, and one, now lost, in a single volume, the fruit of midnight discussions with Emperor Julian while on campaign in Gaul around 358. In his largest enterprise, he assembled passages taken verbatim from a variety of authors

[9] Philostorgius, *Ecclesiastical History*, VIII. 10 (from Philostorgius, *Kirchengeschichte*, 3rd ed., ed. Joseph Bidez (Berlin: Akademie-Verlag, 1981), p. 111; Spanish doctors, *Vitae patrum Emeritensium* IV.11, ed. J.-P. Migne (Patrologia Latina, 80) (Paris: J.-P. Migne, 1850); and Cyneheard, *S. Bonifatii et Lullii Epistolae*, ed. M. Tangl (Berlin: Weidmann, 1916), p. 247. MacMullen, *Christianity and Paganism in the Fourth to Eighth Centuries*, examines social explanations for this cultural change.

to form a coherent mosaic of argument on topics ranging from healthy town planning to the best medicinal honey. His successors, Aetius (fl. 530) and Paul of Aegina (fl. 630), especially in his Book VI, on surgery, included some material taken from less familiar authors, but increasingly Galenic writings came to dominate. Other views were either excluded or silently amalgamated with Galen's, while his own hesitations were edited away. The resulting synthesis was thus tighter both doctrinally and verbally. It admitted no room for doubt.

A similar pattern can be seen in the choice of medical books for copying. Although papyri from the late Empire show that a variety of authors were still being read, alternatives to Galen tended to survive only where he had not dealt at length with a particular topic, for example gynecology, or where they had fortuitously become part of the Galenic corpus (e.g., *Whether the foetus has a soul*) and were copied along with it. A few authors, notably the Hippocratic writers, Dioscorides, and Rufus of Ephesus, survived in part because of Galen's recommendation, but even copies of Rufus became scarce in Byzantium.[10]

The reasons for this triumph of Galenism are many and varied. Changes in education, economics, and medical life in general made it difficult for anyone to follow Galen's own example. He was born into a very wealthy family, with an immense private library and a staff of shorthand writers and copyists. He had traveled widely and had enjoyed an unusually long medical education (over ten years) at the most prestigious medical centers. But he was also hugely talented, a philosopher and scholar as well as a physician, and possessed abundant energy and manual dexterity. His forceful rhetoric also imposed its own vision of Galen as always correct and always better prepared than any other doctor. Opportunities to repeat what he had done, let alone to surpass them, were few and far between. No wonder that some in late antiquity saw him as the last man to master the whole of medicine in its vast unity; they could only specialize in one part of medicine and, as one professor put it, follow where Galen had reaped what had been sown by Hippocrates.

Galen's own pagan piety, most famously expressed in the last book of *On the Use of Parts*, also helped him to survive. His belief in the personal intervention of the god Asclepius in his own life was taken as an unfortunate misunderstanding, trivial when compared with the convincing proofs that he had produced for the world as controlled and organized by some power. His constant references to the order and beauty of a world purposefully set in being by Nature or a Creator allowed his views to be taken over in support of a Christianized view of the world. Readers of the *Ecclesiastical History*

[10] For Galen, see Owsei Temkin, *Galenism: Rise and Decline of a Medical Philosophy* (Ithaca, N.Y.: Cornell University Press, 1973). For Rufus, see Alexander Sideras, "Rufus von Ephesos und sein Werk im Rahmen der antiken Medizin," in *Aufstieg und Niedergang der römischen Welt*, ed. Wolfgang Haase, Teil II, Principat, d. 37, 2 (Berlin: De Gruyter, 1994), pp. 1077–253.

of Eusebius would have seen how some Christians had almost adored him, and Bishop Nemesius of Emesa (fl. 390), in writing what has been termed the first Christian anthropology, drew heavily upon Galen. It was not only physicians who came to regard him as "the most divine" Galen.[11]

But the sheer size of his oeuvre, over 350 known titles, presented difficulties – and still does, even today. To acquire manuscript copies even of the medical half of his writings involved a huge expenditure of time, effort, and money. Besides, in an age when indexes and other finding aids were limited or nonexistent, navigating through the vast Galenic corpus in search of particular topics, let alone a passing comment, was not easy.

Three strategies were followed. One, found in the so-called Alexandrian summaries, involved a series of extracts from individual treatises, keeping only their main points. The second brought together into a coherent block opinions on one topic scattered throughout the Galenic corpus. So, for example, the pseudonymous Galenic treatise *On Pulses, for Antonius*, perhaps from the sixth century, collects a variety of passages on pulsation. A more striking example can be found in the treatise on urines ascribed to Magnus of Emesa, probably to be identified with Magnus of Nisibis, a city in the same region, who dominated the medical life of Alexandria in the late fourth century. Students flocked to hear him from all over the Eastern Mediterranean. Although others, like his colleague Ionicus, might be more expert than he was in practical bandaging and surgery, he had the power of rhetoric to aid his diagnoses. He could convince the sick that they were on the way to recovery purely by the force of his words, and he was reputed to be able to defeat even Death by argument. His treatise on urines takes scattered incidents and observations from all over the Galenic corpus and works them into a well-organized, elegant, and didactically effective presentation. It is Galenic in the sense that it depends on Galen, but it is new in that it turns uroscopy, the examination of urines, from an occasional part into a significant element in the diagnostic process. By 500, in both East and West, uroscopy came to occupy a central place in diagnosis, and its rules were available in a variety of forms, including diagrams.[12]

The third method of coping with the sheer size of the Galenic corpus was to select a number of works as especially important. This procedure,

[11] Temkin, *Galenism*, chap. 2. Nemesius of Emesa, *On the Nature of Man*, in *Cyril of Jerusalem and Nemesius of Emesa*, trans. William Telfer (The Library of Christian Classics, 4) (London: SCM Press, 1955), cited at least sixteen tracts of Galen, including the lost *On Demonstration* and the recently rediscovered *On movements hard to explain*.

[12] For Magnus and Ionicus, see Eunapius, *Lives of the Philosophers*, trans. Wilmer Cave Wright (Loeb Classical Library) (London: W. Heinemann, 1961), pp. 531–9. On uroscopy, see G. Baader, "Early Medieval Latin Adaptations of Byzantine Medicine in Western Europe," *Dumbarton Oaks Papers*, 38 (1984), 254–6; Gundolf Keil, *Die urognostische Praxis in vor- und frühsalernitanischer Zeit* (Medizinische Habilitationschrift, University of Freiburg im Breisgau: 1970); and Faith Wallis, "Signs and Senses: Diagnosis and Prognosis in Early Medieval Pulse and Urine Texts," *Social History of Medicine*, 13 (2000), 265–78.

the formation of a canon, was already familiar in Greek and Roman schools, with their canons of poets, tragedians, orators, and historians laid down for study. Selection served a didactic purpose. It laid down what works should be studied and explained by a teaching commentary, written or oral. Galen had already commented on some Hippocratic treatises, and it was only to be expected that as his own writings became authoritative they should also be subjected to the same process. When and how this occurred cannot be determined from the scanty evidence at our disposal, but Galen had come to occupy a position alongside Hippocrates by 400, and by the time of the earliest surviving medical commentaries, from the end of the fifth century, a Galenic canon of "Sixteen Books" seems to have been already established at Alexandria.

This offered a structured syllabus, beginning with books deliberately written by Galen for beginners and offering a general overview of medicine and diagnosis before moving on to what would today be classed as anatomy, physiology, pathology, diagnosis, therapeutics, and, although this is less certain, preventive medicine. This order also revealed the growing split between medical theory and practice, which were usually treated separately in medical textbooks from 400 onward. Galen's own insistence on the need for a doctor to be philosophically aware became interpreted as a demand for preparatory training in (usually Aristotelian) philosophy. At least some of the teachers of medicine in sixth-century Alexandria were also involved in the teaching of philosophy.[13]

This syllabus of recommended readings of Galen, supplemented by one of Hippocratic texts (largely those selected by Galen), gradually came to define formal medicine, and by implication irregular medicine, first in the Greek world and then in those areas that emulated it intellectually. If, as seems likely, this canon was created at Alexandria, this would explain its diffusion to Ravenna, the center of Byzantine administration in Italy, where in 572 Leontius was "doctor at the Greek school" and where the Latin lecture notes of Agnellus correspond closely to those of John of Alexandria. Via the medium of translation into Syriac around 520 by an Alexandria-trained doctor and theologian, Sergius of Resaena, it passed to the Near East, the Islamic world, and ultimately into late-medieval Europe.[14]

13 Owsei Temkin, *The Double Face of Janus* (Baltimore: Johns Hopkins University Press, 1977), pp. 167–97; Albert Z. Iskandar, "An Attempted Reconstruction of the Late Alexandrian Medical Curriculum," *Medical History*, 20 (1973), 235–58; and, for the Latin West, Augusto Beccaria, "Sulle tracce di un antico canone latino di Ippocrate e di Galeno, I," *Italia medioevale e umanistica*, 2 (1959), 1–56; "Sulle tracce di un antico canone latino di Ippocrate e di Galeno, II," *Italia medioevale e umanistica*, 4 (1961), 1–75; "Sulle tracce di un antico canone latino di Ippocrate e di Galeno, III," *Italia medioevale e umanistica*, 14 (1971), 1–23.

14 Innocenzo Mazzini and Nicoletta Palmieri, "L'Ecole médicale de Ravenne: Programmes et méthodes d'enseignement, langue, hommes," in *Les écoles médicales à Rome: Actes du 2ème Colloque international sur les textes médicaux latins antiques, Lausanne, septembre 1986*, ed. Philippe Mudry and Jackie Pigeaud (Université de Lausanne, Publications de la Faculté des Lettres, 33) (Geneva: Librairie Droz,

A medical orthodoxy, already strengthened by the gradual disappearance of non-Galenic alternatives, grew still further with the provision of scholarly aids – commentaries and summaries of the main outlines of each book and argument. Galen's own injunctions to empirical observation and independent thought were replaced by a reliance on a series of set positions. Alexander of Tralles (fl. 550), no opponent of Galen, vigorously attacked contemporary Galenists who preferred to follow the book of orthodoxy, even to the detriment of their patients, instead of, like him, offering a variety of effective cures adapted to individual circumstances.[15]

Below this formal medicine one could consult a multitude of books giving instructions on the use of magical recipes, medical horoscopes, or physiognomy in treating illness. Medical astrology, or iatromathematics, seems to have been originally a Hellenistic Egyptian specialty, and many of the surviving texts bear the names of Hermes Trismegistus or Pharaoh Nechepso. But Galen, in his treatise *On Critical Days*, appeared to give it partial approval, and it gained support from the widespread conviction that the whole universe was in some sense linked together in ways that only the initiate could fully understand. Treatises on the properties of stones or, like the *Koiranides*, incorporating a variety of magical remedies, charms, and even party tricks alongside more sober investigations of natural phenomena can also be found in Greek.[16] Medicine also entered into the "Problems literature," which treated natural science in question-and-answer form. Here one could find juxtaposed explanations of why salty seawater became sweet on boiling and why eye inflammations gave a sensation of something rough. The theologian Anastasius of Sinai in the early sixth century discussed a similar range of medical and scientific questions in his *Moral Questions*, giving a particular Christian slant to his discussion of the cause of plague and the ways to avoid it.[17]

Natural science in the Greek world was also assimilated into the Christian tradition of the *Hexaemera*, expositions of Creation as revealed in the book of Genesis, like that offered by St. Basil around 360. Information on the natural world was also available in authors such as Aelian, whose *Variegated History*,

1991), pp. 285–310; and Dimitri Gutas, *Greek Thought, Arab Culture: The Graeco-Arabic Translation Movement in Baghdad and Early ʿAbbāsid Society (2nd–4th/8th–10th c.)* (New York: Routledge, 1998).

15 Alexander of Tralles, *Medical Writings*, vol. 2, pp. 83, 155.

16 No good modern survey of this literature exists, but see Georg Luck, *Arcana Mundi: Magic and the Occult in the Greek and Roman Worlds* (Baltimore: Johns Hopkins University Press, 1985); Hans Dieter Betz, ed., *The Greek Magical Papyri in Translation, Including the Demotic Spells*, 2 vols. (Chicago: University of Chicago Press, 1986–1987); David Bain, "Salpe's ΠΑΙΓΝΙΑ: Athenaeus 322A and Plin. H.N. 28.38," *Classical Quarterly*, n.s. 48 (1997), 262–8; and Alain Touwaide, "Iatromathematik," in *Der neue Pauly; Enzyklopädie der Antike*, ed. Hubert Cancik and Hans Schneider (Stuttgart: Verlag J. B. Metzler, 1999), vol. 5, cols. 873–5.

17 Cassius Iatrosophistes, *Problems*, in *Physici et medici graeci minores*, ed. Johann Ludwig Ideler, 2 vols. (Berlin: Reimer, 1841–1842), vol. 2, 64–5; and Anastasius of Sinai, *Moral Questions* (Patrologia Graeca, 89), Question 114, p. 765; cf. his discussion of medical prognosis in Question 20, p. 521.

little snippets of interesting facts, was widely read for centuries. Although writings on veterinary medicine and horticulture were still being produced, and new discoveries made, the surviving treatises are little more than extracts from much earlier tracts, such as that by the veterinarian Apsyrtus (fl. 150–250).

A tradition of high-quality illustration survived into late antiquity in Constantinople and Egypt. Fragments of illuminated herbals on papyri show the quality of some botanical drawing, and the beautiful Vienna manuscript of Dioscorides' *Materia medica*, copied in Constantinople around 512 for a princess, includes detailed pictures of birds, fishes, and animals alongside the plants described by the great pharmacologist almost five hundred years earlier. A similar, although weaker, tradition of pharmacological illustration survived in the West, notably in the so-called Apuleius Herbal. One of the rare manuscripts of Anglo-Saxon medicine, the eleventh-century British Library Cotton Vitellius C III is an illustrated translation of the Latin *Medicine from Quadrupeds* by Sextus Placitus (ca. 450).[18]

LATE LATIN TEXTS ON MEDICINE AND NATURAL SCIENCE

Although the same pattern of the coexistence of medical and scientific texts of various levels of sophistication and audience applied across the world of late antiquity, in general, medicine in the late antique West differed from its equivalent in the East in many ways. Galenism did not gain much hold. Only a handful of works by Galen and Hippocrates were translated into Latin, along with portions of Oribasius and Paul of Aegina, in all probability in the region of Ravenna. Indeed, by 800 Galen was most famous in Latin Europe as the inventor of a drug, the *Hiera Galeni*, or as the author of oracular precepts on uroscopy.[19]

In North Africa around 400, Methodist medicine appeared to be flourishing. Methodist theory, which went back to the first century, rejected humors in favor of explanations of the common features of illness in terms of the relationship between atoms and pores, expressed as fluidity, stricture, or a mixed state. The substantial treatise *On Acute and Chronic Diseases* by Caelius Aurelianus may be little more than a translation into Latin of a Greek book by Soranus of Ephesus around the year 100, but it is translated with flair and intelligence and shows traces of independent thought. Another North African, Cassius Felix, in 447 wrote a neat summary of therapeutics

[18] Alfred Stückelberger, *Bild und Wort: Das illustrierte Fachbuch in der antiken Naturwissenschaft, Medizin und Technik* (Mainz: Philipp von Zabern, 1994); and Linda E. Voigts, "Anglo-Saxon Plant Remedies and the Anglo-Saxons," *Isis*, 70 (1979), 250–68.

[19] Baader, "Early Medieval Latin Adaptations of Byzantine Medicine in Western Europe," pp. 251–3.

according to the principles of the "logical sect." If one also adds the similar treatise that passes under the name of Vindicianus, the abundant information offered by St. Augustine on contemporary learned physicians and surgeons and their contacts with Alexandria, and the geographical and astronomical information collected by yet another African, Martianus Capella (ca. 470), the medical and scientific life of North Africa in the fifth century appears remarkably vibrant.[20] Similar observations can be made about Italy, where the two most important surviving works of ancient veterinary medicine, by Pelagonius and Vegetius, were written toward the end of the fourth century and where the senator Macrobius included a long section on the physiology of eating and drinking in his *Saturnalia*, written around 430. Although they take much of their information from other authors, they organize it into a coherent and effective whole, and the veterinary authors in particular often describe their personal experiences in practice. The senator Cassiodorus, expounding to King Theodoric in around 520 the duties of the count of the *archiatri*, the chief of the doctors of Rome, describes a paragon of medical and scientific virtue, capable of judging all medical disputes, treating the king, and deciding on the capabilities of lesser colleagues, just as in Constantinople.[21]

But signs of decline are also apparent. Cassiodorus himself (ca. 490–590), when advising the monks of his monastery at Vivarium in southern Italy to aid the sick with medicines and with hope in God, also recommended to them a small selection of essential medical texts – Gargilius Martial's *On Gardens*, some Latin versions of Hippocrates and Galen's *Method of Healing, for Glaucoma*, an anonymous compendium, an (illustrated ?) Dioscorides, Caelius Aurelius's *On Medicine*, Hippocrates' *On Herbs and Cures*, and a few other books, a meager harvest from classical antiquity.[22]

To identify these treatises is instructive. First, they are largely texts of practical medicine; what theory they contain is presented dogmatically and with little room for discursive, Galenic argument. Second, although at first sight the authors are well known, the names may well hide a variety of suppositious works. Five or six tracts from the Greek Hippocratic corpus were available in Latin by this date, including *Aphorisms*, *Prognostic*, and *Airs, Waters and Places*, but *On Herbs and Cures* is not among them, unless it was a compilation of extracts from *On diet*. Early-medieval Latin manuscripts

[20] Anna Maria Urso, *Dall' autore al traduttore: Studi sulle passiones celeres e tardae di Celio Aureliano* (Lessico e Cultura, 2) (Messina: EDAS, 1997); and Guy Sabbah, "Observations préliminaires à une nouvelle édition de Cassius Félix," in *I Testi di medicina latini antichi. Problemi filologici e storici. Atti del I Convegno Internazionale, Macerata – S. Severino M., 26–28 aprile, 1984*, ed. Innocenzo Mazzini and Franca Fusco (Università di Macerata, Pubblicazioni della Facoltà di lettere e Filosofia, 28) (Rome: Giorgio Bretschneider, 1985), pp. 285–312.

[21] J. N. Adams, *Pelagonius and Latin Veterinary Terminology in the Roman Empire* (Studies in Ancient Medicine, 11) (Leiden: Brill, 1995), especially pp. 1–65 and 103–58; and Cassiodorus, *Variae*, VI.19, ed. T. Mommsen (Berlin: Weidmann, 1894).

[22] Cassiodorus, *Institutes*, I.31, ed. Sir Roger Mynors (Oxford: Clarendon Press, 1937).

transmitted, under the heading of Galen's *Method of Healing, for Glaucoma*, a translation of its two genuine books along with others that were certainly not his. "Caelius Aurelius" may refer to Caelius Aurelianus's *On Acute and Chronic Diseases* or to a popular compendium on fevers ascribed to one Aurelius. Likewise the treatise of Dioscorides may not be a Latin version of his famous herbal but the tract *On Feminine Herbs* that goes under his name and survives in at least seven manuscripts written before 900. The attribution of short practical works to famous names of the past, or to none, is typical of early-medieval Latin medical manuscripts.[23]

Third, the combination of guides to diagnosis and therapy with an herbal is common. The so-called *Lorsch Book of Medicine*, written about 795 in the Benedictine Abbey of Lorsch (in southwest Germany), contains brief introductory texts on anatomy, the humors, and prognostics (as well as a short version of the Hippocratic Oath), and ends with a long series of recipes and practical advice and a letter on diet dedicated to the Frankish king Theodoric.[24]

Cassiodorus's list is no chance assemblage. It offers an effective, if not always elegantly expressed, distillation of the medicine of the past. Gargilius Martial's book on gardens, for example, contained useful medical as well as horticultural information, and was a far from uncritical compilation from earlier learned writers, including Galen. Cassiodorus's choice of books may have constituted, as one scholar dismissively called it, "a literature for barbarians," but, equally, it preserved much practical therapy of antiquity (and no small amount of its basic theory).[25]

This was to be the pattern of medical book collections for the next five hundred years. Leaving aside the Ravenna commentaries on Galen and Hippocrates, which were present in a few major Benedictine monasteries in Italy but rarely those north of the Alps, discussions of medical theory in the manner of Galen are replaced by, at best, summaries of basic facts (e.g., the number of bones), or by bald definitions, necessary for establishing the framework for medical practice. Handy guides to prognosis are prominent, especially indications of imminent death (of prime concern to clerics intent on saving the patient's immortal soul), as well as advice on uroscopy and

[23] See the possible permutations in Guy Sabbah, Pierre-Paul Corsetti, and Klaus-Dietrich Fischer, *Bibliographie des Textes Médicaux Latins* (Mémoires du Centre Jean Palerne, 6) (Saint-Etienne: Publications de l'Université de Saint-Etienne, 1987).

[24] Ulrich Stoll, *Das 'Lorscher Arzneibuch', ein medizinisches Kompendium des 8. Jahrhunderts (Codex Bambergensis Medicinalis 1)* (Sudhoffs Archiv, Zeitschrift für Wissenschaftsgeschichte, Beiheft 28) (Stuttgart: Franz Steiner Verlag, 1992); and Gundolf Keil and Paul Schnitzer, eds., *Das Lorscher Arzneibuch und die frühmittelalterliche Medizin: Verhandlungen des medizinhistorischen Symposiums im September 1989 in Lorsch* (Geschichtsblätter Kreis Bergstrasse, Sonderband 12) (Lorsch: Verlag Laurissa, 1991).

[25] Hans Diller (echoing the earlier comment of Valentin Rose), *Die Überlieferung des hippokratischen Schrift De aere, aquis, locis* (Philologus, Supplementband 23) (1932), p. 50.

taking the pulse. But most medical manuscripts before 1100 are largely recipe lists, often well organized and well suited to the needs of the community.[26]

These recipe books were not transmitted haphazardly. Some early manuscripts of the *Euporista* (*Easily Obtainable Remedies*) by Theodore Priscian, a North African author of the early fifth century, show important additions and corrections, which indicate that the book was updated for use. But the most important source of earlier pharmacological information, and indeed scientific information in general, was the Elder Pliny's *Natural History*. But its thirteen books on medicine were cumbersome, and expensive to buy or copy. Hence it is not surprising that several different selections of its recipes, the so-called *Pliny's Physic*, were made from the fourth century onward.[27]

Pliny's influence is felt in various ways. The encyclopedia tradition that he represents was continued particularly by Isidore of Seville (d. ca. 640), whose *Etymologies* – Book IV, on diseases and cures, and Book XI, on anatomy – became prime sources of medical learning for succeeding generations. They provided a definition of terms, and to understand the meaning of these terms was also to comprehend the essence of the world as created by God.[28] In England, Bede (d. 735) based his *On the Nature of Things* largely on information from Isidore and Pliny, as did Rhabanus Maurus, writing in his *On the universe* at the German monastery of Fulda a century later. The natural world was increasingly viewed through Plinian spectacles. Its geography depended either directly on the *Natural History* or indirectly on such interpreters as Solinus (ca. 200), whose *Collection of Wonders*, based almost entirely on Pliny and Pomponius Mela, was the most popular geographical compendium until the twelfth century.

This emphasis on the wonders of nature fitted well with a Christian view of the universe, as well as being more accessible to an audience that might hear the stories but not be able to read them. A treatise like the *Physiologus*, a Christianized exposition of the marvelous properties of fifty plants, birds, and animals as reported by an earlier "natural scientist," enjoyed phenomenal success, first in Greek (in several recensions) and then in a variety

[26] The fundamental study is Augusto Beccaria, *I codici di medicina del periodo presalernitano (secoli IX, X e XI)* (Rome: Edizioni di Storia e Letteratura, 1956); supplemented by Ernest Wickersheimer, *Les manuscrits latins de médecine du Haut Moyen Âge dans les bibliothèques de France* (Documents, Études et Répertoires publiés par l'Institut de Recherche et d'Histoire des Textes, 11) (Paris: Editions du Centre Nationale Scientifique, 1966). See also Baader, "Early Medieval Latin Adaptations of Byzantine Medicine in Western Europe." For the importance of prognostics, see Frederic S. Paxton, "*Signa Mortifera*: Death and Prognostication in Early Medieval Monastic Medicine," *Bulletin of the History of Medicine*, 67 (1993), 631–50.

[27] Cornélia Opsomer, *Index de la Pharmacopée du Ier au Xe siècle*, 2 vols. (Alpha-Omega, Reihe A, CV) (Hildesheim: Olms-Weidmann, 2001), lists the texts and shows the continued use of drugs.

[28] Danielle Jacquart and Claude Thomasset, *Sexuality and Medicine in the Middle Ages* (Cambridge: Polity Press, 1985), pp. 8–16.

of translations, including Latin. It informed, instructed, and also pointed the way to God.[29]

MEDICINE AND NATURAL SCIENCE IN AND OUT OF THE MONASTERY

It is this Christian context that determined the direction of medicine and natural science until the twelfth century. The Benedictine Rule favored acts of compassion and charity; at St. Gall (Switzerland) an idealized plan of the monastery around 820 shows an elaborate infirmary, and Abbot Othmar, a decade later, founded a poor house and a leper hospital in the town.[30] His monastic library also contained several books of medicine, as did those of Nonantola (northern Italy) or Reichenau (southern Germany), centers that conserved and copied earlier learning for the benefit of the church. But one should not exaggerate this interest in learned medicine. Out of roughly 9,000 codices surviving from the ninth century, barely 100 can be classed as medical. Collections of recipes, antidotes, and agglomerations of handy rules predominate. Longer and more exacting texts like the Lorsch copies of Caelius Aurelianus or the Latin Dioscorides are rarely mentioned and still less used. Other areas of the natural sciences are even less familiar, except for computational astronomy, which owed its importance to the church's need to know the date of Easter.[31]

The Carolingian Renaissance of the ninth century did not change this picture of medicine and science. Although many Latin writings from classical antiquity were rediscovered, the only medical text among them was a 1,107-line poem based on Celsus and Pliny by the otherwise unknown Q. Serenus (about 200). Whether it became popular because of its medical or, more likely, its poetical merits remains an open question. The didactic *Problems* literature also reappears, and the Platonic tradition represented by the *Timaeus*, especially in Calcidius's Latin, became important again, but John Scotus Eriugena's *Periphyseon (On the Division of Nature)*, written around 864, shows how far this tradition might diverge from a direct investigation into the natural phenomena themselves. Rather, the natural world was used as a way to gain understanding of God and His purposes. Greek texts, including Aristotle's zoological works, remained inaccessible, except for what had already been available in Latin in the sixth century.[32]

29 Roger French, *Ancient Natural History: Histories of Nature* (London: Routledge, 1994), pp. 276–86.

30 Park, "Medicine and Society in Medieval Europe," p. 71.

31 Figures are based on Beccaria, *I codici di medicina del periodo presalernitano*; Wickersheimer, *Les manuscrits latins de médecine du Haut Moyen Âge dans les bibliothèques de France*; and Leighton Reynolds, ed., *Texts and Transmission: A Survey of the Latin Classics* (Oxford: Clarendon Press, 1983).

32 For Serenus, see Reynolds, *Texts and Transmission*, pp. 381–5.

This Carolingian revival led to the wider spread of literacy and the renewed copying of medical texts, whether by themselves or as part of miscellanies. Scholars like Walahfrid Strabo (d. 849), who composed a charming poem on gardens and medicinal plants, or Pardulus of Laon a century later, were familiar with medical learning and practice, and the cathedral libraries of Rheims and Laon contained medical manuscripts that were studied and possibly used in teaching. Medicine was becoming intellectually respectable again.[33] By 1050, most major monasteries or cathedral schools owned a few manuscripts of medicine, and Monte Cassino and Chartres were exceptional in the size and quality of their collections. At Chartres, medical teaching, based on the Hippocratic *Aphorisms* and *On the Concord of Hippocrates* by Heribrand, "an expert in pharmacology, pharmacy, and botany," is attested in 991, albeit characterized by book learning rather than the practical expertise associated with the later medical teaching at Salerno.[34]

How far this literate medicine spread beyond the cloister and into the square is hard to determine. Glosses show readers or scribes attempting to identify in their native languages the plants and herbs they read about in their Latin texts. Besides, it should not be assumed that the *laece* ("leech") was inferior to the *medicus* or that a text written in the vernacular was any less learned or more representative of popular practice than one in Latin. The most extensive surviving vernacular medical literature from this period, that written in Anglo-Saxon, is far from being a mass of ignorant mumbo-jumbo, "a final pathological disintegration of the great system of Greek medical thought," as Grattan and Singer called it. Anglo-Saxon knowledge of plant remedies was both wide and effective, and authors were aware of the difficulties of identifying Mediterranean flora from British. When Bald and Cild compiled their *Leechbook* around 900, perhaps at Winchester, they adapted the best continental practical medicine to an English environment. They included remedies obtained from Ireland or Irish scholars, and apparently simplified some of their Latin recipes by removing some of the more exotic ingredients. Nonetheless, pepper, silk, aloe, and Eastern herbs like zedoary and galingale, perhaps obtained though the medium of Arab traders in northern Europe, are all mentioned in Anglo-Saxon texts. There are traces, too, of Greek words, derived through intermediaries like Isidore and the late-antique Latin medical authors rather than from any direct acquaintance with Greek books. Anglo-Saxon medicine used chants and charms, and a few diseases were explained as "elf-shot" (the result of darts hurled

[33] John J. Contreni, "Masters and Medicine in Northern France during the Reign of Charles the Bald," in *Charles the Bald: Court and Kingdom*, ed. Margaret T. Gibson and Janet L. Nelson (BAR International Series, 101) (London: BAR, 1981), pp. 333–50.

[34] Loren C. MacKinney, "Tenth-Century Medicine as Seen in the *Historia* of Richer of Rheims," *Bulletin of the Institute of the History of Medicine*, 2 (1934), 347–75; Paul Oskar Kristeller, *Studi sulla Scuola medica salernitana* (Naples: Istituto Italiano per gli Studi Filosofici, 1986), pp. 19–21; and Klaus-Dietrich Fischer, "Dr. Monk's Medical Digest," *Social History of Medicine*, 13 (2000), 239–51.

by mischievous elves) or involving a great worm. But this is only a small part of what survives, and is not unique to the Anglo-Saxons. Many recipes come from much earlier sources and are found in other regions and in Latin learned texts like the *On Medicaments* of Marcellus, written in Gaul around 403 and well known to the Carolingians. Even some of the most derided remedies turn out, on closer inspection, to have at least some pharmaceutical value. Indeed, far from being a superstitious backwater, medicine in late Anglo-Saxon England may, in its practical texts, have been the equal of that anywhere else in Western Europe, including the rising Salerno.[35]

But to look at early-medieval medicine and natural science in a teleological perspective, as a botched transmission of something greater from the past (by which is usually meant Aristotle and Galen) or as a descent into irrationality that could only be reversed by Salerno and the scholastics, is to miss the point. In some areas, like southern Italy, medicine and medical practitioners of a recognizably Galenic sort appear to flourish; elsewhere, as in Carolingian Gaul or Germany, they do not. But to suggest an explanation for this divergence solely in terms of economic and social structures, although tempting, may be to mistake the absence of records for the absence of activity, or to repeat the truism that medical practitioners are most often found where there are wealthy patients.[36]

What disappears between 300 and 1000 are the intellectual and professional structures that ordered the medical life of classical antiquity. The early-medieval West knew nothing of civic doctors and, for the most part, formalized teaching centers. State involvement in the control of medicine was replaced by that of the church, but only in part and with theological considerations uppermost in mind. Sophisticated theoretical discussion was unlikely in the absence of educational establishments where this could take place, and even after the Carolingian revival of learning these were few. Argument took second place to the practical imperatives of caring and curing.

Above all, this period in the West shows a major shift in ideas of health and healing that were not always the result of the triumph of Christianity.

35 James H. G. Grattan and Charles Singer, *Anglo-Saxon Magic and Medicine* (London: Oxford University Press, 1952), p. 94; Voigts, "Anglo-Saxon Plant Remedies and the Anglo-Saxons"; Malcolm L. Cameron, *Anglo-Saxon Medicine* (Cambridge Studies in Anglo-Saxon England, 7) (Cambridge: Cambridge University Press, 1993); Robert A. Banks, "The Uses of Liturgy in the Magic, Medicine, and Poetry of the Anglo-Saxons," in *The Timeless and the Temporal*, ed. Edward Maslen (London: Queen Mary and Westfield College, 1993), pp. 19–40; and Audrey Meaney, "The Practice of Medicine in England about the Year 1000," *Social History of Medicine*, 13 (2000), 221–37. For Marcellus, see Robert and Cornélia Halleux-Opsomer, "Marcellus ou le mythe empirique," in Mudry and Pigeaud, *Les écoles médicales à Rome*, pp. 159–78.

36 Patricia Skinner, *Health and Medicine in Early Medieval Southern Italy* (The Medieval Mediterranean: Peoples, Economies, and Cultures, 400–1453) (Leiden: Brill, 1997). For Gaul and Germany, see Loren C. MacKinney, *Early Medieval Medicine with Special Reference to France and Chartres* (Baltimore: Johns Hopkins University Press, 1937); and Park, *Medicine and Society*, pp. 67–9.

It is a world in which the natural and the supernatural coexisted and interpenetrated, when Elpidius, a royal doctor, was convinced that demons were throwing stones at him even within his own house to make him fall ill. Only the exorcism of a saint could effect a cure. Galen's ordered universe explicable solely in terms of logical causation is replaced by one with a variety of overlapping, and to us at times contradictory, explanations. Diseases could be caused by overeating or overexertion – or by the breath of a dragon or a monster in a lake. To be struck down by "elfshot" was both to fall ill and to be the victim of an assault by the arrows of elves, and evil consequences to the body followed moral as well as physical disorder. Likewise, remedies could lie in the application of a drug, a change of diet, exorcism, magic, the swallowing of dust from a saint's tomb, or the personal intervention of a saint or holy man. These explanations cross boundaries of class and literacy. The great Welsh bard Taliesin tells how King Maelgwn Gwynedd in 547 saw through a church keyhole the Yellow Plague, a beast from the marsh, with golden eyes, teeth, and hair, and promptly died. Neither king, nor poet, nor people would have doubted the truth of this explanation.[37]

This variety of explanations is mirrored by a variety of overlapping healers. The doctor operates in the sphere of the natural, yet at the same time he can assist the Christian saint in his supernatural cures. His limitations are transcended by the power of the saint. At the same time, the boundaries between the doctor and the saint's arch-opponent, the enchanter, are easily overridden. The doctor's failure is taken as proof of his incompetence or lack of faith, his natural remedies as magic, his confident belief in his therapeutics as a step on the road to heresy. A very narrow gap separated the doctor's belief that successful bloodletting could only be carried out on certain "Egyptian" days from that of the enchanter appealing to the cosmic sympathy of the heavens.[38]

This is not to say that it was impossible to differentiate one belief from the other, or that patients and doctors had no clear idea of what might be appropriate in any one case. In his report of the life of St. John of Beverley (d. 721), Bede (672/673–735) makes it quite obvious that he (and John) were well aware of the complementary roles, and limits, of both types of healing. If doctors failed to heed their own rules for sound practice in bleeding, it was not surprising that a patient developed nasty complications. The sign of the cross might make a dumb man speak, but his instruction in language required only good teaching, and the same man's skin disorder and dandruff

[37] For Elpidius, see Anonymous, *Vita S. Caesarii*, I.41; and Peregrine Horden, "Diseases, Dragons, and Saints: The Management of Epidemics in the Dark Ages," in *Epidemics and Ideas: Essays on the Historical Perception of Pestilence*, ed. Terence Ranger and Paul Slack (Past and Present Publications) (Cambridge: Cambridge University Press, 1992), pp. 45–76.

[38] Valerie J. Flint, "The Early Medieval 'Medicus', The Saint – and the Enchanter," *Social History of Medicine*, 2 (1989), 127–46; and Valerie J. Flint, *The Rise of Magic in Early Medieval Europe* (Oxford: Clarendon Press, 1991), pp. 203–330.

were left by John to a *medicus*. Herebald, whose skull was broken after a fall in a horse race, was cured through John's prayers as well as through the bonesetting skills of a *medicus*. Ironically, later retellings emphasized the miraculous by omitting the doctors.[39]

Another case history reveals a different hierarchy of explanation. In 576, Leunast, archdeacon of Bourges (central France), sought help for his eye cataracts from several physicians; when they did not succeed as well as he had hoped, he made a pilgrimage to the nearby shrine of St. Martin of Tours. Although this brought some relief, he also had himself bled by a Jewish doctor in Bourges. He became totally blind, fit punishment, remarked Bishop Gregory of Tours (539–594), for seeking the help of a Jew after he had received God's grace through the saint. Gregory condemns the inappropriate use of secular healing (and Jewish at that), not secular healing in itself.[40]

Such fluidity of explanation and healing practices is a prominent feature of the history of medicine in early-medieval Europe. Although espoused and encouraged by Christianity, this fluidity is not an essentially Christian feature. Rather, its presence in classical antiquity is masked in our sources by a stronger "naturalizing" trend of explanation, represented by Seneca or Galen, and it is the weakening of that natural alternative as much as any new growth in irrationality that explains the changed appearance of medicine and natural science. Classical antiquity had marveled at the world and had sought answers to some of its problems. From the mid-eleventh century onward, some of its answers were discovered anew, some of the older structures, social as well as intellectual, were rebuilt, and new perspectives were introduced. The explanations offered in late antiquity and early-medieval Europe did not thereby disappear. They were merely downgraded as more academic ones took their place.

39 Bede, *Ecclesiastical History of the English People*, ed. Bertram Colgate and Sir Roger Mynors (Oxford Medieval Texts) (Oxford: Clarendon Press, 1969), pp. 456–9, 464–9.

40 Nancy G. Siraisi, *Medieval and Early Renaissance Medicine* (Chicago: University of Chicago Press, 1990), p. 11.

14

TRANSLATION AND TRANSMISSION OF GREEK AND ISLAMIC SCIENCE TO LATIN CHRISTENDOM

Charles Burnett

The translation of scientific texts from Greek and Arabic in the twelfth and thirteenth centuries is both a symptom and a cause of one of the greatest shifts in Western science, comparable in importance with the parallel importation of scientific works into Arabic in Baghdad in the ninth century and the Scientific Revolution in seventeenth-century Europe. Details concerning the translators and the texts translated can be found elsewhere;[1] the purpose of this chapter is to describe the translation movement in general and to point out its salient features for the history of science.

THE COURSE OF THE TRANSLATIONS

In the late tenth century, an interest in mathematical learning in the circles of Abbo of Fleury and Gerbert d'Aurillac resulted in the discovery and development of some Arabic techniques concerning the astrolabe and astrology and the use of Arabic numerals on the counters of the abacus.[2] Not until

[1] For translations from Arabic, see Moritz Steinschneider, *Die hebräischen Übersetzungen des Mittelalters* (Berlin: Kommissionsverlag des Bibliographischen Bureaus, 1893); and Moritz Steinschneider, *Die europäischen Übersetzungen aus dem Arabischen* (Sitzungsberichte der Kaiserlichen Akademie der Wissenschaften in Wien, philosophisch-historische Klasse, 149, no. 4 [1904] and 151, no. 1 [1905], Vienna: In Kommission bei Carl Gerold's Sohn). There is no comparable comprehensive list of translations from Greek, except in regard to philosophical texts, for which see Robert Plasnau, *The Cambridge History of Medieval Philosophy*, 2 vols. (Cambridge: Cambridge University Press, 2009), vol. 2, pp. 793–814. Some other authors are included in the ongoing series *Catalogus translationum et commentariorum* (Union académique internationale, 1960–). Both traditions are discussed in Charles Homer Haskins, *Studies in the History of Mediaeval Science*, 2nd ed. (Cambridge, Mass.: Harvard University Press, 1927). See also Geneviève Contamine, ed., *Traduction et traducteurs au Moyen Âge* (Paris: Éditions du CNRS, 1989); and Jacqueline Hamesse and Marta Fattori, eds., *Rencontres de cultures dans la philosophie médiévale: traductions et traducteurs de l'Antiquité tardive au XIVe siècle* (Louvain-la-Neuve: Université Catholique de Louvain, 1990).

[2] André Van de Vyver, "Les plus anciennes traductions latines médiévales (Xe–XIe s.) de traités d'astronomie et d'astrologie," *Osiris*, 1 (1936), 658–89; Paul Kunitzsch, "Les relations scientifiques entre l'Occident et le monde arabe à l'époque de Gerbert," in *Gerbert l'européen*, ed. Nicole Charbonnel and Jean-Eric Iung (Aurillac: La Haute-Auvergne, 1997), pp. 193–203; and David Juste,

the last years of the next century, however, can one say that a translation "movement" was inaugurated. It was then that the translations of Greek and Arabic medical works began in Salerno, where there was a thriving medical school. The archbishop, Alfano, translated Nemesius's *On the Nature of Man* under the title *Premnon phisicon*, and five short essential texts on medicine, later called the *Articella*, were put together from Greek and Arabic sources. Constantine the African (d. before 1098/1099) brought from Qairawān in Tunisia an Arabic medical corpus, which he proceeded to translate in Salerno and then in the Benedictine monastery of Monte Cassino. The principal text of this corpus was a comprehensive medical manual, the *Pantegni* ("The Complete Book of the Medical Art") of ʿAlī ibn al-ʿAbbās al-Majūsī (in Latin "Haly Abbas").[3] Constantine did not finish translating this large work of twenty books and substituted sections from other texts; the translation was continued by his pupils in the early twelfth century. In 1127, Stephen the Philosopher produced a more accurate translation of the whole text, but Constantine's version remained the more popular. An interest in medicine and in natural philosophy as its concomitant appears to have centered on Stephen's hometown, Pisa, which had a quarter in Antioch.[4] Pisa also had a quarter in Constantinople, where James of Venice (attested in 1136) translated several texts of Aristotle's *libri naturales* from Greek, and Burgundio of Pisa (1110–1193), a Pisan notary, translated some major texts of Galen and Aristotle's *On Generation and Corruption*.[5] Other texts of the *libri naturales* were translated by Henricus Aristippus in Sicily and three anonymous (but probably Italian) translators.

From the beginning of the twelfth century, we find an advance in the interest in mathematics (and especially astronomy) that had been initiated by Abbo and Gerbert. This advance may have been due to Petrus Alfonsi, who, having converted from Judaism to Christianity in Huesca in Aragon in 1106, brought some astronomical tables and other texts to France and England. These were put into Latin by Walcher, abbot of Great Malvern (d. 1135), and Adelard of Bath (fl. 1106–49).[6] After Petrus, a group of scholars was

Les Alchandreana *primitifs: Étude sur les plus anciens traités astrologiques latins d'origine arabe (Xe siècle)* (Leiden: Brill, 2007).

3 Charles Burnett and Danielle Jacquart, eds., *Constantine the African and ʿAlī ibn al-ʿAbbās al-Maǧūsī: The Pantegni and Related Texts* (Leiden: Brill, 1994).

4 See Charles Burnett, "Antioch as a Link between Arabic and Latin Culture in the Twelfth and Thirteenth Centuries," in *Occident et Proche-Orient: contacts scientifiques au temps des croisades (Actes du colloque de Louvain-la-Neuve, 24 et 25 mars 1997)*, ed. Isabelle Draelants, Anne Tihon, and Baudouin van den Abeele (Turnhout: Brepols, 2000), pp. 1–78, reprinted with corrections in Charles Burnett, *Arabic into Latin in the Middle Ages* (Farnham: Ashgate Variorum, 2009), article IV.

5 On James of Venice, see Lorenzo Minio-Paluello, "Iacobus Veneticus Grecus: Canonist and Translator of Aristotle," in *Opuscula*, ed. L. Minio-Paluello (Amsterdam: Adolf M. Hakkert, 1972). On Burgundio, see Gudrun Vuillemin-Diem and Marwan Rashed, "Burgundio de Pise et ses manuscrits grecs d'Aristote: Laur. 87.7. et Laur. 81.18," *Recherches de théologie et philosophie médiévales*, 64 (1997), 136–98.

6 Charles Burnett, "The Works of Petrus Alfonsi: Questions of Authenticity," *Medium Ævum*, 66 (1997), 42–79.

working in the valley of the Ebro and Languedoc; these included Hermann of Carinthia (fl. 1138–43), Hermann's colleague Robert of Ketton (fl. 1141–56) and pupil Rudolph of Bruges (fl. 1144), Raymond of Marseilles (fl. 1141), and Hugo of Santalla (fl. 1145), who devoted themselves to the science of the stars, both translating from Arabic and writing original works. Collaborating with them was the Jewish polymath Abraham ibn Ezra (1092–1167), who provides a link with the Pisan translators since we know that he was in the neighborhood of Pisa in the 1140s and contributed toward adapting the astronomical tables of ʿAbd-al-Raḥmān ibn ʿUmar al-Ṣūfī (d. 986) for the meridian of that city.[7]

In the middle of the twelfth century, Toledo became the center for the translation of scientific works, and the separate streams of mathematical, medical, and philosophical translations were united there. Here an archdeacon, Dominicus Gundissalinus (fl. 1161–81), and a "so-called master" (*dictus magister*), Gerard of Cremona (1114–1187), superintended a systematic translation of Arabic works on mathematics, medicine, and the philosophy of Aristotle and the Arabic Peripatetic tradition. Gundissalinus, a Christian theologian, was particularly interested in psychology and cosmology; he was thus responsible for the translation of works on these subjects by Avicenna (Ibn Sīnā, ca. 980–1037), Algazel (al-Ghazālī, 1058–1111) and Avicebron (Ibn Gabirol, ca. 1021–1065). Gerard attended to texts on the other subjects, including perhaps earlier translations made by John of Seville.[8]

Although inspired, perhaps, by the near-contemporary translations of Aristotle made from Greek by his colleagues in Italy and Constantinople, Gerard could have found the plan for his translations in *The Classification of the Sciences* by Alfarabi (al-Fārābī, ca. 870–950), which he translated. In the sections on logic and natural science, Alfarabi describes their parts and summarizes the relevant text of Aristotle for each part. Gerard translated the last of Aristotle's logical texts (The *Posterior Analytics*; the other logical texts had already been translated) and embarked on the texts on natural science. This Aristotelian program, which included commentaries, was continued in Toledo by Alfred of Shareshill (late twelfth century) and Michael Scot (left Toledo after 1217, died before 1236), and culminated in the translations of the Large Commentaries of Averroes (Ibn Rushd, 1126–1198) on Aristotle's *libri naturales* by Michael and by Theodore of Antioch, and of Averroes's Middle Commentaries on the logical texts by William of Luna in

[7] Charles Burnett, "A Group of Arabic–Latin Translators Working in Northern Spain in the Mid-twelfth Century," *Journal of the Royal Asiatic Society*, 109 (1977), 62–108; Charles Burnett, *The Introduction of Arabic Learning into England* (London: The British Library, 1997), pp. 47–58.

[8] Charles Burnett, "The Coherence of the Arabic–Latin Translation Programme in Toledo in the Twelfth Century," *Science in Context*, 14 (2001), 249–88 at p. 269, reprinted with corrections in Charles Burnett, *Arabic into Latin in the Middle Ages* (Farnham: Ashgate Variorum, 2009), article VII.

Naples around 1260 and Hermann the German in Toledo between 1240 and 1256.[9]

The thirteenth century is marked by a high level of intellectual activity and an exchange of intellectual goods between scholars writing in Arabic, Hebrew, Greek, and Latin – the full extent and complexity of which is only now being realized.[10] Two powerful secular patrons stand out. The first is Emperor Frederick II Hohenstaufen (d. 1250), whose power base was in Sicily and southern Italy and who attracted Michael Scot, Theodore of Antioch, John of Palermo, and the Jewish scholar Jacob Anatoli to his entourage. Here, and in Naples, where he had founded a university in 1224, these five scholars translated under his patronage works of Averroes and Avicenna and, in all probability, the great Jewish work on scholastic theology, Maimonides' *Guide for the Perplexed* (originally written in Arabic).[11] Frederick II's initiative was continued by his son Manfred (r. 1258–66), who employed the translator Bartholomew of Messina.

The second leading patron was Alfonso X, king of Castile and León (r. 1256–84), under whose patronage Jewish scholars translated a wide range of Arabic works on astronomy, astrology, and magic into Castilian, and Italian notaries translated some of these into Latin. The philosophical translations sponsored by Frederick were soon assimilated into the European universities, and the astronomical tables prepared at Alfonso's court were the basis for the standard tables used in Europe for the rest of the Middle Ages (the "Alphonsine Tables"; see North, Chapter 19, this volume).[12]

The end of the thirteenth century and beginning of the fourteenth witnessed the translation of great *summae* on philosophy, astrology, and medicine, such as Averroes's *Destruction of the Destruction of the Philosophers*, Abenragel's *Judgments of the Stars*, and al-Rāzī's *Continens*. By the early fourteenth century, the major works on mathematics, medicine, and philosophy had been translated; no new translations of great importance from Arabic appeared until the Renaissance, when, in addition to the desire to revise or replace translations made in the Middle Ages because of the perceived "barbarity" of their Latin style, an interest in Arabic poetry emerged for the first time.[13]

The supremacy of Toledo as a scientific center from the mid-twelfth to the mid-thirteenth century ensured the ascendancy of translations from Arabic

9 Ramon Gonzálvez Ruiz, *Hombres y libros de Toledo* (Madrid: Fundación Ramón Areces, 1997), pp. 586–602. For a similar continuation in the translation of Avicenna's *Shifā'*, see Marie-Thérèse d'Alverny, "Notes sur les traductions médiévales d'Avicenne," *Archives d'histoire doctrinale et littéraire du moyen âge*, 19 (1952), 337–58.

10 See articles in *Micrologus*, 2 (1994), entitled *Sciences at the Court of Frederick II*.

11 For the possibility that the translator was John of Palermo, see Gad Freudenthal, "Pour le dossier de la traduction latine médiévale du *Guide des Égarés*," *Revue des Études Juives*, 147 (1988), 167–72.

12 Julio Samsó, *Islamic Astronomy and Medieval Spain* (Aldershot: Variorum, 1994).

13 Harry A. Wolfson, "The Twice-Revealed Averroes," *Speculum*, 36 (1961), 373–92; and Johann Fück, *Die arabischen Studien in Europa bis in den Anfang des 20. Jahrhunderts* (Leipzig: Harrassowitz, 1955).

over those from Greek. After the mid-thirteenth century, there was a shift back to translating directly from Greek. This interest in Greek is marked in the studies of Roger Bacon (ca. 1220–ca. 1292) and Robert Grosseteste (ca. 1175–1253). With respect to Aristotle's *libri naturales*, it led to the replacement of earlier Arabic–Latin translations by new translations from Greek, the correction of earlier Greek translations, and the completion of the corpus, first by Grosseteste and then, most comprehensively, by William of Moerbeke (ca. 1215–1286). Similarly, Nicholas of Reggio (ca. 1280–ca. 1350), the personal physician of Robert I in Naples, replaced and supplemented the earlier Arabic–Latin translations of Galen by translating over fifty of his works from the original Greek.

Jews played an important role both in circulating and in translating scientific works in Christendom. But in many ways their contribution to Christian learning was a by-product of the high level of their own scientific culture, which merits more space than is given to it here (see Langermann, Chapter 6, this volume). The language of this culture was at first Arabic; however, through the efforts of Abraham bar Ḥiyya (ca. 1065–ca. 1136) and Abraham ibn Ezra in the second quarter of the twelfth century, Hebrew scientific prose was established, and over the next two centuries a massive collection of Arabic texts in philosophy, mathematics, and medicine was translated into Hebrew.[14] It was, then, to the Jews that Latin-reading scholars turned in the Renaissance for new knowledge of texts by Avicenna, Averroes, and other Arabic authorities, and Jewish scholars turned these from Hebrew into Latin.

GOALS

The most obvious results of the translation of scientific texts can be seen in the curriculum of the medieval schools and universities. One should not underestimate the volume of writings that were translated but disappeared from view simply because they were not taken up into these curricula, or the translations of nonuniversity subjects, like alchemy, some parts of astrology, various technological sciences, and Hermetic literature, whose diffusion is difficult to trace.[15] Nevertheless, partly because of the clearer nature of the evidence, this chapter concentrates on works relevant to the official scientific learning of the time.

The motivation for the translations was the perceived lacunae in Latin scientific education, as Burgundio of Pisa and the biographers of Gerard of

[14] Mauro Zonta, *La filosofia antica nel medioevo ebraico* (Brescia: Paideia, 1996); and Gad Freudenthal, "Les sciences dans les communautés juives médiévales de Provence: leur appropriation, leur rôle," *Revue des études juives*, 152 (1993), 28–136.

[15] Charles Burnett, *Magic and Divination* (Aldershot: Variorum, 1996). For alchemical texts, see Robert Halleux, "The Reception of Arabic Alchemy in the West," in *Encyclopedia of the History of Arabic Science*, ed. Roshdi Rashed, 3 vols. (London: Routledge, 1996), vol. 3, pp. 886–902.

Cremona both state.[16] There were three principal areas in which the Latins were felt to be particularly lacking. The first was mathematics, especially geometry and astronomy. These were two of the seven liberal arts, which had formed the framework for the rhetorically based Latin education since late antiquity and had been revived first by Gerbert d'Aurillac at the turn of the millennium and then more generally in the twelfth century. The five other liberal arts were well represented, mostly thanks to the translations of Boethius (d. 529), who had sought to provide Latin readers with a complete curriculum of Greek studies. But geometry and astronomy lacked comprehensive textbooks.

For geometry, Boethius had only translated a small portion of Euclid's *Elements*. Stephen the Philosopher lamented the poor knowledge of geometry among the Latins, and John of Salisbury reckoned that the only place where the study flourished was in (Islamic) Spain.[17] Thus, a pioneer in the twelfth-century translating movement, Adelard of Bath, devoted a text to the description of the seven liberal arts (his *On the Same and the Different*) but also made the first translation of the *Elements* from Arabic.

For astronomy, no translation by Boethius has survived, but the Latins were aware that the most important Greek text was the *Almagest* of Ptolemy. It is specifically for this text that Gerard of Cremona is said to have gone to Toledo, while the slightly earlier translator of the *Almagest* from Greek had rushed to Sicily from Salerno because he heard that a copy of the *Almagest* had just been brought there from Constantinople.[18] How the new translations filled gaps in the traditional curriculum can be seen in the important "Library of the Seven Liberal Arts" assembled by Thierry of Chartres (d. 1151) in the early 1140s. This *Heptateuchon* included both Euclid's *Elements* and astronomical tables by the ninth-century Arabic mathematician al-Khwārizmī.

The second lacuna was in physics (i.e., the investigation of the workings of nature). In this case, we are dealing with a subject that was *not* one of the rhetorically based seven liberal arts but had been part of the curriculum in philosophy in ancient Alexandria and continued to be taught (with interruptions and revivals) in Byzantium and (more continuously) in the Islamic world. In the West, the interest in physics pre-dates the knowledge of the texts of this curriculum and is based on "natural questions" and late antique Latin works such as Calcidius's translation of Plato's *Timaeus* and

16 Peter Classen, *Burgundio von Pisa* (Sitzungsberichte der Heidelberger Akademie der Wissenschaften, philosophisch-historische Klasse, Jahrgang 1974, 4. Abhandlung) (Heidelberg: Carl Winter Universitätsverlag, 1974), p. 84. For Gerard, see Charles Burnett, "Coherence of the Arabic Latin Translation Programme in Toledo in the Twelfth Century," p. 275.

17 Burnett, "Antioch as a Link between Arabic and Latin Culture in the Twelfth and Thirteenth Centuries," p. 58.

18 Haskins, *Studies in the History of Mediaeval Science*, pp. 191–3.

Macrobius's commentary on the *Somnium Scipionis* of Cicero.[19] However, during the course of the twelfth century, the fact that Aristotle, aside from being an authority on logic, was also the authority on natural science became known. Burgundio of Pisa acquired, or had copied for him, a collection of the Greek texts of Aristotle's *libri naturales*, and, on the testimony of John of Salisbury, called Aristotle "the Philosopher" (an early instance of the shift of this title away from Plato to Aristotle).[20] Burgundio, James of Venice, and William of Moerbeke were directly aware of the corpus of Aristotle's natural science from contemporary Greek scholars who were teaching and writing commentaries on it, and Gerard of Cremona knew the same texts in Arabic and as described by Alfarabi in the *Catalogue of Sciences*, which he translated.

The third area was medicine. On this subject, Galen was the master, and the titles of the sixteen basic texts of his that formed the curriculum at Alexandria were known. Constantine the African listed them in his preface to the *Pantegni* and translated two of them from Arabic; several more were translated by Burgundio of Pisa and Gerard of Cremona.[21]

GREEK OR ARABIC?

If the goal of the translators was to restore the ancient learning of Euclid, Ptolemy, Aristotle, and Galen, they had two sources: the centers of Greek and Arabic learning. The Byzantine Greeks on the whole had preserved the ancient texts without substantially altering them; there had been little scientific development, and a twelfth-century commentary on a work by Aristotle could easily be taken for a second-century commentary on the same work. Among the Greeks, therefore, the Latins sought and could find their copies of Ptolemy's *Almagest*, Aristotle's *libri naturales*, and Galen's works, just as Arabic scholars had done in the ninth century. For the Latins, manuscripts that the Greeks could provide were more important than scholarly expertise. For the most part, the interpretations of the ancient texts were ancient themselves: those of Proclus and Themistius, Alexander of Aphrodisias, and John Philoponus. It is significant that the sack of Constantinople by the Latins in 1204, unlike that by the Turks in 1453, did not result in a diaspora of Greek scholars and a consequent resurgence of Greek humanism in the West.

Among the Arabs, the Latins could and did find these same Greek works, but they were also confronted with the results of a tradition of scholarship that

19 For knowledge of Aristotelian physics before the translations, see Thomas Ricklin, *Die "Physica" und der "Liber de causis" im 12. Jahrhundert* (Freiburg im der Schweiz: Universitätsbibliothek, 1995).

20 John of Salisbury, *Metalogicon*, IV.7.6–8, ed. John Barrie Hall and Katherine S. B. Keats-Rohan (Corpus Christianorum Continuatio Mediaevalis, 98) (Turnhout: Brepols, 1991), p. 145.

21 Danielle Jacquart, "Les traductions médicales de Gérard de Crémone," in *Tolède, XIIe–XIIIe siècles*, ed. Louis Cardaillac (Paris: Autrement, 1991), pp. 177–91.

not only had absorbed new elements from other cultures (particularly those of India and Persia) but had also developed, refined, and changed the learning of the ancients.[22] Thus, the astronomical models of Ptolemy had to contend with those of Indian astronomers, and his measurements of the movements of the planets were repeatedly corrected and often replaced ever since the official "testing" sponsored by the caliph al-Maʾmūn in the early ninth century. Aristotle's *libri naturales* together with the ancient commentaries on them were transmitted, but Alfarabi, Avempace (Ibn Bājja, late eleventh century to ca. 1139), and Averroes wrote new commentaries, and Avicenna recast the whole of Aristotle's philosophy.[23] Perhaps even more radical was the replacement of the original works of Galen by new texts on medicine, each generation of Arabic doctors trying to improve on the work of their predecessors.[24]

Thus Latin scholars were confronted by a dilemma: whether to concentrate on restoring the wisdom of the ancients or to introduce the "novelties" promoted by the Arabic scholars with whom they came into contact. The criticism of "newness" in respect to Arabic science is apparent in the words of Adelard of Bath and Stephen the Philosopher, both of whom sought to justify their Arabic studies.[25] The suspicion that somehow the Arabs had corrupted and defiled the ancient tradition lasted throughout the Middle Ages and reached its culmination in the Renaissance,[26] though the arguments between those medievals who saw scientific knowledge as somehow static and only requiring recovery and preservation and those who saw it as progressively advancing cut across linguistic boundaries.[27]

Arabic learning, then, differed from the Greek in that it resided in masters as much as in books. Adelard and Stephen the Philosopher both referred to their Arabic *magistri*, which they encountered in the principality of Antioch. Other translators benefited from the diaspora of Jewish scholars who had cultivated Arabic learning, especially after the expulsion of the Jews from Islamic Spain by the Almohads in 1160 (in much the same way as Renaissance Italian scholars benefited from the exile of Jewish scholars from Spain in 1491). In some cases, translators appeared to have used those scientific works

22 For the Arabic transformation of early astrological literature, see David Pingree, *From Astral Omens to Astrology: From Babylon to Bīkāner* (Rome: Istituto Italiano per l'Africa et l'Oriente, 1997).

23 Dimitri Gutas, *Avicenna and the Aristotelian Tradition* (Leiden: Brill, 1998).

24 Manfred Ullmann, *Islamic Medicine* (Edinburgh: Edinburgh University Press, 1978).

25 See Adelard's preface to his *Questions on Natural Science*, in *Adelard of Bath, Conversations with His Nephew*, ed. Charles Burnett (Cambridge: Cambridge University Press, 1998), pp. 82–3. For Stephen, see Burnett, "Antioch as a Link between Arabic and Latin Culture in the Twelfth and Thirteenth Centuries," p. 47.

26 See the debate in Marco degli Oddi's preface to *Aristotelis omnia quae extant opera... Averrois Cordubensis in ea opera commentarii*, ed. Bernardus Salviatus, 11 vols. (Venice: Iuntas, 1550–1552), vol. 1, fols. 5r–6v.

27 Dag N. Hasse, "Aristotle versus progress: The decline of Avicenna's 'De anima' as a model for philosophical psychology in the Latin West," in *Was ist Philosophie im Mittelalter?* ed. Jan Aertsen and Andreas Speer (Miscellanea Mediaevalia, 26) (Berlin: W. de Gruyter, 1998), pp. 871–80.

promoted by their Arabic masters; examples include Constantine the African, who transmitted the medical tradition of his masters in Qairawān, and Gundissalinus, who translated works of Avicenna, Algazel, and Avicebron, the favored authors of his collaborator, Avendauth (Abraham ibn Dāūd). Gerard of Cremona, although translating several important works by Arabic scholars (especially in the field of medicine), appears to have made a more deliberate effort to recover the ancient texts from among the Arabs. But his translating activity coincided with a reaction against "modern" developments also on the part of a group of Arabic scholars in Córdoba, who tried to restore a pure Aristotle both for natural science and astronomy.[28]

But if scholars were searching for the ancient Greek texts, why did they make the effort to search for them in Arabic at all? The tradition of scholarship in Arabic and the availability of texts (discussed further later in this chapter) provide part of the answer. But there was also a sense that the language was merely contingent – a clothing that could be changed – and that, as long as translators had been faithful, the underlying "truth" would be preserved. Nevertheless, sometimes a translator went out of his way to give the impression that an Arabic–Latin translation had been made from the Greek, such as the translator of Ptolemy's *Almagest* in MS Dresden, Db 87, who gives the work its correct Greek title "*megali xintaxis*" but substitutes other Greek terms that either do not match the Greek original or are completely fabricated. At other times, translations of the same work from Greek and Arabic would be compared, for example in the proemium to a compendium on Euclid, *Elements XIV–XV*, in the pseudo-Grosseteste's *Summa philosophiae*, and, frequently, by Albert the Great.[29]

SOURCES

For translations to be made, either the Latin scholar must go in search of the texts or the texts must be sent or brought to the Latin center of learning. A book may be among diplomatic exchanges, such as the copy of Ptolemy's *Almagest* brought from Constantinople to Palermo by Henricus Aristippus, acting as ambassador for William I, king of Sicily, shortly before 1160.[30] Petrus Alfonsi, when he traveled to Northern Europe after his baptism in Huesca in 1106, brought books with him.

[28] A. I. Sabra, "The Andalusian Revolt against Ptolemaic Astronomy," in *Transformation and Tradition in the Sciences*, ed. Everett Mendelsohn (Cambridge: Cambridge University Press, 1984), pp. 133–53.

[29] For Euclid, see John E. Murdoch, "Euclides Graeco-Latinus: A Hitherto Unknown Medieval Latin Translation of the *Elements* Made Directly from the Greek," *Harvard Studies in Classical Philology*, 71 (1966), 249–302 at p. 285; and Robert Grosseteste, *Summa philosophiae*, in *Die philosophischen Werke*, ed. Ludwig Baur (Beiträge zur Geschichte der Philosophie und Theologie des Mittelalters, 9) (Münster: Aschendorff, 1912), pp. 275–643. For Albert the Great, see his *De anima*, ed. C. Stroick (Münster: Aschendorff, 1968), p. 8, line 60ff., where he compares the "translatio Arabica" and the "translatio Graeca."

[30] Haskins, *Studies in the History of Mediaeval Science*, pp. 160–1.

As for the masters of the sciences, scholars might be part of the booty captured by pirates, as stated in one story concerning how Constantine the African arrived in Salerno (MS London, British Library, Sloane 2426, fols. 8r–v). Other scholars arrived in Latin centers as the result of persecution or religious differences, such as the Arabic-speaking Christians (Mozarabs) who left Islamic Toledo for the Christian North of Spain in considerable numbers in the ninth century. After its reconquest, Toledo was the natural place of refuge for the Jews and Mozarabs who were driven out of Islamic Spain in the mid-twelfth century by the Almohad regime.[31] The flourishing study of the works of Averroes outside Islamic territory in the early thirteenth century may have been partly caused by intolerance toward the philosophical sciences among the later Almohad jurists.

But while enforced exile and the closing of doors brought about some contacts, the opening up of the Mediterranean to the Latins through conquest and trade brought more. It is no coincidence that the translation movement took off soon after the reconquest of Toledo, which opened up the heart of Islamic Spain (1085), the Norman conquest of Sicily with its Greek and Arabic-speaking population (1072–91), and the fall of Antioch, which opened up the Islamic and Greek cultures of the Eastern Mediterranean (1098). The attempts at reunifying the Greek and Latin churches also brought East and West closer together and resulted in scientific translations as well as theological writings. Political leaders exchanged scholars as well as books. For example, the Ayyubid sultan al-Malik al-Ṣāliḥ (1232–9) sent Frederick II one of the most distinguished Muslim scholars, Sirāj al-Dīn al-Urmawī (d. 1283), to help him interpret Arabic logic.[32]

More important over a longer period were the quarters set up by the Pisans and Venetians in cities throughout the Mediterranean, which, aside from their commercial function, offered opportunities for Latin scholars to work in the midst of Arabs and Greeks. Pisan quarters were set up in Antioch in 1108 and Constantinople in 1111. Whereas James of Venice is attested in the Pisan quarter of Constantinople in 1136, Leonard (Fibonacci) of Pisa acquired his knowledge of Arabic mathematics in the Pisan depot of Bougie (Algeria). Pilgrim routes also provided avenues along which intellectual goods could travel, and it is perhaps not a coincidence that among the places where translations are attested to have been made are Toulouse, León, Astorga, and Ponte do Lima, all important stages on pilgrim routes to Santiago. Finally,

[31] Bernard Septimus, *Hispano-Jewish Culture in Transition: The Career and Controversies of Ramah* (Cambridge, Mass.: Harvard University Press, 1982).

[32] For Frederick II's extensive communication with Islamic scholars, usually through political channels, and the possibility that the sons of Averroes were actually at his court (as was reported already by 1287), see Charles Burnett, "The 'Sons of Averroes with the Emperor Frederick' and the Transmission of the Philosophical Works by Ibn Rushd," in *Averroes and the Aristotelian Tradition: Sources, Constitution and Reception of the Philosophy of Ibn Rushd (1126–1198) (Proceedings of the Fourth Symposium Averroicum, Cologne, 1996)*, ed. Gerhard Endress and Jan Aertsen (Leiden: Brill, 1999), pp. 259–99.

from the second quarter of the thirteenth century onward, the Franciscan and Dominican houses founded in intellectual centers and important cities throughout Europe and the Mediterranean provided opportunities for the spread of texts. The Dominicans had a *studium* in Tunis from 1240 onward, and it has been suggested that the Franciscans in Constantinople could have been one source for Grosseteste's Greek manuscripts.[33]

In the case of mathematical translations made from Arabic into Latin, a significant role may have been played by a single royal library, that of the Banū Hūd in Saragossa. Yūsuf al-Muʾtaman ibn Hūd, king from 1081 to 1085, had written a comprehensive book on geometry, *al-Istikmāl*, which exploited a large collection of Greek and Arabic works on the subject. In 1110, the Banū Hūd were driven out of Saragossa by the Almoravids and settled in the fortress of Rueda de Jalón in Aragon. But they took their library with them, for Hugo of Santalla specifically says (in a rare example of the mention of a source) that his patron, Bishop Michael of Tarazona, acquired the Arabic manuscript of a work on astronomical tables in this library ("*in Rotensi armario*").[34] In 1140, the last of the Banū Hūd, Abū Jaʿfar Aḥmad III Sayf al-Dawla (d. 1146), exchanged his property in Rueda de Jalón for a house in the cathedral quarter in Toledo.[35] It may be no accident that only after this move are Greek and Arabic mathematical works translated from Arabic in Toledo.

PATRONS

Translation was a time-consuming and expensive enterprise, and needed sponsorship. The copying of *The Complete Book of the Medical Art* (*Regalis dispositio*) of Haly Abbas occupied three scribes (including the translator himself) for a period of nearly one year.[36] Vincent of Beauvais mentions specifically Aristotle's *De animalibus* (translated by Michael Scot) and the *Canon* of Avicenna (translated by Gerard of Cremona) as examples of large works that would cost a great deal of money to buy or have copied.[37]

What drove and financed the translating movement? Often it is unclear whether a dedication is an appeal for sponsorship or implies that such

[33] A. Carlotta Dionisotti, "On the Greek Studies of Robert Grosseteste," in *The Uses of Greek and Latin*, ed. A. Carlotta Dionisotti, Anthony Grafton, and Jill Kraye (London: Warburg Institute, 1988), pp. 19–40.

[34] Haskins, *Studies in the History of Mediaeval Science*, p. 73.

[35] "Hūdids," in *Encyclopedia of Islam*, 2nd ed., 12 vols. plus Index vol. (Leiden: Brill, 1960–2009), vol. 3, p. 543; and Angel González Palencia, *Los mozarabes de Toledo en los siglos xii y xiii*, 4 vols. (Madrid: Instituto de Valencia de Don Juan, 1926–1930), vol. 1, pp. 151–3.

[36] This is the inference from the dates of writing added at the end of the various books of Stephen the Philosopher's translation; see Burnett, "Antioch as a Link between Arabic and Latin Culture in the Twelfth and Thirteenth Centuries," pp. 67–9.

[37] Vincent of Beauvais, *Speculum naturale, Prologus*, chap. 4, in *Speculum maius*, 4 vols. (Douai: Balthazar Beller, 1624), vol. 1, col. 4.

sponsorship already exists. But we do have some clear examples of translators looking for audiences or employment.[38] Hermann of Carinthia and his colleague Robert of Ketton each appealed to leading intellectual figures of the day in France: Hermann to Thierry of Chartres, the compiler of the *Heptateuchon*, and Robert to Saint Bernard of Clairvaux, to whom he promised a comprehensive book on the science of the stars. The most blatant appeal for employment was that of Petrus Alfonsi. Newly converted to Christianity and leaving his native Aragon, he promised to the "Peripatetics of France" lectures on astronomy that would "rouse to life the knowledge of this art which has disappeared among the Latins." Avendauth, similarly displaced, tried to win support (presumably from the archbishop of Toledo) for his project of translating Avicenna's *Shifā'*. Some texts may have played a role as prestige objects that enhanced the possessor's reputation or symbolized his power in a way similar to the astrolabes and celestial and terrestrial spheres dedicated to royal patrons. This is implicit in the wordplay in Robert's dedication to St. Bernard – "I promise to your celestial highness, who penetrates every heaven, a celestial gift that embraces in itself the whole of science"[39] – and may apply especially to astronomical tables drawn up for the meridian of the king's capital or archbishop's seat.

On the other hand, we have examples of patrons commissioning translations. Bartholomew of Messina prefaced his translations with the remark that they were done "on the command of King Manfred."[40] Peter the Venerable paid Hermann and Robert to translate the Qur'ān and other Islamic texts, and economic support is also suggested in Avendauth's dedication of his translation of the *De anima* to the archbishop of Toledo, completed "by your support" ("*tuo munere*"), and in the colophon to Jacob Anatoli's dedication to Frederick II of the Hebrew translations of Averroes's Middle Commentaries on Aristotle's logical texts (he will "complete the art [of logic] with the help of Him who aids all men, and who put it into the mind of our Lord, the Emperor Frederick, who loves knowledge and those who study it, to nourish me").[41] Some translators clearly worked to satisfy the scientific interests of their patrons. Good examples are Michael Scot and Theodore of Antioch, who translated works on zoology and falconry for Frederick II, whose interest in every aspect of birds of prey was notorious.

But higher ideals, in particular the benefiting of universities, must have motivated some of the patronage. The translations of Averroes's Middle Commentaries on Aristotle's *Organon* and his Large Commentaries on Aristotle's *Libri naturales* must be seen in the context of Frederick II's foundation

38 Charles Burnett, "Advertising the New Science of the Stars *circa* 1120–50," in *Le XIIe siècle*, ed. Françoise Gasparri (Paris: Le Léopard d'Or, 1994), pp. 147–57.

39 J.-P. Migne, ed., *Patrologia Latina*, 221 vols. (Paris: J.-P. Migne, 1844–1864), vol. 189, col. 660.

40 Gudrun Vuillemin-Diem, "La Liste des oeuvres d'Hippocrate dans le Vindobonensis Phil. Gr. 100," in *Guillaume de Moerbeke: Recueil d'études à l'occasion du 700e anniversaire de sa mort (1286)*, ed. Josef Brams and Willy Vanhamel (Leuven: Leuven University Press, 1989), p. 160.

41 Zonta, *La filosofia antica nel medioevo ebraico*, p. 74.

of the new university in Naples in 1224. A well-known letter advertising the new translations of "Aristotle and other philosophers from Greek and Arabic in the field of the whole trivium and quadrivium" was either sent by Frederick II to the masters and students of Bologna, or more likely, addressed to the masters and students of Paris by the emperor's son Manfred, who had himself commissioned translations of several works that were falsely attributed to Aristotle.[42] Gonzalo Pérez, known as Gudiel, who ended up as archbishop of Toledo (1280), commissioned and collected translations of Aristotle's *libri naturales*, Averroes's commentaries, and parts of Avicenna's philosophical encyclopedia, and, in 1293, founded the University of Alcalà de Henares.[43]

The universities themselves commissioned some translations. An example of this is given by the note in one manuscript of the translation of Averroes's preface to his Large Commentary on the *Physics*, which asserts that "Theodore of Antioch had translated [this work] on the request of the students of Padua" (MS Erfurt, Amplon. F 352, fol. 104v). It is striking, however, that, with the exception of Naples, which was a special case, the universities were not the locus of translation activity. Nor were the translators, on the whole, university teachers, as we shall see in the next section.

There is more evidence that translation was supported by the cathedral schools. Whether Thierry, as chancellor of Chartres cathedral, actually commissioned translations is unclear. But the frequent examples of cooperation between an archdeacon and a translator in Spain suggest more than mere personal interest. We have Robert, archdeacon of Pamplona, working with Hermann of Carinthia; Dominicus Gundissalinus, an archdeacon at Toledo, collaborating with the translators Avendauth and Johannes Hispanus; and Mauritius, another archdeacon at Toledo, joining forces with Mark of Toledo (who translated the Qur'ān and a medical work by Galen). In each case, the archdeacon appears as a person of higher rank than his translator colleague.

The translator Hugo of Santalla is attested as a "magister" in the cathedral of Tarazona in 1145, but it was not necessarily the schoolteachers in the cathedral who were the translators. Gerard of Cremona, Mark of Toledo, Michael Scot, and Hermann the German are all attested as canons, and the wording of a reference to Michael's vacant canonship in 1229 suggests that some people were unhappy about the fact that a canonship was occupied by someone who had no pastoral responsibilities.[44] It has been suggested that the archbishops of Toledo were directly responsible for sponsoring and promoting a "school of translators" as a quasi-university.[45] The evidence is not decisive, however, since only two translations are dedicated to archbishops.

[42] René Antoine Gauthier, "Notes sur les débuts (1225–1240) du premier 'Averroïsme'," *Revue des sciences philosophiques et théologiques*, 66 (1982), 321–74.

[43] Francisco J. Hernandez, "La Fundación del Estudio de Alcalá de Henares," *En la España Medieval*, 18 (1995), 61–83.

[44] Ibid., p. 68.

[45] See the fundamental article by Valentin Rose, "Ptolemäus und die Schule von Toledo," *Hermes*, 8 (1874), 327–49, and the objections of Lorenzo Minio-Paluello, "Aristotele del mondo arabe a quello latino," in *L'occidente e l'Islam nel alto medioevo*, 2 vols. (Settimane di studio del Centro Italiano di

Nevertheless, cathedral clergy appear to have played the leading role in the Toledan translation movement, and the locus of this activity must have been in the Cathedral quarter and the adjacent Frankish quarter of the city, the only districts in which foreigners and Latin learning were dominant – also, apparently, where the remains of the library of the Banū Hūd were located, as we have seen.

In the course of the thirteenth century, the papal curia itself became an important locus for translators and scientists. David of Dinant, who knew some of Aristotle's *libri naturales* in their Greek originals and whose works were condemned in Paris in 1215, was a chaplain to Pope Innocent III; subsequent popes supported Michael Scot, and by the second half of the century the curia in Viterbo had become a veritable hotbed of translating activity, as it was for science in general.[46]

TRANSLATORS

It has already been noted that translators were not, on the whole or primarily, school or university teachers, though Theodore of Antioch and Hermann the German may have spent part of their time as teachers in the universities of Naples and Palencia, respectively.[47] Even the language teachers at the *Studia linguarum* set up by the Dominicans and Franciscans from 1250 onward specifically for the training of missionaries to Jews and Muslims are rarely attested as translators of scientific works.[48]

Some translators, however, apparently had received training in the schools, especially those to whose names the term "magister" is commonly prefixed, such as the "magister Johannes" who collaborated with Gundissalinus. A few could be called amateurs in the sense that they found time in the midst of their professional engagements to learn the requisite language and attempt translations themselves. A good example of this is Robert Grosseteste, whose progress in learning Greek can be followed through the grammar written for him, his transcriptions of Greek texts, and his translation of part of Aristotle's *On the Heavens.*[49] In cases of commissioned translations, however, or those

Studi sull' Alto Medioevo, 12) (Spoleto: Centro Italiano di Studi sull'Alto Medioevo, 1965), vol. 2, pp. 603–37.

46 Agostino Paravicini Bagliani, *Medicina e scienze della natura alla corte dei papi nel duecento* (Spoleto: Centro Italiano di Studi sull' Alto Medioevo, 1991).

47 For translators as teachers, see Charles Burnett, "The Institutional Context of Arabic–Latin Translations of the Middle Ages: A Reassessment of the 'School of Toledo'," in *The Vocabulary of Teaching and Research between the Middle Ages and Renaissance*, ed. Olga Weijers (Turnhout: Brepols, 1995), pp. 214–35.

48 An exception is Rufinus of Alexandria, who collaborated with his Arabic teacher at the Dominican school in Murcia, Dominicus Marrochinus, on a translation of Hunayn ibn Ishāq's *Medical Questions.*

49 Dionisotti, "Greek Studies of Robert Grosseteste."

dedicated to a patron, we are entitled to consider the translators as engaging in a profession.

Burgundio of Pisa, Moses of Bergamo, and James of Venice were at least temporarily employed as official interpreters in the discussions between the Eastern and Western churches in Constantinople in 1136. James may have remained in the Byzantine court in the capacity of an interpreter.[50] At least two translators, Cerbanus, a translator of theological texts, and Leo Tuscus, who translated the book on the interpretation of dreams (*Oneirocriticon*) of Ahmed ben Sirin, were employees of the Byzantine emperor, the latter being described specifically as his interpreter.[51]

Some of these interpreters were also lawyers. Legal documents had to be accurately translated; and, as we shall see, the whole concept of the "literal interpreter" comes from the context of the Roman law courts of late antiquity. James of Venice gave advice as an expert in canon law to an archbishop of Ravenna in 1148, and Burgundio was primarily a notary, while the Italian notaries of Alfonso X were responsible for translating Castilian scientific texts into Latin. In Toledo, where a predominantly Arabic- or Romance-speaking population was ruled by an elite who used Latin in their official documents, the translation of legal cases and documents (either orally or in written form) was routine. One would therefore expect a close connection between forensic interpreters and the translations made there.

Other professions that had an interest in a particular aspect of Arabic science were those involving finance. Arabic methods of calculation were adopted in Henry I's exchequer, to which Adelard of Bath may have been attached; Stephen the Philosopher was a treasurer in Antioch; and Leonard of Pisa learned the language of the Arabs and their methods of calculation while accompanying his father on trade expeditions.

A commonly occurring designation of the translators, however, is "*philosophi.*" In his legal pronouncement of 1148, James of Venice is described as "*philosophus*"; Avendauth describes himself as "*israelita philosophus,*" and an early manuscript of a translation of Gerard of Cremona refers to him as "*summus philosophus.*"[52] In some cases, this adjective can be shown to indicate not only an expertise in the secular sciences of the ancients but also an official position. Two translators at the court of Frederick II – John of Palermo and Theodore of Antioch – were described as the emperor's "philosophers" ("*imperialis philosophus*"). The position was probably analogous to that of

[50] Fernand Bossier and Jozef Brams, eds., *Aristoteles Latinus VII. Physica*, 2 vols. (Leiden: Brill, 1990), vol. 1, p. xix.

[51] Antoine Dondaine, "Hugues Éthérien et Léon Toscan," *Archives d'histoire doctrinale et littéraire du moyen âge*, 19 (1952), 67–134.

[52] For further examples, and discussion, see Charles Burnett, "Master Theodore, Frederick's Philosopher," in *Federico II e le nuove culture, Atti del XXXI Convegno storico internazionale, Todi, 9-12 ottobre 1994* (Spoleto: Centro Italiano di Studi sull'Alto Medioevo, 1965), pp. 248–54, reprinted with corrections in Charles Burnett, *Arabic into Latin in the Middle Ages* (Farnham: Ashgate Variorum, 2009), article IX.

the *ḥakīm* in Islamic society – beyond giving medical and astrological advice to their patrons, they contributed to the political ideal of the combination of rulership and philosophy.[53] It has been suggested that the same term, when applied to James of Venice, may refer to the equivalent Byzantine title of "*philosophos.*"[54]

One thing that lawyers, treasurers, and "*philosophi*" had in common was that they belonged to the new secular professional classes that burgeoned at the same time as the translation movement became active. This might suggest that the existence of an educated lay professional class in itself was responsible for the flourishing of translations. The connections are undoubtedly there, but it must also be remembered that simultaneous with the translations of secular science and philosophy was a radical upsurge in the translation of Christian texts from Greek and a new interest in the Hebrew text of the Old Testament and Jewish interpretations of the Scriptures that transformed theology and biblical scholarship.[55] Both lay and ecclesiastical circles were profoundly affected.

TECHNIQUES

Different terms were used for the transmission of Greek and Arabic texts into Latin. Adelard of Bath regularly used the phrase "*ex arabico sumptus*" ("taken from Arabic"), John of Seville sometimes wrote "*interpretatus*" ("interpreted by") and at other times "*translatus ex arabico in latinum*" ("translated from Arabic into Latin"). Petrus Alfonsi described himself as the "translator" of the *Zīj* of al-Khwārizmī, and says that he "arranged [it] into the Latin language" ("*in latinam linguam digessi*"), whereas Gerard "Latinized" (*latinare*) the *Almagest.* The verb "*transferre*" soon became the normal term; it was, for instance, used for the translations of William of Moerbeke. The process by which this translation occurred could, however, take a variety of forms.

In some cases, it involved the fixing in Latin of a text transmitted orally. This appears to be the case in the "translations" of Arabic works on astronomy made by Petrus Alfonsi and Walcher of Malvern in the West Midlands of England. Petrus had left his books "across the sea," and Walcher's translations

53 The statement that "states are blessed either if they are handed over for philosophers to rule or if their rulers adhere to philosophy" is found in two works addressed to rulers or rulers-elect: al-Majūsī's preface to his *Complete Book of the Medical Art*, I.1 (Lyons: Jacobus Myt, 1523), fol. 1r, and Adelard of Bath's *On the Use of the Astrolabe*, in Burnett, *Introduction of Arabic Learning into England*, pp. 32, 46. For the context, see Sarah Stroumsa, "Philosopher-King or Philosopher-Courtier? Theory and Reality in the Falsifa's Place in Islamic Society," in *Identidades marginales*, ed. Cristina de la Puerte (Madrid: CSIC, 2003), pp. 433–59.

54 Bossier and Brams, *Aristoteles Latinus VII. Physica*, vol. 1, pp. xvii–xviii.

55 These translations are not the subject of this chapter, but it must be borne in mind that some of the Greco-Latin translators mentioned here, such as Burgundio of Pisa, Leo Tuscus, and Henricus Aristippus, also translated or promised to translate Christian writings.

are not of a text but of Petrus's words, as is clear from the title of one of them: "The opinions of Peter the Jew, called 'Alphonsus', concerning the lunar node, which the lord Walcher, prior of the church of Malvern, translated into the Latin language."[56] Abraham ibn Ezra, too, dictated his views ("*sententia*") on the astrolabe to an unnamed Latin scholar,[57] and the same process probably occurred in respect to Abraham's explanation of how to use the Pisan Tables, which appears in at least three different forms in Latin manuscripts. That whole books could be dictated by scholars (especially those of Jewish origin) is not implausible, and we have examples even in the Renaissance.[58] However, most medieval translations imply the existence of a written text in the original language, even if this text has been interpreted orally for a Latin writer.

The source text need not be written in a conventional book form. It may be a schematic diagram of a sheep's shoulder blade with the significance of each of the parts written on it,[59] or an astrolabe with the names of prominent fixed stars and various figures inscribed on it.[60] Hindu-Arabic numerals at first seem to have spread through Europe on the beads of the "Gerbertian abacus."[61] Nor is the text in the book merely the author's words. It may include diagrams and illustrations, as well as marginalia, and its layout may influence the choices of the translator.[62] It matters, then, whether the Latin translator has confronted a text in the original manuscript directly or has merely relied on an oral interpretation of that text. Both situations occurred.

Oral interpretation is best attested in respect to Arabic texts. A well-known description of the process is given by Avendauth, in the dedication of his translation of Avicenna's *De anima* to the archbishop of Toledo: "Here you have the book translated from Arabic: I took the lead and translated the words one at a time into the vernacular language, and Archdeacon Domenicus turned them one at a time into Latin."[63] Since this translation was made in Toledo, the "vernacular language" could have been either the Arabic dialect

56 Burnett, "Works of Petrus Alfonsi," pp. 45–7.

57 Burnett, *Introduction of Arabic Learning into England*, p. 58.

58 For example, when Rabbi Saul ben Simon published *Tsori ha-Yagon* in Cremona in 1557, he indicated that he had lost the original manuscript and had reproduced the text from memory; see Raphael Jospe, *Torah and Sophia: The Life and Thought of Shem Tov ibn Falaquera* (Cincinnati: Hebrew Union College Press, 1988), p. 35.

59 For examples of these diagrams, see the article on "Scapulimancy" in Burnett, *Magic and Divination*, article XII.

60 Such an astrolabe has been carefully copied in Paris, Bibliothèque Nationale de France, MS Lat. 7412, fols. 19v–23v; see Paul Kunitzsch, "Traces of a Tenth-Century Spanish-Arabic Astrolabe," *Zeitschrift für Geschichte der Arabisch-Islamischen Wissenschaften*, 12 (1998), 113–20.

61 Guy Beaujouan sees this as an explanation of the fact that Hindu-Arabic numerals are sometimes reversed or turned ninety degrees with respect to their Arabic models; see his "Étude paléographique sur la 'rotation' des chiffres et l'emploi des *apices* du Xe au XIIe siècle," *Revue d'histoire des sciences*, 1 (1948), 303–13.

62 For examples, see John E. Murdoch, *Album of Science: Antiquity and the Middle Ages* (New York: Scribner, 1984).

63 Simone Van Riet, ed., *Avicenna Latinus, Liber de Anima, I–II–III* (Leiden: Brill, 1972), pp. 95*–98* and 4.

spoken in Toledo or a Romance language. According to Roger Bacon, many people understood the spoken forms of Greek, Arabic, and Hebrew, even if they could not read and write the languages.[64] Adelard of Bath made mistakes that imply a mishearing of Arabic words that sounded alike rather than a misreading of an Arabic text.[65] Other translators are also said to have had the help of Arabic speakers, whether Mozarabs (such as the "Galippus" who helped Gerard of Cremona) or Jews (such as the "Salomon Avenraza" whom Alfred of Shareshill mentions as his teacher, and the "Abuteus levita" who helped Michael Scot).[66] Characteristics of texts resulting from translation by this method are the absence of any annotation in Arabic script and the presence of transcriptions of phrases in the original languages or of vernacular words that might have arisen during the conversation with the interpreter.[67] Some idea of the kind of language that these interpreters put the Arabic texts into may be gained from the astronomical notes in a mixture of completely ungrammatical Latin and Italian written into a manuscript by an unknown scholar in Lucca in or soon after 1160.[68]

In the course of the thirteenth century, the vernacular interpretations themselves were written down. An item in the 1273 inventory of Gonzalo Pérez "Gudiel" is described as "the exemplar in Romance from which it [i.e., the Latin translation of the *Nicomachean Ethics*] was translated."[69] The Castilian versions sponsored by Alfonso X were ends in themselves. It is noticeable that Jewish scholars predominated both as interpreters for Latin scholars in the twelfth and early thirteenth centuries and as translators into Castilian for Alfonso X.

It is clear, however, that some Latin scholars read the Arabic texts directly. Hugo of Santalla complains about "the variety of diacritical marks on the [Arabic] letters, or the lack of marks," and the Latin reviser of Abū Maʿshar's *De magnis coniunctionibus* (Gerard of Cremona?) refers to a word that he cannot read in the original.[70] Hugo transcribes Arabic letters in one of his

64 Roger Bacon, *Opus tertium*, ed. John S. Brewer (London: Longmans, 1859), pp. 33–4.

65 Examples are given in Charles Burnett, "Some Comments on the Translating of Works from Arabic into Latin in the Mid-twelfth Century," in *Orientalische Kultur und europäisches Mittelalter*, ed. Albert Zimmermann (Miscellanea Mediaevalia, 17) (Berlin: W. de Gruyter, 1985), pp. 161–71, especially pp. 166–7.

66 Marie-Thérèse d'Alverny, "Les traductions à deux interprètes, d'arabe en langue vernaculaire et de langue vernaculaire en latin," in her *La Transmission des textes philosophiques et scientifiques au Moyen Âge* (Aldershot: Variorum, 1994), article III.

67 Burnett, *Introduction of Arabic Learning into England*, pp. 40–4.

68 Charles Burnett, "The Transmission of Arabic Astronomy via Antioch and Pisa in the Second Quarter of the Twelfth Century," in *The Enterprise of Science in Islam: New Perspectives (Dibner Institute, November 1998)*, ed. Jan Hogendijk and Abdelhamid I. Sabra (Cambridge, Mass.: MIT Press, 2003), pp. 23–51.

69 Ruiz, *Hombres y libros de Toledo*, p. 435.

70 Preface to Jafar Indus, *Liber imbrium*, Madrid, Biblioteca Nacional, MS 10063, fol. 43r, ed. C. Burnett in "Lunar Astrology," *Micrologus*, 12 (2004), pp. 43–133 (see p. 88); and Charles Burnett and Keiji Yamamoto, ed. and trans., *Abū Maʿšar on Historical Astrology*, 2 vols. (Leiden: Brill, 2000), vol. 2, p. 308.

translations, and a diagram is labeled with words in Arabic script by the main scribe of a collection of texts produced by the scholars of the valley of the Ebro.[71] No Arabic manuscript has yet been identified as a copy used by the Latin translator; the nearest approximation to this is an Arabic manuscript of the *Organon* into which a twelfth-century Latin scribe has inserted Latin equivalents of logical terms and some chapter headings. He may have been attempting to learn Arabic by reading texts that were familiar to him.[72]

The situation in regard to Greek is different both in that it seems to have been normal for Latin translators to deal directly with Greek texts and because some of them have been identified. For example, Burgundio of Pisa's hand appears in the margins of Aristotle's and Galen's texts in several manuscripts of the Laurentian Library in Florence and that of William of Moerbeke in Vienna, Österreichische Nationalbibliothek, phil. gr. 100.[73] These annotated manuscripts show in detail how the translators approached their texts, including, as they do, variant readings and corrections to the Greek, letters of the Latin alphabet (in the case of Burgundio's manuscripts) indicating the logical order of the Greek words, and summaries and explanations of the subject matter. They demonstrate, too, the translators' competence in writing Greek; their knowledge of the language was not merely aural. Nevertheless, the variety of annotations in Burgundio's Greek manuscripts indicates the presence of a second scholar working alongside him, whose competence in Greek was probably superior to that of Burgundio himself.

In translating both Arabic and Greek, then, the Latin scholars' principal resource was the native speakers of the languages. Evidence of the use of dictionaries and grammars is less easy to find. The earliest Arabic grammar composed in a Western language is the one for the Arabic dialect of Granada written in 1505 by Pedro de Alcalà. Its purpose was to aid the conversion of the Arabic-speaking population, and much the same aim may underlie the two extensive Arabic–Latin/Latin–Arabic glossaries of the Middle Ages: the "Leiden glossary" and the *Vocabulista in Arabico* edited by Celestino Schiaparelli (Florence, 1871). Potentially more helpful are the glossaries of technical terms attached to certain texts of astronomy, astrology, and medicine.[74]

[71] Hugo's translation of 'Umar ibn al-Farrukhān, *Caesarian Questions*, Vienna, Österreichische Nationalbibliothek, MS 2428, fols. 90r–91r, 93r, which also includes in the margin of 91r the word "al-Shām" ("Syria") written in Arabic letters. The diagram is in Oxford, Bodleian Library, MS Digby 51, fol. 88v.

[72] Istanbul, Topkapi Sarayi, Ahmet III, MS 3362, discussed in Burnett, "Antioch as a Link between Arabic and Latin Culture in the Twelfth and Thirteenth Centuries," p. 75. For Arabic manuscripts that are very close to those used by the translator, see Charles Burnett, "Manuscripts of Latin Translations of Scientific Texts from Arabic," *Digital Proceedings of the Lawrence J. Schoenberg Symposium on Manuscript Studies in the Digital Age*, 1, no. 1 (2009), article 1 (http://repository.upenn.edu/ljsproceedings/vol1/iss1/1/).

[73] See, respectively, Vuillemin-Diem and Rashed, "Burgundio de Pise et ses manuscrits grecs d'Aristote," pp. 157–75; and Vuillemin-Diem, "La Liste des oeuvres d'Hippocrate dans le Vindobonensis Phil. Gr. 100."

[74] For astronomy, see Paul Kunitzsch, *Glossar der arabischen Fachausdrücke in der mittelalterlichen europäischen Astrolabliteratur* (Nachrichten der Akademie der Wissenschaften in Göttingen I.

The most extensive of these is the *Synonyma* of Simon of Genoa, a doctor at the papal court at the very end of the thirteenth century, who explained in detail the Greek and Arabic words in the medical vocabulary of the Latins; he included some notes on Greek and Arabic phonology.[75] These technical glossaries, however, were primarily for the use of readers unfamiliar with terms that had simply been transliterated from Arabic or Greek. Nevertheless, the consistency with which terms are translated suggests either the presence of some written lists or an extraordinary memory on the part of the translators. This is particularly the case with the translations or revisions of Gerard of Cremona, which use word-for-word equivalents of Arabic terms, even across subject boundaries. For the Jewish scholars in the early thirteenth century who translated texts from Arabic and also helped their Christian colleagues, we have more evidence of linguistic aids, especially in regard to philosophical translations.[76] Latin scholars, too, may have had some linguistic tools that are now lost since Simon of Genoa refers to a kind of dictionary that he calls "*Liber de doctrina arabica*" ("book on Arabic learning").[77]

It is not surprising that there is more evidence of language aids for Greek, which was more important than Arabic for understanding theology and classical antiquity. Moses of Bergamo discussed points of Greek grammar, and Robert Grosseteste arranged for a Greek grammar to be written for him by John of Basingstoke. An extant comprehensive Greek–Latin dictionary may be a copy of a glossary revised by Grosseteste from a south Italian exemplar.[78]

For the style of their translations, medieval translators had two contrasting role models to follow, Cicero and Boethius. Cicero, in his *On the Best Kind of Orator*, defended his decision to translate "not as an interpreter, but as an orator," in which he followed the sense, not the words.[79] By "interpreter" Cicero meant the professional interpreter employed in the law courts. Boethius, on the other hand, defended his decision to abandon good-sounding Latin in favor of the literal translation of the "faithful interpreter," claiming that such a method was necessary in scientific writings in which knowledge of things was sought.[80]

philologisch-historische Klasse, Jahrgang 1982, Nummer 11) (Göttingen: Vandenhoeck und Ruprecht, 1983).

75 Danielle Jacquart, "La coexistence du grec et de l'arabe dans le vocabulaire médical du latin médiéval: l'effort linguistique de Simon de Gênes," in her *La Science médicale occidentale entre deux renaissances (XIIe s.–XVe s.)* (Aldershot: Variorum, 1997), article X.

76 For example, the Hebrew–Italian glossary and a Hebrew–Latin wordlist edited by Giuseppe Sermoneta, *Un glossario filosofico ebraico-italiano del XIII° secolo* (Rome: Edizioni dell'Ateneo, 1969).

77 Jacquart, "La coexistence du grec et de l'arabe dans le vocabulaire médical du latin médiéval," p. 279.

78 London, College of Arms, MS Arundel 9; see Dionisotti, "Greek Studies of Robert Grosseteste," pp. 24–5.

79 Cicero, *De optimo genere oratorum*, V.14.

80 Boethius, *In Isagogen Porphyrii Commentorum Editio secunda*, ed. Georg Schepps and Samuel Brandt (Vienna: F. Tempsky, 1906), chap. 1, p. 135. See Werner Schwarz, "The Meaning of *Fidus interpres* in Mediaeval Translation," *Journal of Theological Studies*, 45 (1944), 73–8.

Those scholars who cultivated a Ciceronian style of translation also tended to write elaborate prefaces and to make adaptations of the material they translated, as Cicero had done. In effect, they made the Greek and Arabic works their own. Among these scholars were Constantine the African, Hugo of Santalla, and Hermann of Carinthia. Hermann pointedly remarked, in a preface addressed to his fellow translator Robert of Ketton, that he departed from the method advocated by Boethius because of the prolixity of the Arabic language. His own translations show not only abbreviation but also substitution (e.g., of familiar geographical names for the obscure Arabic ones and of examples from classical Latin texts for those in Arabic), and above all a compact Ciceronian style. This approach might result from the influence of the humanistic ideals of the French schools, which also led John of Salisbury to criticize the literal translator James of Venice as being inadequately trained in grammar and to wish to replace James's obscure translation of the *Posterior Analytics* with a more readable one.[81] The Ciceronian model became fashionable again in the Renaissance.[82]

It was, however, Boethius's example that became the norm in the Middle Ages in translations from both Greek and Arabic. Literal interpretation, of course, could be merely the first stage toward a literary (or other) version of a text; for example, the author of the *Liber denudationis* (a work against Islam) claims to be following the sense and abbreviating a preliminary literal translation that does not survive.[83] But, in most cases, the literal method was deliberately chosen as that most appropriate to the task.

This approach was professionally followed in the Greek–Latin translations of Burgundio of Pisa and James of Venice and the Arabic–Latin translations of Stephen the Philosopher and Gerard of Cremona. Stephen criticizes Constantine the African, the earlier translator of the same work that he is translating, for "missing out many necessary things, changing the orders of many things, and setting out some things in the wrong way,"[84] and promises to remedy all these faults in his new translation. Burgundio, in the preface to one of his translations, writes what is virtually a blueprint for literal translation, and an account of the history of its practice.[85] For both Burgundio and Stephen, the adoption of the literal method is necessary not only for

81 John of Salisbury, Letter 201, in *Letters*, ed. W. J. Millor, rev. Christopher N. L. Brooke (Oxford: Clarendon Press, 1979), vol. 2, p. 294; and John of Salisbury, *Metalogicon*, IV.6, p. 145.

82 Brian P. Copenhaver, "Translation, Terminology and Style in Philosophical Discourse," in *The Cambridge History of Renaissance Philosophy*, ed. Charles B. Schmitt (Cambridge: Cambridge University Press, 1988), pp. 77–110.

83 Thomas E. Burman, *Religious Polemic and the Intellectual History of the Mozarabs* (Leiden: Brill, 1994), p. 384.

84 Burnett, "Antioch as a Link between Arabic and Latin Culture in the Twelfth and Thirteenth Centuries," p. 27.

85 Classen, *Burgundio von Pisa*; also discussed in Charles Burnett, "Translating from Arabic into Latin in the Middle Ages: Theory, Practice, and Criticism," in *Éditer, traduire, interpréter: essais de méthodologie philosophique*, ed. Stephen G. Lofts and Philipp W. Rosemann (Louvain-la-Neuve: Peeters, 1997), pp. 55–78.

scientific reasons (i.e., to preserve the "truth" of the original text) but also for moral reasons. It is a mark of presumptuousness to dare to alter anything in the original text, and a nonliteral translation becomes the composition of the translator, no longer that of the original author. According to the author of the anonymous commentary on the *Timaeus* sandwiched between Henricus Aristippus's translations of Plato's *Phaedo* and *Meno*, "Cicero, in writing the *Republic*, had excerpted many things from Plato, but he made them his own through change of order and sense. [Plato's *Republic*] was untouched as far as translation was concerned."[86] Among the consequences of the literal method is the fact that the translators tend not to add anything of their own by way of prefaces and even neglect to mention their own names. Burgundio also states that literal translators add in the margins alternative translations of words that have two senses in the original language and rephrasings clarifying the sense of passages whose literal translation renders them obscure; they keep the text pure. Burgundio rigorously put his preaching into practice.[87]

Since Burgundio of Pisa stands at the center of the translation movement of the mid-twelfth century, it is not surprising to find that his pronouncements are representative of, if not prescriptive for, the translators of the latter half of the twelfth century and later. Among these is Gerard of Cremona; the fact that he originates from northern Italy like Stephen the Philosopher, Burgundio, and James of Venice and is roughly contemporary with them is probably significant. He wrote no prefaces to his translations, and, according to his students, did not even put his name on them. Thus, he left no account of his translation method. Yet, a consistent method appears to have been established for translating works in Toledo, probably under his supervision. This consisted in either making new translations or revising previous ones to ensure that a stringent and consistent word-for-word terminology was applied throughout,[88] and in adding an apparatus of marginal notes, giving alternative translations of words, rephrasing, and clarifying obscure points, in a way remarkably similar to that exemplified in Burgundio's translations.

The sheer volume and success of the Toledan translations guaranteed that the consistent terminology used in them became the scientific norm. It must be emphasized, too, that the literalness was not restricted to the letters of the text. Where there were diagrams and illustrations, these, too, were reproduced with extreme accuracy. The Toledan translation of the surgery of Albucasis (al-Zahrāwī) reproduced to the last detail the many surgical

86 Oxford, Corpus Christi College, MS 243, fol. 137va.

87 Fernand Bossier, "L'Élaboration du vocabulaire philosophique chez Burgundio de Pise," in *Aux origines du lexique philosophique européen*, ed. Jacqueline Hamesse (Louvain-la-Neuve: FIDEM, 1997), pp. 81–116.

88 Danielle Jacquart, "Note sur la traduction latine du *Kitāb al-Manṣūrī* de Rhazes," in Jacquart, *La Science médicale occidentale entre deux renaissances*, article X; and Abū Maʿšar al-Balḫī [Albumasar], *Liber introductorii maioris ad scientiam judiciorum astrorum*, 9 vols. (Naples: Istituto Universitario Orientale, 1995–1997), vol. 4, pp. 217–76.

instruments depicted in the text, even to the extent of preserving the turbans of the Muslim surgeons and the layout of the Arabic original.[89]

Although literal translation continued to be the norm throughout the thirteenth century, the annotations became more detailed, and Greek and Arabic explanations of the canonical texts were added as comprehensively as possible. Grosseteste's version of Aristotle's *Ethics*, with its panoply of Byzantine commentaries, is matched by Hermann the German's translation of Aristotle's *Rhetoric*, with the relevant commentaries of Averroes, Alfarabi, and Avicenna.

FROM *TRANSLATIO STUDII* TO *RESPUBLICA PHILOSOPHORUM*

Literal translation was perceived as representing the words of the original author so accurately as to make the original texts redundant. Such a view is already apparent in the words of Boethius that, "through the integrity of a completely full translation, no Greek literature is found to be needed any longer."[90] In his advertisement of new translations of Aristotle and his commentators to the students of Paris, King Manfred both emphasized that the "virginal integrity" of the original Greek and Arabic texts has been preserved by representing single words with single words and regarded the Latin versions as providing new clothes for texts that have long languished in their unchanged Greek garb.[91] Boethius's aim for a "transfer of learning" (*translatio studii*) from Greek into Latin in the sixth century was cut short by his untimely death and lack of immediate successors, but it was renewed and largely fulfilled in the twelfth to thirteenth centuries. The rationale of a *translatio studii* is that scientific culture passes from one people to another; once in possession of that culture, the recipients have no need to return to its source. Thus, once Latin scholars had acquired the wisdom of the Greeks and the Arabs, there was no need to learn the Greek or Arabic language to understand this wisdom. Roger Bacon was swimming against the current when he insisted that the only way to truly know theology was to learn Hebrew and Greek and, to know science, Arabic and Greek.[92] Most Western scholars, like Albertus Magnus and Thomas Aquinas, thought it sufficient to compare different Latin translations of the texts they were studying. They were confident that the translation preserved the "truth" of the original text

[89] Eva Irblich, "Einfluss von Vorlage und Text im Hinblick auf kodikologische Erscheinungsformen am Beispiel der Überlieferung der 'Chirurgie' des Abū'l Qāsim Khalaf Ibn 'Abbās al-Zahrāwī von 13. Jahrhundert bis 1500," in *Paläographie 1981*, ed. Gabriel Silagi (Münchener Beiträge zur Mediävistik und Renaissance-Forschung, 32) (Munich: Arbeo-Gesellschaft, 1982), pp. 209–31.

[90] Boethius, *In Isagogen Porphyrii Commentorum Editio secunda*, chap. 1.

[91] Gauthier, "Notes sur les débuts (1225–1240) du premier 'Averroïsme'," pp. 323–4.

[92] Roger Bacon, *Opus maius*, ed. John Henry Bridges, 2 vols. (Oxford: Clarendon Press, 1897), vol. 1, p. 66.

and, in the case of an Albertus Magnus, became so used to "translation Latin" (which, to those accustomed to classical Latin, like John of Salisbury, appeared obscure) that they were able to interpret the original author's meaning in a remarkably accurate way.

But it is probably more accurate to say that we are dealing here not so much with a transfer of learning as with the internationalism of scientific learning. Already Adelard had compared the world to a body in which different parts have been assigned different functions. In the same way, different parts of the world are fertile with different disciplines and what "the [world] soul is unable to effect in a single part of the world, she brings about within its totality."[93] In the next century, Theodore of Antioch went from Christian Antioch to Muslim Mosul, where he studied the works of Alfarabi, Avicenna, Euclid, and Ptolemy with the foremost Islamic scholar, Kamāl-al-Dīn ibn Yūnus (1156–1242); subsequently he studied medicine in Baghdad before serving a Seljuk ruler of Konya in Asia Minor, an Armenian regent, and finally a Christian emperor in Sicily. Another student of Kamāl-al-Dīn, al-Urmawī, also spent time at the same emperor's court, writing a book of logic for him. A Jewish scholar, Juda ben Salomon ha-Cohen, corresponded in Arabic from Toledo with Emperor Frederick II's "philosopher" on questions concerning geometry, and Frederick himself sent questions on mathematics and philosophy to Arabic scholars throughout the Mediterranean. Later in the same century, opinions of Thomas Aquinas were incorporated into an Arabic *apologia* for Christianity written by a Mozarab in Toledo, and an anonymous scholar wrote a Greek introduction to Aristotle's *libri naturales* in Sicily or southern Italy on the basis of Averroistic texts being taught in Paris.[94] Meanwhile, it has been suggested that astronomical information from the observatory of Maragha in the Mongol realms arrived at the Spanish court of Alfonso X.[95] That such scholarly exchanges and intellectual traffic were possible testifies to the fact that, at least by the second quarter of the thirteenth century, the Jewish and Islamic worlds shared with Christendom a common knowledge of science and philosophy; a commonwealth of scholars had come into being that transcended political and religious borders. That such a state could come about resulted, in no small measure, from the achievement of translators who raised the scientific culture of each linguistic group to the point where all shared the same level of excellence.

93 Adelard of Bath, *De eodem et diverso*, in Burnett, *Adelard of Bath*, pp. 70–1.

94 Burman, *Religious Polemic and the Intellectual History of the Mozarabs*, pp. 157–91; and Marwan Rashed, "De Cordoue à Byzance," *Arabic Sciences and Philosophy*, 6 (1996), 215–62.

95 Samsó, *Islamic Astronomy and Medieval Spain*, chap. XIII, pp. 32–3.

15

THE TWELFTH-CENTURY RENAISSANCE

Charles Burnett

THE IDEA OF A RENAISSANCE

Of all generalizations in medieval European history, that of a "twelfth-century Renaissance" has probably generated the most discussion, ever since Charles Homer Haskins launched it in 1927. If a cultural renaissance was taking place in Latin Europe, it apparently was not noticed by its neighboring civilizations. A twelfth-century Arabic physician in Syria, Usama ibn Munqidh, could speak disparagingly about the competence of Latin doctors, and the royal patron of scholars in Constantinople, Anna Comnena, would naturally refer to the Latins as "barbaroi."[1] Nor was the twelfth century seen by Western scholars post-factum as a period of spectacular advance: Petrarch (1304–1374) defined it as part of the "dark ages." Not before the mid-nineteenth century was the existence of a "twelfth-century renaissance" hinted at, and it was Haskins who developed the theme to the greatest extent, and, ever since the publication of his *The Renaissance of the Twelfth Century*, historians have felt obliged to discuss the concept, either in the course of general history or as a topic for a separate book.[2]

[1] Bernard Lewis, *The Muslim Discovery of Europe* (London: Weidenfeld and Nicolson, 1982), pp. 222–3; and Anna Comnena, *Alexiade*, X.8.5, ed. Bernard Leib, 4 vols. (Paris: Les Belles Lettres, 1937–1976), vol. 2, p. 218. For the Byzantines' lack of mention of the material and cultural expansion of Europe in the twelfth century, see Paul Magdalino, *The Empire of Manuel I Komnenos, 1143–80* (Cambridge: Cambridge University Press, 1993), chap. 5.

[2] The contribution of Charles Homer Haskins – *The Renaissance of the Twelfth Century* (Cambridge, Mass.: Harvard University Press, 1927) – was commemorated in Robert L. Benson and Giles Constable, with Carol D. Lanham, eds., *Renaissance and Renewal in the Twelfth Century* (Oxford: Clarendon Press, 1982). For the extensive literature on the concept of a twelfth-century renaissance preceding this commemorative volume, see Gerhart B. Ladner, "Terms and Ideas of Renewal," in Benson and Constable, *Renaissance and Renewal in the Twelfth Century*, pp. 1–33, especially pp. 31–2. Of books subsequent to Benson and Constable's volume, important contributions are the chapters on the renaissance of the twelfth century in Marcia L. Colish, *Medieval Foundations of the Western Intellectual Tradition, 400–1400* (New Haven, Conn.: Yale University Press, 1997), pp. 175–82; Alain de Libera, *La philosophie médiévale* (Paris: Presses Universitaires de France, 1993), pp. 310–12; Françoise Gasparri, ed., *Le XIIe siècle: Mutations et renouveau en France dans la première moitié du XIIe siècle*

There is no doubt that scholars in the late eleventh and early twelfth centuries were aware that something new was being introduced into Western culture. Adelard of Bath, in the 1120s, represents his patron as requesting "some new item of the studies of the Arabs," and Stephen ("the disciple of Philosophy"), in introducing a new cosmology, defends novelty against his critics.[3] One such critic was Hermann of Tournai (late eleventh century), who was alarmed to see teachers like Reimbert of Lille read "their own new inventions" into the texts of Porphyry and Aristotle.[4]

A sense of change was certainly in the air, and historians have given several accounts of the symptoms of, and the reasons for, this change. The eleventh century saw dramatic increases in both agricultural production and population. Consequently, people were no longer tied to the land and could devote themselves to crafts, trade, and scholarship. Communities of craftsmen and other categories of the "middle class" coalesced into towns. The growth of the population, its increasing diversity, and new towns required social and economic regulation, which led to greater literacy and numeracy.[5] Moreover, culture became progressively more secular and spilled over into the vernacular languages. At the same time, since its split from the Eastern church in 1054, the Western church was consolidating itself both spiritually and administratively. A church hierarchy was established, confirming the pope as the supreme leader of a single organization. The effects of this organization percolated to every level of society, especially as a result of the reforms of Pope Gregory VII (1073–85), which, among other things, required every parish priest to know sufficient Latin to say the mass correctly.[6]

Moreover, the late eleventh century was a period of territorial expansion and state-formation – developments that stimulated the drawing up of law codes and ecclesiastical structures (e.g., in the new administrative regions of the Crusader States and the reconquered regions of Spain). Travel from one

(Paris: Le Léopard d'Or, 1994); Jacques Verger, *La Renaissance du XIIe siècle* (Paris: Les Éditions du Cerf, 1996); and R. N. Swanson, *The Twelfth-Century Renaissance* (Manchester: Manchester University Press, 1999).

3 Adelard, *Questions on Natural Science*, in Adelard of Bath, *Conversations with His Nephew*, ed. and trans. C. Burnett with the collaboration of Italo Ronca et al. (Cambridge: Cambridge University Press, 1998), p. 83; Stephen the Philosopher, introduction to *Liber Mamonis*, in Charles Burnett, "Antioch as a Link between Arabic and Latin Culture in the Twelfth and Thirteenth Centuries," in *L'Occident et le Proche-Orient au temps des croisades: traductions et contacts scientifiques entre 1000 et 1300*, ed. Anne Tihon, Isabelle Draelants, and Baudoin van den Abeele (Turnhout: Brepols, 2000), pp. 1–78, reprinted with corrections in Charles Burnett, *Arabic into Latin in the Middle Ages* (Farnham: Ashgate Variorum, 2009), article IV.

4 Constant J. Mews, "Philosophy and Theology, 1100–1150: The Search for Harmony," in Gasparri, *Le XIIe siècle*, pp. 159–203 at p. 163.

5 On literacy, see Brian Stock, *The Implications of Literacy: Written Language and Models of Interpretation in the Eleventh and Twelfth Centuries* (Princeton, N.J.: Princeton University Press, 1983). On numeracy, see Alexander Murray, *Reason and Society in the Middle Ages* (Oxford: Clarendon Press, 1978).

6 Harold J. Berman, *Law and Revolution: The Formation of the Western Legal Tradition* (Cambridge, Mass.: Harvard University Press, 1983); and Giles Constable, *The Reformation of the Twelfth Century* (Cambridge: Cambridge University Press, 1996), pp. 94–119.

end of Christendom to the other became possible as never before, and long journeys were often undertaken for the sake of pilgrimage or for diplomatic or ecclesiastical reasons (such as visits to the papal curia) – or, indeed, for sheer adventure. Scholars were particularly keen to travel and received special privileges to make their peregrinations more secure.[7] Political contacts between the Norman dominions of England, northwest France, Sicily, and Antioch also provided opportunities for travel, as did the connections of the provinces of Spain with France and of Portugal with England. An epitome of civil law appearing in Provençal in the 1140s (*Lo codi*) was translated into Latin soon afterward, probably in Pisa, and became an important source for the evolution of civil law in the Kingdom of Jerusalem.[8] The larger geographical horizons of twelfth-century Europe opened up larger intellectual horizons, and St. Paul's Abbey in Antioch (Syria) or the cathedral of Syracuse (Sicily) could be considered as worthwhile a destination for a scholar as the cathedral of Laon or the Abbey of St. Victor.[9] A common scholarly language and a universal religion made possible a cultural community that spread from Stavanger to Palermo, from Lisbon to Edessa.[10]

THE SYSTEMATIZATION OF ADMINSTRATION AND LEARNING

Both the secular and the sacred spheres in twelfth-century Europe were the beneficiaries of systematizing tendencies. Civil law provides a striking example, which can be taken as archetypal of the trends of the time. The early Middle Ages had inherited fragments of Roman law, which had adapted and blended with local traditions in different ways in different places. The recovery of a manuscript of Justinian's corpus of Roman law (the *Digest*, also known by its Greek name *Pandectae*) in Italy in the third quarter of the eleventh century (the "Florentine codex") marked a watershed. The new text, which together with Justinian's *Institutiones* and *Codex* made up the *Corpus iuris civilis*, quickly came to be seen as the ideal law and therefore the gauge and measure for all civil law henceforth. The *Digest* was arranged by the decrees of the Roman praetors, to each of which several different

7 For example, Emperor Frederick I (Barbarossa)'s privilege, drawn up between 1155 and 1158, which placed under imperial protection all who studied and traveled for the sake of studying throughout the whole kingdom; see Winfried Stelzer, "Zum Scholarenprivileg Friedrich Barbarossas (Authentica "Habita")," *Deutsches Archiv*, 34 (1978), 123–65.

8 Swanson, *Twelfth-Century Renaissance*, p. 72.

9 The words of Fulcher of Chartres (1120s) are significant: "For we who were Occidentals have now become Orientals.... He who was of Rheims or Chartres has now become a citizen of Tyre or Antioch." See Fulcher of Chartres, *A History of the Expedition to Jerusalem, 1095–1127*, III. 37, ed. H. S. Fink, trans. F. R. Ryan (New York: W. W. Norton, 1969), p. 271.

10 This community of culture is well expressed in David Knowles, *The Evolution of Medieval Thought*, 2nd ed. (London: Longman, 1988), pp. 72–84.

lawyers added comments. A copy of the Florentine codex was introduced into Bologna before the end of the eleventh century and became the basis for successive glosses by Bolognese lawyers, each generation of which introduced a greater level of systematization.[11]

Canonists or church lawyers sought to bring the same level of systematization to the decrees of popes; this led to the digests of Ivo, bishop of Chartres (d. 1116) and author of the *Panormia*, and Gratian (his *Decretum* of the 1140s). But systematization can be observed in other spheres, too. It could take the form of ordering the statements of (often conflicting) authorities by topic, as in Peter Abelard's *Sic et Non* (first draft 1121) and Peter Lombard's *Sentences* (ca. 1150) for theology, Abraham ibn Ezra's *Foundations of Astronomical Tables* (1154) for different astronomical opinions, and the *Book of the Nine Judges* (ca. 1175) for astrology. Or an author might provide his own synthesis of previous learning, such as Hugh of St. Victor for religion in his *On the Sacraments of the Christian Religion* (1137) and Constantine the African's *Pantegni* (before 1087)[12] for medicine. Several of these texts aspired to all-encompassing coverage, employing in their titles either the Greek "*pan*" ("all") or the Latin "*summa.*"[13]

THE RECOVERY OF ROMAN AND GREEK CULTURE

The ready adoption of Roman law points to another feature of the eleventh to twelfth centuries, a Romanization of society and culture as a whole, which can be observed in many spheres. Classical Latin texts were rediscovered, read, or copied again in large numbers.[14] Already in the late tenth century, Gerbert d'Aurillac had established a curriculum in logic, consisting of a coherent body of classical texts by Boethius,[15] and during the eleventh century the quadrivium began to be taught on the basis of the genuine works of Boethius (for arithmetic and music), Euclid (the first four books of his *Elements* in Boethius's translation, although without their proofs), and the

11 Berman, *Law and Revolution*, pp. 120–64.

12 Constantine describes himself as the "unifier" of material from the books of many authors ("*ex multorum libris coadunator*"; MS British Library, Add. 22719, fol. 4v). The *Pantegni* is primarily Constantine's adaptation of an Arabic compendium, *The Complete Book of the Medical Art*, by ʿAlī ibn al-ʿAbbās al-Majūsī, but Constantine replaced many of al-Majūsī's chapters with material from other medical works. Later in the twelfth century, a similarly "unifying" book, Avicenna's *Canon of Medicine* (translated before 1187) began to replace Constantine's *Pantegni*.

13 See Richard H. Rouse and Mary A. Rouse, "*Statim inveniri*, Schools, Preachers, and New Attitudes to the Page," in Benson and Constable, *Renaissance and Renewal in the Twelfth Century*, pp. 201–25, especially p. 206.

14 A convenient résumé of the number of authors rediscovered in the eleventh and twelfth centuries is given in the introduction to Leighton Durham Reynolds, ed., *Texts and Transmission* (Oxford: Clarendon Press, 1983).

15 Aberrant texts such as the *Categories* attributed to Apuleius and Augustine dropped out of use; see Margaret T. Gibson and Lesley Smith, eds., *Codices Boethiani*, 4 vols. (London: Warburg Institute, 1995–2009), vol. 1, pp. 2–4.

supposed works of Ptolemy (his "Astrolabe" and introduction to planetary tables, the "*Preceptum Canonis,*" translated into Latin in 534–5). In the Benedictine abbey of Monte Cassino alone, monks in the eleventh century copied several hitherto totally unknown texts, including works of Apuleius, Seneca's *Dialogues*, Tacitus's *Annals*, and Varro's *De lingua latina*, even as one of them wrote the first "textbook" on writing good Latin, Alberic of Monte Cassino's *Ars dictaminis* (late eleventh century).

This increased use of classical texts went hand in hand with an attempt to imitate classical style in writing Latin.[16] The inculcation of good Latin style was also achieved by commenting on the classical *auctores*. A new tradition in doing this can be seen in the late-eleventh-century *Glosule* ("Little glosses") to Priscian's *Grammatical Institutes*. This practice culminated in the teaching method of Bernard of Chartres and his pupils and colleagues, William of Conches, Thierry of Chartres, and Bernardus Silvestris, who commented on Plato's *Timaeus*, Virgil's *Aeneid*, Cicero's *De inventione*, Pseudo-Cicero's *Rhetorica ad Herennium*, Macrobius's *Dream of Scipio*, and Martianus Capella's *Marriage of Mercury and Philology*. These writers cultivated a self-consciously literary style that is especially manifest in their elaborate prefaces. The admiration for classical texts went as far as the evocation of nonexistent authorities such as "Plutarch's *Education of Trajan*" in John of Salisbury's *Policraticus*.

Romanism was also expressed in architecture, from the "romanesque" style of church-building to Abbot Desiderius's (1058–87) reuse of Roman columns in his rebuilding of the Monte Cassino basilica. The city of Rome, too, was revived, and the pope was invested with new powers whose resemblance to those of the ancient emperors was not coincidental.[17]

But awareness of the Latin classics involved the acknowledgment that the Romans were indebted to the Greeks, and an unmediated interest in Greek culture, whether that of the past or of contemporary Byzantium, also characterizes the period. Abbot Desiderius used Byzantine artists to construct his new basilica. Thierry of Chartres added a Greek veneer to his "Library of the seven liberal arts" (*Heptateuchon*) by writing titles in Greek letters.[18] Indeed, the use of Greek or pseudo-Greek titles for original works became popular. Aside from the already-mentioned works with "Pan-" in

[16] Different authors chose different classical styles; see Janet Martin, "Classicism and Style in Latin Literature," in Benson and Constable, *Renaissance and Renewal in the Twelfth Century*, pp. 537–68. All these styles may be distinguished from the monastic, "new Christian" sermon style used in homiletic and Christian texts by authors such as Peter the Venerable and Bernard of Chartres.

[17] The Greek envoy of John Comnenus at the court of the German emperor Lothair II in 1137 is claimed to have said that "the Roman pontiff was no bishop, but an emperor." See Kenneth M. Setton, "The Byzantine Background to the Italian Renaissance," *Proceedings of the American Philosophical Society, section C*, 100 (1956), pp. 1–76 at p. 24.

[18] A practice extended by Hermann of Carinthia, his pupil, to the title of a text translated from Arabic; see Charles H. Haskins, *Studies in the History of Mediaeval Science*, 2nd ed. (Cambridge, Mass.: Harvard University Press, 1927), p. 45.

their headings, one may mention *Dragmaticon*, *Heptateuchon*, *Metalogicon*, *Policraticus*, and the popularity of the term *Isagoge* for "introduction."[19] But scholars also attempted to learn more about Greek language and literature, as witnessed by the endeavors of John of Salisbury to collaborate with Greek scholars, the discussion of Greek grammar in the works of Hugo Eterianus (d. 1182), and the translations made from Greek, which reached a crescendo in the mid-twelfth century.[20]

THE WIDENING BOUNDARIES OF *PHILOSOPHIA*

The whole compass of the secular arts of Greece and Rome was regarded as "*philosophia*." Boethius had provided a striking and popular image of its integral nature at the beginning of his *Consolation of Philosophy*, in which Philosophia is personified as a woman with a gown made up of theory and practice, who complains about those who tear off only portions of her gown. The significance of this image is underlined by William of Conches in his commentary on the *Consolation of Philosophy* and Adelard's emphasis in his introduction to *philosophia* – his *On the Same and the Different* – that none of the seven liberal arts could be studied without all the others.[21] *Philosophia* was not confined to the seven liberal arts but embraced the whole of what was knowable, and it is a characteristic of the scholars who restored the seven liberal arts that they also included in their works discussions of topics that originally fell outside the purview of those arts, such as medicine, the natural sciences, and cosmology. Both Adelard and his follower Daniel of Morley state that the man who does not know about the world in which he lives is unworthy to inhabit it and should be cast out (if that were possible).[22] In contemporary divisions of medicine and the natural sciences, the science of the microcosm (the "small world" of man) is related to the science of the macrocosm (the "large world" of the universe) as two branches of natural science.[23] William of Conches sees no incongruity in discussing these

19 Porphyry's *Isagoge* to logic had set a precedent, which was followed in translations of texts on medicine (the *Liber ysagogarum* of Johannitius, translated from the Arabic in the late eleventh century), astrology (the *Ysagoga minor* of Abū Maʿshar, translated by Adelard of Bath), and theology (the *Ysagoge in theologiam* of Odo).

20 Haskins, *Studies in the History of Mediaeval Science*, especially pp. 141–54, 194–222.

21 William of Conches, *Glosae super Boetium*, ed. Lodi Nauta (Turnhout: Brepols, 1999), p. 34 ("They tear apart the arts who think that one of them alone, or some of them without others can make a perfect philosopher, saying that any of the arts is superfluous"); and Adelard of Bath, *On the Same and the Different*, in Adelard, *Conversations with His Nephew*, pp. 6–7 ("one could not see any of [the maidens personifying the seven liberal arts] unless one looked at them all at once") and pp. 70–1 ("the one I had read would bring no benefit if what remained was lacking").

22 Charles Burnett, *The Introduction of Arabic Learning into England* (London: The British Library, 1997), pp. 31, 63.

23 For example, in the division of science in the anonymous text "Ut testatur Ergafalau," in *Adelard of Bath: An English Scientist and Arabist of the Early Twelfth Century*, ed. Charles Burnett (London: The Warburg Institute, 1987), p. 145.

subjects in the context of commentaries on classical texts (the *Consolation of Philosophy* and the *Timaeus*)[24] and applies the term "*philosophia*" to a systematic description of the microcosm and macrocosm in his *Dragmaticon philosophiae*. The attainment of knowledge of the whole of the natural sciences was a possibility to be striven for by human beings and could even be described as a way of becoming like God.[25]

This humanistic interest in cosmology as an extension of *philosophia* is particularly noticeable among scholars associated with the cathedral school at Chartres in the early twelfth century. A brilliant succession of scholars at this cathedral school revived the reputation of the school established early in the previous century by Bishop Fulbert (1006–28). The first of these was Bernard (chancellor 1119–26), who famously compared the endeavor of his contemporaries to "dwarfs standing on the shoulders of giants" (referring to the classical authors of the past) and whose teaching was contemporary with that of Ivo, the bishop who codified canon law. His pupils and successors as teachers included William of Conches, Thierry of Chartres, and Gilbert of Poitiers, the first two being particularly interested in cosmological questions. As all three scholars encouraged translations of works from Greek and Arabic, some of the earliest manuscripts of these translations were brought to, and copied in, the cathedral.[26] Thierry's *Heptateuchon*, from the early 1140s, is the best testimony to the spirit and ideals of the *philosophia* of the first half of the twelfth century. Lost since World War II,[27] it contained copies of the traditional Latin texts on the seven liberal arts alongside newly rediscovered texts (Boethius's translation of the *Topics* and *Sophistical Refutations* of Aristotle) and newly made translations from Arabic, such as Adelard's translations of Euclid's *Elements* and the astronomical tables of al-Khwārizmī.[28]

Other like-minded scholars had associations with Chartres. Bernardus Silvestris of Tours dedicated to Thierry his *Cosmographia*, which used the prosimetron form of Boethius's *Consolation of Philosophy* to describe the macrocosm and microcosm. Raymond of Marseilles not only introduced instructions for the use of the astronomical tables of az-Zarqallu (the "Tables of Toledo") with 165 lines of verse, replete with classical allusions, but also wrote a defense of the science of the stars in an elaborate style, with many quotations from Ovid, Virgil, and especially Lucan; although he wrote these texts in Marseilles, the earliest manuscript of his astrological work appears to

24 See Nauta, introduction to William of Conches, *Glosae super Boetium*, pp. xliii–lxii.

25 William of Conches, *Glosae super Boetium*, p. 20: "Philosophy is seen by Boethius to be rising above his head because philosophy makes man ascend above the nature of man, i.e. deifies him."

26 Charles Burnett, "The Contents and Affiliation of the Scientific Manuscripts Written at, or Brought to, Chartres in the Time of John of Salisbury," in *The World of John of Salisbury*, ed. Michael Wilks (Studies in Church History, Subsidia 3) (Oxford: Blackwell, 1984), pp. 127–60.

27 Not only is it now no longer extant, except in the form of photographs, but in addition one of its two volumes had lost 104 folios at an early stage in its existence.

28 The *Elements* are without proofs, and the astronomical tables without instructions for use, which suggests that the texts were assembled as a literary collection rather than for scientific use.

have been in Chartres (the now lost MS 213 of the Bibliothèque municipale), alongside the earliest copies of Adelard's translations.

A similar spirit was present in Monte Cassino and Salerno: Alberico of Monte Cassino established rules for Latin prose-composition; Alfano, the archbishop of Salerno (1058–85), wrote both accomplished poetry and original works on medicine, and, before the middle of the twelfth century, a certain "Marius Salernitanus" was writing books on the elements (*De elementis*), on man (*De humano proficuo*), and (probably) on the cosmos (*Alcantarus de philosophia mundi*).[29]

The same humanistic spirit imbues the works of those seeking to extend the boundaries and fill in the details of *philosophia*. Hermann of Carinthia and his colleague Robert of Ketton not only wrote elaborate prefaces to their translations from Arabic, but Hermann also put the astrology of Abū Maʿshar into fine Sallustian prose, substituting examples from Aratus and Boethius where they served to illustrate the argument, while Robert of Ketton wove the short exhortatory phrases of the Qurʾān into the rolling periods of Ciceronian prose. Hermann boasted of being a pupil of Thierry of Chartres, but the same concern for good Latin and the integrity of learning can be seen in translators not associated with Chartres. In his *Pantegni*, for example, Constantine the African adapted the style and contents of the Arabic medical text that was his source to the expectations of a literate Latin audience by beginning with a discussion of the place of "literary medicine" ("*medicina litteralis*") in the Platonic division of letters into logic, ethics, or physics.[30] Even Stephen of Antioch, who provided a more literal translation of the same Arabic text in the 1120s, wrote elaborate literary prefaces. In these, he avowed his own intention to progress from the cures of the body, to which medicine is devoted, to the cures of the soul, to which philosophy is dedicated,[31] and alluded repeatedly to Boethius's *Consolation of Philosophy*. In his original work, the *Liber Mamonis*, he cited Lucan and Cicero and criticized Macrobius as the chief, but unworthy, representative of cosmological knowledge for his contemporaries in France.

THE RISE OF SPECIALIZATION

Thus, from the eleventh century through the first half of the twelfth century, we see a new intellectual self-confidence developing, a belief that man can

29 For the relationship between these works, see Charles Burnett, "The Works of Petrus Alfonsi: Questions of Authenticity," *Medium Ævum*, 66 (1997), 42–79, especially pp. 56–61.

30 Danielle Jacquart, "Le sens donné par Constantin l'Africain à son œuvre: les chapitres introductifs en arabe et en latin," in *Constantine the African and ʿAlī ibn al-ʿAbbās al-Maǧūsī: The Pantegni and Related Texts*, ed. Charles Burnett and Danielle Jacquart (Leiden: Brill, 1994), pp. 71–89 at p. 84. This discussion is entirely lacking in the Arabic text of al-Majūsī.

31 Burnett, "Antioch as a Link between Arabic and Latin Culture in the Twelfth and Thirteenth Centuries."

assimilate the whole of the natural sciences, and the concomitant belief that the same classical Latin style (whether that of Cicero, Seneca, or Sallust) can be applied to every branch of human knowledge. In the second half of the twelfth century, however, this integrity of the sciences started to break down in the face of increasing specialization and the establishment of technical vocabularies and styles that could be understood only by specialists in the various fields. Adelard of Bath implied that there are specialists in different fields when he ended his *On the same and the different* with an elaborate metaphor concerning the necessity of traveling great distances to find different bits of knowledge because not everything can be found in the same place. One of the first specialists was perhaps Adelard's colleague Petrus Alfonsi, who, in a letter addressed to the "Peripatetics of France" apparently written shortly after 1116, promises lectures in astronomy of a character heretofore unheard of in the Latin world. John of Salisbury, in his *Metalogicon* (1159), described his experience of learning different subjects from different teachers in Paris and Chartres, but he was critical of what he saw as the current trend of devoting all one's attention to logic, law, or medicine and failing to aim for a general knowledge of philosophy. He was nostalgic for the time of scholars such as Thierry, whose breadth of learning reminded him of Aristotle's.[32]

The introduction of so many new and more advanced texts in most fields of learning during the course of the twelfth century made it simply impossible for any one person to master them all. Although the seven liberal arts remained the ideal basis of Western university education, a divide began to open up between the elementary textbooks used by the students and the advanced books usable only by specialists. Greater specialization went hand in hand with the development of technical vocabularies and styles of Latin appropriate to each science. The literary translations of the first half of the century gave way to literal translations. Only Henricus Aristippus, translating Plato, the most literary of philosophers, was still attempting literary translations in the 1160s, and these were largely ignored by his contemporaries. In the case of translations of Greek and Arabic scientific works, it was regarded as obvious that the translation should be strictly literal, that there should be consistency in the use of technical vocabulary, and that, if Latin could not provide an appropriate equivalent, a calque or a transliteration of the original word should be accepted. When astronomers, physicians, and other professionals started to compose original works, they used not only the accepted technical terms but also the style of the translations. It became

[32] John of Salisbury, *Metalogicon*, I.5.10–11. The same criticism is voiced by Dominicus Gundissalinus in his introduction to his *On the division of philosophy*, which begins with the nostalgic reference to a former "happy age" that produced so many scholars compared with the present age, in which men study rhetoric and pursue worldly ambitions rather than the sciences. See Dominicus Gundissalinus, p. 373: *De divisione philosophiae*, ed. Ludwig Baur (Beiträge zur Geschichte der Philosophie des Mittelalters, 4.2–3, (Münster: Aschendorff, 1903), p. 3.

both unnecessary and inappropriate to quote Ovid, Cicero, or Lucan when writing about astronomy, and a rhetorical style was held in suspicion. John of Salisbury's complaints bear witness to the change. He criticizes not only specialization but also the literal style of translating, and he accuses its practitioner, James of Venice, of inadequate training in Latin grammar.[33] But it was James of Venice's ungrammatical translation of Aristotle's *Posterior Analytics* that became canonical, rather than the translation apparently commissioned by John of Salisbury, with its preface and elegant Latin style.[34]

Although John of Salisbury respected Aristotle for the comprehensiveness of his philosophy, it was precisely the words of Aristotle that justified the separation of the sciences. According to the *Posterior Analytics*, each science had its own premises and "we cannot ask every question of each individual expert . . . but only such questions as fall within the scope of his own science."[35] And it is perhaps no coincidence that the shift from an all-embracing scientific humanism in the first half of the twelfth century to specialization in the second half occurs at the same time as a shift from regarding Plato as the prince of philosophers to placing Aristotle in that role.[36]

Concomitant with specialization was the establishment and definition of new disciplines. Particularly significant are the sciences of law and theology, to both of which higher university faculties were devoted. The *Corpus iuris civilis* was fully recovered in the third quarter of the eleventh century, but it was the twelfth-century glossators and commentators on the law who made a "science" out of the material.[37] During the same century, the Bible was subjected to a similar analysis, which led to the production of the *Glossa ordinaria*.

The university system, with its separate faculties, was a cause and symptom of the specialization and professionalization in the arts and sciences. The cathedral, monastic, and court schools, which had been the centers for education until the mid-twelfth century, gave way, at the level of higher education, to the new systems of teaching that became institutionalized in the thirteenth century as universities. The beginnings of this new system can be seen in Bologna, where experts in the newly recovered *Corpus iuris civilis* attracted so many students that the students formed themselves into guilds on the basis of national origins and drew up regulations for their own and their teachers' conduct. It is probably not by chance that "*societas*" (the relationship of a master to his pupils) and "*universitas*" (the general body of students

[33] John of Salisbury, *Letter 201*, in *The Letters of John of Salisbury*, ed. W. J. Millor and C. N. L. Brooke, 2 vols. (London: T. Nelson, 1955; Oxford: Clarendon Press, 1979), vol. 2, p. 294.

[34] The latter translation is edited in John of Salisbury, *Aristoteles Latinus*, IV.1–4, ed. Lorenzo Minio-Paluello and Bernard G. Dod (Bruges: Desclée de Brouwer, 1968); see p. xliv for the preface.

[35] Aristotle, *Posterior Analytics*, 1.12, ed. and trans. H. Tredennick (London: William Heinemann, 1960), pp. 77–9.

[36] See later in this chapter.

[37] This is demonstrated at length in Berman, *Law and Revolution*, pp. 120–64.

and teachers) were originally terms from Roman law. In Paris and Oxford, the universities developed because of regulatory initiatives proposed by the teachers themselves. In Paris, at first, several independent masters set up schools and competed for students (as Peter Abelard testified most eloquently in his autobiography – *Historia calamitatum*), and it was, perhaps, precisely their mutual criticism and intellectual rivalry that sharpened wits, advanced new theories, and stimulated the introduction of new authorities.[38] This situation contrasts with the early-medieval monastic and cathedral schools, which emphasized instead the continuity of the tradition of learning. The Augustinian canons regular took a more vigorous attitude toward teaching, manifest especially in the school of the House of St. Victor in Paris in the period of Hugh (d. 1141).[39] But when this house closed its doors to outsiders after Hugh's death, the schools of the university became the main center of learning in Paris (see Shank, Chapter 8, this volume).[40]

THE REFINEMENT OF LANGUAGE

When we consider what is important in the changes of the twelfth century, specifically for the development of scientific discourse, we must bear in mind first of all that this discourse was in Latin. Communication in any branch of the sciences required mastery of Latin, honed into a form that could express ideas and technical points clearly and precisely. The principal medieval guide to the intricacies of the Latin language was Priscian's *Grammatical Institutes* ("*Priscianus maior*"). In the preface to the *Glosule*, the anonymous commentator already heralds a new approach to words and their relationship to ideas, stating that "grammar comes first, because one ought to know how to make appropriate joining of words before truth or falsehood [i.e., dialectic] or the decoration of eloquence [i.e., rhetoric] is learned."[41] The *Glosule* not only inaugurated a new interest in, and tradition of, commenting on Priscian but also prompted reflection on the nature of the significant utterance (*vox*) and its relationship to reality (*res*). The adequacy of human language to reflect reality has always been a basic concern of philosophers, and one with special implications for the natural sciences. Are the patterns that we see in cosmology, for example, merely human conventions that we impose on the

[38] That the sophists of fifth-century B.C. Greece also had to compete in the marketplace for their pupils has been adduced as the primary reason for the rapid advance in science in Greece at that time. See G. E. R. Lloyd, *The Revolutions of Wisdom: Studies in the Claims and Practice of Ancient Greek Science* (Berkeley: University of California Press, 1987).

[39] See C. Stephen Jaeger, "Humanism and Ethics at the School of St. Victor in the Early Twelfth Century," *Mediaeval Studies*, 55 (1993), 51–79.

[40] See Stephen C. Ferruolo, *The Origins of the University: The Schools of Paris and Their Critics, 1100–1215* (Stanford, Calif.: Stanford University Press, 1985), pp. 27–44.

[41] Margaret Gibson, "The Early Scholastic 'Glosule' to Priscian, 'Institutiones Grammaticae': The Text and Its Influence," *Studi medievali*, 20 (1979), 235–54.

universe, or do they reflect the realities of creation? On a practical level, the desirability of a one-to-one relation between a word and what it signifies is undoubted, for it encourages a consistent scientific terminology. The first thing to be established was that the human word was an adequate tool for describing reality, which was already the preoccupation of St. Anselm in his *On the Grammarian* (ca. 1060–3).

While Priscian and the more elementary manuals that preceded study of his work taught correct Latin, the widening range of classical Latin authorities enriched the Latin writer's vocabulary. The vogue for glossing classical texts was a way to define meanings precisely.[42] The very title of Martianus's fifth-century work *The Marriage of Philology and Mercury* called attention to the importance of correct language or "philology" – Mercury standing for the sciences. Horace's *Epistles* and *Satires* in particular were taught for their ability to inculcate a good command of Latin.[43] It is no accident that the leading scholars of the early twelfth century were described as "grammarians." Bernard of Chartres and his pupil William of Conches were both primarily "*grammatici.*" "Grammar" had a wider sense than it does now, including what we may term literary criticism, but the use of the word does indicate the emphasis placed on language during this period.

The sources that Latin writers used to broaden their terminology included the works of classical writers such as Boethius for arithmetic, geometry, and music, Firmicus, Calcidius, and Macrobius for the science of the stars, and Pliny for medicine – but also new translations from Greek and Arabic.[44] The struggle to find the most appropriate Latin vocabulary can be seen when one studies the techniques of the translators.[45] In astronomy, for example, a wide range of terms was at first used for the same concept, some being transliterations, others calques, yet others classical terms, but by the end

42 According to William of Conches in his *Glosae super Platonem*, the correct term for a commentary that explains the individual words is a *glosa*: "[A] *commentum* deals only with the *sententia* [the meaning]. It says nothing about the *continuatio* [context] or the *expositio* of the *littera*. But a *glosa* takes care of all these factors." See William of Conches, *Glosae super Platonem*, ed. Edouard Jeauneau (Paris: Vrin, 1965), p. 67; and also Nikolaus M. Häring, "Commentaries and Hermeneutics," in Benson and Constable, *Renaissance and Renewal in the Twelfth Century*, p. 179.

43 See S. Reynolds, *Medieval Reading: Grammar, Rhetorica and the Classical Text* (Cambridge: Cambridge University Press, 1996).

44 For the development of Latin terminology in different subject areas, see F. A. C. Mantello and A. G. Rigg, eds., *Medieval Latin Studies: An Introduction and Bibliographical Guide* (Washington, D.C.: Catholic University of America Press, 1996). While scholars were establishing Latin as a language for science, Hebrew was being adapted for this purpose by Abraham bar Ḥiyya (d. ca. 1136) and Abraham ibn Ezra (1089–1164). See Y. Tzvi Langermann, "Science in the Jewish Communities of the Iberian Peninsula," in his *The Jews and the Sciences in the Middle Ages* (Aldershot: Variorum, 1999), article I, especially pp. 10–16.

45 For some medical examples, see Danielle Jacquart, "L'enseignement de la médecine: quelques termes fondamentaux," in *Méthodes et instruments du travail intellectuel au Moyen Âge: Études sur le vocabulaire*, ed. Olga Weijers (Turnhout: Brepols, 1990), pp. 104–20; and Danielle Jacquart, "De l'arabe au latin: l'influence de quelques choix lexicaux (*impressio, ingenium, intuitio*)," in *Aux origines du lexique philosophique européen*, ed. Jacqueline Hamesse (Louvain-la-Neuve: Fédération Internationale des Instituts d'Études Médiévales, 1997), pp. 165–80.

of the twelfth century, a relatively uniform vocabulary had emerged. For example, for the small circle on which the body of the planet was supposed to be mounted in the Ptolemaic system, called an "epicycle," Adelard of Bath (ca. 1126) transliterated the Arabic term "elthedwir" (i.e., Arabic "al-tadwīr" from the root "d-w-r" = "roundness"), whereas his contemporary Stephen the Philosopher invented a Latin calque on the Arabic ("*circulus rotunditatis*"). Not until mid-century did the term used by Calcidius and Martianus Capella – "*epicyclus*" – achieve regular use. Several alternative terms are sometimes given for a single word in the original, either as doublets within the text or as marginal glosses. This served not only to achieve a closer approximation to the original sense but also to enrich Latin vocabulary.

The main concern of translators from Greek and Arabic from about the second quarter of the twelfth century onward was for terminological accuracy, as can be seen in Burgundio of Pisa's translations of Galen and of Aristotle's works on the natural sciences,[46] and in Gerard of Cremona's translations of Arabic texts on philosophy, mathematics, and medicine.

From the mid-eleventh century, aids for studying texts were also devised, including Papias's *Elementarium doctrinae erudimentum*, the first alphabetically ordered dictionary, preceding by a few decades the first alphabetically ordered herbals.[47] The *accessus ad auctores* added at the beginning of books asked the questions: "[W]hat is the subject-matter, the method of dealing [with the subject-matter], the aim, the reason for writing the book, and its usefulness? To which part of philosophy does it belong, who is its author, and what is its title?"[48] Latin terms were further defined by providing them with glosses of vernacular equivalents.[49] Such combined efforts of grammarians and translators ensured that Latin could be used with clarity and precision in any branch of human knowledge.

THE DEVELOPMENT OF METHODS OF SCIENTIFIC ARGUMENT

Even more striking than developments in grammar is the extraordinary interest in logic (dialectic) from the late eleventh century onward. An early-twelfth-century manuscript from Saint-Évroul contains a poem (perhaps

46 Fernand Bossier, "L'Élaboration du vocabulaire philosophique chez Burgundio de Pise," in Hamesse, *Aux origines du lexique philosophique européen*, pp. 81–116.

47 Lloyd W. Daly, *Contributions to a History of Alphabetization in Antiquity and the Middle Ages* (Brussels: Latomus, 1967); and the articles collected in John Riddle, *Quid pro quo? Studies in the History of Drugs* (Aldershot: Variorum, 1992).

48 Richard Hunt, "The Introductions to the *Artes* in the Twelfth Century," in *Studia mediaevalia in honorem R. J. Martin O.P* (Bruges: De Tempel, 1948), pp. 85–112.

49 Tony Hunt, *Teaching and Learning Latin in Thirteenth-Century England*, 3 vols. (Cambridge: Brewer, 1991).

written in the mid-eleventh century) that expresses the importance of dialectic: "[dialectic] defines and discerns, divides and asserts, powerful to think, an invincible conqueror, which the lamp of Manlian light [i.e., of Boethius] illuminates."[50] Although late in the tenth century Gerbert d'Aurillac had already established a curriculum based on Boethius's translations and commentaries (the *Vetus logica*, "Old Logic"), it was a series of teachers in France at the turn of the eleventh to twelfth centuries who made dialectic the most advanced intellectual discourse of the time. This trend culminated in the series of commentaries on, and summaries of, the *Vetus logica* by Peter Abelard (d. 1142). Boethius's previously neglected translations of Aristotle's *Sophistical Refutations* and *Topics* were rediscovered, and the first works of Aristotle to be sought in their original Greek were his logical works: the *Topics*, *Prior* and *Posterior Analytics*, and *Refutations*.[51]

The most relevant of these works of the "New Logic" for scientific speculation was the *Posterior Analytics*, in which Aristotle explained how "demonstration" proceeded from "premisses" which "must be true, primary, immediate, better known than, and prior to, the conclusion, which is further related to them as effect to cause."[52] Although the argument of this text was summarized by John of Salisbury in his *Metalogicon* (1159), the *Posterior Analytics* does not seem to have been much read in the twelfth century; it did not, for instance, receive a commentary until the thirteenth. Nevertheless, the demonstrative method that it describes became well known in the twelfth century, partly because of Boethius's summary of it in his second commentary to the *Isagoge* of Porphyry but especially through two works that exemplified the method: Boethius's own *De hebdomadibus* and Euclid's *Elements*. *De hebdomadibus*, in which Boethius demonstrates theological truths by reason rather than authority, was read and commented upon in the schools in the twelfth century and became the principal model on which Alain de Lille and Nicolas of Amiens, toward the end of the century, developed their own schemes for using deductive arguments in theology. Boethius describes the method as using indubitable premises ("common notions of the mind"), which "anyone accepts as soon as he hears them." He gives as an example an axiom from Euclid's *Elements*: "If you take two equal quantities away from two equal quantities, the remaining quantities are equal." Aristotle's *Posterior Analytics* had also cited geometry as the best example of the demonstrative method in practice.[53]

50 Mews, "Philosophy and Theology, 1100–1150," p. 192.

51 For the testimony of Robert of Torigni (between 1157 and 1169) concerning the translations of James of Venice, see Lorenzo Minio-Paluello, *Opuscula: The Latin Aristotle* (Amsterdam: Hakkert, 1972), p. 189.

52 Aristotle, *Posterior Analytics*, 1.2.71b20–21, trans. G. R. G. Mure, in *The Works of Aristotle*, ed. W. D. Ross (Oxford: Oxford University Press, 1928).

53 Charles H. Lohr, "The Pseudo-Aristotelian *Liber de causis* and Latin Theories of Science in the Twelfth and Thirteenth Centuries," in *Pseudo-Aristotle in the Middle Ages: The Theology and Other Texts* (London: The Warburg Institute, 1986), pp. 53–62.

One of the most conspicuous and distinctive features of twelfth-century thought is the rapid assimilation of Euclid's *Elements* and engagement with its methodology. In the early years of the twelfth century, only Boethius's translations of the first four books, without the all-important proofs, were known. Fifty years later, at least five versions of the entire work existed, three based on Arabic texts and one from Greek. As significant as the translations themselves are the examples of numerous interpretations of the text, in the form of either reworkings of the literal translations or the addition of marginalia and further theorems or proofs. In fact, this engagement with the *Elements* exemplifies two kinds of methodology. The first, which follows a literal interpretation of Euclid's text, is that of demonstration by analysis or synthesis within each theorem, using geometrical figures; its conclusions are expressed in the form: "This is what we wished to demonstrate." The second appears to be a development that occurred within the Latin schools, whereby the emphasis is placed on logical continuity from one theorem to the next by showing how theorems can be proved by means of previous theorems.[54] The proofs of Euclid's *Elements* were immediately used by Latin writers in other mathematical contexts – for example, by Hermann of Carinthia in 1143, to ascertain whether the planets shone by their own light, and by the author of the *Parvum Almagestum*.[55]

The dominance of logic in the schools of the early twelfth century entailed the recognition of the leading position of Aristotle, logic's prime exponent. As the century advances, one can see a shift toward regarding Aristotle as the leading philosopher as well. In the early Middle Ages, Cicero and Seneca were the "Philosophers," and Plato was honored as much as Aristotle, if not more.[56] Aristotle's growing preeminence can be seen in the *Metalogicon* of John of Salisbury, who attributes to Burgundio of Pisa the statement that Aristotle, more than any other philosopher, merits the name "Philosophus."[57] The theme of the *Metalogicon* (whose title signifies "After" or "Beyond" Logic) is that Aristotle's logic is an *organon* (i.e., a "tool") in the strict sense of the word, in that it provides the means of argument for any subject; John criticizes the contemporary tendency to pursue logic as an end in itself.[58] It is significant that he appeals to the authority of Burgundio, who was not a logician but

[54] The difference between the two methods is described in detail in Charles Burnett, "The Latin and Arabic Influences on the Vocabulary Concerning Demonstrative Argument in the Versions of Euclid's *Elements* Associated with Adelard of Bath," in Hamesse, *Aux origines du lexique philosophique européen*, pp. 175–201.

[55] Richard Lorch, "Some Remarks on the *Almagestum parvum*," in *Amphora: Festschrift for Hans Wussing on the Occasion of his 65th Birthday*, ed. S. S. Demidov, Menso Folkerts, David E. Rowe, and Christoph J. Scriba (Basel: Birkhäuser Verlag, 1992), pp. 407–37, reprinted in Richard Lorch, *Arabic Mathematical Sciences* (Aldershot: Variorum, 1995).

[56] For Adelard of Bath writing early in the twelfth century, Plato was the "philosophorum princeps." See Adelard of Bath, *De eodem et diverso*, in Adelard, *Conversations with His Nephew*, p. 3.

[57] John of Salisbury, *Metalogicon*, IV.7.6–8.

[58] See Klaus Jacobi, "Logic (ii): The Later Twelfth Century," in *A History of Twelfth-Century Western Philosophy*, ed. Peter Dronke (Cambridge: Cambridge University Press, 1988), pp. 227–51.

a pioneer in collecting and translating the Greek texts of Aristotle's works on the natural sciences, metaphysics, and ethics. The rediscovery of the full range of Aristotle's works, encouraged no doubt by the renewed interest in his logic, is a momentous episode in the history of Western European science. The consequence was the eclipse, or the driving underground, of alternate conceptions of science, such as those of Plato, expressed in the *Timaeus*; those of the Middle Platonists, such as Apuleius and Calcidius; and those of the Neopythagorean (or numerological) tradition, expressed in Nicomachus's arithmetic, the glosses on Calcidius,[59] and the cosmology of Abbo of Fleury. The flowering of interest in these alternative conceptions in the first half of the twelfth century did not lead to the translation and study of new texts by Plato or the Platonists as the interest in logic had led to the translation and ready assimilation of new works by Aristotle.

Burgundio's advocacy of Aristotle is not an isolated instance but can be set within an interest in the natural sciences and cosmology in southern Italy dating at least from the late eleventh century. At that time, Alfanus, archbishop of Salerno, and Constantine the African raised the medical learning of Salerno and Monte Cassino, respectively, to a more theoretical level by translating works that included discussion of the nature of the elements (Nemesius's *On the Nature of Man* and ʿAlī ibn al-ʿAbbās al-Majūsī's *Complete Book of the Medical Art*). By the mid-twelfth century, several texts had been translated or composed in Italy and Sicily on the nature of the elements and on cosmology, including the *Salernitan Questions* (in several versions), Aristotle's *On Generation and Corruption* and the fourth book of his *Meteorology*, Marius's *On the Elements*, *Alcantarus de philosophia mundi*, and the *Tractatus compendiosus*.[60] All of these adhere to an Aristotelian cosmology (in which, for example, the heavenly regions are made of a fifth essence) based on Aristotle's genuine works of natural science and the pseudo-Aristotelian *Problemata* and *De elementis*. The *Tractatus compendiosus* refers explicitly to Aristotle as the "follower of the truth."[61]

Aristotle, then, came to epitomize the method most appropriate for speculation about the natural world. In this form of argumentation, which twelfth-century commentators (following Boethius) referred to as "*ratio*" ("reason") and contrasted with the demonstrative method of mathematics, premises are inferred from the experience of the senses. They are not axiomatic but based on opinion. The resulting arguments are "probable" ("*probabilis*"), and their validity must be judged on the basis of their reasonableness ("*ratio*").[62] Hugh

59 This is fully explored in Anna Somfai, "The Transmission and Reception of Plato's *Timaeus* and Calcidius's *Commentary* during the Carolingian Renaissance" (PhD diss., Cambridge University, 1998).

60 See note 29.

61 *Tractatus compendiosus*, quoted in Piero Morpurgo, *L'idea di natura nell'Italia Normannosveva* (Bologna: CLUEB, 1993), p. 95.

62 See John of Salisbury, *Metalogicon*, II.14.

of St. Victor defined natural science ("*physica*") in terms of its methodology: "Natural science speculates by investigating the causes of things in their effects, and [deriving] the effects from their causes."[63]

The necessity of looking into causes of natural things had also been advocated by Plato in his *Timaeus*, the one philosophical text on cosmology that was available to the Latins before the introduction of Aristotle's *libri naturales*. Plato had stated that everything that comes into being necessarily arises from a certain cause (*Timaeus*, 28A) and proceeded to build up a cosmos piece by piece using "probable arguments." The *Timaeus*, with Calcidius's commentary, which in previous centuries had been mined for its numerological and astronomical data, became a "best-seller" in the early twelfth century and inspired commentaries from Bernard of Chartres, William of Conches, and several anonymous authors. The deliberate use of its methodology can be seen clearly in Adelard of Bath's *Questions on Natural Science*, written in the early years of the twelfth century. At the beginning of this text, Adelard purports to follow the "rational arguments" (*ratio*) of the Arabs rather than the "authority" of the French schools. However, before long, Plato takes over the role of the rational guide, and the most substantial quotations in the text come from the *Timaeus*. Throughout his work, Adelard invites his readers to look for causal connections, causal necessity, the intention of the causes, and so on.[64]

Although the *Timaeus* eventually yielded to the more systematic cosmology of Aristotle, it did encourage the search for causes. On the physical level, this led to interest in the nature of the elements of the universe. Witness the proliferation of texts "On the elements," both in Italy and Spain, and discussions of the nature of the elements in the *Dialogi* of Petrus Alfonsi, the commentaries and original works of William of Conches, and the *De essentiis* of Hermann of Carinthia. On the metaphysical level, it led to definitions of the basic principles of things. Adelard, using Calcidius's terms, called these the "*initium vel initia*" ("the beginning or beginnings"). Hermann of Carinthia called them the five "essences" ("*essentiae*"): cause, movement, place, time, and state ("*habitudo*"). Petrus Alfonsi, following the ninth-century Arabic medical author al-Rāzī, lists them as God, soul, matter, time, and place. Several cosmological works contain "causes" or "principles" in their title (e.g., "*De rerum causis*," the perhaps more authentic title for Adelard's *Questions on Natural Science*, and the *Liber Hermetis Mercurii Triplicis de VI rerum principiis*). This trend culminates in the translation from

[63] Hugh of St. Victor, *Didascalicon*, II.30, ed. C. H. Buttimer (Washington, D.C.: Catholic University of America Press, 1939), p. 46.

[64] Charles Burnett, "Scientific Speculations," in Dronke, *History of Twelfth-Century Western Philosophy*, pp. 151–76, especially pp. 169–70; and Andreas Speer, *Die entdeckte Natur: Untersuchungen zu Begründungsversuchungen einer "scientia naturalis" im 12. Jahrhundert* (Leiden: Brill, 1995), pp. 27–75.

Arabic of the cento from Proclus's metaphysics, called in Latin the *Liber de causis*, which, from the time of its translation (ca. 1150) until the early thirteenth century, substituted for Aristotle's *Metaphysics*.

In psychology, the concern for causes is manifested in the interest in the soul, the source of movement in animate bodies. Introduction of the medical writings of Nemesius and al-Majūsī, both of which devoted chapters to the soul, prompted discussion among theologians as to how the soul as a physiological entity shared by all animate creatures differed from the immortal soul unique to man. This problem was partly solved by distinguishing the former as preeminently "spirit." A lively debate concerning the nature of the soul led to a proliferation of works on the soul and the spirit before the mid-twelfth century. In addition to original works on the soul by William of Saint Thierry, Hugh of St. Victor, Alcher of Clairvaux, and others, the earliest philosophical text to be translated from Arabic was Qusṭā ibn Lūqā's *On the Difference between the Soul and the Spirit* (1151) and the first complete part of Avicenna's philosophical encyclopedia (the *Shifā*ʾ) to be translated was the book on the soul, both works becoming immediately popular.[65] The Platonic theory that the cosmos as a whole was an animal and moved by a "soul of the world" ("*anima mundi*") was even more controversial than medical theories on the human soul,[66] but in this case the triumph of Aristotelianism soon thwarted its success.

Finally, on the level of politics and human actions and sufferings, the search for causes encouraged the adoption of the scientific explanations of astrology. In medieval cosmology, all changes of coming-into-being, growth, decay, and death were regarded as mediated and finely tuned by the motions of the heavenly bodies. People therefore naturally believed that the same celestial causation influenced political events and the course of each person's life and, consequently, that it was possible to determine the appropriate time for commencing or avoiding particular activities. What is remarkable about the burgeoning of scientific astrology in the twelfth century is not so much the strength of its theoretical basis as the fact that it gave the individual practitioner the ability to predict which movements of which heavenly bodies resulted in which effects. Openness to belief in the validity of astrology is already apparent in extracts from Pseudo-Clementine's *Recognitiones* included in Pseudo-Bede's *De mundi terristris celestisque constitutione*, as well as copies of Firmicus Maternus's *Mathesis* and some popular Hebrew-Arabic astrology (known collectively as the *Alchandreana*) made in the eleventh century. But it was in the twelfth century that textbooks covering the whole of astrology were consciously assembled. In their comprehensiveness and bulk, these collections were comparable to the corpus of Aristotle's *Libri naturales*

65 See Dag-Nikolaus Hasse, *Avicenna's* De anima *in the Latin West* (London: The Warburg Institute, 2000).

66 Tullio Gregory, *Anima mundi: La filosofia di Guglielmo di Conches e la scuola di Chartres* (Florence: Sansoni, 1955).

and Avicenna's *Canon of Medicine*. Examples are the *Book of the Nine Judges* and the collection of translations of Arabic texts by Sahl ibn Bishr, Abū Maʿshar, Māshāʾallāh, and others made in Toledo later in the century.

THE POTENTIAL OF MAN

The case of astrology exemplifies another prerequisite for the advancement of the natural sciences: a belief in the human ability to discover the secrets of nature. For to believe that one has the means to make predictions about specific events in the future is surely a mark of supreme confidence in the human ability to fathom the natural world. This confidence is already apparent in Adelard's *Questions on Natural Science*, in which he refuses to resort to divine action as an explanation without first trying to go as far as human knowledge allows.[67] It is also manifest in the resurgence of interest in Hermetic writings. The *Asclepius*, the one text of the *Corpus Hermeticum* translated in the late-classical period, began to be copied again in the early twelfth century. Here we read how man is a "great wonder" (*magnum miraculum*), whose nature links the divine to the terrestrial; he can use the power of words, which are spirits declaring a person's whole wish; and he can create gods by making "statues" or "likenesses" out of plants, stones, and spices (which have in them a natural power of divinity) and implanting life in them by calling up the souls of demons or angels (*daemones vel angeli*). These man-made gods offer protection, the power to control things, and the ability to divine the future.[68] Knowledge of the Hermetic *Asclepius* may have spurred scholars to look for more Hermetic works. They found among the Arabs not only writings that provided the details of how to make the statues or "*imagines*" (usually translated "talismans") discussed in the classical text but also confirmation of man's potential. For example, in a text called in Latin *Liber Antimaquis*, we read of man that "having power over all creatures, he knows all knowledge, he can do all things, see all things, hear all things, eat all things, drink all things. This is because he is the form of forms, embodying in himself the whole of creation, as the microcosm in relation to the macrocosm."[69]

That man can manipulate nature to produce new forms and change the course of events is the premise for a large body of magical and alchemical texts, of which the earliest were Adelard of Bath's translation of a work

[67] Burnett, "Scientific Speculations," in Dronke, *History of Twelfth-Century Western Philosophy*, pp. 151–76, especially pp. 169–70; and Speer, *Die entdeckte Natur*, pp. 27–75.

[68] Brian P. Copenhaver, *Hermetica: The Greek Corpus Hermeticum and the Latin Asclepius in a New English Translation, with Notes and Introduction* (Cambridge: Cambridge University Press, 1992).

[69] Translation from *Liber Antimaquis*, ed. Charles Burnett, in *Hermes Latinus*, VI, ed. P. Lucentini (Turnhout: Brepols, 2001), pp. 209–10. The Arabic original from which this text was translated was known to Hermann of Carinthia.

on talismans by Thābit ibn Qurra and Robert of Ketton's translation of an alchemical text, the *Liber Morieni*. The significance of these texts for the development of experimental science has been documented by Lynn Thorndike.[70] However, the potential of man could also be expressed in terms of the powers of an attribute uniquely his – the intellect (*nous*). Again, Adelard had aptly described this faculty, in his *On the Same and the Different*, when he had stated that:

> The supremely good Creator ... adorned the soul with mind, which the Greeks call "nous." The soul uses this with clarity when she is in her pure state, lacking any disturbance from outside. She reaches not only realities in themselves, but also their causes, and the beginnings of their causes, and from present conditions understands those to come, a long time in advance. She grasps what she is, what is the mind by which she understands, what the reason by which she enquires.[71]

Richard and Mary Rouse have stated that:

> twelfth-century scholarship is characterized by the effort to gather, organize, and harmonize the legacy of the Christian past as it pertained to jurisprudence, theological doctrine, and Scripture.... In a certain sense the "twelfth century" can be said to close with the achievement of these goals and the emergence of a new mode of scholarship characterized by efforts to penetrate these great mosaics of the twelfth century, to gain access to the whole works of authority, and to ask fresh questions of them.[72]

What the Rouses have said about jurisprudence, theological doctrine, and Scripture applies equally to every branch of the natural sciences – to mathematics, physics, medicine, and astrology. Scholars of the twelfth century collected, organized, and harmonized the texts both from the indigenous Latin tradition and from new Greek and Arabic sources. But these texts themselves, and the other work that the scholars did, provided the language, methodology, and grounds for man's self-confidence, which led to the development of original scientific thought in the West in the thirteenth century and beyond.[73]

70 Lynn Thorndike, *A History of Magic and Experimental Science*, 8 vols. (New York: Columbia University Press, 1923–1958).

71 Adelard, *Conversations with His Nephew*, pp. 17–19.

72 Rouse and Rouse, "*Statim invenire*, Schools, Preachers, and New Attitudes to the Page," p. 201.

73 For the developments in nonliterary fields that were equally important in science (i.e., in "the mechanical arts" or technology), see Ovitt, Chapter 27, this volume.

16

MEDIEVAL ALCHEMY

William R. Newman

THE ORIGINS OF MEDIEVAL EUROPEAN ALCHEMY

The term "alchemy" derives from the Arabic *al-kīmiyā*, which in turn comes from the Greek *chymeia* or *chēmeia*, probably derived from *cheō*, and meaning in its root sense "the art of fusing [metals]."[1] Therefore "alchemy," in its primitive sense, concerns the treatment and transmutation of metals, a meaning reflected in many medieval definitions of the art.[2] Nonetheless, the theory and practice of medieval alchemy were not confined to the transmutation of metals. In reality, medieval alchemy concerned itself with all aspects of chemical and mineral technology and theory and had close links to pharmacy as well.

An ordinary codex of medieval alchemy will usually contain recipes for the synthesis and refining of numerous chemicals of commerce, such as sal ammoniac (ammonium chloride), sal alkali (usually sodium carbonate), alum, saltpeter (potassium nitrate), and sal tartari (potassium carbonate), as well as directions for making such pigments as vermilion, ceruse, minium (lead oxide), and verdigris, often with instructions for using these pigments in the manufacture of artificial gemstones. Such manuscript books also often contain recipes for alloying and soldering metals and for making elaborate glues, glazes, and lutes. Alchemical codices contain the most detailed descriptions of assaying and purifying metals that have been transmitted in any medieval documents. They describe a host of processes for purifying minerals by means of sublimation, distillation, solution, coagulation, precipitation, crystallization, and other operations. Finally, from the early

[1] Robert Halleux, *Les textes alchimiques* (Turnhout: Brepols, 1979), p. 45; Otto Lagercrantz, "Das Wort Chemie," *Kungliga Vetenskaps Societetens Arsbok, 1937* (Uppsala: Almqvist & Wiksell, 1938), pp. 25–44; and J. Gildemeister, "Alchymie," *Zeitschrift der Deutschen Morgenländischen Gesellschaft*, 30 (1876), 534–8.

[2] See Halleux, *Textes alchimiques*, pp. 43–7.

fourteenth century on, alchemical recipes form the primary locus for experimentation with the newly discovered mineral acids.

Alchemy in the Middle Ages was therefore a comprehensive field of endeavor concerning itself with every branch of chemical and mineral technology. At the same time, however, it embodied an attempt to link these technological practices to a theoretical framework. The characteristic (though by no means exclusive) structure of the medieval alchemical treatise is the explanatory *theorica* followed by an artisanal *practica*. A typical *theorica* will usually contain a recitation of Aristotle's theory of the elements, followed by a description of the subterranean generation of metals from sulfur and mercury.

Along with this introductory material, one will often find a defense of alchemy against its detractors, above all the Persian philosopher Avicenna (Ibn Sīnā, d. 1037). In many instances, this will be followed by a recounting of such laboratory processes as sublimation and distillation in terms of the fixed (nonvolatile) and unfixed (volatile) components of minerals, often expressed in the terminology of a corpuscular theory that we shall describe presently. Medieval alchemy was not therefore intended to be a "mechanical art" or pure technology; rather, its practitioners viewed it as an integration of theory and practice, descended from natural philosophy and subordinated to it.[3]

Although alchemy originated in the Hellenistic culture of late antiquity, the Greek alchemical corpus was almost entirely unknown in the Latin Middle Ages, and the systematic translation of it was begun only in the sixteenth century, by Domenico Pizzimenti.[4] It is to Islamic civilization that we must turn for the immediate sources of Latin alchemy. By the time of alchemy's arrival in the Latin West – usually identified with Robert of Ketton's mid-twelfth-century translation of Morienus's *On the Composition of Alchemy* from Arabic into Latin – the discipline had evolved considerably from its Greek roots.[5]

Alchemical writings were probably already being translated from Greek into Arabic – in some cases by means of Syriac intermediaries – as early as the eighth century. Already at this early stage, Arabic authors were making important modifications to Greek alchemical theory. One early example is found in the influential *Book of the Secret of Creation*, attributed to Bālīnas or pseudo-Apollonius of Tyana. Bālīnas's book, probably composed in the eighth century (though possibly based on a Greek original), already postulates

3 Chiara Crisciani, "The Conception of Alchemy as Expressed in the *Pretiosa margarita novella* of Petrus Bonus of Ferrara," *Ambix*, 20 (1973), 165–81.

4 For Pizzimenti, see Joseph Bidez, *Catalogue des manuscrits alchimiques grecs*, 8 vols. (Brussels: Maurice Lamartin, 1928), vol. 6. But see Robert Halleux and Paul Meyvaert, "Les origines de la *mappae clavicula*," *Archives d'histoire doctrinale et littéraire du moyen âge*, 72 (1987), 7–58.

5 Julius Ruska, "Zwei Bücher *De Compositione Alchemiae* und ihre Vorreden," *Archiv für Geschichte der Mathematik, der Naturwissenschaften und der Technik*, 11 (1928), 28–37; and Lee Stavenhagen, "The Original Text of the Latin *Morienus*," *Ambix*, 17 (1970), 1–12.

that the known metals are actually compounds of sulfur and mercury.[6] This fundamental doctrine, probably based on the observation that most of the then-known metals would form amalgams with mercury and that the common sulfide-ores of metals tend to deposit sublimed sulfur in the flues of refining furnaces, was accepted in altered form until the end of the eighteenth century.

By the late ninth and early tenth centuries, the Arabs were composing a large number of original and important alchemical treatises. The most significant of these early endeavors were the texts going under the names of Jābir ibn Hayyān and Muammad ibn Zakariyyā al-Rāzī. As Paul Kraus showed in the 1930s and 1940s, the vast corpus of about 3,000 works ascribed to Jābir ibn Hayyān (supposedly died ca. 812) cannot really go back to the eighth century, when this sage is supposed to have been active.[7] In fact, only one work in the corpus, the *Book of Mercy*, could be that early. Jābir's other books develop a complex theory called "the Balance," which incorporates arithmological notions with the belief that all bodies are composed of four "natures" (hot, cold, wet, and dry), which can be manipulated in order to produce new materials.

Jābir also adopts a preexistent theory that terrestrial substances consist of an "occult" or "internal" and a "manifest" or "external" component. Hence silver, for example, which is cold and dry "externally," contains the opposite qualities, hot and wet, "internally." Alchemical processes can induce the occult qualities to become manifest, thus transmuting one substance into another. Jābir's theory of the occult and the manifest was to have a long life in the Western world, where it offered an important alternative to the Galenic and scholastic concept of occult qualities as permanently unknowable entities.[8] The highly speculative Jabirian corpus also contains an important emphasis on the fractional distillation of substances in order to isolate their "natures" before recombining them. This theme would have a huge impact in the Latin West, being transmitted not only by Jābir's *Seventy Books* (translated by Gerard of Cremona as the *Liber de septuaginta*) but also by the highly influential *Book of the Soul in the Art of Alchemy*, a spurious work attributed to Avicenna that became a favorite source of Roger Bacon's, among others.[9]

[6] Paul Kraus, *Jābir ibn Hayyān: Contribution à l'histoire des idées scientifiques dans l'Islam*, 2 vols. (Cairo: Institut français d'archéologie orientale, 1942–1943), vol. 2, p. 1.

[7] Ibid., vol. 1, pp. i–lxv.

[8] Ibid.

[9] The *Liber de septuaginta* (TK 813) was edited in Marcelin Berthelot, "Geber – le livre des soixante dix," *Mémoires de l'Académie des Sciences*, 49 (1906), 310–63. Pseudo-Avicenna's *Liber de anima in arte alkimie* (TK 796) is described in Julius Ruska, "Die Alchemie des Avicenna," *Isis*, 21 (1934), 14–51. Note that "TK" followed by a number signifies an incipit found in Lynn Thorndike and Pearl Kibre, *A Catalogue of Incipits of Mediaeval Scientific Writings in Latin* (Cambridge, Mass.: Mediaeval Academy of America, 1963).

Unlike the corpus of Jābir, the extant alchemical work of Muhammad ibn Zakariyyā al-Rāzī (865–925) avoids elaborate speculation. Rāzī's *Book of Secrets* is a straightforward technical treatise devoted to the production of transmutational elixirs. One of the masterpieces of alchemical literature, the work provides a comprehensive discussion of minerals, alchemical apparatus, and processes. Rāzī's *Book of Secrets* was translated into Latin as the *Book of Ebubachar Rasy*, and it exists in other Latin versions as well.[10] Numerous pseudonymous works attributed to Rāzī were also translated into Latin. The most important of these is the *Book of Alums and Salts*, a highly technical treatise existing in several different Latin versions, whose immense popularity eclipsed the genuine work of Rāzī.[11] Although spurious, the *Book of Alums and Salts* was inspired by Rāzī in its practical orientation and avoidance of speculation. This cataloging of practical information and recipes came to characterize the Razian tradition in the medieval West, where it spawned a veritable genre of books on the refining and classification of salts.

After Jābir and Rāzī, the third Arabic writer to play a major part in Western alchemy was the Persian philosopher Avicenna. Avicenna was, in fact, an opponent of alchemy. He attacks it in a section of his *Book of the Remedy* dealing with minerals, which was translated around 1200 by Alfred of Sareshel as *On the Solidification and Conglomeration of Stones*.[12] This little Latin treatise acquired great prestige, in part because it was attached to the Latin translation of Book IV of Aristotle's *Meteorology* and appeared to be his. Many Latin alchemical treatises begin with a rebuttal of the dictum found in this treatise, "let the artificers of alchemy know that the species of metals cannot be transmuted," which came to be known by its introductory phrase, *sciant artifices*. Partly as a response to Avicenna's attack, medieval alchemy became a vehicle for advocating the power of human art to replicate natural products.

Despite Avicenna's clear rejection of alchemy, however, alchemical texts were attributed to him, such as *The Book of the Soul in the Art of Alchemy*. In addition, there is the *Letter to Hasen*, a remarkable little work written in a philosophical style, which contains valuable ruminations about the nature of alchemical processes.[13] The author of the *Letter to Hasen* derides the unphilosophical character of alchemical recipe books and argues that the

10 Julius Ruska, "Übersetzung und Bearbeitungen von al-Razis Buch Geheimnis der Geheimnisse," *Quellen und Studien zur Geschichte der Naturwissenschaften und der Medizin*, 4 (1935), 153–239.

11 Two versions of the *Liber de aluminibus et salibus* (TK 1388) have been edited. See Robert Steele, "Practical Chemistry in the Twelfth Century," *Isis*, 12 (1929), 10–46; and Julius Ruska, *Das Buch der Alaune und Salze* (Berlin: Verlag Chemie, 1935).

12 E. J. Holmyard and D. C. Mandeville, *Avicennae de congelatione et conglutinatione lapidum* (Paris: Geuthner, 1927).

13 The Latin title of the *Letter to Hasen* (TK 1036) is often given as *De re recta* or, probably more correctly, *De re tecta*. See H. E. Stapleton et al., "Two Alchemical Treatises Attributed to Avicenna," *Ambix*, 10 (1962), 41–82; and Georges C. Anawati, "Avicenne et l'alchimie," *Oriente e Occidente nel Medioevo* (Rome: Accademia Nazionale dei Lincei, 1971), pp. 285–341.

only laboratory processes that can be effective in transmutation are those that are modeled closely on principles derived from observations of natural processes. This would have a major impact on Albertus Magnus and his follower, the Latin author of the *Summa perfectionis* (*Sum of Perfection*), who went under the name of "Geber."

Among the most important Arabo-Latin alchemical translations was the short but influential *Emerald Tablet of Hermes*, a list of oracular utterances that begins "True, without doubt, certain, and most true, that which is below is just as that which is above."[14] The earliest form of the *Emerald Tablet* is found in Bālīnas's *Book of the Secret of Creation*. The text claims to have been inscribed by Hermes on a massive emerald and buried in his tomb between his hands. This "Hermes" is the "Hermes Trismegistus" of late antiquity, a syncretic divinity of Greco-Roman Egypt who was often claimed as the founder of alchemy. The content of the *Emerald Tablet*, with its emphasis on the parallelism between the celestial and terrestrial worlds, thus reflects late-antique Platonic and Hermetic interests.[15] The work is also found in the fabulously successful *Secret of Secrets* of pseudo-Aristotle, originally an Arabic "Mirror of Princes," which grew over time into a sort of "occult encyclopedia." Since the *Secret of Secrets* may have been the most popular secular book of the Middle Ages, currently extant in over six hundred Latin manuscript copies, it obviously gave the *Emerald Tablet* a very wide distribution.[16] The work exercised an immediate influence in the Latin West, where it formed the basis of numerous commentaries, such as the work *On the Perfect Magistery*, an early-thirteenth-century compilation attributed sometimes to Aristotle and sometimes to Rāzī.[17]

Already in *On the Perfect Magistery* one finds the notion that alchemy is a "terrestrial astronomy," an offshoot of the *Emerald Tablet*'s celestial–terrestrial parallelism. The leading idea behind this trope, still evident in the sixteenth century in the work of Tycho Brahe, is that the planets have earthly cognates in the form of the seven known metals.[18] The force behind this analogy lay in the fact that alchemists since late antiquity had used the planets as *Decknamen* ("cover-names") for the metals. In the Middle Ages, this association grew stricter: Gold became the Sun, silver the Moon, copper Venus, iron Mars, tin Jupiter, lead Saturn, and quicksilver Mercury.[19]

14 Julius Ruska, *Tabula smaragdina* (Heidelberg: Winter, 1926), p. 2.

15 Ibid., pp. 6–38.

16 Charles B. Schmitt and Dilwyn Knox, *Pseudo-Aristoteles Latinus* (London: Warburg Institute, 1985), pp. 54–75; and Steven J. Williams, *The Secret of Secrets: The Scholarly Career of a Pseudo-Aristotelian Text in the Latin Middle Ages* (Ann Arbor: University of Michigan Press, 2003).

17 Pseudo-Aristotle, *De perfecto magisterio* (TK 344); and J. J. Manget, *Bibliotheca chemica curiosa* (Geneva: Chouet et al., 1702), vol. 1, pp. 638–59. See also Julius Ruska, "Pseudepigraphe Rasis-Schriften," *Osiris*, 7 (1939), 31–94, especially pp. 45–56.

18 Owen Hannaway, "Laboratory Design and the Aim of Science: Andreas Libavius versus Tycho Brahe," *Isis*, 77 (1986), 585–610.

19 Robert Halleux, *Le problème des métaux dans la science antique* (Paris: Les Belles Lettres, 1974), pp. 149–60.

It is important to note, however, that the planetary *Decknamen* were only a tiny sample of the huge store of obscure synonyms employed by alchemists to veil their processes.[20] Although some scholars have seen this "terrestrial astronomy" as evidence of an integral linkage between alchemy and astrology, the alchemical texts and recipes of the Middle Ages seldom refer to astrology at all, and when they do, their references are usually cursory and sometimes unsympathetic.

THE THIRTEENTH CENTURY

Although alchemy was known in the West from the mid-twelfth century, serious attempts to assimilate the huge mass of Arabic writings on the subject were not widespread until the first half of the following century. From this time, one begins to see commentaries on and elaborations of the Arabic translations, such as the important work *On the Perfect Magistery* attributed sometimes to Aristotle and sometimes to Rāzī, and a closely related text also going under the name of these two authors, the *Light of Lights*.[21] Unlike these works, the *Art of Alchemy* attributed to Michael Scot stems from an identifiable location, namely the court of Frederick II von Hohenstaufen, the Holy Roman Emperor (r. 1215–50). The *Art of Alchemy* contains repeated references to Elias of Cortona (1180–1253), minister general of the Friars Minor from 1232 to 1239, presaging an important link between the Franciscans and alchemy that was to reach its full fruition in the following century.[22] The text is a striking representative of the Razian genre of alchemical works devoted to salts. Not content merely to catalog them, the author gives workable tests for distinguishing salts from one another. He discriminates between saltpeter and washing soda, which had both gone under the name "niter" (*nitrum*), by observing their different properties when placed on a glowing coal, an early predecessor of nineteenth-century blowpipe analysis.[23] The presence of this test in the early thirteenth century reveals that alchemists were keen to employ experimental evidence in the classification and understanding of natural substances, a trait that becomes more evident as the century progresses.

[20] E. O. von Lippmann, *Entstehung und Ausbreitung der Alchemie*, 3 vols. (Berlin: Springer, 1919-1954), vol. 1, p. 11 et sparsim; and Julius Ruska and E. Wiedemann, "Beiträge zur Geschichte der Naturwissenschaften, LXVII: Alchemistische Decknamen," *Sitzungsberichte der physikalisch-medizinischen Sozietät in Erlangen*, 56 (1924), 17–36.

[21] For the *Light of Lights* (*Lumen luminum*, TK 290), see Ruska, "Pseudepigraphe Rasis-Schriften," pp. 56–67.

[22] The *Art of Alchemy* (*Ars alkimie*, TK 281) is edited in S. Harrison Thomson, "The Texts of Michael Scot's *Ars alchemie*," *Osiris*, 5 (1938), 523–59. See also Charles H. Haskins, "The 'Alchemy' Ascribed to Michael Scot," *Isis*, 10 (1928), 350–9.

[23] J. R. Partington, *A History of Greek Fire and Gunpowder* (Cambridge: W. Heiffer and Sons, 1960), pp. 87–9. See also Robert Multhauf, *The Origins of Chemistry* (London: Oldbourne, 1966), p. 33; and Marcelin Berthelot, *La chimie au moyen âge* (Paris, 1893; repr. Osnabrück: Otto Zeller, 1967), vol. 1, p. 98.

ALBERTUS MAGNUS

In the mid-thirteenth century, Latin alchemy underwent a new and powerful treatment in *On Minerals* by Albertus Magnus (ca. 1200–1280). Drawing on translations of Arabic alchemical texts such as the *Letter to Hasen* attributed to Avicenna, the *Seventy Books* of Jābir, and the pseudo-Razian *Book of Alums and Salts*, as well as Arnoldus Saxo's encyclopedic *On the Ends of Natural Things* and the *Questions* of Nicholas the Peripatetic, Albert subjected alchemy to a probing, and yet sympathetic, critique.[24] Although Albert had no intention of writing an alchemical treatise as such, *On Minerals* proved enormously influential in setting the agenda of subsequent alchemical works. In *On Minerals*, for example, Albert takes a hint from Nicholas the Peripatetic that many terrestrial substances contain a "two-fold unctuosity" or sulfur, of which one part is extrinsic and flammable, the other intrinsic and noncombustible.[25] Albert goes on to say that evidence of this double sulfur is found in most metals. Upon calcining (that is, oxidizing) silver, for example, one notices a sulfurous stench – this is the extrinsic, flammable sulfur passing off into the atmosphere. If the calx is then reduced, the silver will be purer than before; it will still contain its intrinsic sulfur, but some of the extraneous type will have been lost. Albert's doctrine of extrinsic and intrinsic sulfurs was refined and generalized by the most significant of the thirteenth-century alchemical treatises, the *Summa perfectionis* of "Geber," which we shall describe shortly.

Although adopting the alchemical theory that metals are composed of sulfur and mercury, Albert rejects the alchemical theory of the occult and the manifest, which he attributes primarily to Hermes. Albert also denies that gold can be the Aristotelian "form" of the other metals, which are, according to "Callistenes" (Khālid ibn Yazīd), pseudonymous author of the *Book of Three Words*, merely imperfect or diseased forms of the most noble metal.[26] Albert argues from experience that metals appear to be "stable"; under normal circumstances, they do not become other metals. Similarly, each metal has its own peculiar set of properties, from which we may conclude that the accidents of the metals are not shared.

Nonetheless, Albert does not deny the possibility of alchemical transmutation, and argues against the Avicennian pronouncement that the species of metals cannot be transmuted (the *sciant artifices*). Conflating "species" with "specific form," Albert employs a well-established scholastic theory, to the effect that one form can be corrupted and replaced by another, to argue that

[24] See Robert Halleux, "Albert le grand et l'alchimie," *Revue des sciences philosophiques et théologiques*, 66 (1982), 57–80. For the articles of Pearl Kibre on Albert and alchemy, see Halleux, *Textes alchimiques*, p. 22.

[25] William Newman, *The Summa perfectionis of pseudo-Geber* (Leiden: Brill, 1991), pp. 216–20.

[26] The *Book of Three Words* (*Liber trium verborum*, TK 76) is described in Ruska, "Pseudepigraphe Rasis-Schriften," pp. 81–7.

the form of a base metal can be removed and replaced with that of a precious metal. Albert suggests that alchemists should act like physicians, first purging the imperfect metals of the extrinsic sulfur and impurities by processes such as washing and sublimation. Once this has been done, the alchemist can strengthen the "elemental and celestial powers" in the metal's substance, primarily by adding the proper ingredients to the base metal. This preparatory work will allow nature itself to perform the transmutation; the alchemist merely provides the correct initial conditions that will permit nature to act. One further condition must still be met, however. The alchemist should design his "artificial vessels" – that is, his apparatus – to imitate the natural vessels provided by caverns and interstices within the earth. The heat of the alchemical furnace will then be able to act on the purged alchemical "first matter" in the same way as the Sun and celestial heat act on unformed matter within the earth to produce genuine precious metals.[27]

The influence of Albert's work on subsequent Latin alchemists would be hard to overstate. In addition to spawning at least thirty pseudo-Albertine alchemical writings, Albert's *On Minerals* was a major source for the pseudo-Jabirian *Summa perfectionis*, a classic text written in the final third of the thirteenth century. Even works attributed to Roger Bacon, such as the pseudonymous *Brief Breviary*, make use of key Albertine notions, such as the concept of multiple sulfurs and mercuries within a given metal, some of which are volatile and extrinsic, others fixed and intrinsic.[28] Before passing to such works of the late thirteenth century, however, let us first consider the alchemical interests of the genuine Roger Bacon, whose writings on the subject are only a decade or so later than those of Albert.

ROGER BACON

The *Opus maius*, *Opus minus*, and *Opus tertium* of Roger Bacon (ca. 1219–ca. 1292), all composed in the 1260s at the behest of Pope Clement IV, reveal the keen interest of this Franciscan friar in alchemy. His *Letter on the Secret Works of Art and Nature* is also deeply concerned with alchemy, but this is largely a compilation from the three *Opera*.[29] Bacon divides alchemy into "speculative" and "operative" branches. He argues that theoretical alchemy teaches how "simple and compounded humors" arise from the elements. Like many Arabic alchemists, Bacon argues that the fractional distillations taught

27 Albertus Magnus, *De mineralibus, in B. Alberti Magni Ratisbonensis... opera omnia*, ed. A. Borgnet, 37 vols. (Paris: Vives, 1890), vol. 5, pp. 1–116. See also Albertus Magnus, *Book of Minerals*, trans. Dorothy Wyckoff (Oxford: Clarendon Press, 1967).

28 For the dependency of the *Brief Breviary* (*Breve breviarium*, TK 180) on Albert's *De mineralibus*, see Newman, *Summa perfectionis of pseudo-Geber*, p. 46, n. 88.

29 William Newman, "The Philosophers' Egg: Theory and Practice in the Alchemy of Roger Bacon," *Micrologus*, 3 (1995), 75–101; and Michela Pereira, "Teorie dell' elixir nell' alchimia latina medievale," *Micrologus*, 3 (1995), 103–48.

by speculative alchemy allow a fundamental analysis of matter into its most basic components. He is much impressed by an example from *On the Soul in the Art of Alchemy*, where pseudo-Avicenna describes an alchemical analysis of milk. Upon its distillation, a clear, tear-like substance "passes over" first. This is the element of water. As the distillation continues, a yellow water follows, and a black, burned substance is left in the flask. The yellow water corresponds to air, the black residue to earth, and the smoky vapors that appear during the process to fire.

In addition to such analyses into the four elements, Bacon claims that the alchemists also teach about the different orders of humors, things that have escaped both the physicians and the natural philosophers. Thus it is the alchemists who have discovered that the four humors found in animals are really compounds formed of four "simple humors" and that these simple humors are in turn compounds formed from the elements. Since alchemy reveals the ultimate nature of matter, Bacon argues, it is also the proper source of knowledge about all the other things that descend from elemental origins, such as precious stones, metals, pigments, salts, sulfurs, oils, and burning pitches. In addition, vegetables, animals, and humans derive from the humors, so that an ignorance of theoretical alchemy necessarily entails ignorance of theoretical medicine, and thus failure in practical medicine.[30]

If Bacon's theoretical alchemy studies the foundation of matter, his practical alchemy teaches the manufacture of precious metals, pigments, and other items of chemical technology. Practical alchemy, moreover, can even surpass nature in the perfection of its products, making a knowledge of it essential to the utilitarian needs of the commonwealth. Practical alchemy is also useful by providing the key to making medicines of a far more perfect sort than those currently available. Bacon envisages two roles for alchemy in medicine, both related to the technology contained in alchemical texts, especially the distillation technology of pseudo-Avicenna's *On the Soul in the Art of Alchemy*. The first role consists in purifying ordinary pharmaceuticals, which Bacon views as having been adulterated by apothecaries. In his work *On the Errors of Physicians*, Bacon is concerned among other things with the removal of toxic components from otherwise beneficial medicines, such as laxatives. He describes several methods of doing this that are explicitly alchemical, such as sublimation and distillation.

The second role for alchemy in Bacon's work concerns a veritable fixation in his mind, namely the prolongation of human life (macrobiotics) by means of substances that have been "reduced to equality" with the aid of alchemy. Influenced above all by the *Secret of Secrets* of pseudo-Aristotle, Bacon argues that alchemical medicines should be made out of the stratified fractions of human blood, mercury, and other ingredients. These should then be exposed

[30] Roger Bacon, *Opus tertium*, in *Opera quaedam hactenus inedita*, ed. J. S. Brewer (London: Longman, Green, Longman and Roberts, 1859), pp. 39–41.

to the influence of the stars and planets, preferably with the aid of burning mirrors to concentrate the beneficial rays. The object is to produce a "body of equal complexion"; that is, a substance whose elemental qualities are perfectly balanced and that is therefore incorruptible. The incorruptibility can then be transferred to the ailing bodies of humans or to a degenerate metal.

Bacon's macrobiotic program based on alchemy represents an important development in the history of the subject. In this area, he foreshadows the remarkable successes of medical alchemy in the following century, associated above all with John of Rupescissa and pseudo–Ramon Lull. Curiously, the alchemical pseudepigrapha circulating under Bacon's name by the end of the thirteenth century do not make much of his medical alchemy, focusing instead on the traditional transmutational themes in his work.[31] Given this lack of documented influence, it is probably a mistake to view Bacon as the progenitor of medical alchemy in the West, as Joseph Needham and others have done.[32]

Despite the high premium that Bacon placed on alchemy, it was not his view of the subject that came to dominate the late thirteenth century but the radically different theory of Albertus Magnus. Although Albert also draws a parallel between medicinal cures and the perfection of base metals, for him this is only a metaphor. In Albert's *On Minerals*, there is never any question of alchemical "medicines" actually being used to cure ailing humans. His medicines are for metals alone. Hence the medical and macrobiotic aspect of Bacon's alchemy is completely absent from Albert's work. Second, Roger's belief that alchemy should operate by producing an "equal body" from the separated and purified humors derived from human blood finds no parallel in the work of Albert. Indeed, Albert's insistence on following the generative methods of nature would seem a priori to exclude such operations from his canon of accepted techniques. Nature does not make the precious metals by separating human blood, so why should the alchemist?

THE *SUMMA PERFECTIONIS* OF "GEBER"

The same emphasis on following the methods of nature can be found in the *Summa perfectionis* attributed to Geber, a work that has been called the "main chemical textbook" of the medieval alchemists.[33] The *Summa* was

31 William R. Newman, "An Overview of Roger Bacon's Alchemy," in *Roger Bacon and the Sciences*, ed. Jeremiah Hackett (Leiden: Brill, 1997), pp. 317–36.

32 Joseph Needham, *Science and Civilization in China* (Cambridge: Cambridge University Press, 1974), vol. 5, pt. 2, pp. 12–15, 74, 123; and Michela Pereira, *L'Oro dei filosofi* (Spoleto: Centro Italiano di Studi sull' Alto Medioevo, 1992), pp. 43–112.

33 George Sarton, *Introduction to the History of Science* (Baltimore: Carnegie Institution, 1931), p. 1043; and Newman, *Summa perfectionis of pseudo-Geber*, pp. 215–23.

apparently written in the last third of the thirteenth century, probably by an obscure Franciscan named Paul of Taranto, who is said to have lectured in Assisi.[34] Although the work is organized in the typically scholastic genre of the *summa* – a comprehensive compendium – the author has consciously set out to appropriate the reputation of the Arab Jābir ibn Hayyān, for the *Summa perfectionis* contains numerous literary borrowings and reworked passages from the latter's *Seventy Books*. The *Summa* became the most authoritative medieval work on the purification and transmutation of metals, exercising a lively influence as late as the seventeenth century.

Among the innovative teachings propounded in the *Summa* is an elaboration of the Albertine and Avicennan notion that alchemy should follow the techniques used by nature in generating metals. But the *Summa perfectionis* also proscribes all "organic" reagents, such as blood, hair, and urine, and restricts the ingredients of the alchemists' transmutative agent, the philosophers' stone, to mercury, sulfur, and "arsenic" (arsenic sulfide), the presumed components out of which nature fashions metals. In practice, however, the *Summa* restricts the role of sulfur and arsenic so much that mercury becomes the essential ingredient of the philosophers' stone. This again is an attempt to model alchemical techniques on the generative methods of nature using the same ingredients as it does. "Geber's" focus on mercury was to have a massive influence on the remainder of the Middle Ages. As Lynn Thorndike remarked, the dominant alchemical theory of the fourteenth century maintained that transmutation should proceed by means of "mercury alone."[35] Thorndike was unaware of the Geberian origin of this idea, however, and failed to see its link with the Albertine and Avicennian claim that alchemists should limit themselves to processes based on the natural generation of metals.

Interestingly, the *Summa* deviates from the model of Roger Bacon in another way, avoiding all attempts to arrive at a proper tempering of elemental qualities; the very discussion of such qualities is largely absent from the work. Instead, the *Summa* propounds a comprehensive corpuscular theory of matter with close links to experiment, whose subsequent influence will still be evident among such seventeenth-century "corpuscularians" as Daniel Sennert, Kenelm Digby, and Robert Boyle.[36] As "Geber" says of sulfur and mercury, "each of these . . . is of very strong composition and uniform substance. This is so because the particles of earth are united through the

[34] William Newman, "New Light on the Identity of Geber," *Sudhoffs Archiv*, 69 (1985), 76–90; Newman, "The Genesis of the Summa perfectionis," *Archives internationales d'histoire des sciences*, 35 (1985), 240–302.

[35] Lynn Thorndike, *A History of Magic and Experimental Science*, 8 vols. (New York: Columbia University Press, 1923–1958), vol. 3, p. 58.

[36] William Newman, "Boyle's Debt to Corpuscular Alchemy," in *Robert Boyle Reconsidered*, ed. Michael Hunter (Cambridge: Cambridge University Press, 1994), pp. 107–18; and Newman, "The Alchemical Sources of Robert Boyle's Corpuscular Philosophy," *Annals of Science*, 53 (1996), 567–85.

smallest [particles] to the aerial, watery, and fiery particles in such a way that none of them can separate from the other during their resolution."[37]

This revealing passage refers to the fact that mercury and sulfur leave little if any residue when they are sublimed. Their resistance to such analysis is due to the "very strong composition" (*fortissima compositio*) of their elementary corpuscles. Already the term *compositio*, literally "setting side-by-side," implies that "Geber" thinks of sulfur and mercury as secondary particles composed of the simpler elementary corpuscles. The sulfur and mercury are of "uniform substance," meaning primarily that they are composed of uniformly sized particles. This homogeneity allows all their particles to sublime at the same rate, so that none is left behind to form a residue.

The *Summa* goes on to describe this corpuscular micro-mechanics in greater detail when analyzing particular alchemical processes. In the case of distillation, for example, small corpuscles (*subtiles partes*) pass off more easily than large ones (*grosse partes*). For this reason, a nonhomogeneous substance such as a mixture of different vegetable oils can be separated into its constituents by means of distillation. Similarly, extreme heating of a base metal in the process of calcination will drive off its tiny, volatile sulfur particles, leaving behind only a fixed, nonvolatile mercury combined with any fixed sulfur that remains. Here "Geber" links his own corpuscular theory to Albertus Magnus's notion of extrinsic and intrinsic metallic principles, explaining the escape of extrinsic sulfur in terms of particle size. Interestingly, "Geber" treats the metallic corpuscles themselves as tertiary particles composed of mercury and sulfur. For this reason, he is able to think of metals as "homogeneous" despite the fact that they are made up of spatially distinct subparticles. The sources for the *Summa*'s corpuscularism are mainly Book IV of Aristotle's *Meteorology*, the "natural questions" literature, and medical theories deriving from the School of Salerno.

The immense influence of the *Summa perfectionis* was not limited to its corpuscular theory and the doctrine of "mercury alone." Basing himself on vague hints from the *Seventy Books* of Jābir, "Geber" also originated the groundbreaking theory of three orders of "medicines" for perfecting metals. According to this theory, transmutative agents occur in three forms: medicines of the first order work only superficial, temporary change, as in the case of artificial gold (actually brass) made of copper and zinc oxide; medicines of the second order produce real change, but only in one quality, as when lead is purged of its sulfurous component by calcination; and medicines of the third order genuinely transmute base metals into gold or silver. Since "Geber" believes that the particles from which the metals are made diminish in size with the perfection of the metal, his major goal is the acquisition of an exceedingly subtle medicine whose particles can penetrate deeply into a base metal and bond therewith. Although he gives no explanation for the fact

37 Newman, *Summa perfectionis of pseudo-Geber*, p. 663.

that these subtle particles do not pass off when the precious metal is heated, the answer probably lies in a vaguely stated concept of "like-to-like." The ultra-pure mercury of the philosophers' stone will bond with the mercury already resident in a base metal, as a result of their mutual affinity.

ALCHEMY IN THE LATE MIDDLE AGES

The *Summa perfectionis* represents the apogee both of alchemy in the high Middle Ages and its interaction with scholasticism. The originality and rigorous disputational form of the *Summa*, with its remarkable degree of integration between theory and practice, was not to be equaled again in the Middle Ages. Indeed, beginning in the final third of the thirteenth century, one begins to see an increasing rejection of alchemy on the part of religious and university authorities. Giles of Rome, for example, probably writing in the 1280s, attacked the possibility of alchemical transmutation in a quodlibetal question. Among the religious orders, alchemy had become a bone of contention and was condemned by the Dominicans in 1272, 1287, 1289, and 1323.[38] The Franciscans and Cistercians also prohibited alchemy in their orders, a compelling testimony to the widespread presence of alchemical practice in medieval monasteries.[39] The anti-alchemical backlash received papal support in John XXII's decretal *Spondent quas non exhibent* ("They promise wealth that they do not deliver"), promulgated in 1317.[40] At the end of the century, the prominent Inquisitor Nicholas Eymerich wrote a comprehensive denunciation of alchemy, based in part on the *Spondent*, in a treatise called *Contra alchimistas* (1396).[41]

Unlike medicine and natural philosophy, alchemy had no institutional home within the universities of the thirteenth century, and the increasing censure led only to its further marginalization from the mainstream of intellectual life. Nonetheless, the fourteenth century did witness a number of rigorous defenses of alchemy by legal writers and alchemists. The well-known canonist Oldrado da Ponte wrote a widely circulated *consilium* in the early fourteenth century that defended the sale of alchemical gold. His views were adopted by an impressive list of legal authorities.[42]

A far more sustained defense of alchemy is found in the work of an otherwise obscure physician, Petrus Bonus, who wrote his lengthy *Precious*

38 Ibid., p. 35.

39 Halleux, *Textes alchimiques*, p. 127; and C. Narbey, "Le moine Roger Bacon et le mouvement scientifique au XIIIe siècle," *Revue des questions historiques*, 35 (1884), 115–66.

40 Francesco Migliorino, "Alchimia lecita e illecita nel Trecento," *Quaderni medievali*, 11 (1981), 6–41; cf. p. 15.

41 Sylvain Matton, "Le traité *Contre les alchimistes* de Nicolas Eymerich," *Chrysopoeia*, 1 (1987), 93–136.

42 Thorndike, *History of Magic and Experimental Science*, vol. 3, pp. 48–51; and Migliorino, "Alchimia lecita e illecita nel Trecento," pp. 29–41.

Pearl between 1330 and 1339.[43] Petrus argues that alchemy is a legitimate offshoot of Aristotelian natural philosophy, to which it is subordinate. Natural philosophy supplies alchemy with the theoretical principles that explain it but are too general to guide alchemical practice by themselves. Alchemy, being a more specific science than natural philosophy per se, can make a more detailed investigation of minerals. At the same time, alchemy offers something that natural philosophy cannot match at all, namely the claim to be a "gift of God" (*donum Dei*) that only the "sons of doctrine," an elite few, can successfully attain. This well-worn motif, present in many Arabic alchemical tracts and transmitted by the *Seventy Books* of Jābir, among others, had already been elaborated by the *Summa perfectionis*. Petrus Bonus develops it much further, however, claiming that the ancient alchemists were also prophets who foretold such events as the Last Judgment and the birth of Jesus Christ.[44]

Petrus Bonus's defense of alchemy is an interesting mixture of Aristotelian disputation and appeal to the special revelatory status of alchemy. As the fourteenth century progresses, one sees ever more appeals to the divine status of alchemy, whereas the clear argumentation of the previous century dwindles. At the same time, the figurative character of alchemy becomes more pronounced, as though in inverse relation to the disputational content of the texts. The fourteenth century sees the transfer of older alchemical conceits into actual visual imagery in such texts as the *Book of Secrets* attributed to one Constantinus and translated into Flemish. As the fourteenth century passes into the fifteenth, alchemical illustrations become ever more elaborate, as in the *Rising Dawn* (*Aurora consurgens*) and the strange vernacular *Book of the Holy Trinity* (*Buch der heiligen Dreifaltigkeit*) composed at the Council of Constance. In these works, one finds the elaborate depictions of the alchemical hermaphrodites, murdered kings and queens, dying crows, copulating couples, and tail-eating dragons that become the favorite *topoi* of early-modern alchemy.[45]

The rationalistic tradition of the *Summa perfectionis* did not die out in the fourteenth century but was elaborated in numerous derivative texts. At the same time, however, one sees an increasingly religious overlay applied to what were originally philosophical and scientific doctrines, such as the "mercury alone" theory and the doctrine of three medicines. This is not surprising if one considers the writers to whom such fourteenth-century texts are ascribed.

43 Petrus Bonus, *Margarita pretiosa*, in Manget, *Bibliotheca chemica curiosa*, vol. 2, pp. 1–80. See Crisciani, "Conception of Alchemy as Expressed in the *Pretiosa margarita novella* of Petrus Bonus of Ferrara," passim; and Thorndike, *History of Magic and Experimental Science*, vol. 3, pp. 147–62.

44 Crisciani, "Conception of Alchemy as Expressed in the *Pretiosa margarita novella* of Petrus Bonus of Ferrara," p. 171; and Bonus, *Margarita pretiosa*, pp. 30, 50.

45 For Constantinus, see Barbara Obrist, *Constantine of Pisa: The Book of the Secrets of Alchemy: Introduction, Critical Edition, Translation and Commentary* (Leiden: Brill, 1990). For the history of alchemical imagery in the Middle Ages, see Obrist, *Les débuts de l'imagerie alchimique (XIV–XVe siècles)* (Paris: Le Sycomore, 1982).

The Catalan physician Arnald of Villanova (1240–1311), who had a long and brilliant medical career under the kings of Aragon, fell under the reformatory influence of the Spiritual Franciscans and became an exponent of the twelfth-century prophet Joachim of Fiore, who had chosen to express himself in the form of elaborate pictorial figures.[46] Thus Arnald composed his own prophetic works, such as *The Coming of Antichrist* (*De adventu antichristi*), which predicted the appearance of the Great Beast in 1378. Although it is highly unlikely that the genuine Arnald of Villanova ever wrote an alchemical text as such, it is no surprise that his alchemically minded followers would have made him the representative of a school that stressed the interaction of alchemy and religion. Several treatises attributed to Arnald, such as the *Exempla* or *Parabolae*, compare the "great work" of the alchemists to the life and death of Jesus. Here a detailed comparison is made between the transformation of mercury and the passions or torments of Jesus. Just as Jesus was first scourged until he bled, made to wear a crown of thorns, nailed to the cross, and finally treated to gall and vinegar, so must mercury be tortured in four stages.[47]

Arnald's peculiar interests – Joachite prophecy, medicine, and the reformatory zeal of the Spiritual Franciscans – were combined with alchemy in a most fruitful way by a follower of the next generation, John of Rupescissa (Jean de Roquetaillade). Rupescissa is famous to medieval historians for his large corpus of Joachite prophecies, mostly written during the 1340s and 1350s. He also wrote two alchemical works, however, the *Book of Light* (TK 255) and *Book of the Consideration of the Quintessence of all Things* (TK 458). The latter contains the first sustained attempt to apply alchemical techniques to medicine and is justifiably famous as a source for the early-modern iatrochemistry of Paracelsus. Its huge popularity is reflected in the fact that well over one hundred manuscripts of the work survive today. Unlike the medical alchemy of Roger Bacon, John's work gives detailed recipes for extracting the virtues and quintessences of numerous minerals, metals, and plants, often with the aid of *aqua ardens* or *aqua vite* (ethyl alcohol) and the mineral acids. Writing under the influence of Arnald of Villanova's possibly genuine work *On the Water of Life*, John argued that specially purified *aqua ardens* was an earthly analogue of the Aristotelian quintessence. Like the quintessence, this earthly "human heaven" (*coelum humanum*) was incorruptible. Not only was it incapable of generating flies by spontaneous generation; it actually preserved dead flesh from putrefaction. Hence it should provide an ideal medicine for prolonging life and warding off disease.

In the *Book of the Quintessence*, John complains that he wasted five years at the University of Toulouse studying the "useless disputes" of

[46] Michael McVaugh, *Arnaldi de Villanova opera medica omnia, Aphorismi de gradibus*, 16 vols. (Granada-Barcelona: Seminarium historiae medicae granatensis, 1975), vol. 2, pp. 75–80. For the alchemical corpus ascribed to Arnald, see Halleux, *Textes alchimiques*, p. 105; Michela Pereira, "*Arnaldo da Villanova e l'alchimia,*" *Arxiu de Textos Catalans Antics*, 14 (1995), 95–174; and Antoine Calvet, *Le Rosier alchimique de Montpellier* (Paris: Université de Paris-Sorbonne, 1997).

[47] MS Venice, San Marco VI, 214, fols. 167v–168r.

philosophy.[48] In 1332, John became a Franciscan, and continued his studies for five more years. Soon he fell under the influence of reformatory currents promoting ecclesiastical poverty, probably inspired by the Beghards. Prophetic visions led him to issue a series of diatribes against the venality of the members of the religious orders, especially the Dominicans, whom he called "heretical followers of Mammon."[49] His intransigent appeals to absolute poverty were not well received within his own order; in December 1344, he was sent to a succession of gruesome prisons, where he spent at least the next twenty years.[50]

But it is John's alchemical works that give us the most information about his plans for the righteous.[51] The *Book of the Quintessence* records a vision that revealed to him the secrets of alchemy after seven years in prison. As a result of his vision, John says, he has written his work of medical alchemy in order to help the righteous – especially Franciscans – during the coming time of tribulation. This is a reference to a recurrent theme in his prophetical works – the notion that between 1370 and 1415 there will be forty-five years of horrible bloodshed resulting from the coming of the Antichrist, followed by a millennium of peace under the rule of a universal monarch. A similar theme is proposed in his alchemical *Book of Light*, where he reveals the secret of the philosophers' stone as a means of saving oneself during the time of tribulation. In the *Book of the Quintessence*, John also points out that people can hide their gold from the forces of the Antichrist by reducing the metal into "earth" with the aid of mercury.

John, then, views alchemy as a means of protecting the righteous downtrodden from the premillennial fury of the Antichrist while the proud dignitaries of the church get their just desserts. This message is accompanied by a full-fledged integration of alchemy and prophecy and an association between the "evangelical men" of his religion and the "sons of doctrine" of Arabic alchemy. In sum, John of Rupescissa has fully assimilated alchemy into the social and reformatory ideology of the Spiritual Franciscans. His eschatological goals, coupled with his disdain for university education, illustrate the great change in outlook between his alchemy and that of "Geber" a mere generation or two earlier.

A similar integration of alchemy with religious themes characterizes the huge corpus ascribed to the Catalan philosopher and Franciscan tertiary (or lay brother) Ramon Lull, which numbers over one hundred forty separate

48 Jeanne Bignami-Odier, "Jean de Roquetaillade," in *Histoire littéraire de la France*, Académie des inscriptions et belles-lettres, 43 vols. (Paris: Imprimerie nationale, 1981), vol. 41, pp. 75–209; cf. pp. 76–7.

49 Ibid., pp. 102–3.

50 Ibid., pp. 77–84.

51 Robert Halleux, "Les ouvrages alchimiques de Jean de Rupescissa," in *Histoire littéraire de la France* (Paris: Imprimerie nationale, 1981), vol. 41, pp. 241–84; and Bignami-Odier, "Jean de Roquetaillade." See also Leah DeVun, *Prophecy, Alchemy, and the End of Time: John of Rupecissa in the Late Middle Ages* (New York: Columbia University Press, 2009).

titles.[52] The oldest of these works, the *Testament*, may have been composed as early as 1332.[53] Since the genuine Ramon Lull died in 1315 and is known to have disparaged alchemy in his works, his authorship of any alchemical work is doubtful at best. The *Testament* of pseudo-Lull, despite its pseudonymous character, betrays the characteristic stamp of what has come to be viewed as the Lullian school of alchemy. Like the genuine philosophical works of Lull, the *Testament* contains circles and other diagrams with letters that symbolize various concepts and substances. This is a version of Lull's famous "combinatory art," which acted as a sort of mnemonic technique as well as being a method for the mechanical combination of concepts. One of the more interesting characteristics of this influential work may be found in its elemental theory. The author of the *Testament* argues that the four elements that we encounter in the world are not pure but corrupted. The Creator fashioned the universe by first making a quintessence ex nihilo.[54] He divided the quintessence into three portions of descending purity. The purest became the realm of the angels, the less pure was relegated to the "heavens, planets, and stars," and the third portion, least pure, became the sublunary elemental world.

A similar ranking occurs in the elemental world, for the Creator divided the substance of the elements into five parts. The purest is called the "quintessence of the elements," followed in descending order by fire, air, water, and earth. According to pseudo-Lull, these elements were originally "pure and clear by reason of the clear part of nature";[55] that is, the quintessence. But as a result of the Fall they have been corrupted, and indeed they grow more corrupt daily as the Last Judgment approaches. When the end of the world arrives, the world will be consumed in a vast conflagration, and all that "is not of the purity of the elements . . . will be poured into the abyss."[56] Only the pure, transparent elements, unmarred by terrestrial putrefaction, will remain, untouched by the fire of the Last Judgment at the end of time. The philosophers' stone itself, which has the power of bestowing incorruptibility, must be made from such incorruptible, central elements. Just as the pure elements will be able to withstand the final conflagration, so will a metal transmuted by the pure elements survive the assayer's fire. The twin themes of a Creator God who makes the world by a process involving alchemical separation and of an eschatology also pictured in terms of alchemy were to have significant repercussions in the early-modern period, when Paracelsus, for example, made them the basis of his own "spagyric" interpretation of the Bible.

[52] Michela Pereira, *The Alchemical Corpus Atributed to Ramon Lull* (London: Warburg Institute, 1989).
[53] Ibid., p. 3.
[54] Michela Pereira and Barbara Spaggiari, *Il "Testamentum" alchemico attribuito a Raimondo Lullo* (Florence: SISMEL-Edizioni del Galluzo, 1999), pp. 12–14.
[55] Ibid., p. 14.
[56] Ibid., p. 16.

The religious focus of the pseudo-Lullian corpus is emphasized in another way by the work *On the Secrets of Nature* (TK 408), which contains a dialogue between the supposed Ramon Lull and a Benedictine monk. The monk convinces Ramon, who is weeping dejectedly, to turn his alchemical talents to curing the afflictions of the diseased.[57] All of this is in fact a framing device allowing the pseudepigrapher to incorporate large chunks of John of Rupescissa's *On the Quintessence* directly into his own work. In fact, pseudo-Lull's *On the Secrets of Nature* served as an extremely important point of diffusion for Rupescissa's text since the mystique of the Lullian art added an additional novelty to Rupescissa's straightforward exposition. The huge vogue of Lullism among writers on natural magic and occult philosophy in the fifteenth and sixteenth centuries, such as Cornelius Agrippa von Nettesheim, served to popularize these works of pseudo-Lullian alchemy still further.

CONCLUSION

Medieval Latin alchemy developed in three stages. The first stage, translation – primarily from Arabic into Latin – was followed by a second period, in which the philosophical fruits of scholasticism were applied to an understanding of alchemical transmutation and its implications for matter theory. In the third stage, alchemy gradually dissociated itself from the university context, as religious motifs and figurative forms of expression grew ever more dominant. All three stages present a variety of historiographical problems, which will remain unsolved until the relevant manuscripts are examined and the texts edited. The most pressing problem concerns the identities of the alchemical authors since the vast majority chose to write under pseudonyms. In most cases, we do not even know the geographical locations in which alchemical texts were composed, much less the identity of their authors or the contexts in which they were writing. The role of alchemy in extra-university settings such as monasteries and courts could provide material for a host of local studies, but only when manuscript and archival sources are exploited fully.

Another problem concerns the evolution and dispersion of alchemical techniques in technologies such as metallurgy and the refining of minerals. Alchemical recipes are also clearly related to the medieval industries of pigment making, assaying, salt manufacture, and pharmaceuticals, but the precise relationship between alchemy and these commercial enterprises remains to be established.

57 Michela Pereira, "Filosofia naturale lulliana e alchimia," *Rivista di storia della filosofia*, 4 (1986), 747–80.

Finally, the link between the alchemy of the Middle Ages and early-modern science casts new light on the old issue of continuity, or the lack thereof, between the two periods. The corpuscular theory of "Geber" and the alchemical medicine of John of Rupescissa are but two examples of medieval innovation in alchemy that would have a remarkably successful afterlife in the early-modern period. One may also point to the extensive alchemical literature denying an essential and necessary difference between products of human art and those of nature. This theme, adopted by Francis Bacon in the early seventeenth century, continued to find striking echoes in the work of Robert Boyle and Isaac Newton. In the realm of experiment, the identity of artificial and natural products had important implications since it meant that alchemists were not inclined to view the artificial constraints of an experimental situation as producing an "unnatural," and hence invalid, result. The role of medieval alchemy in nurturing an incipient experimentalism points to the need to revise the traditional identification of medieval science with a supposedly nonexperimental, academic natural philosophy.[58]

[58] This is a major theme of William R. Newman, *Promethean Ambitions: Alchemy and the Quest to Perfect Nature* (Chicago: University of Chicago Press, 2004); see particularly chapter 5, pp. 238–89.

17

CHANGE AND MOTION

Walter Roy Laird

Throughout the Middle Ages, motion and change were seen as the fundamental and immediate expressions of the innate natures of physical things. To understand their causes and effects was to grasp the nature of physical reality and to approach an understanding of higher realities such as spiritual substances and God. Philosophical treatments of motion and change in the Middle Ages arose out of teaching and commenting on Aristotle's natural works (the *libri naturales*) by masters in the medieval universities. Although banned several times at Paris early in the thirteenth century, by mid-century these works came to constitute almost the entire program in natural philosophy (*physica* or *philosophia naturalis*), which, along with logic, moral philosophy, and metaphysics, made up the core of the medieval arts curriculum. Beyond the faculty of arts, theories of motion and change came to be applied to theological problems such as grace and the sacraments, and they were in turn influenced by theological and other considerations. Motion and change were matters of intense interest to medieval natural philosophers, logicians, mathematicians, and theologians alike, and the scope of their studies extended far beyond the narrow limits of what in the seventeenth century would become dynamics and kinematics. Within medieval treatments of motion and change, there emerged a number of scientific and mathematical advances of great interest and lasting importance in the history of science, including a rule on the relation between force and speed, impetus theory, the quantification of qualities, the mean-speed theorem, and the graphical representation of qualities and motion. The medieval science of motion, therefore, was not merely an imperfect foreshadowing of the physical science of Galileo and Newton but a comprehensive program of natural philosophy, the purpose of which was a complete understanding of the natural world – the world of motion and change – through its principles and causes.

Aristotle's *libri naturales* covered the whole of physical reality and all the particular kinds of natural bodies, with their various motions and changes, including the heavens, elemental bodies, atmospheric and terrestrial phenomena, and living things. Of these, by far the most important within the medieval arts curriculum was his *Physica* (*Physics*), in which Aristotle gave an account of the general principles of nature, motion, and change. Here he discussed the nature of motion and change and their kinds, as well as time, place, void, continuity, speed, the relation between movers and moved bodies, self-movers, natural and forced motions, projectiles, and the first motion and first mover. Of the books on particular branches of natural philosophy, *De caelo* (*On the Heavens*) treated the nature and motions of the celestial realm and contained Aristotle's most extensive treatment of natural and forced motions, the motions of heavy and light bodies, and the nature and motions of celestial bodies. Together, *Physics* and *De caelo* thus provided the framework and most of the material for medieval discussions of motion and change.[1]

CHANGE AND MOTION

Change and motion are as puzzling to the mind as they are obvious to the senses. Most philosophers before Aristotle recognized their existence in the sensible world but were unable to explain satisfactorily what they were or their relation to other existing realities. The pre-Socratic philosopher Parmenides argued against the reality of all change on the grounds that nothing can come from nothing and that what already exists cannot come to be. Aristotle, for his part, dismissed such denials of the obvious and sought instead to account for motion and change through the discovery of the general principles of natural things. A principle of something is its source or origin, that from which it comes or what makes or constitutes it. In the first book of his *Physics*, Aristotle concluded that three principles underlie the existence and motion of natural things: the most general pair of contraries, called form and privation, and the subject that underlies them, called matter. Form is whatever qualifies a thing, makes it what it actually is – for example, white or hot. Privation, by contrast, is the exact absence of that form, what the thing could be but is not – black or cold. And matter is the underlying subject that receives or is subjected to different forms. The idea of form and privation thus breaks the Parmenidean deadlock by allowing a thing both to be in some sense (to be white or hot) and not to be in another (not to be black or cold). Motion, then, is the transition of a subject from privation to form or

[1] On motion and change in general in the Middle Ages, see John E. Murdoch and Edith D. Sylla, "The Science of Motion," in *Science in the Middle Ages*, ed. David C. Lindberg (Chicago: University of Chicago Press, 1978), pp. 206–64; and James A. Weisheipl, *The Development of Physical Theory in the Middle Ages* (1959; repr. Ann Arbor: University of Michigan Press, 1971). Many of the original texts are edited and translated, with commentary and introductions, in Marshall Clagett, *The Science of Mechanics in the Middle Ages* (Madison: University of Wisconsin Press, 1961), pp. 163–625, and Edward Grant, *A Source Book in Medieval Science* (Cambridge, Mass.: Harvard University Press, 1974), pp. 228–367.

from form to privation that occurs successively – over time – rather than all at once.[2]

Aristotle took it as obvious that all motion and change proceeds from something (some condition or place) toward something. The goal to which the motion and change of a natural thing is directed is determined precisely by its *nature*. The nature of a seed is to grow into a particular sort of plant; the nature of a heavy body is to descend. The nature of a thing is thus its internal source or principle of change or motion toward its natural goal. This nature is often associated with its matter (as the nature of a tree is sometimes said to be wood), though it is more properly to be associated with the form, which makes the thing what it is and gives it its proper propensity to move.[3]

In addition to matter and form, which medieval commentators often called the internal causes of a thing, Aristotle had distinguished two external causes: the efficient cause and the final cause. The efficient cause is what effects the change or impresses a form on matter; in Aristotle's example, the sculptor is the efficient cause of the statue. The final cause is the purpose or end – that for the sake of which. Thus the final cause of the statue is to be venerated in a temple. Natural things, as well as artificial things such as statues, can be accounted for by all four causes, though in natural things the formal, efficient, and final causes often coincide. The final cause also often determines the others in that ends often determine means, and for this reason Aristotle considered the final cause to be the most important in nature. As a result, his physics is often called *teleological* (from the Greek *telos*, meaning end or goal). According to Aristotle, nature – the internal source or principle of motion in natural things – acts for a purpose.[4]

Aristotle distinguished the motion and change that happens by nature from what is forced on a body by an external agent in opposition to its natural tendencies. Such motion was called violent or forced or compulsory, obvious examples of which are heavy bodies forced upward and projectiles propelled by a sling or bow. Unlike natural motions, which tend to be slower at the beginning and faster at the end, violent motions are usually faster at the beginning, when the external cause is strongest, and slower at the end, when the natural tendency of the body has taken over. In addition to natural and violent motions, Aristotle discussed motions and changes that are the result of willed or deliberate action, which medieval commentators called "voluntary" motions, thus distinguishing, according to cause, three kinds of motion: natural, violent or compulsory, and voluntary.[5] The medieval natural philosopher sought to distinguish these kinds of motions from one

2 Aristotle, *Physics*, I.7.189b30–190b17; and Aristotle, *Metaphysics*, XI.9.1065b5–1066a34.

3 Aristotle, *Physics*, II.1.192b8–193b21.

4 Ibid., II.3–9.194b16–200b8.

5 James A. Weisheipl, "Natural and Compulsory Movement," *New Scholasticism*, 29 (1955), 50–81, reprinted in James A. Weisheipl, *Nature and Motion in the Middle Ages*, ed. William E. Carroll (Washington, D.C.: Catholic University of America Press, 1985), pp. 25–48.

another and to determine their proper and immediate causes. This will become especially apparent later in this chapter in the discussions of motion in a void, projectiles, falling bodies, and celestial motions.

Whatever their causes, motion and change are of several different kinds, so one must ask whether these are fundamentally different or fundamentally the same, and whether motion itself is some reality apart from the subject in motion, on the one hand, or the goal or terminus of the motion on the other. For Aristotle, motion or change – both words translate his *kinesis* and henceforth will be used interchangeably – was not a reality itself but an actuality occurring in relation to quantity, quality, and place.[6] Thus he distinguished three kinds or species of motion: change of size or quantity (augmentation and diminution in the case of living things, rarefaction and condensation in the case of nonliving things), change of quality (or alteration), and change of place (local motion). A fourth candidate – substantial change (the transformation of one kind of thing into a completely different kind of thing) – Aristotle eliminated as a kind of motion since it occurs suddenly rather than over time. His medieval commentators called substantial change mutation (*mutatio*), and some at least included it in their discussions of motion in general.

To confuse the matter, Aristotle had said in his *Categories* that being heated and being cooled were examples of *passio* (being acted upon), which implied that all motion is a single kind of thing – a *passio* – distinct from the body in motion and its terminus.[7] The eleventh-century Arabic commentator Avicenna (Ibn Sīnā, 980–1037) took up this implication, holding that motion in general was a *passio* with four species or kinds (local motion and qualitative, quantitative, and substantial change), and distinct both from the body in motion and its terminus. Averroes (Ibn Rushd, 1126–1198), who was known in the Latin West as "The Commentator" for his exhaustive commentaries on Aristotle, tried to resolve the apparent contradiction in Aristotle by conceding that although motion was a *passio* formally, it was of four kinds materially, and in this respect motion belongs to the same species as its terminus.[8]

The question of the nature of motion was among the many physical problems first treated in a thorough, scholastic manner in the Latin West by Albert the Great (Albertus Magnus, ca. 1200–1280), in his commentary on *Physics*, and his discussion largely set the terms for later debate. Albert was especially concerned with distinguishing the truth of Aristotle's views from what he saw as distortions and misinterpretations introduced by Avicenna and Averroes, whose commentaries had accompanied the works of Aristotle in their reception in the Latin West. Aristotle had defined motion as

[6] Aristiotle, *Physics*, V.1–2.224a21–226b17; and Aristotle, *Metaphysics*, XI.11–12. 1067b1–1069a14.

[7] Aristotle, *Categories*, 9.11b1–5.

[8] Ernest J. McCullough, "St. Albert on Motion as *Forma fluens* and *Fluxus formae*," in *Albertus Magnus and the Sciences*, ed. James A. Weisheipl (Toronto: Pontifical Institute of Mediaeval Studies, 1980), pp. 129–53 at pp. 129–35.

"the actualization of what exists potentially insofar as it is potential," a definition that gave medieval commentators far less trouble than it later gave René Descartes (1596–1650), who famously declared it to be incomprehensible. It was obvious to Aristotle that motion was some kind of actuality since a body in motion is actually different from one at rest. But the crucial characteristic of motion is that, while it is occurring, it is incomplete. Motion is an actuality, but an incomplete actuality (what we might call a process), for before the motion has begun, the body is merely potentially in a new place, and once it has arrived in that new place its actuality is complete. Only while that potentiality is actually being realized precisely as potentiality is there motion. Motion is thus an incomplete actuality, the actualization of what exists potentially insofar as it is potential. Medieval commentators called this definition the formal definition of motion to distinguish it from what they called its material definition: that motion is the actuality of the mobile body precisely as mobile.[9]

For Albert, as for his student Thomas Aquinas (ca. 1224–1274), motion was not a univocal term – having exactly the same meaning in its various applications and referring to a single kind of being – but an analogous term, in that it is applied by analogy to more than one kind of being. Thus each different motion is in fact the same in essence as its terminus or end, differing from it only in perfection; for example, the motion of becoming white is in its essence the same as whiteness – its goal and end – except less perfect or complete (i.e., less white). According to Albert, then, motion is the continuous going forth of a form (*continuus exitus formae*) from mover to mobile, a form (e.g., whiteness) that flows rather than a static one. Thus Albert gave what he called the "total definition" of motion as the simultaneous activity of both mover and moved body, and in this way he encompassed the external causes – the efficient and, at least implicitly, the final cause. Although the phrase was not used by Albert, motion under the view that it was identical in essence or definition to its terminus came to be called a *forma fluens* (a form flowing), the view almost universally accepted among later commentators, in contrast to a *fluxus formae* (the flow of a form), which later came to characterize Avicenna's position. But the phrases *forma fluens* and *fluxus formae*, however convenient for modern historians, were not used by medieval commentators with sufficient consistency and precision to delineate accurately their various positions concerning the nature of motion.[10]

The view of motion as a *forma fluens* was pushed to its extreme form by the relentless logical analysis of terms and the spare, nominalist ontology of the Oxford logician and theologian William of Ockham (1290–1349). For

9 Aristotle, *Physics*, III.1–3.200b12–202b29.

10 McCullough, "St. Albert on Motion as *Forma fluens* and *Fluxus formae*," pp. 135–53.

Ockham, there are no realities beyond absolute things (*res absolutae*) – individual substances and the qualities that inhere in them. Quantity, relation, place, time, action, passion, and the other Aristotelian categories are not absolute things but merely convenient ways of speaking that can be reduced to statements about absolute things. Likewise motion is not some kind of reality beyond the mobile body and the accidents or places that the body acquires successively without pausing in any. Although later natural philosophers did not always accept this conclusion, in general they formulated the question of the nature of motion in the same way as Ockham had.[11]

PLACE AND TIME

The genius of Aristotle's definition of motion is that it appeals only to prior and more general metaphysical principles – actuality and potentiality – and does not depend on preexisting or subsisting space and time, in which or through which bodies are supposed to move. Rather, three-dimensional extension is the first, essential, and fundamental property of all bodies. The essence of a body is to be extended, and to be extended is to be a body. Space, according to Aristotle and many of his medieval commentators, was a mental abstraction of the extension of bodies, a result of considering the three-dimensionality of a material body abstracted from its matter and all other forms. But the three-dimensionality of a body no more exists separately from that body than does its weight, though both can be thought of as though they did. The science that treats weight as separate from material bodies is statics or, as it was known in the Middle Ages, the science of weights (*scientia de ponderibus*); the science that treats extension in general as separate from material body is geometry, whereas stereometry treats specifically three-dimensional extension. To maintain that space is a subsisting, three-dimensional reality somehow separate from bodies not only results in certain physical absurdities (for example, that each part of space is inside another and so on to infinity) but also fails to explain how a body can move from one place to another. In Book IV of *Physics*, after rejecting this naive but common idea of empty space, Aristotle defined place (*ho topos*, Latin *locus*) in such a way as to avoid these absurdities while allowing for local motion. The place of something, he concluded, is the innermost, motionless surface of the body immediately surrounding it. The immediate and proper place of Socrates, for example, is

[11] James A. Weisheipl, "The Interpretation of Aristotle's *Physics* and the Science of Motion," in *The Cambridge History of Later Medieval Philosophy*, ed. Norman Kretzmann, Anthony Kenny, and Jan Pinborg (Cambridge: Cambridge University Press, 1982), pp. 521–36; and Anneliese Maier, "The Nature of Motion," in *On the Threshold of Exact Science: Selected Writings of Anneliese Maier on Late Medieval Natural Philosophy*, ed. and trans. Steven D. Sargent (Philadelphia: University of Pennsylvania Press, 1982), pp. 21–39.

the surface of the air touching him and of the ground touching the soles of his feet. Place is thus not some empty, separate three-dimensionality but the surface of a real body, which can be left behind and subsequently be occupied by some other body (for example, the air that flows in when Socrates walks away). Nor is place a mere mathematical abstraction; it is the real physical relation between bodies expressed as their surfaces in contact. This is crucial for Aristotle's physical theory because it explains the natural distinction between up and down that he attributes to place as well as the curious power that place has in the natural motions of elemental bodies.[12]

According to Aristotle, each of the four primary bodies or elements (earth, water, air, and fire) has its own natural place, the place to which it is naturally moved and in which it is naturally at rest. The place of earth is at the center of the cosmos, and that of fire is immediately below the sphere of the Moon. Between them are the concentric places of water and air. Fire has the natural tendency, called lightness or levity, to rise to its natural place at the circumference of the cosmos, whereas earth has the natural tendency, called heaviness or gravity, to sink to the center. The two other elements have a tendency either to rise or to fall depending on where they find themselves, as well as a certain susceptibility in general to be moved. Mixed bodies – those composed of several elements – have the tendency of the element predominant in them. The place of an elemental body is determined by the body that surrounds it and determines in turn whether it will be at rest (if it is in its natural place) or in motion (if it is not) and in which direction. Place, then, has within itself the distinction of up and down and the power or potency to actualize a body's natural propensity for rest or for motion up or down. Without place, there could be no distinction between natural and forced motions. Aristotle's definition of place was defended by almost all medieval commentators against what was taken to be the unlearned view that place was a three-dimensional reality somehow distinct from the bodies in it – what we now call "space."[13] The possibility of the separate existence of such a three-dimensional, empty space is best discussed later in the context of arguments concerning the void.

Time, similarly, is not an existing reality in which or through which motion occurs. For Aristotle, time was rather "the number of motion with respect to before and after." Time is a number or measure of motion because

[12] Aristotle, *Physics*, IV.1–5.208a27–213a11.

[13] Ibid., IV.1.208b8–27; Aristotle, *De caelo*, IV.3–5.310a14–3313a14; Pierre Duhem, *Theories of Infinity, Place, Time, Void, and the Plurality of Worlds*, ed. and trans. Roger Ariew (Chicago: University of Chicago Press, 1985), pp. 139–268; Edward Grant, "Place and Space in Medieval Physical Thought," in *Motion and Time, Space and Matter: Interrelations in the History of Philosophy and Science*, ed. Peter K. Machamer and Robert G. Turnbull (Columbus: Ohio State University Press, 1976), pp. 137–67, reprinted in Edward Grant, *Studies in Medieval Science and Natural Philosophy* (London: Variorum, 1981); and Edward Grant, "The Concept of *Ubi* in Medieval and Renaissance Discussions of Place," *Manuscripta*, 20 (1976), 71–80, reprinted in Grant, *Studies in Medieval Science and Natural Philosophy*.

it is by time that we measure one motion to be longer or faster than another. Unlike whiteness or heat, which are internal to the body, time is external to the moved body, similar to distance in local motion, which is the other external measure of motion with respect to longer and faster. But whereas distance measures only local motion with respect to here and there (i.e., with respect to place), time measures all motions with respect to before and after, now and then. And just as distance or space is the metric abstraction arising from the extension of bodies, time is the metric abstraction arising from the motion of bodies. Motion is thus prior to and more fundamental than time.[14]

MOTION IN A VOID

Whether there could be motion through a void space and what its characteristics might be were subjects of considerable controversy among medieval commentators on Aristotle's *Physics*. Although most followed Aristotle in asserting that actually existing void spaces are impossible by nature, the general belief that an omnipotent God could create a void provided the occasion for considering hypothetical motion through such a void and for examining the nature of motion in general, the role of the medium, and the relation between moving powers and resistances.

The ancient atomists Democritus and Leucippus had posited the existence of an empty space or void to allow their atoms to move rather than be fixed motionless against each other. The force of Aristotle's arguments in Book IV of *Physics* against the existence of the void was to show that positing a void not only could not serve the purpose intended but would also result in certain absurdities concerning motion. Natural motion, he argued, requires that there be a place in which the body is now located and a place toward which it naturally tends. But if there were a void space, places and directions in it would be undifferentiated, so that there could be neither natural nor violent motion. Nor would there be any reason for a body to come to rest in any particular place, so in a void all bodies would be at rest; or, if they were moved, having no natural tendency or any medium to counteract their motion, they would be moved indefinitely unless stopped by a more powerful body, which Aristotle took to be self-evidently absurd. Again, since according to Aristotle some violent motions depend on the medium (notably projectile motion, discussed later), these also would be impossible in a void. If, as the atomists asserted, the void were the cause of motion by drawing a body into itself, then a body surrounded by a void would tend to be moved equally in all directions; that is, it would not be moved at all. Furthermore, the ratio of

[14] Aristotle, *Physics*, IV.10–14.217b29–224a17. See Duhem, *Theories of Infinity, Place, Time, Void, and the Plurality of Worlds*, pp. 295–363.

speeds of a given body moving through two different resisting media is as the ratio of the rarity of the media, but between a medium however rare and the void there is no ratio, which implies that the speed of motion in a void would be infinite. Finally, since bodies with greater weight divide a medium more readily and thus move more swiftly, in a void all bodies, whatever their weights, will move at equal speeds, which Aristotle took to be impossible.[15] As for interstitial voids – the supposed microscopic voids between the atomists' particles – they are not needed to explain condensation and rarefaction. Like temperature and weight, rarity and density for Aristotle were a kind of qualitative accident changed by alteration, though in this case their intensity is measured by size alone rather than in degrees or pounds. Thus, even by the atomists' own assumptions, the void is neither sufficient nor necessary to explain either local motion or increase and decrease.[16]

Some of Aristotle's later commentators, however, used the hypothetical void to analyze the nature and conditions of motion itself. In particular, Aristotle had argued that, if the void were the sole cause of motion, there could be no ratio between the speed of a motion in a void and that of a motion in a medium. The implications of this – that motion in a void would be instantaneous and consequently that the temporal successiveness of motion depended on a resisting medium – were to have a long history among medieval commentators. Averroes reported that Avempace (Ibn Bajja, d. 1138), perhaps influenced by the sixth-century Greek commentator John Philoponus (d. ca. 570), asserted that motion does not require a resisting medium in order to have a finite speed. As an example, Avempace cited the motions of the celestial spheres, which despite the absence of anything to resist their motion still move with finite speeds. Rather, he argued, the speed of a body is determined by the ratio of the body moved to the cause of its motion, any medium serving only to slow the motion further by subtracting speed. This slowing or retardation, and not the speed itself, is thus proportional to the resistance of the medium.[17]

15 Aristotle, *Physics*, IV.6–9.213a12–216b21.

16 Ibid., IV.9.216b22–217b28.

17 Edward Grant, "Motion in the Void and the Principle of Inertia in the Middle Ages," *Isis*, 55 (1964), 265–92, reprinted in Grant, *Studies in Medieval Science and Natural Philosophy*; James A. Weisheipl, "Motion in a Void: Aquinas and Averroes," in *St. Thomas Aquinas, 1274–1974: Commemorative Studies*, ed. A. A. Mauer, 2 vols. (Toronto: Pontifical Institute of Mediaeval Studies, 1974), vol. 1, pp. 467–88, reprinted in Weisheipl, *Nature and Motion in the Middle Ages*, pp. 121–42. See also Duhem, *Theories of Infinity, Place, Time, Void, and the Plurality of Worlds*, pp. 369–414; Ernest A. Moody, "Ockham and Aegidius of Rome," *Franciscan Studies*, 9 (1949), 417–42, reprinted in Ernest A. Moody, *Studies in Medieval Philosophy, Science, and Logic* (Berkeley: University of California Press, 1975), pp. 161–88 at pp. 166–70; and Ernest A. Moody, "Galileo and Avempace: The Dynamics of the Leaning Tower Experiment," *Journal of the History of Ideas*, 12 (1951), 163–93, 375–422, reprinted in Moody, *Studies in Medieval Philosophy, Science, and Logic*, pp. 203–86 at pp. 226–9.

Averroes himself advanced a different opinion regarding motion in a void. He took Aristotle to require that every body be moved by a mover in constant contact with it (a *motor conjunctus*, or conjoined mover). In the case of living things, the soul, which is the form of a living thing, is the mover of the matter. In nonliving things, the form is also the mover, but it cannot move the matter directly since otherwise there would be no distinction between living and nonliving things. Instead, Averroes argued, the form somehow moves the medium, which in turn moves the matter of the body. Therefore, in the absence of a medium, inanimate natural motion could not occur. But even if it did, such motion would be instantaneous, for matter alone is purely passive and thus can offer no resistance to form, which is pure actuality. Nevertheless, Averroes concluded that, despite the absence of a resisting medium, the motion of the celestial spheres is not instantaneous because the spheres themselves offer the resistance to their movers necessary for finite speeds.[18]

Although Averroes's views on motion in a void exerted considerable influence through the high Middle Ages, on this point as on many others Thomas Aquinas offered an important alternative to the Averroist interpretation. According to Aquinas, motion in a void would not be instantaneous since a body, even with all its other forms removed (including heaviness and lightness), simply by virtue of being extended, has a natural inertness (*inertia*) to being moved. Note that this inertness does not arise from heaviness or some other property of the body since these have been explicitly removed from the body – still less does it arise from mass or quantity of matter in Newton's sense. Rather, it arises from the purely spatial extension of the body. Similarly, the magnitude through which a body is moved also has spatial extension, and since its parts are in sequence, so are the parts of the motion itself. Thus, for Aquinas, motion takes a finite time even in the absence of a medium and all other resistance, which would serve only to subtract from the body's natural speed. This argument came to be known as the "incompossibility of termini" (*incompossibilitas terminorum*) or "separation of termini" (*distantia terminorum*) since the mere separation of places (*termini*) and the impossibility of the same body's occupying them simultaneously were seen to be sufficient causes of the finite successiveness of motion. Nevertheless, most natural philosophers in the fourteenth century, including Aegidius of Rome, Walter Burley, and John of Jandun, followed the opinion of Averroes against Avempace. William of Ockham, however, saw no contradiction between the separation of termini as the cause of the successiveness of motion and Avempace's view that speeds decrease in proportion to the resistance of the medium. Ockham's resolution of the problem in effect distinguished between motion considered purely in terms of distance and time, and motion considered in relation to

[18] Weisheipl, "Motion in a Void," pp. 131–4; and Moody, "Galileo and Avempace," pp. 229–41.

its causes – a distinction that would be made explicit by later scholars at Oxford.[19]

Natural philosophers sometimes turned Aristotle's arguments concerning the void on their head, so that the absurd consequences of positing a void were simply accepted as hypothetically true and then analyzed and explained. In his *Treatise on the Ratios of Speeds in Motions* (*Tractatus de proportionibus velocitatum in motibus*) of 1328, the Oxford theologian Thomas Bradwardine (d. 1349) included a theorem stating that in a void all mixed bodies of similar composition would in fact move at equal speeds whatever their sizes, which Aristotle had deemed absurd. A mixed body was one compounded of several or all of the elements; because the elements are never actually found pure and unmixed, all actual physical bodies are mixed bodies. Since each of the elements has a natural tendency to be moved up or down, a mixed body, according to Aristotle in *De caelo*, will be moved with the tendency of the predominant element.[20] In refining these suggestions of Aristotle, medieval commentators usually identified the predominant element as the motive power and the elements with the contrary tendency as the resistive power. Depending on where the body was, one element of the body could be indifferent to motion and thus would contribute to neither motive nor resistive power. For example, a body located in air and consisting of three parts of fire and two parts each of earth, air, and water has in air a motive power downward since the downward power of the earth and water combined overcomes the upward power of the fire, whereas the air in its own element tends neither up nor down. In water, however, the body has an upward tendency since the upward motive power of the fire and air together exceeds the downward power of the earth, whereas the water this time has no tendency up or down.

There was considerable debate over whether the resulting speed depended on the excess of motive over resistive power (found by subtracting the resistive power from the motive) or on the ratio of the motive to the resistive power. Bradwardine, as will be shown at length in the next section, asserted that speeds follow the ratio of motive to resistive powers. And since mixed bodies of the same composition (that is, with the same ratio of elements) have the same ratio of motive to resistive powers, their speeds will be the same in a void. Bradwardine acknowledged that this conclusion was not at all obvious. He noted that it seems to contradict the common Aristotelian opinion that bodies move naturally at speeds in proportion to their magnitudes. But he explained that because the intensities of the motive forces are in the same ratio as the intensity of the resistances, they produce equal *intensities* of

19 Weisheipl, "Motion in a Void," pp. 134–42; Grant, "Motion in the Void and the Principle of Inertia in the Middle Ages," pp. 268–72; and Moody, "Galileo and Avempace," pp. 241–56.

20 Aristotle, *De caelo*, I.2.268b30–269a6.

motion (that is, equal speeds) despite the difference in their absolute sizes or *extensions*.

This distinction between the intensity and extension of motion would become a significant part of the latitude of forms subsequently pursued at Oxford and Paris in the next generation. Albert of Saxony (ca. 1316–1390) arrived at the same conclusion and went on to contrive a number of paradoxical results concerning the speeds of mixed and pure elemental bodies moving in various media and the void, including a case in which natural motion would be slower at the end than at the beginning. As for a hypothetical body composed of a single element moving in a void, Bradwardine had declined to assign it any definite speed, whereas Albert asserted that its speed would be infinite.[21]

BRADWARDINE'S RULE

In Book VII of *Physics*, Aristotle had treated in general the relation between powers, moved bodies, distance, and time, but his suggestions there were sufficiently ambiguous to give rise to considerable discussion and disagreement among his medieval commentators. The most successful theory, as well as the most mathematically sophisticated, was proposed by Thomas Bradwardine in his *Treatise on the Ratios of Speeds in Motions*. In this tour de force of medieval natural philosophy, Bradwardine devised a single simple rule to govern the relationship between moving and resisting powers and speeds that was both a brilliant application of mathematics to motion and also a tolerable interpretation of Aristotle's text.

In what is known as the ship-haulers argument, Aristotle had argued that if a certain power (e.g., a certain number of men) can move a certain body (e.g., a ship) a certain distance in a certain time, that same power can also move half that body the same distance in half the time or twice the distance in the same time. Similarly, twice that power can move the same body twice the distance in the same time or the same distance in half the time. But half the power does not necessarily move the same body half the distance in the same time or the same distance in twice the time, for the power may be too weak to move the body at all.[22]

[21] H. Lamar Crosby, Jr., ed. and trans., *Thomas of Bradwardine: His Tractatus de Proportionibus, III.2*, Theorem XII (Madison: University of Wisconsin Press, 1961), pp. 116–17; and Edward Grant, "Bradwardine and Galileo: Equality of Velocities in the Void," *Archive for the History of Exact Sciences*, 2 (1965), 344–64, reprinted in Grant, *Studies in Medieval Science and Natural Philosophy*, pp. 345–55.

[22] Aristotle, *Physics*, VII.5.249b27–250b7. See Israel E. Drabkin, "Notes on the Laws of Motion in Aristotle," *American Journal of Philology*, 59 (1938), 60–84.

It was not Aristotle's purpose here to state a precise quantitative rule governing powers and motions. Rather, this argument was one of the preliminaries leading to his proof in Book VIII of *Physics* that there must be a first mover and a first mobile body. Medieval commentators, however, considered the ship-haulers argument to be his definitive statement of the relationship between movers, resistances, and speeds, and they tried to put his rule into more definite form. They identified powers and moved bodies as moving powers and resisting powers, which, since both were powers, could then be put into ratios with each other. They thought of speed not in the modern sense as a ratio of distance to time (since such dissimilar quantities cannot form ratios) but as the intensity of motion, measured in degrees. (A motion with two degrees of speed, for example, is twice as fast as a motion with one degree of speed and will therefore cover twice the distance in the same time or the same distance in half the time.) Several interpretations of the relationship between motive and resisting powers and the resulting speeds had been proposed by Aristotle's ancient and medieval commentators in order to reconcile the rules of motion in Book VII of *Physics* with other passages and with common observation. The goal was to come up with a single general rule that expressed the relationship between moving and resisting powers and speed and at the same time precluded motion when the moving power is less than or equal to the resisting power, a goal that Bradwardine's rule achieved brilliantly.

Bradwardine wrote his *Treatise on the Ratios of Speeds in Motions* in the form of a deductive argument, beginning with definitions and suppositions and proceeding to theorems and proofs. In the first of its four chapters, he laid the necessary mathematical groundwork – the medieval theory of ratios and proportions, which until then had been applied mainly to music theory. The crux of Bradwardine's treatment is what it means for a ratio to be a *multiple* of – or, in general, greater than – another ratio. A ratio is a *multiple* (e.g., *double* or *triple*, in a technical sense that applies only to ratios) of another ratio when it can be compounded from that smaller ratio. For example, the ratio 4:1 is *double* 2:1 since 2:1 compounded twice with itself or *doubled* – (2:1)(2:1), or $(2{:}1)^2$ – yields the ratio 4:1; similarly, 8:1 is *triple* 2:1. And in general this is what Bradwardine meant when he said that one ratio is *greater than* another – that the larger is compounded from the smaller.[23]

Before using the theory of compounded ratios in his own rule, Bradwardine considered and rejected four alternative opinions on the relationship between powers, resistances, and speeds. The first is that speeds follow the excesses of motive powers over resistances ($V \propto [M-R]$, where $V =$ speed, $M =$ motive power, and $R =$ resistance), the opinion attributed to Avempace and others and discussed previously in connection with motion in a void. The second opinion, which Bradwardine attributed to Averroes, is that

[23] Crosby, *Thomas of Bradwardine*, I.3, pp. 76–81.

speeds follow the ratio of the excesses of the motive over the resisting powers to the resisting powers ($V \propto [M-R]/R$). The third opinion is that the speeds follow the inverse of the resistances when the moving powers are the same ($V \propto 1/R$ when M is constant) and follow the moving powers when the resistances are the same ($V \propto M$ when R is constant), which was the usual interpretation of Aristotle's rule. The fourth opinion is that speeds do not follow any ratio because motive and resistive powers are quantities of different species and so cannot form ratios with each other. Bradwardine rejected all of these because they are either inconsistent with Aristotle's opinion or yield results contrary to common experience.[24]

Bradwardine's own rule is that the ratio of speeds follows the ratios of motive to resistive powers. This deceptively simple formulation appears at first to be no different from the third rejected opinion until one understands it in the light of medieval ratio theory just explained. Speed is tripled, for example, when the ratio of motive to resistive power is *tripled* – in the technical sense that applies to ratios. For example, if a motive power of 2 moves a resistance of 1 at a certain speed, a motive power of 8 will move the same resistance with a speed three times the first speed since the ratio 8:1 is *triple* the ratio 2:1; that is, 8:1 is compounded of 2:1 with itself three times (i.e., $8:1 = (2:1)(2:1)(2:1)$ or $(2:1)^3$). And, in general, speeds are doubled or tripled or halved as the ratios between motive and resistive powers are *doubled* or *tripled* or *halved* in the technical sense that applies to ratios.[25]

Furthermore, Bradwardine's rule precludes the possibility of motion when the motive power becomes equal to or less than the resisting power because no ratio of greater inequality of motive to resisting power (e.g., 8:1) can be compounded of ratios of equality (e.g., 1:1 or 2:2) or of lesser inequality (e.g., 1:2), for no matter how many times one compounds ratios like 1:1 or 1:2 with themselves, one will never arrive at a ratio like 8:1. This means effectively that speed approaches zero as the ratio of motive to resisting power approaches equality; when the motive power is less than the resisting power, there will be no speed at all.[26] Thus, by applying medieval ratio theory to a controversial passage in Aristotle's *Physics*, Bradwardine devised a simple, definite, and sophisticated mathematical rule for the relation between speeds, powers, and resistances.

The rest of the *Treatise* consists of a series of theorems in which Bradwardine applied this rule to various combinations of motive and resistive powers. Where the ratios were *multiples*, he could determine the speeds as multiples, but in the many more cases where they were not *multiples*, he was content with proving only whether the resulting speed would be greater than, less than, or equal to some multiple of the original speed. He did not resort to

[24] Ibid., II, pp. 86–111.

[25] Ibid., III.1, pp. 110–13; and Anneliese Maier, "The Concept of the Function in Fourteenth-Century Physics," in Sargent, *On the Threshold of Exact Science*, pp. 61–75.

[26] Crosby, *Thomas of Bradwardine*, II, pp. 81–6, 114–15.

the actual measurement of powers or speeds to confirm his results, beyond his earlier appeals to common experience to disprove the alternative rules. Nevertheless, his rule led him to the remarkable conclusion, mentioned earlier, that all mixed heavy bodies of similar composition will move naturally at equal speeds in a vacuum.[27]

Bradwardine's rule was widely accepted in the fourteenth century, first among his contemporaries at Oxford, where Richard Swineshead and John Dumbleton applied it to solving sophisms, the logical and physical puzzles that were just beginning to assume an important place in the undergraduate arts curriculum (See Ashworth, Chapter 22, this volume.). Soon thereafter it appeared at Paris, in the works of Jean Buridan and Albert of Saxony, and, by mid-century, at Padua and elsewhere. It circulated not only in copies of Bradwardine's own work but even more widely in an abbreviated version dating probably from the mid-fourteenth century.[28]

In one of the more interesting applications of Bradwardine's rule to a physical problem, Richard Swineshead (fl. 1340–55), known to posterity as "the Calculator" from the title of his influential *Book of Calculations*, set out to determine whether a natural body acts as a unified whole or as the sum of its parts. He imagined a long, uniform, heavy body falling in a vacuum (to eliminate the complication of a resisting medium) down a tunnel through the center of the earth. The motive power is represented by the length of the bar above the center and still falling, and the resisting power is represented by the length of the bar that has already fallen past the center and so is now rising on the other side. Using Bradwardine's rule – that the speed of the bar follows the ratio of motive to resisting power – he examined each successive speed as the distance between the center of the bar and the center of the world is successively halved. He found that the speed of the bar decreases in such a way that its center will never reach the center of the world in any finite time. But Swineshead in the end rejected this conclusion because of the unacceptable consequence that the nature of the whole bar as a heavy body would be thwarted since its center could never coincide with its natural goal, the center of the world. Instead, he argued that all the parts of the body would assist the whole in achieving its goal. Despite the subtlety of the analysis, then, for Swineshead the physical argument trumped the mathematical.[29]

Elsewhere in his *Book of Calculations*, Swineshead applied Bradwardine's rule to the analysis of motions through variously resisting media, devising rules to compare not only speeds but the rate of change of speeds when either

[27] Ibid., III.1, pp. 112–17; the conclusion mentioned is Theorem XII, pp. 116–17.

[28] Clagett, *Science of Mechanics in the Middle Ages*, pp. 421–503, 629–52.

[29] Michael A. Hoskin and A. G. Molland, "Swineshead on Falling Bodies: An Example of Fourteenth-Century Physics," *British Journal for the History of Science*, 3 (1966), 150–82; A. G. Molland, "Richard Swineshead and Continuously Varying Quantities," in *Acts of the Twelfth International Congress on the History of Science, Paris, 1968*, 12 vols. in 15 (Paris: Albert Blanchard, 1971), vol. 4, pp. 127–30; and Molland, Chapter 21, this volume.

the moving or the resisting power is continuously increased or decreased with time. He also considered more complex cases, such as a body moved by a constant power through a medium the resistance of which increases uniformly with distance (rather than with time). Since the speed of the body at any instant depends on where it is in the medium but where it is in the medium depends in turn on its speed and the time it has traveled, to solve this problem Swineshead effectively had to reduce the distance dependency of speed to time dependency.[30]

The only other significant extension of Bradwardine's rule – of more interest to the history of mathematics than of the science of motion – was by Nicole Oresme. As mentioned earlier, where the ratios were not simple *multiples* of other ratios, Bradwardine could calculate only whether the resulting speeds were larger or smaller than some multiple. Oresme, however, generalized what he called the ratio of ratios to include improper fractional and even irrational ratios of ratios, which extended the mathematical rigor and generality of Bradwardine's rule but without changing its basic application to motion.[31]

FALLING BODIES AND PROJECTILES

Two of the most controversial problems in the medieval science of motion concerned the acceleration of falling bodies and the continued motion of projectiles. The ancient commentators had already criticized Aristotle's treatments, and medieval commentators in turn tried to solve these problems in ways that were consistent both with general Aristotelian principles and with everyday experience. In the effort to explicate Aristotle, however, they devised solutions that went far beyond what could reasonably be attributed to Aristotle, producing some of the most interesting innovations in medieval physical thought. And, in the process, Aristotle's distinction between natural and violent motions tended to become blurred, implicitly calling into question a fundamental tenet of Aristotelian natural philosophy.

Falling bodies and projectiles came up in the course of Aristotle's arguments for a first or prime mover in Book VIII of *Physics*. The main premise of the argument was that everything that is moved is moved by another. To establish this premise, Aristotle distinguished three classes of motion according to the mover's relationship to the thing moved: the motion of living (self-moving) things, forced (compulsory or violent) motion, and natural (or spontaneous) motion. He then showed that in each case the moved body is distinct from its mover. This distinction is clearest in the case of forced or

[30] John E. Murdoch and Edith Dudley Sylla, "Swineshead, Richard," *Dictionary of Scientific Biography*, XIII, 184–213, especially pp. 201–4.

[31] Nicole Oresme, *De proportionibus proportionum and Ad pauca respicientes*, ed. and trans. Edward Grant (Madison: University of Wisconsin Press, 1966), especially pp. 24–36.

violent motion, for there the cause of the motion is obviously different from the body moved. The horse moves the cart, and the bowstring moves the arrow. Where motion is imposed on a body from without, the mover must be in continuous contact with the body in order to endow it with motion. The difficulty with violent or forced motion was not in distinguishing the mover from the moved but rather in determining why the motion continues once the mover is no longer in contact with the moved body. This is the problem of projectile motion, which will be treated later in the chapter.

In the case of living things, since they are self-moving, there would seem to be no distinction between mover and moved. But here Aristotle distinguished self-movements proper – voluntary movements that arise from appetite – from involuntary movements such as growth, nutrition, generation, heart-beat, and the like that arise from the nature of the animal as a certain kind of living thing. Properly speaking, animals move themselves by walking, running, crawling, swimming, flying, and such, motions the efficient cause of which is the animal itself but understood in the sense of "part moving part." That is, in every case of an animal's moving itself, the motion is caused by one part of the animal first moving another part and then being moved in turn. In this way, Aristotle preserved the distinction between mover and moved in animals that move themselves and thus preserved the generality of the dictum that everything that is moved is moved by another.

But the most difficult cases, Aristotle admitted, are the involuntary movements of living things and the natural motions of inanimate bodies. These movements arise from the nature of the body, nature being the source of motion in those things that tend to move by virtue of what they are. In such cases, the immediate efficient cause of the motion is the generator of the body itself; that is, whatever brought the body into being as a heavy – or as a living – thing in the first place. Once a body has been endowed by its generator with a certain nature, it acts spontaneously in accordance with that nature, no further efficient cause being necessary. A heavy body, for example, will spontaneously fall by nature unless it is impeded by some other body. Thus the only other efficient cause that Aristotle admitted for natural motion is the cause that removes an impediment, although this is only an incidental rather than an essential cause of motion. For Aristotle, inanimate elemental bodies are not to be thought of as self-moving. Since they are homogeneous (that is, indefinitely divisible into parts of similar composition), there is in them no distinction of parts and therefore no first part that could move the other parts. And since they are not moved part by part, they are not self-moving.[32]

The nature of a body, which is the source and principle of its motion, is expressed in its formal and final causes; for example, the form of a heavy

[32] Aristotle, *Physics*, VII.1.241b34–242a49 and VIII.4.254b7–256a3; and Weisheipl, "The Principle *Omne quod movetur ab alio movetur* in Medieval Physics," *Isis*, 56 (1965), 26–45.

body makes it heavy and causes it to seek the center of the cosmos, its goal or final cause. For this reason, Avicenna identified the form of a natural body or its natural goal as efficient causes of its natural motion. Averroes, although allowing that the generator gives a body its form and all of its characteristic activities, including local motion, nevertheless argued that Aristotle required a mover conjoined to the body at all times, even in natural motion. Since only in living things can the form be the direct mover of the body, Averroes suggested that in inanimate natural bodies the form first moves the medium, which in turn moves the body itself, like a man rowing a boat. Thus, for Averroes the medium plays the same role in natural motion as it did for Aristotle in the continuation of forced, projectile motion.

Although both Albertus Magnus and Thomas Aquinas argued against Averroes's interpretation of Aristotle, the tendency to look for an efficient cause of natural motion within the moved body proved irresistible to many subsequent thinkers. One notable exception was Duns Scotus, who denied that everything moved is moved by another, asserting instead that all natural and voluntary motions are in fact self-motions.[33] However, many later commentators tended, like Averroes, to look within the naturally moving body itself for an efficient cause of its motion, similar to the cause of violent motion, especially of projectiles. Consequently, they posited a cause of forced motion, often called *impetus*, that inhered almost like an artificial, contrary nature in a moving projectile. Impetus was also sometimes invoked to explain the acceleration of freely falling heavy bodies. Thus, in the fourteenth century, the causes of natural and violent motion tended to be assimilated to each other; by the seventeenth century, the distinction between natural and violent motions would break down entirely in the physics of Galileo and Descartes. We shall return to the cause of the acceleration of falling bodies after discussing impetus theory.

PROJECTILE MOTION AND THE THEORY OF IMPETUS

In the course of his argument for a first motion and a first mover in Book VIII of *Physics*, Aristotle considered the possibility that the first motion (which he identified with the rotation of the outermost cosmic sphere) could be like that of a projectile in that it might continue not through the continued action of a first mover but rather as a stone continues to be moved through

[33] James A. Weisheipl, "The Specter of *Motor coniunctus* in Medieval Physics," in *Studi sul XIV secolo in memoria di Anneliese Maier*, ed. A. Maierù and A. P. Bagliani (Rome: Edizioni di Storia e Letteratura, 1981), pp. 81–104, reprinted in Weisheipl, *Nature and Motion in the Middle Ages*, pp. 99–120; and James A. Weisheipl, "Aristotle's Concept of Nature: Avicenna and Aquinas," in *Approaches to Nature in the Middle Ages: Papers of the Tenth Annual Conference of the Center for Medieval and Early Renaissance Studies*, ed. Lawrence D. Roberts (Medieval and Renaissance Texts and Studies, 16) (Binghamton, N.Y.: Center for Medieval and Early Renaissance Studies, 1982), pp. 137–60.

the air after it has left the hand or the sling that has thrown it. He rejected this possibility, however, by asserting a theory of projectile motion that would later be accepted by few of even his most loyal followers. According to Aristotle, the thrower imparts not only motion to the projectile and to the medium surrounding it but also a certain power of moving to the medium alone, either air or water, which has a natural propensity to receive such a power of moving. Each part of the medium so affected thus becomes a mover of the projectile by passing both motion and the power of moving to its adjacent parts. In passing from one part of the medium to the next and then to the projectile, both the motion and the power of moving are weakened, in part at least by the natural tendency of the projectile as a heavy or light body to be moved down or up. Eventually, the power of moving is so weak that the last part of the medium can pass on motion but not the power of moving. When this last motion of the medium ceases, so does the forced motion of the projectile.[34]

In Aristotle's defense, it must be said that his main purpose was not to explain projectile motion but rather to show that it was an unsuitable candidate for the first motion, for he had shown earlier that the first motion must be absolutely continuous, which in his theory the motion of a projectile is not since it is caused by a succession of different movers (i.e., the different parts of the medium). Nevertheless, his views on projectile motion received devastating criticism from John Philoponus, who in the sixth century first advanced most of the arguments that would be leveled against it over the next millennium. In general, Philoponus argued that a medium can act only to resist the motion of a projectile; what keeps it moving is an "incorporeal motive force" received from the projector. Although Philoponus's criticisms and alternative theory were not known directly in the Latin West, they probably influenced several Islamic philosophers, including Avicenna, who argued that projectiles receive from the projector an inclination – in Arabic, *mayl* – that inheres in the body and is not a motive force but only its instrument. But since these passages of Avicenna were not fully translated into Latin, the theory of *mayl* had little influence in the Latin Middle Ages, though it formed part of a parallel tradition in Arabic that lasted into the seventeenth century.[35]

In the Latin West, the first alternative to Aristotle's theory of projectile motion appeared, surprisingly, in the context of sacramental theology. In his commentary on Peter Lombard's *Sentences*, the Franciscan theologian Peter John Olivi (d. 1298) discussed how the sacraments could continue to be effective long after they had been instituted by Christ. By way of analogy, Olivi described a theory of projectile motion wherein the projector imparts to the projectile a similitude or *species* of itself as mover. Olivi was apparently

[34] Aristotle, *Physics*, VIII.10.266b27–267a15.
[35] Clagett, *Science of Mechanics in the Middle Ages*, pp. 505–14.

drawing on the species theory of Roger Bacon (ca. 1220–ca. 1292), which explained all causation by means of species or images of a cause imposed on other things.[36] In a later work, however, Olivi rejected this theory of projectile motion on the grounds that, since motion is a real mode of being in itself that needs nothing to sustain it, the continuation of a projectile needs no cause besides the original projector. This later position is identical to that taken by William of Ockham, though for exactly opposite reasons. Ockham not only rejected all species as unnecessary in the explanation of causes and their effects, but he also claimed that local motion was nothing over and above a body and the places it occupies successively without stopping in any. Since the continued motion of a projectile is not a new effect, it has no need of a cause at all. Around the same time, another theologian, Franciscus de Marchia, again drew the analogy between sacramental causality and projectile motion. According to Franciscus, in his lectures on the *Sentences* given at Paris in 1319–20, the mover gives to the projectile – rather than to the medium, as Aristotle thought – a certain moving power that remains for a time in the projectile and causes its motion to continue.[37]

Whether informed by these theological treatments or not, the philosopher Jean Buridan, twice rector of the University of Paris between 1328 and 1340, presented the canonical version of what he called *impetus* in his questions on Aristotle's *Physics* and *De caelo*. But first he mustered a decisive set of arguments against the view that the medium is somehow responsible for the continuing motion of projectiles. First, a top or a smith's wheel will continue to rotate in place long after it is pushed, despite the fact that air cannot get behind it, and continues to rotate even when one obstructs the air by placing a cloth next to its edge. Again, a lance pointed at both ends can be thrown as far as one pointed only in front, though the point behind presents less area for the air to push against. Moreover, air is so easily divided that it is difficult to see how it could move a thousand-pound stone thrown from a siege engine. Furthermore, ships will continue to move without there being any wind detected behind them, and no matter how hard one beats the air behind a stone, little motion is produced compared with throwing it. And finally, if the air moved projectiles, then a feather could be thrown farther than a stone since it is more easily moved by air, but the contrary is observed. Therefore, Buridan concluded, a projectile is not moved by the air but by a certain *impetus* or motive force (*vis motiva*) impressed on it by the mover in proportion to the mover's speed and to the amount of prime matter in the projectile. The air, along with the natural tendency of the heavy body to move downward, serves only to diminish the impetus by resisting the motion, so that the forced motion becomes slower and slower until it is completely overcome. This is why larger and denser bodies can be thrown

[36] See Lindberg and Tachau, Chapter 20, this volume.
[37] Ibid.

farther, for a greater impetus can be impressed on them by virtue of their greater quantity of prime matter, and why a greater resistance of air and natural tendency is necessary to overcome their motion. Buridan took pains to point out that impetus is not the same as the motion it produces, for motion is present only successively, one part after another, whereas impetus is a quality or accident present all at once in its subject. Impetus is thus a real motive power in the body it moves, impressed on it by the mover and inhering in it in the same way that a magnet impresses a quality on a piece of iron that causes it to be moved.[38] Using this idea of impetus, Buridan went on to explain a variety of physical phenomena, including the rebound of bodies from obstacles, the vibration of strings, the swinging of bells, and, most notably, the motion of the celestial spheres and the acceleration of falling bodies, to which we now turn.

ACCELERATION OF FALLING BODIES

According to Aristotle, natural motion (such as the descent of a heavy body) is faster at its end than at its beginning. He implied in *Physics* and *De caelo* that the reason for this in the case of falling bodies is that the closer a heavy body comes to the center of the cosmos (its natural place), the heavier it becomes and the faster it will fall.[39] Aristotle's ancient commentators gave several alternative explanations of the acceleration of falling bodies, attributing it either to the medium through which they fall or to some power temporarily resident in the body itself. Simplicius (sixth century), in his commentary on *De caelo*, recorded that, according to the astronomer Hipparchus, whatever power had lifted or sustained the heavy body continues to act on it as it falls, resulting in an increase in speed as this power gradually wears off. Others (Simplicius did not identify them) suggested that a lesser quantity of medium beneath a falling body offers less support, so that the lower the body gets, the faster it goes. In the end, however, Simplicius agreed with Aristotle, although he acknowledged that Aristotle's explanation implied that a body should weigh more when it is in a lower place unless perhaps the difference is insensible.

Philoponus, for his part, implicitly denied Aristotle's explanation of the acceleration of falling bodies when he argued in his commentary on *Physics* that two different weights fall with almost the same speed and not with speeds proportional to their weights. One medieval Arabic commentator, Abu ʿl-Barakat (d. ca. 1164), under the influence of Philoponus's explanation of projectile motion, argued that a projectile, after it has ceased moving upward

[38] Clagett, *Science of Mechanics in the Middle Ages*, pp. 532–40; and Maier, "The Significance of the Theory of Impetus for Scholastic Natural Philosophy," in Sargent, *On the Threshold of Exact Science*, pp. 76–102 at pp. 86–7.

[39] Aristotle, *De caelo*, 1.8.277a27–33; and Aristotle, *Physics*, IV.8.216a13–16.

and begins to fall, still retains some of its upward, violent inclination, or *mayl*, as it begins to fall, and because this violent *mayl* continues to weaken while the natural *mayl* increases as the body falls, the body speeds up. Averroes, in contrast, although agreeing with Aristotle's explanation, added that a falling body also speeds up because as it falls it heats the air through which it passes, which then becomes more rarefied and therefore offers less resistance.[40]

Drawing on Averroes and perhaps on Simplicius, whose commentary on *De caelo* had been translated by William Moerbeke in 1271, Jean Buridan offered a number of arguments against both Aristotle and those who attributed the acceleration of falling bodies to the medium. Against Averroes, he noted that bodies do not fall faster in the summer, when the air is hot, than in the winter, and that a man does not feel the air heat up as he moves his hand through it. Against Aristotle, he argued, as Philoponus had, that a stone is not lighter the higher up it is. When dropped from a high place, it moves much more quickly through the last foot of its fall than a stone dropped to the same level from the height of a foot; nor does a stone dropped from a high place through some distance fall faster than a stone dropped from a low place through an equal distance. For similar reasons, he rejected the opinion that the greater the amount of air beneath a falling body, the greater the resistance, suggesting instead that the air higher up should be rarer and thus less resistant. He also suggested that air closer to the ground should in fact offer more resistance to a falling body, not less, because there is less room for it to be pushed out of the way. Finally, he dismissed the suggestion that acceleration is caused by an insensible difference in weight between a body in a high place and one in a low place on the grounds that since the differences in speed and impact are very sensible, they cannot result from an insensible difference in weight.[41]

For all of these reasons, Buridan concluded that both the weight of the stone and the resistance of the medium are constant throughout the fall and thus do not contribute to acceleration. He suggested instead that, as a heavy body falls, it is moved not only by its gravity or weight (which remains constant and thus by itself would produce only a constant speed) but also by a certain impetus that the body acquires from its motion. At first, the body is moved only by its gravity, so it moves only slowly; then it is moved both by its gravity and by the impetus acquired from its first motion, so it moves more swiftly; and so on, its speed continually increasing as more and more impetus is added to its natural gravity. Buridan noted that this impetus could properly be called "accidental gravity," which placed him in agreement with both Aristotle and Averroes that increasing gravity is the cause of acceleration. Having rebutted earlier explanations with reasons drawn from a careful analysis of the observed behavior of falling bodies,

40 Clagett, *Science of Mechanics in the Middle Ages*, pp. 541–7.

41 Jean Buridan, *Questions of the Four Books on the Heavens* II, qu. 12, trans. in Clagett, *Science of Mechanics in the Middle Ages*, pp. 557–60.

Buridan thus applied his theory of impetus, originally devised to explain the continued violent motion of projectiles, to explain the acceleration of naturally falling bodies.[42]

Buridan's use of impetus theory to explain the acceleration of falling bodies was widely accepted in the late Middle Ages, by Albert of Saxony and Nicole Oresme, among many others. Albert applied it to a body that falls through a hole passing through the center of the earth, suggesting that its impetus would carry it past the center of the earth until overcome by the body's natural tendency; it would then again descend to the center and continue to oscillate about the center until its impetus was entirely spent. In adopting the impetus theory of falling bodies, Oresme introduced a modification of his own. According to him, impetus arises not from the speed of the falling body but from its acceleration; this impetus in turn causes more acceleration, and so on. He also repeated the now familiar example of a body falling through the center of the earth, which he compared to the swinging of a pendulum, in order to argue that impetus cannot properly be weight or gravity since in these cases it causes heavy bodies to rise, whereas weight can cause them only to fall. All motions, Oresme concluded, whether natural or violent, possess impetus – with the notable exception of the celestial movements.[43]

Compared with the cause of the acceleration of falling bodies, the mathematical rule that governed their actual acceleration received almost no attention in the Middle Ages. In general, medieval natural philosophers tended to assume that the speed of falling bodies increases with the distance and the time of fall, without distinguishing the two and without specifying the exact relationship. Nicole Oresme, however, suggested several alternatives for the rule governing the increase of speed of falling bodies, including the correct one – that equal speeds are acquired in equal times – though he did so only to determine whether the speed would increase indefinitely or approach some finite limit. Besides Oresme, only one other late scholastic philosopher asserted what is now known to be the correct rule to describe the acceleration of falling bodies: the sixteenth-century Spaniard Domingo de Soto, who applied to falling bodies the mean-speed theorem for uniformly accelerated motions, described in the next section.[44]

THE OXFORD CALCULATORS AND THE MEAN-SPEED THEOREM

In the fourteenth century, motion and change provided material for the logical and physical puzzles known as *sophismata*, which came to assume a

[42] Ibid., pp. 560–4.

[43] Ibid., pp. 552–3, 565–71.

[44] Ibid., pp. 541–56; and William A. Wallace, "The Enigma of Domingo de Soto: *Uniformiter difformis* and Falling Bodies in Late Medieval Physics," *Isis*, 59 (1968), 384–401.

significant place in the arts curriculum of the medieval university. *Sophismata* gave rise to an extensive body of literature in the form of handbooks for undergraduate students, in which the quantitative treatment of qualities and the comparison of speeds of motion resulted in a remarkable series of theoretical insights, including the idea of instantaneous velocity, the mean-speed theorem, the distance rule for accelerated motions, and the graphical representation of qualities and motions.[45] In this context, a distinction was drawn between considering the speed of motion with respect to its causes (that is, with respect to its powers and resistances) – as in Bradwardine's rule and impetus theory – and considering it with respect to its effects (that is, with respect to distances and times). This distinction is roughly equivalent to the modern distinction between dynamics and kinematics.

For motion considered with respect to its effects, Aristotle was again the main starting point; in Books VI and VII of *Physics*, he had discussed the comparison of speeds and the continuity and infinity of magnitude, time, and motion, partly to refute Zeno's paradoxes of motion but ultimately to show in Book VIII that the only possible infinite motion is the rotation of a sphere moved by an immaterial mover. In Book VI, he had defined the faster as what traverses a greater magnitude in an equal time or traverses an equal or greater magnitude in less time.[46] Then, in Book VII, he asked whether every change or motion is comparable with every other change or motion. Clearly, a change of place cannot be directly compared with an alteration, for although their times may be compared, a distance cannot be compared with a quality. Similarly, circular motions cannot be compared with straight ones because, for Aristotle, circular and straight were two different species in the genus of line, and since they are different kinds of things, they cannot be compared with each other. Strictly speaking, then, the speeds of circular and straight motions were for Aristotle incomparable.[47]

Apart from Aristotle, the other possible inspiration for the earliest medieval attempts to compare motions was *On the Motion of the Sphere* of Autolycus of Pitane (fl. 310 B.C.). Translated into Latin several times in the twelfth and thirteenth centuries and circulated widely with astronomical and other mathematical texts, *On the Motion of the Sphere* may have been known to Gerard of Brussels, who wrote his *Book on Motion* sometime between 1187 and 1260. Drawing on Archimedes' *On the Measure of the Circle* and a derivative of his *Sphere and Cylinder*, Gerard, like Autolycus, tried to find the point on a rotating line (for instance) such that, if the whole line were moved uniformly and rectilinearly at the speed of this point, its motion would be equivalent to its motion of rotation; that is, it would sweep out

45 Edith Dudley Sylla, "Science for Undergraduates in Medieval Universities," in *Science and Technology in Medieval Society*, ed. Pamela O. Long (Annals of the New York Academy of Sciences, 441) (New York: New York Academy of Sciences, 1985), pp. 171–86.

46 Aristotle, *Physics*, VI.2.232a23–27.

47 Ibid., VII.4.248a10–249b26.

an equal area in the same time. The point he chose for a rotating line was its midpoint, and he applied the same technique to the rotation of surfaces and volumes.[48] In a similar way, Thomas Bradwardine, in the fourth chapter of his *Treatise on the Ratios of Speeds in Motions*, reduced circular motions to their rectilinear equivalents, except that for him the speed of a rotating line was to be measured not by its midpoint but by its fastest point.[49] The solution to *sophismata* that involved the comparison of speeds of rotating bodies often hinged on whether the overall speed of a rotating body should be measured by its midpoint or by its fastest point – in either case, Aristotle's reservations concerning the comparison of rectilinear and circular motions came to be effectively ignored.

A number of scholars at Oxford in the 1330s and 1340s – notably William Heytesbury, Richard Swineshead, and John Dumbleton – avidly took up the study of motion with respect to its effects, developing quasi-mathematical techniques, known as calculations (*calculationes*), to deal with it and its associated problems. Since several of these masters, including Bradwardine, were associated with Merton College, Oxford, they are often referred to as the Merton school or the Mertonians, though the more inclusive "Oxford calculators" is preferable.[50] These scholars considered local motion within the more general context of what they called the intension and remission of forms – how formal qualities and properties in general are intensified and diminished and how they are to be compared as they change. Duns Scotus had suggested that a body became whiter, for example, not by exchanging its existing form of whiteness for an entirely new form of a higher degree or intensity but by an addition of whiteness to the existing form, resulting in a higher degree of whiteness. In this way, qualitative forms were quantified and their quantities expressed in degrees or intensity. Since the range of intensity was called "latitude," the theory concerning the quantity of qualities came to be called the "latitude of forms."

The mathematical techniques developed in connection with the latitude of forms were applied to a variety of natural phenomena. For example, the hotness of a body can be measured in two ways – by its intensity (what we call its temperature) and by its extension (the total quantity of heat in the body, which we measure in calories). Thus a one-pound and a two-pound block of iron at the same temperature have the same intensity of heat, but the larger block has twice the extension of total quantity of heat. Now motion itself fell under the same consideration, so that among the Oxford calculators speed came to be seen as the intensity of motion; that is, as an intensional quality with a latitude that could be intensified or remitted (diminished).

48 Clagett, *Science of Mechanics in the Middle Ages*, pp. 163–97.

49 Crosby, *Thomas Bradwardine*, IV, pp. 124–41.

50 James A. Weisheipl, "Ockham and some Mertonians," *Mediaeval Studies*, 30 (1968), 163–213; and Edith Dudley Sylla, "The Oxford Calculators," in Kretzmann, Kenny, and Pinborg, *Cambridge History of Later Medieval Philosophy*, pp. 540–63.

The intensity or speed of motion was distinguished from its extension in the same way that the intensity of a quality was distinguished from its extension. By analogy to hotness and other qualities, the motion of a body came to be measured both intensively as its speed at any instant and extensively as its total time and distance.[51]

Building on these ideas, William Heytesbury devoted Part VI of his comprehensive *Rules for Solving Sophisms* to problems concerning the three species of change: local motion, or change of place; augmentation, or change of size; and alteration, or change of quality. In the section on local motion, he distinguished uniform motion, which he defined as motion in which equal distances are traversed at equal speeds in equal times, from nonuniform or difform motion. Motion can be difform with respect to the subject, as for example the motion of a rotating body, since various points on the body move at various speeds; or with respect to time, as when a body is speeding up or slowing down; or with respect to both, as when a rotating body is altering its speed. Difform motion with respect to time can be uniformly difform (in modern terms, uniformly accelerated or decelerated), which Heytesbury defined as motion in which equal intensities of speeds are acquired or lost in equal times. Difformly difform (nonuniformly accelerated or decelerated) motions can vary in an infinite number of ways, which Heytesbury and his successors attempted in part to classify. In all cases, however, the speed of a motion, as well as its uniformity or difformity, is to be determined from the speed of the body's fastest-moving point. In the case of a motion difform with respect to time, speed at any instant is measured by the distance the fastest-moving point would have traveled if it had moved uniformly at that speed for some period of time.[52] Now Aristotle had insisted that there can be neither motion nor rest in an instant because motion and rest by definition occupy some finite time.[53] Heytesbury has thus defined, in defiance of Aristotle and common sense, instantaneous speed – an important step in the abstract treatment of motion that would culminate in the new science of Galileo and Newton.

The concept of instantaneous speed set the stage for the most significant contribution of the Oxford calculators to the kinematics of local motion – the mean-speed theorem. Its earliest known statement is found in Heytesbury's *Rules for Solving Sophisms*: a body uniformly accelerated or decelerated for a given time covers the same distance as it would if it were to travel for the same time uniformly with the speed of the middle instant of its motion,

[51] Clagett, *Science of Mechanics in the Middle Ages*, pp. 199–219; Edith Sylla, "Medieval Quantifications of Qualities: The "Merton School," *Archive for History of Exact Sciences*, 8 (1971/1972), 9–39; and Edith Sylla, "Medieval Concepts of the Latitude of Forms: The Oxford Calculators," *Archives d'histoire doctrinale et littéraire du moyen âge*, 40 (1973), 223–83.

[52] Curtis Wilson, *William Heytesbury: Medieval Logic and the Rise of Mathematical Physics* (Madison: University of Wisconsin Press, 1960), pp. 117–22; and Clagett, *Science of Mechanics in the Middle Ages*, pp. 235–7.

[53] Aristotle, *Physics*, VI.10.241a15–26.

which is defined as its mean speed. Thus a body accelerating uniformly for two hours will cover the same distance as would a body moving uniformly for two hours at the speed reached by the first body after one hour. From this Heytesbury drew several conclusions, including what is called the distance rule: that the distance covered in the second half of a uniformly accelerated motion from rest is three times the distance covered in the first half.

Heytesbury gave the mean-speed theorem without proof, but if he was the author of a related work on motion entitled the *Proofs of Conclusions*, he is responsible for a common proof repeated by a number of his successors, including Richard Swineshead and John Dumbleton. In this proof, a uniformly accelerated motion is divided at its middle instant into two halves, and each instant in the first half corresponds to an instant in the second half. Since the speed at each instant in the slower half is less than the mean speed by exactly as much as the speed at the corresponding instant in the faster half exceeds the mean speed, the same overall distance is covered as would be covered by a uniform motion at that mean speed.[54]

Like Bradwardine's theorem, the methods and results of the other Oxford calculators spread to the continent over the next generation, appearing most notably at the University of Paris in the works of Albert of Saxony, Nicole Oresme, and Marsilius of Inghen. The most significant addition to the Oxford subtleties was Nicole Oresme's graphical representation of qualities and speeds, including a graphical proof of the mean-speed theorem. Although the method of representing qualities graphically perhaps occurred earliest in Giovanni di Casali's *On the Speed of Alteration*, dated 1346 in one manuscript, Oresme, at about the same time or somewhat later, developed a graphical method of considerable power and flexibility. Whereas the Oxford calculators had represented the intensities of qualities and speeds with lines of various lengths, Oresme, in his *On the Configuration of Qualities*, represented them with figures of two or more dimensions. A horizontal line – called the subject line – represented the subject or body, from each point of which was raised a vertical line representing the intensity of some quality at that point (see Figures 17.1 and 17.2). The line joining the tops of the intensity lines – called the line of summit or line of intensity – defined what Oresme called the configuration of the quality. In the case of a quality uniform over the length of the subject line, the configuration is rectangular (Figure 17.1); in the case of a quality uniformly increasing from zero at one end of the subject line to some finite intensity at the other, it is triangular (see Figure 17.2), and so on. Oresme also used the subject line to represent the time of a motion and the intensity lines to represent its speed at each instant. A uniform motion is thus represented by a rectangle, and a uniformly accelerated motion by a right triangle. Oresme called the area of such a figure the total velocity of the

54 Wilson, *William Heytesbury*, pp. 122–3; and Clagett, *Science of Mechanics in the Middle Ages*, pp. 255–329.

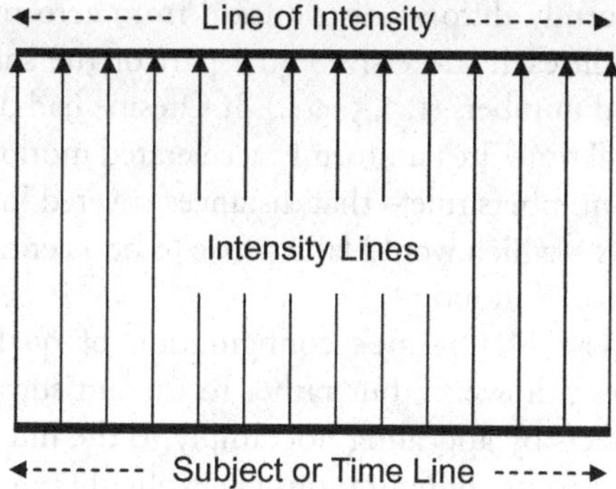

Figure 17.1. The configuration of a uniform quality or motion.

motion since it is composed of all the infinite number of velocity lines within it. He also recognized that the total velocity was dimensionally equivalent to the distance traversed by the motion.

From this the mean-speed theorem and the distance rule immediately follow, for it is evident that the area of a right triangle, representing the total intensity of a quality distributed uniformly difformly from zero to some intensity, is equal to that of the rectangle on the same base with half the altitude, representing a uniform quality at the mean degree. And since the area of the triangle representing the slower half of the motion is one-fourth the area of the whole triangle representing the whole motion, the distance rule – that the distance covered in the faster half of the motion is three times that covered in the slower – is also confirmed. In his *Questions on Euclid's Elements,* Oresme in effect generalized the distance rule, showing that for

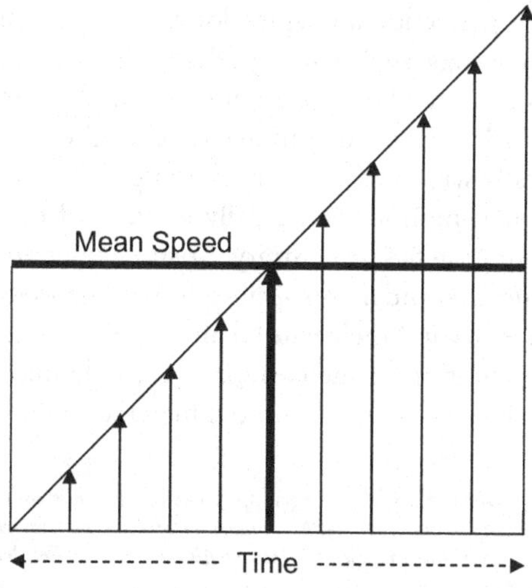

Figure 17.2. The configuration of a uniformly difform motion and the mean speed.

qualities in a uniformly difform distribution from zero to some intensity, the total intensity in each successive equal part of the subject follows the sequence of the odd numbers (1, 3, 5, etc.). If Oresme had demonstrated this specifically for equal times in a uniformly accelerated motion, he would have expressed the odd-numbers rule – that distances covered in equal times vary as the odd numbers – which would later prove to be a central proposition in Galileo's new science of motion.[55]

The main purpose of Oresme's configuration of qualities was not to develop a kinematics, however, but rather to explain surprising and unexpected physical effects by appealing not simply to the intensity of a quality in a subject but also to its configuration. He explicitly recognized that these configurations were imaginary or hypothetical, and in devising them he did not rely on any actual measurements. His extension of configuration theory to motion seems to have been a legacy from the Oxford calculators, where motion was one of many topics in natural philosophy that provided material for the paradoxes and logical puzzles that made up the *sophismata* of the arts curriculum. And, in general, natural philosophers from the fourteenth century on were more concerned with classifying types of uniform and difform motions and compiling solutions to various hypothetical situations than with discovering the way things occur in nature.

CELESTIAL MOVERS

Within the Aristotelian cosmos, the sublunary realm – the spheres of the four elements, located below the Moon – was the realm of motion and change, augmentation and diminution, growth and decay. Above the Moon, by contrast, was the celestial realm, occupied by bodies as far surpassing elemental bodies in perfection and splendor as they are distant from them. In *De caelo*, Aristotle suggested that the celestial spheres should be thought of as living bodies composed of a celestial material called ether or, as it was later known, the fifth essence, or quintessence. Unlike the four elements, which move naturally with rectilinear motions up or down, the ether moves naturally with circular motion, as the daily rotation of the heavens reveals. And since circular motion has no contrary, Aristotle inferred that the celestial spheres were ungenerated and incorruptible, fully actualized and without any potential for change beyond their natural and eternal circular motions.[56]

The existence of motion in the cosmos as a whole implied for Aristotle the existence of realities that lie beyond the material. From the premise that

55 Clagett, *Science of Mechanics in the Middle Ages*, pp. 331–418; and Marshall Clagett, *Nicole Oresme and the Medieval Geometry of Qualities and Motions: A Treatise on the Uniformity and Difformity of Intensities known as Tractatus de configurationibus qualitatum et motuum* (Madison: University of Wisconsin Press, 1968).

56 Aristotle, *De caelo*, I.1–2.269a18–b17; II.2.285a29–30; and II.12.292a19–22.

everything that is moved is moved by another, he proved in Book VIII of *Physics* that there must be a first mover and a first motion – not first in time, for Aristotle held that the world and its motion were eternal, but first in priority. This first motion, he argued, must be the uninterrupted rotation of a single body, the first or prime mobile (*primum mobile*), moved by a single mover, the first or prime mover (*primum movens*), acting on it with absolute uniformity. The prime mobile must be a physically real celestial sphere, above the sphere of the fixed stars, which receives its own perfectly uniform and regular motion from the first or prime mover and somehow communicates that motion to the sphere of the fixed stars below and from there to the spheres of the planets, Sun, and Moon. The prime mover must itself be immaterial and unmoved, moving the prime mobile as its final cause (its goal or purpose).[57] In his *Metaphysics*, Aristotle suggested that the lower planetary spheres, each of which has its own proper motion, are also moved by unmoved movers, each acting as an external final cause.[58] For Aristotle, then, the existence and persistence of motion in the world implied the existence of an eternal and immaterial unmoved mover as its first cause, an argument that Aquinas adopted as the first of his famous five proofs for the existence of God.[59]

How the prime mover moves the prime mobile, how that motion is passed on to the spheres below, and how each of the lower spheres has, in addition, its own proper motion gave rise to much controversy among medieval commentators. That the celestial spheres were living bodies possessing animating souls was greeted with skepticism, especially after the opinion was condemned in 1277. If, on the other hand, they were inanimate, how could they be moved by nature, since they were already located in their natural places? Furthermore, since their circular motion was infinite (that is, without any definite end or goal), it could not be natural since it would be incapable of fulfillment and would thus be in vain; and nature, according to Aristotle, does nothing in vain. Faced with these perplexities, many took up Aristotle's suggestion of an external, unmoved mover for each celestial sphere, though there was no agreement on whether these unmoved movers – called celestial intelligences and sometimes identified as angels – moved the heavenly bodies directly or through the mediation of purely intellective (rather than animating) souls.[60] One alternative to celestial intelligences was to suggest, as Buridan did, that once the celestial spheres had been created and set in

57 Aristotle, *Physics*, VIII.1–10.250b11–267b26; and Aristotle, *Metaphysics*, XII.7.1072a19–1073a13.

58 Aristotle, *Metaphysics*, XII.8.1073a14–1074b14.

59 Thomas Aquinas, *Summa theologiae*, prima pars, qu. 2, art. 3; and Thomas Aquinas, *Summa contra gentiles*, I.13.

60 James A. Weisheipl, "The Celestial Movers in Medieval Physics," *The Thomist*, 24 (1961), 286–326, reprinted in Weisheipl, *Nature and Motion in the Middle Ages*, pp. 143–75; and Edward Grant, *Planets, Stars, and Orbs: The Medieval Cosmos, 1200–1687* (Cambridge: Cambridge University Press, 1994, repr. 1996), pp. 469–87, 514–48.

motion by God, in the absence of any resistance their impetus would continue to move them perpetually. This alternative was taken up by very few, however. Only Albert of Saxony seems to have adopted impetus as the mover of the heavens, most others preferring some version of celestial intelligences. Nicole Oresme, for one, objected that the celestial material was incapable of receiving impetus, perhaps because it was neither heavy nor light.[61]

The many remarkable developments that took place within medieval treatments of motion arose mainly out of the reading and interpretation of Aristotle's works, although they often took very un-Aristotelian turns and sometimes challenged fundamental tenets of Aristotelian natural philosophy. At the core of that natural philosophy was the premise that motion and change are the characteristic expressions of the natures of physical bodies, where nature – the internal source of motion and change within natural things – always acts for a purpose, toward a definite end or goal that fulfills or perfects the body moved. Medieval natural philosophers examined the nature and reality of motion in the light of both Aristotelian philosophy and newer philosophical positions, such as Ockham's nominalism, and they employed hypothetical cases, such as motion in a void, as a way of scrutinizing motion and discovering its essential characteristics. Dissatisfied with Aristotle's explanation of projectile motion, natural philosophers devised impetus theory and then applied it to naturally falling bodies, a move that tended to blur Aristotle's fundamental distinction between natural and violent motions. Similarly, the suggestion that the celestial spheres could be moved by impetus implies a blurring of the distinction between the celestial and elemental realms.

Aristotle's application of mathematics to motion had been rudimentary at best. Medieval logicians and natural philosophers brought to bear on motion new mathematical methods and techniques of great sophistication, which in turn yielded a number of conceptual insights. The first of these was Bradwardine's rule, with its application of ratio theory to motion, implying a recognition of the idea of a functional dependence between powers, resistances, and speeds. With Bradwardine and the Oxford calculators in general, motion came to be considered in abstraction from physical bodies and their natural tendencies, and speed came to be regarded as a magnitude that could be measured both intensively and extensively. Once they had grasped the paradoxical notion of instantaneous speed, the Oxford calculators and their followers could compare the speeds of various motions, both real and hypothetical, in a variety of sophisticated ways, most notably by applying the mean-speed theorem. With Oresme's powerful method of graphing qualities and motions, results such as the mean-speed theorem and the distance rule became immediately obvious. Motion and change also provided material for

61 Clagett, *Science of Mechanics in the Middle Ages*, pp. 536, 561, 570; and Grant, *Planets, Stars, and Orbs*, pp. 548–52.

purely logical inquiries that used new quasi-mathematical methods of analysis to treat problems concerning infinity, beginnings and endings, maxima and minima, and first and last instants, which, among other things, yielded new mathematical insights into infinity and continuity.[62]

The study of motion in the Middle Ages, then, was not a slavish and sterile commentary on the words of Aristotle, but neither was it a failed attempt at the experimental science of motion that Galileo, Descartes, and Newton would establish in the seventeenth century. Rather, from a critical examination of the best sources available at the time, medieval logicians, philosophers, and theologians undertook to explain motion and change and the many puzzles surrounding them, developing in the process a series of new insights and analytic techniques that yielded a number of notable results. Part of the measure of their success – but only part – is that some of these insights and results had to be rediscovered later by Galileo and others in the course of the Scientific Revolution.

[62] See John E. Murdoch, "Infinity and Continuity," in Kretzmann, Kenny, and Pinborg, *Cambridge History of Later Medieval Philosophy*, pp. 564–91 at pp. 585–90.

18

COSMOLOGY

Edward Grant

Cosmology was not an independent discipline during the late Middle Ages but an undifferentiated part of the broad domain of natural philosophy. During the early Middle Ages, from approximately 500 to 1150, Plato's *Timaeus*, which was available in an incomplete Latin translation by Chalcidius (fl. fourth or fifth century), was the primary source of cosmological knowledge. During the twelfth and thirteenth centuries, however, a Greco-Islamic storehouse of natural philosophy and science reached Western Europe in the form of Latin translations from Greek and Arabic. Among these translations, Aristotle's works on natural philosophy were of the utmost importance since they provided the primary basis for late-medieval cosmology. Among the natural books of Aristotle, only his treatise *On the Heavens* (*De caelo*) can be regarded as a cosmological work, although significant cosmological ideas appear in his *Physics*, *Metaphysics*, and *Meteorology*. Aristotle devotes the first two books of *De caelo* to the celestial region and the last two to the four terrestrial elements that exist below the Moon.

Apart from commentaries on Aristotle's *De caelo*, cosmological ideas and discussions were also embedded in translations on natural philosophy, astronomy, astrology, and even medicine. Particularly noteworthy is the first chapter of Claudius Ptolemy's (ca. 100–ca. 170) *Almagest*, which contained widely cited arguments about the immobility of the Earth. Ptolemy also wrote a work titled *Hypotheses of the Planets*, in which he showed how the eccentric and epicyclic celestial orbs might be arranged in the heavens. Although ideas from the latter work would have a large impact on medieval cosmology, the work itself was not translated into Latin during the Middle Ages, leaving the manner in which it influenced the West a mystery.[1]

1 For a German translation of the two books of *Hypotheses of the Planets*, see Ptolemy, *Opera quae exstant omnia*, vol. 2: *Opera astronomica minora*, ed. J. L. Heiberg (Leipzig: Teubner, 1907), pp. 69–145. For part of Ptolemy's treatise that is missing from the Greek text and German translation, see Bernard R. Goldstein, "The Arabic Version of Ptolemy's Planetary Hypotheses," *Transactions of the American Philosophical Society*, n.s. 57, pt. 4 (1967), 1–55.

In addition to the basic works just cited, a number of authors within Islamic civilization, not all of whom were Muslims, made significant contributions to cosmological thought. In this group are Averroes (Ibn Rushd, 1126–1198), Avicenna (Ibn Sīnā, 980–1037), Alhacen (Ibn al-Haytham, 965–ca. 1039), al-Farghani (d. after 861), al-Farabi (ca. 870–950), a Sabean star worshiper named Thābit ibn Qurra (836–901), and the Jewish author Moses Maimonides (1135–1204). To this Arabic corpus should be added the Latin literature of the early Middle Ages, contributed by a group known collectively as the Latin encyclopedists, who are noteworthy because they preserved some cosmological ideas and interpretations from the ancient world during the darkest period of Western intellectual life (see Burnett, Chapter 14, this volume).

The natural books of Aristotle became the core of the curriculum in the arts faculties in the medieval universities, the earliest of which were in existence by around 1200. The commentaries on these Aristotelian works, which constitute the major source of our knowledge of medieval cosmology, took two forms. The first and original form was a straightforward textual commentary in which the text is explained sequentially, section by section. The second type is called the *questio* method and is the more significant. Each Aristotelian treatise was divided into a series of questions, with each question in turn discussed according to a stylized format that allowed a thorough consideration of the issues. Jean Buridan (ca. 1295–ca. 1358), perhaps the greatest arts master in the Middle Ages, divided his *Questions on "De caelo"* into fifty-nine questions, devoted overwhelmingly to the first two books on the celestial region, reflecting its greater importance. Among Buridan's questions on the celestial region are the following:[2]

Whether the sky [or heaven] has matter.
Whether the world is perfect.
Whether an infinite body is possible.
Whether it is possible that there are several worlds.
Whether there is something beyond the sky [or heaven].
Whether the whole earth is movable.
Whether the sky is always moved regularly.
Whether the [planets and] stars are self-moved or moved by the motions of their spheres.
Whether the earth is spherical.

Buridan's questions on Aristotle's *On the Heavens* were typical of the kinds of questions on that treatise posed by natural philosophers. Works by authors other than Aristotle also contributed to the medieval understanding of cosmology. Included within this category are the commentaries of Averroes

[2] The titles of these questions are drawn from my translations of the titles of all fifty-nine questions in *A Source Book in Medieval Science*, ed. Edward Grant (Cambridge, Mass.: Harvard University Press, 1974), pp. 203–5. I have altered the translation of the eighth question in the sequence.

on the works of Aristotle, John of Sacrobosco's *Treatise on the Sphere* and the commentaries on it, and the *Commentaries on the "Sentences"* of Peter Lombard (d. ca. 1160). The four books of the *Sentences* (or *Opinions*) of Peter Lombard – God, the Creation, the Incarnation, and the Sacraments – served as the basic theological textbook from around 1200 to 1700. Cosmological issues arose most frequently in the second book, which treated the Creation.

Our knowledge of medieval cosmology is derived primarily from the hundreds of questions, and, to a lesser extent, commentaries, that medieval natural philosophers and theologians have left us on all of these treatises, especially those on Aristotle's natural books. Responses to these questions varied among scholastic authors between the thirteenth and sixteenth centuries. Occasionally, new questions were proposed. Some were prompted by the Condemnation of 1277 issued by the bishop of Paris. A number of natural philosophers and theologians at the University of Paris took cognizance of condemned articles in formulating their questions and responses. Changes in medieval Aristotelian cosmology were thus under way long before the assault on the geocentric world began in the sixteenth century.

It might seem that a major source of cosmological discussion would be astronomical treatises. In the ancient and medieval periods, however, cosmology was a subject studied primarily by natural philosophers, not mathematical astronomers. Following a well-established Greek tradition, the roles of astronomer and physicist were distinguished, sometimes explicitly and sometimes simply by the way astronomers and natural philosophers treated the celestial region. The astronomer sought to account for astronomical phenomena by appropriate, though not necessarily true, hypotheses from which one could readily predict planetary positions, as well as eclipses, conjunctions, and oppositions. By contrast, the task of the natural philosopher was to determine the essential nature and structure of the heavens: what they were made of, how they moved, how they were arranged and why, the influences they might exert on earth, and so on. Problems such as these were rarely treated in astronomical works, where they would have been regarded as inappropriate.

During the Middle Ages, cosmology was primarily a subdivision of Aristotelian natural philosophy. Moreover, natural philosophy, or physics, was considered superior to traditional astronomy since the former concerned itself with the real operations and structure of the world, whereas the latter need not and often did not. Despite this alleged difference, a number of astronomers in the late Middle Ages included cosmological problems in their astronomical treatises and also sought physical causes for celestial phenomena. During the sixteenth century, physics and astronomy became more widely integrated, as more and more astronomers came to realize that mathematical astronomy needed to expand its horizons. In the first quarter of the seventeenth century, Johannes Kepler was the most noteworthy astronomer to call for the necessary integration of natural philosophy and astronomy.

The terms *mundus* and *universum* were used rather indifferently to refer to the world, or cosmos, as a whole. That world was regarded as a large, finite, unique sphere that was everywhere filled with matter, thus making the formation of vacua impossible. From the outermost convex surface of the world to its geometrical center, cosmic matter was arranged in a vast hierarchy of nested orbs, which contained planets, elements, and compound bodies. But the nested orbs of the cosmos were not a seamless continuum. In Aristotle's cosmology, the world was divided into two radically different parts, one celestial, the other terrestrial. The properties and activities of these two domains differed dramatically. Before describing these two parts, it will be well to consider one monumental problem about the world taken as a single totality: Was it created or eternal?

IS THE WORLD CREATED OR ETERNAL?

From Holy Scripture (Gen. 1:1–2; John 1:2–3 and 17:5), Christians learned that the world was created supernaturally and would eventually be destroyed supernaturally. Despite the absence of any explicit statement in Jewish, Christian, or Muslim scriptures that the world was created "from nothing" (*ex nihilo*), creation from nothing had been widely assumed since the second century. Its appeal was obvious in that a God able to create a world from nothing would seem, prima facie, a more powerful deity than one who, like the Demiurge in Plato's *Timaeus*, did not and could not.[3]

In the fourth canon promulgated at the Fourth Lateran Council in 1215, creation from nothing was made Christian doctrine. Aristotle's natural books, introduced in the twelfth and thirteenth centuries, proclaimed the contrary message that the world could not have had a beginning and could never come to an end. In Aristotle's philosophy, it was impossible for matter or substance to come from nothing. The clash of these opinions was a major problem in the thirteenth century, especially at the University of Paris. The issue climaxed in 1277, when the bishop of Paris, following the recommendations of a committee of theologians, condemned 219 articles, of which many – at least 27 – were directed against the eternity of the world. The following condemned articles convey something of the flavor of the controversy:[4]

> 90. That a natural philosopher ought to deny absolutely the newness [that is, the creation] of the world, because he depends on natural causes and natural reasons. The faithful, however, can deny the eternity of the world because they depend upon supernatural causes.

[3] See Edward Grant, *Planets, Stars, and Orbs: The Medieval Cosmos, 1200–1687* (Cambridge: Cambridge University Press, 1994), pp. 89–90.

[4] The articles are cited from Grant, *Source Book in Medieval Science*, pp. 48–9.

93. That celestial bodies have eternity of substance but not eternity of motion.

94. That there are two eternal principles, namely the body of the sky and its soul.

98. That the world is eternal because that which has a nature by [means of] which it could exist through the whole future [surely] has a nature by [means of] which it could have existed through the whole past.

Throughout the Middle Ages, the creation of the world from nothing was an article of faith. Different interpretations were, however, still possible. Three opinions found varying degrees of support. Some scholars, such as St. Bonaventure (1221/1222–1274), Richard of Middleton (fl. second half of the thirteenth century), and Henry of Ghent (d. 1293), argued that the creation of the world was capable of rational proof. Others, such as Nicole Oresme (ca. 1320–1382), insisted that in terms of reason the world should be regarded as eternal because no natural agent could be invoked to account for its beginning. Nevertheless, we must accept on faith that the world was created by God from nothing.[5] In the middle was Thomas Aquinas (1225–1274), who argued that no rational proof was possible for either side but insisted that God could have willed the existence of the world without also causing it to have a temporal beginning, thus producing, if He wished, a world that was both created and eternal.[6] This opinion, which sought to reconcile the Aristotelian view of an uncreated world with the Christian belief in a created world, was the most popular of the three.

THE TWO PARTS OF THE WORLD: CELESTIAL AND TERRESTRIAL

In his *Meteorology* (Book I, Chapter 2), Aristotle speaks clearly of a world that is divided into two radically different parts. The first, or celestial, part consists of "one element from which the natural bodies in circular motion are made up."[7] That element was a special incorruptible ether, or fifth element, from which all the planets, stars, and orbs were composed. It ranged from the concave surface of the lunar orb upward to the outermost celestial orb, which by the late Middle Ages extended beyond the sphere of the fixed stars, the

[5] See *The Questiones super De celo of Nicole Oresme*, ed. and trans. Claudia Kren (PhD diss., University of Wisconsin, 1965), bk. 1, qu. 10, pp. 148–50.

[6] For a discussion of these issues, see St. Thomas Aquinas, Siger of Brabant, and St. Bonaventure, *On the Eternity of the World ("De Aeternitate Mundi")*, translated from the Latin with an introduction by Cyril Vollert, Lottie H. Kendzierski, and Paul M. Byrne (Milwaukee, Wis.: Marquette University Press, 1964), pp. 4–17.

[7] Aristotle, *Meteorologica*, I.2.339a.11–12, trans. H. D. P. Lee (London: William Heinemann; Cambridge, Mass.: Harvard University Press, 1962), p. 7.

limit of Aristotle's cosmos, all the way to the empyrean heaven, a Christian creation perceived as an abode for God and the elect.

The second, or terrestrial, region consists of "four other physical bodies produced by the primary qualities [that is, hot, cold, dry, and moist], the motion of these bodies being twofold, either away from or towards the centre. These four bodies are fire, air, water and earth."[8] The terrestrial region extends downward from the concave surface of the lunar orb to the center of the Earth, an Earth assumed to be spherical. By contrast to the celestial region, the part of the world below the lunar orb was a realm of incessant change. Although the celestial and terrestrial regions differed radically, Aristotle assumed that the two regions were continuous, and, because the heavens were nobler and more perfect, he further argued that the motions of the celestial region governed the movements and changes of bodies in the terrestrial region.[9] These two distinct parts, celestial and terrestrial, made up the physical world. Let us examine them more closely.

As we saw, all motions and activities of physical bodies in the terrestrial region were governed from the celestial region. By virtue of the heat from the Sun's rays, elements continually interact, forming compound bodies that eventually degenerate and release their elements to form new compounds in a never-ending cycle of generation and corruption. If the Sun failed to provide its life-giving heat, the four elements would sort themselves out into four static, concentric rings in the following order: earth, water, air, and fire. This fails to occur because the elements continually combine and dissociate, thereby producing generation and corruption. Change in general, or generation and corruption, the coming to be and passing away of things, occurs in all physical substances because the matter in each substance possesses a form or quality that is replaceable by its contrary. For example, fire possesses the properties of hotness and dryness. If the hotness were replaced by its contrary quality, coldness, the fire would be transformed into earth, with the properties of dryness and coldness. Generation and corruption, and therefore all change, involves the possession of one, and the expulsion of the other, of a pair of contrary qualities.[10] When an element, say earth, is transformed into fire, it moves toward the region of the sublunar world in which fire predominates, namely just below the concave surface of the lunar sphere. Although the world never sorts itself out into four static concentric rings, it does retain a structure of natural places that corresponds to the following arrangement: earth, water, air, and fire.

When compared with the size of the firmament, the Earth was regarded as no more than a point. Nevertheless, it had a measurable size. At least three estimates for its circumference circulated during the late Middle Ages, the

[8] Ibid., I.2.339a.12–16.
[9] Ibid., I.2.339a.22–24.
[10] See Grant, *Planets, Stars, and Orbs*, pp. 193–4.

smallest of which, 20,400 miles (of Arabic origin), was used by Christopher Columbus (1451–1506) in seeking patronage for his voyage of discovery to America.[11] The Earth was unanimously regarded as roughly spherical. Aristotle himself provided proofs of this claim. In Aristotle's natural philosophy, the round Earth lay immobile at the center of the world. In the Middle Ages, this simplistic model yielded to one in which the Earth's status became more complex.

Two major problems about the Earth affected cosmological discussions: (1) Was the Earth, located at the center of the world, the center of all planetary orbits? And (2) was the Earth really immobile? Aristotelian cosmology, as reinforced by the ideas of his commentator Averroes, required that every celestial orb rotate around a physical body situated at its center. In a system in which all physical orbs were concentric, the Earth fulfilled this essential function. The realities of astronomy, however, made this untenable. Observed variations in the brightness of each planet made it apparent that the distance of a planet from the Earth varied, from which it followed that the Earth could not be the physical center of the planetary orbits. Aristotle's Earth-centered concentric astronomy was replaced by Ptolemy's system of eccentric orbs in which the Earth, though still situated at the center of the world, was no longer the center of planetary motions.

But if the Earth lay at the center of the world, was it really immobile? In the thirteenth century, John of Sacrobosco simply assumed that the "earth is held immobile in the midst of all" because "the center is a point in the middle of the firmament. Therefore, the earth, since it is heaviest, naturally tends toward that point."[12] By the fourteenth century, however, Jean Buridan attributed real but minute motions to the Earth. Buridan assumed that, under the influence of geological changes caused ultimately by solar heat, the Earth's density was differentially altered, thus causing continuous shifts of its center of gravity. Since the Earth always sought to rest in its natural place at the geometric center of the universe, it continually shifted its position until its center of gravity coincided with the geometric center. In this manner, the Earth oscillated perpetually around the geometric center of the world.

Buridan, and, a few years later, Nicole Oresme, contemplated an even more daring concept of terrestrial motion when they considered whether the daily rotation of the heavens might not derive from a real daily axial rotation of the Earth. For different reasons, both eventually rejected this alternative but in the process proposed cogent arguments for believing in axial rotation, some of which appear in Copernicus's *De revolutionibus orbium caelestium*. On one momentous point they agreed: If the earth did actually rotate daily on its axis while the heavens remained immobile, the astronomical phenomena would

[11] Ibid., pp. 620–1.

[12] Lynn Thorndike, *The "Sphere" of Sacrobosco and Its Commentators* (Chicago: University of Chicago Press, 1949), p. 122.

be saved just as well as if the Earth were immobile and the heavens moved with a daily rotation. Their arguments included appeals to the relativity of motion and economy of effort, the latter exemplified by the much smaller velocity required for the Earth to complete a daily rotation.

Phenomena such as comets, meteors, and even the Milky Way posed special problems for Aristotle, who argued (in his *Meteorology*) that they did not occur in the celestial region but were located in the upper atmosphere between the Earth and Moon, in the natural places of air and fire. He assumed that the fiery and airy regions were carried around by the celestial revolutions and that under certain conditions parts of the dry and warm exhalations of the upper atmosphere might ignite. These burning exhalations formed a variety of phenomena – comets, shooting stars, and the Milky Way – that appeared celestial but were not. They were phenomena that Aristotle regarded as corruptible and transient, though recurrent. By confining them to the upper atmosphere, Aristotle reserved incorruptibility for the celestial region. Medieval natural philosophers such as Thomas Aquinas, Albertus Magnus (ca. 1200–1280), Aegidius of Lessines (ca. 1235–ca. 1304), Geoffrey of Meaux (fl. 1310–48), and others, agreed with Aristotle on the sublunar location and corruptible nature of comets and the other phenomena discussed. Only the significance of comets for human existence varied from author to author.

When natural philosophers made the mental leap beyond the invisible concave surface of the lunar sphere, leaving behind the upper atmosphere, they entered another world, one that bore little resemblance to the region below. To terrestrial observers, the planets and stars appeared to be self-moved, in the manner of fish in water or birds in air. In the ancient world, Plato assumed self-motion for the planets, attributing the motion of each planet to its own soul. Aristotle, however, denied self-movement and argued for the existence of transparent, invisible spheres to which the celestial bodies were attached or in which they were enclosed. The spheres were invisible, but the celestial bodies they carried were readily observable. Why? Although planets, stars, and orbs were all composed of the same transparent ether, stars and planets were visible because each celestial body was a region of highly concentrated ether that was capable of receiving light and becoming self-luminous or, for those who considered the celestial bodies opaque, of reflecting light from the Sun. Medieval natural philosophers largely agreed that the planets, and probably the stars, received their light from the Sun, with a few further assuming that the planets were also weakly self-luminous.

ARISTOTLE AND PTOLEMY

The natural rotary motion of the celestial spheres, or orbs, produced the observable circular motion of every visible celestial body. Although the terms "sphere" (*sphera*) and "orb" (*orbis*) were used interchangeably in the Middle Ages, Pierre d'Ailly (1350–1420), among others, recognized an important distinction between them when he observed that "properly speaking" an orb is enclosed by two surfaces, one concave and the other convex, but a sphere is contained solely by a single convex surface.[13] Following d'Ailly's distinction, we may say that the multitude of fixed stars were attached to a single sphere, whereas each planet was carried around by its own orb. Not only was each planet assumed to possess its own orb, but each of its motions – daily motion, sidereal motion, motion in latitude, and so on – was also assumed to possess its own orb. Ptolemy and Aristotle each invoked a plurality of orbs to account for the resultant celestial position of each planet. All told, Ptolemy employed as many as forty-one orbs, and Aristotle assigned as many as fifty-five. Both systems nested their respective orbs one within the other but differed radically in the way they configured the latter. Aristotle's were concentric with respect to the Earth, whereas Ptolemy's were eccentric (that is, had centers other than the Earth), as we saw earlier (from this point, the terms "orb" and "sphere" will be used interchangeably).

Medieval natural philosophers had to choose between Aristotle's system of purely concentric orbs, which assumed that the Earth was at rest in the geometric center of the world and therefore could not account for the variation in planetary distances from the Earth, or go with Ptolemy's system of eccentric orbs, which could account for the variation in planetary distances but achieved this by having the orbs move around centers other than the Earth, thus violating Aristotle's principle that the Earth is the sole center of all planetary motions.

To reconcile the two systems, medieval natural philosophers adopted a compromise that had in fact already been made by Ptolemy (in his *Hypotheses of the Planets*). In medieval terminology, they distinguished between a "total orb" (*orbis totalis*) and a "partial orb" (*orbis partialis*). A total orb is a concentric orb whose geometric center is the center of the Earth (in Figure 18.1, which represents the Moon's orbs, the total orb consists of the convex circumference *ADBC* and the concave circumference *OQKP*, both concentric to the Earth at center *T*); a partial orb is any eccentric orb whose geometric center lies outside the center of the world. Between the concave and convex surfaces of a concentric sphere lay at least three partial orbs, one of which, the eccentric deferent (represented in Figure 18.1 by *b'* enclosed by circumferences *AGFE* and *HNKM* around center *V*, a center other than the Earth) carried an epicycle (enclosed by circumferences *KLFI* and *ΘRYS*) within which a planet (the Moon, *O* in this figure) is enclosed.

Most medieval natural philosophers adopted this compromise system, for although they now had to assume the rotation of partial eccentric orbs around

13 Cited from Pierre d'Ailly, *14 Questions on the Sphere of Sacrobosco*, in Grant, *Planets, Stars, and Orbs*, p. 115, n. 37.

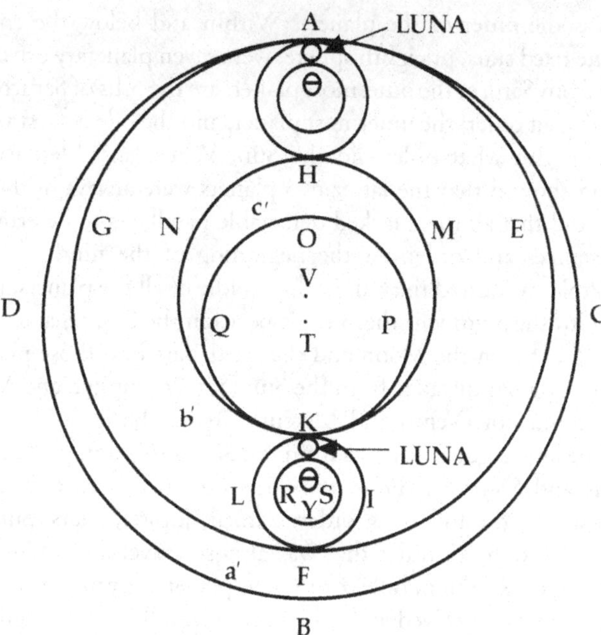

Figure 18.1. A representation of the Moon's concentric, eccentric, and epicyclic orbs as described in Roger Bacon's *Opus Tertium*. See Pierre Duhem, ed., *Un fragment inédit de l'Opus tertium de Roger Bacon, précédé d'une étude sur ce fragment* (Quaracchi: Collegium S. Bonaventurae, 1909), p. 129.

their own geometric centers, thus denying to the Earth its former glorified status as sole center of planetary motions, they could console themselves with the thought that, in the larger scheme, each total concentric, planetary orb, which contained all of its planet's partial eccentric orbs, had the Earth as its center. In this sense, one could still speak of a concentric world system in which the Earth was at the center of all planetary motions. This compromise was adopted because it gave the appearance of uniting Aristotle's cosmology with Ptolemy's astronomy.

THE NUMBER OF ORBS AND THE ORDER OF THE PLANETS

Since each total, or concentric, orb is composed of at least three partial eccentric orbs, there are obviously many more partial orbs than total ones. When inquiring about the number of spheres in the celestial region, however, natural philosophers counted only total concentric orbs, ignoring partial orbs. Because each planet had its own total concentric orb, at least eight orbs from the Moon to the fixed stars were required for the celestial bodies.

What was the order of the planets? Within and below the enveloping sphere of the fixed stars, or eighth sphere, were seven planetary orbs. Moving downward from Saturn, the outermost planet, are the orbs of Saturn, Jupiter, and Mars, in that order; the innermost planet, and the one nearest the Earth, is the Moon. But what order did the Sun, Venus, and Mercury follow? Observation showed that the latter two planets were always in the vicinity of the Sun and that all three lacked detectable parallaxes to determine their relative distances and order. At the beginning of the ninth book of his *Almagest*, Ptolemy located the Sun in the middle of all the planets, justifying the move with the argument that in this position the Sun separated the five planets lying between the Moon and the fixed stars into those planets that could be any angular distance from the Sun (Saturn, Jupiter, and Mars) and those that could not (Venus and Mercury; in this he ignored the Moon, which can be any angular distance from the Sun but is nonetheless grouped with Venus and Mercury). From the outermost planet to the innermost, Ptolemy assumed the following order: Saturn, Jupiter, Mars, Sun, Venus, Mercury, and Moon, an order that was almost universally adopted in the late Middle Ages. As the middle, or fourth, planet counting from above or below, the Sun was perceived as "a king in the Middle of his kingdom" or a "heart in the middle of an animal."[14]

The relationship of the Sun, Venus, and Mercury had been given a strikingly different arrangement in the fifth century, when Martianus Capella assumed that Venus and Mercury circled the Sun rather than the Earth, a system that came to be called "Capellan." Anyone who subscribed to such an arrangement could not, of course, assume a fixed order of the planets since the Sun, Mercury, and Venus would be related differently when the latter two were above the Sun than when they were below it. In the fourteenth century, Jean Buridan incorporated the Capellan system into Ptolemaic eccentric and epicyclic astronomy. He may have been the first to present an unequivocal description of heliocentric orbits for Venus and Mercury within a system of eccentrics and epicycles, an arrangement he regarded as "probable."[15]

The seven planets and the fixed stars required eight concentric orbs. No visible celestial bodies existed beyond the fixed stars. The latter, however, played a crucial role in determining what went on above and beyond. Medieval natural philosophers assumed with Aristotle that every simple celestial motion required a separate sphere to produce it. Because a minimum of three simple motions was attributed to the sphere of the fixed stars, at least two more orbs were required. As its proper motion, the eighth sphere of the fixed stars moved with the motion that accounted for precession of the equinoxes,

[14] For references to those who used these metaphors, see Grant, *Planets, Stars, and Orbs*, p. 227 and n. 28.

[15] Ibid., pp. 313–14.

moving one degree in 100 years and thus completing a revolution in 36,000 years. The two other motions associated with the sphere of the fixed stars were the daily motion from east to west, which was assigned to a ninth sphere, and an irregular motion of the equinoxes that oscillated between east and west, known as "trepidation," which was actually a nonexistent motion "discovered" by the ninth-century Arab astronomer Thābit ibn Qurra. Trepidation was usually assigned to a tenth orb. The total number of orbs was, however, a function of the number of motions assigned to the fixed stars; occasionally, only the daily rotation was assigned to it, and the total number of orbs would then be eight. If precession were also assigned, nine orbs would be required, and if trepidation were added, there would be at least ten celestial spheres.

THE THEOLOGICAL SPHERES

Passages in Genesis concerning the second day of Creation convinced many theologians that the sphere of the fixed stars and one or more of the concentric spheres beyond should be interpreted as manifestations of the biblical Creation account. Of fundamental importance was the firmament, which (according to Gen. 1:7) divided "the waters from the waters." Although various interpretations of the firmament had been proposed – some, for example, viewed it as a single heaven that embraced all the planets and the fixed stars, whereas others equated it with the region of air – the most popular identified it solely with the eighth sphere of the fixed stars.

The waters above the firmament came to be called *crystalline*, a term that was applied to fluid waters as well as to waters congealed like ice or crystal. Defenders of each view could be found. For St. Jerome (ca. 347–419?) and the Venerable Bede (672–735), the waters were hard and crystalline, whereas for St. Basil (ca. 331–379) and St. Ambrose (ca. 339–397) they were fluid and soft. Whether hard or soft, the crystalline sphere was usually located above the firmament and was identified with the ninth or tenth transparent celestial orb, and sometimes with both.

Beyond all the mobile spheres of the cosmos lay a sphere called the empyrean heaven (*caelum empyreum*), which only emerged as a distinct cosmic entity in the twelfth century. In his famous theological treatise, the *Sentences*, Peter Lombard identified the empyrean sphere with the invisible heaven created on the first day. The empyrean heaven had no astronomical functions. It was a purely theological creation and seems to have acquired quasi-doctrinal status in Latin Christendom. Theologians such as Anselm of Laon (d. 1117), Peter Lombard, and many others who followed regarded the empyrean heaven as a place of dazzling luminosity, receiving its light directly from God. It was soon viewed as the dwelling place of God and the angels as well as the abode of the blessed. Despite its radiant state, the empyrean heaven transmits none of the dazzling light that perpetually fills it.

The empyrean is noteworthy because it was always regarded as immobile, and therefore a suitable container of the mobile orbs that moved with circular motion within it. The idea of an immobile heaven as a container was contrary to Aristotelian physics and was therefore opposed by some natural philosophers, for example Jean Buridan and Albert of Saxony.[16] Although the empyrean heaven was created for theological reasons and was initially assumed to have no physical impact on the world, over the centuries scholars persistently inquired whether the empyrean heaven could influence terrestrial change as did all mobile celestial orbs. Opinions varied from denials of influence, to modest influences that barely extended to the inferior planets, all the way to the exalted claim by the anonymous author of the *Summa philosophiae* that "the immobility of the empyrean heaven is more the universal cause of every transmutation of generable and corruptible things than the *primum mobile* and the other inferior spheres."[17] By the middle of the seventeenth century, however, few scholastic theologians believed that the empyrean heaven influenced terrestrial events, and most eliminated it from their cosmologies.

CELESTIAL MOTIONS AND THEIR CAUSES

Since uniform circular motion was the only kind of motion deemed appropriate for celestial bodies, natural philosophers were expected to identify the cause of such motions. Aristotle had already furnished two responses that one might make: external and internal. In his *Physics* and *Metaphysics*, Aristotle argued that each orb has an external, immaterial, spiritual mover, or intelligence, which is separate and distinct from it. This intelligence does not move its orb by pushing or pulling, or by exerting itself in any way. Rather, Aristotle characterized it as an "unmoved mover" because it caused its orb to move without being in motion itself, a task it achieved "by being loved," as Aristotle expressed it. The nature of this extraordinary love relationship between intelligence and celestial orb was left vague. Of the fifty-five intelligences, one for each orb, the first was accorded special status because it was associated with the outermost moving sphere, or sphere of fixed stars, that enclosed the world. Indeed, Aristotle identified the first unmoved mover with the deity, although all of the other unmoved movers had identical properties. Aristotle's internal explanation of celestial motion appears in his treatise *On the Heavens*, where he assumed that the celestial ether, by its very nature, moved with uniform circular motion.

16 Ibid., pp. 375–6.
17 Ibid., p. 379.

In the Christian Middle Ages, the external explanation with intelligences, which by then were usually equated with angels, was clearly the more popular. Aristotle's first unmoved mover was now called the Prime Mover, an expression synonymous with God. Although God could, of course, move the celestial orbs directly as an efficient cause, He chose rather to assign this task to one of his creations, an angel, or intelligence. The explanation by which angels were assumed to move their orbs was shaped by an article (no. 212) condemned in 1277 when the bishop of Paris decreed it an excommunicable offense to hold "That an intelligence moves a heaven by [its] will alone,"[18] an opinion held by many, including Thomas Aquinas. The underlying rationale for the condemnation of this article was that only God, not angels, could move by will alone. A new approach was devised by Richard of Middleton (fl. second half of the thirteenth century), Godfrey of Fontaines (d. 1306), and Hervaeus Natalis (ca. 1260–1323). They assumed that, in addition to intellect and will, an intelligence, or angel, possessed a third power, a motive force (*virtus motiva*). Responding to the command of the intellect, the will, in turn, commands the motive force to move the orb. The latter, it must be emphasized, is not an impressed force. In producing the uniform circular motion of its orb, an angel was assumed to be in direct contact with it. Since an angel performs its actions voluntarily, the celestial motions were assumed to be voluntary. The angelic *virtus motiva* became a common explanation to account for celestial motions.

Although most medieval natural philosophers assumed that angels, or intelligences, external to the celestial orbs caused them to move, a few explained those motions by a cause internal to the orbs. One causal candidate was the soul of the orb. But this proved unpopular and controversial since a soul implied that the orbs were alive. One expression of the hostility against souls in orbs was the condemnation of 1277, which denounced the existence of souls in the heavens, as well as the idea of intelligences acting as forms. The appeal to internal causes was expressed in different ways. Already in the thirteenth century, John Blund (ca. 1175–1248) and Robert Kilwardby (d. 1279) argued that each celestial orb possessed a natural, intrinsic capability for self-motion. According to James Weisheipl, Kilwardby held that "To each planet and orb God gave an innate natural inclination to move in a particular way in rotational motion; to each he accorded an innate order, regularity and direction without the need of a distinct agency like a soul, an angel or Himself here and now producing the motion."[19]

The most significant use of internal movers was proposed by Jean Buridan, who applied his famous impetus theory (which he employed to explain such terrestrial motions as that of a projectile; see Laird, Chapter 17, this volume)

[18] Ibid., p. 528. For the details, see ibid., pp. 528–35.

[19] James A. Weisheipl, "The Celestial Movers in Medieval Physics," *Thomist*, 24 (1961), 316.

to the celestial orbs. Since the Bible makes no mention of intelligences moving celestial bodies, Buridan felt free to abandon intelligences and propose his own interpretation:

> God, when He created the world, moved each of the celestial orbs as He pleased, and in moving them He impressed in them impetuses which moved them without his having to move them any more except by the method of general influence whereby he concurs as a co-agent in all things which take place.... And these impetuses which he impressed in the celestial bodies were not decreased nor corrupted afterwards, because there was no inclination of the celestial bodies for other movements. Nor was there resistance which would be corruptive or repressive of that impetus.[20]

In the absence of external resistances and contrary tendencies, the impressed impetus of an orb would remain constant and move its orb with uniform circular motion forever.

Medieval natural philosophers accepted the Aristotelian principle that each sphere could have only one proper motion. But this did not mean that an orb might not move with two motions. Indeed, the planets did move with two simultaneous motions: for example, the Sun, which moves east to west with a daily motion, while also moving west to east along the ecliptic. How could this happen? Nicole Oresme explained it by assuming that two intelligences operated simultaneously. The Sun's west to east motion was its proper motion, caused by its own intelligence. The Sun's daily east to west motion, however, was caused by another intelligence, probably the *primum mobile*, which produced the daily motion for all the planets. Oresme realized that the resultant motion caused by the two intelligences was a spiral line. Although Oresme fails to explain how the external intelligence causes the daily motion of the Sun and planets, his famous successor at the University of Paris, Pierre d'Ailly, considered the problem and concluded that the first movable sphere, or the *primum mobile*, caused the daily east to west motion of the planets by transmitting a certain "virtue" (*virtus*) to them that induces them to follow it in its own east to west daily motion. D'Ailly therefore agrees with Oresme that the west to east motions are natural and proper. But, surprisingly, he regards the daily motion, caused, as we saw, by the virtue transmitted to the planets by the *primum mobile*, as neither natural nor violent but a motion that lies outside of nature (*praeter naturam*).

Oresme and d'Ailly were aware that they might seem to have violated Aristotle's principle of one motion per orb. They resolved this difficulty by re-defining the concept of "one motion." The latter was understood to mean only "proper motion" caused by the orb's own intelligence, thus allowing a sphere to have an additional "improper," or "accidental," motion that

[20] Translated by Marshall Clagett, *The Science of Mechanics in the Middle Ages* (Madison: University of Wisconsin Press, 1959), p. 536, par. 6.

originated from a source – ultimately an intelligence – external to the orb. In this way, simultaneous motions were integrated into the Aristotelian system.

DIMENSIONS OF THE WORLD

What were the dimensions of the finite cosmos we have just described? The dimensions were measured only from the Earth to the sphere of the fixed stars (that is, from the innermost body of the world to the outermost bodies, the fixed stars). Thus the measurements did not include the ninth and tenth orbs and the empyrean heaven beyond, none of which contained a celestial body. Cosmic dimensions were rendered even more uncertain by a difficulty with the eighth sphere.

In a tradition stemming from Ptolemy, celestial spheres were thought to be nested within one another. It was routinely assumed that the concentric spheres or spherical shells that bore the planets were packed together tightly, so as to leave no empty space between consecutive spheres. In other words, the innermost surface of the sphere of Jupiter (its concave surface), for example, contains, and is contiguous with, the outermost surface of the sphere of Mars (that is, its convex surface), and so on for any two successive concentric planetary orbs. By this technique, natural philosophers and astronomers made certain that no extraneous matter or void spaces could lie between the surfaces of any two planetary orbs.

In measurements provided by Campanus of Novara, the distance from the Earth's center to the outermost surface of Saturn's sphere is 73,387,747 miles.[21] On the basis of the nested-sphere pattern described earlier, the innermost surface of the sphere of the fixed stars would be assumed contiguous with the outermost surface of the sphere of Saturn, from which we may infer that the inner surface of the sphere of fixed stars was generally assumed to be some 73 million miles from the Earth's center. But what of the outer surface? Were all the visible stars arrayed on the inner surface? Or were they distributed at different distances, so that the eighth sphere was of considerable thickness, like the planetary spheres? If the latter, some stars might be much farther away than 73 million miles. In his *Commentary on De caelo*, Albertus Magnus suggested two ways to explain the six magnitudes of brightness attributed to the fixed stars:

1. They are equidistant from the earth and on the same spherical surface, but they are of six different sizes and therefore of six different degrees of brightness; or

[21] The measurements cited here appear in Campanus of Novara's *Theorica planetarum*. They are taken from the tables that appear in Grant, *Planets, Stars, and Orbs*, pp. 436–7. The tables in Grant's work represent a rearrangement of data in tables produced in Francis S. Benjamin, Jr. and G. J. Toomer, eds., *Campanus of Novara and Medieval Planetary Theory: Theorica planetarum* (Madison: University of Wisconsin Press, 1971), pp. 356–63.

2. all stars are of the same size and brightness but located at six different distances from the Earth.

Albertus rejected the second possibility because he believed it would require six separate orbs, one for carrying each star around the heavens. Because he assumed that each of the six orbs would move with a different velocity and there was no empirical evidence that fixed stars moved with differing speeds, he opted for the first and traditional opinion. But Albertus did not ask whether the six magnitudes of fixed stars might be distributed at different distances within a single thick orb. That possibility was not seriously entertained until the seventeenth century, when some scholastics adopted it, and even Galileo suggested that the fixed stars and the planets were distributed over varying distances within a single all-embracing sphere.[22]

Among the planets, the Sun's distance measurements were the most anomalous compared with modern values. From the inner surface of the solar sphere, whose inner surface, representing its nearest distance from the Earth, is approximately 3,900,000 miles, to its outermost surface, representing its farthest distance from the Earth, is approximately 4,200,000 miles. These figures differ radically from the modern value of a mean distance of approximately 93,000,000 miles. The medieval value of 35,000 miles for the Sun's diameter should be contrasted with the modern value of 864,000 miles. Anomalous measurements abound, making the cosmos seem idiosyncratic rather than the beautifully crafted product of an infinite Creator. The thicknesses of the planetary spheres seem discordant. The sphere of Venus, for example, is approximately nine times thicker than those of the Sun and Moon, the planets immediately above and below. With the spheres of the superior planets, we witness an enormous increase in thickness over the Sun and inferior planets. The diameters, and therefore the sizes, of the planets also seem to vary in strange ways. Were medieval natural philosophers aware of these anomalies? Or better, did they even regard them as anomalies? There is no evidence that they did. It is more likely that medieval scholars regarded their cosmos as a thing of beauty and harmony. They probably would have agreed with C. S. Lewis's judgment that "the human imagination has seldom had before it an object so sublimely ordered as the medieval cosmos."[23]

EXISTENCE BEYOND THE COSMOS

Although Aristotle had denied the existence of matter, place, time, and vacuum beyond our world, the urge to contemplate the possibility that something might exist outside the cosmos proved irresistible. Scholastics

[22] See Grant, *Planets, Stars, and Orbs*, pp. 440–1.

[23] C. S. Lewis, *The Discarded Image: An Introduction to Medieval and Renaissance Literature* (Cambridge: Cambridge University Press, 1964), p. 121.

focused on topics that Aristotle had discussed, namely the possible existence of other worlds and the possibility of an infinite void space.

Aristotle and the Creation account in Genesis were in agreement on the uniqueness of our universe. But the manner in which Aristotle derived the world's uniqueness was offensive because he claimed to demonstrate that the existence of another world was impossible, and to argue that creation of other worlds was impossible, even for God, was viewed as a restriction on God's absolute power to do as He pleases. For example, article 34 of the Condemnation of 1277 declared it an excommunicable offense to hold "that the first cause [God] could not make several worlds." As a consequence, scholars in the fourteenth century were compelled to concede that God could create other worlds, although they were free to deny that He had actually done so. Before and after the condemnation, however, some scholars chose to argue that although God could create other worlds, He probably had not done so. Thomas Aquinas, for example, argued that if God created other worlds they would be either similar or dissimilar. If the former, they would be superfluous; if the latter, none of them would be perfect because none could incorporate all the natures of sensible bodies. Only by combining all of these imperfect worlds could we obtain a perfect world, a state of affairs that could be achieved more economically by a single world. In the fourteenth century, Jean Buridan declared that "it must be believed on faith that beyond this world God could form and create however many other spheres, and other worlds, and other finite magnitudes He wishes."[24] But Buridan believes that there is no space, magnitude, or other world beyond our world not only because Aristotle has given natural arguments against such possibilities but also because God Himself probably chose not to make other worlds. For if he wished to create additional creatures like those in our world, he could simply increase the size of our world to double, or one hundred times, its present size. Nicole Oresme was even convinced that there was an infinite space beyond our world in which other worlds could subsist, which led him to conclude "that God can and could in His omnipotence make another world besides this one or several like or unlike it. Nor will Aristotle or anyone else be able to prove completely the contrary." In accordance with the Bible, however, Oresme insists that "there has never been nor will there be more than one corporeal world."[25]

On the assumption that God could create other worlds identical to ours, natural philosophers considered the physical consequences of simultaneously

[24] Jean Buridan, *Acutissimi philosophi reverendi Magistri Johannis Buridani subtilissime questiones super octo libros Aristotelis diligenter recognite et revise a Magistro Johanne Dullaert de Gandavo antea nusquam impressa* (Paris, 1509; facsimile, Frankfurt am Main: Minerva, 1964), bk. 3, qu. 15, fol. 57v, cols. 1–2. See also Grant, *Planets, Stars, and Orbs*, p. 163.

[25] See Nicole Oresme, *Le Livre du ciel et du monde*, ed. Albert D. Menut and Alexander J. Denomy, translated with an introduction by Albert D. Menut (Madison: University of Wisconsin Press, 1968), bk. 1, chap. 24, pp. 177–9.

existing worlds. Some of the derived consequences were in conflict with basic Aristotelian principles. For example, although Aristotle had argued that vacua of any kind were impossible, the existence of void spaces between these hypothetical and discontinuous worlds seemed a plausible and even compelling inference. Nicole Oresme, for example, not only found this acceptable but also concluded that if such worlds were like ours, they would obey the same physical laws. Each world would be spherical and have its own "up," "down," "center," and "circumference." Since all such worlds would be coequal, it followed that no single center or circumference would be privileged, a conclusion that implied rejection of a unique center and circumference of the universe, on which Aristotle had founded much of his physics and cosmology.

If other worlds, and therefore matter, could not exist beyond our world, perhaps an infinite void space might. Although Aristotle had rejected this possibility in the third book of his *Physics*, the existence of infinite space became linked to theological discussions about God and His whereabouts. The idea that God might, in some sense, occupy an infinite space in which our world is located seems not to have played any significant role until the fourteenth century, when Thomas Bradwardine (ca. 1290–1349), in a treatise titled *In Defense of God Against the Pelagians*, proposed the following five brief corollaries that established the existence of a real infinite void space:

1. First, that essentially and in presence, God is necessarily everywhere in the world and all its parts;
2. And also beyond the real world in a place, or in an imaginary infinite void.
3. And so truly can He be called immense and unlimited.
4. And so a reply seems to emerge to the questions of the gentiles and heretics "Where is your God?" "And where was God before the [creation of the] world?"
5. And it also seems obvious that a void can exist without body, but in no manner can it exist without God.[26]

Bradwardine not only identified God's infinite immensity with infinite void space but insisted that infinite space was dimensionless because God "is infinitely extended without extension and dimension." Although it is virtually impossible for us to conceive of a dimensionless space, scholastic theologians were driven to that conclusion because they identified infinite void space with God's omnipresent immensity. Thus, to attribute tridimensionality to that space would be the same as attributing tridimensionality to God, who would then be physically extended, and therefore also divisible, just like any body. But God was spirit, not body. In the Middle Ages, it would

[26] Thomas Bradwardine, *De causa Dei contra Pelagium et de virtute causarum ad suos Mertonenses libri tres... opera et studio Dr. Henrici Savili Collegii Mertonensis in Academia Oxoniensis custodis, ex scriptis codicibus nunc primum editi* (London, 1618; repr. in facsimile, Frankfurt am Main: Minerva, 1964), p. 135. See also Grant, *Planets, Stars, and Orbs*, pp. 173–7.

have been unthinkable to conceive of God as extended and divisible. Bradwardine's ideas were influential. Many scholastic theologians discussed the same theme, especially in the sixteenth and seventeenth centuries (Bradwardine's *De causa Dei* was printed in 1618 in London). Through Bradwardine, and perhaps others, scholastic ideas about God and space played a significant role in the history of spatial conceptions between the late sixteenth and eighteenth centuries, the period of the Scientific Revolution. From the assumption that infinite space is God's immensity, medieval and early-modern scholastics derived most of the same properties of space – except for tridimensionality – as would be conferred upon it by the likes of Henry More, Isaac Newton, Otto von Guericke, and Samuel Clarke.

The ideas and concepts described, and others not mentioned here, transformed Aristotle's cosmology and natural philosophy. By the end of the Middle Ages, *Aristotelian cosmology* differed radically from *Aristotle's cosmology* in the form in which it had been received in the twelfth and thirteenth centuries. It is no exaggeration to proclaim that the assault on Aristotle's cosmology was begun and developed by medieval Aristotelians. They not only made cosmology a viable and vital subject for many centuries but posed many of the crucial problems (Is there an infinite space beyond the world? Can vacua exist? Does the Earth rotate? Are terrestrial and celestial matter the same or different?), and a number of thought-provoking and provocative hypothetical problems, that would be resolved by sixteenth- and seventeenth-century natural philosophers such as Copernicus, Galileo, Kepler, and lesser lights. Even in the seventeenth century, numerous scholastic natural philosophers sought to adjust to the newly emerging cosmology derived from Copernicus and Tycho Brahe in which celestial orbs had been largely replaced by a fluid medium. Scholastic cosmology was far from a congealed body of doctrine. It was rather a body of varied opinion in which a genuine effort was made to incorporate aspects of the new cosmology. To understand the dramatic changes in the seventeenth century, it is essential to study the fate of medieval cosmology, which existed, and even evolved, alongside the gradually developing new Copernican cosmology for approximately 150 years.

19

ASTRONOMY AND ASTROLOGY

John North

To humanity at large, astronomy has always had an appeal linked more to the vastness of the world we inhabit than to the details of any scientific analysis of it. It is a striking fact of medieval academic life that every student in the arts was obliged to pursue a course in astronomy. Yet the texts used for this purpose were cast in an ostensibly literary form, far removed from the mathematical form that was essential to the ancient scientific eminence of astronomy. The most widely circulated of all, the *Treatise on the Sphere* by John of Sacrobosco (d. 1236 or 1246), must have been for many students a bitter pill to swallow, but it was one sweetened by quotations from the poetry of Ovid, Virgil, and other classical writers. It was studied as part of an evolving curriculum that taught the mathematical sciences, including the traditional four (quadrivium) that made up the more advanced part of the seven liberal arts, but also because it had a practical value. Every cleric, for example, needed to understand the church calendar as a guide to the festivals of the church year, and the universities were in turn intended to serve the church. Even the study of the calendar took the student into mathematical waters deeper than were necessary to understand the simple cosmological patterning of the Aristotelian universe. Nevertheless, in many respects, astronomy was still in the service of that cosmology, and both served theology as needed. The interrelations of those three arms of medieval learning were complex, and changed materially with time, but by the thirteenth century we find something happening that the early church fathers could not have contemplated, namely theologians trying to reach an understanding of the habitation of God by analogy with the stellar and planetary heavens of Aristotelian philosophy. It was in this spirit that Albertus Magnus, for example, wrote of an eleventh heaven in which Father, Son, and Holy Spirit had their place.[1] At a time when such theology was placing the Trinity, no

[1] Albertus Magnus, Commentary on Peter Lombard's *Sentences*, bk. 2, distinction II.h., art. 7–8; *Opera omnia*, ed. Auguste Borgnet, 38 vols. (Paris: Vivès, 1890–1899), vol. 27, pp. 56–8.

less, in the setting of the cosmology of the schools, it was natural enough that some scholars should wish to delve deeper into the precise movements of the cosmos and not content themselves with its crude topology.

Astronomy at a lower theoretical level entered into the daily experience of many people outside the universities. Almost all of those who lived in centers of population rich enough to support churches decorated through painting, carving, or stained glass would have been familiar with astronomical symbolism of some sort. From the late thirteenth century onward, this encompassed church clocks and mechanical contrivances originating in the desire to drive around a wheel with a daily motion – the wheel of the sky – and so produce a model of the universe. This universe was most readily conceived in broad symbolic forms, the basic cosmological symbols being those of the Aristotelian cosmology described in the previous chapter. Students could easily grasp, at least superficially, the Aristotelian system of concentric spheres, but they also learned that the geometrical models needed to understand the intricacies of planetary motion went far beyond that system and owed more to Ptolemy than to Aristotle. The scholar tended to think of the Aristotelian motions as fundamental and in a way real and force-driven, and of the Ptolemaic motions as in some sense geometrical embroidery on the basic reality. Little did the scholar realize that Ptolemy himself felt the same way about the primacy of the Aristotelian description and the need to adapt the eccentrics, epicycles, and the rest to that much older account.

Many beliefs draw their force from the simplicity of the language in which they are expressed. The crudest of symbols are the most persistent but are also the most fickle, in that they tend to take on whatever meaning is convenient, regardless of their authors' intentions. Thus Aristotle's doctrine of the mean came to be read as a doctrine that the center of the world was the best of all places. In the account of the (real or fictive) fourteenth-century voyages of Sir John Maundeville, at the beginning of the prologue, the notion was stretched to explain that Christ chose the Jews because their country was in the very middle of the world. In the intellectual world of the astronomers, this type of reasoning was much less readily invoked. Here mathematics was master and had to be learned in the form of arithmetic and geometry, two other parts of the traditional quadrivium. Much of this learning was by rote. The expert astronomer who wished to pass beyond this level had to master techniques of what we now describe as spherical trigonometry, something taught within technical astronomical treatises that were the subject of higher study alone. Scholars in this class were relatively few. They might not have had the social standing of the senior theologians within the universities, but they certainly comprised an intellectual elite, and of course many individuals qualified on both counts. The most original English astronomer of the Middle Ages, Richard of Wallingford (ca. 1292–1336), was also abbot of St. Albans, England's premier monastery. An astronomical and astrological writer of high reputation later in the same century was Pierre d'Ailly (1350–1420), who went from doctor in theology through various preferments – including chancellor of the University of Paris and bishop of Cambrai – to cardinal, no less.[2] And there are many similar examples.

Astronomers also used another type of reasoning that was, however, less taxing than mathematical astronomy and more open to the use of simple human analogy. It offered one of the chief rewards for mastering the harder technicalities and was one of the chief motives for doing so. Astrology in one form or another had, from even before its codification in the Hellenistic period, gone hand in glove with mathematical planetary and stellar astronomy. The liaison continued throughout the Middle Ages and beyond in the highest reaches of academia. The intrinsic difficulties of astronomy gave a certain respectability to astrological subjects, and they in turn were allowed a place at the fringes of theology, metaphysics, and philosophy – notably through the continuing discussion of astral determinism in its various forms. Although astrology had other implications, it was its relevance to human life that underlay its perceived value and made possible its role of great importance in art and literature. The simple cosmological picture of the spheres of "Aristotelian" cosmology lent itself readily to the provision of overall structure to Dante's *Divine Comedy*, but there, and even more in the poetry of Chaucer, evidence is to be found for the use also of astrological color and even deep poetic structures based on astrology. We who regard astrology as a valueless pursuit need to bear constantly in mind the high value then placed on it and manifested in such ways.

It is often said that astronomy and astrology were treated as a single discipline until after the end of the Middle Ages, but this remark needs much qualification. Those who make the point often have the ulterior purpose of proving that astronomy was driven entirely by social pressures coming from elsewhere. The fact that the one was often a motive for studying the other and that many ignorant writers confused the two (as some do today) does not mean that their practitioners did not distinguish them. The university curriculum, with its purely astronomical introductory course (the doctrine of the sphere and the theory of planetary motions), shows that they did so. One hardly needed to read a text carefully to know the category into which it fell; the type of illustration, the type of material, and the very method of argument made it desirable that the two genres have different labels. Many surviving treatises even opened with a careful distinction between the meanings of the two words. It is hoped that the distinction will become more obvious as we proceed. For the time being, it will be enough to note that, although both relied heavily on the use of Greek and Arabic authorities, astronomy developed criteria for the acceptance of

[2] John North, *God's Clockmaker: Richard of Wallingford and the Invention of Time* (London: Continuum, 2005); and Laura Ackerman Smoller, *History, Prophecy, and the Stars: The Christian Astrology of Pierre d'Ailly* (Princeton, N.J.: Princeton University Press, 1994).

those authorities – conformity with observed planetary and stellar positions – whereas astrology was and remained shamefully eclectic and uncritical. There were some who, especially in the fourteenth century, tried to put astrological weather prediction to the test and to derive or improve meteorological laws, but for the most part there was blind acceptance of even inconsistent bodies of doctrine, deriving for the most part from Eastern Islam and the ancient world. For this reason we begin with astronomy, the scientific objectives of which were generally clear and on which the science of astrology relied heavily.[3]

PLANETARY ASTRONOMY

The importance of Spain in providing translations of key classical and Arabic texts is discussed by Burnett (Chapter 14, this volume). From the eighth century onward, Spain was more or less under Muslim control, and it remained partially so until the fifteenth century, but the great period of its influence was over by the end of the thirteenth. Eleventh-century authors had written extensive treatises on the three most educationally useful astronomical instruments – the sphere, the astrolabe, and the equatorium – and they had introduced astronomical tables that made use of al-Khwārizmī's, thus setting fashions that in due course caught the European imagination. The system of reckoning using Hindu-Arabic numerals was introduced along this astronomical route, in which the change from the old Roman numerals was more or less complete by the early fourteenth century. Compendia of astronomical tables, known by the generic Arabic term *zīj*, were of numerous different sub-types. Most had a limited life, and very few achieved much fame, but the two most pervasive were closely associated with the city of Toledo. The first, the "Toledan Tables," had instructions for their use written by the renowned scholar Ibn al-Zarqāllu (d. 1100).[4]

The Toledan Tables superseded all others in Spain and Europe generally until the early fourteenth century. Like most *zījes* before and after them, they were a compilation of elements from such ninth-century Eastern authors as al-Khwārizmī and al-Battānī, and they drew also on a program of new observation that was meant to verify such older material. They contained a little astrological material. Star coordinates were updated for precession, and in later Latin recensions they were usually updated yet again for the time of copying. They were overshadowed by the most famous Spanish *zīj* of all, produced under the patronage of the Christian king Alfonso X of Leon and Castile in the 1270s. The old Toledan Tables, in Latin translation,

[3] For a survey of ancient and medieval astronomy, see John North, *Cosmos* (Chicago: University of Chicago Press, 2008), chaps. 4, 8–10; and James Evans, *The History and Practice of Ancient Astronomy* (New York: Oxford University Press, 1998).

[4] Gerald J. Toomer, "A Survey of the Toledan Tables," *Osiris*, 15 (1968), 1–174.

had carried Islamic astronomy well and truly to the heart of Europe, but they were introduced alongside the al-Khwārizmī *zij* and did not become the dominant force at once. The Alfonsine Tables took their place in Paris, Oxford, and elsewhere in the 1320s, half a century after they had been drafted and in modified form. In some places, their arrival was much later still.

After the first universities emerged, and especially after their rise in social and intellectual importance at the beginning of the thirteenth century, masters wrote new introductory texts. The literary style of that by John of Sacrobosco has already been mentioned. Sacrobosco was perhaps an Oxford master, more probably a Paris master, and he certainly taught at the university in Paris. His book dealt with only elementary spherical astronomy and geography, and included virtually nothing on planetary theory. Sacrobosco produced other popular texts, on arithmetic, on the *computus* (calendar calculation), and on a simple type of quadrant, but until well into the seventeenth century his name in university circles was known chiefly for his work "on the sphere." At much the same time, another work of the same sort was written by Robert Grosseteste (ca. 1168–1253), quasi-chancellor of the University of Oxford.

These works of "spherical astronomy" (as the subject is still called) needed to be supplemented by texts on planetary theory, and this gave rise to a type of book known by the generic title *Theorica planetarum* (*Theory of the Planets*). John of Seville's translation from Arabic to Latin of a treatise by the ninth-century writer al-Farghānī was an example of this type of work, one that was used in the schools from the twelfth century onward, and Roger of Hereford (fl. second half of the twelfth century) wrote another. The most widely circulated Western example, by an unknown author, is often referred to simply by its opening Latin words: "*Circulus eccentricus vel egresse cuspidis....*" Despite some technical shortcomings, the work helped to stabilize astronomical vocabulary in the West, and its diagrams helped the student to master the essentials of the Ptolemaic planetary models. Unfortunately, it gave no real understanding of how those models had been derived in the first place. Ptolemy's *Almagest* could have done that, for it was twice translated into Latin in the twelfth century, once from Greek and once from Arabic, but it was a long, difficult, and costly work, unsuited to a university course in the arts faculty. Another translation of the *Almagest* was prepared in 1451. Even for more advanced study, Ptolemy's long work had been replaced by astronomical digests in the West as in Islam.

OBSERVATION AND CALCULATION

The character of the medieval university might seem to have been unfavorable to the growth of astronomy, regarded as a science related to the *observed* world, since there was as yet no very clear sense of the importance of systematically extending scientific knowledge through new observations. Early-medieval attitudes toward knowledge were strongly influenced by the techniques used in discussing Holy Scripture; that is, as an inheritance, to be purified and restored to its original form, then analyzed and commented upon before being transmitted to later generations. Fortunately, as better astronomical material became available and as scholars developed their own skills, prompted in part by astrological ambitions, the situation slowly changed. Roger Bacon (ca. 1220–ca. 1292) was one who helped to advance a sense of scientific empiricism, although he was never deeply immersed in astronomy. For clearer signs of change, we should look instead to a scholar like William of St. Cloud, who flourished at the end of the thirteenth century and was in some way connected with the French court. In 1285, he observed a conjunction of Saturn and Jupiter. He compiled an accurate "almanac," giving the calculated positions of the Sun, Moon, and planets at regular intervals between 1292 and 1312. There was nothing unusual in this, but William also introduced into his almanac an account of the observations and planetary tables (of Toledo and Toulouse) on which the almanac was based, as well as the corrections he found necessary to make to them. In connection with his work on almanacs, with their references to solar and lunar eclipses, William considered the projection of the Sun's image on a screen through a pinhole aperture. (Casting an image in this way into a darkened room, a camera obscura, was an alternative to doing the same with a simple lens.) This would avoid damaging the eyes, William said, as had happened in the case of so many who had observed the eclipse of 4 June 1285.

Roger of Hereford had mentioned the pinhole technique in the twelfth century and, following a suggestion by William, the Jewish scholar Levi ben Gerson (1288–1344) actually used pinhole images in 1334 to yield a figure for the eccentricity of the Sun's orbit. Levi is another who keenly felt the need for an empirical approach to astronomy. He observed the Sun at summer and winter solstices using a combination of the "Jacob's staff," an instrument of his own invention, and the camera obscura. Kepler observed a solar eclipse in 1600 in almost the same way. The derivation of the eccentricity of the solar model depends on the fact that the diameter of the image is in inverse proportion to the Sun's distance, so that the connection between the quantities observed and the geometry of the eccentric circular orbit is very direct. After the end of the thirteenth century, the practical side of astronomy began to grow rapidly, and it continued to do so steadily. To take yet another example from William of St. Cloud, by observation he found that the positions of the stars implied that the theory (wrongly) ascribed to Thābit ibn Qurra was about 1° in error. For this reason, he favored a *steady* precessional motion, going against the general trend of belief at that time.

Distinct from the instruments that made possible this mild empiricism were instruments designed chiefly for *calculation* and yet others used mostly for *demonstration* in a schoolroom situation, such as the simple armillary

Figure 19.1. Mid-thirteenth-century English brass astrolabe with silvered plates or tympans. Courtesy of the Adler Planetarium, Chicago (M-26).

sphere. (This was hardly more than a three-dimensional model made of rings, *armillae*, representing the main circles of the celestial sphere – the equator, the ecliptic, the tropics, and so forth.) All three functions – observation, calculation, and demonstration – overlapped considerably; indeed, the most important instrument of all, the astrolabe (Figures 19.1 and 19.2), combined all three in some measure. An essential part of the astrolabe was its "alidade" (*H*), a rule pivoted about the center of the instrument and carrying a vane at each end, pierced with holes through which a star might

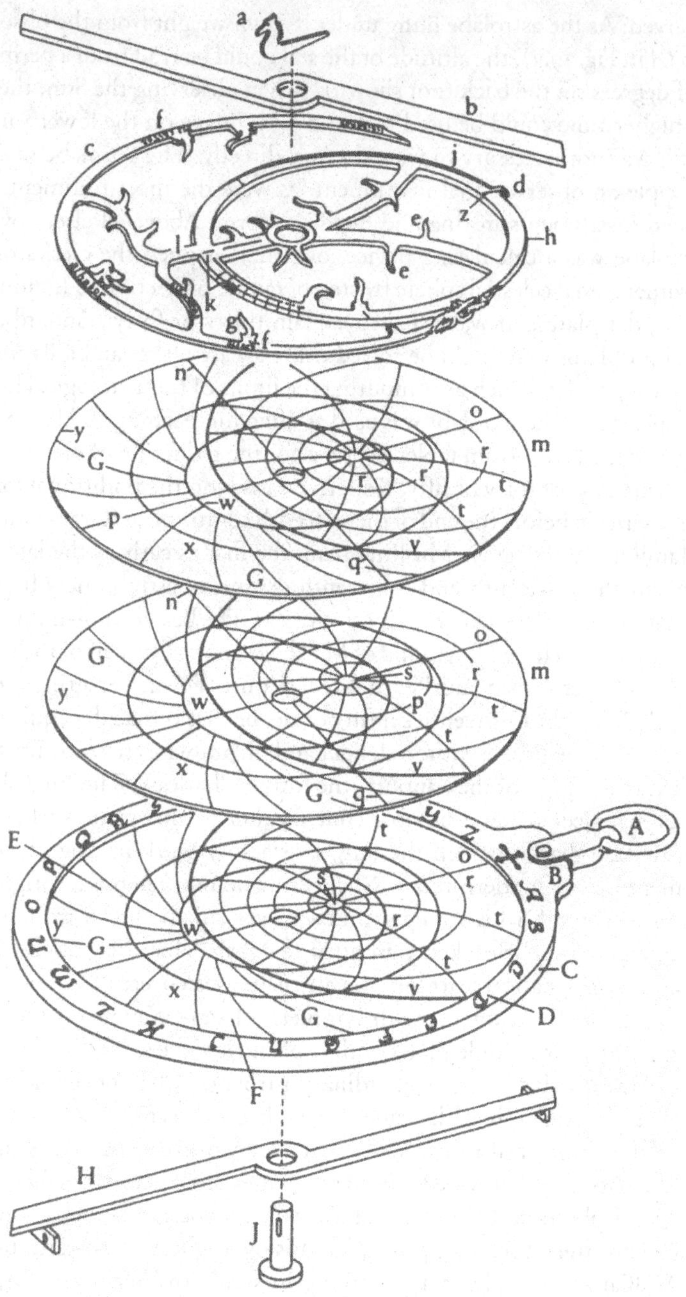

Figure 19.2. An "exploded" view of the astrolabe. (a) Horse (wedge with a horse-head design); (b) rule; (c) rete; (e) star pointers; (h) ecliptic circle; (k) lines marking the signs of the zodiac; (m) climates; (r) almucantars (circles of equal altitude); (s) zenith; (t) lines of equal azimuth; (v) horizon line; (w) Tropic of Cancer; (x) equator; (y) Tropic of Capricorn; (A) ring by which the thumb holds the astrolabe; (C) mater (mother); (G); hour angle lines; (H) alidade with sighting holes; (J) pin (to be locked by horse-head wedge).

be observed. As the astrolabe hung under its own weight from the observer's thumb (*A* in Fig. 19.2), the altitude of the star could be read from a peripheral scale of degrees on the back (not shown). When observing the Sun, the hole in the higher vane could be used to cast a solar image on the lower vane H, so that it was not necessary to view the Sun directly. The astrolabe was thus in principle an observational instrument, as were the most rudimentary of graduated instruments in ring and quadrant form. Above all else, however, the astrolabe was a calculating device, one that mapped the circles on the three-dimensional celestial sphere by stereographic projection to (graduated) circles on flat plates.[5] It was usual to explain thirty or forty standard astronomical problems that could be solved using an astrolabe, and a handful of astrological problems such as computing the limits of the astrological houses. One suspects that the astrolabe was used as often for teaching and for the processing of data drawn from tables as it was for the reduction of observations.

There are at least forty or fifty Western treatises on the traditional form of astrolabe written before the end of the sixteenth century, but they fall into just three families. All came via Muslim Spain, the first as early as the late tenth century. In the thirteenth and fourteenth centuries, astronomers began to blend Iberian astrolabe texts with local variants. The better-known European writers in Latin included Raymond of Marseilles in the first half of the twelfth century, and Sacrobosco and Pierre de Maricourt (Petrus Peregrinus) in the thirteenth. With the fourteenth century came the first vernacular equivalents. In 1362, Pélerin de Prusse wrote a slender and incomplete astrolabe treatise in French at the request of the dauphin, the future Charles V. The English poet Geoffrey Chaucer made a notable contribution to this genre of literature with a treatise that remained the only satisfactory work in English on the instrument before modern times. Chaucer made no pretense to originality. His book, written for his young son Lewis, was chiefly drawn from a work that was at the time mistakenly ascribed to Māshā'allāh (fl. 762–815). As a symbol of mathematical learning, the astrolabe was coveted by nonexperts, who fell for fake instruments with symmetrical star-pointers on their retes (see Figure 19.3) or latitude plates with concentric circles.

One of the disadvantages of an ordinary astrolabe is that it needs a separate plate for each geographical latitude. In the late thirteenth century, universal astrolabes (universal in the sense that they worked at any geographical latitude whatsoever) provided inexpensive alternatives to astrolabes of the simple planispheric type. In Spain, al-Zarqāllu had designed such a universal astrolabe, but the underlying principle was slow to filter northward. Indeed, when William the Englishman, working in southern France in 1231, tried to reproduce the universal astrolabe from the written source only, he made several mistakes. The instrument, often known by the Arabic word *saphea*

[5] A stereographic projection is one from a pole of the sphere onto the equatorial plane (or a plane parallel to it). In this projection, circles on the sphere are transformed into circles on the plane.

Figure 19.3. Fifteenth-century Eastern brass astrolabe body, Iran. Fake, identical tympans and symmetrically located, unnamed stars on rete. Courtesy of the Adler Planetarium, Chicago (A 91).

(plate), was not well understood in the Latin world until long after Ibn Tibbon's Hebrew translation of al-Zarqāllu's text in 1263, and even then it remained an instrument for experts rather than ordinary students.[6]

[6] This will be mentioned again in context later.

Unlike the *saphea*, an ordinary astrolabe for use at five different latitudes might have the inside of the "mater" inscribed with circles for one latitude and two plates, each inscribed on both sides, for the four others. The extra metal for the plates added somewhat to the cost, and the work needed for the plates was roughly five times as much as that to inscribe the plate (*saphea*) of a universal instrument. Later, the instrument of universal design was further modified in the interests of economy – but not of simplicity. The compound diagram (of lines in stereographic projection appropriate to the rete and the plate of an ordinary astrolabe) was folded along a main axis and then folded again along the axis at right angles to the first, giving a morass of lines, points, and graduated scales, many of them double. To make use of this "astrolabe-quadrant," one first had to be trained in the principles of the *saphea*, but to the skilled astronomer, the merits of the astrolabe-quadrant were not insignificant. Its accuracy was roughly comparable with that of an instrument four times as large in area. Like the traditional instrument, it could be used for observation. To measure altitudes, small sighting vanes with pinholes could be fitted to one edge of the quadrant, and the quadrant tilted until the star (or whatever was under observation) was seen through the two holes. The vertical was settled by means of a hanging plumb-bob on the thread, and the altitude was read off the peripheral scale.

The earliest known description of such an astrolabe-quadrant was written between 1288 and 1293 in Hebrew by Jacob ben Machir ibn Tibbon, a scholar known in Latin as Profatius Judaeus. As mentioned earlier, he was responsible for translating into Hebrew an Arabic treatise on the universal astrolabe. He was born in Marseilles (ca. 1236) and died in Montpellier (1305), but his family had come from Granada. It has often been assumed that Ibn Tibbon's "new quadrant" must have been based on an Islamic prototype. Whether this is true or not, he called it the "Israeli quadrant." His work was turned into Latin (in 1299, by Armengaud), and the Latin was expanded by the Danish astronomer Peter Nightingale (fl. 1290–1300),[7] through whose text the instrument became well known to Latin astronomers. The astrolabe-quadrant was often made in a very rudimentary form, its lines drawn on parchment or paper glued to a wooden quadrant. From the fifteenth century onward, it became popular, especially in this wooden form, in the Ottoman Empire; indeed, it was used in Turkey even into the twentieth century by those who organized religious life by traditional astronomical means.

Instruments were designed for far more than the numerical solution of problems in the astronomy of the celestial sphere. The most important class of calculating instrument after the astrolabe was the equatorium, for evaluating planetary positions for any time by simulating (usually with metal disks) motions in the epicycles, equants, and deferent circles of models of Ptolemaic type. Al-Zarqāllu wrote original treatises on this type of instrument,

[7] This important scholar also goes by such names as Peter Philomena and Peter of Dacia.

but outside Iberia, the first important text on such a device was by Campanus of Novara, who was consciously following in the wake of his Spanish precursors. Roughly speaking, this type of equatorium directly simulated the Ptolemaic motions of the planetary circles, giving each circle its own disk (with graduations to set the angular motions in the circles). Occasionally instruments of this ostensibly simple and intuitive type were developed with great subtlety, as when two Paris-trained astronomers, Guido de Marchia (ca. 1310) and Johan Simon of Zeeland (ca. 1418), turned a simple eccentric construction (lacking an epicycle) into an epicyclic construction with a concentric deferent circle. (To have a single central pivot was of course mechanically desirable.) Another interesting construction was that of the Augustinian canon Rudolf Medici (1428), who used a double epicycle in a way not unlike that adopted later by none other than Copernicus. Rudolf's equatorium was no pocket instrument; in the form of a triptych, it was more than a meter square when closed. Like Copernicus – but for different reasons – he rejected the Ptolemaic equant but made use of nonuniform graduations on his disks.

The most notable instance of a very different and subtler type of equatorium was the albion ("all by one") designed by Richard of Wallingford. Disks were now used to perform the intermediate calculations that would normally require planetary tables. The results were then combined, again by resorting to circular scales on the instrument – rather as one may add or subtract numbers by sliding two rulers alongside one another. The nonanalog technique, which modern readers would certainly find difficult at first, was one that would have come fairly naturally to astronomers accustomed to the use of tables. A nonuniform graduation of certain scales was required for some of the calculations.

Once the intuitive connection between a scale and a circular orbit was broken (that is, once the scales were regarded simply as repositories of mathematical information), astronomers began to think much more freely about mathematical functionality than they had before. There was hardly a problem of classical astronomy that could not be solved to a reasonable degree of accuracy on the albion. With its subsidiary instruments, it could provide knowledge of parallax, velocities, conjunctions, oppositions, and eclipses of the Sun and Moon. First in England and later in Southern Europe, it remained in vogue in various anonymous forms until the sixteenth century. At least seven treatises were derived from it, and astronomers began to abstract its subsidiary instruments, especially its parallax and eclipse instruments. The Viennese master John of Gmunden (d. 1442) produced what was perhaps the most often copied version (ca. 1430). Johannes Regiomontanus (1436–1476) drew from it to produce a rather careless manuscript edition, and John Schöner (1477–1547) abstracted its eclipse instrument. The most striking printed work to use it was the *Astronomicum Caesareum* (1540) by Peter Apian of Ingolstadt (1495–1552). This book, dedicated to Emperor

Charles V and his brother, was one of the most lavishly illustrated and colorful scientific works of the first four centuries of printing. It contains several equatoria, each with moving disks of paper, but mostly of the simple types that simulated the Ptolemaic planetary models.[8]

THE ALFONSINE TABLES

The drive to improve calculational techniques gathered much momentum in the first decades of the fourteenth century. John of Lignères (fl. second quarter of the fourteenth century), for example, wrote treatises on a new type of armillary, the saphea, the Campanus-type equatorium, and a "directorium" – a calculating instrument related to the astrolabe but of specifically *astrological* use. Above all, like many of his near contemporaries (John of Murs and John of Saxony, for example), he took a strong interest in the refinement and reorganization of astronomical tables, in particular the various versions of the Alfonsine Tables. Astronomical tables were not merely for astrological use. Thus John of Murs compiled tables for the conjunctions and oppositions of the Sun and Moon (for 1321–96), which had a direct bearing on the subject of ecclesiastical calendar reckoning.[9] Paris, beginning in the 1320s, was the single most important point of diffusion of the "Alfonsine Tables." It has been argued that they were a Parisian creation, and not truly Alfonsine at all, but in the last analysis this question reduces to one of definition.

The Alfonsine Tables are only one aspect of the important scholarly activity of this thirteenth-century ruler, Alfonso X of Leon and Castile, "Alfonso the Wise" (1221–1284).[10] Alfonso encouraged the translation from Arabic into Castilian of many philosophical and scientific writings, a task already begun under the patronage of his father, King Ferdinand III of Castile. Translation from Arabic into Latin had long been prevalent at Toledo, but Alfonso established a school that included Christian and Jewish savants, as well as a Muslim convert to Christianity, and he presided over this group, revising their works and writing parts of the introductions to them. The names of fifteen collaborators are known from the complete collection of Alfonsine books, which includes treatises on the movement of the eighth sphere (precession), the universal astrolabe in its various guises, the spherical astrolabe, water

8 For a short history of the equatorium and a full account of the albion, see John D. North, *Richard of Wallingford*, 3 vols. (Oxford: Clarendon Press, 1976). For a comprehensive survey of the history of the equatorium, see Emmanuel Poulle, *Equatoires et horlogerie planétaire du XIIIe au XVIe siècle*, 2 vols. (Geneva: Droz, 1980).

9 Along with Firmin de Bellaval, he was invited by Pope Clement VI to Avignon to advise on calendar reform (1344–5).

10 For a broad survey of the scientific work of the court, see Mercè Comes, Roser Puig, and Julio Samsó, eds., *De astronomia Alphonsi Regis: Proceedings of the Symposium on Alfonsine Astronomy Held at Berkeley in August 1985* (Barcelona: Instituto Millàs Vallicrosa, 1987); and Julio Samsó, *Islamic Astronomy and Medieval Spain* (Aldershot: Variorum, 1994).

and mercury clocks, a simple quadrant, sundials, and equatoria. In view of discrepancies between new observations and predictions based on the old Toledan Tables, Alfonso had instruments constructed and observations made at Toledo. Two Jewish scholars are given as the compilers of the new tables, namely Yehuda ben Moses Cohen and Isaac ben Sid. They made observations at Toledo for more than a year, but the king moved his court often, and much work was no doubt also done at Burgos and Seville.

Little is known of the fortunes of all this activity before John of Lignères and his pupils assembled the essential ingredients of what was to become the most popular version of the tables, the edition composed by John of Saxony in 1327. The various Parisian versions of the tables were in use in England and Scotland very soon after they were written. Around 1340, William Rede of Merton College (Oxford) adapted them to the local meridian. There are versions for other English towns, but Oxford's key academic position ensured that Rede's became widely known in England, and two much more significant revisions were made in Oxford. The first dates from about 1348, and was perhaps by William Batecombe, and the second was by the Merton College astronomer John Killingworth (ca. 1410–1445). These were meant to be used in calculating a full planetary almanac (ephemeris), and a superb illuminated copy survives that was made for Humfrey, duke of Gloucester. Killingworth's work contains evidence of a type of theorizing that he does not explicitly spell out for us, but one that few could reproduce today without recourse to the differential calculus. Much the same could be said of the thought processes of various earlier astronomers whose solar and lunar velocity tables have been the subject of recent investigation.[11]

Briefly expressed, the aim of all these revisions was to speed up the tedious business of calculation. The 1348 tables went much further in this respect than those of John of Lignères; they were used in the northern Netherlands – where they were known simply as "the English tables" – and as far afield as Silesia and Prague. In fifteenth-century Italy, Mordechai Finzi, assisted by an anonymous Christian from Mantua, made a Hebrew translation of them. Giovanni Bianchini, the most notable Italian astronomer of the mid-fifteenth century, was influenced by them in producing a similar set of tables much used by such leading contemporaries as Georg Peuerbach (1423–1461) and Johannes Regiomontanus (1436–1476). Through this intermediary and a fourteenth-century Eastern European version (the *Tabulae resolutae*) that Copernicus had studied while a student at Cracow, they provided the core of Johannes Schöner's tables of the same name (printed in 1536 and 1542). These in turn were widely used for many decades, though their origins had long been forgotten.

[11] For an introduction to this type of material, see Bernard R. Goldstein, José Chabás, and José L. Mancha, "Planetary and Lunar Velocities in the Castilian Alfonsine Tables," *Proceedings of the American Philosophical Society*, 138 (1994), 61–95.

Such high-quality materials as these were not the fare of most astronomers, who found the simpler versions of John of Saxony, William Rede, and the like more acceptable. Even simplified, however, they had an important educational function that transcended their ostensible purpose. Like the Toledan Tables before them, the canons explaining the use of the tables helped to focus scholars' attention on the underlying spherical astronomy, and this in due course led to a new type of specialized trigonometrical text. The growth of this type of literature can be observed through commentaries on the canons to the older Toledan Tables – for instance that written by John of Sicily (fl. 1291) on Gerard of Cremona's translation of Ibn al-Zarqāllu's canons.

The first significant separate treatise on spherical trigonometry to be written in Christian Europe was Richard of Wallingford's *Quadripartitum*. Developing his work at Oxford on the basis of the *Almagest*, the Toledan canons, and a short treatise perhaps by Campanus of Novara, Richard found time to revise it while he was abbot, taking into account a work by the twelfth-century Sevillian astronomer Jābir ibn Aflaḥ (one of two famous men known in the West as Geber). Works such as this offered exact solutions of problems in the geometry of the sphere, for instance those involving spherical triangles. Again, calculation proved tedious. Working with an actual solid (or armillary) sphere could give approximate answers, but such instruments were difficult to construct accurately. Richard of Wallingford therefore designed what he called a "rectangulus," incorporating a system of seven straight rods pivoted in three dimensions to solve the same sort of problem. Like the armillary, it could in principle be used for observation, giving coordinates directly. The fact that the rods were straight meant that great accuracy of construction and graduation was in principle possible. Whether or not this particular device came to anything is of far less interest than the sheer vitality of the astronomical activity of the early fourteenth century in Paris, Oxford, and associated centers, especially in the matter of devising new calculational techniques and the speed at which the results of this burst of activity spread across Europe.

CRITICS OF THE OLD ASTRONOMY

By 1300, Europe had two dozen universities, most of them small and recent. By 1500, some ninety had been founded, with sixty already in existence.[12] With this expansion in learning came a new wave of enthusiasm for

[12] For a general account of the university movement in Europe, see Hilde De Ridder-Symoens, *A History of the University in Europe*, vol. 1: *Universities in the Middle Ages* (Cambridge: Cambridge University Press, 1992).

astronomy that in some of the new universities – for example, those in Vienna and Prague – outstripped that in the older centers. One of the most notable of the masters displaced from Paris to Vienna in the religious turmoil of the period was Henry of Langenstein (ca. 1325–1397), who was at the same time one of the most notorious of Ptolemy's critics.[13] Like several other Western astronomers of the late Middle Ages, he followed the example set by Muslim philosophers and used physical arguments to criticize the planetary schemes found in the *Almagest.* He was not only an adherent of homocentrism but was willing to abandon the almost sacred principle of uniform circular motion. In a work written in 1364 in Paris, he vented his feelings in petty academic criticism of the standard university text mentioned earlier, *Theorica planetarum.* The circles of Ptolemaic astronomy cannot, he argued, be physically real mechanisms existing in the heavens. They were for Henry mere mathematical constructions, justified only by the predictions based upon them. He was dissatisfied with Ptolemy's account in various other respects. He disliked the equant, for example, and the irregularities introduced into the theory of planetary longitude by the complicated Ptolemaic theory of planetary latitude. These were points at which Copernicus, nearly two centuries later, would attack the old astronomy with great success.

Astronomy seemed to revive in the middle of the fifteenth century. New attention to the *Almagest* is evident in the works of Giovanni Bianchini and George of Trebizond, both of whom translated and commented on Ptolemy's masterwork. This was in part a result of the multiplication of texts made possible by the invention of printing – texts that were being called for in ever larger numbers by the new university populations. Both Georg Peuerbach and Johannes Regiomontanus were important to the humanist movement, but at the less exalted level of textbook writer they were perhaps even more telling in the long run. Peuerbach, an Austrian scholar, followed John of Gmunden, a master who had taught the subject for two decades at the University of Vienna. John had died (in 1442) before Peuerbach's arrival, but he had bequeathed to the university the substantial collection of invaluable manuscripts and instruments he had assembled. Peuerbach received the master's degree at Vienna in 1453, but both before and after this time he traveled throughout France, Germany, and Italy. He became court astrologer first to Ladislaus V, king of Bohemia and Hungary, and then to the king's uncle, Emperor Frederick III. At Vienna, he taught the classics in the new humanist style, and there completed his famous textbook *Theoricae novae planetarum* (*New Theories of the Planets*, 1454), a replacement for the thirteenth-century text. After Regiomontanus printed it posthumously

[13] He was also known as Henry of Hesse. The Viennese context can be studied through Michael H. Shank, "*Unless You Believe, You Shall not Understand*": *Logic, University, and Society in Late Medieval Vienna* (Princeton, N.J.: Princeton University Press, 1988).

around 1472, it went into nearly sixty editions before falling into disuse in the seventeenth century. It made popular the solid spheres akin to those found in Ptolemy's *Planetary Hypotheses* and various Eastern works.

Intimately bound up with Peuerbach's career was that of Johannes Müller of Königsberg, better known as Regiomontanus, who probably studied under Peuerbach at Vienna and graduated master in November 1457. Within two years, he had joined Peuerbach in a program of observation of the planets, eclipses, and comets. Both were typical in writing on astrology. In 1460, their careers took a new direction when Cardinal Bessarion, papal legate to the Holy Roman Empire, arrived in Vienna. Bessarion, who was Greek, persuaded Peuerbach, who knew no Greek, to produce an improved abridgement of the *Almagest* as a foil against George of Trebizond's treatments of Ptolemy, which Bessarion despised. Peuerbach was about halfway through the work when he died in 1461. Regiomontanus completed the task within another two years, but the resulting *Epitome of the Almagest* was first printed only in 1496, twenty years after his premature death. The first part of the work relied on a widely used medieval *Abbreviated Almagest*, but the whole was much more comprehensive and, until recent times, offered the best available commentary on Ptolemy. Copernicus himself would use the *Epitome* heavily.[14]

In 1471, after working for the king of Hungary and drafting a long attack on George of Trebizond's commentary on the *Almagest*, Regiomontanus settled in Nuremberg, one of the chief commercial centers of Europe. There he could obtain fine instruments, and he set up a printing press that issued Peuerbach's *New Theory of the Planets* as its second publication. Regiomontanus followed it with his own planetary almanac for the period 1474–1506 – the first printed work to exploit a new and potentially vast astrological market for precomputed planetary positions. Regiomontanus had planned to print many other works in the mathematical sciences, but he died on a visit to Rome in 1476. A program of astronomical observation that he had begun was continued by an able colleague, Bernhard Walther (1430–1504), over the period 1475 to 1504. For the orbit of Mercury, Copernicus later used some of Walther's observations.

Regiomontanus's reputation by the time of his death was considerable. He was a great humanist scholar-printer, an excellent expositor of existing astronomical knowledge, and made a number of useful additions to spherical trigonometry. His *De triangulis omnimodis*, first published in 1533, made use of the "cosine law" as well as the "sine law" for spherical triangles, although the laws were not new. In astrology, the name of Regiomontanus became

14 Noel M. Swerdlow and Otto Neugebauer, *Mathematical Astronomy in Copernicus's De Revolutionibus*, 2 vols. (Berlin: Springer-Verlag, 1984), vol. 1, pp. 51–2; and Michael H. Shank, "Astronomia tra corte e università," in *Il Rinascimento italiano e l'Europa*, vol. 5: *Le scienze*, ed. Antonio Clericuzio and Germana Ernst (Treviso: Angelo Colla, 2008), pp. 3–20.

attached to a method of calculating the "houses," although it had been used in Spain long before.[15]

ASTROLOGY

That astrology was a motive force for much of the work done under the name of astronomy cannot be doubted. If one examines the eclectic astrological summaries of such men as Pietro d'Abano, Cecco d'Ascoli, and Andalò da Nigro, or their Arabic models, one does not immediately see where mathematics would be needed, but of course the basic need of the expert astrologer is for precise planetary positions, and this was what astronomy claimed to provide.[16] Such "data" were needed for meteorological astrology, the astronomical "prognostication of times,"[17] no less than for the prognostication of human fortune. Indeed, at least one Oxford scholar, William Merle, had an empirical outlook on meteorology that would be hard to match in the human branch of the subject.[18] Similar empirical concerns are evident in some observations of comets, with classifications of their position, color, and so forth, where the focus of interest was not on the comets as such but on the disasters they portended. Robert Grosseteste, Roger Bacon, and Albertus Magnus (ca. 1200–1280) were to a greater or lesser extent aware of Arabic astrological theories of cometary influence, and all had their ideas as to how comets might be introduced into the chain of natural causes. For them and most of their Christian contemporaries, comets were signs of divine Providence.

By the mid-twelfth century, astrology in various forms had become a recognized part of Western intellectual life, even though it continued to be haunted by the debate over determinism.[19] The founding and rapid expansion of the universities that was then beginning helped to give legitimacy to the subject, which gradually edged its way into the higher reaches of university curricula, for example in the faculty of medicine. By the fourteenth century, the place of astrology in the university curriculum was generally secure. In Paris in 1331, for example, John of Saxony added a commentary to John of Seville's translation of al-Qabīsī (Alchabitius, fl. ca. 950), and this was read by students of medicine, together with Ptolemy's

[15] For a general history of the mathematical principles behind the calculation of horoscopes, see John D. North, *Horoscopes and History* (London: Warburg Institute, 1986).

[16] For a specimen of this kind of work, see Graziella Federici Vescovini, ed., *Il "Lucidator dubitabilium astronomiae" di Pietro d'Abano: Opere scientifiche inedite* (Padua: Programma e 1 + 1, 1988).

[17] Compare the French, Spanish, and Italian words for the weather (*temps*, *tiempo*, and *tempo*, respectively).

[18] He compiled a journal of weather observations covering January 1337 through January 1344.

[19] Theodore Otto Wedel, *The Mediaeval Attitude Toward Astrology, Particularly in England* (New Haven, Conn.: Yale University Press, 1920), is still a useful guide to medieval concerns about the question of foretelling the future and its implications for free will and hence human responsibility.

Quadripartitum (the Latin title of a four-part astrological work known in Greek as *Tetrabiblos*) and a very popular collection of a hundred aphorisms known as *Centiloquium* that was mistakenly ascribed to Ptolemy. Between 1346 and 1348, John Ashenden, a fellow of Merton College, Oxford, composed a summary of astrology (*Summa iudicialis de accidentibus mundi*) on a truly grand scale. In view of its size, it had a surprisingly wide currency – some sections were even translated into Middle Dutch. Although numerous native European treatises were composed at this time – notably the popular *On the Science of the Stars* (*De astrorum scientia*) by Leopold of Austria – the classic Arabic works continued to hold center stage. Astrology was practiced in many possible styles, as it had been in Islam, ranging from the flamboyant and impressionistic to the desiccated and formal – the latter being favored in the northern universities of Oxford and Paris. As mentioned earlier, the two greatest European poets of the fourteenth century, Dante at the beginning and Chaucer at the end, both wove much astrology into their writings. It is a curious fact that their styles mirror those of the two poles of astrological practice.[20] Dante wrote for the most part in architectonic generalities. His Hell – where we meet the astrologers Cecco d'Ascoli and Guido Bonati, not to mention the magician Michael Scot – and his Paradise were both structured with great attention to Aristotelian cosmology and its mirror image. There was much astrological color in Dante's imagery, and no doubt he drew heavily on the mammoth work of Guido, even though he condemned Guido to always looking backward as a penalty for having tried to see into the future. Dante's use of astrology was quite unlike that found for it by Chaucer, who in a whole series of works performed precise calculations silently, but with extraordinary expertise, to pattern plot and imagery alike.

The rapid rise in astrology's respectability during the fourteenth century led to its increasing use in literature, architecture, book illustration, and the visual arts generally. It therefore comes as no surprise to find horoscopes, for example, placed in church windows, in portrait paintings, and on tombstones. The great astronomical clocks from the end of the thirteenth century and later, such as those in churches at St. Albans, Wells, and Strasbourg, like their civilian counterparts – for instance the famous astrarium of Giovanni Dondi, who taught astronomy at the University of Padua – were filled with astrological imagery and purpose. The oldest for which we have details – as it happens, the most complex mechanism known from the whole of the Middle Ages – was that by Richard of Wallingford, abbot of St. Albans, and yet it contained a wheel of fortune, a device with positively pagan antecedents. In this respect, it was by no means rare.

[20] For Dante, the most comprehensive sources are the astronomical entries by Emmanuel Poulle in the *Enciclopedia Dantesca*, 6 vols. (Rome: Istituto della Enciclopedia Italiana, 1970–1978); and M. A. Orr [Mrs. John Evershed], *Dante and the Early Astronomers* (London: Gall and Inglis, 1913). For the work of Chaucer, see John D. North, *Chaucer's Universe*, 2nd ed. (Oxford: Clarendon Press, 1990).

COURT ASTROLOGY AND PATRONAGE

A very high proportion of the astrologers whose names we know were connected in some way with royal courts. It had already been so in Islam in both the East and the West. In Sicily, second only to Spain in its importance for the introduction of Greek and Arabic learning into Europe, Emperor Frederick II (1194–1250) not only supported the sciences but is said to have employed Michael Scot as court astrologer and magician. From this time onward, many European courts seem to have had advisers with a knowledge of the newly arriving sciences, even if they were not always astrologers as such. To take but one famous example, Vincent de Beauvais (ca. 1190–ca. 1264), a Dominican who held a position in the Cistercian abbey of Royaumont and served Louis IX as royal chaplain, librarian, and tutor to the royal children, took a critical view of both demonology and astrology, but it is quite obvious that he believed in both of them. When he condemned them, it was with arguments that had been circulating since the time of the church fathers. Skeptics such as Vincent would not have protested so much had there been nothing to protest.

Certainly by the end of the thirteenth century, astrology was an accepted part of court life. One can see this from its increasing use in illuminated ecclesiastical calendars of a sort that only the wealthy could afford. As time went on, these included more and more astrological information. One such calendar was prepared by William of St. Cloud in 1296 for Queen Marie of France. In England, Richard of Wallingford, who often acted as host to the royal household, wrote on nativities in the margins of a calendar, perhaps for Philippa, wife of Edward III. At the end of the fourteenth century, two English friars composed royal astrological calendars that were both mentioned and used by the poet Geoffrey Chaucer.[21] The lists of comparable European material grow exponentially from about this time and contain increasing numbers of works in the various vernaculars.

This kind of activity represents a relatively modest astrology, free from the worst excesses of what we now judge to have been superstition. Indeed, since kings could generally afford the services of the most intelligent of their subjects, they occasionally had to suffer writings on the iniquities of scientific determinism. It is virtually impossible to find an educated man who did not believe in the reality of celestial influence, but many were appalled by the implications of the idea, notably for the limitation it seems to place on human freedom of action and hence on human responsibility. This was an important issue raised in at least two works with an anti-astrological flavor written for royal consumption, one by Thomas Bradwardine for Edward III of England and one by Nicole Oresme for Charles V of France. The works

[21] The Franciscan John Summer and the Carmelite Nicholas of Lynn. See North, *Chaucer's Universe*, pp. 91–5.

in question were very different in character but had in common authorship by men high in the ecclesiastical hierarchy – Bradwardine died archbishop of Canterbury and Oresme bishop of Lisieux. Later, even this theme was democratized. Philippe de Mézières claimed to be writing in French so that the *common people* might learn how to avoid astrological nonsense. He still regarded certain kinds of astrological prediction as perfectly possible, even though spiritually dangerous. Of course, he was no democrat in the modern sense. He believed that kings and princes should study astrology since it was, he said, a noble science, and nobles and princes should know more of what is good and beautiful than anyone else. He listed several kings who knew the subject well, and repeated the well-worn story that Aristotle, the supposed author of the *Secret of Secrets*, had advised Alexander the Great to do nothing in the world without first consulting an astrologer.

The main centers of astrological learning in the fourteenth century included, of course, not only Paris and Oxford but also the universities of northern Italy, notably Bologna and Padua, for Italian princes were no less active in patronizing astrology than their northern cousins. They began their wars, received foreign embassies, laid foundation stones, and dedicated their churches, all with respect to the state of the heavens. Guido Bonati (ca. 1210–post-1296) advised nobles and warlords, in particular the Ghibelline leader Guido da Montefeltro, on the precise times of arming, mounting, and departure of his army. The times of the crowning of popes were calculated by astrologers well into the early-modern period. Pope Leo X (formerly Giovanni de' Medici) was particularly beholden to astrologers, and had patronized them long before he became pope. Many supposedly correct predictions were made concerning his period of office.

It became a habit in the late Middle Ages and afterward for astrologers to collect the horoscopes of the famous. One manuscript in the Bibliothèque Nationale in Paris contains a triple collection of about sixty figures in all, including princes and other persons of importance. Among them are horoscopes of Henry VI of England, possibly done by Jean Halbout of Troyes, minister general of the Trinitarians.[22] Other horoscopes in the collection are for John II ("the Fearless"), duke of Burgundy; John I, count of Alençon; and other nobles; and one for John Fastolf – who gave his name at least to Shakespeare's character Falstaff. Of great interest is that the collection as a whole was assembled in an attempt to find an astrological explanation for political events. There are plenty of horoscopes surrounding the assassination in 1407 of the duke of Orléans at the instigation of John II of Burgundy. Eclipses of the Sun and Moon in that year, and a conjunction of Jupiter and Saturn

[22] Emmanuel Poulle, "Horoscopes princiers des XIVe–XVe siècles," *Bulletin de la Société Nationale des Antiquaires de France* (12 February 1969), 63–77. For much on astrology at the English court, see Hilary M. Carey, *Courting Disaster: Astrology at the English Court and University in the Later Middle Ages* (London: Macmillan, 1992). For Burgundian and French affairs, see Jan R. Veenstra, *Magic and Divination at the Courts of Burgundy and France* (Leiden: Brill, 1998).

in 1405, were related to the possibility of a dynastic change such as the one that had taken place around 988, when the nobles of France had given the crown to Hugh Capet. It was calculated that a Saturn–Jupiter conjunction had taken place then, just as on 1 June 1325. On the second occasion, the conjunction was supposed to have heralded the accession of Philippe VI of Valois – although this did not come about until 1328. Other astrological phenomena with suitably delayed actions were introduced. By the time of the last cluster of horoscopes in this series (1437), Charles VII had made his triumphant entry into Paris, and so the astrologer thought it desirable to make a thorough analysis of Charles's birth horoscope.

It is clearly somewhat misleading to speak of "astrologers in the service of kings" since astrology was not a full-time activity. Every medieval king had his medical advisers, and every university-trained physician had some knowledge that one would describe as astrological. Every king therefore had his astrologer, in this simple sense. Some were more intensely motivated than others. Simon de Phares (1444–post-1499), for example, was astrologer to Charles VIII of France and author of a long biobibliographical account of astrology from its beginnings to his own day. This contained no fewer than 1,226 people said to have been astrologers, astronomers, or lovers of the science of the heavens, although many seem to have existed only in Simon's imagination.[23] Astrology, however, covered a multitude of sins, from the timing of bloodletting to meteorology, from casting nativities to making prognostications about wars, from analyzing history through the occurrence of great conjunctions to predicting the end of the world from the sighting of comets. In a sense, doing any of these things amounted to the exercise of power – bloodletting was no doubt for some patients almost as terrifying as the idea that the end of the world was at hand. (Indeed, in Descartes' case, the two amounted to the same thing.)

POPULAR ASTROLOGY

The introduction of printing helped to accelerate a movement toward the democratization of astrology that had been gathering momentum for a century or more before the press. One symptom of this trend was the growing production of almanacs, calendars of a sort already mentioned as having long been available to the learned and the rich. A general demand for these works was created by giving them a prophetic content, and they were issued from most of the main centers of printing in Europe. For example, at least eight editions of the almanacs of Johannes Engel were issued, in Latin or German, between 1484 and 1490. Less ephemeral was a work Engel

[23] Jean-Patrice Boudet, *Lire dans le ciel: La bibliothèque de Simon de Phares, astrologue du XVe siècle* (Brussels: Centre d'Études des Manuscrits, 1994).

published that caught the popular imagination through its inclusion of a whole series of illustrated horoscopic figures, one for each degree of the ecliptic in the ascendant.[24] Engel kept dark the fact that the latitude for which the calculations were made was 45°, and so was quite inappropriate to his German clientele (he did name the person responsible for computing the houses, Pietro d'Abano, whose latitude it was). This work, which eventually earned the condemnation of the Paris faculty of theology, was much imitated, and effectively created a new genre of astrological literature – horoscopy without tears. It was also, unfortunately, astrology without accuracy. Rather more difficult to use, but more accurate, were the written and printed tables that Regiomontanus prepared for a whole range of latitudes. Just as princely horoscopes had earlier been collected together, likewise a similar type of popular printed literature came into being, books including horoscopes of famous men and women. Martin Luther might have vehemently opposed astrology, but that did not prevent the astrologers from adding his horoscope to a collection that included the horoscopes of Christ and Muḥammad.

Popular interest in astrology grew spectacularly in the fifteenth and sixteenth centuries, and held its own at higher academic levels, too. One can detect a slight disillusionment, but nothing more. The best critics of astrology were mostly lapsed believers, able to make use of their "inside knowledge." Nicole Oresme's attack on astrology, mentioned earlier, was the product of a clear mathematical mind, and made the point that since the planetary motions are incommensurable, strict repetitions of planetary configurations are impossible, so that the principal axiom of the astrologer must be abandoned. Such limited skeptics as Oresme – who did not doubt the reality of celestial influence – made few substantial inroads into academic astrology, and even fewer into popular beliefs, which, as in so many other respects, were largely based on trust in the authority of superiors. The decline of astrology as a serious scientific pursuit came long after the end of the Middle Ages, with the growth of new systems of thought that seemed to destroy its plausibility. And no science carried greater responsibility for those new systems of thought than astronomy itself.

APPENDIX: THE PTOLEMAIC THEORY OF PLANETARY LONGITUDE AS APPLIED IN THE MIDDLE AGES

In the Latin Middle Ages, the methods used to calculate planetary positions – primarily in longitude (that is, measured along the ecliptic, the Sun's apparent path relative to the stars), more rarely in ecliptic latitude – were those presented a millennium earlier by Ptolemy in his *Almagest* and elsewhere.

[24] Such material is well illustrated in Stefano Caroti, *L'Astrologia in Italia* (Rome: Newton Compton, 1983).

The intervening period had seen a certain skepticism about the mechanism of the heavens, which in turn had resulted in certain modifications of the Ptolemaic picture. The great strength of Ptolemy's geometrical schemes, however, lay in their predictive success, so that the revisions made to the basic parameters from time to time were generally minor. More substantial changes had been made in the intervening period, chiefly by Islamic astronomers, to the theory of "the motion of the eighth sphere," the sphere of the fixed stars. (Assigning responsibility for this apparent drift to an instability in the coordinate system rather than in the stars themselves, we now call this phenomenon the "precession of the equinoxes.") The movement results in a slow apparent drift in the ecliptic longitudes of the stars but not in their latitudes. Some of these models for the movement of the eighth sphere were highly complex, and made use of a three-dimensional scheme that yielded a nonlinear change in longitude.

The strength of astronomy in the Latin Middle Ages lay not in devising fundamentally new planetary models, or materially different parameters for them, but in elaborating the old techniques for handling the ancient models. It is not possible to explain in detail here what new techniques were invented, but to begin to understand the nature of the problem we shall need a general idea of how planetary longitudes were derived (using tables and equatoria) given the various planetary models and the parameters governing their scales and fundamental motions.[25]

We begin with the Sun's simple eccentric model shown in Figure 19.4. T is the Earth and C is the center of the eccentric circle on which the Sun (S) moves with constant speed (that is, on a radius rotating at constant angular velocity about C). To an observer at T, the Sun will appear to be at the point S' on the ecliptic, known as the Sun's "true place." The direction of the remotest part of the eccentric from T (the apogee A, or aux, as it was then called) may be taken as a datum. The value of the constantly increasing angle $\overline{\gamma}$, measured from this datum line, can be easily calculated on the basis of two items of information, namely the value of $\overline{\gamma}$ at any convenient date and time (known as the radix date or root date) and the constant rate of increase of $\overline{\gamma}$, the Sun's mean angular motion. Simple tables allow $\overline{\gamma}$ to be found for any time and date, the first step being that of establishing the interval between the date sought and the radix date. From here it would not be a difficult matter to find the direction of the Sun as seen from the Earth, T, as long as we can calculate the value of the angle marked ξ. The latter is known

[25] On the evolution of astronomical tables, important works include Otto Neugebauer, *The Astronomical Tables of al-Khwārizmī* (Copenhagen: Historisk-filosofiske Skrifter, Danske Videnskabernes Selskab, 1962); Bernard R. Goldstein, *Ibn al-Muthannā's Commentary on the Astronomical Tables of al-Khwārizmī: Two Hebrew Versions* (New Haven, Conn.: Yale University Press, 1967); and Emmanuel Poulle, ed., *Les tables alphonsines, avec les canons de Jean de Saxe* (Paris: Editions du CNRS, 1984). A general survey of the history of the Alfonsine Tables appears in John D. North, "The Alfonsine Tables in England," in his *Stars, Minds, and Fate* (London: Hambledon, 1989).

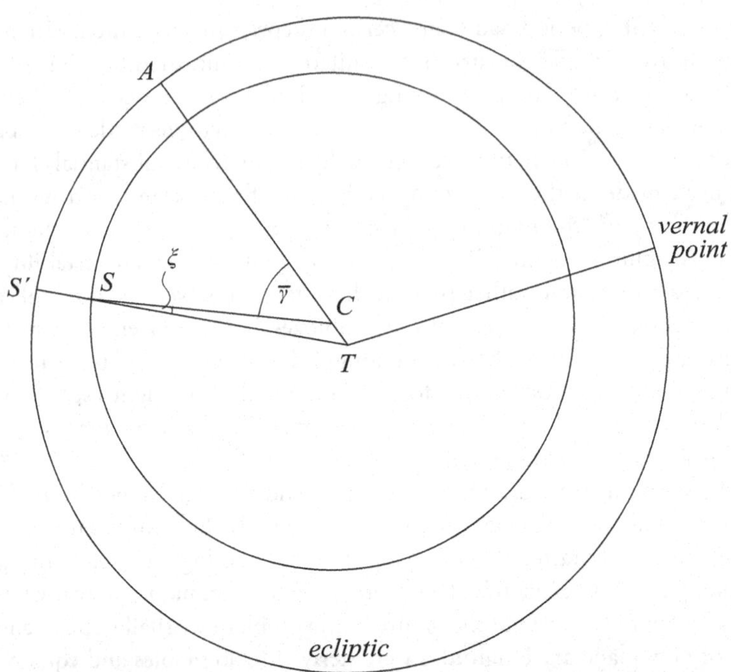

Figure 19.4. Ptolemaic eccentric model for the Sun.

as the "equation of the center." The required angle *ATS* is simply $\overline{\gamma}$ minus ξ, in the case shown in Figure 19.4. The answer will obviously depend on the eccentricity; that is, not on the absolute distance *TC* but on the ratio of the distance *TC* to the radius *CS* of the eccentric circle. To save the astronomer the trouble of performing the calculation on every occasion, Ptolemy drew up a table of the equations of center. Later revisions of the value assigned to the Sun's eccentricity led to new tables being calculated. It should be obvious that ξ is a simple function of $\overline{\gamma}$ and that the table encapsulates this function.

One further step in the calculation is needed, to take account of the slow drift in the direction of the line of apogee (*TA*). This is partly owing to an intrinsic movement of the apogee with reference to the equinoctial point from which, ultimately, planetary longitudes are measured. There is also a contribution from the precessional movement mentioned earlier.

For the planets Venus, Mars, Jupiter, or Saturn, all of the corresponding models involve an epicycle carried around an eccentric circle (see Figure 19.5 for the general scheme). The "true center" of the epicycle (γ) is calculated with the help of tables as a function of $\overline{\gamma}$ in much the same way as for the Sun, with a similar "equation of center," although there is a fundamental difference in the way in which the planetary tables of that equation of center are calculated. The reason for this is that Ptolemy's data did not allow him

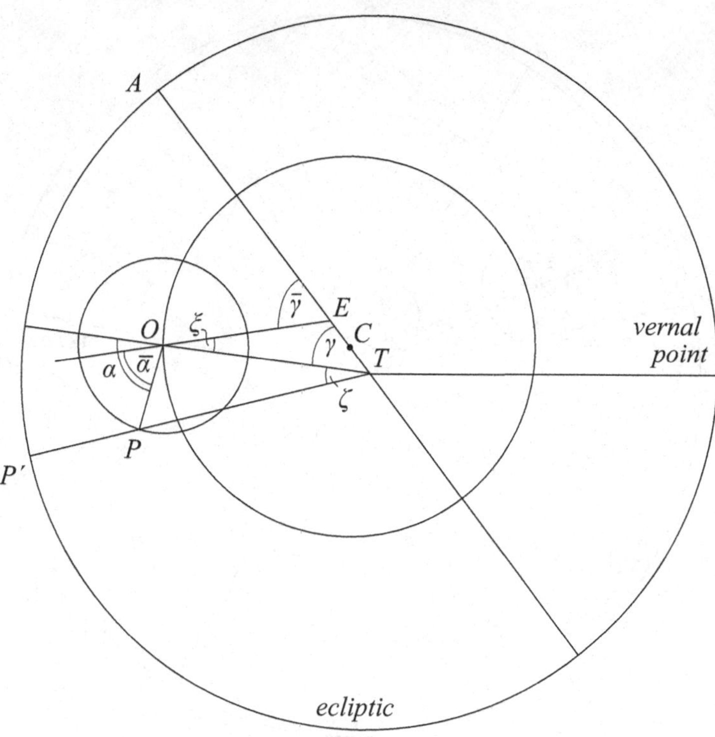

Figure 19.5. Ptolemaic model for Venus and the superior planets.

to have the center of the epicycle move uniformly around the carrying circle (the deferent). He thus supposed its motion to be uniform not around *C* but around the "equant" point (*E* in Figure 19.5); that is, he made the line *EO* rotate at constant velocity. He took the center of the deferent circle *C* to be midway between the equant point (*E*) and the Earth (*T*).

The planet is at point *P* on the epicycle, and the "true place" of the planet is at *P′*. The epicycle radius *OP* rotates at constant speed relative to the (constantly moving) line *EO*. The angle between them ($\overline{\alpha}$) is known as the mean argument, and the name "equation of the argument" is reserved for the angle marked ζ. It should be clear that the final longitude of the planet relative to the apogee line can be found by combining the angles $\overline{\gamma}$, ξ, and ζ by addition or subtraction. (Whether one adds or subtracts will depend on the relative positions of the lines.) A table is drawn up for the equation of the argument ζ, but this will be more complex in character than that for the equation of center. It was conventional (following Ptolemy) to express ζ as a function of $\overline{\alpha}$, $\overline{\gamma}$, and ξ, and to make use of certain rather clever interpolation methods when extracting ζ from the table in a particular case. The final stages in the calculation of the planets' true places required adjustments for any apogee and precessional movements, as before.

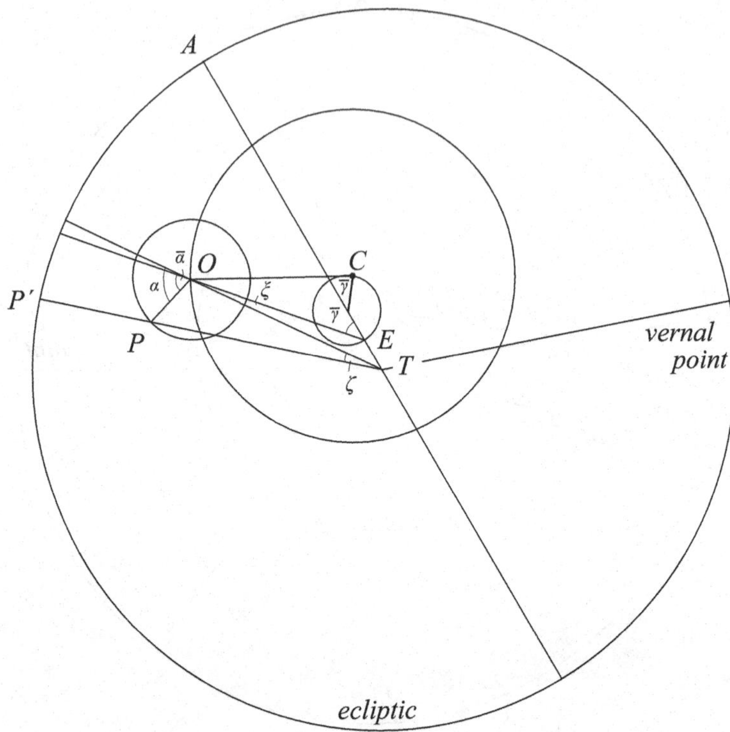

Figure 19.6. Ptolemaic model for Mercury.

In his effort to improve their fit with observations of position and speed, Ptolemy made his models for the motion of Mercury and the Moon even more complicated. For Mercury, he conceived that the center C of the deferent circle was carried around a small circle at the center of the figure (see Figure 19.6) in such a way that the carrying radius made with the line of apogee TA an angle $\overline{\gamma}$ equal to AEO on the other side of the apogee line. Despite these complications, the resulting tables were not strikingly more difficult to use than the tables for the other planets.

Ptolemy took the Moon, like Mercury, to have a moving deferent circle, but with a movement generated in a different way. In fact, the entire line of apogee, passing through the equant point (E' in Figure 19.7), the Earth (T, fixed), and the center of the deferent circle (C), rotates in the direction opposite that of the general planetary rotation. (In common parlance, it is said to be retrograde rather than direct.) The key to this movement of the line of apogee is that it brings the Sun (or more strictly the mean Sun) into the lunar calculation. It is a movement such that the line to the mean Sun bisects the angle ATO. The use of lunar tables of equations differs in some small respects from that of the planetary tables, but the broad principles are the same; that is, a final longitude is found by simple addition or subtraction

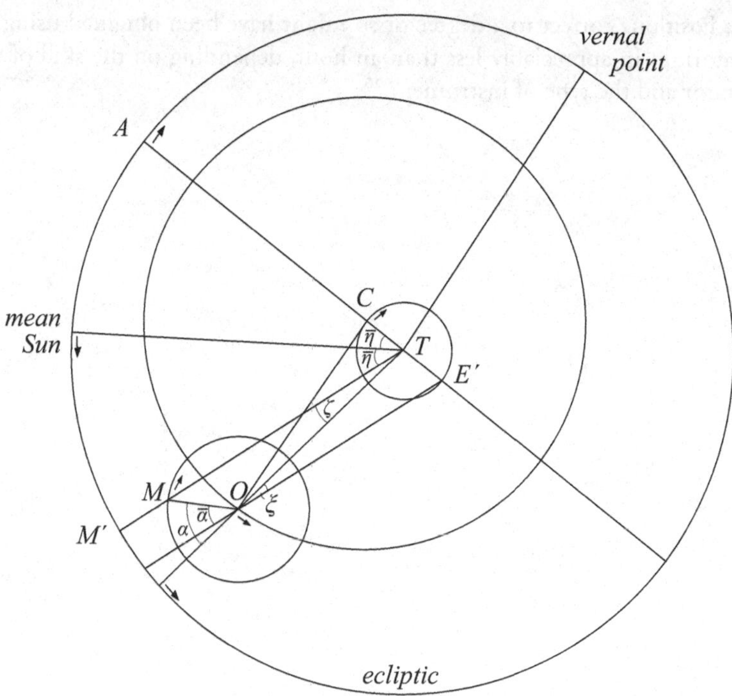

Figure 19.7. Ptolemaic model for the Moon.

of an equation of the argument and a mean motion. (The equation of center, ξ, is not added or subtracted at the *final* stage in the case of the Moon, but it is needed in the course of the calculation to help provide the true argument $\overline{\alpha}$.)

The genius of the medieval calculators is evident from the way in which they managed to reduce the number of stages in these Ptolemaic calculations. They achieved their ends with double- and triple-entry tables; that is, tables that made one angle a function of two or three others. These new tables were of necessity much longer than the old versions, and were therefore more costly, but they helped astronomers to mass-produce planetary almanacs (ephemerides) with less effort.

There is no room here for a complete set of calculations of even a set of planetary positions for a single date and time by any of these methods, old or new. In an attempt to convey a feeling for the sheer energy required for this sort of activity, however, we give the barest details of the intermediate stages – that is, the key quantities named in this section as they would have been found by an astronomer casting, for example, a single horoscope for a specific time and date. Using standard tables, the full set of planetary longitudes would probably have taken even the most skilled calculator an hour or two. A university student might have spent a day or more on the task. Positions correct to a degree or so might have been obtained using an equatorium in appreciably less than an hour, depending on the skill of the operator and the type of instrument.[26]

[26] For a fuller account of the procedures to be followed in using and composing tables of planetary longitude, see North, *Richard of Wallingford*, vol. 3, Appendix 29, pp. 168–200.

20

THE SCIENCE OF LIGHT AND COLOR, SEEING AND KNOWING

David C. Lindberg and Katherine H. Tachau

Sophisticated theories of light, color, and vision were produced in the ancient Greek world – committed to writing and transmitted in treatises by Aristotle, Euclid, Ptolemy, Galen, and others. Although most of these works were unavailable to Latin-reading scholars in early-medieval Europe, fragments of Greek optical theory did circulate in a variety of encyclopedic works and philosophical or theological treatises. It was not until the thirteenth century, inspired by the wholesale translation of Greek and Arabic scientific literature, that the scholarly discipline called *perspectiva* emerged in Europe as the object of university lectures and a recognizable piece of the intellectual equipment of the educated university graduate.

Perspectiva came to be represented in the thirteenth century by influential standard texts that treated the physics and mathematics of light, color, and vision. Also included were discussions of the anatomy and physiology of the eye and visual apparatus, inquiries into the psychological faculties and cognitive processes that made vision possible, and analyses of phenomena associated with light as an astrological or cosmological agent. Roger Bacon, writing in the seventh decade of the century, could plausibly declare that without a knowledge of *perspectiva* one could "know nothing of value in philosophy." How *perspectiva* achieved this exalted status in medieval Europe, and how it managed to embrace theories ranging from cognitive processes to the radiation of cosmic power, are the subjects of this chapter.[1]

[1] The quotation is from Roger Bacon, *Opus minus*, in *Opera quaedam hactenus inedita*, ed. J. S. Brewer (London: Longman, Green, Longman, and Roberts, 1859), p. 327. In what follows, we draw upon and extend research that we have published elsewhere, especially David C. Lindberg, *Theories of Vision from Al-Kindi to Kepler* (Chicago: University of Chicago Press, 1976); and Katherine H. Tachau, *Vision and Certitude in the Age of Ockham: Optics, Epistemology, and the Foundations of Semantics, 1250–1345* (Leiden: Brill, 1988). Portions of this chapter bear a close resemblance to David C. Lindberg and Katherine H. Tachau, "Perspectiva: la scienza della luce, del colore e della visione," in *Storia della scienza*, vol. 4: *Medioevo, Rinascimento*, ed. Sandro Petruccioli (Rome: Istituto della Enciclopedia Italiana, 2001), pp. 397–406, which we were writing at about the same time and (owing to the publisher's oversight) appeared without Lindberg's name.

The physical transmission of ancient Greek optical knowledge is well understood. The Greek sources were transplanted first to the Islamic world and later to medieval France and England, where the first shoots of perspectival knowledge appeared in the early-thirteenth-century *œuvre* of Robert Grosseteste. Flowering at mid-century in the Parisian works of Grosseteste's admirer Roger Bacon, it was disseminated from Paris and the papal court at Viterbo (to which Bacon had dispatched his major writings) through treatises on *perspectiva* written by several younger scholars who were inspired by Bacon and influenced by the same sources that had shaped his perspectival theories.

If we are to understand the range of meaning of perspectivist science for medieval intellectuals, we must explore three of its aspects: first, its account (both mathematical and physical) of light and vision; second, its exploration of cognitive processes, including theories of the interior senses and the anatomical structure of the brain; and third, the metaphysical, cosmological, and astrological speculation concerning light in which all of this was embedded.

GREEK BEGINNINGS

It seemed clear to Aristotle (384–322 B.C.) that perception, no matter which sense we have in mind, requires contact. This contact is obvious in taste and touch, but sight seems to be a long-range phenomenon that violates that requirement. Thus one of the central problems confronting Greek philosophers attempting to understand how we perceive the world visually was to explain how *invisible* contact occurs between a visible object and the organs of sight. Ancient authors produced a variety of answers, all of them requiring either that the eye emit something in the direction of the visible object (extramission) or receive something from the visible object (intromission); an intervening medium might or might not play a role in conveying this something (visual images or visual power) between object and eye.

The intromission alternative was most influentially formulated by Aristotle, who undertook a detailed analysis of the nature of light, its relationship to the transparent medium through which it was propagated, and the causal mechanism by which it provoked visual sensation. He concluded that light does not consist of independent entities ("substances" such as atoms). Rather, he argued, light is a state of the transparent medium that is achieved when the presence of a self-luminous body, such as the Sun, moves the transparency of that medium from potentiality to a state of actuality. Having thus become actually transparent, that medium may be further endowed with the quality of color, owing to the presence of a colored body. The medium transmits this color instantaneously to the eye of an observer, where coloring of the interior humors produces the appropriate sensation. Some version or variation of this intromission theory survived within Aristotelian traditions until the seventeenth century.[2]

One of Aristotle's purposes was to refute earlier philosophers' mistrust of sense perception. His refutation rested on the understanding that the "coloring of the interior humors" makes the eye temporarily similar to the colored body, thereby, through its assimilation, providing the sense organ with direct experience of what the world is really like. Moreover, in human beings and other animals, according to Aristotle, "sensations and images remain in the sense-organs even when the sensible objects are withdrawn," which enables the process of assimilation to continue into the heart and brain. When we observe an object in the outside world, he argued, the visual images impressed upon the eye by the illuminated medium, along with impressions from other exterior senses (such as odors received in the nose) and sensible features of the object (such as its motion or rest) common to more than one exterior sense, are brought together in a "common sensation." Aristotle's readers understood him to have been speaking of an interior faculty or capacity, the "common sense," and also to have described other internal senses. These were "phantasy," so called from the "appearances" (phantasmata, or phantasms) or residual perceptual impressions of sensed objects that it manipulates, and "memory," a treasury where these images of what the senses have perceived may be stored for later use. Not least among such uses is the higher work of the rational (or intellectual) soul.[3]

Aristotle barely touched on the mathematics of light and vision, preferring to deal with natures and causes. However, a tradition of analyzing these phenomena mathematically soon sprang from the work of Euclid (fl. 300 B.C.) and his successors.[4] Euclid devoted his attention almost entirely to the geometry of radiating light, which he presented in his *Optica* and *Catoptrica*. In these works, he established the equal-angles law of reflection, presented

[2] Lindberg, *Theories of Vision from Al-Kindi to Kepler*, pp. 6–9. On all of the ancient theories of vision, also see the excellent account by A. Mark Smith, trans., *Alhacen's Theory of Visual Perception: A Critical Edition, with English Translation and Commentary of the First Three Books of Alhacen's "De Aspectibus,"* 2 vols. (Philadelphia: American Philosophical Society, 2001), vol. 1, pp. xxvi–xliv.

[3] Aristotle, *De anima*, III. For the quoted line, see III.2.425b. See also Harry A. Wolfson, "The Internal Senses in Latin, Arabic, and Hebrew Philosophical Texts," *Harvard Theological Review*, 28 (1935), 69–133.

[4] On this mathematical tradition, see Lindberg, *Theories of Vision from Al-Kindi to Kepler*, chap. 1; Albert Lejeune, *Euclide et Ptolémée: deux stades de l'optique géométrique grecque* (Louvain: Université de Louvain, 1948); Albert Lejeune, *Recherches sur la catoptrique grecque* (Brussels: Académie Royale de Belgique, Classe des lettres et des sciences morales et politiques, Mémoires, 1957), vol. 2, fasc. 2; A. Mark Smith, "Ptolemy's Theory of Visual Perception: An English Translation of the Optics, with Introduction and Commentary," *Transactions of the American Philosophical Society*, 86, pt. 2 (1996); Sylvia Berryman, "Euclid and the Sceptic: A Paper on Vision, Doubt, Geometry, Light and Drunkenness," *Phronesis*, 43 (1998), 176–96; A. Mark Smith, "Ptolemy, Alhacen, and Kepler and the Problem of Optical Images," *Arabic Sciences and Philosophy*, 8 (1998), 9–44, especially pp. 20–1; and A. Mark Smith, "Ptolemy's Search for a Law of Refraction: A Case-Study in the Classical Methodology of 'Saving the Appearances' and Its Limitations," *Archive for History of Exact Sciences*, 26 (1982), 221–40.

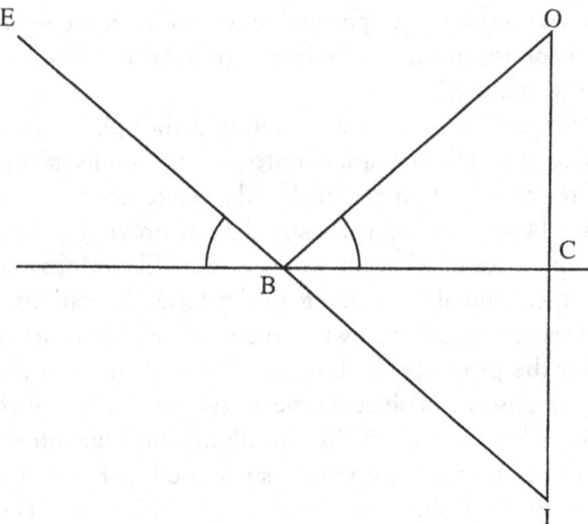

Figure 20.1. Vision by reflected rays according to Euclid and Ptolemy. *E* is the observer's eye and *O* the visible object. An observer at *E* judges the image, *I*, to be located where the linear extension of the visual ray *EB* intersects the perpendicular *OCI*, drawn vertically from the object to the reflecting surface.

rules for locating the image of an object observed in a mirror (Figure 20.1), and offered a qualitative understanding of the phenomena of refraction – all applicable to both the light emanating from external sources, such as the Sun, and the light emanating from the eye. Integral to this mathematical analysis was Euclid's extramissionist theory of vision, which declared that radiation reaches outward from the observer's eye in the form of visual rays (rectilinear rays bundled as a cone, with its apex in the observer's eye) and perceives visible objects simply by virtue of encountering them. The visual cone became the foundation of a theory of mathematical perspective according to which (for example) objects that intercept rays high in the visual cone appear high and those intercepting rays low in the visual cone appear low, and the perceived size of a visible object is proportional to the angle formed in the eye by the visual rays that touch its periphery (Figure 20.2).

Borrowing Euclid's extramissionist approach and his geometrical analysis, Ptolemy (fl. 127–48) was able to set them in a theory of considerably greater mathematical sophistication that, moreover, broadened to embrace notions of the physical nature of the visual radiation and the process of perception. Included in his *Optica* was an impressive analysis of reflection in spherical mirrors, both concave and convex, and an investigation, both experimental and quantitative, of the refraction of light as it passes from one transparent medium to another. More than any earlier scholar, Ptolemy brought together in his work the shared geometrical foundations of optics and astronomy. Most notably, he provided a mathematical understanding of the problems

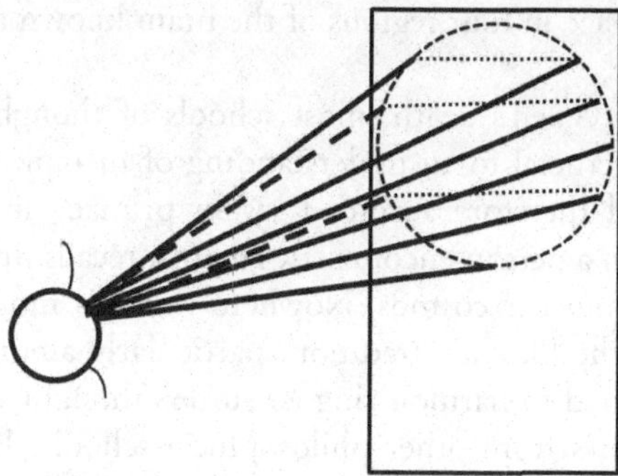

Figure 20.2. Euclid's visual cone. Objects intercepting rays high within the cone of visual rays will appear high within the visual field of the observer. Objects intercepting rays low within the visual cone will appear low within the observer's visual field.

that the reflection and refraction of light posed for the observation of celestial phenomena and thereby a means of correcting for the resulting perceptual distortions.

A third tradition of ancient optical thought, concerned with the anatomy and physiology of the eye and optic nerves, was codified by the physician Galen (d. after 210), a great but eclectic scholar who studied and contributed to Platonic, Stoic, and Aristotelian philosophy.[5] Galen provided a detailed (and largely accurate) description of the tunics and humors of the eye, including the aqueous and vitreous humors, the crystalline humor (or lens), cornea, iris, retina, and optic nerves. His theory of vision, heavily influenced by Stoic thought, maintained that the crystalline lens is the principal organ of vision, endowed with visual power by pneuma conveyed from the brain by the optic nerves. This pneuma emerges from the eye and transforms the surrounding transparent medium into an extension of the optic nerve and an instrument of vision, capable of perceiving the objects it touches. Perceptions, in turn, as "impressions" conveyed to the brain, provide the requisite basis of other psychological processes, which are carried out in distinct regions of the brain. Galen himself evidently thought that the psychic pneuma carries out these processes – the completion of perception and, with it, consciousness, the storage of sensed images, memory – in the substance of the brain itself. Those who built on Galen's work attributed to him the view that these

[5] On Galen's philosophical formation and contributions, see John Whittaker, "Plotinus at Alexandria: Scholastic Experiences in the Second and Third Centuries," *Documenti e studi sulla tradizione filosofica medievale*, 8 (1997), 159–90, especially pp. 172–85. On Galen's theory of vision, see Lindberg, *Theories of Vision from Al-Kindi to Kepler*, pp. 9–11; Glenn Lesses, "Content, Cause, and Stoic Impressions," *Phronesis*, 43 (1998), 1–25; and Fernando Salmón, "The Many Galens of the Medieval Commentators on Vision," *Revue d'histoire des sciences*, 50 (1997), 397–419.

operations took place in four regions of the brain known as its "cavities" or "ventricles."[6]

By the time of Galen's death, most schools of thought considered the workings of light crucial to an understanding of metaphysical and cosmological reality and therefore accorded vision primacy in accounts of the processes by which a person encounters, records, recalls, interacts with, and comes to understand the cosmos. Nowhere was this more completely the case than within the Platonic tradition, particularly among Neoplatonists, who were committed to harmonizing Aristotle's thought with Plato's while also absorbing views from other philosophical schools. Plotinus (d. 270), generally regarded as the founder of Neoplatonic philosophy, explained the relationship between the singularity of the One, the ground of all being, and the multiplicity of created things in terms of radiating light, arguing that just as rays of light emanate from the Sun, so a lesser form of being emanates from the One through an overflowing of its essence (a theory referred to by historians as "emanationism"). Plotinus also declared light to be the form that endows matter with corporeality or dimensionality. And, in a doctrine that would have important repercussions in the later Middle Ages, he maintained that everything that exists radiates images or likenesses of itself into surrounding bodies. Also significant for later thinkers was Plotinus's acceptance of Galen's anatomical and physiological evidence that the starting point of sense perception (including visual perception), of the subsequent sensory evaluation or "judgment" of what the eyes have seen, and of such other functions as imagination lay in the brain and required its use of the body's nerves.[7]

One of the few pieces of the Greek optical achievement available to late-Roman and early-medieval authors was this Neoplatonic metaphysics and epistemology of light. Scholars in Latin Christendom knew the theories of Plotinus indirectly through the writings of Augustine of Hippo, the pseudo-Dionysius, and other fathers of the Christian church. No patristic author was more influential in Western Christendom than Augustine, for whom vision required activity on the part of the observer, exercised through the extramission of visual rays (borrowed from Plotinus and Stoic sources) and the choice of objects within the visual field on which to focus attention. Augustine admitted that the viewer was not only an agent but also a

6 J. Leyacker, "Zur Entstehung der Lehre von den Hirnventrikeln als Sitz psychischer Vermögen," *Archiv für Geschichte der Medizin*, 19 (1927), 253–86; and Walter Pagel, "Medieval and Renaissance Contributions to Knowledge of the Brain and Its Function," in *The History and Philosophy of Knowledge of the Brain and Its Functions*, ed. F. N. L. Poynter (Oxford: Blackwell, 1958), pp. 95–114.

7 On light in the Platonic and Neoplatonic traditions, see David C. Lindberg, "The Genesis of Kepler's Theory of Light: Light Metaphysics from Plotinus to Kepler," *Osiris*, ser. 2, 2 (1986), 5–42 at pp. 9–14; and David C. Lindberg, *Roger Bacon's Philosophy of Nature: A Critical Edition, with English Translation, Introduction, and Notes, of De multiplicatione specierum and De speculis comburentibus* (Oxford: Clarendon Press, 1983), pp. xxxv–xlviii. See also Teun Tieleman, "Plotinus on the Seat of the Soul: Reverberations of Galen and Alexander in *Enn.* IV, 3 [27], 23," *Phronesis*, 43 (1998), 306–25.

recipient of the exterior light that made vision possible and of an image (or "species") of the seen object in the viewer's eye, transmitted from there to the memory for storage. For Augustine, as for Plato, the *corporeal* vision of physical objects "by the [extramitted] visual rays that shine through the eyes and touch whatever we see" served as a model for understanding how the "mind's gaze" (*acies mentis*) actively reaches toward and achieves intellectual knowledge or spiritual understanding of *intelligible* objects such as God.[8] Augustine's frequent references to the spiritual implications of both sensory and intellectual sight would offer encouragement to thirteenth-century readers interested in the mathematical and empirical investigation of light and vision.

THE ISLAMIC CONTRIBUTION

Although scholars in early-medieval Christian Europe had only fragmentary access to the Greek traditions just sketched, the major Greek works were translated into Arabic after the advent of Islam and stimulated scholarly activity within the Islamic world in each of the areas of Greek achievement. Galen's works, for example, were translated by scholars such as the Nestorian Christian Ḥunayn ibn Isḥāq (d. 877), a physician whose interpretation of Galen's views on the anatomical location of the various faculties of the interior senses had a powerful influence in both Islam and (later) European Christendom, and whose own treatment of the anatomy and physiology of the eye and visual pathway made Hellenistic ophthalmology widely available.[9]

Scholars in the Islamic world also had better access to the teachings of Plotinus and his systematizer, Proclus. Their emanationism was widely diffused among scholars who recovered and furthered Aristotle's thought, from Alkindi to Ibn Sīnā (known as Avicenna to those who would read his works in Latin). Indeed, somebody in the Islamic world created the distillation of Plotinus and Proclus that later circulated in the Latin West under the name of Aristotle as the *Liber de causis*.[10]

8 Margaret Miles, "Vision: The Eye of the Body and the Eye of the Mind in Saint Augustine's *De Trinitate* and *Confessions*," *The Journal of Religion*, 63 (1983), 125–42.

9 See Lindberg, *Theories of Vision from Al-Kindi to Kepler*, pp. 18–57. On Hunayn's contributions to visual theory, see also Bruce S. Eastwood, "The Elements of Vision: The Micro-cosmology of Galenic Visual Theory According to Ḥunayn Ibn Isḥāq," *Transactions of the American Philosophical Society*, 72, pt. 5 (1982). Hunayn's views became known to Western readers under the name of Constantine the African, the eleventh-century translator of Hunayn's *De oculis* and *Introduction (Isagoge) to Galen's Tegni*. On the *De causis*, which bound Aristotelian to Neoplatonic (meta)physics, see Cristina d'Ancona Costa, *Recherches sur le livre De causis* (Paris: Vrin, 1995). On the translations, see Newman, Chapter 16, this volume.

10 Christina d'Ancona Costa, *Recherches sur le livre De causis*. See also Gerhard Endress, "L'Aristote arabe: Réception, autorité et transformation du premier maître," *Medioevo: Rivista di storia della filosofia medievale*, 23 (1997), 1–42; and Pinella Travaglia, *Magic, Causality and Intentionality: The Doctrine of Rays in al-Kindī* (Florence: Sismel/Edizioni del Galluzzo, 1997).

Scholars also undertook synthetic efforts. The philosopher Alkindi (or al-Kindī, d. 873) was the first Arabic writer to master the entirety of Greek treatments of light and vision as well as the science of the stars. And he emphasized that students of the stars and students of sight investigated the same fundamental phenomenon: the radiation of power itself, including light. After Alkindi, many other savants in the Islamic world augmented a growing body of literature on light and vision. The physician and polymath Avicenna (980–1037) was especially prolific, and the wide diffusion and influence of his writings into early-modern times would be difficult to overstate. Two works in which Avicenna detailed these subjects were particularly widely known in Latin translation. The first, a section from his *Book of Healing* (*Kitāb al-Shifā'*), circulated as a commentary on Aristotle's *On the Soul* (*De anima*); it included lengthy treatises on vision and on the interior senses, in which Avicenna extensively refuted extramissionist accounts. The second, his *Canon of Medicine* (*Liber canonis* or *Kitāb al-Qānūn*), long a standard textbook for physicians in the West, brought together his thinking on these topics and ophthalmology. Ibn Rushd (1126–1198), known in the Latin West as Averroes, also contributed significant treatments of light, vision, and the internal senses.[11]

The geometrical tradition of Euclid and Ptolemy was taken up by a number of skillful Islamic mathematicians, including Alkindi, Ahmad ibn ʿĪsā (ninth century), Qustā ibn Lūqā (d. 912), and Ibn al-Haytham. Of all authors writing about light and vision in Arabic, Ibn al-Haytham, known in the West as Alhacen (965–1040/1041), was by far the most significant, for his great optical treatise *Kitāb al-Manāzir* (*Book of Optics*) marks a solid synthesis of several of the distinct traditions of optical thought. Moreover, it was extraordinarily influential, profoundly shaping 300 years of Western optical thought after its translation into Latin (as *De aspectibus*) around the beginning of the thirteenth century.[12]

As Alhacen recognized, the major ancient optical traditions were at odds not merely about theoretical matters such as the nature of light and the directionality of the vision-causing rays but also the very criteria that a theory needed to satisfy in order to be judged successful: mathematical criteria for the Euclideans; physical or causal criteria for Aristotle, Aristotelians, and

[11] Lindberg, *Theories of Vision from Al-Kindi to Kepler*, pp. 43–56; Dimitri Gutas, *Avicenna and the Aristotelian Tradition: Introduction to Reading Avicenna's Philosophical Works* (Leiden: Brill, 1988); and Simone van Riet, ed., *Avicenna Latinus: Liber de Anima*, 2 vols. (Leiden: Brill, 1968–1972).

[12] On the spelling of Alhacen's name (which in medieval Europe was always spelled with a "c" rather than the "z" invented by Friedrich Risner in his 1572 printing of Alhacen's *Optica*), see David C. Lindberg, *Roger Bacon and the Origins of Perspectiva in the Middle Ages* (Oxford: Oxford University Press, 1996), p. xxxiii n. 75. On Alhacen's achievement, see *The Optics of Ibn al-Haytham, Books I–III on Direct Vision*, ed. and trans. A. I. Sabra, 2 vols. (London: The Warburg Institute, 1989); Smith, *Alhacen's Theory of Visual Perception*; and Lindberg, *Theories of Vision from Al-Kindi to Kepler*, chap. 4.

Neoplatonists; and anatomical and physiological criteria for Galen and the physicians. Refusing to cast his lot with any one set of criteria and the visual theory it spawned, Alhacen set out to merge the three approaches by devising a single comprehensive theory of vision capable of satisfying all three sets of criteria simultaneously. Those who followed Alhacen's lead could no longer be satisfied with a geometrical account of vision that overlooked the nature of light and the anatomy of the visual apparatus, or one that examined physical or physiological issues while ignoring the mathematics of light and vision. Their optical theories had to be comprehensive in scope in order to meet comprehensive measures of success.

However, delivering on this promise proved a formidable challenge. The challenge was not experimental, although at every point Alhacen took empirical data seriously as measures of theoretical adequacy. The problems were theoretical, and theoretical effort was required to solve them. By some means unknown to us (but most probably by reading the texts of predecessors, both Greek and Islamic), Alhacen had gained a full mastery of all of the major ancient optical traditions; he knew his Aristotle, his Euclid, his Ptolemy, his Galen, and his Alkindi. His project required him to submit the theoretical claims on which these traditions were founded to careful scrutiny and meticulous criticism. He was obliged to identify errors, adjudicate rival claims, craft compromises, and (above all) construct arguments. The goal was to demonstrate the mutual compatibility of the core achievements of the Aristotelian Neoplatonists, the Euclideans, and the Galenists.

A brief look at Alhacen's intromission theory of vision will serve to illustrate this point. In order to establish that vision occurs by the reception in the eye of rays issuing from the visual object, Alhacen appealed to relevant empirical data, such as afterimages and the ability of bright light to injure the eye, which in his view demonstrate that in the act of vision the eye is the recipient of action from outside. He argued further (following Alkindi) that the forms of light and color issue in all directions from every point of visible objects and that it is the nature of transparent bodies to receive and transmit these forms (Figure 20.3). Alhacen (like Avicenna) therefore insisted that visual radiation emanating from the eye would be redundant and should not be supposed to exist.

But if vision were to occur by intromission of rays from the visible object, how could one salvage the mathematical analysis of Euclid and Ptolemy, which depended on the visual cone, consisting of rectilinear rays emanating from an apex in the eye to a base in the visual field? Is it possible to have mathematical analysis without positing the visual cone, and to have the visual cone without visual rays emanating from the eye? Alhacen answered the first question in the negative, the second in the affirmative; the visual cone must be posited, but this can occur within an intromission theory. Alhacen advanced the following demonstration. Take an arrow, observed

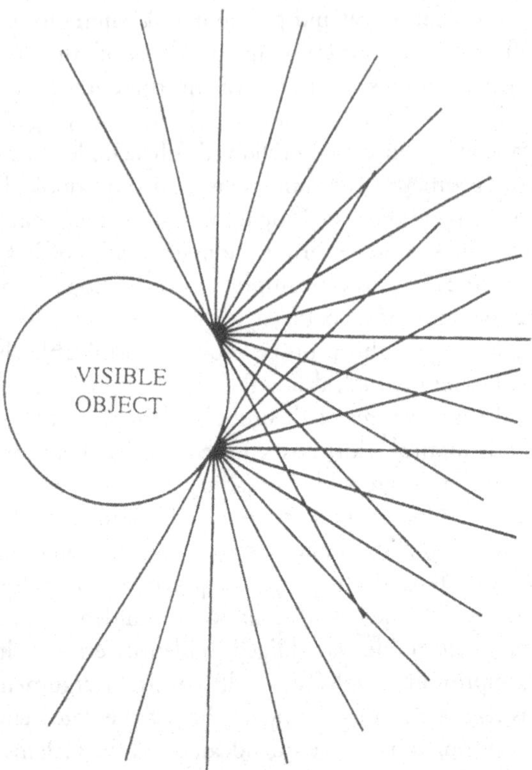

Figure 20.3. A representation of al-Kindi's (d. ca. 866) theory of independent and incoherent radiation in all directions from each point on the surface of the visible object. Radiation from only two points is shown for the sake of simplicity.

by the eye, as in Figure 20.4. Alhacen argued (here borrowing a conclusion from Alkindi) that objects radiate not as coherent wholes but as incoherent collections of points, each point on the surface radiating in all directions independently of the others. As a result, every point of the eye receives rays from every point of the object. And it would seem to follow from this that any attempt to observe an object would end in total confusion. Clarity of vision requires order, and order is achieved only if each point of the sensitive organ of the eye (which Alhacen, following Galen, identified as the crystalline humor or lens) receives radiation from a single point of the visible object. In short, a one-to-one correspondence must be established between points in the visual field and points in the eye, and this is precisely what the visual cone of the extramissionists does.

Yet, Alhacen believed, an intromissionist can obtain the same result. Of the infinity of rays emanating from the fletching, center, and head of the arrow in Figure 20.4, only one from each of these three points falls on the

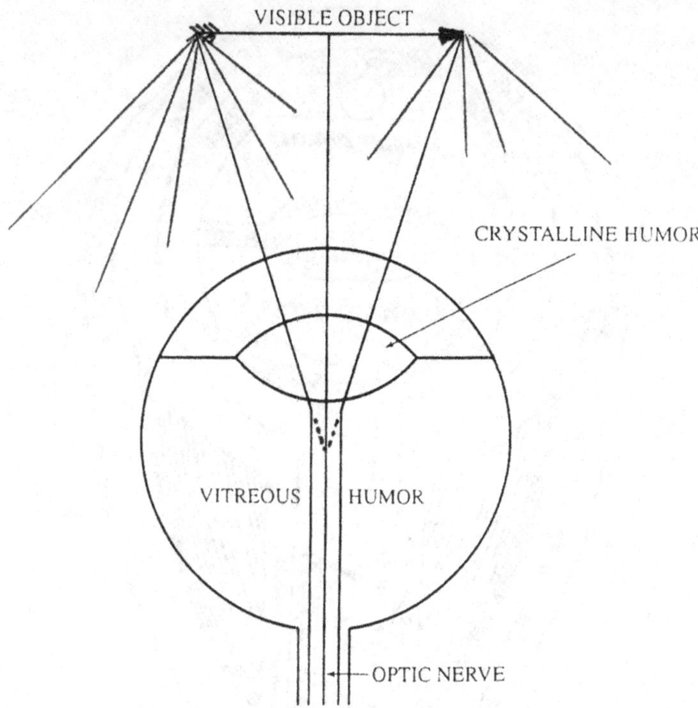

Figure 20.4. The eye and visual cone according to Alhacen's intromission theory. Rays from the visible object falling obliquely on the outer surface of the eye are refracted, thereby weakened, and thus eliminated from serious consideration. The unrefracted, vision-causing rays, one from each point on the surface of the visible object (three are shown), are perpendicularly incident on the cornea and front surface of the crystalline lens/humor. Passing into the crystalline lens, they give rise to visual perception. Emerging through the back side of the crystalline lens, they pass through the vitreous humor, and from there into the optic nerve, which carries them to the optic chiasma in the front of the brain, where rays from the two eyes join and the act of vision is "completed."

eye perpendicularly and enters without refraction. All others are refracted and thereby weakened, and only the lone unrefracted, unweakened ray is powerful enough to stimulate the visual power in the crystalline humor. The same is true of radiation issuing from every point of the visible object. But the collection of unrefracted rays, one from each point on the surface of the object, forms a cone of rays, directed toward an apex at the center of the eye. The rays of this cone fall on the front surface of the crystalline lens configured exactly as are the points of the object from which they originated, and clarity of vision is explained. Moreover (and this is a crucially important point), with the visual cone come all of the mathematical

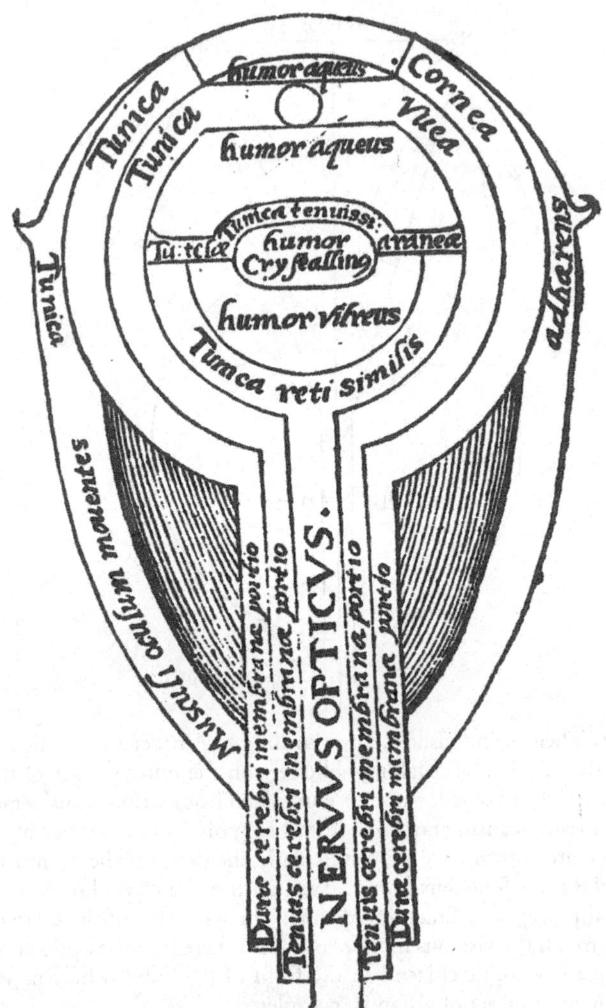

Figure 20.5. Anatomy of the eye as conceived by the editor of Alhacen's great optical treatise, *Opticae Thesaurus*, published in its early-thirteenth-century Latin translation. Basel, 1572.

capabilities of the mathematical tradition of Euclid and Ptolemy. Alhacen has thus demonstrated that an intromission theory need not be limited to questions about the nature of light and the causes of vision but can also support all of the mathematical analysis that had previously been associated exclusively with Euclidean extramissionism. By adding discussions of the anatomy and physiology of the eye to this account of vision, Alhacen provided the beginnings of a theory that could satisfy the criteria of all three ancient traditions (Figure 20.5).

THE BEGINNINGS OF *PERSPECTIVA* IN THIRTEENTH-CENTURY EUROPE

The intellectual life of Europe was dramatically enriched and transformed in the twelfth and thirteenth centuries by the translation of Greek and Arabic books into Latin (see Burnett, Chapter 14, this volume).[13] Among the translated works appearing piecemeal from late in the eleventh century to early in the thirteenth were nearly all of the Greek and Arabic optical sources. The translation into Latin of Aristotle's works, especially his *On the Soul* and *On Sensation*, with Avicenna's and Averroes's commentaries, powerfully reinforced attention to the physical and psychological aspects of vision. The *Liber de causis* attributed to Aristotle began to circulate at about the same time. Averroes's medical compendium, the *Kitāb al-Kulliyāt* (Latin *Colliget*), and Alkindi's *On Stellar Rays* reached readers of Latin during the 1220s. By the early thirteenth century, Neoplatonic philosophy (which had absorbed aspects of Aristotelian and Galenic thought) was reaching European universities from two directions, one Arabic, the other Patristic.

The effect of this influx of new sources was not only to augment but also to complicate optical knowledge in the West, for the new materials were widely divergent in both theoretical content and methodology. The first signs that the newly translated works were exercising a significant influence on the content and methodology of indigenous Western optical thought appeared in the writings of the distinguished scholar, later bishop of Lincoln, Robert Grosseteste (ca. 1169–1253).

Much of Grosseteste's education in the liberal arts and medicine occurred in the schools of the twelfth century, where Neoplatonic philosophy loomed large, but a crucial portion of his career was also spent at Hereford, then an intellectual milieu at the forefront of scientific and philosophical learning, with ready access to newly translated Arabic works. Grosseteste also acquired a theological education, perhaps at Paris before 1224. Ever interested in the latest discoveries and ideas, and their applications to Christian learning, Grosseteste was among the first Latin scholars to bring together an extensive reading of Patristic literature, Neoplatonic sources (including the pseudo-Dionysius, whose work he retranslated from Greek to Latin), most of Aristotle's writings (Grosseteste translated the *Nicomachean Ethics*), the commentaries of Avicenna and Averroes, and Euclid's and Alkindi's books on geometrical optics.[14]

[13] See also Lindberg, *Theories of Vision from Al-Kindi to Kepler*, pp. 209–13; and David C. Lindberg, "Roger Bacon and the Origins of Perspectiva in the West," in *Mathematics and Its Applications to Science and Natural Philosophy in the Middle Ages: Essays in Honor of Marshall Clagett*, ed. Edward Grant and John E. Murdoch (Cambridge: Cambridge University Press, 1987), pp. 249–68.

[14] On Grosseteste's career, see James McEvoy, *The Philosophy of Robert Grosseteste* (Oxford: Clarendon Press, 1982); Joseph Goering, "When and Where did Grosseteste Study Theology?" in *Robert Grosseteste: New Perspectives on His Thought and Scholarship*, ed. James McEvoy (Turnhout: Brepols,

What Grosseteste learned from his reading of Euclid and Alkindi was the power of geometry when applied to questions about nature in general and light and vision in particular. He issued a manifesto to that effect at the beginning of a little treatise entitled *On Lines, Angles, and Figures*:

> The usefulness of considering lines, angles, and figures is very great, since it is impossible to understand natural philosophy without them. They are useful in relation to the universe as a whole and its individual parts. They are useful also in connection with related properties, such as rectilinear and circular motion. Indeed, they are useful in relation to activity and receptivity, whether of matter or sense – and this is so whether in the sense of sight . . . or in the other senses. . . . Now all causes of natural effects must be expressed by means of lines, angles, and figures, for otherwise it is impossible to grasp their explanation.[15]

This declaration is reminiscent of Alkindi's claim that everything radiates its force in all directions into suitable recipients, "so that every place in the world contains rays from everything that has actual existence."[16] For optics, this theory meant that every point of a visible object radiates its power or image in all directions. Alkindi's doctrine of the universal radiation of force, appropriated by Grosseteste, emerged in the latter's writings as a doctrine that he called the "multiplication of species." According to Grosseteste, everything in the universe acts on its surroundings through the emanation (or "multiplication," as he preferred) of its species (or likeness) in all directions. In choosing to employ the term "species" for what he also called "power" or "force" (*vis* or *virtus*), Grosseteste and those who followed him were conveying the notion of "what is visible," as in the Latin "*speculum*" (mirror) and such English words as "spectacle," "speculate," "aspect," and "perspective." Grosseteste was thus treating the radiation of light as the paradigmatic case – in part because (owing to its visibility) it reveals the workings of other forms of radiation that are invisible but also because of its cosmic role in the creation of the universe of concentric spheres.[17] It follows, as one of the fundamental assumptions of Grosseteste's natural philosophy, that such radiation is implicated in all forms of natural causation.

Grosseteste believed that radiation naturally exerts more force on recipients that it reaches along short, straight lines than on those it reaches by means of

1995), pp. 17–52; and N. M. Schulman, "Husband, Father, Bishop? Grosseteste in Paris," *Speculum*, 72 (1997), 330–46.

[15] Robert Grosseteste, *On Lines, Angles, and Figures*, trans. David Lindberg, in *A Source Book in Medieval Science*, ed. Edward Grant (Cambridge, Mass.: Harvard University Press, 1974), p. 385.

[16] Quoting from pp. 69–71 of Marie-Thérèse d'Alverny and F. Hudry, "Al-Kindi, De radiis," *Archives d'histoire doctrinale et littéraire du moyen âge*, 41 (1974), 139–260.

[17] Grosseteste believed that God created an original point of light, which diffused itself in all directions, generating matter and giving rise to the cosmos as we know it. The Plotinian origins of this thesis are indisputable, although sometimes overlooked, because Grosseteste frequently supports it with Aristotle's arguments. See also Lindberg, *Roger Bacon's Philosophy of Nature*, pp. xliv–lxvi; and Richard C. Dales, "Robert Grosseteste's Views on Astrology," *Mediaeval Studies*, 29 (1967), 357–63.

longer lines; likewise, the reflection or refraction of rays by a denser medium – unless that medium has the polished, uniform smoothness of a mirror – weakens their impact upon objects they subsequently strike. Sounds that are magnified by reflection from smooth bodies are dissipated by striking rough surfaces; a planet's influence on mundane objects waxes or wanes with its motion through the heavens because the angle of its rays' incidence on any given object is altered; and a hot object warms cold objects in its vicinity more than distant ones by radiating and so sharing its heat. Hence – and, for Grosseteste, crucially – if we are to understand the diversity of effects within the recipient (whether that recipient is exterior matter or a sense organ), we must investigate this radiation by means of "lines, angles, and figures," in short, through the application of geometrical methods. In his short treatise *On the Rainbow*, Grosseteste attempted to exemplify the application of such mathematical methods to a specific natural phenomenon.[18]

But the mathematical analysis of light and vision did not exhaust Grosseteste's interests in visual phenomena. Grosseteste was among the earliest medieval Christian authors to show real appreciation for Avicenna's treatise *On the Soul*, with its important treatments of the psychological aspects of visual perception. This work provided Grosseteste with a convincing delineation of the interior senses (derived from Nemesius of Emesa) and an understanding of the process by which visual perception is "completed" in the brain. According to Grosseteste, what we see with our eyes when we look at a visible object are light and colors – the latter identified as the original light of the universe incorporated into transparent material bodies (thus literally "embodied"). On the object's surface, colors are its visible aspects or form, its "species," and when they are suffused with rays of light, they generate rays in turn, multiplying (or reproducing) themselves rectilinearly in all directions. Sight occurs as these visible "species" interact with a visual spirit made "of the same nature as the sun's light" that emanates from the eye.[19]

When, as a result, the eye has received this form, the act of vision is complete, but the work of the interior senses – by which we know what we see – has not yet begun. Grosseteste enumerated four such senses, locating them in three cavities or cells of the brain. In the first cell, the forms of an object sensed by the different external senses are brought together by the "common sense" and are preserved by the "imagination"; in the middle cell resides the "estimative sense," which Grosseteste defined (following Avicenna) as the power that, for example, enables the lamb to judge the danger presented by a wolf. The basis for the lamb's judgment is "intentions" rather than

[18] See especially Grosseteste, *On Lines, Angles, and Figures* and *On the Rainbow*, in Grant, *Source Book in Medieval Science*, pp. 385–91.

[19] Robert Grosseteste, "De operationibus solis," in James McEvoy, "The Sun as *res* and *signum*: Grosseteste's Commentary on Ecclesiasticus ch. 43, vv. 1–5," *Recherches de théologie ancienne et médiévale*, 41 (1974), 38–91 (the quoted passage is at pp. 69–71).

forms, and these are stored by the brain in the "memory," within the brain's posterior cavity. Like earlier Neoplatonists, Grosseteste also held that we possess, in addition, an intellect able to experience insights regarding intelligible objects (such as mathematical objects, forms, universals, or truths) using real processes precisely like those of corporeal vision.[20]

Grosseteste's theories no doubt held intrinsic appeal for many of his readers. But we should not overlook the official positions that he held within the university and church, for these contributed to the dissemination of his ideas. Grosseteste was lecturer to the Oxford Franciscans and subsequently bishop of Lincoln, offices that positioned him to advocate a synthesis of the Greco-Arabic learning newly available in translation with the wisdom found in Patristic writings. At about the same time, Grosseteste's close friend William of Auvergne, who had similar inclinations, became bishop of Paris – a position in which he was even better situated to promote the emerging synthesis.[21] By the time of their deaths (Auvergne's in 1249, Grosseteste's in 1253), Albert the Great and Roger Bacon (the first Latin scholars to boast a comprehensive mastery of the new Aristotle) were already hard at work, immersed in the available literature on light and vision.

Albert (ca. 1200–1280), sometime professor in the Dominican priory at Cologne and the University of Paris, was a towering figure in thirteenth-century scholarship. Generations of medieval scholars mined his commentaries on the writings of both the pseudo-Dionysius and Aristotle. Lecturing on the entire corpus of works attributed to Aristotle, Albert commented at length on Aristotle's theories of light and color. He was also acquainted with the optical treatises of Euclid, Alkindi, and Alhacen and influenced in his understanding of Aristotelian psychology by the works (including the *Canon of Medicine*) of Avicenna, whose classification of five internal senses Albert generally shared. Confronted with the alternative methodologies of Aristotle, on the one hand, and the Euclidean mathematicians, on the other, Albert cast his lot firmly with Aristotle. In his paraphrase of Aristotle's *Metaphysics*, Albert wrote: "We must beware of the error of Plato, who said that natural things are based on mathematical things and mathematical things on divine things, just as the third cause is based on the second and the second on the first; and therefore he said that the principles of natural things are mathematical, which is altogether false."[22] Albert certainly did not intend to forbid the mathematical analysis of radiation; indeed, he engaged in such

[20] Robert Grosseteste, *Commentarius in posteriorum analyticorum libros*, II.6, ed. Pietro Rossi (Florence: Olschki, 1981), p. 404; and McEvoy, *Philosophy of Robert Grosseteste*, pp. 297–312.

[21] See Steven Marrone, "Metaphysics and Science in the Thirteenth Century: William of Auvergne, Robert Grosseteste and Roger Bacon," in *Medieval Philosophy*, ed. John Marenbon (London: Routledge, 1998), pp. 204–24.

[22] Albertus Magnus, *Opera omnia*, ed. Bernhard Geyer, 40 vols. (Münster: 1951–), vol. 16, pt. 1, p. 2, lines 31–35. In addition to James A. Weisheipl, ed., *Albertus Magnus and the Sciences: Commemorative Essays* (Toronto: Pontifical Institute of Mediaeval Studies, 1980), see Lindberg, *Theories of Vision from Al-Kindi to Kepler*, pp. 104–7, on Albert's optics.

analysis on a limited scale himself. But he firmly believed that the central issues in the study of light and vision had to do with physical natures and must be approached through natures and causes.

THE BACONIAN SYNTHESIS

Nobody contributed more to the development of the science of *perspectiva* in the West than Roger Bacon (ca. 1214/1220–ca. 1292). On the methodological issues that separated Grosseteste and Albert the Great, Bacon dismissed Albert's opinion nearly as enthusiastically as he praised Grosseteste's. Although Bacon did not study under Grosseteste, and may never even have met him, he did have access to Grosseteste's library, left to the Franciscan convent in Oxford, and was clearly inspired by his example. However, Bacon was also powerfully moved by sources that had been unavailable to Grosseteste, principally the optical works of Ptolemy and Alhacen, where the promise of geometrical optics had been much more completely fulfilled.[23]

Thus Bacon, even more than Grosseteste, became an apostle of the application of mathematical method as the gate and key to all subjects, claiming, for example, that "no science can be grasped without this science [mathematics], and . . . nobody can perceive his ignorance in other sciences unless he is excellently informed in this one. Nor can things of this world be known, nor can man grasp the uses of body and things, unless he is imbued with the mighty works of this science."[24] Despite Bacon's rhetoric in this passage, he actually advocated a scientific methodology that coupled a Neoplatonic conviction that the universe is mathematically structured and can be mathematically deciphered with a more Aristotelian expectation that scientific understanding must also be empirically grounded in (or tested against) what the senses experience.

It is, of course, one thing to promote a methodological program but quite another to practice it. Did Bacon consistently practice the mathematization of light and vision? The answer is a qualified and complicated "yes." Following Alhacen's lead, he applied geometrical analysis to optical phenomena wherever promising and possible, given the conceptual framework and the

[23] Lindberg, *Roger Bacon's Philosophy of Nature*; Lindberg, *Roger Bacon and the Origins of Perspectiva in the Middle Ages*; and David C. Lindberg, "Roger Bacon on Light, Vision, and the Universal Emanation of Force," in *Roger Bacon and the Sciences: Commemorative Essays*, ed. Jeremiah Hackett (Leiden: Brill, 1997), pp. 243–75.

[24] Roger Bacon, *Communia mathematica*, in *Opera hactenus inedita Rogeri Baconi*, 16 fascicules, ed. Robert Steele and Ferdinand M. Delorme (Oxford: Clarendon Press, 1905–1940), fasc. 16, p. 7. For similar remarks, see Roger Bacon, *Opus maius*, pt. 4.1.1, in *The Opus Majus of Roger Bacon*, ed. John H. Bridges, 3 vols. (London: Williams and Norgate, 1900), vol. 1, pp. 97–8. On Bacon and the application of mathematics to optical phenomena, see Lindberg, *Roger Bacon and the Origins of Perspectiva in the Middle Ages*, pp. xlii–lii; and Lindberg, "Roger Bacon and the Origins of *Perspectiva* in the West," pp. 258–64.

mathematical techniques available to him, thereby pushing the mathematization of light and vision as far as it would go before the seventeenth century. Yet his mathematical analysis, though generally successful, was not flawless, and there were aspects of optical theory that simply were not vulnerable to a mathematical assault. The Aristotelian side of Bacon that aspired to nonmathematical causal knowledge was also a player in the game.

Several examples will serve as illustrations. First, the magnitude of Bacon's commitment to a mathematical analysis is superficially apparent from a glance at his *Perspectiva*, which contains fifty-one geometrical diagrams, or his *On the Multiplication of Species*, which contains thirty-nine, all fully integrated into the argument of their respective treatises. Second, Bacon's works reveal a complete mastery of the geometry of reflection and image-formation in plane, convex, and concave mirrors. Bacon understood, of course, that in reflection the angles of incidence and reflection are equal, and he knew that the image of an object seen by reflection is situated where the backward, rectilinear extension of the reflected ray reaching the eye intersects the perpendicular drawn from the visible object to the reflecting surface (Figure 20.1). He revealed an impressive understanding of the magnification and diminution of images seen by reflection or refraction, and the concept of focal point or focal plane is implicit in his analysis of convex spherical mirrors.[25]

We can see the depth of Bacon's commitment to the mathematization of optical phenomena in a third example – Bacon's remarkable supposition (following Alhacen) that the visual apparatus and the very act of vision will submit to geometrical analysis. According to Bacon, all of the tunics and humors of the eye (cornea, crystalline lens, aqueous and vitreous humors, and retina) are defined or enclosed by spherical surfaces, the centers of which are situated on a straight line running from the center of the pupil at the front to the opening into the optic nerve at the back. He believed, as Alhacen had taught, that only rays incident on the eye perpendicularly, which enter it without refraction, are capable of stimulating the eye's visual capabilities. These perpendicular rays form a cone or pyramid extending from the visual object as base toward an apex (which the rays never actually achieve) at the center of the observer's eye (Figure 20.4). The rays that make up this visual cone pass without refraction through the cornea and front surface of the crystalline lens (which are concentric, so that a ray perpendicular to the one will be perpendicular to the other. At the rear surface of the crystalline lens, they are refracted in such a way as to be projected through the opening of the optic nerve, which conducts them to its point of union with the other optic nerve (our optic chiasma). There the completion of vision occurs, as the species from the two eyes join to form a single image. That image, in turn, continues to multiply itself into

[25] Lindberg, *Roger Bacon and the Origins of Perspectiva in the Middle Ages*, pp. xliv–xlv.

the three chambers of the brain that house the five inner senses defined in Avicenna's *On the Soul.* Although Bacon's theory contains much more detail, a striking feature of his quest to understand the act of vision is his willingness (following Alkindi, Grosseteste, and especially Alhacen) to extend mathematical analysis to something so apparently unmathematical as human anatomy.[26]

Mathematical analysis was only one prong of Bacon's methodological campaign. The other – his Aristotelian side, reinforced by Alhacen's example – was the practice of what he called "experimental science," the aspect of his scientific work on which his reputation as one of the most original of medieval scholars primarily rests. Among Bacon's many proclamations on the subject of experimental science were arguments for its epistemological authority as the mistress to whom the other sciences, her handmaidens, owed obedience.[27] But is the reputation deserved? And how thoroughly did *perspectiva* become an experimental science in Bacon's hands?

We do not have space here to probe deeply into the vexed question of the meaning of "experiment" in Bacon's work, but on any definition his theories of light and vision were *not* primarily the products of experimental investigation. Bacon proceeded as natural scientists have nearly always proceeded, by borrowing from the past and constructing a theoretical framework from inherited materials. His overarching goal was not originality but truth, and he had every reason to suppose that each of the available authorities – Plato, Aristotle, Euclid, Ptolemy, Galen, Augustine, Ḥunayn ibn Isḥāq, Alkindi, Avicenna, Alhacen, and Grosseteste – possessed some portion of the truth. None was wrong, but some were incomplete, and others had been misunderstood.

Nevertheless, it does not follow that Bacon imitated his guides uncritically. He recognized that authorities had to be interpreted, criticized, and corrected, disputes had to be adjudicated, choices made, and compromises worked out, and all of this could be done without depriving them of their authoritative status. What emerged was Bacon's considered opinion on the matters at issue.

In the end, any theory had to pass the tests of reason and experience. Bacon exhibited no reluctance toward bringing observational experience to bear on his theories of light and vision. Indeed, a few of his observations can be confidently attributed to contrived experiments employing instruments. The clearest evidence of such an experiment is the value of 42° for the maximum elevation (also radius) of the rainbow, first expressed by Bacon and therefore, we may reasonably assume, the product of his own measurements, probably using an astrolabe.

[26] Ibid., pp. xlvi–xlviii, 21–55; and A. Mark Smith, "Getting the Big Picture in Perspectivist Optics," *Isis*, 72 (1981), 568–89.

[27] See especially Bacon, *Opus maius*, vol. 2, pp. 167–222.

That said, historians have no adequate window on Bacon's actual scientific practices. The evidence at our disposal is limited to texts that Bacon wrote (occupying about two feet of bookshelf in their nineteenth- and twentieth-century editions), and, apart from a few exceptional cases like the elevation of the rainbow, what they reveal are the functions that empirical data perform in his scientific *arguments.* A handful of his optical observations are intended simply to discover or verify factual data, with no direct bearing on optical theories. Most, however, serve some theoretical function, such as confirming, refuting, or challenging a theoretical belief. The best of many examples is Bacon's attempt to devise a test of Alhacen's claim that, although vision is primarily the result of rays incident perpendicularly on the eye, which enter without refraction, rays that are nonperpendicular (and therefore refracted) also play some role. To prove this point, he proposes that we place a thin straw in our visual field between our eye and a visible object. Since the straw must intercept perpendicular rays from some parts of the object, and since, despite the presence of the straw, we perceive the entire object, it is evident that nonperpendicular rays are (under such circumstances) visually efficacious.[28]

Finally, complicating any taking of Bacon's methodological temperature is the fact that his optical writings were not research monographs but polemical *persuasiones,* meant to reveal to a nonspecialist audience the awesome possibilities of optical knowledge through a survey of its contents. In that effort, Bacon displayed his remarkable ability to synthesize the disparate sources available to him. For example, his sources presented him with a variety of apparently incompatible theories of the nature of light, including Plato's "visual fire," Aristotle's actualization of the potential transparency of the medium, Alkindi's universal propagation of force, Alhacen's "forms" of light and color, and Grosseteste's "species." Bacon's solution was to overlook the differences. Plato's hypothesis of visual fire (which had experienced something of a revival in the twelfth century) was too vague to cause any trouble, and Bacon found that he could successfully merge the others. He took Grosseteste's species, assigned them all of the properties and functions of Alhacen's forms, claimed that this was what Aristotle had in mind, and agreed that they radiate in all directions from every point of every visible object, as Alkindi had proposed. Anything that can be described in so few words may seem simple, but in fact Bacon devoted great effort and a treatise of 40,000 words, entitled *On the Multiplication of Species,* to the project. In this treatise, he developed the ideas of Alkindi and Grosseteste in exquisite detail – defining species as likenesses of the agents from which they issue,

[28] Bacon's experimental practice is treated more fully in Jeremiah Hackett, "Roger Bacon on *Scientia experimentalis,*" in Hackett, *Roger Bacon and the Sciences,* pp. 277–315. On perspectival experimentation, see David C. Lindberg, "Bacon on Light, Vision, and the Universal Emanation of Force," in Hackett, *Roger Bacon and the Sciences,* pp. 265–72; and Lindberg, *Roger Bacon and the Origins of Perspectiva in the Middle Ages,* pp. lii–lxvii.

maintaining that all natural causation requires them, working out in detail the mathematics and physics of their propagation, and specifying their role as agents of visual perception.[29]

But in what direction are the species responsible for vision propagated? Here the alternatives were stark. Either vision occurs by extramitted rays, as Euclid, Ptolemy, Galen, Hunayn ibn Ishāq, and Alkindi all maintained; by intromitted rays, as Aristotle, Alhacen, Avicenna, and Averroes insisted; or by some mechanism involving a two-way transmission, as in the theories of Plato, St. Augustine, and Grosseteste.[30] Bacon was a convert to the intromissionist gospel, thanks to Aristotle, Avicenna, and Alhacen. He joined them in arguing that intromitted species are necessary for vision and its principal cause; the senses (as Aristotle pointed out in a passage quoted by Bacon) are initially passive recipients, whereas the species of the visible object are the agents.

However, convinced that our senses are active as well as passive, Bacon was unwilling to dismiss the extramissionist position. Moreover, extramissionists were able to advance persuasive arguments and relevant empirical evidence, such as the (alleged) ability of cats to see in the dark. Bacon escaped the dilemma by noticing that Alhacen, Avicenna, and Averroes had proved the necessity, and therefore the existence, of intromitted rays without ever demonstrating the nonexistence of extramitted ones. This allowed Bacon to concede that intromitted rays are the immediate cause of vision (as the intromissionists insist) while acknowledging that extramitted rays also exist (as the extramissionists maintain) and perform the ancillary function of "ennobling" the medium and the incoming species, thus rendering the latter capable of "completing its action" on an animated body such as the eye. According to Bacon's understanding of Galen, the optic nerves are hollow tubes descending from the brain that can accommodate the two-way traffic of extramission and intromission.[31]

THE RAINBOW AND ITS COLORS

Rainbows – spectacular natural phenomena as well as symbols of God's promise to Noah – could hardly be overlooked by medieval scholars claiming an interest in light and vision. Their starting point was Aristotle's *Meteorologica*, where the rainbow was attributed to the reflection of solar light

[29] For a critical edition, English translation, and analysis of this treatise, see Lindberg, *Roger Bacon's Philosophy of Nature*. A shorter analysis appears in Lindberg, "Roger Bacon on Light, Vision, and the Universal Emanation of Force," pp. 245–50.

[30] An argument can be offered for classifying Galen and Ḥunayn with those who defended a two-way transmission since they required the air (which has been transformed by visual spirit into an extension of the optic nerve) to return its impressions to the eye and brain.

[31] Lindberg, *Roger Bacon and the Origins of Perspectiva in the Middle Ages*, pp. lxxxiii–lxxxvi.

in a cloud, with colors emerging as a result of variable weakening of that solar light by reflection and transmission through the nebular mists. Aristotle also treated the geometry of the rainbow, concluding (for example) that a straight line connects the Sun, the observer, and the center of curvature of the rainbow.[32]

One of the first medieval scholars to take a serious interest in the rainbow after Aristotle's works became available in translation was Robert Grosseteste. In a short work entitled *De iride*, composed early in the thirteenth century, he dismissed Aristotle's reflection theory on the grounds that reflection in a hollow cloud could not explain the rainbow's shape. He urged instead that the rainbow is the product of multiple refractions in the cloud and its mists – considered holistically rather than as a collection of discrete drops of moisture.[33]

Grosseteste's *De iride* devoted fewer than 1,000 words to the rainbow. Roger Bacon, writing some thirty to forty years later and doubtless aware of Grosseteste's little treatise, allocated about 8,000 words to the subject in his *Opus maius*. The context of Bacon's discussion was the power of what he called "experimental science," with the rainbow as his primary example. One observational result, mentioned earlier, was the universal 42° radius of the rainbow. As for the rainbow's cause, Bacon completely discounted Grosseteste's argument on behalf of *holistic* refraction, insisting that the key was to be found in the *individual* droplets of moisture in the cloud (perhaps following Albertus Magnus on this point). The causal importance of these droplets was evident, Bacon argued, from the fact that if the observer of a rainbow walks toward the rainbow, it recedes from him; if he withdraws, it follows him; and if he moves sideways, the rainbow does the same. From this Bacon concluded that if there are multiple observers, each sees a different rainbow. And that can be true only if, for each of them, the rainbow is produced by a different set of droplets. As for what happens in these droplets, Bacon did not see how refraction could produce the known phenomena of the rainbow. He argued instead for reflection from small drops "infinite in number" descending from a cloud, from each of which "reflection occurs as from a spherical mirror."[34]

Early in the fourteenth century, a Dominican friar, Theodoric of Freiberg (d. ca. 1310), followed Bacon in opting for the importance of individual drops. In one of the most remarkable experimental investigations of the Middle Ages, Theodoric projected rays of light through crystalline spheres and water-filled flasks, meant to simulate the passage of sunlight through a drop of moisture in a rain cloud. He concluded that the primary bow resulted from refraction as a solar ray entered the raindrop, an internal reflection at

32 Aristotle, *Meteorologica*, III.4.373a–b.

33 For a translation of this treatise, see Grant, *Source Book in Medieval Science*, pp. 388–91.

34 Bacon, *Opus maius*, vol. 2, p. 192. See also David C. Lindberg, "Roger Bacon's Theory of the Rainbow: Progress or Regress?" *Isis*, 57 (1966), 235–48.

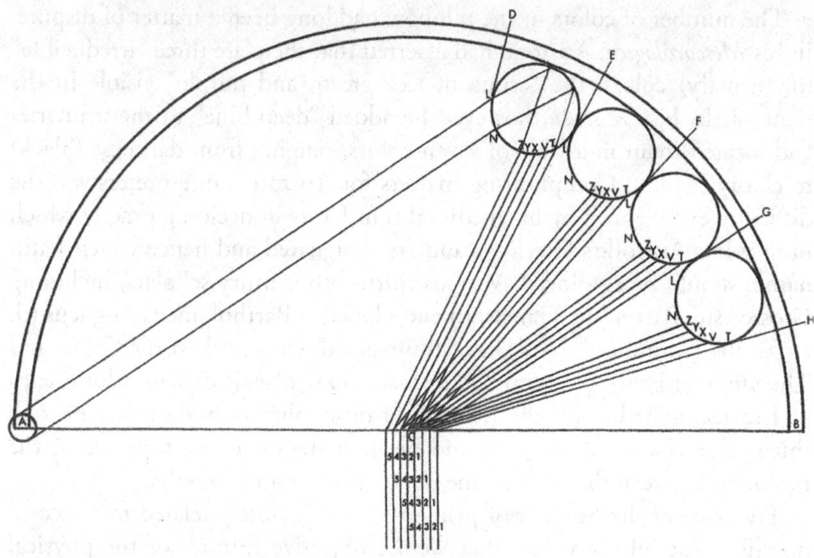

Figure 20.6. Theodoric of Freiberg (d. ca. 1310) on the rainbow. The drawing shows four small circles, representing droplets of moisture within a cloud responsible for the rainbow. The Sun (*A*) is on the horizon at the lower left; the observer is at *C*, the center of the large semicircle. The paths of two solar rays are represented in the uppermost droplet, refracted as the light enters it (*L*, *N*) and again as it leaves it (*Z*, *T*), following a total internal reflection at the back of the same droplet. This happens to be exactly how the rainbow is formed.

the back of the raindrop, and a second refraction as it emerged from the raindrop (Figure 20.6). To explain the secondary rainbow, Theodoric added a second internal reflection within the raindrop.[35] Whether Descartes (with whom this theory became a permanent piece of meteorological knowledge, still valid) borrowed the idea from Theodoric is open to conjecture.

COLORS, APPEARANCES, AND THE KNOWABILITY OF THE WORLD

Theorizing about the rainbow might seem from the foregoing account to have been concerned exclusively with reflection and refraction, treated geometrically wherever possible. But the rainbow also provided ancient and medieval scholars with an occasion and an opportunity for discussing the nature and number of colors and, in the process, for raising important perceptual and epistemological questions.

[35] William A. Wallace, "Dietrich von Freiberg," *Dictionary of Scientific Biography*, IV, 92–5.

The number of colors in the rainbow had long been a matter of dispute. In his *Meteorologica*, Aristotle had asserted that there are three "irreducible" (or primary) colors, the "crimson, leek-green, and purple" visible in the rainbow. In his *De sensu*, however, he added "deep blue" to the primaries and located them in a scale of seven colors, ranging from darkness (black) to clarity (white). Complicating matters for Aristotle's interpreters was the difficulty experienced by his medieval translators in deciding exactly which pure colors Aristotle's Greek vocabulary designated and hence which Latin names should be employed. Various thirteenth-century scholars, including Grosseteste, Bacon, the Franciscan encyclopedist Bartholomeus Anglicus (d. 1272), the Dominicans Vincent of Beauvais (d. 1264), Albert the Great, and Theodoric of Freiberg, attempted to make sense of Aristotelian color theory and to reconcile it with the theories of other ancient authorities. Bacon's three-value color scale – white and black at the extremes, ruby-red at the midpoint between them – became popular but not universal.[36]

The issue of the number of primary colors is closely related to a second question. Are all the colors that we see objective features of the physical world, or are some created within our perceptual apparatus? If, as Bacon had pointed out, no two observers see the same rainbow, it follows that the colors that had seemed to be objective and extramental were in fact at least partially subjective phenomena – differing, in his view, from the "true" colors produced when solar rays pass through crystal. Bacon also pointed to the iridescent colors visible in peacocks' tails and pigeons' necks and shifting spectra projected upon a wall, which he believed had "more an appearance of true color than its existence."[37] Nonetheless, he regarded such phenomena as exceptional.

Peter Auriol (d. 1322), a Franciscan belonging to the generation after Bacon, had quite a different opinion. He formulated the critical question as follows:

> Those who ask concerning the colors of a rainbow, or the colors that are in the neck of a dove, or an image that appears in a mirror, or a candle appearing somewhere other than in its true location, whether these have real being or only intentional being, mean to ask whether these have only *subjective* and fictitious or apparent being, or whether they have real and fixed being externally in the nature of things, independent of any apprehension.[38]

36 Aristotle, *De sensu et sensato*, 3–4; Aristotle, *Meteorologica*, III.2.371b34–372a11 and also III.5.375a1–11; Roger Bacon, *De sensu et sensato*, in Steele and Delorme, *Opera hactenus inedita Rogeri Baconi*, fasc. 16, pp. 74–5.

37 Roger Bacon, *De multiplicatione specierum*, I.3, lines 178–207, in Lindberg, *Roger Bacon's Philosophy of Nature*, pp. 54–6 (see p. 54, line 182, for the quoted words). See also Roger Bacon, *Perspectiva*, III, dist. 1, chap. 5, in Lindberg, *Roger Bacon and the Origins of Perspectiva in the Middle Ages*, p. 278, lines 295–318; and Bacon, *Opus maius*, VI ("Scientia Experimentalis), chap. 8, in Bridges, *The Opus Majus of Roger Bacon*, vol. 2, pp. 190–2.

38 Auriol, *Scriptum in primum librum Sententiarum*, quoted in Tachau, *Vision and Certitude in the Age of Ockham*, pp. 96–7 n. 35.

Auriol's controversial answer – that these appearances have no extramental existence – made him famous for generations of theologians. What Bacon and other perspectivists had thought the observer's perceptual powers sometimes do, Auriol held, they always do in the act of vision, veridical or not – they actively form purely subjective conceptions that appear like, and thereby represent, that object. Contents of the mind are not *unreal* entities, Auriol insisted, but they have a kind of existence that is "diminished," "intentional," or "apparent." Auriol was without followers, apart from the Franciscan theologian Bernard of Arezzo (fl. 1335). A broad philosophical movement prepared to accept and exploit skeptical arguments such as Auriol's had to await the sixteenth century.[39]

THE DIFFUSION OF *PERSPECTIVA* AFTER ROGER BACON

Bacon produced his optical works in haste and semisecrecy – not because anybody considered the subject matter dangerous but because in their very composition Bacon had broken prohibitions of the Franciscan Order against writing books on any subject without permission. He then dispatched them to Pope Clement IV, who had asked to see them. We do not know whether Clement approved of them or even read them, but in the long run they inspired an optical tradition that gave rise to commentaries and competing texts and entered the mainstream of university education, becoming the intellectual property of generations of European scholars.

The first fruits of Bacon's influence are seen in two optical texts written in the 1270s, about a decade after Bacon composed his *De multiplicatione specierum* and *Perspectiva*. Probably the first was an enormous volume entitled *Perspectiva*, written by a Silesian cleric named Witelo (d. after 1281), who probably learned of Bacon's work through his affiliation with the papal court. Witelo's *Perspectiva*, consisting of about 400,000 words and containing nearly 600 geometrical figures (ten times the 44,000 words and 51 figures of Bacon's *Perspectiva*), was meant as an exhaustive, systematic account of the new discipline, borrowing freely from Bacon, Alhacen, and other sources in the perspectivist tradition.[40]

Whereas Witelo aimed for completeness, the second text inspired by Bacon was a teaching text, written for brevity and accessibility by an English Franciscan and future archbishop of Canterbury, John Pecham (ca. 1230/1235–1292). After teaching theology at Paris, Pecham was appointed to do the same at

39 On Auriol's theory of knowledge and apparent being, see Tachau, *Vision and Certitude in the Age of Ockham*, pp. 85–112.

40 See *Vitellonis opticae*, printed with Alhacen's *Optica* under the title *Opticae thesaurus*, ed. Friedrich Risner (Basel: Eusebius Episcopius, 1572), reprinted with an introduction by David C. Lindberg (New York: Johnson Reprint, 1972).

the papal curia in Viterbo, where he likely encountered Witelo in person as well as Bacon's writings (if he had not already seen them in Paris). Written probably for Pecham's own students, his text, the *Perspectiva communis*, became in the long run the most widely used textbook on the subject in the universities of the Middle Ages and Renaissance.[41] Along with the works of Euclid, Alkindi, Alhacen, Bacon, and Witelo, it conveyed the basics of the discipline (especially its mathematical aspects) to generations of scholars until the early decades of the seventeenth century. Evidence of the strength of this optical tradition in the intervening centuries is found in the writings of Jean Buridan, Dominicus de Clavasio, Nicole Oresme, Henry of Langenstein (d. 1397), Biagio Pelacani da Parma (d. 1416), and many others.[42]

Among the scholars whom Bacon influenced, Pecham, Auriol, and Theodoric of Freiberg were not unusual in being theologians. Through its application to issues of sensation and cognition, *perspectiva* spoke to epistemological questions salient among theologians. Fellow Franciscans such as Roger Marston, Guillaume de la Mare, Matthew of Aquasparta, and Peter Olivi (all at Paris) absorbed (or criticized) perspectivist tenets in the thirteenth century, as did John Duns Scotus and William of Ockham in the fourteenth. Others who revealed the influence of *perspectiva* were Nicole Oresme, Henry of Langenstein, and Robert Holcot.[43]

The influence of the medieval perspectivist tradition did not end here but continued well into the seventeenth century. Alhacen's *De aspectibus* was printed in 1472, Bacon's *Perspectiva* in 1614, Witelo's *Perspectiva* in 1535, 1551, and 1572, and Pecham's *Perspectiva communis* in ten editions between 1503 and 1627. A measure of the influence of these four authors is the frequency of citations of their books in sixteenth- and seventeenth-century scientific writings by Leonardo da Vinci, Regiomontanus, Giambattista della Porta, Francesco Maurolico, Tycho Brahe, Michael Maestlin, Johannes Kepler, Galileo Galilei, William Gilbert, Christopher Scheiner, Ismaêl Boulliau, Thomas Harriot, Willebrord Snell, René Descartes, Francesco Maria Grimaldi, and Isaac Barrow (to name only the best known). Most remarkably, Kepler's theory of the retinal image – the most important optical innovation of the first half of the seventeenth century – represents the culmination, rather than a repudiation, of the medieval perspectivist tradition. Such a claim is not intended to question Kepler's originality or diminish its luster but rather to identify the foundation on which he built and the source materials that he employed

41 David C. Lindberg, ed. and trans., *John Pecham and the Science of Optics: Perspectiva communis* (Madison: University of Wisconsin Press, 1970).

42 David C. Lindberg, *A Catalogue of Medieval and Renaissance Optical Manuscripts* (Toronto: Pontifical Institute of Mediaeval Studies, 1975); and Lindberg, *Theories of Vision from Al-Kindi to Kepler*, pp. 116–46.

43 Tachau, *Vision and Certitude in the Age of Ockham*, passim.

in his creation of a new theory of how we see.[44] Tracing the influence of the medieval perspectivist tradition shows that it contained the theoretical and factual resources that later scholars exploited (not without alteration, of course, but nonetheless recognizably) to fuel the optical innovations of seventeenth-century optical practitioners.

44 See especially Lindberg, *Theories of Vision from Al-Kindi to Kepler*, pp. 188–208.

21

MATHEMATICS

A. George Molland

Mathematicians have perennially been regarded as an odd lot, partly for exoteric and partly for esoteric reasons. The exoteric side usually concerns readily perceived or at least reported eccentricities, such as Thales falling into a well while contemplating the stars, Archimedes streaking through Syracuse shouting "Eureka," or G. H. Hardy's skirmishes with God. The esoteric side relates to mathematicians' supposed possession of arcane knowledge that veers between being absolutely useless and equipping them with extraordinary powers. Galileo spoke of "the superhuman Archimedes, whose name I never mention without a feeling of awe," and a little later Robert Burton said of the fourteenth-century schoolman Richard Swineshead that "he well nigh exceeded the bounds of human genius." This brings us to the Middle Ages, when the perceived oddness or threat of mathematicians was more usually located among astrologers (and by extension magicians) than among pure mathematicians. Indeed, since Roman times, the word *mathematici* had applied especially to such practitioners, and in the thirteenth century Roger Bacon still strove at length to convince his audience of the difference between false mathematics (deterministic astrology) and true mathematics, which was of the utmost use to the commonwealth and had applications in many other sciences.

The criterion of usefulness must constantly be borne in mind when we consider medieval mathematics. Although few expressed this point as forthrightly as Roger Bacon, who railed at Euclid's multiplication of useless conclusions,[1] the surviving records display few signs of delight in doing pure mathematics for its own sake. The implicit emphasis was almost always on

[1] Compare A. George Molland, "Roger Bacon's Knowledge of Mathematics," in *Roger Bacon and the Sciences: Commemorative Essays*, ed. Jeremiah Hackett (Leiden: Brill, 1997), pp. 151–74. Good general accounts of medieval mathematics are not numerous. Among more recent works, we may mention A. P. Juschkewitsch, *Geschichte der Mathematik im Mittelalter* (Leipzig: Teubner, 1964); Michael S. Mahoney, "Mathematics," in *Science in the Middle Ages*, ed. David C. Lindberg (Chicago: University of Chicago Press, 1978), pp. 145–78; and Ivor Grattan-Guinness, ed., *Companion Encyclopedia of*

the service that mathematics could provide in other fields, such as practical mensuration, commerce, natural philosophy, astronomy, music (the last two being regarded as parts of mathematics), theology, and even the playing of games. The Roman Empire can, with only slight unfairness, be described as overwhelmingly lowbrow in its attitude toward mathematics. Perhaps its main contribution was the treatises of the *agrimensores* or land surveyors, which were transmitted to the early Middle Ages, but these exhibit only low levels of mathematical ability. This stance toward mathematics was reinforced by the oft-repeated etymology of geometry as the measurement of the Earth, necessitated in ancient times by the periodic flooding of the Nile River. The eighth-century bishop and encyclopedist Isidore of Seville pointed out that its scope soon broadened to include the measurement of the sea and the heavens, but this did little to enhance its intellectual content.

BOETHIUS AND THE EARLY MIDDLE AGES

If pure geometry received short shrift in the early Middle Ages, the situation was different with regard to arithmetic. This was partly a result of a historical accident embodied in the person of Boethius. Born in the late fifth century into an aristocratic and Christian Roman family, Boethius was from an early age immersed in Greek culture, and very soon formed a project of rendering much of it, including the works of Plato and Aristotle, into Latin. Unfortunately, and largely because of an involvement in politics that resulted in his execution in 523 or 524, his achievement was truncated, but it did include some mathematics. According to his slightly younger but longer-lived contemporary Cassiodorus, Boethius did render Euclid into Latin, but the portions of his translation that can be identified with any confidence are skeletal, consisting mainly of some definitions and enunciations of propositions, and usually found fleshed out with inferior later material. Thus, until the twelfth century, medieval pure geometry had an ignominious history. However, Boethius also produced an especially influential Latin work on arithmetic, which was very closely based on the Greek *Introduction to Arithmetic* of the neo-Pythagorean Nicomachus of Gerasa (late first century) and also a substantial treatise on music, a science that was then regarded as intimately dependent upon arithmetic (because of the association of the principal musical consonances with simple whole number ratios). It may be more illuminating to the modern reader to describe both Boethius's *Arithmetic* and the mathematical parts of his *Music* as dealing with number theory rather than arithmetic as we generally know it, for calculation

the History and Philosophy of the Mathematical Sciences, 2 vols. (London: Routledge, 1994), vol. 1, pp. 167–285.

finds virtually no place there. Instead we are treated especially to disquisitions on the properties of whole numbers – even, odd, prime, figurate (e.g., square), and so on – and those of numerical ratios. To some of the purely mathematical features we shall refer later, but here we must emphasize the cosmological and epistemological resonances of the treatises, which made them admirable companions to Plato's works, especially his *Timaeus*, which was widely studied in twelfth-century Europe.

For Boethius, it was a firm precept that whoever neglected the quadrivium (the mathematical sciences of arithmetic, music, geometry, and astronomy) lost the whole teaching of philosophy, for these sciences guided us from matters of sense to the more certain things of intellect. Moreover, "All things whatsoever that have been constructed from the primeval nature of things seem to have been formed on the rationale of numbers, for this was the principal exemplar in the mind of the Creator,"[2] a sentiment well in accord with the assertion in the Wisdom of Solomon that "Thou hast ordered all things in number, weight, and measure."[3] The elements were bound together in a mathematically determined harmony, and the heavenly bodies moved in accord with a music of the spheres. Mankind's propensity for music also helped bolster the somewhat nebulous Pythagorean and Platonic idea that the human soul itself was a harmony. The effect of all this was to produce a picture of a mathematically determined universe, which the soul's nature rendered accessible to human reasoning. The details of this outlook were overwhelmed by the dominance of Aristotelianism from the thirteenth century onward, but the general emphasis on the necessity of mathematics for a proper understanding of the natural world still flourished among many important thinkers, who often cited Boethius in support of their orientation.

SEMIOTIC CONSIDERATIONS

Mathematics, both pure and applied, constantly employs signs and symbols. In all ages, but perhaps especially the Middle Ages and the twentieth century, much of the community of mathematicians and mathematical controversialists has been very conscious of the fact, although not often to the extent of concurring with Bertrand Russell's notorious definition of the subject as that "in which we never know what we are talking about, nor whether what we are saying is true."[4] And even when contemplating the work of less reflective practitioners, we often get a firmer grasp of what essentially was going on by thinking semiotically. A few general reflections are thus in order.

[2] Boethius, *Arithmetica*, I.2, in Boèce, *Institution arithmétique*, ed. and trans. Jean-Yves Guillaumin (Paris: Les Belles Lettres, 1995), p. 11.

[3] Wisd. of Sol. 11:21.

[4] Bertrand Russell, *The Collected Papers of Bertrand Russell* (The McMaster University Editions, 36 vols.) (London: Allen and Unwin, 1983–), vol. 2, p. 366.

Boethius may be read as asserting that arithmetic was the language of nature (a role that Kepler and Galileo later assigned to geometry), and it also provided for him (and for informed common sense) at least part of the language of geometry. Boethius grounded this view in ontological priority:

> If you take away number, from where are triangle and quadrangle or whatever is considered in geometry, which are all denominatives of numbers? But if you remove quadrangle and triangle, and all geometry is consumed, three and four and the other names of numbers will not perish. Again, when I speak of any geometric form, a name of numbers is immediately bound up with it; when I speak of numbers, I have not yet named any geometric form.[5]

Numbers acted as symbols, but also needed to be symbolized by numerals themselves, either written, spoken, or otherwise concretely expressed. In like manner, diagrams symbolized geometrical objects, but less satisfactorily, since they could not have the exactness that pure geometry demanded (for instance, how does one physically draw a line of "breadthless length," as Euclid defines it?) or the generality that geometrical theorems characteristically possessed by speaking, for instance, of *any* triangle of a certain type. Also, diagrams themselves still had to be described by means of words, letters, and other means.

An especially significant word in the passage quoted from Boethius was "denominatives" (*denominativa*), for this term and its cognates played an important role in medieval mathematical discussions. Although the term appeared in both rhetoric and logic, with its own nuances in each case, in mathematics denomination is best described as the giving of a meaningful name to an object, a name like Humpty Dumpty ("my name means the shape I am") as opposed to Alice ("with a name like yours, you might be any shape almost").[6] Discussions in terms of denomination reflected, at least in the late Middle Ages, a concern with the logical structure of mathematics (see Ashworth, Chapter 22, this volume), and arose partly, but only partly, from the deep divide between the discrete and the continuous, something that is easily obscured from the modern reader by later radical extensions of the number concept.

For classical Greece and the Middle Ages, numbers (the proper subject matter of arithmetic) were, strictly speaking, simply collections of indivisible units – corresponding to our positive integers. By contrast, geometry was the science of continuous quantity or, more narrowly, spatially extended continuous quantity, and a hallmark of such quantities was that they were infinitely divisible. This was to be interpreted in the potential sense, meaning that (in thought at least) they could always be divided into smaller and smaller

[5] Boethius, *Arithmetica*, I.1, p. 9.

[6] Lewis Carroll, *Through the Looking Glass* (New York: M. F. Mansfield and A. Wessels, 1899), chap. 6.

parts without end. This incursion of the infinite meant that they could be very slippery to deal with, especially since they did not have any, so to speak, built-in markers indicating their size. For example, given a number, we can immediately say that it is 46 rather than 45 or 52, but in order to express the size of one straight line we need to compare it with another, and for practical and some theoretical purposes this was done by means of units, such as feet and inches, so that we could, for instance, say that a line is six feet long. Philosophically, Aristotle had assimilated this approach to the counting of discrete collections by saying that we treated units as if they were indivisible. In this way, numbers could be used as part of a language for the description of lines, weights, speeds, and so forth. Also, geometrical figures themselves could perform important semiotic functions, as when in the fourteenth century Nicole Oresme used graphic-like diagrams to symbolize variations of speeds and qualities. (For more details, see Laird, Chapter 17, this volume.)

FROM BOETHIUS TO THE TWELFTH-CENTURY RENAISSANCE

Although Boethius did not deal much with calculation, this activity is necessary for any remotely civilized society, and even plays a large role in the paradigmatic geometric activity of measuring fields. We should thus expect it to figure prominently in the early Middle Ages. Scraps of literary evidence do indeed show that calculation then reached levels of sophistication that counterbalance the banality of most other mathematical texts of the period. We may take as an example finger-reckoning,[7] a technique that the early Middle Ages received from the Romans. To us, this is almost a sign of mathematical illiteracy, but, as a text by the Venerable Bede (ca. 673–735) in particular makes clear, the medieval version was far more intricate. By using different flexings of joints and differing relations of index finger and thumb, one could readily represent numbers up to 9,999. From Roman times, additions were performed with finger-reckoning, which could also aid in multiplication, although it was not clear that the latter development had taken place by Bede's time. Just like semaphore, it embodied a sign language, and, much more clearly than Roman notation, this language employed a decimal place-value principle, with different sets of fingers representing units, tens, hundreds, and thousands.

More familiar to the modern reader, but often misconstrued as to its form, was the medieval abacus. We tend to think of these instruments as comprising beads strung on wires, but such were typically found only in the

7 On this and related matters, see Karl Menninger, *Number Words and Number Symbols: A Cultural History of Numbers*, trans. Paul Broneer (Cambridge, Mass.: MIT Press, 1969).

East and in Russia. In the West, the counters were pebbles or scratch marks in the sand, or artificial disks on specially constructed boards (sometimes spheres on grooved boards). A decimal place-value system was again strongly in evidence, with different columns being used to represent units, tens, hundreds, and so on. Details vary, but in a typical Roman abacus, the "vertical" columns would be divided by a horizontal bar. On each line, there was one counter above the bar whose potential value was five times that of each of the four counters below the bar. Different positionings of the counters then readily represented in each column values from zero to nine, which, depending on the column, were interpreted as the number of units, tens, hundreds, and so on in the whole given number. By moving counters around, a skilled operator could swiftly perform addition, subtraction, and (somewhat less swiftly) multiplication and division.[8]

Such abaci presumably continued to be used in the early Middle Ages, but the literary evidence is sparse, and ironically the first substantial testimony concerns an idiosyncratic instrument whose conception probably rested on a misunderstanding. It is associated with one of the few names regularly mentioned in histories of early-medieval mathematics, Gerbert of Aurillac,[9] who from 999 until his death in 1003 would reign as Pope Sylvester II. Perhaps his spanning of the millennium as well as his reputation for learning helped foster his fame as a magician! In any event, Gerbert's abacus (if he was really its inventor) was characterized by its having at any time no more than one counter on each of its columns. However, the counters were individualized by having marked on them symbols for the numbers from one to nine, written in characters known as ghubar or West Arabic numerals. Gerbert probably became acquainted with these numerals during time that he spent in Spain. In form, they are ancestors of our own "Arabic" numerals, and when they are positioned on the vertical columns of an abacus, a number can be read off horizontally, with an empty column being interpreted as zero for that power of 10. However, for calculatory purposes, they were not as efficient as the more usual system of unmarked counters, whose value depended upon position alone, since they demanded continually changing one counter for another with a different numeral marked on it.

THE TWELFTH CENTURY

Gerbert's level of mathematical sophistication was, despite his renown, rather low, but it was better than that of most of his contemporaries, and few surpassed him in Christian Europe until the twelfth century. By then there

8 For examples of the use of the abacus, see, for instance, Mahoney, "Mathematics," pp. 147–8. For the use of a later-medieval form of abacus ("reckoning on the lines"), see Menninger, *Number Words and Number Symbols*, pp. 349–62.

9 On Gerbert, see, for example, Dirk J. Struik, "Gerbert," *Dictionary of Scientific Biography*, vol. 5, 364–6.

were so many intellectual stirrings in different fields that it has become commonplace to speak of a renaissance of learning. Much of the revival had a strongly Platonic flavor, and the large part of Plato's *Timaeus* that was available in Calcidius's Latin translation came in for close scrutiny and provided a strong case for the necessity of mathematics in understanding nature. The relevant mathematics was of the Boethian type discussed earlier, and this meant that his *Arithmetic* and *Music* were much studied. Also, his theological works gave insight into the axiomatic method, in which the desired conclusions were reached by deduction from a relatively small number of first principles.[10] (This was before relevant geometrical works became available in adequate form.) But with the new intellectual vitality there came a justified feeling of inferiority with respect to the neighboring Arabic culture. Several scholars traveled in Spain and other Muslim lands, and many translations were made – mainly from Arabic, even if the original work had been Greek, but also some directly from the Greek (see Burnett, Chapters 14 and 15, this volume).

Mathematics as well as other subjects benefited from this movement, and we may mention two authors in particular. The first is Euclid. As we saw, only small fragments of a Boethian translation of the *Elements* were extant in the earlier Middle Ages, but now the complete work became available in a bewildering array of versions, including straight translation, précis, and elaboration. Traditionally three versions were ascribed to the learned and much-traveled Englishman Adelard of Bath, and in a seminal article a modern commentator labeled them Adelard I, II, and III.[11] Adelard I was a straight translation from the Arabic and Adelard III an elaboration on Euclid's text. Adelard II, the most used in the Middle Ages, was more of a précis, with much abbreviated proofs; its modern editors regard it as being the work not of Adelard but of Robert of Chester,[12] who, together with Hermann of Carinthia (who himself produced a Latin version of the *Elements*), also translated the Qur'ān into Latin. In the thirteenth century, Adelard II's enunciations were used in a very substantial version of the *Elements* (including many of his own comments) by the notable mathematician Campanus of Novara. This became the standard text of Euclid for the late Middle Ages, and in the sixteenth century it still vied with new versions made directly from the Greek.[13]

[10] Compare Mechthild Dreyer, *Nikolaus von Amiens: Ars fidei catholicae – Ein Beispielwerk axiomatischer Methode* (Beiträge zur Geschichte der Philosophie und Theologie des Mittelalters, Neue Folge, 37) (Münster: Aschendorff, 1993); and Gillian R. Evans, "Boethian and Euclidean Axiomatic Method in the Theology of the Later Twelfth Century: The *Regulae Theologicae* and the *De Arte Fidei Catholicae*," *Archives Internationales d'Histoire des Sciences*, 30 (1980), 36–52.

[11] Marshall Clagett, "The Medieval Latin Translations from the Arabic of the Elements of Euclid, with Special Emphasis on the Versions of Adelard of Bath," *Isis*, 44 (1953), 16–42.

[12] Hubert L. L. Busard and Menso Folkerts, eds., *Robert of Chester's (?) Redaction of Euclid's Elements: The So-Called Adelard II Version* (Basel: Birkhäuser, 1992).

[13] A translation directly from Greek was made in the twelfth century, but it never acquired much influence. See John E. Murdoch, "Euclides Graeco-Latinus: A Hitherto Unknown Medieval Latin

The second set of noteworthy translation events concerns Muhammad ibn Mūsā al-Khwārizmī, a ninth-century writer who did much to promote the acceptance of Indian numerals (constructed on the same principle as our own) among the Arabs and later gave his name to the techniques known in the Latin West as algorismus. His original Arabic work, *Book on Indian Reckoning*, is lost, but the Latin versions[14] were important factors in getting "Hindu-Arabic" numerals and calculations made with them accepted in Christian Europe. The form of the numerals was basically that of the ghubar numerals that Gerbert had used, but importantly they now included a round symbol for zero instead of an empty space, and they were no longer tied to the columns of an abacus. Al-Khwārizmī also provided rules for addition, subtraction, multiplication, and division. The operations, which involved frequent erasures and changes of one numeral into another, were often performed with figures scratched in sand. This art of algorism was made more readily available to the West in a work by the thirteenth-century writer Johannes de Sacrobosco[15] that became a prescribed text in several universities, institutions that began to flourish at about the same time (see Shank, Chapter 8, this volume). It was potentially a very useful introduction to the art, but in its pristine form (shorn of elaborations and commentaries) it had rather little numerical exemplification. Also, as so often happens, we cannot say how much influence it exerted on students who lacked prior mathematical inclinations.

Al-Khwārizmī was also the author of another significant work, which became available in three separate Latin translations in the twelfth century. This was the *Kitāb al-jabr w'al-muqābalah*, and once again a name was appropriated for use in Latin, and later in English, namely algebra. This work, whose title may be translated as *Book of Restoration and Opposition*, used ordinary language and numerical signs to formulate and solve problems without employing the special symbols that we regard as so characteristic of algebra. We may surmise that this subject had less general influence than algorism and was instead regarded as a matter for those more mathematically accomplished.

Translation of the *Elements* Made Directly from the Greek," *Harvard Studies in Classical Philology*, 71 (1967), 249–302; and H. L. L. Busard, ed., *The Mediaeval Latin Translation of Euclid's Elements Made Directly from the Greek* (Boethius: Texts and Essays on the History of the Exact Sciences, 15) (Wiesbaden: Franz Steiner Verlag, 1987).

14 Four (none a direct translation) are edited in Muhammad ibn Mūsā al-Khwārizmī, *Le Calcul Indien (Algorismus)*, ed. André Allard (Paris: Blanchard; Brussels: Société des Études Classiques, 1992).

15 The most useful edition is contained in F. Saaby Pedersen, ed., *Petri Philomenae de Dacia et Petri de S. Audomaro Opera Quadrivialia*, 2 vols. (Copenhagen: Gad, 1983), vol. 1, pp. 165–201. Parts are translated into English in Edward Grant, ed., *A Source Book in Medieval Science* (Cambridge, Mass.: Harvard University Press, 1974), pp. 94–101. A Middle English version may be found in Robert Steele, *The Earliest Arithmetics in English* (Early English Text Society, Extra Series, 118) (London: Humphrey Milford, Oxford University Press, 1922), pp. 33–51. On the limited evidence about Sacrobosco himself, see Olaf Pedersen, "In Quest of Sacrobosco," *Journal for the History of Astronomy*, 16 (1985), 175–221.

DOING MATHEMATICS: LEONARDO OF PISA

In continuing our account of medieval mathematics, we should pay regard to a bifurcation of styles. There is often a temptation to assert that medieval writers were more interested in talking and thinking *about* mathematics than actually doing it and that they veered toward philosophy of mathematics rather than actual mathematical practice. There is quite a lot of truth in this view, and the former type of mathematics will be the concern of the next section. However, the view is only partial, and this section will consider evidence (somewhat Italocentric) for actual mathematical practice. We start with a man who has been regarded as the greatest of medieval mathematicians.

Leonardo Fibonacci of Pisa came from the Bonacci family, well established in Pisa for many years before his birth around 1170. When his father was appointed secretary or chancellor in charge of the Pisan custom-house at the trading port of Bugia (Bougie) in present-day Algeria, he had the young Leonardo join him and arranged for him to be instructed in mathematics for a short time with a view to a future (presumably) mercantile career. Leonardo later recalled this time with palpable excitement:

> There, when introduced by wonderful teaching in the art through the nine figures of the Indians, knowledge of this art above other ones so much pleased me, and I so much believed in it, that I learned through much study and the practice of disputation what was studied in it, with its various methods, in Egypt, Greece, Sicily, and Provence, to which places I afterwards traveled in the course of business. But all this, both algorism and Pythagoras's arches, I reckoned as an error in comparison with the Indians' method. Wherefore, binding myself more strictly to this method, studying more carefully in it, adding some things from my own thinking, and introducing also some things from the subtleties of Euclid's art of geometry, I labored to compose as intelligibly as I could the main features of this book.[16]

This passage comes from the introduction to the "second edition" (1228) of Leonardo's *Liber abbaci*, probably his most influential work, which was dedicated to the famous sage and putative magician Michael Scot, who practiced as court astrologer to the Emperor Frederick II. The title of the work is confusing, for "abbacus" (spelled with two *b*s) has nothing to do with the calculatory device that we have already discussed. The latter "abacus" (spelled with one *b*) is almost certainly what Leonardo refers to as "Pythagoras's arches" (*arcus Pictagore*), for that description was regularly applied to the semicircular figures placed at the top of the columns in Gerbert's and other

[16] My translation from the text in Richard E. Grimm, "The Autobiography of Leonardo Pisano," *Fibonacci Quarterly*, 11 (1973), 99–104.

abaci. Also disparaged was algorism, which seems surprising, given that the name was usually applied to calculations with Hindu-Arabic numerals in the tradition of the Latin versions of al-Khwārizmī. A modern scholar has put the difference between the two traditions as follows:

> The algorisms [that is, books on algorism] teach the Arabic "dust-board" method of calculation which requires the repeated erasure and shifting of numerals as the calculation is performed. They are also comparatively short treatises, confined to outlining the basic methods with few or no practical examples. The abbacus books, in contrast, use more modern methods of calculation in which all the digits are retained and laid out in a distinctive pattern, and also provide hundreds of practical examples and problems which illustrate these methods.[17]

The characteristic milieu for abbacus arithmetic was an Italian mercantile community. After Leonardo, most of the treatises were written in Italian, their authors often known as *maestri d'abbaco*. Rules for performing multiplication quickly by means of pen and ink, without the need for frequent erasures, were laid down. Skills were often facilitated by proceeding according to certain geometrical patterns, such as those of crosses, bells, and cups. For instance, to borrow a simple example from a modern commentator,[18] multiplying 432 by 543 by *cross*-multiplication uses the following figuration:

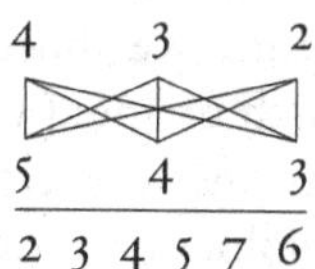

The result on the bottom line comes from the following sequence of multiplications along the vertical and diagonal lines:

$2 \times 3 = 6$; write 6, no carry
$(2 \times 4) + (3 \times 3) = 17$; write 7, carry 1
$1 + (2 \times 5) + (3 \times 4) + (4 \times 3) = 35$; write 5, carry 3
$3 + (3 \times 5) + (4 \times 4) = 34$; write 4, carry 3
$3 + (4 \times 5) = 23$; write 23

Leonardo himself also gave much attention to problems of an algebraic nature, although, like that of his Arabic predecessors, his algebra looks very different from ours and was for the most part expressed in ordinary language. The following is the procedure for dealing with one class of what we would

[17] Warren Van Egmond, "Abbacus Arithmetic," in Grattan-Guinness, *Companion Encyclopedia of the History and Philosophy of the Mathematical Sciences*, vol. 1, pp. 200–9 at p. 201.

[18] Ibid., pp. 203–4. Several other examples of such techniques may also be found in this article.

call quadratic equations; the particular example would now be symbolized as $x^2 + 10x = 39$, where the problem in the end is to find x^2 rather than x:

> When . . . you wish to find the quantity of the amount (*quantitatem census*) which with given roots is equal to a given number, you should proceed thus: Take the square of half of the roots, and add it to the given number, and take the root of what comes. From this take away the number of half of the roots, and what remains will be the root of the sought amount. For example, the amount (*census*) and ten roots are equal to 39, and so a half out of the roots is 5, which multiplied by themselves make 25, which added to 39 make 64. If from the root of these, which is 8, there be taken away half of the roots, namely 5, there will remain 3 as the root of the sought number of the amount, wherefore the amount is 9.[19]

Although well adapted to actual problem solving and conveying a precise meaning with regard to the principal matter at hand, this passage, like others in his work, can seem very confusing to the modern reader (the main reason for quoting the text here) and was probably correspondingly difficult for the medieval student to assimilate.

Despite the greatness of Leonardo's achievement, his original work has received little intensive scholarly analysis, and accounts of it are often given in modern mathematical notation, with the attendant danger of an understanding marred by anachronism. We must accordingly be selective and sketchy. Leonardo was very much oriented toward solving problems; some were traditional, others were of his own creation or proposed to him by other people. One of these was John of Palermo, mathematician at the court of Emperor Frederick II, a vibrant intellectual center with which Leonardo maintained close contact. One of Leonardo's problems that became famous was that of two rabbits placed in a walled enclosure. Given certain assumptions about breeding habits, it was asked how many rabbits the enclosure would contain at the end of a year or any arbitrary period of time. This gave rise to a series of numbers in which each term (after the second) was the sum of its two immediate predecessors. Besides rendering 754 rabbits at the end of the year, this "Fibonacci series" (as it is now known) was later shown to have many other interesting properties. Series often found their place in Leonardo's work, and he has also been greatly admired for his versatility in solving so-called Diophantine equations (those that forbade surds or "irrational numbers" as solutions), which were often indeterminate (that is, not yielding a single or a determinate number of solutions).

[19] My translation from Guillaume Libri, *Histoire des sciences mathématiques en Italie depuis la renaissance des lettres jusqu'à la fin du 17ᵉ siècle*, 4 vols. (Paris: J. Renouard, 1838–1841; repr. Hildesheim: Olms, 1967), vol. 2, pp. 358–9. The preferred edition, Baldassarre Boncompagni, ed., *Scritti di Leonardo Pisano*, 2 vols. (Rome: Tipografia delle scienze matematiche e fisiche, 1857–1862), is less accessible.

Leonardo was undoubtedly a mathematician of originality, but of course he still depended upon other sources – to an extent that awaits further elucidation. For instance, besides his use of the work of al-Khwārizmī and Abū Kāmil, particular problems (but not, it appears, their methods of solution) can be shown to derive either directly or indirectly from al-Karajī, ʿUmar al-Khayyāmī, and Diophantus of Alexandria.[20] Leonardo's own algebraic influence, though small among academics, was great among more practical mathematicians, who characteristically wrote in the vernacular. Later in the Middle Ages, his work was mediated to other parts of Europe. In Germany, the subject flourished, and its practitioners became known as cossists, after the Italian *cosa* for "thing" or unknown quantity, and naturally they also dealt with other associated areas of mathematics, notably methods of calculation. In France, although activity was not as vigorous, it did result in the production in the late fifteenth century of a huge treatise (still not completely published) by Nicolas Chuquet, which included a substantial section on commercial arithmetic. However, the outstanding monument to the tradition came from the country of greatest activity in the form of the vast *Summa of Arithmetic, Geometry, Ratio, and Proportion* (1494) of Luca Pacioli, teacher, traveler, friar, and – fittingly – friend of Leonardo da Vinci and other practical luminaries of the Italian Renaissance.

CONSIDERING MATHEMATICS: JORDANUS DE NEMORE AND THE UNIVERSITIES

Archimedes was undoubtedly a practical man, but his extant treatises are models of the most austere, unembellished theory in mathematics. As the famous Greek biographer Plutarch said, it was "not possible to find in geometry more difficult and weighty questions treated in simpler and purer terms."[21] As such, to the medieval mind, his works called out for commentary and explication, and five huge volumes (bound as ten) have been devoted to the Latin *fortuna* of these and associated treatises.[22] A productive modern literary critic would not wish to be dubbed inactive, but just as he or she does not usually write novels or poetry, so the medieval Archimedean did not compose new mathematics in the Archimedean spirit but aimed at deeper understanding of the existing texts. A small portion of the latter were translated from the Arabic in the twelfth century, and the (almost) complete corpus by William of Moerbeke directly from the Greek later in

[20] Kurt Vogel, "Fibonacci, Leonardo, or Leonardo of Pisa," *Dictionary of Scientific Biography*, vol. 4, 603–13; and Roshdi Rashed, "Fibonacci et les mathématiques arabes," *Micrologus*, 2 (1994), 145–60.

[21] Plutarch, *Marcellus*, XVII.5, quoted in Ivo B. Thomas, *Selections Illustrating the History of Greek Mathematics* (London: Heinemann, 1939), p. 31.

[22] Marshall Clagett, *Archimedes in the Middle Ages* (vol. 1, Madison: University of Wisconsin Press, 1964; vols. 2–5, Philadelphia: American Philosophical Society, 1976–1984).

the thirteenth century. The aim was to probe inward rather than expand outward. The approach paralleled that adopted for other authoritative texts, such as the Bible and the works of Aristotle, and meshed well with the commentary and the disputation, common modes of teaching in the nascent universities (see Shank, Chapter 8, this volume).

However, some apparent exceptions to this general rule must be noted. In standard histories of mathematics, the name of Jordanus de Nemore is often linked with that of Leonardo of Pisa as another medieval mathematician of undoubted originality. Recently he has been compared to Melchisedec, the biblical priest and king who was "without father, without mother, without descent,"[23] but this comparison must apply to image more than reality. On the side of descent, Jordanus's treatises generated a respectably large number of manuscript copies, but they did not inspire further work in the same style so much as elaborations and explications in a spirit that accorded with Archimedes' works. On the side of ancestry, we cannot allow generation ex nihilo, as much on a priori grounds as detectability, of specific Archimedean and other influences, but the problem of the immediate matrix remains acute.

The difficulty is made more acute by the fact that the person of Jordanus remains very shadowy.[24] A similar situation obtains in the case of two other (probably contemporary) mathematicians, Johannes de Tinemue and Gerard of Brussels, each of whom (unlike Jordanus) is famous for a single work, Johannes for a geometrical treatise closely related to Archimedes' *On the sphere and the cylinder* and Gerard for a highly mathematical treatment of the kinematics of extended bodies.[25] Many hypotheses have been proposed about the identities and careers of these men, but to me most of these seem to founder on the obstacle that the Latin cultural ambience at the time (probably late twelfth century) does not appear adequate to allow the speedy growth of works of such sophistication and tightness of traditional geometrical form, and so one is led to infer nourishment from outside. This would plausibly arise from their use of Greek or Arabic texts now no longer extant, and it may be that the works of Jordanus, John, and Gerard were nearer to translations than to independent compositions arising in late-twelfth- or early-thirteenth-century Europe, in much the same manner in which Boethius used Nicomachus of Gerasa.[26]

23 Heb. 8:3. The comparison was made independently by A. George Molland, "Ancestors of Physics," *History of Science*, 14 (1976), 54–75 at p. 64, reprinted in A. George Molland, *Mathematics and the Medieval Ancestry of Physics* (Aldershot: Variorum, 1995); and Barnabas B. Hughes, "Biographical Information on Jordanus de Nemore to Date," *Janus*, 62 (1975), 151–6.

24 Flimsy attempts to identify him with Jordanus of Saxony, the second master-general of the Dominicans, have now been almost universally abandoned.

25 Johannes's treatise, *De curvis superficiebus*, is edited with English translation in Clagett, *Archimedes in the Middle Ages*, vol. 1, pp. 439–557, and Gerard's *Liber de motu* in Clagett, *Archimedes in the Middle Ages*, vol. 5, pp. 1–151.

26 My own conjecture (not necessarily better than any of the others, and certainly not well corroborated) is that all three were Flemings who had (in the Near East) been in close contact with Greek texts, possibly in the aftermath of the Fourth Crusade in the years 1206–16, when Henry

The Jordanus corpus comprises several works that have the notable feature, consonant with the Latin Euclidean and Archimedean traditions, that they mostly appear in two versions, one pristine and the other explicated and elaborated, presumably by a writer or writers later than Jordanus himself. Jordanus is probably best known in the history of science for his works on statics, or, to use the medieval term, the theory of weights. These adopt a mathematical and axiomatic approach that is different from that of Archimedes and has somewhat anachronistically been compared to the principle of virtual velocities. Jordanus is also responsible for a large work on arithmetic in the Boethian spirit, but written in a more axiomatic style, and a sophisticated geometrical treatise known in its pristine version as *Book of the Philotechnist* (*Liber philotegni*) and in its elaborated form as *Book on triangles*.[27] It attempts to develop definitions of basic geometrical objects from philosophical notions of continuity, again not an approach that one would readily expect to see as an independent product of late-twelfth- or early-thirteenth-century Latin culture.

A similar situation holds with regard to Jordanus's treatise *On given numbers* (*De numeris datis*).[28] This is often described as a work on algebra, but it reads very differently from the algebraic writings of al-Khwārizmī and Leonardo of Pisa. Instead it focuses on showing that various problems formulated in general terms have determinate solutions. It parallels quite closely Euclid's *Data* (*Given Things*), and even the titles resonate. But whereas Jordanus's work concerned numbers (whole or fractional) and their relationships, Euclid treated the more complicated subject of geometrical objects and their interrelations. Similar motivations seem to lie behind both works, but, although Euclid's *Data* had been translated into Latin, I do not think that it could have been sufficiently assimilated to allow Jordanus to make the requisite transformation by his own efforts.

In my account, Jordanus, Johannes, and Gerard are themselves exemplars of a transmissory and commentatorial spirit rather than creators of new knowledge. Indeed they occasionally show signs that their grasp of their putative sources (while very good) is not completely firm. As noted, a similar urge toward the propagation and deeper understanding of existing knowledge is also characteristic of the medieval universities, but we should not suppose that these institutions were hotbeds of mathematical activity or that Archimedes and Jordanus were staple fare for the average undergraduate. Even the statutes do not aim that high, typically prescribing Boethius's *Arithmetic* and textbooks by Sacrobosco and making such demands as that

of Flanders was the Latin emperor of Constantinople, and would thus have prefigured another Fleming, William of Moerbeke, who in 1260 was translating from the Greek at Nicea. See Clagett, *Archimedes in the Middle Ages*, vol. 2, p. 4.

27 Clagett, *Archimedes in the Middle Ages*, vol. 5, pp. 143–477.

28 Jordanus de Nemore, *De numeris datis*, ed. and trans. Barnabas B. Hughes (Berkeley: University of California Press, 1981).

the first six books of Euclid's *Elements* be heard "adequately." Moreover, the works of many schoolmen strongly suggest that these requirements were interpreted very laxly. Nevertheless, the universities probably provided a congenial environment for small groups of enthusiasts to study mathematical texts in a more private fashion.[29] Also, even if, unlike in ancient Greece or the modern world, mathematics was not taken very seriously for its own sake, something more than a rudimentary knowledge of it was necessary in order for a university teacher to properly fulfill his functions. In theology and some other areas of study, important symbolic value was often attached to particular numbers and geometrical objects, but this rarely demanded much mathematical manipulation. More exacting were the works of Aristotle, which formed the basis of so much university education.

Aristotle had little time for numerology, but mathematics did infiltrate his thinking to an extent that is often unrecognized (especially when he is compared to Plato). First of all, from a structural point of view, geometry had provided Aristotle with a model of what he regarded as a properly constituted science, one in which all relevant conclusions could be deduced from a relatively small number of first principles (axioms, postulates, etc.). Accordingly, geometrical examples abound in his work on scientific methodology known as the *Posterior Analytics*. But besides providing a model, mathematics could to an extent dominate other sciences. Although Aristotle was concerned with maintaining the independence of the particular sciences, each retaining its own special methods, there were cases where subordination was legitimate, so that the superior science gave explanations of facts provided by the inferior science.[30] And often the superior science was more mathematical than the inferior. Thus optics was subordinate to geometry, and the study of the rainbow was subordinate to optics. This principle ceded considerable power to mathematics in sciences such as astronomy, optics, music, and mechanics (conceived as the science of machines) for providing explanations. In the Middle Ages, there were many attempts to extend the principle to other areas, notably motion and the quantification of qualities, and in this way medieval writers anticipated, though in a different form, the mathematical natural philosophies of the seventeenth century. (See later in this chapter and particularly Laird, Chapter 17, this volume.)

But even though mathematics could have this explanatory power, there were still many problems as to how exactly it related to the physical world. Aristotle did not acknowledge an independent realm of mathematical objects but held instead that the mathematician considered natural bodies, but in

[29] Compare A. G. Molland, "The Quadrivium in the Universities: Four Questions," *Miscellanea Mediaevalia*, 22 (1994), 66–78.

[30] See several studies by Steven J. Livesey, notably "William of Ockham, the Subalternate Sciences, and Aristotle's Theory of *Metabasis*," *British Journal for the History of Science*, 18 (1985), 127–45, and "The Oxford Calculatores, Quantification of Qualities and Aristotle's Prohibition of Metabasis," *Vivarium*, 24 (1986), 50–69.

abstraction from their sensible qualities. This meant that, although mathematical objects were rooted in the physical world, the mind still had a considerable role in constituting them, and often there seemed to be a lack of a precise match between the mathematician's reasoning and the state of the outside world. For instance, the geometer dealt with figures possessing an exactness and generality denied to their physical counterparts, so that Aristotle, like Plato, insisted that the geometer was talking not about his diagrams but about what they symbolized. Also, the geometer moved his figures around in a way that had no obvious physical counterpart, and his reasonings seemed to demand an infinitely extended space, whereas the Aristotelian world was finite, with nothing, not even space, existing beyond the outermost heaven. The resulting tensions allowed for different emphases on the role of mathematics in the understanding of the physical world. Some, like Averroes (Ibn Rushd) and Albertus Magnus, minimized its relevance and stressed its purely mental aspects. But others, such as Robert Grosseteste, Roger Bacon, and Thomas Bradwardine, actively sought means to bolster and extend the beneficent role of mathematics in natural philosophy.

RATIOS AND PROPORTIONS

A striking feature of medieval mathematics, especially in its theoretical genre, is the amount of discussion that was conducted in terms of ratios (*proportiones*) and their equalities or proportions (*proportionalitates*). This characteristic had strong ancient roots, notably in Euclid and Boethius, but to modern eyes is open to misinterpretation, for we tend immediately to identify ratios with fractions; that is, numbers in a more extended sense than was strictly allowable in the Middle Ages. For medieval writers, a ratio was emphatically not a number but a certain relation in respect to size between two quantities that may be susceptible to numerical description or denomination. In the case of ratios between (whole) numbers, this was relatively easy. For instance, the ratio of 6 to 3 is double (*dupla*) and that of 18 to 6 is triple (*tripla*); descriptions for more complicated cases could also be adapted from Boethius (such as the cumbersome expressions *triplex superquadripartiens quintas* for 19:5).

With continuous quantities, more serious problems arose because of incommensurability. To take the most famous example, it had been known since the time of the ancient Pythagoreans that the diameter of a square is incommensurable with its side; that is, there is no smaller line that fits an integral number of times into each of them. Thus the ratio of these two lines cannot be represented by the ratio of two integers, and hence is irrational. (In more modern terms, it is as $\sqrt{2}$ to 1, and $\sqrt{2}$ is an irrational number, or surd; that is, it cannot be expressed as a fraction in which both the numerator and denominator are whole numbers.) In Greek times, the discovery of incommensurability resulted in a sharp separation between arithmetic and geometry, so that, although numbers were used in geometrical arguments, geometry could certainly not be reduced to arithmetic. Book 5 of Euclid's *Elements* contains a much-admired general definition of the equality of ratios, which is usually ascribed to the earlier mathematician Eudoxus of Cnidos:

> Magnitudes are said to be in the same ratio, the first to the second and the third to the fourth, when, if any equimultiples whatever be taken of the first and third, and any multiples whatever of the second and fourth, the former multiples alike exceed, are alike equal to, or alike fall short of, the latter equimultiples respectively taken in corresponding order.[31]

In more modern terms, $a{:}b = c{:}d$ if and only if, for all positive integers m and n, $ma > nb$ when and only when $mc > nd$, $ma = nb$ when and only when $mc = nd$, and $ma < nb$ when and only when $mc < nd$. This admirably rigorous definition is rather cumbersome, and it is not surprising that it reached the Latin Middle Ages in garbled form. In the thirteenth century, Roger Bacon fulminated against its obscurity, and at first glance it seems that even Campanus of Novara completely misunderstood the force of the definition.[32] But the medieval semiotic style was not happy with such a stark definition of equality. It sought, possibly under Aristotelian influence, something that would be more informative about the size of the geometrical objects involved. If this were done, it would then be possible to define the equality of ratios in terms of their having the same denomination, a term whose general meaning has been discussed above[33] but that in this context has caused particular confusion by signifying both the giving of a meaningful name to an object and the name itself.

The denomination of ratios between commensurable quantities of any kind was relatively easy, for one could immediately adopt the principles used in arithmetic. By definition, this could not apply to incommensurable quantities, but in the Middle Ages there dawned a recognition that the situation was not altogether hopeless. This resulted from a development of syntactical relations used in the mathematical theory of music. It had long been agreed, in geometry as well as in music, that the result of combining the ratio of a to b with that of b to c, where a was greater than b and b greater than

[31] Thomas L. Heath, *The Thirteen Books of Euclid's Elements*, 2nd ed. (Cambridge: Cambridge University Press, 1926; repr. New York: Dover, 1956), vol. 2, p. 114.

[32] John E. Murdoch, "The Medieval Language of Proportions: Elements of the Interaction with Greek Foundations and the Development of New Mathematical Techniques," in *Scientific Change*, ed. Alistair C. Crombie (London: Heinemann, 1963), pp. 237–71. Despite his garbling of the definition, the rest of his version of Book V of the *Elements* shows that Campanus did acquire a good understanding of how the criterion actually operated; see A. G. Molland, "Campanus and Eudoxus; or, Trouble with Texts and Quantifiers," *Physis*, 25 (1983), 213–25, reprinted in Molland, *Mathematics and the Medieval Ancestry of Physics*.

[33] Section entitled "Semiotic Considerations," pp. 514–16.

c, was the ratio of a to c. To us it seems (as it did to some medieval writers) natural to interpret this act of composition as one of multiplication, for $a/b \times b/c = a/c$, but in the musical tradition it was regarded as addition, so that adding 3:2 to 4:3 gave rise to 2:1, not 17:6, and this does seem more appropriate when the ratios represent musical intervals. With hindsight, extensions of the principle to the multiplication and division of ratios by whole numbers readily suggest themselves, but historically this took longer to develop. In a passage that has often been overlooked, Robert Grosseteste in the early thirteenth century spoke, albeit somewhat circumspectly, of the ratio of the diagonal to the side of the square as being half of the double ratio (*medietas duple proportionis*),[34] but it was only in the fourteenth century that the new principle infiltrated mainstream mathematics to a significant degree. A key figure in effecting this was Thomas Bradwardine, a scholar and ecclesiastic of diverse talents, who in 1349 became archbishop of Canterbury and shortly afterward died of the plague. His early textbook on pure geometry, known as the *Geometria speculativa*, was appropriately described by the editor of its first printed version (1495) as "gathering together all the geometrical conclusions that are most needed by students of arts and the philosophy of Aristotle."[35] Of its four parts, one is devoted to ratios, proportion, and associated problems of measure, and denomination plays an explicit and crucial role. Bradwardine asserts that, whereas rational ratios are immediately denominated by number, an irrational ratio is only mediately (*mediate*) so denominated. He gives as an example the diagonal and side of a square, and says that in the same way "other species of this ratio receive denomination by number."[36]

Bradwardine also raised the topic in his *Tractatus de proportionibus* of 1328, in which he used the addition view of the composition of ratios to produce a simple, although empirically very vague, law relating speeds of motions to forces and resistances. This was then applied by successors in a variety of sophisticated, if not sophistical, arguments. (See Laird, Chapter 17, this volume, for more details.) Bradwardine did not go into great detail concerning his strategy for denominating irrational ratios, and for exemplification usually restricted himself to talk of halving ratios. It was left to the great French scholar Nicole Oresme later in the fourteenth century to develop the theory systematically and to assert that "an irrational ratio is said to be mediately denominated by some number when it is an aliquot part or aliquot parts of some rational ratio, or when it is commensurable with some

34 Robert Grosseteste, *Commentarius in Posteriorum Analyticorum Libros*, ed. Pietro Rossi (Unione Accademica Nazionale, Corpus Philosophorum Medii Aevi, Testi e Studi, 2) (Florence: Olschki, 1981), p. 120.

35 See A. George Molland, "An Examination of Bradwardine's Geometry," *Archive for History of Exact Sciences*, 19 (1978), 113–75 at p. 120, reprinted in Molland, *Mathematics and the Medieval Ancestry of Physics*.

36 Thomas Bradwardine, *Geometria Speculativa*, ed. A. George Molland (Stuttgart: Franz Steiner, 1989), pp. 88–9.

rational ratio, which is the same."[37] This is from Oresme's treatise *On ratios of ratios*, in which he self-consciously treats ratios as continuous quantities that had ratios between them. This immediately leads to the supposition that one ratio can be incommensurable with another, and, although he says that he cannot prove it, Oresme thinks that there are irrational ratios that are incommensurable with any rational one and thus are not susceptible to numerical denomination even in its newly extended form.

CONCLUSION

We may conclude by drawing together two leading themes. One that has raised its head at various points (for instance, with the abacus and algorism) is that of symbolism and the number concept. Many modern expositors have used exponential notation to reveal the "meaning" of late-medieval treatments of ratios, so that, for instance, two-thirds of the ratio 5:4 becomes $(5/4)^{2/3}$. In a sense this is accurate, but it replaces medieval semiotics with a modern one and so, to the extent to which its language is vital to the nature of mathematics, distorts the historical picture. But it also prompts us to see a certain chafing at the restrictiveness of the number concept, which was manifest in other contexts. Fractions are an obvious example. They could be, and were, treated as successive operations of division and multiplication performed on continuous quantities, or on unity treated as if it were a continuous quantity. But they did have numerical symbols of their own, and there was a tendency, especially in the more practical traditions, to treat them as if they were numbers in their own right. A similar situation arose in algebra, where there was a similar temptation to treat irrational, negative, and even imaginary solutions of equations as if they were not just symbols but again numbers in their own right. However, the fuller maturation of the process whereby mathematical symbols helped extend the range of the symbolized is more properly the subject of later volumes of this series.

The second theme, which has appeared spasmodically throughout this chapter, is that of usefulness.[38] This is a teleological concept, and there were a variety of purposes for which mathematics could seem fitted. On the social, economic, and political fronts, lack of evidence and of specialized studies necessitates vagueness and conjecture. Numeracy, although to a lesser extent than literacy, certainly seems to have been an increasingly valuable qualification for the upwardly mobile, and had obvious potential applications

[37] Nicole Oresme, *De proportionibus proportionum* and *Ad pauca respicientes*, ed. Edward Grant (Madison: University of Wisconsin Press, 1966), pp. 160–1.

[38] For two stimulating but not unproblematic discussions, see Alexander Murray, *Reason and Society in the Middle Ages* (Oxford: Clarendon Press, 1978), chaps. 6–8, and Alfred W. Crosby, *The Measure of Reality: Quantification and Western Society, 1250–1600* (Cambridge: Cambridge University Press, 1997).

in trading, accountancy, and taxation – although it is difficult to assess how much these fields were affected by new mathematical developments. Practical geometry continued to be used for surveying and other ends but, apart from its treatises acquiring a more theoretical tinge, seems to have undergone scant medieval development – with one important exception.[39] This is where it extended into the construction and use of astronomical instruments. Here we approach the better-documented areas of the applications of mathematics in other sciences. Most discussion of this has appropriately been left to the chapters on the relevant sciences – optics, astronomy, and motion – but here we have touched on some of the more general intellectual uses of mathematics. Roger Bacon would probably have continued to complain about the neglect of the subject during the fourteenth and fifteenth centuries. Nevertheless, the ground was being well prepared for the many variegated and fruitful eruptions of both pure and applied mathematics that occurred during the Renaissance and the Scientific Revolution.

[39] Compare Stephen K. Victor, *Practical Geometry in the High Middle Ages: Artis cuiuslibet consummatio and the Pratike de geometrie* (Memoirs of the American Philosophical Society, 134) (Philadelphia: American Philosophical Society, 1979); and Hervé l'Huillier, "Practical Geometry in the Middle Ages and the Renaissance," in Grattan-Guinness, *Companion Encyclopedia of the History and Philosophy of the Mathematical Sciences*, vol. 1, pp. 185–91.

22

LOGIC

E. Jennifer Ashworth

Medieval logic is crucial to the understanding of medieval science for several reasons.[1] At the practical level, every educated person was trained in logic, which provided not only a technical vocabulary and techniques of analysis that permeate philosophical, scientific, and theological writing but also the training necessary for participation in the disputations that were a central feature of medieval instruction. At the theoretical level, medieval logicians made several contributions. First, they discussed logic itself, its status as a science, its relation to other sciences, and the nature of its objects. Here it is important to note that medieval thinkers took a science (*scientia*) to be an organized body of certain knowledge that might include theology, logic, and grammar as well as mathematics and physics. Second, they discussed the nature of a demonstrative science and scientific method in general. Third, they provided a semantics that allows one to sort out the ontological commitments carried by nouns and adjectives. The discussion of connotative terms is particularly important here since it allowed logicians to analyze such terms as "motion" without postulating the existence of anything other than ordinary objects and their qualities. Fourth, they provided particular logical strategies that allow one to sort out the truth-conditions for scientific claims. Particularly important here are supposition theory, the distinction between compounded and divided senses, and the analysis

[1] For full information about medieval logic, see Catarina Dutilh Novaes, *Formalizing Medieval Logical Theories: Suppositio, Consequentiae and Obligationes* (Dordrecht: Springer, 2007); Dov M. Gabbay and John Woods, eds., *Handbook of the History of Logic 2: Mediaeval and Renaissance Logic* (Amsterdam: Elsevier/North-Holland, 2008); Klaus Jacobi, ed., *Argumentationstheorie: Scholastische Forschungen zu den logischen und semantischen Regeln korrekten Folgerns* (Leiden: Brill, 1993); Norman Kretzmann, Anthony Kenny, and Jan Pinborg, eds., *The Cambridge History of Later Medieval Philosophy* (Cambridge: Cambridge University Press, 1982); Norman Kretzmann and Eleonore Stump, trans., *The Cambridge Translations of Medieval Philosophical Texts, vol. 1: Logic and the Philosophy of Language* (Cambridge: Cambridge University Press, 1988); and Mikko Yrjönsuuri, ed., *Medieval Formal Logic: Obligations, Insolubles and Consequences* (Dordrecht: Kluwer Academic Publishers, 2001).

of propositions containing such syncategorematic terms as "begins" and "ceases."

BACKGROUND: TEXTS AND INSTITUTIONS

Medieval logic has to be seen against the background of the split of the Roman Empire into the Latin-speaking, Catholic West and the Greek-speaking, Orthodox East. Despite the efforts of Boethius to translate the logical works of Aristotle into Latin, and to supplement them with commentaries and monographs, only a handful of logic texts were available in Latin after about 550, and, in any case, there were few institutions to teach from them. The real story of medieval logic begins with Gerbert of Aurillac, who taught in the cathedral school at Reims from about 972 before becoming a bishop and finally pope.[2] He taught the *Logica vetus* (the "Old Logic"; i.e., Porphyry's *Isagoge* and Aristotle's *Categories* and *On Interpretation*), Cicero's *Topics*, and a good deal of Boethius. As Western Europe became more settled, educational institutions began to flourish. Better monastic schools and cathedral schools developed, and in the twelfth century, especially in Paris, schools grew up around individual masters such as Peter Abelard and Adam of Balsham (*Parvipontanus*), among others. They used the same basic curriculum as Gerbert had, but added the *Book of Six Principles* (*Liber sex principiorum*; i.e., the last six categories, about which Aristotle had less to say), attributed to Gilbert of Poitiers. At the same time, the philosophical study of grammar became increasingly important, with a focus on the *Grammatical institutions* of Priscian.[3]

Three important developments took place from about 1150 on: the foundation of universities and the *studia* of the new religious orders; the recovery of the rest of Aristotle's logical works, along with other texts; and the development of new areas of logic. By the beginning of the thirteenth century, the universities at Oxford, Paris, and Bologna were taking shape as organized institutions.[4] Gradually, other universities were founded, along with the *studia* of the new religious orders, especially the Dominicans and the

[2] For the early period, see John Marenbon, *Early Medieval Philosophy (480–1150): An Introduction* (London: Routledge, 1983).

[3] For twelfth-century grammar and logic, see Peter Dronke, ed., *A History of Twelfth-Century Philosophy* (Cambridge: Cambridge University Press, 1988).

[4] For logic teaching at the University of Oxford, see P. Osmund Lewry, "Grammar, Logic and Rhetoric 1220–1320," in *The History of the University of Oxford*, vol. 1: *The Early Oxford Schools*, ed. J. I. Catto (Oxford: Clarendon Press, 1984), pp. 401–33; and E. Jennifer Ashworth and Paul Vincent Spade, "Logic in Late Medieval Oxford," in *The History of the University of Oxford*, vol. 2: *Late Medieval Oxford*, ed. J. I. Catto and Ralph Evans (Oxford: Clarendon Press, 1992), pp. 35–64. For Bologna, see Dino Buzzetti, Maurizio Ferriani, and Andrea Tabarroni, eds., *L'Insegnamento della logica a Bologna nel XIV secolo* (Studi e Memorie per la Storia dell'Università di Bologna, Nuova Serie, 8) (Bologna: L'Istituto per la Storia dell'Università, 1992).

Franciscans, and all these institutions offered training in the liberal arts.[5] Two things should be noted. First, a degree in the arts was required for all advanced (graduate) study in theology, medicine, and law (though, especially in Italy, medicine and law, unlike theology, could be pursued at the undergraduate level). Second, logic formed a very large part of the arts curriculum, especially during the first two years. One reason for its importance was the large place given to disputation at all levels of university teaching; logic clearly offers training in the techniques of debate and inculcates quickness in analyzing and responding to arguments. However, university teaching also involved the close study of texts, especially those of Aristotle, and many commentaries were written by medieval logicians.

Aristotle's *Topics* and *Sophistical Refutations* were known by the 1130s, and the entire *Logica nova* (the "New Logic"), including the *Prior Analytics* and *Posterior Analytics*, was known by 1159, when John of Salisbury referred to all four works in his *Metalogicon*. In the second half of the twelfth century, people began to translate Arabic logic, including writings by Avicenna (Ibn Sīnā), Farabi, and Ghazali. In the 1230s, several logic commentaries by Averroes (Ibn Rushd) were translated, though they were less successful than the Arabic works translated earlier. Some Greek commentators were also translated. For instance, Themistius's *Paraphrase of the Posterior Analytics* and some extracts from Philoponus on the *Posterior Analytics* became known in the late twelfth century. The important thing about all these texts is not just their advanced content but the fact that they provide a full logic curriculum for an organized institution.

Although the writings of Aristotle were always central to the logic curriculum, there were matters that he did not discuss. This left room for a considerable number of new developments, all of which have roots in the second half of the twelfth century. The most prominent is the so-called *Logica moderna* or *modernorum* (a label coined by twentieth-century authors). Also known as terminist logic, this includes supposition theory and its ramifications. Treatises on supposition theory deal with the type and range of reference that subject and predicate terms have in propositions, and they have as a corollary the treatises on syncategoremata, which deal with all the other terms in propositions, including "every," "not," "and," "except," and so on. The material dealt with in treatises on syncategoremata was also dealt with in treatises on sophismata, but in the fourteenth century much of it also appeared in treatises on the Proofs of Terms. Three other important developments are found in treatises on insolubles or semantic paradoxes, obligations or the rules one is obliged to follow in a certain kind of disputation, and consequences or valid inferences. The latter only became the subject of separate treatises around 1300.

[5] See Alfonso Maierù, *University Training in Medieval Europe*, trans. and ed. D. N. Pryds (Leiden: Brill, 1994).

Although supposition theory constitutes the most important new development, it was not always privileged. Paris in the second half of the thirteenth century and Bologna and Erfurt at the beginning of the fourteenth were centers of modist logic and tended not to use supposition theory.[6] Modist logicians were more interested in the sense than the reference of terms, and they made much use of the postulated relationships between the modes of being of things (*modi essendi*), the modes of understanding in the mind (*modi intelligendi*), and the modes of signifying of words (*modi significandi*). They were also characterized by their lengthy discussions of the rational nature of logic and its relation to second intentions, or higher-order concepts. Another rival to supposition theory as a tool of analysis was provided by treatises on Proofs of Terms, a title covering a variety of analytic techniques, which were particularly important in the late fourteenth century.

Another new form of writing was the comprehensive textbook. At least six survive from the thirteenth century, including those by William of Sherwood, Peter of Spain, and Roger Bacon.[7] In the fourteenth century, we find those by William of Ockham, Jean Buridan, and Albert of Saxony.[8] Some universities, especially Oxford and Cambridge, preferred to use loose collections of brief treatises on various topics; a good example of such a collection is the *Logica parva* of Paul of Venice, who studied at Oxford.[9]

Nearly all of these writings on logic show traces of the important disputes between nominalists and realists about the status of common natures (or universals, though the name "universal" was often restricted to concepts). Nominalists held that common natures, such as humanity and horse-ness, existed only as concepts in the mind, whereas realists held that they had a special kind of existence, different from that of concepts and from that of individual humans or horses. However, it would be a mistake to suppose that any particular part of logical theory could be equated with nominalism, realism, or any other "ism." Such realists as Walter Burley and John Wyclif employed supposition theory; the nominalist Ockham was happy to accept

[6] For modism, see Michael A. Covington, *Syntactic Theory in the High Middle Ages* (Cambridge: Cambridge University Press, 1984); Costantino Marmo, *Semiotica e linguaggio nella scolastica: Parigi, Bologna, Erfurt 1270–1330. La semiotica dei Modisti* (Rome: Istituto Storico Italiano per il Medio Evo, 1994); and Irène Rosier, *La grammaire spéculative des Modistes* (Lille: Presses Universitaires de Lille, 1983).

[7] For William of Sherwood, see Norman Kretzmann, trans., *William of Sherwood's Introduction to Logic* (Minneapolis: University of Minnesota Press, 1966). For selections from Peter of Spain, see Kretzmann and Stump, *Logic and the Philosophy of Language*, passim.

[8] Some of Ockham's *Summa logicae* has been translated. See Michael J. Loux, trans., *Ockham's Theory of Terms: Part I of the Summa Logicae* (Notre Dame, Ind.: University of Notre Dame Press, 1974); and Alfred J. Freddoso and Henry Schuurman, trans., *Ockham's Theory of Propositions: Part II of the Summa Logicae* (Notre Dame, Ind.: University of Notre Dame Press, 1980). For a complete annotated translation of Buridan's *Summulae*, see Gyula Klima, trans., *John Buridan: Summulae de Dialectica* (New Haven, Conn.: Yale University Press, 2001).

[9] For a translation with useful introductory material, see Alan R. Perreiah, trans., *Paulus Venetus: Logica Parva* (Washington, D.C.: The Catholic University of America Press; Munich: Philosophia Verlag, 1984).

the modist view that logic dealt with second intentions, and although some logicians eagerly embraced the use of mathematical language, others, such as Buridan, did not.

THE NATURE OF LOGIC

The purpose of logic had nothing to do with the setting up of formal systems or the metalogical analysis of formal structures. Instead, it had a straightforwardly cognitive orientation. Everyone accepted the view that logic is about discriminating the true from the false by means of argument. This is why logic was essential to the speculative sciences, since it provided the instruments for finding truth and for proceeding from the known to the unknown.

How logic should be classified had always been a subject of discussion.[10] In the earlier period, it was called both a linguistic science (*scientia sermocinalis*) and a rational science (*scientia rationalis*). There were considerations supporting both titles. On the one hand, the Stoics had divided philosophy into natural, moral, and rational, and the last was equated with logic, which could then, as Boethius pointed out, be seen as both an instrument and a part of philosophy. On the other hand, logic, along with rhetoric and grammar, was one of the three liberal arts constituting the trivium, which made it seem a linguistic discipline. Some logicians, such as William of Sherwood, preferred to call logic just a linguistic science, but many others in the thirteenth century, including Robert Kilwardby, called it both linguistic and rational.[11]

In the late thirteenth and fourteenth centuries, the notion of logic as a rational, rather than a linguistic, science became predominant. This move was partly associated with the rediscovery of the *Posterior Analytics* and the new emphasis on demonstrative science, and it raised certain problems about the nature of logic. If a science (*scientia*) consists of universal necessary propositions, if it proceeds by demonstration, and if it deals with being (*ens*), how can the study of fallacies or individual arguments count as science? Modist logicians and others agreed that logic did count as science, both because it dealt with the universal, necessary principles governing fallacies and other logical phenomena and because the notion of being included not only real beings but also beings of reason, which owe their existence to the mind's activity.[12]

10 For more on these topics, see James A. Weisheipl, "Classification of the Sciences in Medieval Thought," *Mediaeval Studies*, 27 (1965), 54–90; and Costantino Marmo, "*Suspicio*: A Key Word to the Significance of Aristotle's *Rhetoric* in Thirteenth Century Scholasticism," *Cahiers de l'institut du moyen-âge grec et latin*, 60 (1990), 145–98.

11 For Kilwardby on the nature of logic, see Kretzmann and Stump, *Logic and the Philosophy of Language*, pp. 264–77.

12 For more about beings of reason, see Gino Roncaglia, "Smiglecius on *entia rationis*," *Vivarium*, 33 (1995), 27–49.

The next question was how logic fitted into the division of science into practical and speculative, along with the associated division of speculative science into natural philosophy, mathematics, and divine science, which, unlike the Stoic classification, offered no obvious place for logic. Some later logicians, including William of Ockham, Buridan, and Albert of Saxony, classified logic as a practical science, but most preferred to think of it as speculative (though of course with practical applications). Logic was then called a supporting science (*scientia adminiculativa*), subordinated to the three important sciences, and seen as a tool directing reason in the acquisition of knowledge. This function of logic provided one reason for calling it rational.

However, the most important reason for calling logic rational had to do with views about its subject matter, and here the influence of Avicenna is crucial, for it was he who had said that logic was about second intentions. Second intentions, often identified with beings of reason (*entia rationis*), are those higher-level concepts that we use to classify our concepts of things in the world, including such notions as genus, species, subject, predicate, and syllogism. Nominalists and realists disagreed about whether second intentions pick out special common objects, including both universals and logical structures, or whether they are just mental constructs reached through reflection on individual things and on actual pieces of discourse or writing, but this did not prevent such nominalists as Ockham from agreeing that logic deals with second intentions and that the syllogism the logician considers is neither a thing in the world nor a piece of writing or speaking. Some people preferred to say that logic was about things in the world as they fall under second intentions, whereas other people preferred to pick out some special second intention such as argumentation or the syllogism as the subject of logic, but there was still a strong consensus that the objects of logic are rational objects.

These remarks have to be balanced by the obvious fact that people reason in ordinary language and that ordinary locutions are often vague, ambiguous, and misleading. Medieval logicians spent an enormous amount of time on the analysis of ordinary Latin, not because they thought that logic was theoretically concerned with ordinary language but because this was the only way to avoid fallacious reasoning. Indeed, the avoidance of fallacy was at the heart of all the new types of logical writing.

DEMONSTRATION AND SCIENTIFIC METHOD

The most important of Aristotle's logical works for the history of science is his *Posterior Analytics*. Its reception was slow and indirect, and the first complete commentary, by Robert Grosseteste, dates from the 1220s. Roger Bacon's *Summule* was the first general textbook to contain a full section on the demonstrative syllogism.[13] People agreed that the demonstrative syllogism, with its necessarily true premises leading to a necessarily true conclusion, is the highest form of argumentation and crucial to any *scientia*. However, there were two areas of dispute. One area was largely epistemological and concerned the status of necessary knowledge. The more properly logical area had to do with methodology. Aristotle had distinguished between two types of demonstration, demonstration of the fact (demonstration *quia*) and demonstration of the reasoned fact (demonstration *propter quid*). He also claimed that in establishing the principles of nature one should start with those things that are more knowable to us and proceed to those things more knowable or intelligible by nature though not to us. It seemed then that, in order to achieve full scientific understanding, one should start with an effect and reason to its cause using the first kind of demonstration and then reverse the procedure by showing how knowledge of the cause includes knowledge of the effect. Such Aristotelians as Pietro d'Abano (1257–1315), known later as "the Conciliator," conflated these two types of demonstration with the two methods that Galen had set out in his *Ars Medica*. These were the method of resolution, in which an object is broken down into its component parts, and the method of composition, in which the components yielded by the resolution are put into their proper order. Logicians such as Paul of Venice were particularly concerned with defending the whole procedure against the charge of circularity. However, the full story of this defense belongs to the Renaissance.[14]

NEW TECHNIQUES: SOPHISMATA AND OBLIGATIONS

Medieval logic is characterized by a new technique, the analysis and solution of sophismata.[15] The word "sophisma" covers two phenomena. First, there is the sophisma sentence itself, which is a logical puzzle intended to introduce or illustrate a difficulty, concept, or general problem. Examples include "Every phoenix exists" and "Socrates is whiter than Plato begins to

[13] For Roger Bacon on demonstration, see secs. 146–240 in Alain de Libera, ed., "Les *Summulae dialectices* de Roger Bacon: III. De argumentatione," *Archives d'histoire doctrinale et littéraire du moyen âge*, 54 (1987), 171–278. For the first part of Roger Bacon's work, see Alain de Libera, ed., "Les *Summulae dialectices* de Roger Bacon: I–II. De termino, De enuntiatione," *Archives d'histoire doctrinale et littéraire du moyen âge*, 53 (1986), 139–289.

[14] See William A. Wallace, "Circularity and the Paduan *Regressus*: From Pietro d'Abano to Galileo Galilei," *Vivarium*, 33 (1995), 76–97.

[15] See Stephen Read, ed., *Sophisms in Medieval Logic and Grammar: Acts of the Ninth European Symposium for Medieval Logic and Semantics, held at St Andrews, June 1990* (Dordrecht: Kluwer Academic Publishers, 1993). See also Sten Ebbesen, "The More the Less: Natural Philosophy and Sophismata in the Thirteenth Century," in *La nouvelle physique du XIVe siècle*, ed. Stefano Caroti and Pierre Souffrin (Florence: Leo S. Olschki, 1997), pp. 9–44.

be white."[16] The origin of these sentences seems to lie in the twelfth-century use of *instantiae*, or counterexamples. Second, there is the technique of the sophismatic disputation, which is used to show that the very same reasoning that supported a plausible thesis could also be used to establish something implausible. By the end of the twelfth century, the sophisma was established in different genres of logical and grammatical writing, which included special treatises devoted to sophismata. Typically, these treatises would start with a sophisma, and, using disputational techniques, either solve it by appealing to logical distinctions and to facts about syncategorematic terms such as "every" or "begins" or dissolve it by showing that different truth-values were possible according to different senses of the sophisma sentence. Various types of and treatises on sophismata survive. It seems that in Paris the sophisma was an element of teaching and had to be debated in the schools; that is, within a fairly formal setting. From the latter part of the thirteenth century, very elaborate sophismata survive that were given solutions by the teaching masters and were often used as a vehicle for the straightforward discussion of interesting logical problems.[17] On the face of it, Oxford sophismata seem to have been quite different. They were the subject of live debate, but this debate was at a strictly undergraduate level and was part of the practical training in logical disputation. These sophismata were not primarily a vehicle for discussing substantive issues, and they seem to disappear as a genre when the curriculum changes. Fourteenth-century Oxford sophismata had two special features. First, they exhibit a new emphasis on the *casus*, or initial hypothesis about the context of the sophisma-sentence, which might itself be identified as the source of the problems in determining the value of that sentence. This notion of *casus* also played a key role in treatises on insolubles and obligations. Second, those produced by the Oxford calculators, such as Richard Kilvington and especially William Heytesbury,[18] are quite different both in content and in presupposition from thirteenth-century sophismata, which had focused on terms and their syntactico-semantic properties, and also from those produced by other fourteenth-century authors, including Walter Burley. Syncategorematic terms and supposition theory continue to have an important role to play, but now logical language is joined by the mathematical language of proportion and the analysis of continuous

[16] According to legend, each new phoenix must be born from the ashes of its predecessor. For a discussion of the phoenix, see Read, *Sophisms in Medieval Logic and Grammar*, pp. 3–23 and 185–201. The second example comes from Richard Kilvington: see Norman Kretzmann and Barbara Ensign Kretzmann, introd. and trans., *The Sophismata of Richard Kilvington* (Cambridge: Cambridge University Press, 1990), pp. 2–3.

[17] See Boethius of Dacia, "The Sophisma 'Every Man Is of Necessity an Animal'," in Kretzmann and Stump, *Logic and the Philosophy of Language*, pp. 482–510.

[18] See John Longeway, *William Heytesbury: On Maxima and Minima. Chapter 5 of Rules for Solving Sophismata, with an anonymous fourteenth-century discussion* (Synthese Historical Library, 26) (Dordrecht: Reidel, 1984).

magnitudes and processes. Examples include "A begins to intensify whiteness in some part of B, and each proportional part in B will without interval be diminished" and "Socrates will move over some distance when he will not have the power to move over that distance."[19] These new sophismata are characterized by a concern with measure (whether of the changes in speed over time or the distribution of a quality such as whiteness that varies in degree), and by a concern with limit-decisions (whether about the beginning and ending of the measurable processes just mentioned or the setting of limits to such powers as the power to lift a weight). At the same time, there is a concern with how to draw comparisons between two or more of the infinite sets of degrees or temporal instants that were appealed to in the analysis both of measurable processes and the limit-decisions that these may involve.

Treatises on obligations were closely related to treatises on sophismata in their provision of new techniques.[20] They discussed the rules that students were obliged to follow (hence the title) in special disputations that served as logical exercises. One type of obligational disputation that was important up until the time of Ockham, though it disappeared in the late fourteenth century, was impossible *positio*, in which the consequences of granting an impossible proposition were explored. Ockham explained the two limitations that were necessary. The rules of inference had to be restricted, to prevent the inference of anything whatsoever, and the proposition itself could not be or imply a formal contradiction of the form "P and not-P."

However, the central type of obligational disputation was possible *positio*. In this, the opponent began the game by positing a proposition, which the respondent had to grant, provided that it was logically possible. This *positum* was normally false, and it was usually trivial in content. Frequently it was preceded by a *casus*, which specified the conditions supposedly obtaining. Next, the opponent had to propose a series of other propositions, which could be either true or false. In replying to these, the respondent had to follow the rules of the game. If a proposition followed from the set formed by the *positum*, the propositions already granted, and the negations of propositions already denied, it had to be granted. If a proposition was inconsistent with the set so determined, it had to be denied. If a proposition was such that neither it nor its negation followed from the set so determined, it was called *impertinens* or irrelevant, and the respondent had to reply to it according to his current state of knowledge. This allowed him a third type of reply – besides "*concedo*" (I grant it) and "*nego*" (I deny it), he could reply "*dubito*"

[19] See Kretzmann and Kretzmann, *Sophismata of Richard Kilvington*, pp. 47–9 and 65–70.

[20] For a discussion of obligations, see Hajo Keffer, *De Obligationibus: Rekonstruktion einer spätmittelalterlichen Disputationstheorie* (Leiden: Brill, 2001) and Dutilh Novaes, *Formalizing Medieval Logical Theories*. For the translation of a text and many references, see E. Jennifer Ashworth, ed. and trans., *Paul of Venice, Logica Magna: Part II, Fascicule 8 (Tractatus de Obligationibus)* (Oxford: Published for the British Academy by the Oxford University Press, 1988).

(I am in doubt about it).[21] The task of the opponent in this game was to lead the respondent to the point of accepting an inconsistent set, either by granting and denying the same proposition or by granting a proposition whose negation would follow from the set of propositions already granted. The task of the respondent, of course, was to avoid falling into these traps. In order to be successful, the respondent needed an excellent grasp of the logical relations between propositions and also an excellent memory. This is why obligational disputations could serve as valuable exercises for young logicians. However, they were also valuable for scientists because they provided a model for the full exploration of hypothetical situations, or reasoning *per imaginationem*. The specification of the initial *casus* was particularly important here.

SIGNIFICATION

Medieval logicians employed the notion of signification as their basic semantic notion.[22] They regarded spoken words, at least so far as ordinary natural-kind terms such as "dog" and "cat" were concerned, as conventional signs that pointed beyond themselves in two ways, to the things that were their significates and to the concepts that were an essential part of the significative process. There was considerable disagreement about the proper description of this process. In the thirteenth century, authors tended to agree that the things in question were universal natures and that the individuals exhibiting these natures were at best secondary significates. As a result, they made a clear distinction between signification and supposition or reference, and when they discussed whether a spoken word primarily signified concepts or things they were concerned only with the relative importance of concepts and universal natures, however construed. With the rise of nominalism in the fourteenth century, the things signified were taken to be individuals, and the issue of whether words primarily signify concepts or things had to do with the relative importance of concepts and individuals. Some authors, such as Jean Buridan, held that words primarily signified or made known concepts, whereas others, following Ockham, held that words signified things alone while being subordinated to concepts.

Ockham also insisted that the concept itself must be regarded as a sign. Although this notion was not new, Ockham made it central, and in so doing made mental language rather than spoken language the paradigm of signification. The doctrine that there is a language of thought that is naturally significant and common to all human beings was present at least

[21] One sort of obligational disputation required that the respondent reply "*nego*" to the *positum* itself and another that he reply "*dubito.*"

[22] For a discussion of medieval semantics, see Jan Pinborg, *Medieval Semantics: Selected Studies on Medieval Logic and Grammar* (London: Variorum Reprints, 1984). For a discussion of signs and signification, see Joël Biard, *Logique et théorie du signe au XIVe siècle* (Paris: Vrin, 1989). For Ockham, see Claude Panaccio, *Ockham on Concepts* (Aldershot: Ashgate, 2004).

from Augustine on, but it was only fully developed in the fourteenth century, first by Ockham and then by Pierre d'Ailly.[23] Mental propositions were thought of as having syntactic structure, and mental terms were thought of as having supposition, or different types of reference, so that the notion of a language system was internalized. As a corollary, philosophical grammar as presented by the modists was devalued and lost its role in logic.

Both nominalists and realists viewed natural-kind terms such as "cat" as relatively unproblematic in that they picked out something unequivocally real, however reality was to be described. Other terms were more problematic. First, there are negative terms (e.g., "non-being"), privative terms (e.g., "blindness"), and pure fictions (e.g., "chimera"). Negations, privations, and fictions were often said to be beings of reason since their existence was due to the mind, though in some cases there could be a foundation in reality. Negations and privations were particularly important for natural philosophers since some terms they used (such as "darkness") were clearly privative, whereas others, including "infinite," "point," and "instant," were thought to include an element of negation. Second, there are transcendental terms, such as "being" and "one," which are so called because they transcend or fall outside Aristotle's ten categories. These terms were often said to be analogical, meaning that they are used in extended senses.[24] This notion of extended senses was applied to at least some of the terms used in natural philosophy, such as "principle" and "cause." Third, there are denominative terms (also called paronyms). These are adjectives such as "just," which refer both to individuals and to an abstract quality, though there was much disagreement over which type of reference was primary and which secondary. Ockham pulled most of this material together with his doctrine of connotative terms, which have both a primary and a secondary reference to individual substances and qualities. Thus the term "father" has a primary reference to men who are fathers, but it connotes their children and can be analyzed without introducing the supposed real relation of paternity. The theory of connotation is important for natural philosophy because it allows an analysis of such words as "vacuum" and "motion" without involving a commitment to a vacuum or motion as positive things in the world.

SUPPOSITION

The most notable new theory that took shape in the twelfth century was supposition theory and its ramifications, particularly ampliation and

[23] For mental language, see Claude Panaccio, *Le discours intérieur: De Platon à Guillaume d'Ockham* (Paris: Editions du Seuil, 1999). For a translation of Peter of Ailly, see Paul Vincent Spade, trans., *Peter of Ailly: Concepts and Insolubles* (Synthese Historical Library, 19) (Dordrecht: Reidel, 1980).

[24] For some discussion of analogical terms, see E. J. Ashworth, "Analogy and Equivocation in Thirteenth-Century Logic: Aquinas in Context," *Mediaeval Studies*, 54 (1992), 94–135; and E. Jennifer Ashworth, *Les théories de l'analogie du xiie au xvie siècle* (Paris: Vrin, 2008).

restriction.[25] This theory explored the different types of reference that a subject or predicate term could have in various contexts. The three main types of supposition were material, simple, and personal. A term was said to have material supposition when it stood for itself or for other occurrences of the same term, as in "Man is a noun." A term was said to have simple supposition when it stood for a universal, as in "Man is a species." Both material supposition and simple supposition gave rise to controversy, but especially the latter, because of the obvious problem of the ontological status of universals or common natures. Finally, a term has personal supposition when it is taken for its normal referents, as when "man" is taken for Socrates, Plato, and so on.

Some logicians distinguished accidental personal supposition from natural supposition, which allowed a term to have pre-propositional reference to all its referents – past, present, and future – whereas others insisted that supposition must be purely propositional and contextual. This debate in turn affects the doctrines of ampliation, whereby the range of reference of a term can be extended, and restriction (the opposite of ampliation). Parisian logicians, such as Jean le Page, writing around 1235, tended to accept natural supposition and to say (like Buridan in the fourteenth century) that terms had natural supposition in scientific propositions (i.e., universal necessary truths), so that no ampliation was necessary. As a corollary, in nonscientific propositions, the supposition of terms was restricted in various ways. For English logicians in the thirteenth century, all supposition was contextual, and the notion of ampliation had to be used when the subject of a proposition was to extend beyond present existent things.

The notion of ampliation was particularly important in the analysis of propositions containing tensed verbs, modal terms, and epistemic terms such as "imagine." Logicians generally held that affirmative propositions with nonreferring terms are false, yet many of the propositions we wish to take as true have terms that refer to nothing currently existent. The doctrine of ampliation allowed reference to extend over past, future, and possible objects. In the late fourteenth century, Marsilius of Inghen argued that one should also allow reference to imaginable objects that were impossible. By allowing this kind of ampliation to occur when such terms as "imagine" were used, he could save the truth of "I imagine a chimera" while still holding that "A chimera is an animal" was false.[26]

The three types of personal supposition most often appealed to are determinate, purely confused (*confuse tantum*), and confused and distributive. These types came to be illustrated by means of the descent to singulars. For instance, to say that the subject of a particular affirmative proposition,

[25] See L. M. de Rijk, *Logica Modernorum: A Contribution to the History of Early Terminist Logic*, 2 vols. in 3 pts. (Assen: Van Gorcum, 1962–1967).

[26] For more about chimeras, see E. J. Ashworth, *Studies in Post-Medieval Semantics* (London: Variorum Reprints, 1985).

"Some A is B," has determinate supposition is to say that one can infer the disjunction of singular propositions, "This A is B, or that A is B, or the other A is B, and so on." To say that the predicate of a universal affirmative proposition, "Every A is B," has purely confused supposition is to say that one can infer a proposition with a disjoint predicate, "Every A is this B or that B or the other B, and so on." To say that the subject of a universal affirmative proposition has confused and distributive supposition is to say that one can infer a conjunction of propositions, "This A is B, and that A is B, and the other A is B, and so on." Some people distinguished between mobile and immobile cases. For instance, no descent is possible from "Only every A is B," and so A has immobile supposition. A fourth type of supposition is collective supposition, as in "Every man is hauling a boat," given that they are doing it together. Here any descent will involve a conjoint subject, as in "This man and that man and the other man are all hauling a boat."

The theory of personal supposition was used to solve a variety of problems. One standard problem had to do with promising (or "owing" in the works of some authors). If I promise you a horse, is there some horse that I promise you, and if not, how is the original sentence to be construed? A wide variety of answers were proposed.[27] Walter Burley suggested that "horse" has simple supposition; Heytesbury took it that "horse" had purely confused supposition and that it did not imply "There is some horse that I promise you" because the new position of "horse" before the verb gave it determinate supposition. Ockham preferred to replace the sentence by a more complex sentence, "You will have one horse by means of my gift." This solution is closer to the type found in treatises on the Proofs of Terms, which focused not on the reference of the term "horse" but on the analysis of the word "promise" in terms of giving a right to an object.

COMPOUNDED AND DIVIDED SENSES

One of the basic tools of propositional analysis was the distinction between compounded and divided senses, which is generally associated with modal logic but has its origin in Aristotle's discussion of the fallacy of composition and division, to which Abelard was one of the first to pay careful attention. The basic point concerns two ways of reading the sentence "A seated man can walk." Interpreted according to its compounded sense, this proposition is *de dicto* (about a dictum or "that" clause) and means "That-a-seated-man-walk (i.e., while seated) is possible." Interpreted according to its divided sense, the proposition is *de re* (about a *res* or thing) and means "A seated man has the power or ability to walk." The proposition is false in the first sense but true in the second. It became standard when considering modal inferences,

[27] For more about promising and requiring horses, see Ashworth, *Studies in Post-Medieval Semantics.*

including modal syllogisms, to distinguish between the compounded and divided senses of premises and conclusion and to work out the logical results of these different readings.

In the fourteenth century, the distinction between the compounded and divided senses was very widely used, as can be seen from the work of William Heytesbury, the first to devote a complete treatise to this topic.[28] He listed nine types of logical problems that could be solved by paying attention to the distinction. These problems were signaled by the presence of modal words but also, for example, by the presence of future-tense verbs, verbs producing confused supposition, terms such as "infinite," and words about acts of will and intellect. This provided another solution to the question of whether "I promise you a horse" implies "There is a horse that I promise you." In the antecedent, the sense is compounded, for the sentence involves an attitude toward a propositional complex. In the consequent, the sense is divided, for the subject of discussion is an actual horse, not a proposition, and one cannot argue here from the compounded to the divided sense. Precisely similar analyses, involving both different types of supposition and different senses, were applied to inferences about matters in natural philosophy, such as "Immediately after this there will be some instant; therefore, some instant will be immediately after this."[29]

SYNCATEGOREMATA; PROOFS OF TERMS

Supposition theory dealt with the subject and predicate terms in propositions, albeit in relation to other terms. Treatises on syncategoremata dealt with all these other terms, such as "all," "some," or "not," that appear in a proposition and exercise some logical function.[30] Treatises on syncategoremata were most prominent in the thirteenth century, and they did not altogether disappear in the fourteenth century. For instance, the late-fourteenth-century English logician Richard Lavenham wrote one. However, two other forms of writing come to the fore. First, there are a lot of short treatises on particular syncategorematic terms, including "know" (*scire*), "begins" and "ceases" ("*incipit*" and "*desinit*"), and terms with the power to produce purely confused supposition. Second, and most important, are the treatises on the Proofs of Terms (or Proofs of Propositions), whose best-known example is the *Speculum puerorum* of Richard Billingham, an Oxford author of the mid-fourteenth century.

[28] For William Heytesbury, "The Compounded and Divided Senses," see Kretzmann and Stump, *Logic and the Philosophy of Language*, pp. 415–34.
[29] Heytesbury in Kretzmann and Stump, *Logic and the Philosophy of Language*, p. 415.
[30] Two works are available in translation. See Norman Kretzmann, trans., *William of Sherwood's Treatise on Syncategorematic Words* (Minneapolis: University of Minnesota Press, 1968); and Peter of Spain, *Syncategoreumata*, ed. L. M. de Rijk, trans. Joke Spruyt (Leiden: Brill, 1992).

In this context, a proof seems to be a method of clarifying a sentence containing a particular sort of term or of showing how one might justify that sentence. There were three groups of terms. Resoluble terms are those whose presence calls for explanation or clarification through ostensive reference, as captured in an expository syllogism (i.e., one with singular terms). Thus "A man runs" is resolved into the expository syllogism "This runs, and this is a man, therefore a man runs." Exponible terms are those whose presence calls for exposition of the sentence in terms of a set of equivalent sentences. For instance, the sentence "Only a man is running," which contains the exclusive term "only," is expounded as "A man is running and nothing other than a man is running." Along with exclusives, the main groups of exponible terms were exceptives (such as "except") and reduplicatives (such as "inasmuch as"), but many other terms, such as "begins" and "ceases" and "infinite," were included. In fact, exponible terms are all terms that had figured prominently in treatises on syncategoremata. Finally, there are "official" or "officiable" terms ("*officiales*," "*officiabiles*"), so called because they performed a function (*officium*). They included any term that governed a whole sentence or that treated a whole sentence as modifiable, such as modal terms and such terms as "know," "believe," "promise," "desire," and "owe." Analysis of sentences containing such terms shows why they are referentially opaque when taken in the compounded sense.

For natural science, the most important syncategorematic terms are "infinite," "begins," and "ceases." "Infinite" was said to have both categorematic and syncategorematic senses, and this distinction was used to prevent such inferences as "*Infinitus est numerus, ergo numerus est infinitus*" (literally, "infinite is number, therefore [a] number is infinite"). In the early thirteenth century, Robert Bacon, O.P., explained that the antecedent uses "infinite" syncategorematically to indicate the infinity of the subject in relation to the predicate. In other words, one can keep adding to the number series. The consequent uses the term in the categorical sense and indicates that there is a number that is actually infinite. Later authors, especially Heytesbury, used more elaborate examples and appealed to a variety of solutions using both supposition theory and the distinction between the compounded and the divided senses.

Discussions of "begins" depended on a basic distinction between two types of object or characteristic, together with the notion of a hidden reference to the past or future. A successive object had no first instant of being, whereas a permanent object had no last instant of non-being. This allows two different expositions of "Socrates begins to be white," depending on whether whiteness is permanent or successive. The first is "Socrates is not now white, and immediately after this he will be white," and the second is "Socrates is white, and immediately before this he was not white." Once more, Kilvington and Heytesbury give much more elaborate examples, including "Socrates is infinitely whiter than Plato begins to be white," but the principle of exposition in terms of equivalent sentences in order to make clear the different sorts of truth-conditions for different senses remained the same.[31]

CONCLUSION

In this short account of medieval logic, we have been forced to omit discussion of some important topics, including insolubles or semantic paradoxes, consequences or valid inferences, and developments in modal logic.[32] Nonetheless, enough has been said to show that medieval logicians developed a rich and sophisticated semantics together with techniques for propositional analysis that enabled them to tackle a wide variety of problems in philosophy, theology, and natural science. It is true that their interests and methods of approach cannot be mapped precisely onto those of the twenty-first century, but the very recognition of these differences can still be philosophically fruitful.

[31] For the example, see Kretzmann and Kretzmann, *Sophismata of Richard Kilvington*, p. 4.

[32] For insolubles, see Paul Vincent Spade, *Lies, Language and Logic in the Late Middle Ages* (London: Variorum Reprints, 1988). For consequences, see the sources listed in note 1. For modal logic, see Simo Knuuttila, *Modalities in Medieval Philosophy* (London: Routledge, 1993); and Paul Thom, *Medieval Modal Systems: Problems and Concepts* (London: Ashgate, 2004).

23

GEOGRAPHY

David Woodward

In the Latin Middle Ages, the scientific questions we now associate with geographical inquiry did not fall easily into any one of the subjects constituting the trivium (grammar, rhetoric, and logic) or the quadrivium (arithmetic, geometry, astronomy, and musical theory). Even the word "*geographia*" was seldom used and did not assume currency until the fifteenth and sixteenth centuries. Works that we might now call geographies reflected the concern with describing the earth as an abode designed and planned by God and were intended to supplement knowledge of the earth provided by the Scriptures.[1]

Medieval geographical representations, in the form of both texts and maps, fulfilled three broad functions: (1) explaining and teaching views of the world in scholarly mathematical geography or religious ideology; (2) documenting, claiming, and planning natural and political resources on a regional or local level; and (3) wayfinding or navigation (on land and sea) using itineraries and charts of various types. This chapter is organized around these three themes.

SCHOLARLY MATHEMATICAL GEOGRAPHY AND THE WORLDVIEW

The intellectual's concepts of the shape, size, and physical nature of the Earth in the Middle Ages rested on classical foundations, particularly the works of Aristotle (384–322 B.C.). The concept of a spherical Earth is traditionally associated with the authority of the Pythagorean school around 500 B.C., but the first explicit proofs are laid out in Aristotle's *De caelo* (*On the Heavens*).[2]

[1] Clarence J. Glacken, *Traces on the Rhodian Shore: Nature and Culture in Western Thought from Ancient Times to the End of the Eighteenth Century* (Berkeley: University of California Press, 1967), pp. 171ff.

[2] Aristotle, *On the Heavens*, trans. W. K. C. Guthrie (Cambridge, Mass.: Harvard University Press, 1986). See especially II.13.294 (p. 227), II.14.297 (p. 249), and II.14.297 (p. 253).

I would like to thank Evelyn Edson and Victoria Morse for their helpful comments.

In this work, Aristotle disposes of the views of Anaximenes, Anaxagoras, and Democritus that the flatness of the Earth caused it to remain at rest, settling on the air beneath it "like a lid." In contrast, Aristotle's a priori view was that the Earth must be spherical because all its parts naturally move toward the center and because the resulting mass must be shaped similarly on all sides. He provides two empirical proofs to back up this statement: in lunar eclipses, the shadow of the Earth on the Moon is always circular; and the maximal height of a star above the horizon changes as we move north.

In the Hellenistic and Greco-Roman periods, the spherical shape of the Earth appears never to have been in doubt, and the concept is described in the works of Eratosthenes (275–194 B.C.), Pliny (23/24–79), Ptolemy (second century), Macrobius (fl. 399–422), and Martianus Capella (fl. 410–39), among others. Even in widely disseminated objects, such as coins of the Roman period, the image of the Earth as a globe appears as a symbol of imperial power. The notion that the early church fathers rejected the Greco-Roman idea of the spherical Earth is almost totally false. To be sure, Firmianus Lactantius (ca. 240–ca. 320) and Severianus, bishop of Gabala (fl. late fourth century, d. after 408), could not reconcile the idea of the spherical Earth with the literal teachings of the Bible (which contains the words "circle" and "four corners" but not "sphere" or "globe"), and Cosmas Indicopleustes (a sixth-century Nestorian Christian residing in Alexandria, who wrote a *Christian Topography*) described a flat, rectangular, four-cornered Earth with a vaulted heaven.[3] But these views were anomalous – Cosmas was the object of contemporary ridicule, whereas the first- to fourth-century views of the Earth's sphericity in Pliny, Pomponius Mela, Macrobius, and Martianus Capella were far more representative of the period. Furthermore, the church had no specific teachings on matters geographical or cosmographic and at worst regarded scientific pursuits as unprofitable and irrelevant to the Christian life. Despite the strangely persistent modern view that medieval scholars routinely believed the Earth was flat, almost nobody expressed such an idea.

To the contrary, medieval Europe had two prevailing estimates of the Earth's circumference: 180,000 and 252,000 stades. The more usual figure, 252,000, is found in Lambert of St. Omer's *Liber floridus* (1100) and Sacrobosco's *De sphaera* (ca. 1220s or 1230s) and was derived from the famous measurement by Eratosthenes, reported by Strabo (63 B.C. – A.D. 20), Pliny,[4] Cleomedes (second century), and Macrobius. The figure of 180,000 stades, which is stated in the *De imagine mundi* of Honorius of Autun (ca. 1130) and the text of the Catalan Atlas (ca. 1375), derives from Posidonius's "correction"

[3] David Woodward, "Medieval *Mappaemundi*," in *History of Cartography*, vol. 1: *Cartography in Prehistoric, Ancient, and Medieval Europe and the Mediterranean*, ed. J. B. Harley and David Woodward (Chicago: University of Chicago Press, 1987), pp. 286–370; and Jeffrey Burton Russell, *Inventing the Flat Earth* (New York: Praeger, 1991).

[4] Pliny the Elder, *Natural History*, II.112.247.

(first century B.C.) of Eratosthenes' figure, which Ptolemy reported in the *Geography*.

The problem of the differing values stemmed from the difficulty of making precise astronomical measurements and the lack of standardized units. The usual practice of expressing the number of units in one degree of the Earth's circumference depended on the size of the "units" and was therefore fraught with misunderstandings. The confusion stemmed partly from the difference between the Roman mile (75 miles to a degree) and the Arabic mile (56 2/3 miles to a degree). Early in the ninth century, the Islamic caliph al-Ma'mūn, apparently frustrated at not understanding Ptolemy's meaning of "stade," reportedly commissioned two independent measurements of a degree of latitude in the desert of Sinjar (between the Tigris and Euphrates rivers), which resulted in a figure between 56 and 57 miles. The purpose of this measurement seems to have been less to establish an empirical value for the size of the Earth than to settle the choice of the values received from antiquity.[5] The value of 56 2/3 was reported by al-Farghānī (Alfraganus) (fl. 813–61), whose ideas found favor in the Latin West. A gross underestimate of the Earth's circumference ($56\ 2/3 \times 360 = 20{,}400$ miles) persisted into the fifteenth century and was the figure adopted by Columbus in his calculations.[6]

The idea of the spherical Earth was presupposed using a prevalent concept in medieval geography – that of five latitudinal zones, in which different climatic conditions were encountered according to the angle of incidence of the Sun's rays on the Earth's surface at different latitudes. The zones consisted of two uninhabitable polar zones, two habitable temperate zones between the tropics and the polar circles, and an uncrossable equatorial zone between the tropics. The southern temperate zone, according to the original Greek concept, was inhabited by the Antipodeans, a race of "opposite footed" people. The roots of this classification can be traced to the works of Aristotle, particularly his *Meteorology*.[7] The system of zones gained currency during the Middle Ages largely through Macrobius's early-fifth-century commentary on Cicero's *Dream of Scipio*, in which he prescribed precise measurements for the zones.[8]

An associated fourfold division of the Earth became known to the medieval world through the works of Macrobius and Martianus Capella. The concept

[5] Raymond P. Mercier, "Geodesy," in *History of Cartography*, vol. 2, bk. 1: *Cartography in the Traditional Islamic and South Asian Societies*, ed. J. B. Harley and David Woodward (Chicago: University of Chicago Press, 1992), pp. 175–88 at pp. 178–81.

[6] George E. Nunn, *The Geographical Conceptions of Columbus: A Critical Consideration of Four Problems* (New York: American Geographical Society, 1924; 2nd ed., Milwaukee: American Geographical Society Collection, 1992), pp. 1–30.

[7] Aristotle, *Meteorologica*, II.5.362a.33, trans. H. D. P. Lee (Loeb Classical Library) (Cambridge, Mass.: Harvard University Press; London: William Heinemann, 1952).

[8] Macrobius, *Commentary on the Dream of Scipio*, ed. and trans. William Harris Stahl (New York: Columbia University Press, 1952, repr. with corrections, 1966), pp. 206–7.

of four parts originated from a large globe reportedly made by the Stoic philosopher Crates of Mallos around 150 B.C. and were known as *oikoumene*, *antoikoi*, *perioikoi*, and *antipodes*, with the antipodes located in the area now associated with South America. Macrobius reports that all four quarters were believed to be inhabited, but that the inhabitants could not communicate with each other. This posed a doctrinal problem for the church in that it implied that there might be men isolated from the words of Christ, a concern expressed by St. Augustine in *The City of God*.[9] Belief in the Antipodeans became a heretical doctrine, highlighted in the debate between St. Boniface and Virgil, bishop of Salzburg.[10]

The five Macrobian zones should be distinguished from an alternative scheme – the seven "climates" dividing the northern temperate zone on the basis of the longest day at different latitudes. This concept, modeled on a Ptolemaic scheme, was employed by Islamic geographers. The exact position of these seven climates and the criteria for bounding them varied.[11]

The study of geography in the Latin West differed fundamentally from that in the Islamic world in one important respect: that the West did not have ready access to Ptolemy's *Geography* until the beginning of the fifteenth century. In the first book of that work, Ptolemy had clearly explained his system of geography and mapmaking in the second century. Central to Ptolemy's ideas – based largely on the methods of Hipparchus for star mapping – was the principle that maps should be plotted using measurements of latitude and longitude, as described in his *Almagest* and *Geography*.

The transmission of the *Geography* differed, in both timing and route, from that of Ptolemy's great astronomical text, the *Almagest*. Transmission rarely involved unqualified adoption. Conservatism in methods of mapmaking abounded in both the Western and Islamic traditions. We do not know what happened to the *Geography* between the second century and the early ninth century, when clues to its existence began to appear. Since the tables of al-Khwārizmī (d. ca. 847), who worked at the court of the caliph al-Maʾmūn (r. 813–33) in Baghdad, were based not only on a text but also probably on a map associated with the *Geography*, we know that either a Greek or, more likely, a Syriac version was available sometime before 833.[12] Al-Kindī

9 Macrobius, *Commentary on the Dream of Scipio*, pp. 205–6. See Eva Matthew Sanford and William M. Green, eds. and trans., *The City of God against the Pagans*, 7 vols. (Loeb Classical Library) (Cambridge, Mass.: Harvard University Press, 1965), vol. 5, p. 51. See also John Block Friedman, *The Monstrous Races in Medieval Art and Thought* (Cambridge, Mass.: Harvard University Press, 1981), p. 47; and Serafin Moralejo, "World and Time in the Map of the Osma Beatus," in *Apocalipsis Beati Liebanensis Burgi Oxomensis II: El Beato de Osma, Estudios*, ed. John Williams (Valencia: V. Garcia, 1992), pp. 151–79.

10 John Kirtland Wright, *The Geographical Lore of the Time of the Crusades: A Study in the History of Medieval Science and Tradition in Western Europe* (American Geographical Society Research Series, 15) (New York: American Geographical Society, 1925; New York: Dover, 1965), p. 57.

11 Gerald R. Tibbetts, "The Beginnings of a Cartographic Tradition," in Harley and Woodward, *History of Cartography*, vol. 2, bk. 1, pp. 90ff.

12 Ibid., p. 95.

(813–874) is credited with making a translation during al-Maʾmūn's reign, but a full translation was probably not available for general scholarly use until after al-Maʾmūn's death in 833.

In Islamic science, tables of geographical coordinates based on Ptolemy were used for astrological purposes to pinpoint the place of birth, but the idea of using these longitude and latitude tables to make maps – the core of Ptolemy's message – was not heeded for maps of the earth (as opposed to maps of the heavens). No surviving Islamic terrestrial maps before the fourteenth century use a grid of longitude and latitude; thereafter, they do so only in a highly stylized fashion. We have only one account of how this could be done – from the work of Suhrab in the tenth century.[13]

In the medieval West, tables of longitude and latitude were likewise available. The Toledan Tables, composed by al-Zarqālī (ca. 1029–ca. 1087), a Spanish Muslim from Córdoba, were the prime model. These tables, in addition to the expected astronomical calculations for the city of Toledo, contain a long list of geographical coordinates based on the prime meridian of the Canary Islands (as in Ptolemy and al-Khwārizmī). They were translated in the twelfth century and adapted for several other cities in the next two centuries, including Marseilles, Hereford, London, Toulouse, Cremona, and Novara. The measurement of longitude by comparing local solar times at different locations was explained on several occasions, and for the first time the length of the Mediterranean was given correctly as 42° of longitude.[14] As in the Islamic world, however, maps were not compiled from these measurements until much later (the fifteenth century for the West). The Toledan Tables, for example, may have been used as a source for two sketch maps possibly prepared by Conrad of Dyffenbach in 1426 as part of a school of astronomy under the leadership of Johannes de Gmunden at the University of Vienna and the prelate Georg Müstinger at the Augustinian monastery of Klosterneuburg.[15]

Ptolemy's *Almagest* was translated in Toledo by Gerard of Cremona in 1175 as part of the influx of new knowledge into Western Europe in the twelfth century, channeled by dozens of translations of Arabic and Greek classics, particularly in philosophy, mathematics, astronomy, and the physical and natural sciences.[16] The importance of the *Almagest* to geography lay in the

[13] Ibid., pp. 104–5.

[14] John K. Wright, "Notes on the Knowledge of Latitudes and Longitudes in the Middle Ages," *Isis*, 5 (1923), 75–98.

[15] Dana Bennett Durand, *The Vienna-Klosterneuburg Map Corpus of the Fifteenth Century* (Leiden: Brill, 1952). However, these manuscripts need to be carefully restudied and dated, in line with the skeptical view of Patrick Gautier Dalché, who questions whether Dyffenbach made the maps in 1426 (see his "Pour une histoire du regard géographique: Conception et usage de la carte au XVe siècle," *Micrologus*, 4 (1996), 77–103, especially p. 85 and n. 3) and severely criticizes other conjectures by Durand. See Patrick Gautier Dalché, *La Géographie de Ptolémée en occident (IVe–XVIe siècle)* (Turnhout: Brepols, 2009), especially pp. 180–3.

[16] Gerald J. Toomer, trans., *Ptolemy's Almagest* (New York: Dutton, 1984). The *Almagest* was an extremely difficult treatise that circulated only among a small number of mathematicians. Only in

transfer of the major geometric divisions of the heavens (the equator, ecliptic, tropics, etc.) to the Earth as well as the description of the Earth's general shape and size and the proportions of land and water on its surface. It also raised the question of which meridian should be used in world mapping as a prime meridian (suggesting Alexandria). A more popular text was the *Tractatus de sphaera* of Sacrobosco (also known as John of Holywood or Halifax, d. 1256), a small introductory textbook drawing on classical and Arabic sources on the position of the Earth with respect to the heavens, illustrated with world maps and diagrams. Thanks to its clarity and brevity, it enjoyed widespread use in multiple manuscript codices and printed editions until the seventeenth century.

When the *Geography* was transmitted to the West, it was not in its modified Islamic form but in its original Greek form, as found in Byzantine manuscripts. During the fourteenth century, Maximos Planudes (1260–1310) had had Greek copies made. As Byzantium was subjected to increasing attacks by the Turks at the end of the fourteenth century, the Byzantine emperor Manuel II Paleologus sent the Greek scholar Manuel Chrysoloras (1335–1415) to Venice to enlist help. He and other scholars brought word of the *Geography*, among other treatises that Italian humanists, particularly in Padua, Ferrara, and Florence, were eager to examine and translate. The translation of the Greek text of the *Geography* was completed by Jacopo d'Angelo da Scarperia in Florence in 1406–7, and manuscript and printed versions, with and without maps, appeared throughout the fifteenth century. The name of Ptolemy became associated with geographical authority, but as empirical observations progressively contradicted his information, he became a historical icon. Furthermore, his method of map compilation using coordinates was not adopted as quickly or as completely as commonly believed.

DESCRIPTIVE GEOGRAPHIES OF THE WORLD AND *MAPPAEMUNDI*

Geographical description in the early Middle Ages was included in general encyclopedias as a backdrop to explain historical events. Paulus Orosius (fl. 414–17), in the *Historia adversum paganos*, a polemical history designed to demonstrate that Christianity had not been responsible for the collapse of the Roman Empire, stated, "[W]hen the theatres of war and the ravages of disease shall be described, whoever wishes to do so may the more easily obtain a knowledge not only of the events and their dates but of their geography as well." Orosius's detailed geographical narrative of the world was widely used

the fifteenth century did it attract broader scholarly attention. Simplified versions of it, such as the *Little Almagest*, *Epitome*, or *On the disposition of the sphere*, possibly based on Geminus's *Introduction to the Phenomena*, were in more common use. See Olaf Pedersen, *A Survey of the Almagest* (Odense: Odense University Press, 1974), pp. 16–19.

throughout the Middle Ages, even as late as the work of Pierre d'Ailly (1410). It was superseded, however, by the *Origins*, or *Etymologies*, of St. Isidore, bishop of Seville (ca. 560–636), comprising twenty books, the thirteenth and fourteenth of which dealt with geography. Encyclopedias remained the most popular format for geographical descriptions in the high- and late-medieval Latin West.[17] Their purpose was to present the world as God's creation and as a stage for establishing Christ's kingdom on earth. These works were firmly grounded in the legends and narratives of classical and biblical literature.

The geographical encyclopedias of the twelfth and thirteenth centuries, such as those of Albert the Great, Bartholomew the Englishman, and Vincent of Beauvais, differed from the general encyclopedias of the early Middle Ages, such as Isidore's, in their attempt to reconcile the traditional sources with empirical observations made by pilgrims, missionaries, and traveling scholars. For example, the Franciscan Bartholomew says in *On the Properties of Things* (before 1260) that he started out merely to describe biblical places but finished by including many others on the basis of his own or his contemporaries' personal travels and observations. Vincent of Beauvais's 3,000-page trilogy, the *Speculum naturale*, *Speculum doctrinale*, and *Speculum historiale* (mid-thirteenth century), draws not only on Pliny and Isidore but also the recently discovered Aristotle, many Arabic writers, and new geographical data gleaned from the Asian travels of John of Plano Carpini and Simon of St. Quentin.[18]

Running through the encyclopedias is a common thread of environmental theory. The four elements (earth, air, fire, and water) were associated with pairs of the "qualities" (cold, hot, wet, and dry), and the mixtures of their counterparts in the human body, the humors (black bile, blood, yellow bile, and phlegm), were held to affect the body's health. The doctrine of the elements and the humors thus emphasized the changes in the body that could be brought about by geographical variations in temperature and moisture. Much of this thinking was based on the classical medical writings of Hippocrates and Galen, transmitted through Isidore or Avicenna (980–1037). In the case of the Dominican Albert the Great (1193–1280), particularly in his treatise on the nature of places (*Liber de natura locorum*), a detailed study of the influences of climate, mountains, bodies of water, and forests on places and peoples of the world was a means to uncover the unity of creation. This theme is echoed in the *Opus maius* of Roger Bacon (ca. 1267), who discusses geographical matters when they are relevant to his general aim of employing a knowledge of the world in the service of Christian missionary activity. Mathematical geography, Bacon says, is useful to the Christian theologian in

17 Roy J. Deferrari, trans., *The Seven Books of History against the Pagans* (Washington, D.C.: Catholic University of America Press, 1964); and Jacques André, ed. and trans., *Etymologies: Isidore de Seville* (Paris: Les Belles Lettres, 1981). See John Williams, "Isidore, Orosius and the Beatus Map," *Imago Mundi*, 49 (1997), 7–32.

18 George H. T. Kimble, *Geography in the Middle Ages* (London: Methuen, 1938; New York: Russell and Russell, 1968), pp. 80–2.

three main ways. First, it leads to an understanding of the "infinite vastness of things in the heavens," to which the Christian's gaze should be fixed. Second, mathematics helps us understand the astronomical matters discussed in the Bible, such as the explanation of the length of the day when the sun stood still (Josh. 10:13) or such phenomena as when the sun retreated ten degrees (II Kings 20:12). Third, it provides the physical basis of history. To understand the Creator, the theologian must know the locations of places (Bacon specifies longitude and latitude) and must be able to visualize their character.[19]

The trend toward positive valuation of experience and observation in geographical description is exemplified by the activities of an Islamic–Norman collaboration at the court of Roger II of Sicily (1095–1154). In the last fifteen years of his life, Roger developed a clearinghouse of Islamic and classical knowledge at his court in Palermo. He had a personal interest in mathematical geography and devised "iron" instruments for measuring longitude and latitude. Around 1138, he invited the Islamic scholar al-Idrīsī (1100–1165) to his court, at first probably for political reasons as a potential puppet ruler rather than as a scholar. Al-Idrīsī had made up for his apparent lack of formal geographical training with extensive travel in Asia Minor, France, England, Spain, and Morocco and garnered an impressive reputation as a geographer. In collaboration with other scholars in Roger's court, over a period of some fifteen years, he completed in 1154 a large world map engraved on a silver disk. To accompany the map, and apparently derived from it, he also completed a geographical treatise called *The book of pleasant journeys into faraway lands* (also known as the Book of Roger). This work contained a small world map and seventy sectional maps that can theoretically be fitted together into a large world map. Al-Idrīsī used the system of seven climates to organize both the treatise and the sectional maps.[20]

Al-Idrīsī's description of the compilation of the silver map and the Book of Roger reveals a great amount of sifting and winnowing of geographical writings, scholarly opinions, and firsthand observations. The Book of Roger was understandably focused on the king's own domains, and – sensing a great deal of disagreement in the usual sources – Roger interviewed scholars and firsthand observers singly and in groups until a consensus was reached.

Al-Idrīsī's sources included Ptolemy's *Geography*, probably a version seen by al-Khwārizmī and later commentators and translators of the *Geography*. In addition, he used Islamic traditions (such as that of Ibn Ḥawqal of the "Balkhī school of geography") and benefited also from a number of contemporary European travelers at Roger's court and possibly from

[19] David Woodward and Herbert M. Howe, "Roger Bacon: Geography and Cartography," in *Roger Bacon and the Sciences: Commemorative Essays 1996*, ed. Jeremiah Hackett (Leiden: Brill, 1997), pp. 199–222.

[20] Al-Idrīsī, *Opus geographicum; sive, "Liber ad eorum delectationem qui terras peragrare studeant,"* issued in nine fascicles by the Istituto Universitario Orientale di Napoli, Istituto Italiano per il Medio ed Estremo Oriente (Leiden: Brill, 1970–1984).

some maritime information of the North African coast derived from Roger's navy.[21]

Despite several allusions to the importance of observation in compiling geographical information in the twelfth and thirteenth centuries, the prevalent taste was to perpetuate the world's marvels and legends, many of which originated in classical times.[22] The stories of monstrous races go back at least to Herodotus (fifth century B.C.). The location of the monstrous races in classical literature had largely been India, and after Alexander the Great's invasion of India in 326 B.C., a body of legend known as the Alexander romances grew out of his travels. Strabo, in the seventeen books of the *Geography*, specifically avoided these myths and legends, but his work was far less influential in the Middle Ages than that of Pliny the Elder's *Natural History* (ca. 77), which emphasized the marvels. Gaius Julius Solinus (third century), a popularizer of Pliny, emphasized the marvels and little else in his *Collection of Memorabilia*, a book that enjoyed widespread interest. Many of the other standard geographical sources of the Middle Ages – Isidore, Aethicus of Istria, the Ravenna geographer Rabanus Maurus, Honorius of Autun, Walter of Metz, Gervase of Tilbury, Bartolomeus Anglicus, Brunetto Latini, Vincent of Beauvais, and Pierre d'Ailly – perpetuated the lore. The monstrous races were necessarily located on the fringes of the civilized world, and the Antipodes was a convenient geographical location in which to place them.[23]

The Crusades (1096–1270) enormously widened the geographical horizons of many classes of people, increased mobility, and fostered a culture of travel. They also prompted a series of Asian expeditions in the last half of the thirteenth century, which we will shortly describe. Maps and travel narratives of the Holy Land and the pilgrimage routes to it were among the most popular geographical subjects. But despite these indirect effects, the general medieval view of the world persisted. Donald Lach, in his fundamental series tracing the role of Asia in the making of Europe, writes that "medieval representations of the world were not altered materially by the geographers of the crusading age. . . . In short, the Crusades themselves changed almost nothing in Europe's pictorial image of Asia."[24]

A key source of new information in thirteenth- and fourteenth-century geographical treatises derived from a series of diplomatic, missionary, and

[21] S. Maqbul Ahmad, "Cartography of al-Sharīf al-Idrīsī," in Harley and Woodward, *History of Cartography*, vol. 2, bk. 1, pp. 156–74.

[22] Friedman, *The Monstrous Races in Medieval Art and Thought.*

[23] Wright, *Geographical Lore of the Time of the Crusades*, p. 287. On the instability of the concept of Gog and Magog, see Scott Westrem, "Against Gog and Magog," in *Text and Territory: Geographical Imagination in the European Middle Ages*, ed. Sylvia Tomasch and Sealy Gilles (Philadelphia: University of Pennsylvania Press, 1998), pp. 54–75.

[24] Donald Lach, *Asia in the Making of Europe*, 2 vols. (Chicago: University of Chicago Press, 1965–1970), vol. 1, pp. 30–3.

mercantile expeditions to the East. East Asia had long been thought to harbor the tribes of the Antichrist, Gog and Magog (although the meanings of these terms were very unstable in the Middle Ages). It was believed that the only thing preventing these tribes from spilling out of Asia and overwhelming Christendom was a great wall and gate that Alexander had constructed in antiquity to hold them back.[25] The devastating raids of Hungary and Poland by the Mongols in 1240–1 renewed fears of such an apocalypse but also raised the possibility that the Mongol emperor Genghis Khan might help to fight a rearguard action against the Muslims to rescue the Holy Land and help to protect the potentially lucrative but dangerous silk routes to China. In 1245, from the Council of Lyon, Pope Innocent IV dispatched the Franciscan John of Plano Carpini to discover the history and customs of the Mongols, to attempt to convert the Grand Khan, and to seek an alliance. John succeeded in reaching the Mongol capital of Karakorum and, upon his return in 1247, recorded his travels in the *History of the Mongols*, which discredited many of the misconceptions about Mongol customs current in the Christian West. Its topographical and geographical information was subsequently incorporated into Vincent of Beauvais's encyclopedia, the *Speculum historiale*.

John of Carpini's narrative was largely superseded by the accounts of other travelers to Asia – William of Rubruck (1253–5) for similar missionary purposes and, for very different mercantile reasons, Marco Polo (1298). Although Vincent of Beauvais and Albertus Magnus ignored William's work, Roger Bacon not only studied it with great care but also discussed it with its author. Bacon used William, for example, to confirm that the Caspian Sea was not an arm of the circular *Oceanus* surrounding the world – a classical opinion – but instead a huge inland sea.

The influence on medieval geographical knowledge of the travels of Niccolò, Maffeo, and Marco Polo (1260–9, 1271–95) is difficult to assess. Some scholars, such as Charles Beazley, regarded Marco Polo's book as the primary source for medieval geography and exploration. Others question the extent to which Marco Polo had seen the things he described.[26] In any event, his book provided a mine of mercantile knowledge of Asia, if limited in its effect on geographical theory and cartography. With the exception of the Catalan Atlas (ca. 1375), information from the narrative did not find its way onto maps until the fifteenth century.

[25] Charles Raymond Beazley, *The Dawn of Modern Geography*, 3 vols. (London: John Murray, 1897–1906; New York: Peter Smith, 1949), vol. 3, p. 15.

[26] For example, Kimble, *Geography in the Middle Ages*, pp. 142–3, comments on his lack of references to tea, the Great Wall, the compressed feet of Chinese women, or the use of fishing cormorants, but points out that none of the medieval European travelers to China mentioned tea, and only Odoric of Pordenone refers to the compressed feet of the women or to the cormorants. More recently, Frances Wood admits that the contents of the book remain a rich source of information about China and the Near East but leans toward the view that Marco never traveled much farther than the Black Sea. See Frances Wood, *Did Marco Polo Go to China?* (London: Secker and Warburg, 1995), p. 150.

Figure 23.1. The Ebstorf map. Mid-thirteenth century. This map represents the earth as the body of Christ, with Jerusalem at its center. A small T-in-O map is found in the margins, illustrating the general structure of the map. Size: 3.56 × 3.58 m. From Walter Rosien, *Die Ebstorfer Weltkarte* (Hanover: Niedersächisches Amt für Landesplanung und Statistik, 1952). By permission of the Niedersächisches Institut für Landeskunde und Landesentwicklung an der Universität Göttingen.

The most revealing graphical manifestation of global geographical knowledge was the *mappamundi*, or "world map," the primary purpose of which was to instruct the faithful in the geographical and historical setting of divinely planned events associated with the creation, salvation, and last judgment of the Earth.[27] Medieval world maps are thus often diagrammatic and formulaic in structure, falling into two broad categories. One type, the T-O map (Figure 23.1), consists of an interpretation of a three-part earth separated by the great water bodies of the Don River, Nile River, and Mediterranean Sea,

[27] Juergen Schulz, "Jacopo de' Barbari's View of Venice: Map Making, City Views, and Moralized Geography before the Year 1500," *Art Bulletin*, 60 (1978), 425–74; Woodward, "Medieval *Mappaemundi*," pp. 286–370.

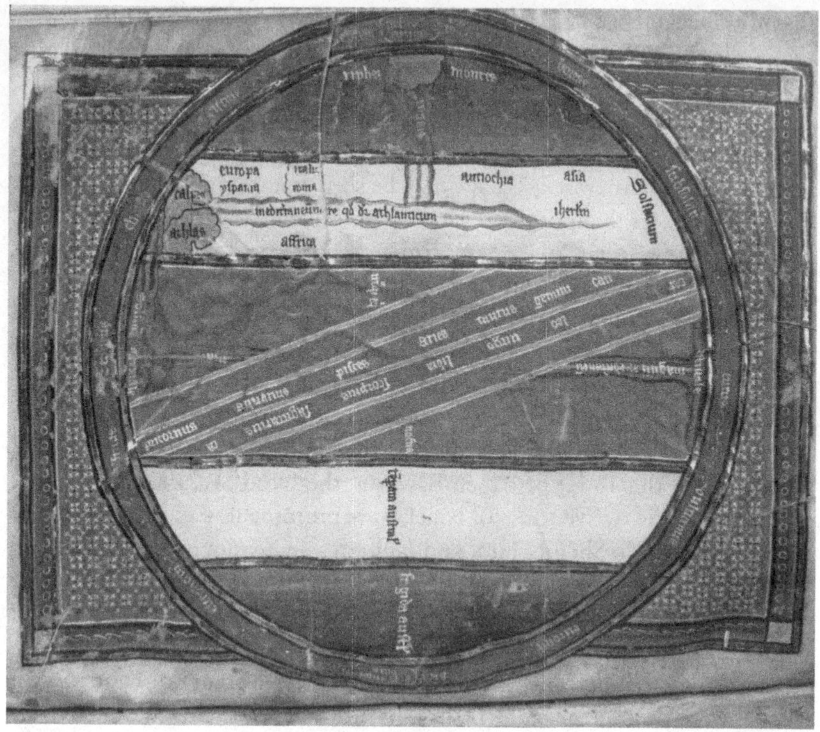

Figure 23.2. Zonal *mappaemundi* by William of Conches, from a twelfth-century manuscript of the *De philosophia mundi*. Diameter: 12.8 cm. By permission of the Bibliothèque Sainte-Geneviève, Paris (MS 2200, fol. 34v).

thus suggesting a T inscribed in an O. This type of map characteristically had east, the Christian sacred direction, at the top, and were often – but by no means always – centered on Jerusalem.

The other type (Figure 23.2), with north or south at the top, followed the zonal, or Macrobian, model and was less concerned with showing the world's divinely planned events than with the goal of demonstrating the Sun's apparent motion through the seasons and its effect on the Earth's climates. Over 150 *mappaemundi* of this type are found in manuscripts of Macrobius's commentary on Cicero's *Dream of Scipio* from the ninth century to the fifteenth and in several other works, such as the *Liber floridus* of Lambert of St. Omer (ca. 1120) and *De philosophia* of William of Conches (ca. 1130).

Although it is generally true that the purpose of the *mappaemundi* was to help the faithful navigate the spiritual and historical rather than the physical world, the occurrence and position of real places on the maps can often be used as evidence for dating or authorship. For example, Armin Wolf has persuasively used evidence about the occurrence of place names on the Ebstorf map that were associated with Otto the Child, concluding that the map must have been made in 1239.[28] Medieval world maps were thus collages of information of uneven veracity, complexity, and secularity, reflecting widely differing ideas of geographical and cosmographical "reality."

Unlike many modern maps, the *mappaemundi* were not representations of a slice through space at a moment in time but a projection of history onto a geographical framework.[29] They may be seen as analogous to the medieval narrative pictures that portray several events separated by time and included within the same scene. Some cities had ceased to exist long before the maps were drawn, but their historical importance merited their inclusion; examples of this are Troy in Asia Minor and Leptis Magna and Carthage in North Africa. The names of classical regions, such as Gallia, Germania, Achaea, and Macedonia, often appear along with cities that had gained contemporary commercial importance, such as Genoa, Venice, Bologna, and Barcelona. The locations of places mentioned in the Old Testament, such as Noah's ark, Mount Sinai, the Tower of Babel, Babylon, the Dead Sea, the river Jordan, Samaria, and the twelve tribes of Israel, were prominent. The descendants of the sons of Noah – Shem, Ham, and Japheth – increasingly appear assigned to their respective continents, Asia, Africa, and Europe.[30] Places in the New Testament marked on the maps included those associated with the life of Christ and the journeys of Paul and the apostles.[31] Pilgrimage sites such as Santiago de Compostela in Spain and the French island of Mont Saint-Michel were prominent, and the associated itineraries were the sources for many of the place names. Not surprisingly, Rome appears on almost every map, reflecting its multiple roles as the old imperial capital of the West, the seat of the papacy, and the city of many churches where indulgences were offered to pilgrims.

LOCAL DESCRIPTIONS AND MEASUREMENTS OF LAND AND PROPERTY

On a different scale, and adhering to a different geographical tradition than that of the scholarly worldview, was the medieval practice of describing land and property in a local setting. The functions of the few documents that survive can be divided into three broad categories: property management in both rural and urban contexts; legal purposes; and engineering, planning,

[28] Armin Wolf, "News on the Ebstorf World Map: Date, Origin, Authorship," in *Géographie du monde au moyen âge et à la Renaissance*, ed. Monique Pelletier (Paris: CTHS, 1989), pp. 51–68.

[29] Evelyn Edson, *Mapping Time and Space: How Medieval Mapmakers Viewed Their World* (London: British Library, 1997); and Paul Zumthor, *La mesure du monde* (Paris: Seuil, 1993).

[30] For a critique of the historical association of the continents with the sons of Noah, see Benjamin Braude, *Sex, Slavery, and Racism: The Secret History of the Sons of Noah* (forthcoming).

[31] Loveday Alexander, "Narrative Maps: Reflections on the Toponymy of Acts," in *The Bible in Human Society: Essays in Honour of John Rogerson*, ed. M. Daniel Caroll Rodas, David J. A. Clines, and Philip R. Davies (Sheffield: Sheffield Academic Press, 1995).

and constructing public works. A transition occurred from qualitative to quantitative measurement, but it is difficult to date precisely when this happened. In the early thirteenth century, surveys began to provide area measurements, and tables exist that give the length of an acre of land for any given width. In the early thirteenth century, *The practice of geometry* of Leonardo da Pisa (Fibonacci) describes how to use a plumb-bob level to find the horizontal area of a slope and shows how a quadrant can be used in surveying. Although we cannot infer from works such as Fibonacci's that the recommended instructions were routinely practiced, their appearance does reflect a rudimentary knowledge of measurement units and techniques needed in land descriptions. Some of these skills came from the continuing Roman tradition of *agrimensores*, or field measurers. Indeed, the Roman surveying manuals were still being copied in the twelfth, thirteenth, and fourteenth centuries, and they were often mentioned in the Middle Ages, notably by Isidore of Seville and Adelard of Bath.[32] Rather than drawing maps from these measurements, the land surveyor of the Middle Ages compiled *terriers*, or written descriptions showing the size, location, and character of fields, water, and other features on an estate. For example, in the *Domesday Book*, the 1086 survey of William I's entire kingdom, there are no maps.[33]

In some areas of Europe, particularly northern France and England, landholdings operated by tenants were under the direct supervision of the lord of the manor and did not require precise or frequent resurveying as the plots changed tenants. However, as the management of agriculture became more complex in the fourteenth and fifteenth centuries, with the breakdown of feudalism and the development of demesne farming (estate managers tilling lands through their own employees), efficient management became a concern. This resulted in a marked increase in written estate surveys for inventory purposes and the appearance of treatises on estate management. One of the best known of these is the thirteenth-century *Seneschalship*, which describes the duties of the medieval land steward.[34] In the sixteenth and seventeenth centuries, as the population and demand for land increased, the concept of land ownership demanded dramatic developments in the technique of surveying and the translation of the surveyor's data into more graphic plans.

In addition to written surveys of towns, including lists of householders, rents payable, and occasional dimensions of plots, city plans start appearing in the fourteenth century. Examples include the plans of Venice and Rome in Paolino Veneto's historical works (ca. 1320) and Petrus Vesconte's plans of

[32] Danielle Lecoq, "La 'Mappemonde' du *De Arca Noe Mystica*," in *Géographie du monde au moyen âge et à la Renaissance*, ed. Monique Pelletier (Paris: CTHS, 1989), p. 15; and Brian Campbell, "Shaping the Rural Environment: Surveyors in Ancient Rome," *Journal of Roman Studies*, 86 (1996), 74–99.

[33] F. M. L. Thompson, *Chartered Surveyors: The Growth of a Profession* (London: Routledge and Kegan Paul, 1968), p. 7.

[34] Thompson, *Chartered Surveyors*, pp. 5–6; Derek J. Price, "Medieval Land Surveying and Topographical Maps," *Geographical Journal*, 121 (1955), 1–10; and H. Clifford Darby, "The Agrarian Contribution to Surveying in England," *Geographical Journal*, 82 (1933), 529–31.

Acre and Jerusalem. Town planning based on orthogonal streets is evident in the fourteenth-century "new towns," as in the plan of Talamone (1306).[35]

The earliest medieval city plan bearing a geographical scale is of Vienna and Bratislava – a mid-fifteenth-century copy of an original drawn around 1422. By the middle of the fifteenth century, Leon Battista Alberti describes the practice of drawing maps of towns to scale. In his *Description of the City of Rome*, he explains how key features in a city can be plotted by sighting their angles from a central prominent location; in later works, *Mathematical Games* (1450) and *Ten Books on Architecture* (1452), he describes other methods of constructing maps to scale, including the concept of triangulation.[36]

Many local maps in the Middle Ages seem to have been made to help document a case or to be presented as evidence in court. A treatise by the Italian jurist Bartolo da Sassoferrato (1313–1357), *On Rivers* (1355), describes how plans might be used to settle disputes over the division of watercourses.[37] In 1395, Jehan Boutillier wrote that maps might be used when presenting cases to the French Parlement. In 1444, the Duke of Burgundy, disputing his duchy's boundary with those of the French king, had maps made to show which towns and villages were included in the duchy so as to guard against royal encroachments. In medieval England, a map of Inclesmoor was drawn between 1405 and 1408 to illustrate a dispute over rights to pasture and peat in the area.[38]

The use of maps and plans in planning and constructing buildings, bridges, aqueducts, water supply systems, canals, mines, tunnels, walls, and fortifications was rare in this period, although a sufficient number of examples demonstrate the use of practical geometry in laying out public works. An early application was in the description and planning of water systems (Figure 23.3).

WAYFINDING AND NAVIGATION WITH ITINERARIES AND CHARTS

The medieval pilgrim or traveler usually found his way, as travelers had done for centuries, by asking directions. Under these circumstances, a map was not necessary. Routes were documented, however, in the form of written itineraries that provided a list of suitable waypoints. Such itineraries had classical roots in the Roman *itineraria scripta* (written itineraries), examples

35 See David Friedman, *Florentine New Towns: Urban Design in the Late Middle Ages* (Cambridge, Mass.: MIT Press, 1988), pp. 50–5.

36 Joan Kelly Gadol, *Leon Battista Alberti: Universal Man of the Early Renaissance* (Chicago: University of Chicago Press, 1969), pp. 175–8.

37 Bartolo da Sassoferrato, *La Tiberiade di Bartole da Sasferrato; del modo di dividere l'alluvioni, l'isole, & gl'alvei* (Rome: G. Gigliotto, 1587).

38 R. A. Skelton and P. D. A. Harvey, *Local Maps and Plans from Medieval England* (Oxford: Clarendon Press, 1986), pp. 489–93.

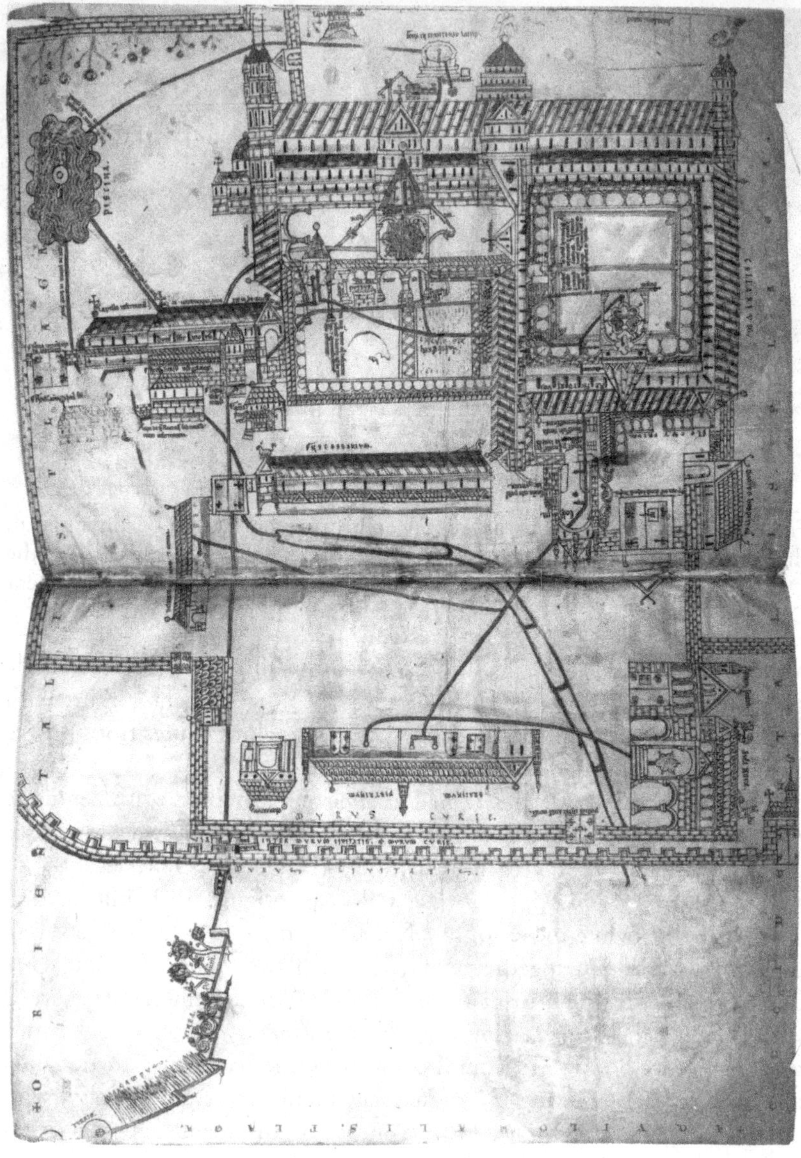

Figure 23.3. Detail from the plan of Canterbury Cathedral, mid-twelfth century, showing the water supply. Size: 45.7 × 55 cm. By permission of the Master and Fellows of Trinity College, Cambridge (MS. R.17.1, fol. 285).

of which include the Antonine itinerary (late third century) and the Bordeaux itinerary (ninth-century copy of a fourth-century archetype), which describe Christian pilgrimage routes.[39] They became common in the high and late

[39] O. A. W. Dilke, "Itineraries and Geographical Maps in the Early and Late Roman Empires," in Harley and Woodward, *History of Cartography*, vol. 1, pp. 234–42.

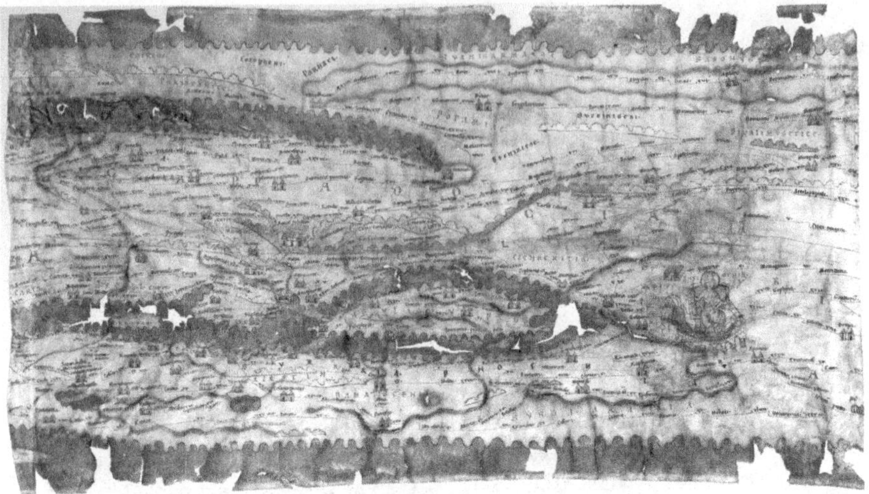

Figure 23.4. One sheet from the Peutinger map of the Eastern Mediterranean, showing Cyprus and Antioch. Size: 33 × 61.4 cm. By permission of the Österreichische Nationalbibliothek, Vienna (Codex Vindobonensis 324, segment IX).

Middle Ages, and their use in pilgrimage was sometimes extended to include mercantile functions; an excellent example is the Bruges itinerary (fourteenth century), which lists the distances between places along routes from Bruges to other parts of Europe.[40]

The Roman *itineraria scripta* had as their rarer graphic equivalents the *itineraria picta* (graphic itineraries), of which the best-known example is the *Tabula Peutingeriana* (Peutinger Map) (Figure 23.4), an eleventh- or twelfth-century manuscript copy of a Roman original with information possibly dating before the eruption of Vesuvius in 79. Its vellum sheets were designed to be fitted together to form a map twenty-one feet long and one foot wide on which roads throughout the Roman Empire are marked with distances between staging posts. Some itinerary maps, like the four versions of Matthew Paris's from the mid-thirteenth century, were designed in a strip format, showing the route from London to Apulia (Italy), with distances noted in days' journeys. The purpose of this manuscript was less as a practical aid to the traveler than as a history of contemporary pilgrimage. Others by Matthew Paris look more like modern maps but have itinerary information inscribed upon them. Examples are his Oxford map of Palestine and his map of Great Britain, which shows the route from Newcastle upon Tyne to Dover, with intermediate places, including Matthew Paris's own monastery at St. Albans.[41]

40 Joachim Lelewel, *Géographie du moyen âge*, 4 vols. and epilogue (Brussels: Pilliet, 1852–1857; repr. Amsterdam: Meridian, 1966), pp. 281–308.

41 Richard Vaughan, *Matthew Paris* (Cambridge: Cambridge University Press, 1958).

In the classical world, the marine equivalent of the land itineraries were sailing directions known as the *itineraria maritima* (maritime itineraries) in Latin or *periploi* in Greek.[42] *Periploi* were coastal descriptions with information about places and approximate distances between selected ports. They were often more qualitative than quantitative. One of the more practical is the Greek *Stadiasmus, or Circumnavigation of the Great Sea*, possibly compiled in the fourth or fifth century. It contains detailed and remarkably consistent distances in stadia along the Mediterranean and Black Sea coasts, as well as between selected ports across open water.[43]

In the Middle Ages, textual sailing directions without graphics (not to be confused with the charts derived from them) became known as *portolani*. The oldest systematic *portolano* for the Mediterranean is the *Book of the Coasts and the Form of our Mediterranean Sea*, which contains place names datable to 1160–1200. Another example is *Lo conpasso de navegare* (*The Sailing Compass*), a 1296 manuscript based on a lost mid-thirteenth-century source.[44]

The portolan chart (not to be confused with the textual portolan from which it was derived) was a nautical chart, usually of the Mediterranean and Black Seas, drawn on vellum and crossed with a pattern of radiating rhumb lines used for laying out courses. The oldest surviving chart, the Carte Pisane, dates from the last quarter of the thirteenth century.[45] Coastal place names are written perpendicularly to the trend of the coasts, and the interior land areas are usually devoid of information, as would be expected given their navigational function. The character of the coastline drawing is usually different from that of the *mappaemundi*, consisting of exaggerated bays and headlands, furnished with symbols locating rocks and shoals. Graphical scales are usually present, but latitude and longitude graduations are missing.

A clue to what was available to the late-thirteenth-century mariner for practical navigation is afforded by a passage in the *Tree of Science*, written in 1295–6 by Ramón Lull (ca. 1233–1315), the Majorcan polymath. In it, Lull describes how sailors measure their mileage at sea, with a *raxon de marteloio* (table of distances traveled along various directions), a chart, sailing directions, magnetic compass, and the polestar.

The word "compass" was used to mean either sailing directions or a pair of dividers for stepping off measurements rather than what we refer to as the

[42] Adolf Erik Nordenskiöld, *Periplus: An Essay on the Early History of Charts and Sailing Directions*, trans. Francis A. Bather (Stockholm: Norstedt, 1897); and Lionel Casson, introd., trans., and comment., *The Periplus Maris Erythraei* (Princeton, N.J.: Princeton University Press, 1989).

[43] Nordenskiöld, *Periplus*, p. 14.

[44] Tony Campbell, "Portolan Charts from the Late Thirteenth Century to 1500," in Harley and Woodward, *History of Cartography*, vol. 1, p. 383; Patrick Gautier Dalché, "Les savoirs géographiques en Méditerranée chrétienne, XIIIe siècle," *Micrologus*, 2 (1994), 75–99; and John H. Pryor, *Geography, Technology, and War: Studies in the Maritime History of the Mediterranean, 649–1571* (Cambridge: Cambridge University Press, 1988).

[45] For an opinion on an earlier origin, see Patrick Gautier Dalché, *Carte marine et portulan au XIIe siècle: Le 'Liber de existencia riveriarum et forma maris nostri Mediterranei' (Pise, circa 1200)* (Rome: École Française de Rome, 1995).

Figure 23.5. The Cortona chart, early or mid-fourteenth century. Size: 47 × 50 cm. By permission of the Pubblica Biblioteca Comunale e dell'Accademia Etrusca, Cortona.

magnetic compass, which they called the "needle." The origin and use of the magnetic compass in Europe has been the subject of much mythology, most recently clarified by Julian Smith, who summarizes the early descriptions by Alexander Neckam, Guyot de Provins, and Jacques de Vitry and points out that the main value of the *Letter on the Magnet* of Peter Peregrinus (1269) was in organizing, rather than originating, knowledge of the compass. By then, the principle of the compass was already widely known, even if the instruments were probably not in common use for navigation.[46]

The association of the magnetic compass with the early portolan charts seems plausible. Not only do they appear around the same time, but the axis of early charts is rotated about 10° or 11° from an east–west line, suggesting that the rhumb lines may have been laid out with a magnetic compass without adjusting for magnetic declination (Figure 23.5). It is perhaps most likely, however, that they were compiled from various sources, including bearings (both wind and magnetic compass) and rough distances gleaned from repeated voyages, written itineraries, or other charts.

[46] Julian A. Smith, "Precursors to Peregrinus: The Early History of Magnetism and the Mariner's Compass in Europe," *Journal of Medieval History*, 18 (1992), 21–74.

CONCLUSIONS

In the Christian Middle Ages, during which the overarching conceptual framework of knowledge was related to understanding God's creation, geography was often cast in a religious and cosmographical context. Theoretical issues in geography, such as knowledge of the size and shape of the Earth, daily and seasonal changes in the relationship of the Earth to the Sun, the hydrological cycle, or the classification of climates and winds, were routinely addressed to confirm the regularity of a divinely created Earth. Geography provided an understanding of how the cycles of day and night and the seasons were tied to the rhythms of human life and worship. Knowledge of how climates varied across the Earth was essential for understanding the physical differences between human races, which were the object of proselytizing. The locations of physical features, such as rivers, lakes, seas, marshes, forests, and mountains, were needed to inform missionary expeditions of benefits that would aid travel and barriers likely to be encountered in their journeys.

It would be misleading, however, to assume that religious thought provided the only context for medieval geography. Mercantile, economic, military, and political aims were especially significant in the case of applied geographic technology (although far from absent in theoretical geography), and the practical arts of navigation, sea charting, land surveying, and their associated instrumentation were more in evidence in the Middle Ages than is commonly supposed.

More often than not, sacred and secular motives were mixed. Roger Bacon's call for an awareness of geography combined a pragmatic interest in the exercise of secular power with attention to the spiritual meaning of physical places and the goals of Christian missionary activity. Even the utilitarian portolan charts often carried the iconography of Christ, the Virgin Mary, or the parting of the Red Sea along with the starkly precise representation of coastlines, rocks, shoals, and sailing courses. In the fourteenth and fifteenth centuries, three fundamentally different geometries of map construction that had hitherto existed side by side started to converge. The world maps of Pietro Vesconte of the 1320s, the fascinating moralized maps of Opicinus de Canistris (1296–ca. 1350), and the Catalan Atlas (ca. 1375) were all mixtures of portolan charts and *mappaemundi*.[47] The map of Fra Mauro (1459) represents the essence of this convergence, including information derived from Ptolemy's *Geography*, portolan charts, and *mappaemundi*. In its circular framework, it is clearly medieval, and the southern orientation shows Arabic influence, but the Mediterranean coasts are modeled on

[47] Pierluigi Tozzi, *Opicino e Pavia* (Pavia: Libreria d'Arte Cardano, 1990); and Peter Barber, "Old Encounters New: The Aslake World Map," in *Géographie du monde au moyen âge et à la Renaissance*, ed. Monique Pelletier (Paris: CTHS, 1989), pp. 69–88.

portolan charts, and the map bears an allusion (and thereby reveals a debt) to the Ptolemaic tradition.

A dominant theme in the study of medieval geography is the constant interplay between text and image. In the early Middle Ages, geographical texts and itineraries were preferred to graphic images as sources, and images were usually derived from texts and not vice versa.[48] By the late Middle Ages, texts were increasingly accompanied by images, which took on a greater degree of independence. The distinction between text and image was blurred. Graphical *mappaemundi* were sometimes called *histories*, and narrative works of a geographical nature containing no maps were sometimes called *mappaemundi*.

Medieval geography owed much to its classical sources and was to some extent controlled by the availability of these texts in translation. Even land surveying was rooted in an extensive Roman tradition. Only the portolan charts appear to have had no classical precursor. It is important, however, to evaluate medieval geography in the context of the sacred and secular needs of medieval society. It was not a cultural backwater between the ancient and modern worlds; nor did it represent a regression in the otherwise progressive march of geographical knowledge. In the course of the period from the eleventh to the fifteenth centuries, a transition occurred in which itineraries were supplemented by maps and in which qualitative geographical descriptions began to include more quantitative measurements. This says much more about the changing needs of the society in which it grew than about the cognitive or practical abilities of medieval people.

[48] P. D. A. Harvey, *Medieval Maps* (London: British Library, 1991), pp. 7–9.

24

MEDIEVAL NATURAL HISTORY

Karen Meier Reeds and Tomomi Kinukawa

In the intellectual world of the medieval Latin West, natural history had no standing. It did not figure in the curriculum of schools or universities. It offered no employment. It brought no honors. Yet, in an age when food and medicine came from the gardens and fields, woods, and rivers that lay just beyond the walls of even the most crowded city, the natural world could not be ignored.[1]

Even though natural history found no home in medieval institutions of learning, its constituency ranged across medieval society. From emperors to falconers, from popes to nuns, from kings to poets, from schoolmasters to artists, there is occasional evidence of interest in the creatures, objects, and workings of the natural world. The motives for that interest ranged from the intellectual to the practical to the decorative, from the spiritual to the magical. This chapter aims to survey the motives and opportunities in the high and late Middle Ages for observing and discussing "natural things" – to use a generic term often employed by medieval writers – and to characterize the ways in which that natural knowledge expressed itself.[2]

"Natural history" takes its name from Pliny the Elder's massive encyclopedia, *Historia naturalis* (completed in 75). Its subject matter, like Pliny's encyclopedia, encompasses virtually all natural phenomena and objects on earth. Natural history stresses the collection of interesting facts about the particulars of nature rather than the formulation of theories about natural processes, although scientific theories and laws may be the ultimate goal or result of the accumulation of many specific natural histories. It relies on the observation, collection, and comparison of individual specimens. In practice, plants, animals, and minerals have always provided the core subjects for

[1] Howard Saalman, *Medieval Cities* (New York: George Braziller, 1968), pp. 24, 40–1; and C. S. Lewis, *The Discarded Image: An Introduction to Medieval and Renaissance Literature* (Cambridge: Cambridge University Press, 1964), pp. 146–7.

[2] Guy Beaujouan, "Motives and Opportunities for Science in the Medieval Universities," in *Scientific Change*, ed. A. C. Crombie (New York: Basic Books, 1963), pp. 219–36, 301–25.

natural history, with weather and exotic human cultures as frequent additions. The study of natural history has often been regarded as the precursor to the sciences of botany, zoology, mineralogy, meteorology, geography, and anthropology.

This description of natural history's scope and methods works quite well for the ancient world – notably, Aristotle's works on animals, Theophrastus's on plants and stones, much of Dioscorides' on materia medica, and much of Pliny's own *Natural History* – and again for the period from the late fifteenth century to the present.[3] But it fails for the Middle Ages. From roughly the eleventh century through the fifteenth, only a handful of people left any systematic written evidence of their observations of plants, animals, rocks, and weather, and even to these few we cannot apply the term "natural history" very strictly. However, if we count other forms of expression of knowledge of nature's particulars, then we begin to get a sense of what part knowledge of natural things played in medieval life and thought.

So this chapter will look at the treatment of beasts, stones, and herbs in medieval scientific writings, but it will also ask what a mystic's visions, a schoolchild's Latin assignment, a theologian's dating of Creation, or a sculptor's carving of oak leaves can tell us about the extent of medieval knowledge of natural things and, more broadly, about attitudes toward the natural world. Finally, it will consider the role of visual images in medieval natural history.

NATURAL HISTORY'S PLACE IN THE MEDIEVAL INTELLECTUAL WORLD

It is appropriate to use medieval classifications of knowledge to understand medieval reasons for looking at the objects of nature (see Cadden, Chapter 9, this volume). Broadly speaking, medieval rankings of the sciences gave highest status to understanding the divine order through the study of the Bible. Next in importance were the theoretical sciences – physics and

[3] Roger French and Frank Greenaway, *Ancient Natural History: Histories of Nature* (London: Routledge, 1994); Peter Dilg, "Die botanische Kommentarliteratur Italiens um 1500 und ihr Einfluss auf Deutschland," in *Der Kommentar in der Renaissance*, ed. August Buck and Otto Herding (Kommission für Humanismusforschung, 1) (Bonn: Deutscheforschungsgemeinschaft, 1975), pp. 225–52; Charles G. Nauert, Jr., "Humanists, Scientists, and Pliny: Changing Approaches to a Classical Author," *American Historical Review*, 84 (1979), 72–85; Charles G. Nauert, Jr., "C. Plinius Secundus (Naturalis Historia)," in *Catalogus Translationum et Commentariorum: Mediaeval and Renaissance Latin Translations and Commentaries; Annotated Lists and Guides*, ed. F. Edward Cranz and Paul Oskar Kristeller (Washington, D.C.: Catholic University Press, 1980), vol. 4, pp. 297–422; Marjorie Chibnall, "Pliny's *Natural History* and the Middle Ages," in *Empire and Aftermath: Silver Latin II*, ed. T. A. Dorey (London: Routledge and Kegan Paul, 1975), pp. 57–78; and Pedanius Dioscorides of Anazarbus, *De materia medica*, Engl. trans. Lily Y. Beck (Altertumswissenschaftliche Texte und Studien, 38) (Hildesheim: Olms, 2005). The shortened medieval versions of Dioscorides generally retained the medicinal uses of natural products and dropped the natural history observations.

mathematics – and the human sciences, which included the tools of language and logic. The practical, "mechanical" arts – weapon-making, farming, healing, trading, cooking, and weaving – were recognized as essential to human life, though ranking lower than the liberal arts. Magic, if acknowledged at all, came at the bottom, with the warning that it was to be avoided.[4] The study of the particulars of nature did not figure explicitly in any medieval division of knowledge, but the disciplines of theology, physics (that is, natural philosophy), rhetoric, medicine, farming, hunting, and magic each provided the occasion for some expression of natural history knowledge.

The *form* of expression was shaped to a large degree by each discipline's expectations about its purpose, method, and audience. It is a truism that the Middle Ages cherished its authorities and preferred the commentary and compilation to original research. Each domain of knowledge claimed a set of authorities for itself and surrounded those key texts with distinctive genres of explication and exposition.

Saint Augustine (354–430) justified natural history to generations of Christians in his very influential treatise *On Christian Doctrine*. Full understanding of the Bible's figurative language was impossible when "we are ignorant of the natures of animals, or stones, or plants, or other things which are often used in the Scriptures for purposes of constructing similitudes."[5] Thus, the admonition "Be ye wise as serpents" (Matt. 10:16) appealed to two "well-known fact[s]": first, that "a serpent exposes its whole body in order to protect its head" from its enemies; and second, that a serpent forces its way through narrow openings and so loses its old skin and "renews its vigor." From serpents Christians should learn not to deny their head, Christ, in order to save their bodies from persecutors of the faith and that, by entering at "the narrow gate" (Matt. 7:13), they could shed their old ways and put on the new (Gen. 8:11).

The first chapter of Genesis – the six days of Creation – offered a particularly rich opportunity for natural history to serve biblical exegesis. Following the example of treatises on the hexaemera (six, *hexa*; days, *hemera*) by the church fathers Ambrose and Basil, medieval theologians used the third day as the starting point for discussing the arrangement of mountains and rivers, the generation of minerals in the earth, the phenomena of rain, hoarfrost, and comets, and the growth of plants, and the fifth and sixth days for describing the kinds and natures of "winged fowl," moving creatures of the waters, "beasts of the earth," and human beings. However, natural history is

[4] James A. Weisheipl, "The Nature, Scope, and Classification of Sciences," in *Science in the Middle Ages*, ed. David C. Lindberg (Chicago: University of Chicago Press, 1978), pp. 461–82, especially pp. 474, 479–80; Elspeth Whitney, "Paradise Restored: The Mechanical Arts from Antiquity through the Thirteenth Century," *Transactions of the American Philosophical Society*, 80 (1990), 1–170, especially chap. 4; and Hugh of St. Victor, *The Didascalicon of Hugh of St. Victor*, trans. Jerome Taylor (New York: Columbia University Press, 1961).

[5] Augustine, *On Christian Doctrine*, XVI.24, trans. D. W. Robertson, Jr. (Indianapolis: Bobbs–Merrill, 1958), pp. 50–1.

not brought to bear on Genesis solely for the sake of what Augustine called "narration or description."[6] When Henry of Langenstein lectured on the third day to his theology students at the University of Vienna in the 1380s to 1390s, for example, he used general knowledge of plant growth to raise, though not settle, the question: In what season of the year did Creation take place? In spring when grass and herbs spring forth from the earth, or "in September when the fruit in trees are ripe and have seeds in them?"[7]

When Henry came to discuss the animals created on the fifth and sixth days, he relied heavily on Aristotle's zoology. The Aristotelian and pseudo-Aristotelian treatises on living things had reached the Latin West through translations from Arabic versions by the early thirteenth century, through translations directly from Greek by the mid-thirteenth century, and through translations (by the 1230s) of loose paraphrases by the Islamic commentator Avicenna. By the mid-1200s, these works had become assimilated into the university arts curriculum as parts of Aristotle's "books on nature" (*libri naturales*). Aristotle's references to natural history in his nonbiological works could also stimulate a receptive reader. Robert Grosseteste, bishop of Lincoln (ca. 1168–1253), picked up on Aristotle's brief examples of cause and effect in nature in the *Posterior Analytics* and even used his own experience to challenge some of Aristotle's explanations. Grosseteste argued, for instance, that broad-leaved plants are deciduous because the sap going to the leaves dries up rather than thickening and congealing as Aristotle had asserted. Moreover, the breadth of the leaf makes it both heavier and more likely to be blown off by the wind. However, although "Aristotle did not understand this," the thickening of sap does account for petals falling off flowers as the sap congeals to form the fruit.[8]

Aristotle's *On the Soul* (*De anima*) was the key text for understanding the nature of living things, not only because it defined the three kinds of soul (vegetative, sensitive, and rational) that governed the functions of life in plants, animals, and humans but also because it set the mode for medieval discussions of them. Philosophical questions about "the reason for something being the way it is" – as Robert Kilwardby put it in the mid-thirteenth century – were applied to broad classes of natural things and took precedence over the investigation of the details of individual kinds.[9]

Although Aristotle's four zoological books – known collectively as *On Animals* (*De animalibus*) – discussed the kinds, parts, movements, behavior, and reproduction of a great variety of individual animals, often drawing upon

[6] Ibid., XXIX, p. 65.

[7] Nicholas H. Steneck, *Science and Creation in the Middle Ages: Henry of Langenstein (d. 1397) on Genesis* (Notre Dame, Ind.: University of Notre Dame Press, 1976), p. 55.

[8] R. W. Southern, *Robert Grosseteste: The Growth of an English Mind in Medieval Europe*, 2nd ed. (Oxford: Oxford University Press, 1992), pp. 161–3.

[9] Whitney, "Paradise Restored," p. 121; and Lynn S. Joy, "Scientific Explanation from Formal Causes to Scientific Laws," in *The Cambridge History of Science*, vol. 3: *Early Modern Science*, ed. Katharine Park and Lorraine Daston (Cambridge: Cambridge University Press, 2006), chap. 3.

Aristotle's own observations, the primary appeal of Aristotelian zoology to medieval readers lay in its explanations of how living creatures function. The short treatises on colors, memory, sleep, length of life, and youth and old age (grouped together as the *Short Works on Nature*, or *Parva naturalia*) and *On Plants* (*De plantis*) – all ascribed to Aristotle in the Middle Ages – reinforced this attitude.[10] *On Plants* referred to about fifty plants by name, but only as examples of the kinds of phenomena that warrant explanation by means of natural philosophy.

Although from the thirteenth century onward most universities required courses on these Aristotelian works for the bachelor of arts degree, they generated very few written commentaries, and the emphasis was always on the causes of general biological phenomena.[11] To judge by Roger Bacon's *Questions regarding On Plants* (*Questiones supra de plantis*) – an apparently verbatim record of his five-week course at Paris in the 1240s – the teacher and his teenage students concentrated entirely on the properties of the vegetative soul. In grafted plants, for example, are the souls of the stock and scion continuous? The references in *On Plants* to the familiar marjoram and cherry and the exotic fig tree and deadly nightshade of Persia evoked no discussion of those plants' special features and properties.[12]

Albertus Magnus (before 1200–1280) was not restricted by university schedules when he agreed to teach all the Aristotelian natural philosophical works to his fellow Dominicans at their order's equivalent of a university in Cologne.[13] His lengthy commentaries on *On Minerals* (*De mineralibus*, ca. 1250), *On Plants* (*De vegetabilibus*, ca. 1259), and *On Animals* (*De animalibus*, completed by the early 1260s) have endowed him with the reputation of premier medieval naturalist.[14] Both for their own content and for the

[10] Nicolaus Damascenus, *De Plantis: Five Translations*, ed. H. J. Drossaart Lulofs and E. L. J. Poortman (Aristoteles Semitico-Latinus) (Amsterdam: North-Holland, 1989). See also Petrus de Alvernia, *Sententia super librum 'De vegetabilibus et plantis'*, ed. E. L. J. Poortman (Aristoteles Semitico-Latinus) (Leiden: Brill, 2003).

[11] Steneck, *Science and Creation in the Middle Ages*, pp. 106, 182 n.2.

[12] Pseudo-Aristotle, "De plantis," in Aristotle, *Minor Works*, trans. W. S. Hett (Loeb Classical Library) (Cambridge, Mass.: Harvard University Press, 1936), pp. 139–233; and Karen Meier Reeds, *Botany in Medieval and Renaissance Universities* (New York: Garland, 1991), pp. 8–9.

[13] Simon Tugwell, ed., *Albert & Thomas: Selected Writings* (Classics of Western Spirituality) (New York: Paulist Press, 1988), p. 3.

[14] Albertus Magnus, *Book of Minerals*, trans. and ed. Dorothy Wyckoff (Oxford: Oxford University Press, 1967), pp. viii, 203–34; John M. Riddle, "Lithotherapy in the Middle Ages: Lapidaries Considered as Medical Texts," reprint III; John M. Riddle and James A. Mulholland, "Albert on Stones and Minerals," reprint VII, in John M. Riddle, *Quid pro quo: Studies in the History of Drugs* (Variorum Collected Studies Series) (Aldershot: Variorum/Ashgate, 1992), pp. 221, 228–30; Albertus Magnus, *De Vegetabilibus Libri VII, Historiae Naturalis Pars XVIII*, ed. Ernst Meyer and Carl Jessen (Frankfurt am Main: Minerva, 1982) (facs. repr. of Berlin edition, 1867); Albertus Magnus, *De animalibus libri xxvi*, ed. Hermann Stadler, 2 vols. (Münster: Verlag der Aschendorffschen Verlagsbuchhandlung, 1920); Albertus Magnus, *On Animals: A Medieval Summa Zoologica*, trans. and annot. Kenneth F. Kitchell, Jr. and Irven Michael Resnick, 2 vols. (Baltimore: Johns Hopkins University Press, 1999); and James A. Weisheipl, "Life of St. Albert," in *Albertus Magnus and the Sciences: Commemorative Essays, 1980*, ed. James A. Weisheipl (Toronto: Pontifical Institute of Mediaeval Studies, 1980), pp. 13–51 at p. 35.

contrast they present with other medieval works touching on natural things, they warrant a closer look.

Albertus's goal was to "compose . . . a complete exposition of natural science . . . from which [my brothers] might be able to understand correctly the books of Aristotle."[15] Albertus worked through all the natural philosophy treatises, beginning with the most universal issues and simplest forms of matter and working up to the most complex, animals and human beings.[16] He systematically paraphrased each text, explained and reorganized the structure where it was confused, checked other authorities, and commented on terms, ideas, and difficult passages, sometimes raising a question and arguing pros and cons of possible answers in standard scholastic style. Avicenna's paraphrases of Aristotle were his model, and he did not hesitate to fill out unfinished or defective treatises with material from other works and substantial digressions of his own.[17]

Albertus's range of sources and observations show how he took advantage of his many journeys across Europe – his administrative duties as head of the Dominicans' German Province required him to visit each of the order's thirty-plus houses from the Lowlands to Latvia on foot – to consult libraries, talk to experienced observers, and closely observe the natural world.[18] *On Minerals* demanded special effort. Although he had heard of a book by Aristotle on stones and had searched hard for a copy, all he could find was an incomplete manuscript. Accordingly, he had to flesh it out with material from Avicenna, encyclopedias, several different lapidaries, and other Aristotelian treatises.[19]

Albertus devoted the great bulk of *On Animals* and *On Plants* and much of *On Minerals* to discussing causes. He used the complete scholastic toolkit of explanations – properties of the four elements, humors, complexions, and their mixtures, the perfection of the sphere's shape, multiplication of forms, the radical or seminal fluid, celestial powers instilled by the stars, the actions of the vegetative and sensitive souls, essence and accident, and so on – while almost completely ignoring theological issues. Typical chapter headings include: "on the cause of the lungs and its natural features (*naturalibus accidentibus*) and the liver and spleen, and their natural workings (*operationibus*)," "why some egg-layers have many eggs and others only a few"; and "into what kinds the genus of plants is divided, and on account of what cause."[20]

15 Benedict M. Ashley, "St. Albert and the Nature of Natural Science," in Weisheipl, *Albertus Magnus and the Sciences*, pp. 73–102, especially p. 78, quoting Albertus Magnus, *Physica*, bk. I, tr. 1, chap. 1.

16 Albertus Magnus, *Minerals*, bk. I, tr. 1, p. 11.

17 Ashley, "St. Albert and the Nature of Natural Science," in Weisheipl, *Albertus Magnus and the Sciences*, p. 79 n. 31; and Weisheipl, "Life of St. Albert," p. 31.

18 Tugwell, *Albert & Thomas*, p. 12.

19 Albertus Magnus, *Minerals*, bk. III, tr. 1, chap. 1, pp. xxxiv–xxxix, 153; and Riddle and Mulholland, "Albert on Stones and Minerals," pp. 230–2.

20 Albertus Magnus, *De animalibus*, vol. 2: Lib. XIII, tr. 1, chap. vi; Lib. XVII, tr. 1. chap. 1; and Albertus Magnus, *De Vegetabilibus*, Lib. I, tr. II, chap. 5; Lib. III, tr. 1, chap. 4.

EXPERIENCE AND THE WORLD OF PARTICULARS

As a philosopher and a teacher, Albertus seems to have taken pleasure in explaining the causes of natural things. He stands alone among contemporary Aristotelian commentators in his readiness to insert examples of particular kinds of animals and plants and in his frequent appeals to his own experience and observations. Aristotle's text, for example, alluded only tangentially to the shape of fruit (*On Plants*, 829b5–20); Albertus spelled out the variety of shapes – spherical, pyramidal, oval, and columnar – since each needed a separate explanation. Moreover, he went beyond familiar cultivated plants to wild plants for his examples. For plants with small, round, spherical fruit, he cited *dracontea* and "field lettuce" as well as grapes.[21]

Albertus's interest in particular things is most striking in the final sections of *On Animals*, *On Plants*, and *On Minerals*. Here he deliberately changed modes of discourse and method, switching away from the universals of natural philosophy to the particulars of the practical sciences. Up to this point, he had mentioned forms and natures of particular kinds for the sake of example; now, he would list and describe the same things and their properties individually.[22]

He justified the move somewhat differently in each work. For minerals, he pointed to Aristotle's unfulfilled intention of describing individual kinds of metals, stones, and gems (*Meteorologica*, III.378b).[23] For plants, he was trying "rather to satisfy the curiosity of students than [the requirements of] philosophy – since philosophy cannot be about particulars."[24] For animals, he wanted his book to appeal to an even wider audience: "[W]e feel we are under obligation to both the learned and the unlearned alike, and . . . we feel that when things are related individually and with attention to detail, they better instruct the rustic masses."[25] For each group, he began with a general description and described the most perfect, noble, and complex things (gems, trees, humans, and quadrupeds) in each class of objects. He apologized for proceeding alphabetically through the members of each class. The alphabetical order was unphilosophical, yet it was convenient and customary among physicians in their books on simple medicines and – he might have added – in the bestiaries, herbals, and lapidaries that he used as sources for much of his information.[26]

[21] Albertus Magnus, *De Vegetabilibus*, Lib. III, tr. 1, chap. 4 and n. ΩΩ; and Karen Reeds, "Albert on the Natural Philosophy of Plant Life," in Weisheipl, *Albertus Magnus and the Sciences*, pp. 341–54.

[22] Jerry Stannard, "Albertus and Medieval Herbalism," in Weisheipl, *Albertus Magnus and the Sciences*, pp. 358–9.

[23] Albertus Magnus, *Minerals*, p. xxxi.

[24] Albertus Magnus, *De Vegetabilibus*, Lib. VI, tr. 1, chap. 1, para. 1.

[25] Lynn Thorndike, *A History of Magic and Experimental Science*, 8 vols. (New York: Columbia University Press, 1923–1958), vol. 2, p. 536; Albertus Magnus, *De animalibus*, vol. 2, Lib. XXII, tr. 1, chap. 1, para. 1; and Albertus Magnus, *Animals*, vol. 2, Lib. XXII, tr. 1, p. 1441.

[26] Albertus Magnus, *Minerals*, bk. II, tr. 2.

Bestiaries, lapidaries, and herbals resemble one another in general structure. Each consists of a list of individual kinds of natural objects, typically arranged alphabetically, with a paragraph or so of description. Each entry begins with the word, the name of the object, often accompanied by an etymology and synonyms in various languages. Then comes a physical description, often relying on comparisons with other objects and distinguishing various subgroups by color, size, or habitat. Finally comes an account of the thing's composition, behavior, and properties – both visible and hidden – coupled with their applications (chiefly medical) to human life. The entries often refer to earlier authors and compilations. There are many surviving manuscripts of each kind of treatise, as well as encyclopedias that incorporate all three. Albertus's own sections on particular plants, animals, or stones can be found as treatises separated from the main body of his work.[27]

Despite these similarities in form, the three genres had different, albeit overlapping, purposes and audiences. Albertus's deviations from his sources help highlight what was customary in each genre – and his own skills as a naturalist.

Herbals, for example, generally confine themselves to medicinally useful plants ("book of simple medicines" is a common title for herbals). Herbals' descriptions of plants are usually brief, their lists of medical uses long. Not only did Albertus swell the number of entries by including interesting plants of no medicinal value (e.g., the dye plant weld), he also provided far more substantial descriptions, with notes on habitats, dimensions, shapes, structures of plant parts, smells, tastes, colors, and varieties.[28] He noticed, for example, that "the flowers of roses grow above their fruit, as do the flowers of cucumber and pomegranate."[29] For medicinal uses, though, he was usually content to quote standard medical works: Matthaeus Platearius, Constantine the African, or Avicenna.[30] In keeping with his overarching emphasis on causes, he explained the therapeutic properties using the language of the four elemental qualities, but, somewhat surprisingly, he rarely mentioned Galen's assignments of degrees of heat and moisture.[31] Albertus concluded with a long digression on the changes domestication caused in plants, drawing his brief accounts of cultivated plants from the Roman agricultural author Palladius.

Lapidaries tend to be terser than herbals, if only because gems, earths, and metals have relatively few external properties and parts to describe. However, lapidaries vary greatly in their purpose, in the number of minerals listed,

[27] John M. Riddle, "Lithotherapy," pp. 39–50; and Riddle and Mulholland, "Albert on Stones and Minerals," pp. 203–34.

[28] Stannard, "Albertus and Medieval Herbalism," in Weisheipl, *Albertus Magnus and the Sciences*, p. 362; and Albertus Magnus, *De Vegetabilibus*, Lib. VI, tr. 2, ca. VIII.

[29] Albertus Magnus, *De Vegetabilibus*, Lib. VI, tr. 1, ca. XXXII.

[30] Stannard, "Albertus and Medieval Herbalism," p. 372.

[31] Riddle and Mulholland, "Albert on Stones and Minerals," p. 207.

and in the details they include or omit.[32] Like herbals, most lapidaries list medical properties and uses. Wearing the stone will often suffice: "Experience shows that jasper reduces bleeding in menstruation. They say that it prevents conception and aids childbirth; and that it keeps the wearer from licentiousness."[33] Some lapidaries emphasize stones with strong religious symbolism: the twelve stones on Aaron's breastplate (Exod. 28) and the precious stones that adorned the foundations of the city of God (Rev. 21).[34]

Although Albertus had commented on the symbolism of precious stones and metals in his earliest known work, a theological treatise on the nature of the Good (ca. 1243–4), he ignored them in *On Minerals*.[35] Although the encyclopedist Thomas of Cantimpré – a student of Albertus at Cologne and apparently Albertus's source for much material, especially on minerals and zoology – had praised coral because the cross formed by its branches terrifies demons, Albertus made no mention of this apotropaic quality.[36] Instead, he focused on the unusual phenomena of minerals, their properties – practical, technical, medicinal, alchemical, and magical – and their causes, augmented by his own observations from his wanderings through German mining regions talking to miners.

The emphasis on Christian symbolism was even more pronounced in many bestiaries. The names of animals, their features, and their behavior provided a springboard for moralizing. The beaver, for example, was a symbol of chastity and was named *castor* because of its trick of castrating itself to save itself from the hunters who sought its musk for medicine. Albertus, however, thought the whole story nonsense: "As has been ascertained frequently in our regions, it is false."[37] His active interest in animals ranged

32 Riddle, "Lithotherapy," pp. 46ff.

33 Albertus Magnus, *Minerals*, p. 100.

34 Riddle, "Lithotherapy," p. 40.

35 Albertus Magnus, *De natura boni*, ed. Ephrem Filhaut (Alberti Magni Opera Omnia . . . Institutum Alberti Magni Coloniense, 25) (Aschendorff: Monasterii Westfalorum in Aedibus Aschendorff, 1974), pp. 89, 227.

36 Albertus Magnus, *Minerals*, p. 81; and Thomas Cantimpratensis [Thomas of Cantimpré], *Liber de natura rerum: Editio princeps secundum codices manuscriptos*, ed. H. Boese (Berlin: Walter de Gruyter, 1973), vol. 1, Lib. XIV, tr. 15, p. 359 (the second volume was never published). For Albertus's use – without acknowledgment and often with considerable editing – of Thomas of Cantimpré's encyclopedia and the mysterious Experimentator and *Liber rerum* frequently cited by Thomas, see Pauline Aiken, "The Animal History of Albertus Magnus and Thomas of Cantimpré," *Speculum*, 22 (1947), 205–25; John Block Friedman, "Thomas of Cantimpré, *De naturis rerum*, Prologue, Book III and Book XIX," in *La science de la nature: théories et pratiques* (Cahiers d'études médiévales, 2) (Montréal: Bellarmin, for the Institut d'études médiévales, Université de Montréal, 1974), pp. 107–54; John Block Friedman, "Albert the Great's Topoi of Direct Observation and His Debt to Thomas of Cantimpré," in *Pre-Modern Encyclopaedic Texts: Proceedings of the Second COMERS Congress*, Groningen, 1–4 July 1996, ed. Peter Binkley (Leiden: Brill, 1997), pp. 379–92; Robbin S. Oggins, "Albertus Magnus on Falcons and Hawks," in Weisheipl, *Albertus Magnus and the Sciences*, pp. 441–62; and Albertus Magnus, *Animals*, vol. 1, p. 40, and Kitchell and Resnick's citations of Thomas of Cantimpré [ThC] passim. On coral, see John Block Friedman, "The Prioress's Beads 'Of Smal Coral'," *Medium Aevum*, 39 (1970), 301–5 at p. 305 n. 6.

37 Albertus Magnus, *De animalibus*, vol. 2, Lib. XXII, tr. 2, cap. 1, no. 22; Albertus Magnus, *Animals*, vol. 2, p. 1467 and n. 143. Thomas of Cantimpré was more specific about his source: "the Poles

from ruefully attesting to the toughness of boots made from ass-hide, to watching a drunken snake wobble along, half-alive, through his cloister, to checking on the number of eggs and chicks in an eagle's nest: "We found this out by visiting the nest of a certain eagle for six straight years.... We gained firsthand experience only by being lowered from the cliff on a rope of very great length."[38]

Albertus had been watching birds of prey since childhood; over half of his section on birds in *On Animals* describes falcons, eagles, and other raptors.[39] To supplement his written sources and his own observations, Albertus apparently interviewed Emperor Frederick II's falconers during his stays in Italy and used a book written by a falconer of Frederick's father, King Roger of Sicily.[40] Frederick's own work *On the Art of Hunting with Birds* (*De arte venandi cum avibus*, ca. 1244–50) – which Albertus may have seen – was even more remarkable in its challenges to received authority and its stress on personal observation and experiment, testing, for example, whether a favorable wind or the availability of food was more important to birds about to migrate.[41] Later practical treatises on hunting, such as Count Gaston Phoebus's *Book of Hunting* (*Livre de chasse*, late fourteenth century), also incorporate direct experiences of animals and terrain.[42]

THE PRACTICE AND USE OF NATURAL HISTORY

Thus far, we have been looking for evidence of interest in natural history in books that conformed to the framework of the medieval Christian scholarly world; that is, in works that were written and read primarily by men who knew Latin, who had received at least a liberal arts education, and who valued their book learning and knowledge of the causes of things. But medieval interest in natural things was no more limited to this one part of society than it was to one category of knowledge. The natural world linked the spiritual, intellectual, and material realms, and interest in it crossed the

(*Poloni*) said the story was false in their beavers"; see Thomas Cantimpratensis, Lib. IV, tr. 14, p. 116.

38 Albert the Great, *Man and the Beasts*, p. 88 n.2; Ephrem Filthaut, "Um die *Questiones de animalibus* Alberts des Grossen," in *Studia Albertina: Festschrift für Bernhard Geyer zum 70. Geburtstag*, ed. Heinrich Ostlender (Münster in Westphalen.: Aschendorffsche Verlagsbuchhandlung, 1952), pp. 112–27, especially p. 117; Albertus Magnus, *De animalibus*, vol. 2, Lib. VI, tr. 1, c. 6; and Albertus Magnus, *Animals*, vol. 1, Lib. VI, tr. 50, p. 547, n. 76; Lib. VI, tr. 54, p. 549; Lib. XXII, tr. 18, p. 1451, n. 38; p. 1710, n. 21.

39 Tugwell, *Albert & Thomas*, pp. 4–5, 97 n.10, 98 n.17.

40 Oggins, "Albertus Magnus on Falcons and Hawks," pp. 441–62.

41 C. A. Willemsen, *Fredericus II: De arte venandi cum avibus. Ms. Pal. lat. 1071, Bibliotheca Apostolica Vaticana* (Graz: Akademische Druck- und Verlagsanstalt, 1969), c. 20; and Frederick II, *The Art of Falconry, Being the De Arte Venandi cum Avibus of Frederick II of Hohenstaufen*, trans. and ed. Casey A. Wood and F. Marjorie Fyfe (Stanford, Calif.: Stanford University Press, 1943), chap. 20.

42 Baudouin Van Den Abeele, *La Littérature cynégétique* (Typologie des Sources du Moyen Age Occidental, 75) (Turnhout: Brepols, 1996), pp. 40–56, 62–3, 76–7.

lines that ordinarily divided the powerful and the humble, the learned and the illiterate, the clergy and the laity, men and women, city-folk and rustics, travelers and homebodies, Christians and heathens.

The natural history content of books written in Latin spread to a broader public by several routes. Morals drawn from bestiary accounts of animal behavior were as effective and popular with preachers and their lay audiences in the fourteenth century as they had been with Augustine and his parishioners in the fourth. Bestiaries were sometimes bound with books on the art of preaching, which themselves included compendia of moralized animal stories.[43] They could be tailored to the special concerns of their immediate audiences. Thus, a bestiary that belonged to an English Franciscan monastery in the late thirteenth century stressed that the beaver's self-castration signified not only the customary virtues of chastity and celibacy but also the casting aside of all worldly goods required of Franciscans.[44]

With the growth of literacy among the urban laity, the demand for vernacular versions of all genres of books grew as well. Deluxe, elaborately illuminated copies of books treating natural things were commissioned by emperors, kings, and nobility. To take just one royal family's natural history shelf, King Henry I of England (r. 1100–35) owned a digest of Pliny's *Natural History* (and a menagerie that included an ostrich and a camel); Henry's first wife, Matilda, had a lapidary and exotic travelers' tales in Anglo-Norman French; his second wife, Adeliza, had a bestiary and an allegorical lapidary in French; and his grandson, King John, perhaps another Pliny.[45] The Latin encyclopedias of the thirteenth century were translated into French, German, Flemish, Occitan, and English for lay readers. Herbals, bestiaries, and lapidaries were abridged and combined with a strong dash of magic, astrology, and "marvels of the world" to create the extremely popular genre known as "books of secrets" and often attributed to Albertus Magnus.[46]

The educated gentry and clergy were expected to have some general knowledge of natural history, if only to read the standard poets in the rhetoric syllabus intelligently. The natural history sections in Isidore of Seville's very widely circulated *Etymologies*, compiled in the seventh century, conveniently

43 Debra Hassig, *Medieval Bestiaries: Text, Image, Ideology* (Cambridge: Cambridge University Press, 1995), p. 175.

44 Ibid., pp. 86–9, 175–6, 187.

45 L. J. Reynolds and N. G. Wilson, *Scribes and Scholars: A Guide to the Transmission of Greek and Latin Literature*, rev. and enlarged ed. (Oxford: Oxford University Press, 1974), p. 100; Edward J. Kealey, *Medieval Medicus: A Social History of Anglo-Norman Medicine* (Baltimore: Johns Hopkins University Press, 1981), pp. 19–21; and Aberdeen Bestiary, "The Aberdeen Bestiary Website," Aberdeen University Library (accessed November 11, 2008): http://www.abdn.ac.uk/bestiary/.

46 Pseudo-Albertus Magnus, *The Book of Secrets of Albertus Magnus of the Virtues of Herbs, Stones and Certain Beasts, Also A Book of the Marvels of the World*, ed. Michael R. Best and Frank H. Brightman (Oxford: Oxford University Press, 1973). For a survey of the genre, see William Eamon, *Science and the Secrets of Nature: Books of Secrets in Medieval and Early Modern Culture* (Princeton, N.J.: Princeton University Press, 1994).

quoted passages from Virgil, Horace, and Ovid. Seven hundred years later, Wenceslaus Brack echoed Isidore in his textbook *Vocabulary of Things* (*Vocabularius rerum*), written for his arts students at Constance. This combination of thesaurus and Latin–German dictionary provided semi-alphabetical lists of cattle, "tiny animals," serpents, insects, fish, birds, herbs, trees, potherbs, stones, and the verbs for animal sounds. Brack's authorities included Albertus on minerals as well as lines from Virgil and Ovid not cited by Isidore. (Brack's comment that an abundance of the dyeplant woad [*sandix, wayd*] was found around Erfurt also seems to be his own addition.)[47] Another schoolmaster, in fifteenth-century England, expected his pupils to have the vocabulary to translate "I saw a goldfinch feeding himself upon a thistle in a thistly place" into a Latin tongue-twister.[48]

Because poetry was easier to memorize than prose, bestiaries, lapidaries, and herbals were often turned into verse or composed in verse to begin with, such as the muchcopied and translated poem *On the Powers of Herbs* (*De viribus herbarum*).[49] The versified allegorical bestiary lent itself to secular social commentary. Richard of Fournival turned the animals in his poem *Bestiary of Love* (*Bestiaire d'amour*, late thirteenth century) into a misogynist's satire of courtly love.[50] Geoffrey Chaucer (d. 1400) and the fifteenth-century Scottish poets William Dunbar and Richard Holland drew heavily on the bestiary attributes of birds in their satires of social classes and occupations.[51] In lyric and religious verses, the delights of nature were a common theme, no less heartfelt for being invoked through generic images of green meadows, leafy trees, sweet-scented blossoms, and singing birds or through allusions to the plants, birds, and gems in the Rosary and Song of Solomon.[52]

47 Wenceslaus Brack (or Brach), *Vocabularius rerum* (n.c.: n.p., n.d. [1490?]), fo. xxviii r°, xxx r°, cited from National Library of Medicine, History of Medicine Division, microfilm C-79–3 of early printed book in Bürgerbibliothek, Bern; and Nina Pleuger, *Der Vocabularius rerum von Wenzeslaus Brack: Untersuchung und Edition eines spätmittelalterlichen Kompendiums* (Berlin: Mouton de Gruyter, 2005), pp. 284, 414.

48 "*Vidi vnum cardoelum/ vel vnum cardoelem passentem se super cardonem in cardeto.*" See Nicholas Orme, "A Grammatical Miscellany of 1427–1465 from Bristol and Wiltshire," *Traditio*, 38 (1982), 300–26 at p. 314, n. 9.

49 Also called *Macer*, after its supposed author. See Thorndike, *History of Magic and Experimental Science*, vol. 1, pp. 612–15.

50 Debra Hassig, "Marginal Bestiaries," in *Animals and the Symbolic in Mediaeval Art and Literature*, ed. L. A. J. R. Houwen (Groningen: Egbert Forsten, 1997), pp. 171–88.

51 Pseudo-Albertus Magnus, *Book of Secrets of Albertus Magnus of the Virtues of Herbs, Stones and Certain Beasts*; Geoffrey Chaucer, "*Parlement of the Foules,*" in *Chaucer's Major Poetry*, ed. Albert C. Baugh (New York: Appleton–Century–Crofts, 1963), pp. 60–73, e.g., lines 319–72; Priscilla Bawcett, "'Nature Red in Tooth and Claw': Bird and Beast Imagery in William Dunbar," in Houwen, *Animals and the Symbolic in Mediaeval Art and Literature*, pp. 93–105; and Regina Scheibe, "The Major Professional Skills of the Dove in *The Buke of The Howlat*," in Houwen, *Animals and the Symbolic in Mediaeval Art and Literature*, pp. 107–37.

52 F. J. E. Raby, ed., *The Oxford Book of Medieval Latin Verse* (Oxford: Oxford University Press, 1959); see, e.g., no. 163, Adam of St. Victor, ca. 1140; no. 179, Peter the Venerable, d. 1155; nos. 194, 196, Walter of Châtillon, d. after 1184; nos. 202, 204, 210, 212, anon., twelfth century; no. 211, Carmina Burana, twelfth–thirteenth century; and no. 271, John of Howden, d. 1275.

An even more intense connection to the natural world and natural things found expression in the writings and deeds of some mystics in the late twelfth and early thirteenth centuries. In her visions, Hildegard of Bingen joyfully beheld the "greenness" (*viriditas*) and moisture that God had instilled into the earth and all created things.[53] In her herbal, *Book of Simple Medicines* (called *Physica* or *Liber de simplicis medicinae*), and in her *Causes and Cures* (*Causae et curae*), a medical book and encyclopedia combined, that vital greenness provided the explanatory force for the healing properties of herbs.[54] Images of greenness, moisture, and "gardens both physical and spiritual" pervaded her letters and sermons to church leaders as she admonished them to weed and water "the garden of all virtues."[55] Saint Francis of Assisi's fellowfeeling for all living creatures – the wolf of Gubbio, a tree cricket, a flock of birds, or wildflowers – was a particularly compelling expression of two currents of medieval spirituality. Francis manifested both the ever-present awareness that "All the world's creatures are as a book and a picture/And a mirror for us" (to use the famous lines of his contemporary, the poet Alan of Lille) and a new emotional desire for direct, concrete, physical experience of God's creation.[56]

Although medieval healers had especially strong reasons to take an interest in natural things, actual knowledge varied markedly with the individual healer's position in society. University-trained physicians claimed, by virtue of their knowledge of Latin, the liberal arts, and medical theory, the right to regulate the work of the surgeons, apothecaries, and herbalists, and to prosecute anyone who practiced medicine without a license.[57] Yet learned doctors

[53] Charles Singer, "The Visions of Hildegard of Bingen," in his *From Magic to Science* (New York: Dover, 1928, repr. 1958; essay originally published in 1917), pp. 199–239, plate II; Hildegard of Bingen, *The Letters of Hildegard of Bingen*, trans. Joseph L. Baird and Radd K. Ehrman (New York: Oxford University Press, 1994), vol. 1, pp. 7–8; and Victoria Sweet, *Rooted in the Earth, Rooted in the Sky: Hildegard of Bingen and Premodern Medicine* (Studies in Medieval History and Culture) (New York: Routledge, 2006), especially chap. 5.

[54] Thorndike, *History of Magic and Experimental Science*, vol. 2, c. xl; Hildegard of Bingen, *Causae et curae*, ed. Paul Kaiser (Leipzig: Teubner, 1903); and *Beate Hildegardis Cause et cure*, ed. Laurence Moulinier and Rainer Berndt (Rarissima mediaevalia, 1) (Berlin: Akademie Verlag, 2003). For Books I and III of *Liber de simplicis medicinae* on healing plants and on trees, see Hildegard of Bingen, *Das Buch Von den Pflanzen*, trans. and comment. Peter Rieth (Salzburg: Otto Müller Verlag, 2007); Hildegard of Bingen, *Das Buch von den Baümen*, trans. and comment. Peter Rieth (Salzburg: Otto Müller Verlag, 2001); and Laurence Moulinier, "Deux jalons de la construction d'un savoir botanique en Allemagne aux XIIe–XIIIe siècles," in *Le Monde végétal (XIIe–XVIIe siècles): Savoirs et usages sociaux*, ed. Alan J. Grieco, Odile Redon, and Lucia Tongiorgi Tomasi (Vincennes: Presses Universitaires de Vincennes, 1993), pp. 89–106.

[55] Hildegard of Bingen, *Letters of Hildegard of Bingen*, pp. 7–8, e.g., Letters 15r, 17.

[56] Raby, *Oxford Book of Medieval Latin Verse*, Selection 242, our translation; Ernst Robert Curtius, *European Literature and the Middle Ages*, trans. Willard Trask (Princeton, N.J.: Princeton University Press, 1952), pp. 316–21; and Lynn White, Jr., "Natural Science and Naturalistic Art in the Middle Ages," in his *Medieval Religion and Technology: Collected Essays* (Berkeley: University of California Press, 1978), pp. 33–9.

[57] Vivian Nutton, "Medicine in Medieval Western Europe, 1000–1500," in *The Western Medical Tradition: 800 B.C. to A.D. 1800*, ed. Laurence I. Conrad et al. (Cambridge: Cambridge University Press, 1995), pp. 139–205, especially pp. 153–72; and Reeds, *Botany in Medieval and Renaissance Universities*, pp. 1–5, 24–5.

evidently had little firsthand acquaintance with the natural substances that they were prescribing.[58] Their books about simple and compound drugs told them what to direct the apothecaries to prepare; the apothecaries, in turn, depended on their own gardens, spice merchants, and herb-gatherers to supply the raw materials.[59] The gap in class and literacy between the physicians and the herb-gatherers was often compounded by a difference in gender. Although both men and women did this work, medieval and early-Renaissance writers assumed that most herbalists were women.[60]

Instances of the learned seeking out the knowledge of the illiterate are therefore all the more striking. Simon Cordo of Genoa, physician to Pope Nicholas IV, said in his massive dictionary of plant names and medical terms, *Synonyms of Medicine* (*Synonyma medicinae*, also known as *Clavis sanationis*, finished around 1292), that "I made myself the companion and student to an old Cretan woman" in order to learn how to recognize herbs, hear their proper Greek names, and learn their virtues.[61] Rufinus, Simon's contemporary and fellow Genoan (although neither seems to have known of the other) was even more open-minded in his herbal, *On the virtues of herbs* (*De virtutibus herbarum*, ca. 1287). Rufinus frequently cited the opinions of herbalists, rustics, and women in Bologna and Naples about the vernacular names and properties of plants, including some that were not described in books, such as *aucheta* (which "makes hair blonde, so women say"). He also proved a keen observer of the shapes, textures, colors, growth patterns, and varieties of plants and their habitats.[62] But Rufinus, though well versed in medical literature, did not claim to be a physician. He had studied and taught arts and astrology at the University of Bologna before becoming a monk; his interest in the ways the stars influenced things on earth had aroused his

[58] See, however, Nancy G. Siraisi, *Medieval and Early Renaissance Medicine: An Introduction to Knowledge and Practice* (Chicago: University of Chicago Press, 1990), fig. 11 (Sächsisches Landesbibliothek, Dresden MS Db 93, fol. 397 recto, c. 1460, Galenic works) at pp. 74–5.

[59] Loren C. MacKinney, "Medieval Medical Dictionaries and Glossaries," in *Medieval and Historiographical Essays in Honor of James Westfall Thompson*, ed. James Lea Cate (Chicago: University of Chicago Press, 1938), pp. 240–68 at fig. 21; and Robert S. Lopez and Irving W. Raymond, eds., *Medieval Trade in the Mediterranean World* (Records of Civilization, Sources and Studies, 52) (New York: Columbia University Press, 1955), Doc. 175.

[60] Marguerite Tjader Harris, ed., *Birgitta of Sweden: Life and Selected Revelations* (Classics of Western Spirituality) (New York: Paulist Press, 1990), p. 30; Andrew Wear, "Medicine in Early Modern Europe, 1500–1700," in *The Western Medical Tradition*, pp. 215–362, especially pp. 236–40; and Walter Pagel, "Religious Motives in the Medical Biology of the XVIIth Century," *Bulletin of the History of Medicine*, 3 (1935), 97–123, especially pp. 120–3.

[61] Lynn Thorndike, assisted by Francis S. Benjamin, *The Herbal of Rufinus* (Chicago: University of Chicago Press, 1945), pp. xii–xiv; Simon a Cordo Januensis, *Synonyma medicinae, sive Clavis sanationis* (Milan: Antonius Zarota, 1473), fol. 2v a; and Danielle Jacquart, "La coexistence du grec et de l'arabe dans le vocabulaire médical du latin médiéval: l'effort linguistique de Simon de Gênes," in *Transfert de vocabulaire dans les sciences*, ed. M. Groult (Paris: CNRS Editions, 1988), pp. 277–90, reprint X, in Danielle Jacquart, *La science médicale occidentale entre deux renaissances (XIIe s.–XVe s.)* (Variorum Collected Studies Series) (Aldershot: Variorum/Ashgate, 1997).

[62] Thorndike, *Herbal of Rufinus*, pp. 50, 152, *herba Gualterii*.

desire to acquire knowledge of herbs.[63] As an amateur, so to speak, he could talk freely with doctors, pharmacists, herb-gatherers, and farmers without infringing upon anyone's prerogatives.

THE DEPICTION OF NATURE

Implicit in the references by Simon Cordo, Rufinus, Albertus Magnus, and Emperor Frederick II to their own experience and to the testimonies of experts was faith, above all, in the evidence of their own eyes. The word, written or spoken, of men and women (no matter how low their estate) who had observed natural things closely also deserved respect – as did the writings of ancient authorities, although even the most highly regarded texts were vulnerable to mistranslation, miscopying, and misunderstanding. Travelers' tales (like those deftly reassembled in the immensely popular *Travels of Sir John Mandeville*) and things known by common report – "they say that" – were worth further investigation but meanwhile had to be regarded with caution; ignorance, fraud, malice, or irony could all lead to deception.[64]

All these claims to authority on the basis of experience relied on the transformation of things seen into words. However, experience and information could also be recorded in visual images. Several different traditions of illustrating the objects of the natural world ran alongside the written and oral traditions of medieval natural history. Many bestiaries included elegant paintings of animals and their attributes. Drawings of plants often accompanied the texts of herbals and occasionally appear on their own with no more than labels.[65] Specimens of gems and minerals are little more than lumps in medieval drawings, but their sources and uses could make memorable

[63] Ibid., pp. 1–2. It is not clear what monastic order Rufinus joined. For mendicant friars' interest in plants and gardens and relations with secular apothecaries, see Angela Montford, *Health, Sickness, Medicine and the Friars in the Thirteenth and Fourteenth Centuries* (Aldershot: Ashgate, 2004), pp. 18–19, 56–9, and chap. 9, "The Hand of Christ: Drugs for the Sick Friar."

[64] Sir John Mandeville, *The Travels of Sir John Mandeville*, ed. and trans. C. W. R. D. Moseley (London: Penguin Books, 1983), Introduction; Albertus Magnus, *Minerals*, bk. II, tr. 1; and Bert Hansen, *Nicole Oresme and the Marvels of Nature: A Study of His De causis mirabilium with Critical Edition, Translation, and Commentary* (Toronto: Pontifical Institute of Mediaeval Studies, 1985), pp. 7–10. See also Friedman, "Albert the Great's Topoi of Direct Observation and His Debt to Thomas of Cantimpré."

[65] For representative illustrations and discussion, see Hassig, *Medieval Bestiaries*, especially pp. 85–6, fig. 17; Wilfrid Blunt and Sandra Raphael, *The Illustrated Herbal* (New York: Thames and Hudson, with the Metropolitan Museum of Art, 1979), especially p. 97; L. F. Sandler, "Jean Pucelle and the Lost Miniatures of the Belleville Breviary," *Art Bulletin*, 66 (1984), 73–96; Minta Collins, *Medieval Herbals: The Illustrative Tradition* (The British Library Studies in Medieval Culture) (London: The British Library, 2000); and Jean A. Givens, Karen M. Reeds, and Alain Touwaide, eds., *Visualizing Medieval Medicine and Natural History, 1200–1550* (AVISTA Studies in the History of Medieval Technology, Science and Art, 5) (Aldershot: Ashgate, 2006). For a bibliography, see chap. 5, "The Natural World," in John B. Friedman and Jessica M. Wegmann, *Medieval Iconography: A Research Guide* (New York: Garland, 1998).

pictures in lapidaries and encyclopedias.[66] To these illustrations in works of reference must be added the images of natural things in architectural ornamentation, tapestries, paintings, nonscientific manuscripts, drawings in artists' model books, maps, and the occasional historical record of an exotic animal.[67]

The picture-and-entry format of medieval bestiaries and herbals tempts us to assume that they were used as field guides; that is, taken out into gardens, meadows, and woods to enable the user to identify animals and plants.[68] This view of their function, however, overlooks both the intrinsic value of books and the content of their text and illustrations. Because expensive illustrated books had to be kept safely indoors, the social divide between those who could read the books and those who did the hands-on work in the field was reinforced. Scenes of hunts, herb-gathering, or preparation of medicines customarily distinguished between the sumptuously dressed nobleman or long-robed physician in charge and the unbearded youths in knee-length jackets who pulled up the plants, pounded the ingredients, or tracked the game.[69]

A bigger objection to the comparison with field guides is that, as a rule, the medieval pictures were not accurate or detailed enough to ensure reliable identifications (Figure 24.1). Even assuming that the illustrations in the original text had been drawn from life (as, Pliny tells us in *Natural History*, XXV, chaps. 4–5, the ancient herbalist Crateuas had done with plants), the process of copying and recopying inevitably resulted in the distortion of details, the schematizing of forms, or the confusion of colors. For bestiaries, this did not matter much. A preacher could moralize from the animals without accurate pictures, indeed without pictures at all. Familiar animals

[66] Riddle, "Lithotherapy," p. 43. See also illustrations of earth-forms and sources of gems in a fifteenth-century French manuscript of Bartholomaeus Anglicus, *Liber de proprietatibus rerum* (French translation by Jean Corbechon, 1372), Bibliothèque Nationale, Paris, BNF, Fr 136, in Bibliothèque Nationale, Web site exhibit, 1997 (accessed November 21, 2008), "The Age of King Charles V (1338–1380): 1,000 Illuminations from the Department of Manuscripts," http://www.bnf.fr/ENLUMINURES/Aaccueil.htm; and Ulrike Spyra, *Das "Buch der Natur" Konrads von Megenberg: Die illustrierten Handschriften und Inkunabeln* (Pictura et Poësis, 19) (Cologne: Böhlau Verlag, 2005).

[67] Brunsdon Yapp, *Birds in Medieval Manuscripts* (New York: Schocken Books, 1982), pp. 14–15, 75–6; Theodore Bowie, ed., *The Sketchbook of Villard of Honnecourt* (Bloomington: Indiana University, 1959), plates 31 and 32; Hassig, *Medieval Bestiaries*, pp. 137–44, 250 n.77, n.80, n.81, n.86; and Nona C. Flores, "The Mirror of Nature Distorted: The Medieval Artist's Dilemma in Depicting Animals," in *The Medieval World of Nature: A Book of Essays*, ed. Joyce E. Salisbury (New York: Garland, 1993), pp. 3–54.

[68] John E. Murdoch, *Album of Science: Antiquity and the Middle Ages* (New York: Charles Scribner's Sons, 1984), pp. xi, 213–14; and Loren C. MacKinney, *Medical Illustrations in Medieval Manuscripts* (Publications of the Wellcome Historical Medical Library) (London: Wellcome Historical Medical Library, 1965), p. 24.

[69] Margaret B. Freeman, *The Unicorn Tapestries* (New York: The Metropolitan Museum of Art, 1976), p. 95; Tony Hunt, *The Medieval Surgery* (Woodbridge: The Boydell Press, 1992), plates 29 and 48; MacKinney, *Medical Illustrations in Medieval Manuscripts*, pp. 27–38, figs. 19–27; Peter Murray Jones, *Medieval Medical Miniatures* (Austin: University of Texas Press and The British Library, 1984), pp. 90–3, figs. 40 and 41; and Blunt and Raphael, *Illustrated Herbal*, pp. 50–1.

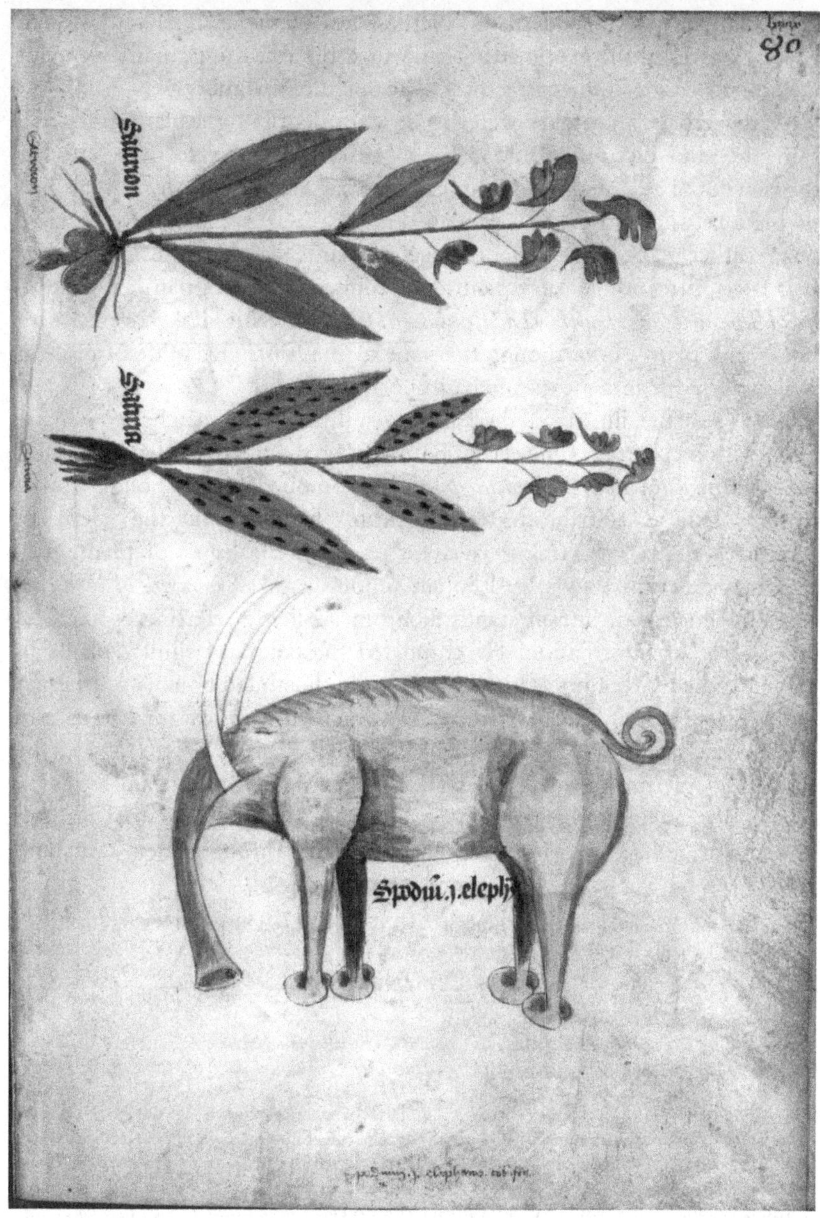

Figure 24.1. The picture of the male orchid (*satirion*) in this album of materia medica images is meant to suggest the use of the plant's roots as an aphrodisiac. The elephant's tusks yielded spodium (burnt ivory), used as a remedy for nosebleeds. From Joannes Platearius, *Compendium Salernitanum*, by permission of The Pierpont Morgan Library, New York, MS M. 873, fol. 80r (mid-fourteenth century).

were recognizable by label, if not picture, and there was rarely any way to check pictures of the exotic creatures against the real thing (many were not real, in any case), so the correctness of the picture was moot.[70]

With herbals, the stakes were higher. Misidentifying a plant could kill someone, and pictures would seem to be the best way to overcome the deficiencies of the descriptions and nomenclature.[71] Yet, many herbals had no pictures, and in those that were illustrated, the artist almost always relied on an earlier manuscript rather than actual plants for models. It is indicative that late-fourteenth-century pictures of Albucasis (Ibn Botlān) at work on his *Handbook of Health* (*Tacuinus sanitatis*) show his desk covered with books, not plants, even though the same manuscripts give pride of place to well-rendered plants shown in charming genre scenes.[72]

Nor were the illustrated herbals generally used to match up a living plant with its portrait. One exception might be the medical/pharmaceutical manuscript known as *Tractatus de herbis*, compiled for a scholarly reader in Italy at the end of the thirteenth century. In illustrating the Salernitan text *Circa instans*, the unknown artist rendered the leaves of plants in a strikingly descriptive style.[73] Although Simon Cordo's *Synonyms of Medicine* was not illustrated, Simon stands alone in explicitly referring to pictures as a source of information. He compared the colors of similar plants in an "illustrated Greek book" and "another old illustrated book."[74] Although herbals were certainly read, used, and augmented – as marginal notes and the incorporation or substitution of local plants prove – in practice, the formalized picture of a plant served more as a bookmark than as a way to identify an unfamiliar specimen.[75] Because the truly essential features of plants were hidden from view, detectable only through smell, taste, and

70 On an elephant and a lion "drawn from life," see Otto Pächt, "Early Italian Nature Studies and the Early Calendar Landscape," *Journal of the Warburg and Courtauld Institutes*, 13 (1950), 13–47, especially pp. 24–5; Hassig, *Medieval Bestiaries*, pp. 141–3; Flores, "Mirror of Nature Distorted"; and Jean A. Givens, *Observation and Image-Making in Gothic Art* (Cambridge: Cambridge University Press, 2005), chaps. 2 and 5, plate V, and figs. 11, 16, 23, 52.

71 Thorndike, *Herbal of Rufinus*, p. 95, *cicuta*; pp. 116–17, *elleborum, elacterium*.

72 Luisa Cogliati Arano, *The Medieval Health Handbook: Tacuinum Sanitatis* (New York: George Braziller, 1976), p. 153, figs. 1, 92, 94, 95.

73 For discussions of the production, ownership, and use of the earliest known manuscript of *Tractatus de herbis* (London, British Library Egerton MS 747, also called *Compendium Salernitanum*), its place in the Salernitan medical tradition, and later Latin and vernacular illustrated versions, see Pächt, "Early Italian Nature Studies and the Early Calendar Landscape," pp. 27–30, plates 7a, 7b, 9b; Minta Collins, "Introduction," in *A Medieval Herbal: A Facsimile of British Library Egerton MS 747* (London: The British Library, 2003), p. 14 (see also fol. 90 verso and 91 recto for a comparison with fig. 1); Jean A. Givens, "Reading and Writing the Illustrated *Tractatus de herbis*, 1280–1526," in Givens, Reeds, and Touwaide, *Visualizing Medieval Medicine and Natural History*, pp. 115–45; and Pseudo-Bartholomaeus Mini de Senis, *Tractatus de herbis* (*MS London, BL, Egerton 747*), ed. Iolanda Ventura (Edizione Nazionale "La Scuola Medica Salernitana," 5) (Florence: SISMEL, 2009).

74 Simon a Cordo Januensis, *Synonyma medicinae*, e.g., entries for *achantos seu achantinos, artemisia, centaurea*.

75 Jones, *Medieval Medical Miniatures*, pp. 78–80; Linda Voights, "Anglo-Saxon Plant Remedies," *Isis*, 70 (1979), 250–68; and John M. Riddle, "Pseudo-Dioscorides' *Ex herbis femininis* and Early Medieval Medical Botany," *Journal of the History of Biology*, 14 (1981), 43–81, especially p. 51.

physiological effects, and hence impossible to paint, neither medical theory nor natural philosophy encouraged pictorial accuracy.[76]

In the context of religious and secular art, however, images of natural things tended toward greater naturalism and descriptiveness.[77] A thirteenth-century illustration of an angel summoning the birds at the Last Judgment, for example, depicts several identifiable kinds of birds.[78] From the fourteenth century onward, the borders of devotional manuscripts became a birdwatcher's delight – the magpie, long-eared and tawny owls, coal tit, and a possible capercaillie – are only a few of the birds that have been named with reasonable certainty.[79]

In a number of thirteenth-century Gothic ecclesiastic buildings in France and England, most notably at the Southwell chapter house of secular canons, stonemasons carved leaves and fruits of common hedgerow plants on the capitals and ceiling bosses. In some cases, it is clear that these sculptors were using local plants as models – two species of oak, one with insect galls, are differentiated clearly – rather than copying an earlier work.[80] Of all the illustrated herbals produced by the mid-fifteenth century, only two from northern Italy – the Carrara herbal (ca. 1400) and its partial copy, the *Codex Roccabonella* (between 1415 and 1449) – rival the roses, peapods, and columbines that decorate the borders of late-medieval Books of Hours in their observation of natural forms and their skill at re-creating those forms on the page.[81] Like the Books of Hours and the sumptuously illustrated copies of Albucasis's *Handbook of Health* (*Tacuinum sanitatis*) from the same period and locale, the Carrara herbal at least was produced with the tastes

[76] Hansen, *Nicole Oresme and the Marvels of Nature*, pp. 214–15, 232–3; Bert S. Hall, "The Didactic and the Elegant: Some Thoughts on Scientific and Technological Illustrations in the Middle Ages and Renaissance," in *Picturing Knowledge: Historical and Philosophical Problems Concerning the Use of Art in Science*, ed. Brian S. Baigrie (Toronto: University of Toronto Press, 1996), pp. 3–39, and review by Karen Meier Reeds in *Technology and Culture*, 39 (1998), 760–2.

[77] Jean Ann Givens, "The Garden Outside the Walls: Plant Forms in Thirteenth-Century English Sculpture," in *Medieval Gardens*, ed. Elisabeth Macdougall (Washington, D.C.: Dumbarton Oaks Research Library and Collection, 1986), pp. 187–98 plus 18 plates; Jean Ann Givens, "The *Leaves of Southwell* Revisited," in *Southwell Minster and Nottinghamshire: Medieval Art, Architecture and Industry*, ed. Jenny Alexander (British Archaeological Association Transactions, 21) (Leeds: Manley, 1998), pp. 60–6, plates XVIII–XXI; and Givens, *Observation and Image-Making in Gothic Art*, introd. and chap. 1, "Gothic Naturalism."

[78] Francis Klingender, *Animals in Thought and Art to the End of the Middle Ages*, ed. Evelyn Antal and John Harthan (Cambridge, Mass.: MIT Press, 1971), pp. 402–13; and Yapp, *Birds in Medieval Manuscripts*, pp. 104–11.

[79] Yapp, *Birds in Medieval Manuscripts*.

[80] Givens, "Garden Outside the Walls," pp. 155–6, 182; Givens, "*Leaves of Southwell* Revisited"; and Givens, *Observation and Image-Making in Gothic Art*, pp. 1–14, 82–101, 106–27, 135–45, 165–8.

[81] London, British Library, Egerton 2020 (*Carrara Herbal*); Venice, Biblioteca Marciana, Cod. lat. 59.2548 (*Codex Roccabonella*); Felix Andreas Baumann, *Das Erbario Carrarese und die Bildtradition des Tractatus de herbis: Ein Beitrag zur Geschichte der Pflanzendarstellung im Übergang von Spätmittelalter zur Frührenaissance* (Berner Schriften der Kunst, 12) (Bern: Benteli Verlag, 1974); and Pächt, "Early Italian Nature Studies and the Early Calendar Landscape," pp. 30–2, plates 7c, 8a, 8b, 8c, 9c, 10a, 10b. On naturalistic flowers, shells, and animals in Books of Hours, see Celia Fisher, *Flowers in Medieval Manuscripts* (Medieval Life in Manuscripts) (London: The British Library, 2004).

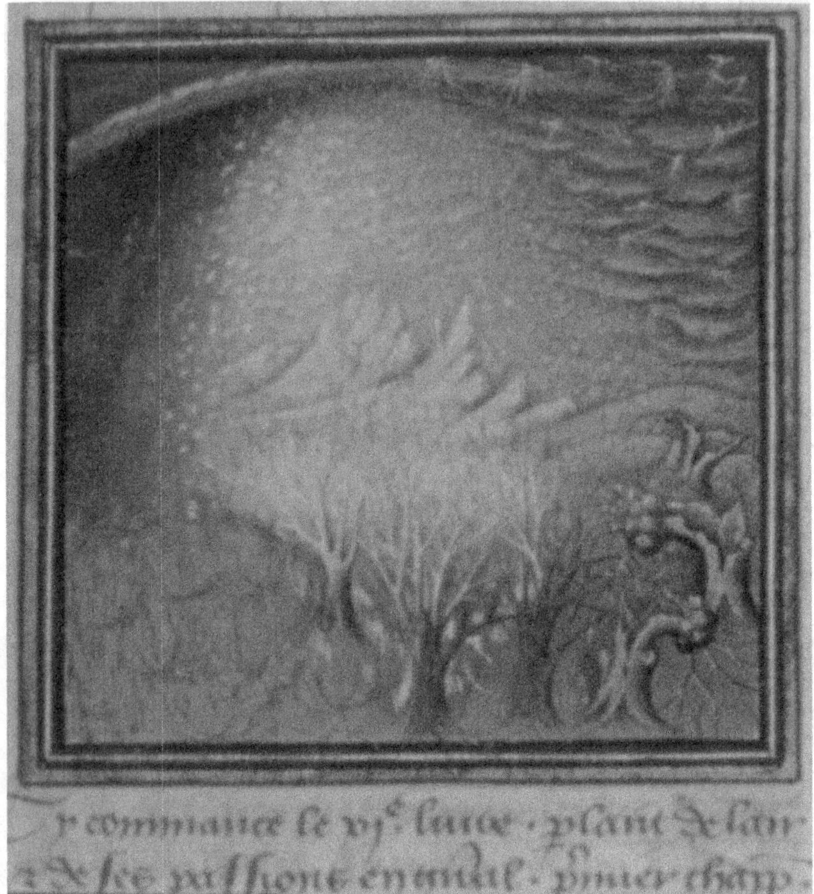

Figure 24.2. In his popular thirteenth-century encyclopedia, *On the Properties of Things*, the Franciscan Bartholomaeus Anglicus employed Aristotle's *Meteorologica* to explain storms, rainbows, meteors, and comets in terms of the four sublunary elements. In this French translation of the encyclopedia, the artist turned these natural phenomena into an apocalyptic vision appropriate to the author's moralizing intentions. Bartholomaeus Anglicus, *Liber de proprietatibus rerum* (fifteenth century). By permission of the Bibliothèque Nationale, Paris, BNF, FR 136, fol. 4 verso.

of a wealthy lay reader very much in mind – in this case Francesco Carrara the Younger, last duke of Padua (1359–1406), whose love of beautiful plants, birdsong, gardens, and herbal infusions was attested by a Paduan chronicler.[82]

[82] Cogliati Arano, *The Medieval Health Handbook*, p. 153; Albucasis, *The Four Seasons of the House of Ceruti*, trans. Judith Spencer (New York: Facts on File, 1983); Baumann, *Das Erbario Carrarese und die Bildtradition des Tractatus de herbis*, pp. 11–12, n. 11; and Cathleen Hoeniger, "The Illuminated *Tacuinum sanitatis* Manuscripts from Northern Italy ca. 1380–1400: Source, Patrons, and the Creation of a New Pictorial Genre," in Givens, Reeds, and Touwaide, *Visualizing Medieval Medicine*

Some art historians have pointed to the carved leaves at Southwell, the illustrated *Tractatus de herbis*, the two exceptional northern Italian herbals, and the flower-strewn borders of prayer books as evidence that the origins of the Scientific Renaissance in the sixteenth century should be sought in the art rather than the science of the late Middle Ages.[83] For the revival of botany, zoology, and anatomy, however, the ability to paint animals and plants "so artfully you would have thought they grew on the pages" was not sufficient.[84]

Pictures – like specimens – need captions and explanations for full understanding.[85] We can admire, for example, the sharp observation of the fifteenth-century artist who captured the very moment a snowstorm descended on a forest. But the illumination's equally impressive meteors, rainbow, and lightning-shattered tree make sense only when read with the encyclopedia article on meteorology that the picture as a whole illustrates (Figure 24.2).[86]

Moreover, neither images nor texts have any influence unless they are seen and discussed. For books in demand by the university, the church, and rich layfolk, the stationers and scriptoria created a reasonably effective system of book production and distribution. Access to books on natural history, though, was a hit-or-miss affair. The lives of Albertus Magnus, Simon Cordo, and Rufinus overlapped in time and place, yet they apparently did not know one another's work. As soon as humanist circles and medical schools provided natural history with intellectual respectability and an institutional home, and as soon as printing made the ancient classics of natural history accessible, the study of natural things through words and pictures began to flourish.[87]

and Natural History, pp. 51–81. Sarah Rozalja Kyle, "The *Carrara Herbal* in Context; Imitation, Exemplarity, and Invention in Late Fourteenth-Century Padua" (Ph.D. diss., Emory University, 2010).

83 Pächt, "Early Italian Nature Studies and the Early Calendar Landscape"; Hall, "The Didactic and the Elegant"; David Topper, "Towards an Epistemology of Scientific Illustration," in Baigrie, *Picturing Knowledge*, pp. 215–49, especially pp. 221–9, 241–7; and Givens, *Observation and Image-Making in Gothic Art*, especially pp. 1–26.

84 Thorndike, *History of Magic and Experimental Science*, vol. 4, p. 599, n. 19, quoting the humanist Pandolfo Collenuccio, circa 1492, referring to the Roccabonella codex.

85 Brian S. Baigrie, "Introduction," in Baigrie, *Picturing Knowledge*.

86 Bartholomaeus Anglicus, *Liber de proprietatibus rerum*.

87 Karen Meier Reeds, "Renaissance Humanism and Botany," *Annals of Science*, 33 (1976), 519–42; Reeds, "Leonardo da Vinci and Botanical Illustration: Nature Prints, Drawings, and Woodcuts ca. 1500"; Claudia Swan, "The Uses of Realism in Early Modern Illustrated Botany," in Givens, Reeds, and Touwaide, *Visualizing Medieval Medicine and Natural History*, pp. 205–37 and 239–49, respectively; Brian W. Ogilvie, *The Science of Describing: Natural History in Renaissance Europe* (Chicago: University of Chicago Press, 2006), especially chap. 3; and Paula Findlen, "Natural History," in Park and Daston, *Early Modern Science*, pp. 435–68.

25

ANATOMY, PHYSIOLOGY, AND MEDICAL THEORY

Danielle Jacquart

Medicine became a university discipline during the first decades of the thirteenth century; that is, at the very beginning of this academic institution. Limited at that time to three main centers, Paris, Montpellier, and, from the 1260s, Bologna, the number of medical faculties increased during the fourteenth and fifteenth centuries all over Europe. Despite this institutional spread, it is likely that, even at the end of the Middle Ages, most ordinary practitioners were not trained at a university. Nevertheless, the model of the literate physician was definitively established. As was also the case for law – the other branch of university learning to be endowed with a practical aim – the entry of medicine into the university framework was the result of its intellectual development during the twelfth century. Indeed, it had been preceded by what may be called "Salernitan medicine," although other places of learning, in France, England, or Germany, participated in this renewal, which tended to closely link everyday practice to theoretical knowledge.[1]

Medicine was among the first fields of learning to be revived during the twelfth century. In this respect, it preceded other disciplines, such as astronomy or mathematics, to which it served in some way as an example. Among the reasons, perhaps related to its practical aim, it may be pointed out that during the early Middle Ages the Greco-Latin background had remained more significant in medicine than in other fields, notably in southern Italy. It was thus not by chance that the first coherent corpus of translations made from Arabic, in the second half of the eleventh century, dealt with medicine and was issued at Monte Cassino, in close connection with Salerno. Apart

[1] For a general survey of university medicine and its twelfth-century background, see Nancy G. Siraisi, *Medieval and Early Renaissance Medicine: An Introduction to Knowledge and Practice* (Chicago: University of Chicago Press, 1990); and Danielle Jacquart, "La scolastique médicale," in *Histoire de la pensée médicale*, vol. 1: *Antiquité et Moyen Âge*, ed. Mirko D. Grmek (Paris: Editions du Seuil, 1995), pp. 175–210, or the English version, Danielle Jacquart, "Medical Scholasticism," in *Western Medical Thought from Antiquity to the Middle Ages*, ed. Mirko Grmek, trans. Antony Shugaar (Cambridge, Mass.: Harvard University Press, 1998), pp. 241–58.

from presenting more detailed descriptions of diseases and treatments than those previously available in Latin, Constantine the African's translations (between 1077 and 1098), unfaithful though they were to the originals, introduced physiological principles, founded on an Arabic interpretation of Galen's works, which provided rational explanations for the healthy functioning of the human body, as well as for many kinds of pathological disorders. More decisively, they allotted to medicine a place in the general classification of the sciences that it had lacked during Greco-Latin antiquity, thus removing it from the range of "mechanical" (i.e., merely practical) arts and defining it more accurately than Isidore of Seville, who in his *Etymologies* (early seventh century) had called it "a second philosophy." As a branch of natural philosophy, medical discipline henceforth had a recognized status.[2]

This recognition went together with the division of medical learning into two parts, theoretical and practical, both belonging to the realm of intellectual knowledge. This partition largely underlay university teaching, even if it was continually questioned and reconsidered, since it covered both a distribution of topics and differences in the method of inquiry (for example, demonstrative reasoning versus sensory experience). In particular, anatomical knowledge did not easily find its place, oscillating between theoretical and practical medicine. On the one hand, the solid constituents of the body (called "members") were among the seven "natural things" that formed, at a theoretical level, the basis of physiology, but, on the other hand, their knowledge involved to a great extent observation, at least on the part of the ancient authors who had elaborated it.[3] To be included in a university framework implied that a scholastic method of teaching and writing would be followed. Academic medical readings were supplied by the numerous translations made from Greek and Arabic throughout the medieval period up to the middle of the fourteenth century. The diversity of opinions that they expressed ensured for university masters a relatively wide range of choices. The failure of scholastic medicine did not lie so much in its static character as in its inability to reach a scientific consensus by reliable means, even about the major components of its doctrine. Moreover, through their duty to train future practitioners and their own involvement in actual practice, university masters were able to detect some inadequacies between bookish

[2] On medical translations from Arabic and their significance, see Danielle Jacquart and Françoise Micheau, *La médecine arabe et l'Occident médiéval* (Paris: Maisonneuve et Larose, 1990; repr. 1996); Charles Burnett and Danielle Jacquart, *Constantine the African and ʿAlī ibn al-ʿAbbās al-Maǧūsī, The "Pantegni" and Related Texts* (Leiden: Brill, 1994). See also the following chapters in this volume: Newman, Chapter 16; and Laird, Chapter 17.

[3] On the divisions of medicine, see Jole Agrimi and Chiara Crisciani, *Edocere medicos, Medicina scolastica nei secoli XIII–XV* (Naples: Guerini e associati, 1988), pp. 21–47; Danielle Jacquart, "L'enseignement de la médecine: quelques termes fondamentaux," in *Méthodes et instruments du travail intellectuel au Moyen Âge*, ed. Olga Weijers (Turnhout: Brepols, 1990), pp. 104–11, reprinted in Danielle Jacquart, *La science médicale occidentale entre deux renaissances XIIe s.–XVe s.* (Aldershot: Variorum, 1997), chap. XII.

knowledge and the data of real life. Many signs of such a crisis appear in fifteenth-century medicine, to which the humanist emphasis on philological faithfulness to ancient Greek and Latin works did not always give the most appropriate answer.

ANATOMICAL KNOWLEDGE: A SLOW RECONSTRUCTION

Anatomical accuracy depends both on a sufficient stock of bookish knowledge, in order to be aware of the different structures of the body, and on an ability to observe reality and to depict it faithfully. These two requirements were slow to be fulfilled during the Middle Ages. Contrary to what has often been stated, it was not so much an aversion to practicing dissection of the human body that slowed the advancement of this field as the poverty of available information. In a rather consistent pattern, physicians resorted to dissections at the very time when their bookish knowledge was becoming more significant.

Before the eleventh century, anatomy had probably been the scantiest part of medical learning. Constantine the African's translations brought new material, but truncated and confused through its Latin transmission. Even when they came to rely on Constantine's *Pantegni*, which provided them with the best anatomical knowledge available at the time, twelfth-century Salernitan anatomists could not go further than recognizing the main organs of the sows that they had dissected. Already summarized in Constantine's Arabic source, ʿAlī ibn al-ʿAbbās al-Majūsī's "Royal book," written at the end of the tenth century in Iran, the extensive Galenic descriptions were hardly recognizable in this early Latin form.[4] It is likely that medieval translators themselves failed to understand Arabic or Greek anatomical terminology – for which, most of the time, there were no Latin equivalents – and to visualize the precise organization of the internal body. It was thus a vicious circle, and it took a long time to escape from it. Whereas Galen's *Anatomical procedures*, which introduced the technique of dissection, became available in Latin only in the Renaissance (1531), the other fundamental Galenic work, *On the usefulness of the parts of the body*, was translated from the Greek in 1317 by Niccolò da Reggio. As to the Latin version from Arabic, which had circulated from the twelfth century, it was based on a mere paraphrase.[5] Before Niccolò da Reggio's translations, the most extensive account of anatomy was provided

[4] On Salernitan anatomy, see George W. Corner, *Anatomical Texts of the Earlier Middle Ages* (Washington: 1927); Morris H. Saffron, "Salernitan Anatomists," *Dictionary of Scientific Biography*, vol. 12, pp. 80–3; and Romana Martorelli Vico, "Gli scritti anatomici della *Collectio Salernitana*," in *La* Collectio Salernitana *di Salvatore de Renzi*, ed. Danielle Jacquart and Agostino Paravicini Bagliani (Florence: SISMEL, 2008), pp. 79–88. The anatomical chapters of al-Majūsī are edited in Arabic and translated into French in Pierre de Koning, *Trois traités d'anatomie arabe* (Leiden: Brill, 1896).

[5] On Latin translations of Galenic works, see Richard J. Durling, "Corrigenda and Addenda to Diels' Galenica": I, *Traditio*, 23 (1967), 461–76; II, *Traditio*, 37 (1981), 373–81.

by Avicenna's *Canon*, which Gerard of Cremona had translated during the second half of the twelfth century.

The partition of this account between the first and the third books of the *Canon* reflected the ambiguous status of anatomy. The first book included a rough enumeration of the main organs, followed by a more detailed description of those anatomical elements that were defined, from Aristotle and Galen, as the homogeneous parts of the body (bones, muscles, veins, arteries, and nerves), in contrast to the composite ones (the head, hand, heart, liver, etc.). In the third book, which reviewed all the diseases affecting one specific composite organ, from head to toe, Avicenna followed the Galenic principle, according to which, as Galen had stated in *On the affected parts* (I.2): "Every activity of the living organism is connected with a separate part of the body whence it arises. Therefore, an activity is necessarily damaged when the part which produces it is affected."[6] Each account of a particular disease and its appropriate treatment was preceded by the description of the affected part of the body in its normal state. If this mode of exposition had the advantage of stressing the importance of anatomical knowledge, it did not incite medieval physicians to acquire it for its own sake. It was mainly valued for its double practical aim: manual intervention and internal treatment of affected parts.

Apart from anatomical treatises of the twelfth and early thirteenth centuries, which depended on the Salernitan tradition, one of the proper places for developing accounts of anatomy was in surgical works. Although their authors practiced surgery themselves, they had generally received a learned medical training. For the medieval period, it is difficult to guess the actual anatomical knowledge of ordinary surgeons. The habit of including a separate section on anatomy in surgical books was inaugurated by William of Saliceto in the successive versions of his *Chirurgia*, dated respectively 1268 and 1275 or 1276. Still relegated to the fourth section of William's book, anatomy was placed by his followers at the beginning in order to stress the necessity for surgeons to get some knowledge of the main parts of the body. These accounts were nevertheless short and usually limited to a summary of Avicenna's *Canon*, as the French surgeon Henri de Mondeville stated in the first decade of the fourteenth century. By contrast, in his *Chirurgia Magna* (1363), Guy de Chauliac was able to quote Niccolò da Reggio's version of Galen's *On the usefulness of the parts of the body*. He also clearly specified the twofold method of acquiring anatomical skill: by reading books and by sensory experience. The necessity of the latter resort was attested by a quotation from Averroes's *Colliget*, a work that, from its translation into Latin in 1285, played a significant part in attributing to anatomy a new status and

[6] Rudolph E. Siegel, *Galen On the Affected Parts, Translation from the Greek Text with Explanatory Notes* (Basel: S. Karger, 1976), p. 23. Gerard of Cremona's Latin translation of Avicenna's *Canon* can be consulted in its printed version (Venice: Paganinus de Paganinis, 1507; repr. Hildesheim: Olms, 1964).

in stressing its experimental side. Averroes had started his book, meant to present all general knowledge necessary to practice medicine, with a section on anatomy.[7]

Considered as the starting point of a new trend that led to the Renaissance, Mondino de' Luzzi's *Anatomia* innovated more in its claim of being based on actual dissections of human cadavers than in its content. Written in 1316, a year before Niccolò da Reggio's translation of Galen's major treatise on anatomy, Mondino's work was less informative than Avicenna's *Canon* and even repeated some errors that most medieval authors avoided, the best known example being the division of the uterus into seven cells that Mondino allegedly saw in female cadavers. More reasonably, fourteenth-century authors preferred to describe two or more "kinds of folds" in the womb. Besides its occasional inaccuracy, Mondino's method of conducting dissection was based on the ancient distinction between homogeneous and composite members; it adopted also a mode of inquiry that followed a scholastic pattern provided by late-Alexandrian commentaries on Galen's treatise *On Sects*. Despite these inadequate features, which persisted well into the following century in anatomical procedures, Mondino's work gave (or testified to) an irreversible impulse to anatomy by referring for the first time since antiquity to actual dissections of human cadavers.[8]

Traditional assumptions regarding the history of medieval anatomical observation have to be revised. Contrary to what is too often stated, ecclesiastical prohibitions do not seem to have played a crucial role in limiting dissections. The striking feature in this respect is that the first prohibitions were issued at the same time as the practice of dissections was mentioned in diverse kinds of sources. Boniface VIII's bull *Detestande feritatis* (1299), intended merely to prevent burial practices involving dismembering and boiling bodies – in particular of French royal persons – in order to convey them from a long distance and to keep them in different sanctuaries, did indirectly affect the practice of dissection.[9] Henceforth anatomists felt obliged to obtain papal dispensations, which made the event more formal.

7 On the surgeons mentioned, see Jole Agrimi and Chiara Crisciani, "The Science and Practice of Medicine in the Thirteenth Century According to Guglielmo da Saliceto, Italian Surgeon," in *Practical Medicine from Salerno to the Black Death*, ed. Luis García-Ballester et al. (Cambridge: Cambridge University Press, 1994), pp. 60–87; Michael McVaugh, *The Rational Surgery of the Middle Ages* (Florence: Edizioni del Galluzzo, 2006); Julius L. Pagel, *Die Chirurgie des Heinrich von Mondeville* (Berlin: Georg Reimer, 1892); and Michael R. McVaugh, *Guidonis de Caulhiaco (Guy de Chauliac) Inventarium sive Chirurgia Magna* (Leiden: Brill, 1997). The Latin translation of Averroes's *Colliget* can be consulted in the printed edition (Venice: apud Juntas, 1562; repr. Frankfurt am Main: Minerva, 1962).

8 Ernest Wickersheimer, *Anatomies de Mondino dei Luzzi et de Guido de Vigevano* (Geneva: Slatkine Reprints, 1977); Roger French, "A Note on the Anatomical Accessus of the Middle Ages," *Medical History*, 23 (1979), 461–8; and Mark Infusino, Dorothy Win, and Ynez Violé O'Neill, "Mondino's Book and the Human Body," *Vesalius*, 1 (1995), 71–6.

9 Elizabeth A. Brown, "Death and the Human Body in the Later Middle Ages: The Legislation of Boniface VIII on the Division of the Corpse," *Viator*, 12 (1981), 221–70; and Katharine Park, "The

One finds references to dissection in regulations concerning their place in university teaching or surgical training, descriptions in medical books, and judicial proceedings involving the violation of graves or the theft of cadavers. The earliest official university regulation seems to have been issued in 1340 in Montpellier, where the masters planned a biennial dissection of a human cadaver. Mondino's work attests that the practice was introduced at Bologna long before its mention in the statutes dated 1405; it was probably the same at Padua, where the first written regulation was issued in the statutes of 1465. No such regulation appears in medieval Paris. It is attested that in 1407 the body of the bishop of Arras was opened in the presence of some university masters, most likely for an autopsy or an embalming. The first official mention of a dissection in Paris is dated 1477, a few years after the faculty of medicine had settled into a permanent building. In Paris, as in Italy or elsewhere, private dissections in a master's home (where lessons also took place) could have been practiced without leaving any trace in any source, if they were conducted in a regular way from a judicial point of view. There is no reason why Guido da Vigevano, who referred in his *Anatomia* (1345) to many actual dissections, could not have practiced some of them in Paris, where he lived at this time, a town in which physicians and surgeons were trained in embalming royal bodies. Moreover, observations of bones provided by ossuaries happened to take place in Parisian teaching, as a few authors attest.[10] The scarcity of direct information in this field does not allow historians to treat medieval anatomy as purely an Italian concern, on the grounds of Renaissance evolution.

Although the weight of ecclesiastical prohibitions must not be overvalued, it remains true that in the Middle Ages, as in most periods or civilizations, the opening of cadavers was a difficult task, fraught with material and psychological obstacles. Medical authors nevertheless recognized the validity of this means of knowing truth and, from the beginning of the fourteenth century, mentioned it frequently.[11] It served mainly as a basis for choosing between the different statements provided by Greek and Arabic sources. In this context, it is never clear whether medieval authors referred to actual dissections that they had conducted or to those of Galen or another predecessor. Whatever the reality of dissections during the Middle Ages, they lacked method and never took the form of theatrical demonstration, which was a Renaissance invention.

Criminal and the Saintly Body: Autopsy and Dissection in Renaissance Italy," *Renaissance Quarterly*, 47 (1994), 1–33.

10 On regulations, see Nancy G. Siraisi, *Medieval and Early Renaissance Medicine*, pp. 86–7. For a reassessment of the Parisian case, see Danielle Jacquart, *La médecine médiévale dans le cadre parisien (XIVe–XVe siècles)* (Paris: Fayard, 1998).

11 See, for instance, an analysis of Arnald of Villanova's statement in Michael R. McVaugh, "The Nature and Limits of Medical Certitude at Early Fourteenth-Century Montpellier," *Osiris*, 2nd ser., 6 (1990), 76–8.

The slow reconstruction of bookish anatomical knowledge found a kind of achievement in Jacques Despars' commentary on Avicenna's *Canon*, written between 1430 and 1453. Printed in Lyon in 1498, it was not without influence on the anatomists of the beginning of the sixteenth century, in particular Berengario da Carpi, who quoted it explicitly. Unlike Italian commentators who preceded him, Jacques Despars commented at length on the sections of Avicenna's *Canon* dealing with bones, muscles, nerves, veins, and arteries. He systematically gave long quotations from Galen's works, thus revealing in detail the incompleteness of Avicenna's account of them. This juxtaposition of many sources pointed out also the extraordinary heterogeneity of anatomical terminology, which was very often misleading. Moreover, his careful reading of Galen's work led Jacques Despars to express some admiration for the structure of the human body and the geometrical harmony that rules it. Whereas the few illustrations reproduced in the 1498 edition are in no way realistic and belong to the medieval tradition of schematic figures, an aesthetic point of view was not completely lacking in this commentary.[12] In the middle of the fifteenth century, the medieval reconstruction of anatomical knowledge had attained its limits. A real shift intervened with the convergence, during the Renaissance, of a reappraisal of bookish knowledge, greater skill in opening bodies, and an interest in realistic descriptions and representations.

HUMORS, "VIRTUES," AND QUALITIES

The Arabic Galenism that underlay medieval Latin physiology took its principles from Aristotelian physics. Like every substance in the sublunary world, the human body was considered as resulting from the mixture of the four elements. In Salernitan commentaries, the actions and passions of these ultimate physical constituents had been discussed at length since this topic reinforced the dependence of medical theory on natural philosophy. At a time when the demonstration of this philosophical root was no longer a main concern of physicians, university masters in medical faculties found in their reading of Avicenna's *Canon* a justification for not lingering on this topic. Avicenna had clearly stated that on some issues, for example the number and nature of elements, "the medical doctor had to trust the physicist." Scholastic debates focused more on the definitions of the other constituents of the human

[12] On Jacques Despars' account and the Parisian mentions of observations of bones, see Jacquart, *La médecine médiévale dans le cadre parisien*. On the use of his commentary by Berengario da Carpi, see Roger K. French, "Berengario da Carpi and the Use of Commentary in Anatomical Teaching," in *The Medical Renaissance of the Sixteenth Century*, ed. Andrew Wear, Roger K. French, and Ian M. Lonie (Cambridge: Cambridge University Press, 1985), pp. 42–74. On the lack of commentaries on this part of Avicenna's *Canon* before Jacques Despars, see Nancy G. Siraisi, *Avicenna in Renaissance Italy, The "Canon" and Medical Teaching in Italian Universities after 1500* (Princeton, N.J.: Princeton University Press, 1987), p. 58.

body on which medical practice could act – namely, besides anatomical parts, complexions, humors, spirits, "virtues," and operations. These entities formed the touchstones of physiological teaching. The immutability of this framework did not preclude multiple variations from one author to another, all the more so since Greek and Arabic sources did not provide uniform views.[13]

Even the doctrine of humors was open to variation, as revealed by such a scholastic question as "Does blood alone nourish?," regularly raised in diverse contexts. It led authors first to define blood and, consequently, to determine the exact nature of the liquid flowing in veins and arteries. Since it was generally admitted that all four humors were present in blood vessels, the interpretation of the Aristotelian statement (*Parts of animals*, II.3) according to which blood was the only provider of "food" (i.e., of reconstitution) to the bodies of animals with blood came into question. If the word "blood" meant only one of the four humors, to the exclusion of the others (phlegm, yellow bile, and black bile), then Aristotle's statement could not be reconciled with a fundamental principle of medical theory according to which each homogeneous part of the body was formed, during gestation, by a specific predominant humor and, consequently, had to be regenerated, during life, by the same specific humor. For instance, black bile was considered the main "food" of bones. In all of these scholastic discussions, one may recognize, albeit differently formulated, the reminiscence of an ancient debate, going back to the ancient Hippocratic treatise *On the Nature of man*, which stated, in opposition to atomistic ideas, the impossibility that the human body is formed of only one element or humor.[14] In the medieval metamorphosis of this debate, the point was to define the word "blood," with all the refinements of scholastic reasoning.

Despite numerous divergences in detail, a kind of consensus led the authors to conclude that, in veins and arteries, blood was always mingled with a small amount of the other humors. For instance, Pietro d'Abano, in his *Conciliator* (1310), admitted that "pure blood" alone nourished the different parts of the body but defined "pure" in such a way as to permit mingling with another substance. Not surprisingly, an analogy with wine

[13] Avicenna dealt with elements in *Canon* I.1.2. Until the fifteenth century, the basic text providing, at an elementary level, the principles of physiological theory was the *Isagoge Iohannitii*. See Gregor Maurach, ed., "*Isagoge Iohannitii*," *Sudhoffs Archiv*, 62 (1978), 148–74, and the English translation in Edward Grant, *A Source Book in Medieval Science* (Cambridge, Mass.: Harvard University Press, 1974), pp. 705–15. On the Salernitan discussions on elements, see Richard McKeon, "Medicine and Philosophy in the Eleventh and Twelfth Centuries: The Problem of Elements," *The Thomist*, 24 (1961), 211–56; and Bernhard Pabst, *Atomtheorien des lateinischen Mittelalters* (Darmstadt: Wissenschaftliche Buchgesellschaft, 1994).

[14] Hippocrates, *La Nature de l'homme* (*Corpus medicorum graecorum*, I.1.3), ed., trans., and comment. Jacques Jouanna (Berlin: Akademie-Verlag, 1975). Among the numerous authors who raised the question *Utrum solus sanguis nutriat*, Taddeo Alderotti and Dino del Garbo can be quoted for Bologna. See Nancy G. Siraisi, *Taddeo Alderotti and His Pupils, Two Generations of Italian Medical Learning* (Princeton, N.J.: Princeton University Press, 1981), p. 328.

was put forward – one could call "pure" a wine with a small amount of water. Another fourteenth-century writer, Turisanus (ca. 1320–30), made a more innovative assumption. When he dealt with the same kind of question, in his commentary on Galen's *Tegni* (ca. 1320–30), he concluded that the humors numbered three – phlegm, yellow bile, and black bile – rather than four; what was then called "blood" was actually the result of a mixture, in the Aristotelian sense of this term. A perfectly balanced mixture of all three humors made perfect blood, whereas the predominance of one humor made "choleric blood," "phlegmatic blood," or "melancholic blood." Turisanus did not even hesitate to question the traditional numerical equivalence between four elements and four humors, claiming that it was groundless. His solution, which viewed blood as a mixture and not a mere association of the other humors, was best suited to reconcile the idea that "blood alone does nourish" with the assumption that the different members were nourished, each by its specific humor. Had it been adopted by Turisanus's readers – which it was not – it would have led to the reconstruction of a large part of medical theory.[15]

Although continuing to be at the center of many explanations of pathological disorders and many therapeutic procedures, notably purgation and bloodletting, the doctrine of four humors was completed by a kind of extension that became particularly significant in the late Middle Ages. It must be recalled that, before the discovery of the circulation of the blood, veins and arteries were viewed as irrigating ducts, subdivided into thinner and thinner vessels. After successively expelling "superfluities" and, for bile and melancholy, segregating their overflow in appropriate receptacles (the gallbladder and spleen, respectively), the remaining humoral liquid was described as restoring the different anatomical elements through a process of assimilation. The interpretation of Avicenna's *Canon* in the light of Galen's *On the natural faculties* led to an emphasis, in this respect, on the function of the "secondary moistures." Primarily generated after the successive digestions of food in the stomach and the liver, humors continued to be "digested" in blood vessels in order ultimately to be able to penetrate the invisible pores disseminated throughout every part of the body. This third digestion transformed humors into "secondary moistures," which in turn underwent a fourth or final assimilation. In Avicenna's *Canon*, this concept of constantly regenerated moistures went along with the definition of another fluid, called "radical moisture," which was innate and intended to maintain life, from birth to death, by preserving the coherence of the smallest parts of the body. Discussions of "secondary moistures" came to supplant the traditional debates about humors. Revealing in this respect is the Paduan master

[15] Pietro d'Abano, *Conciliator* (Venice: apud Juntas, 1565; facs. Padua: Antenore, 1985), diff. XXX; and Turisanus [or Pietro Torrigiano], *Plusquam commentum*, II.44 (*Digressio de sanguine an sit quartus humor*) (Venice: apud Juntas, 1557).

Giacomo da Forlì's commentary (1414) on Avicenna's *Canon*, which did not raise the old question "Does blood alone nourish?" but replaced it with another one, treated at length: "Is it necessary to define, besides four humors, four secondary moistures which nourish the members?"[16] A better knowledge of Galen's works helps to explain this evolution, but it may also be related more generally to an increasing interest, at the end of the Middle Ages, in all kinds of transformations of matter, an interest to which the development of alchemy also testifies. Perhaps it is not too risky to suggest also that this medical concern about the transformation of blood into bodily substance was not completely extraneous to theological discussions on transubstantiation.

Arabic Galenism had combined pneumatism with humoral theory. Three pneumas or spirits were described as flowing in the three kinds of bodily ducts (veins, arteries, and nerves) and as conveying three immaterial forces or virtues. Emanating from the faculties of the soul, these forces found their seat in the organs defined as "principal": the liver for natural virtue, the heart for vital virtue, the brain for psychical virtue. Whether the testicles had to be considered as a fourth principal organ was a matter of debate. The virtues themselves governed all the functions ("operations") necessary for the preservation of the individual (digestion, growth, generation, pulsation, sensation, etc.), which were performed by anatomical elements naturally devised for this purpose. This explanatory system remained, as a whole, unchanged during the medieval period, but its different components underwent many interrogations and divergences. Two main topics of controversy arose: the nature and the number of spirits, and the appropriate hierarchy among virtues.

Constantine the African's translations had transmitted quite a clear representation of the generation of the spirits, defined as the finest substances of the body. At the second stage of digestion, which takes place in the liver and generates humors, the resulting vapor gives rise to the natural spirit that flows in the veins. When the natural spirit reaches the heart, it is again refined because of the heat of this organ; moreover, by passing with the arterial blood through the lungs, it receives some airy substance and becomes vital spirit. A part of this second spirit is itself refined at the base of the brain, in particular in the tiny and intertwined vessels that form, according to Galen, the so-called *rete mirabile* ("admirable network"). There the third stage of refinement takes place, which generates psychical (or animal) spirit,

[16] Giacomo da Forlì, *Quaestiones super Canonem* (Venice: apud Juntas, 1520), qu. XL. Avicenna dealt with this matter in *Canon*, I.1.4.1–2. For Galen, see Galen, *De facultatibus naturalibus*, ed. and trans. A. J. Brock (Cambridge, Mass.: Harvard University Press, 1991). For a state of the problem in the thirteenth century, see Joan Cadden, "Albertus Magnus' Universal Physiology: The Example of Nutrition," in *Albertus Magnus and the Sciences, Commemorative Essays 1980*, ed. James A. Weisheipl (Toronto: Pontifical Institute of Mediaeval Studies, 1980), pp. 321–39. On the concept of "radical moisture," see Michael R. McVaugh, "The *humidum radicale* in Thirteenth-Century Medicine," *Traditio*, 30 (1974), 259–83; and Arnald of Villanova, *Tractatus de humido radicali*, ed. Michael R. McVaugh, pref. and comm. Chiara Crisciani and Giovanna Ferrari (Arnaldi de Villanova Opera medica omnia, V. 2) (Barcelona: Universidad de Barcelona, 2010).

conveyed through nerves to all the parts of the body. Apart from the materialist or, on the contrary, animist temptation that this representation could afford – denounced by some theologians in the twelfth century – Constantine the African's explanations left many questions unsolved.[17] It could seem illogical to imagine that a cruder matter, like natural spirit, was able to generate two such noble substances as vital and psychical spirits. Since Avicenna's *Canon* did not offer any precise account of spirits, defining them only as conveyors of virtues, their nature was described by university masters in different ways, from a subtle vapor, for most authors, to something analogous to light according to others, for example Turisanus. But the most difficult point was that Galen, and even some Arabic authors, like Averroes, had called into question the existence of natural spirit.

Questions about spirits were closely related to a more fundamental issue, in which several ancient controversies converged. The first difficulty lay in a numerical discrepancy. Although the definition of three forces or virtues conveyed by three spirits was originally patterned to fit with the idea of a tripartite soul (vegetative, sensitive, and rational), as it was stated by Plato and Aristotle, in medieval philosophical views the upper level (rational) of the soul could not have any bodily medium. Physicians thus had to subsume three physical forces under only two faculties of the soul (vegetative and sensitive). Whereas Constantine's *Pantegni* had placed reason in the brain (specifically in the middle ventricle), Avicenna's *Canon* had restricted the role of this organ to sensation, in a manner more compatible with theological concerns. It was therefore difficult to assign a precise status to vital virtue, which belonged both to the vegetative faculty of the soul, by its function of maintaining life, and sensation, since it was supposed to manage the emotions. This question was also connected with an ancient controversy involving the heart, which was held to be the seat of this virtue. It led inevitably to a reopening of the debate about the primacy of this organ in the body and, consequently, to the contrast of Aristotelian views with Galenist ones, the former according primacy to the heart, the latter to the brain. This controversy had been an occasion for Avicenna to put forward a kind of medical instrumentalism, stating that "qua physician, the physician need not consider which of these two opinions is the true one, that being the task of the philosopher."[18] As for the question of primacy of that one principal organ, medieval discussants developed a compromise that conceded much to the Aristotelian view. The image of a feudal king could help in this

[17] See Charles Burnett, "The Chapter on the Spirits in the *Pantegni* of Constantine the African," in Burnett and Jacquart, *Constantine the African and ʿAlī ibn al-ʿAbbās al-Maǧūsī, The "Pantegni" and Related Texts*, pp. 99–120. On Galen's *rete mirabile*, see Plinio Prioreschi, "Galenicae Quaestiones Disputatae Duae: rete mirabile and pulmonary circulation," *Vesalius*, 2 (1996), 67–78.

[18] See Michael R. McVaugh, "Nature and Limits of Medical Certitude at Early Fourteenth-Century Montpellier," p. 79. On this question of the primacy of the heart, see also *Il cuore, The Heart*, ed. Nathalie Blancardi et al. (Micrologus, 11) (Florence: SISMEL, 2003).

matter, allowing them to consider heart as "more principal" without unduly restricting the power of the brain and psychical virtue. But the primacy attributed to the heart led them consequently to reconsider the origin of spirits. As Avicenna had suggested, vital spirit was viewed as generated first and became "natural" and "psychical" in the appropriate organs because of their specific qualities or complexion.

If the image of a feudal king was used by some authors, others preferred to resort to an analogy with the sun. Turisanus developed this analogy very fully, stating that just as sunlight reflected from the Moon is altered by the nature of the Moon, so the nature of the spirit coming from the heart changes when it flows within the brain.[19] Turisanus is also famous for another assumption, his rejection of the Galenic idea that heartbeat and pulsation were caused by vital virtue. Relying on Aristotle, who described a process of ebullition or boiling in his short treatise *On youth and old age*, Turisanus opted for a mechanistic explanation of pulsation, without any connection to a faculty of the soul.[20] It does not follow that Turisanus was a forerunner of modern physiology. In fact, his solution had, from a medieval point of view, the advantage of giving a less ambiguous status to vital virtue. On the one hand, the movement of the spirit generated and heated in the heart was the only thing responsible for pulsation. On the other, the luminous nature of this spirit enabled it to convey the virtue emanating from a faculty of the soul, which remained in the sole domain of sensation. Like his proposal for limiting the number of humors to three, Turisanus's explanation of pulsation did not convince many followers among university masters, even if it was considered an Aristotelian view.

All these debates about humors, spirits, and virtues, which attacked some main theoretical principles, did not have a great effect on the whole explanatory system because the latter was firmly based on the concept of complexion. Human beings shared with all the other animate or inanimate beings of the sublunary world a constitution involving a balance of the qualities of hot, cold, moist, and dry. Inherited from Galenism, complexion was defined by Avicenna as "the quality which results from the mutual action and passion of the opposite qualities of the elements."[21] Each individual was characterized, as a whole, by an innate balance, which was subjected to a range of variations throughout life, according to age, activities, environment, and pathological disorders. Apart from this primary individual complexion, to each part of the body, in its normal state, was attributed a tendency toward a specific complexion, common to the human species. For instance, brain was

19 Turisanus, *Plusquam commentum* (Venice: apud Juntas, 1557), fol. 35r. The analogy with the feudal king is, for instance, used by Jacques Despars in his commentary on Avicenna's *Canon*, I.1.5 (Lyon: Iohannes Trechsel, 1498).

20 On Turisanus's explanation of pulsation, see Danielle Jacquart, "Coeur ou cerveau? Les hésitations médiévales sur l'origine de la sensation et le choix de Turisanus," in *Il cuore/The Heart*, pp. 73–95.

21 Avicenna, *Canon*, I.1.3.1.

described as cold and moist, bone as cold and dry, whereas blood had the very qualities of life, namely warmth and humidity. Explanations of health and disease thus involved a complex system of interactions between qualities. On this matter, medical authors shared with natural philosophers the same kinds of problems regarding the definition of these qualities, their actions and passions, the conditions of their permanence or change within mixed bodies, their intensity, and so forth. Additional difficulties were raised by the living and sensitive features of the human body.

FROM HEALTH TO DISEASE

"Medicine is the science of health, disease and the neutral state." Despite its Galenic origin, this definition was not widespread in university medicine. Except in commentaries on Galen's *Tegni*, in which this sentence occurs, the question of a neutral state did not seem crucial since the overwhelming notion of complexion implied for the healthy state a wide latitude, which could just as well be called "neutral." Whereas a perfectly balanced complexion was considered as an ideal, attained only by exceptional men – women being excluded from this ideal – the multitude of factors that acted on the functioning of the human body made it difficult to view the passage from health to disease either as a linear evolution or as a sudden change, except perhaps when a traumatic event, like an injury or a fall, was the cause of the pathological disorder.[22]

At the center of causal explanations was a category of factors that were considered common to health and disease. They formed the "nonnaturals," a concept of Galenic origin that Arabic authors, and after them scholastic physicians, developed at length. Six factors were generally listed: air, food and drink, sleep and wakefulness, motion and rest, evacuation and repletion, and the passions of the mind; baths, exercise, or sexual practices were occasionally added. This theory was highly developed, for it allowed learned practitioners to interfere in every aspect of the ordinary lives of their patients. Apart from its scientific grounds, it has to be connected with the process of "medicalization" that historians have observed in the late Middle Ages and with the increasing concern for the preservation of health, notably illustrated by the writing of numerous regimens. The usual pattern of these writings followed the enumeration of the six nonnaturals, but it can be noted that from the middle of the fourteenth century they focused more and more on dietetics, supplying even some culinary advice.[23] This dietetic concern had many social

22 Per-Gunnar Ottosson, *Scholastic Medicine and Philosophy* (Naples: Bibliopolis, 1984), pp. 127–78.

23 On the concept of "nonnaturals," the "medicalization" of society, and the regimens of health, see Luis García Ballester, "On the Origin of the 'Six Non-natural Things' in Galen," in *Galen und das hellenistische Erbe*, ed. Jutta Kollesch and Diethard Nickel (Stuttgart: Franz Steiner Verlag, 1993), pp. 105–15; Danielle Jacquart and Nicoletta Palmieri, "La tradition alexandrine des *Masa'il fi t-tibb* de Hunain ibn Ishaq," in *Storia e ecdotica dei testi medici Greci*, ed. Antonio Garzya and Jacques

and cultural implications, but it was also related to discussions about the transformation of foods into "secondary moistures" and their assimilation in bodily substance. The idea that "one is what one eats" is then to be understood in its literal sense.

Beyond the framework of regimen of health and the concern for the quality of food, two nonnaturals, air and the passions of the mind, deserve particular attention. From the Hippocratic tradition, best illustrated in the well-known treatise *Airs, waters and places*, the influence of the surrounding air upon human bodies was clearly established. In its larger sense, in use among medical authors, the word "air" could embrace climatic features of seasons and lands and geographical exposure to the winds, as well as astral influences. Astrology was unequally involved in medical writings, but it is worth noting that the theory of nonnaturals could supply a rational basis for resorting to it. Despite this possibility, Avicenna was reluctant to espouse any medical application of astrology. His statement that astral influences – which otherwise he considered as not precisely knowable by human science – constitute remote causes, beyond the reach of the physicians' art, undoubtedly restrained some university masters' enthusiasm. At the end of the Middle Ages, physicians nevertheless had to take into account the astrological vogue that pervaded society.[24] Pietro d'Abano was probably the author who tried hardest to combine Galenic medicine with astrological explanations. He even attempted to closely relate the variations of complexions and the phases of periodic fevers to the Ptolemaic theory of epicycles.

Epidemics that raged after the Black Death gave university masters recurrent opportunities for dealing with air. Their Arabic sources explained the epidemic feature of plague as a corruption of air. It was then debated, in a scholastic manner, whether the air was corrupted in its substance or only in its qualities, and what kinds of climatic events could bring about this corruption. The external agents of this corruption were identified as some vapors, arising from fetid places and brought by winds to every region. When breathed by patients, the corrupted air vitiated their vital spirit, thus severely damaging their health. If some particular predispositions of bodies, related to their complexion, age, sex, and diet, explained for the most part the selective character of epidemics, contagion was also regularly mentioned as a cause,

Jouanna (Naples: M. D'Auria editore, 1996), pp. 217–36; Michael R. McVaugh, *Medicine before the Plague: Practitioners and Their Patients in the Crown of Aragon 1285–1345* (Cambridge: Cambridge University Press, 1993); Pedro Gil Sotres, "The Regimens of Health," in Grmek, *Western Medical Thought from Antiquity to the Middle Ages*, pp. 291–318; Marilyn Nicoud, *Les régimes de Santé au Moyen Âge: Naissance et diffusion d'une écriture médicale (XIIIe–XVe)* (Rome: École française de Rome, 2007).

24 Pietro d'Abano's major account of the use of astrology in medicine is in *Conciliator*, diff. X. For a more detailed analysis of this account and a summary of Avicenna's views, see Danielle Jacquart, "L'influence des astres sur le corps humain chez Pietro d'Abano," in *Le corps et ses énigmes au Moyen Age*, ed. Bernard Ribémont (Caen: Paradigme, 1993), pp. 73–86; Danielle Jacquart, "Médecine et astrologie à Paris dans la première moitié du XIVe siècle," in *Filosofia, scienza e astrologia nel Trecento europeo*, ed. Graziella Federici Vescovini and Francesco Barocelli (Padua: Il Poligrafo, 1992), pp. 121–34.

all the more since in Avicenna's *Canon* (I.2.2.1.9) plague was listed among those diseases passing from one man to another. For the medieval period, there is no reason to oppose an "aerist" theory to a "contagionist" one. Contagion from person to person was viewed as a transmission of corrupted air through breath, clothes, or pieces of furniture. Although it was commonly mentioned among predisposing factors, contagion was not particularly emphasized. The fact that the recognized pathological agent was air explains why plague and leprosy were considered to be the only diseases that putrefied the blood, a humor wet and hot like air.[25]

Among nonnaturals, the passions of mind also played an important part in medical explanations. Under the words "accidents of the soul" – which literally translated an Arabic expression – were included all the emotional states, such as fear, joy, sadness, anger, and anxiety, that had a psychological root and a visible physical expression. These "accidents" made plain the link existing between body and soul. One of the most detailed accounts of this psychosomatic process was given by Arnald of Villanova (d. 1311) in his *Speculum medicine* and his *De parte operativa*. The Arabic theory of the internal senses and their cerebral localizations provided the framework within which medical explanations of psychological states developed. By passing through the three ventricles of the brain, located at the front, in the middle, and at the back of this organ, the psychical virtue had the power of catching the images perceived by the external senses, organizing them, evaluating whether they would bring some good or not, and storing them in the memory. Although the psychical virtue was an immaterial instrument, its conveyor, the spirit, was a material substance and so could be affected in its quality of coldness, hotness, wetness, or dryness by the perceived impressions. The alteration of psychical spirit was described as the result of a previous alteration of the vital one from which it originated. According to the quality of the impression made by the perceived image, the instrument of the emotions – that is, the vital virtue located in the heart – affected the movement of the blood responsible, for instance, for the blushing or paleness of the face. A continuous chain of reactions, in both directions, thus linked psychological to physical states.[26]

Everyday emotions, as well as psychiatric disorders and psychotherapeutic means, found their justification in this explanatory system. Avicenna had been reluctant, from a philosophical point of view, to locate reason in the

[25] Works dealing with the medieval conception of plague are numerous; one can find a statement of the question and a bibliography in Jon Arrizabalaga, "Facing the Black Death: Perceptions and Reactions of University Medical Practitioners," in García-Ballester et al., *Practical Medicine from Salerno to the Black Death*, pp. 237–88.

[26] Pedro Gil-Sotres, "Modelo teórico y observación clinica: Las pasiones del alma en la psicologia medica medieval," in *Comprendre et maîtriser la nature au Moyen Age, Mélanges d'histoire des sciences offerts à Guy Beaujouan* (Geneva: Librairie Droz, 1994), pp. 181–204; and Mary E. Wack, *Lovesickness in the Middle Ages: The "Viaticum" and Its Commentaries* (Philadelphia: University of Pennsylvania Press, 1990).

middle ventricle of the brain; nevertheless, he could not avoid it when he dealt with severe psychiatric diseases, such as melancholy or mania. A contradiction thus existed between the statements that he put in the first and third books of his *Canon*. When dealing with mental disorders, Latin commentators had to set their own limits according to their appreciation of theological requirements. They generally showed a tendency to widen their own field by emphasizing somatic causes or implications of irrational behavior.[27]

Whereas "nonnatural" factors took their name from their nonmembership among the internal components of the body, Galen called pathological disorders "against nature." Their classification rested on their causes and symptoms, as well as on their effect upon one part of the body or upon its whole. Medieval classifications of disease had reached a level of complexity that makes difficult any clear summary or any comparison with the modern subject. It associated entities with names inherited from earliest antiquity, like pleurisy or epilepsy, with diseases defined by one major symptom or one presumed cause. In Avicenna's *Canon* (Book III), diseases affecting one particular part of the body are systematically described according to principles that Galen had put forward in his diverse works dealing with pathology. Different criteria were taken into account – for example, an observed dysfunction, which could involve diminution, change, or disappearance of a function or "operation," or the intervention of any humoral excess that could putrefy or not – all these disorders being generally explained by a complexional alteration or "malice."[28]

In the field of pathology, the most coherent body of medieval medical knowledge was formed by the description of fevers, considered as morbid entities. University medicine inherited from Greek and Arabic authors a system of classification of fevers according to the bodily substance that was primarily affected: spirits for "ephemeral fevers," humors for "putrid or humoral fevers," solid parts for "hectic fever." As for "pestilential fevers," their main feature was to be epidemic. Humoral fevers had the peculiarity of being periodic, with regular attacks: a phlegmatic origin caused a fever daily; a choleric origin, every third day (tertian fever); a melancholic origin, every fourth day (quartan fever). Since infectious diseases that were so described did not always have a regular periodicity, the phenomena were saved by imagining many kinds of composite fevers, which stimulated scholastic subtlety. In contrast to other humoral fevers, blood fever was considered continuous.[29]

27 Danielle Jacquart, "Avicenne et la nosologie galénique: l'exemple des maladies du cerveau," in *Perspectives arabes et médiévales sur la tradition scientifique et philosophique grecque*, ed. Ahmad Hasnawi, Abdelali Elamrani-Jamal, and Maroun Aouad (Leuven: Peeters-IMA, 1997), pp. 217–26.

28 For an example of the implications of this concept, see Luis García-Ballester and Eustaquio Sanchez Salor, *Commentum supra tractatum Galieni De malicia complexionis diverse* (Arnaldi de Villanova Opera medica omnia, 15) (Barcelona: Edicions de la Universidad de Barcelona, 1985).

29 Apart from Avicenna's *Canon* (Book IV, fen. 1), Isaac Israeli's *De febribus* (Lyon: Barthélemy Trot, 1515) provided the most complete account of fevers.

If they did not much improve the general classification of fevers, university masters did deepen their reflections on the nature of fever, which was a favorite topic of discussion. This question was more generally related to the medieval problem of the intensity of qualities, but it also had its properly medical roots. The point was to decide whether fever was a totally "extraneous" heat, as Galen seemed to claim, or a mixture of "natural" and "extraneous" heats, as Averroes's *Colliget* had put it.[30] In order to understand the difficulties that scholastic authors had to face, it suffices to recall that the intensity of fever was defined not by the intensity of its heat but by the importance of the dysfunctions that it caused. Apart from the fact that medieval people had no other means for measuring intensity of heat than touch, it has to be stressed that, for any one kind of fever, the intensity of heat was supposed to vary according to the innate complexion of the patient. This is clearly stated by Gentile da Foligno (d. 1348), for instance, who opted for a definition of fever involving an addition of extraneous heat to the natural one. In dealing with the same problem, Pierre de Saint-Flour, a Parisian master around 1350, made an original assumption, imagining a double scale or "latitude." To natural heat was attributed a "major latitude," which covered the whole life, divided into four degrees according to ages. Within this latitude, he defined a "minor latitude" or "latitude of health," also divided into four degrees. In the first one, in which heat was faint, the functions were performed imperfectly; the second degree was equivalent to perfect health; in the third, functions were damaged but in an imperceptible manner; and in the fourth, functions were patently damaged. The final stage of this fourth degree corresponded to the final stage of the fourth degree of the "major latitude," which was marked by death.[31] This imaginative explanation attested the influence of philosophical debates on the evaluation of increase and decrease in the intensity of a quality, as was often the case in medical scholastic debates about fever and indeed about any topic dealing with the notion of complexion.

FROM THEORY TO PRACTICE

The late-Alexandrian method of teaching and commenting had passed on to Arabic as well as Latin physicians a division of medical topics and purposes that remained stable for centuries. Few were those, like Turisanus, who called

[30] On the background and fate of this question, see Ian M. Lonie, "Fever Pathology in the Sixteenth Century: Tradition and Innovation," in *Theories of Fever from Antiquity to the Enlightenment*, ed. Walter F. Bynum and Vivian Nutton (London: Wellcome Institute for the History of Medicine, 1981), pp. 19–44.

[31] Gentile da Foligno, Comments on Avicenna's *Canon*, IV.1.1 (Venice: heredes O. Scoti, 1520); Pierre de Saint-Flour, *Colliget florum*; and Julius Pagel, *Neue litterarische Beiträge zur mittelalterlichen Medizin* (Berlin: Georg Reimer, 1896), p. 19. On the philosophical debates on increase and decrease in the intensity of a quality, see Laird, Chapter 17, this volume.

its prime foundation into question by claiming that the subdivision into theory and practice had no authentic Galenic origin. What was at stake here was the validity of recognizing a purely speculative side of medicine. This was clearly stated, for instance, in Constantine the African's preface to *Viaticum*, where the translator distinguished three categories among his possible readers: those who intended to practice; those who were interested in medical science in itself, and consequently mainly in its theoretical part; and, finally, those who were interested in both of these aims. During the twelfth century, theoretical medicine had truly delivered to some philosophers, like William of Conches, a body of knowledge about the natural world that could supply the information lacking in this field before the main translations of Aristotle's or Avicenna's philosophical works. In some Salernitan commentaries, such as Bartholomew's, medicine was almost equivalent to natural philosophy since its study was meant to include the science of animals, herbs, and stones, as well as the human body. The context of university medicine was quite different, and the treatment of the usual questions on the definition of medicine as science or art, and its subdivision into theory and practice, tended to emphasize its practical aim. As Turisanus put it, medicine was an active science, despite its subordination to a speculative one, namely natural philosophy.[32]

Often defined as a science on its theoretical side, since it was based on demonstrable principles, and as an art on its practical side, medicine could not help but possess an ambiguous status, which was inherent in both its foundation and its aim. A single author like Averroes seemed to leave two contradictory statements. Whereas in his *Colliget* he described medicine more as *ars* than as *scientia*, in his commentary on Avicenna's *Cantica*, he stressed more the demonstrable character of medicine in both its theoretical and its operational aspects. Apart from the actual practical application that constituted the ultimate goal of medicine, the respective roles of reasoning and experience in the acquisition of medical knowledge were the difficult points to resolve. In his tract *De erroribus medicorum*, Roger Bacon blamed the ordinary physicians for not resorting to experience sufficiently, lost as they were in infinite useless scholastic debates. When Roger Bacon wrote this work, around 1260–70, medicine had been introduced into the university for only a short time, and the medical professors' major concern probably was the adaptation of their teaching methods to this institution's rules.[33]

Although medieval medicine never resembled the experimental science that Roger Bacon had dreamed of, there was a place even in a university

[32] Mark D. Jordan, "Medicine as Science in the Early Commentaries on Johannitius," *Traditio*, 43 (1987), 121–45; and Siraisi, *Taddeo Alderotti and His Pupils*, pp. 118–46.

[33] Roger Bacon, *De erroribus medicorum*, ed. A. G. Little and E. Withington (Opera hactenus inedita Rogeri Baconi, 9) (Oxford: Clarendon Press, 1928), pp. 150–79. On Averroes's definitions, see Michael R. McVaugh, "Nature and Limits of Medical Certitude at Early Fourteenth-Century Montpellier," pp. 69–70.

framework for thorough reflections on the role of experience. One of the favorite occasions was provided by commentaries on Hippocrates' first aphorism, in which it was stated that "experience is dangerous" or "treacherous" (depending on which Latin translation one employed). Galen's commentary had emphasized the notion of "danger" – the human body is not like wood or leather and needs to be respected; moreover, in living phenomena it is difficult to connect an effect to a precise cause. In the same vein, the Italian master Ugo Benzi claimed in the second decade of the fifteenth century that the ever-changing state of physiological processes impeded the conducting of reliable experiences. In commenting on this first Hippocratic aphorism, Italian masters mostly understood "experience" as "experimental pharmacology." It was thus usual to see it from a historical viewpoint – experience was used by the first inventors of medicine, before its doctrine was fixed. The pharmacological emphasis was also inferred from the reading of Avicenna's *Canon* and the notion of specific form (I.2.2.1.15). This theory went along with the complexional action of drugs. Whereas the effect of a substance through its complexional qualities could supposedly be known by reasoning, the effect of its specific form, independent of the balance of primary qualities, was known only through experience. The usual example was the power of scammony to attract bile, in comparison with the action of the magnet, which in the thirteenth century both Robert Grosseteste in his commentary on Aristotle's *Physics* and Pierre de Maricourt in his *Letter on the magnet* (1258) had quoted. Even though they repeated the seven rules that Avicenna had set for conducting experimental testing of drugs, it is not certain that medieval masters were ready to experiment themselves. In this respect, Gentile da Foligno stressed the conjectural character of any measurement involving the evaluation of the intensity of qualities. In the footsteps of the Arabic author al-Kindī, Arnald of Villanova around 1300 had set forth a system based on mathematics in order to evaluate the final complexion of a compound drug. Since the intensity of each quality was evaluated according to four degrees, each substance being possessed of two qualities, the interaction of several substances within a compound had a final result that depended on a complex calculation. Despite all such efforts, appreciation of the intensity of heat or humidity continued to rest on either bookish knowledge or inexact sensory perception.[34]

Because of the multiple variable factors that physicians had to take into account before prescribing, any practical endeavor had a kind of experimental

[34] Chiara Crisciani, "History, Novelty, and Progress in Scholastic Medicine," *Osiris*, 2nd ser., 6 (1990), 118–39; Alistair C. Crombie, *Styles of Scientific Thinking in the European Tradition*, 3 vols. (London: Duckworth, 1994), vol. 1, pp. 339–51; Danielle Jacquart, "L'observation dans les sciences de la nature: possibilités et limites," *Micrologus*, 4 (1996), 55–75; Arnald of Villanova, *Aphorismi de gradibus*, ed. Michael R. McVaugh (Arnaldi de Villanova Opera medica omnia, 2) (Barcelona: Universidad de Barcelona, 1975); and Michael McVaugh, "The 'Experience-Bases Medicine' of the Thirteenth-Century," *Early Science and Medicine*, 14 (2009), 105–30.

character. This was clearly perceived by academic authors. When repeating Aristotle's definition of experience stated at the beginning of *Metaphysics*, that the art was born from a multitude of experiences, they understood this in terms of both an experience acquired through long habit and an experimental process. Both of these concerns were associated with the method of healing named from Gerard of Cremona's translation of the Galenic work dealing with it, "*ingenium sanitatis*." A peculiarity of the Arabic translation from Greek had led the notion of "method" to be rendered by the polysemic word *ingenium*. The same word was used in Gerard of Cremona's translation of al-Farabi's classification of the sciences, with the acceptation of "ingenious devices" that applied theoretical principles. It is not impossible that medieval interpretations of the Galenic work *On the Method of Healing* were partially arrived at in the light of al-Farabi's *scientia ingeniorum*. The notion of *ingenium sanitatis* then served to characterize the passage from general principles to particular cases, from theory to practice, from observed signs on a patient's body to a set of prescriptions.

Thus, the Montpellier master Bernard of Gordon wrote the treatise *De decem ingeniis*, or *De ingenio sanitatis* (early fourteenth century), which listed ten *ingenia* able to cure diseases. Under this word were contained all the criteria that the practitioner had to deal with; they involved both his theoretical knowledge and his observation of the actual situation offered by the patient and his environment. The first *ingenium* was drawn from the essence of the disease and involved eleven criteria related to medieval ways of characterizing an illness: its simple or composite character, its cause, the "nonnaturals" playing a part in it (including the complexion of the person who prepared the food), the physician's behavior, the observation of any tumor or abscess, and the mode of therapeutic action usually prescribed for this kind of disease. The nine other *ingenia* were related to the patient's complexion, the natural faculties and the operations performed by the body, and finally the nature of the affected parts (their principal character, their position, their relation to other parts, and their sensitivity). Bernard of Gordon thus gave a precise picture of the various questions that a learned practitioner had to answer before undertaking his treatment.[35] Fourteenth- and fifteenth-century works testified to a major concern on the part of university masters for the reliability of their science or art. Several factors converged to heighten this concern: the profound influence of Galen's *Method of Healing*, Aristotelian epistemology relating to the knowledge of particulars, and the increasing social need for learned practitioners. Gentile da Foligno clearly summarized the requirements of the medical art, founded on a twofold approach: a "scientific" one, taught and based on demonstration, and an experimental or experiential one,

[35] Danielle Jacquart, "De l'arabe au latin: l'influence de quelques choix lexicaux (*impressio, ingenium, intuitio*)," in *Aux origines du lexique philosophique européen*, ed. Jacqueline Hamesse (Louvain-la-Neuve: FIDEM, 1997), pp. 165–80; and Luke E. Demaitre, *Doctor Bernard of Gordon: Professor and Practitioner* (Toronto: Pontifical Institute of Mediæval Studies, 1980).

acquired through training and the experience of numerous particular cases. The development, principally in northern Italy, of the genre of *consilia* may be related to this attempt to register as many particular cases as possible.[36] Even if some of these letters of advice – which offered, along with a set of prescriptions, a description of the patient's state and an identification of his disease – were addressed to particular patients, they were more than mere consultations. The fact that they were copied several times in manuscripts and sometimes gathered in collections does not suggest that they served as models to be followed but instead that they provided examples of actual or purportedly actual cases. Since "life is short," the experience of one practitioner could not suffice and, in order to fulfill the requirement of Aristotle's definition, the multitude of other practitioners' experiences contributed to the building of the art. From the thirteenth to the fifteenth century, medical theory had revealed, through the scholastic method of reading sources and reasoning, some failures that to a modern mind seem irreparable. The solidity of the notion of complexion, together with its flexibility, ensured the durability of the whole explanatory system. The difficulty of measuring intensive qualities and the changing character of living beings were seen as impediments to applying an experimental approach outside the observation of dissected bodies, the latter not yet going further than a verification of bookish knowledge. Although it is hard to find in medieval writings the descriptions of real cases without any deformation imposed by theoretical presuppositions, nevertheless scholastic medicine developed, in parallel with its "dissection" of doctrinal principles, an awareness of the specificity of its object, which meant dealing with a wide variety of particular situations.

[36] Jole Agrimi and Chiara Crisciani, *Les "consilia" médicaux* (Typologie des sources du Moyen Age occidental, 69) (Turnhout: Brepols, 1994).

26

MEDICAL PRACTICE

Katharine Park

This chapter deals with medical practice in Western Europe, the institutions and circumstances that shaped it, and their evolution during the period from about 1050 to 1500. I have taken "practice" in its broadest sense, to refer to the varied activities engaged in by medieval Europeans of all classes in order to manage illness and to repair or maintain health. It includes approaches that spanned what we would call religion, magic, and science – the boundaries between those categories were less distinct in the Middle Ages – and included activities as varied as domestic nursing, faith healing, and the founding and administration of hospitals, as well as the work of the men and (occasionally) women who practiced physic and surgery in accordance with the principles of learned or text-based medical knowledge.[1]

If medicine as a learned discipline was shaped primarily by the relationship between academic writers and their students and colleagues,[2] historians of medical practice focus rather on the relationship between healers and the sick. The distinction is both important and fairly recent. Earlier historians of medieval medicine tended to rely disproportionately on the perspective of learned surgeons and, especially, physicians, as well as on the atypical example of Paris. This resulted in a picture of medical practice that overemphasized the importance of elite healers and attributed to them (and to university culture in general) an exaggerated role in health care institutions and practices, prompting premature claims about the emergence of medicine as a "profession" in the late Middle Ages. More recent historians have recast the problem, focusing less on the power and authority of academically trained doctors over their patients and other practitioners than on the "negotiation,"

[1] Medical practice should not be confused with practical medicine, or *practica*, which was the part of the learned discipline of medicine that dealt with diseases and their treatment. On the distinction between practice and *practica*, see Geneviève Dumas and Faith Wallis, "Theory and Practice in the Trial of Jean Domrémi, 1423–1427," *Journal of the History of Medicine*, 54 (1999), 56.

[2] The best general account of the culture of learned medicine is Jole Agrimi and Chiara Crisciani, *Edocere Medicos: Medicina nei secoli XIII–XV* (Naples: Guerini, 1988).

"exchange," "contract," or "encounter" between patients and healers. This reformulation has two clear benefits. It allows the historians to move beyond narrow (if important) questions of organizational structure to look at the activities that actually constituted medical practice, and it foregrounds the relative parity of the parties involved.[3]

Indeed, for much of the period covered by this chapter, European society lacked stringent licensing procedures, effective public regulation, and clear disciplinary boundaries for medicine. Under these circumstances, anyone could be a medical practitioner, provided he or she could attract clients – a situation that gave patients a large degree of power and choice. This situation began to change in the late Middle Ages, which saw the first effective attempts to control and limit medical practice in some urban centers. This should not be seen simply as an attempt by formally trained practitioners to monopolize the business of healing, as the impetus for even this development seems to have come at least as much from patients eager for some way to identify those practitioners that deserved their commerce and their trust.[4] "Authority and trust are social products that are reconfigured differently in every generation," in the words of Michael R. McVaugh,[5] and the history of the medical profession and medical practice is as much the history of those reconfigurations as of any collection of institutions or techniques.

This chapter presents an overview of this history insofar as it is currently known.[6] The caveat is important, for much work remains to be done, particularly in the form of local studies based on archival sources.[7] Only studies of this sort can provide an accurate sense of the dynamics that shaped medieval health care and the enormous geographical and chronological variety of the institutions, laws, and practices that constituted it. Above all, these studies reveal by their gaps and silences the challenges of understanding and describing the health care offered and received by the vast majority of medieval Europeans, who did not live in major cities and whose experiences

3 For example, Michael R. McVaugh, "Bedside Manners in the Middle Ages," *Bulletin of the History of Medicine*, 71 (1997), 203, 205, 209; and Gianna Pomata, *Contracting a Cure: Patients, Healers, and the Law in Early Modern Bologna*, trans. by the author, with the assistance of Rosemarie Foy and Anna Taraboletti-Segre (Baltimore: Johns Hopkins University Press, 1998), especially chap. 1.

4 Michael R. McVaugh, *Medicine before the Plague: Practitioners and Their Patients in the Crown of Aragon, 1285–1345* (Cambridge: Cambridge University Press, 1993), especially chap. 7 and conclusion.

5 McVaugh, "Bedside Manners in the Middle Ages," p. 222.

6 Useful overviews include Katharine Park, "Medicine and Society in Medieval Europe, 500–1500," in *Medicine in Society*, ed. Andrew Wear (Cambridge: Cambridge University Press, 1991), pp. 59–90; Vivian Nutton, "Medicine in Medieval Western Europe, 1000–1500," in *The Western Medical Tradition, 800 BC to AD 1800*, ed. Lawrence I. Conrad et al. (Cambridge: Cambridge University Press, 1995), chap. 5; and Nancy G. Siraisi, *Medieval and Early Renaissance Medicine: An Introduction to Knowledge and Practice* (Chicago: University of Chicago Press, 1990), especially chap. 2, and literature cited therein.

7 For example, McVaugh, *Medicine before the Plague*, on Aragon; Pomata, *Contracting a Cure*, on Bologna; Katharine Park, *Doctors and Medicine in Early Renaissance Florence* (Princeton, N.J.: Princeton University Press, 1985); and Joseph Shatzmiller, *Jews, Medicine, and Medieval Society* (Berkeley: University of California Press, 1994), on Provence.

and practices, largely divorced from literate culture, were never recorded in documents and texts.

"BETWEEN DOCTORS AND HOLY SHRINES," 1050–1200

There is little evidence for a sharp break between early-medieval patterns of healing and those that dominated the eleventh and twelfth centuries, though the increase in literacy and the rising production of written documents means that we are much better informed about the latter than the former. In particular, both periods saw the coexistence of religious, magical, and naturalistic modes of healing, as is evident in the story of an English nun from shortly after 1200, when one of the canons of the Gilbertine order compiled a list of the miracles performed by their late founder, Gilbert of Sempringham. One of these miracles involved the nun Mabel of Stotfold, who had injured her foot. According to the report, her fellow nuns "tried all kinds of remedies, putting the foot both in traction and in plaster." When she eventually consulted a doctor (*medicus*), after two years of increasing pain, he "stated that there was no alternative to amputating her foot, which was, she said, as black as her veil." Understandably apprehensive, Mabel instead "requested that her measurements should be used for a candle intended for Master Gilbert" – it was common practice to offer the saint a candle with a wick as tall as the patient – "and when this had been made, she was taken along with the candle into the church." There the prioress "wrapped her foot in the liturgical towel which had lain upon Master Gilbert's breast when he was about to die." Shortly afterward, Gilbert appeared to her in a dream and blessed her, and she woke completely healed.[8]

It may seem counterintuitive to begin an account of medical practice with a miraculous cure, but hagiographical documents – saints' lives, lists of miracles, and proceedings of canonization inquests – are among the best sources for the history of health care in the eleventh and twelfth centuries. The majority of saintly miracles in this period consisted of supernatural cures, by holy people both living and (especially) dead, and the petitioners to whom they were granted came from all walks of life.[9] Many accounts of such miracles, like that granted to Mabel of Stotfold, detail the varied measures taken by the sick to heal themselves before they resorted to the

8 *The Book of Saint Gilbert*, ed. and trans. Raymonde Foreville and Gillian Keir (Oxford: Clarendon Press, 1987), p. 287.

9 On saintly healing in this period, see Ronald C. Finucane, *Miracles and Pilgrims: Popular Beliefs in Medieval England* (London: J. M. Dent, 1977), chaps. 4 and 5; Pierre-André Sigal, *L'homme et le miracle dans la France médiévale (XIe–XIIe siècle)* (Paris: Edition du Cerf, 1985), especially chap. 5; Constanze Rendtel, *Hochmittelalterliche Mirakelberichte als Quelle zur Sozial- und Mentalitätsgeschichte und zur Geschichte der Heiligenverehrung* (Inaugural-Dissertation, Freie Universität Berlin, 1982) (Düsseldorf: [n.p.], 1985); and Thomas Head, *Hagiography and the Cult of the Saints: The Diocese of Orléans, 800–1200* (Cambridge: Cambridge University Press, 1990), especially pp. 165–87.

inconveniences of a pilgrimage and the extreme measures of a supernatural cure. Thus the miracle books offer arguably the most comprehensive picture of the attitudes toward disease and healing of medieval Christians – and even, on occasion, members of the Jewish minority[10] – as well as the types of medical practice available to them.

As Mabel's story indicates, although medieval Christians made a clear mental distinction between natural and supernatural healing, few saw the two as incompatible or opposed. Rather, both formed part of a single world of health care, in which patients and their families moved back and forth between secular and saintly healers: "between doctors and holy shrines," as the account of another of Gilbert's miracles specified.[11] Although the monastic apologists who drew up the miracle lists sometimes disparaged the efforts of the former to the glory of the latter, most laypeople and members of religious orders saw doctors and saints as collaborators (sometimes literally) in the work of healing. Thus, when a doctor found himself unable to extract an arrowhead from the cheekbone of a patient in the mid-eleventh century, Saint Faith shifted the point so that the human practitioner could remove it with ease.[12]

Writing on the early Middle Ages, Peter Brown has emphasized the plurality of healing practices and the choice of therapeutic systems available to the sick. He notes that no one system had final authority; rather, "social and cultural criteria" dictated which illnesses might be taken to particular kinds of healers.[13] (In the eleventh and twelfth centuries, the illnesses taken to saints' shrines were for the most part chronic conditions that failed to respond to domestic or secular medical care, such as lameness, paralysis, blindness, and the like.)[14] Brown divided his healers into two main groups: the saints, whose power to heal was invested in them personally by God and who required the sick to enter into a quasi-feudal relationship of dependence, and what he called the "diffuse resources of the neighborhood."[15] These included family members, cunning men and women, and occasionally parish priests. Their authority stemmed not from any special relationship with the divine but from their learned ability to mobilize the powers invested by God in the environment – plants and animals, springs, the heavens – using natural remedies, often catalyzed by prayers, incantations, and charms.[16]

[10] For Jewish recourse to Christian shrines, see Shatzmiller, *Jews, Medicine, and Medieval Society*, pp. 121–3.

[11] Foreville and Keir, *Book of Saint Gilbert*, p. 325.

[12] *The Book of Sainte Foy*, trans. Pamela Sheingorn (Philadelphia: University of Pennsylvania Press, 1995), pp. 231–2. On the relations between supernatural and naturalistic healing, see Stephen R. Ell, "The Two Medicines: Some Ecclesiastical Concepts of Disease and the Physician in the High Middle Ages," *Janus*, 68 (1981), 15–25.

[13] Peter Brown, *The Cult of the Saints* (Chicago: University of Chicago Press, 1981), pp. 114–15 (quotation at p. 115).

[14] Sigal, *L'homme et le miracle dans la France médiévale*, pp. 235–51.

[15] Brown, *Cult of the Saints*, pp. 116–20 (quotation at p. 120).

[16] We know these remedies only indirectly, through written sources, most of them monastic, or through archaeological excavations. See, for example, Lea Olsan, "Latin Charms of Medieval England: Verbal

Much more rarely, depending on location, the healing resources of the neighborhood might include those described in Latin texts as *medici*, or doctors. Unlike the villagers described earlier, most doctors trained by apprenticeship and practiced medicine for a fee and as a regular occupation. In the late eleventh and twelfth centuries, such men – outside southern Italy, there are few references in this period to *medicae*, or female doctors – gravitated to the relatively few population centers able to support their practices. They appear with some regularity in miracle lists, as I have already mentioned, where they are recorded as treating mainly townspeople and the well-to-do;[17] the most successful found posts as court doctors in the employ of high nobles, kings, and queens.[18] Villagers and the poor had infrequent access to their services; Mabel of Stotfold was treated by her fellow nuns for two years before she consulted the doctor who wished to amputate her foot. Historians know relatively little about *medici* in this early period and the degree to which they cultivated specialized skills, though the existence of a vocabulary that distinguished between different types of doctors – physicians, surgeons, barbers, herbalists, bleeders, leeches – suggests that some kind of differentiation was already in progress.[19] The miracle accounts report them as performing a variety of surgical procedures (lancing abscesses, amputating members, extracting teeth, cutting for stone, and so forth), as well as administering what one miracle collection refers to as "potions, pills, decoctions, plasters, and oils."[20] There is no sign of licensing in the eleventh and twelfth centuries; practitioners built up a clientele through local reputation and word of mouth.

A final group of *medici* included monks, nuns, and other dependents of monastic institutions. This group (together with court doctors the only ones to work in a predominantly literate environment) is the best documented of

Healing in a Christian Oral Tradition," *Oral Tradition*, 7 (1992), 116–42 (on oral formulas); Audrey Meaney, "Women, Witchcraft, and Magic in Anglo-Saxon England," in *Superstition and Magic in Anglo-Saxon England*, ed. Donald Scragg (Manchester: Manchester Centre for Anglo-Saxon Studies, 1989), especially pp. 9–12 (on excavated objects); and in general Karen Louise Jolly, "Magic, Miracle, and Popular Practice in the Early Medieval West: Anglo-Saxon England," in *Religion, Science, and Magic: In Concert and in Conflict*, ed. Jacob Neusner, Ernest S. Frerichs, and Paul Virgil McCracken Flesher (New York: Oxford University Press, 1989), pp. 166–82.

17 Rendtel, *Hochmittelalterliche Mirakelberichte als Quelle zur Sozial- und Mentalitätsgeschichte und zur Geschichte der Heiligenverehrung*, p. 63. On the situation in southern Italy, see Monica H. Green, *Making Women's Medicine Masculine: The Rise of Male Authority in Pre-Modern Gynecology* (Oxford: Oxford University Press, 2008), chap. 1.

18 See, for example, Edward J. Kealey, *Medieval Medicus: A Social History of Anglo-Norman Medicine* (Baltimore: Johns Hopkins University Press, 1981), chap. 3, on physicians in the court of Henry I.

19 Jole Agrimi and Chiara Crisciani, *Medicina del corpo e medicina dell'anima: Note sul sapere del medico fino all'inizio del secolo XIII* (Milan: Episteme, 1978), p. 62 n. 31; Kealey, *Medieval Medicus*, p. 34; and Danielle Jacquart, *Le milieu médical en France du XIIe au XVe siècle* (Geneva: Droz, 1979), p. 262.

20 Miracles of St. Thomas of Canterbury (late twelfth century), cited in Benedicta Ward, *Miracles and the Medieval Mind: Theory, Record, and Event*, rev. ed. (Philadelphia: University of Pennsylvania Press, 1987), p. 245 n. 37.

any group of healers in this period.[21] Medicine was taught as a book discipline in monastic and cathedral schools, as part of general learned culture. Important centers of this kind of learning in the eleventh and twelfth centuries included schools at Reims and Chartres in northern France and the abbeys of Monte Cassino in central Italy and Bury St. Edmunds in England. The library catalogues of these and other monasteries included a range of Greek medical texts in Latin translation.[22] A smaller number of priests and members of religious orders had more specialized training, through reading and (occasionally) apprenticeship, which allowed them to treat a range of patients: fellow monks and nuns, laypeople who came to the monasteries in search of treatment, and sometimes high-status clients as well. Some of these religious, like the French abbot Fulbert of Chartres or the German abbess Hildegard of Bingen, limited themselves to dispensing occasional medical advice and medicines to their correspondents (and presumably also to members of their communities),[23] whereas others had more advanced training, in several schools or with several masters, which gave them lucrative skills and sometimes prestigious positions as court physicians. For example, Baldwin, who studied at Chartres and later became abbot of Bury St. Edmunds, was medical adviser to both Edward the Confessor and William the Conqueror.[24] Everything we know about the work of such monk-practitioners suggests that it was largely naturalistic, based on principles and information derived from Greek and Roman texts, though often combined with spiritual guidance and advice.[25]

Monasteries also served as local medical centers, dispensing medicines, medical knowledge, and medical services to the broader community. Some abbeys had extensive herb gardens and engaged in the exchange of medicinal plants and seeds, and monastic herbals and antidotaries contain evidence of

[21] Well-documented discussions include Anne F. Dawtry, "The *Modus Medendi* and the Benedictine Order in Anglo-Norman England," in *The Church and Healing*, ed. W. J. Sheils (Oxford: Blackwell, 1982), pp. 25–38; Johannes Duft, *Notker der Arzt: Klostermedizin und Mönchsarzt im frühmittelalterlichen St. Gallen* (St. Gall: Ostschweiz, 1972); and David N. Bell, "The English Cistercians and the Practice of Medicine," *Cîteaux; Commentarii Cistercienses*, 40 (1989), 139–73. Loren C. MacKinney, *Early Medieval Medicine, with Special Reference to France and Chartres* (Baltimore: Johns Hopkins University Press, 1937), is still useful; see especially chap. 3.

[22] See, for example, Rodney M. Thomson, "The Library of Bury St. Edmunds Abbey in the Eleventh and Twelfth Centuries," *Speculum*, 47 (1972), 617–45; and Herbert Bloch, *Monte Cassino in the Middle Ages*, 3 vols. (Cambridge, Mass.: Harvard University Press, 1986), vol. 1, pp. 98–110.

[23] Fulbert of Chartres, *The Letters and Poems of Fulbert of Chartres*, ed. and trans. Frederick Behrends (Oxford: Clarendon Press, 1976), nos. 24, 47, and 48; and MacKinney, *Early Medieval Medicine*, pp. 133–9. Hildegard also wrote her own medical treatises and engaged in miraculous healing; see Florence Eliza Glaze, "Medical Writer: 'Behold the Human Creature'," in *Voice of the Living Light: Hildegard of Bingen and Her World*, ed. Barbara Newman (Berkeley: University of California Press, 1998), pp. 125–48.

[24] Dawtry, "*Modus Medendi* and the Benedictine Order in Anglo-Norman England," pp. 27–8, 31.

[25] On the spiritual services often combined with clerical medical practice, see Faye Marie Getz, *Medicine in the English Middle Ages* (Princeton, N.J.: Princeton University Press, 1998), pp. 4–5, 13–15. For an example of spiritual advice in medical language, see Fulbert of Chartres, *Letters and Poems*, pp. 119–29.

direct botanical observations.[26] In addition to their own often impressively structured infirmaries, some had hospitals or clinics for laypeople, as is clear from one final miracle attributed to Gilbert of Sempringham. A doctor (*medicus*) from Castle Donnington had long suffered from tertian fever. At length, despairing of curing himself, he went to the local hospital, "to search among the powerful herbs and roots that were kept there for a means of restoring his health." Instead, the warden of the hospital gave him water in which Gilbert's staff had been washed, which cured him on the spot.[27]

Accounts of this sort underscore the fluid and pluralistic nature of eleventh- and twelfth-century medical practice, where supernatural and naturalistic forms of healing coexisted easily with incantations and charms. In the absence of exclusionary ideas as to what constituted acceptable and effective healing practice, the authority of healers – saintly and secular – depended on their reputations, and ultimately on the satisfaction of their patients, who exercised to the limit their freedom of choice.

URBANIZATION AND THE TRANSFORMATION OF MEDICAL PRACTICE, 1200–1350

Later-medieval health care retained many of the characteristics described previously, particularly in rural areas. Beginning in the thirteenth century, however, demographic, economic, social, and political changes transformed medical practice in the towns and cities that had sprung up along trade routes and around important administrative centers. Urbanization generated an unprecedented market for medical services, both because city inhabitants were numerous and relatively wealthy and because the dense concentration of people in medieval towns, living in crowded and often unsanitary conditions, supported a host of infectious illnesses.[28] At the same time, the rise of a commercial economy, based on the sale of goods and services, created a template for the commercialization of medical practice, and the centralization of political authority and the increasing ambitions of both secular and ecclesiastical governments set the stage for early forms of licensing and control.

The effect of urbanization was to support a larger, more diverse, and more specialized community of medical practitioners. This change is visible even in

[26] Linda E. Voigts, "Anglo-Saxon Plant Remedies and the Anglo-Saxons," *Isis*, 70 (1979), 250–68, especially pp. 259–66. See also Audrey Meaney, "The Practice of Medicine in England about the Year 1000," *Social History of Medicine* 13 (1993), 221–37.

[27] Foreville and Keir, *Book of Saint Gilbert*, p. 307; Latin on p. 306. On monastic infirmaries, see Bell, "English Cistercians and the Practice of Medicine," pp. 161–71; and Dieter Jetter, *Das europäische Hospital: von der Spätantike bis 1800* (Cologne: DuMont, 1986), pp. 34–46; and the literature cited therein.

[28] On the health conditions in medieval cities, see Ynez V. O'Neill, "Diseases in the Middle Ages," in *The Cambridge World History of Human Disease*, ed. Kenneth F. Kiple (Cambridge: Cambridge University Press, 1992), pp. 272–8.

the arena of saintly healing, where the rising population enlarged the pool of supplicants to any given saint. This facilitated the appearance of specialized cults, such as those of St. John (for epilepsy) or St. Maur (for gout).[29] At the same time, the appearance of conditional vows – I'll fulfill my vow if and only if you heal me – expressed a new sense of a contractual, even commercial, relationship between saint and petitioner.[30] The faithful no longer offered their entire person to the saint but substituted more specific and limited commitments, in the form of objects, charitable works, and, increasingly, monetary offerings; one witness at the canonization inquest of St. Nicholas of Tolentino (1325) testified that he had promised the saint "as much money as he would give to a doctor who cured him [of the same complaint]."[31] As this last remark suggests, religious and secular systems of healing retained a strong relationship to one another; subject to the same market forces, they evolved in roughly parallel, though increasingly autonomous, ways.

It is virtually impossible to establish absolute numbers for secular medical practitioners in this period. In addition to virtually insurmountable problems of documentation (particularly acute for female healers),[32] the continuing indeterminacy of the definition of medical practitioner means that any attempt to fix the group's boundaries must be largely arbitrary. But it is clear that the thirteenth and early fourteenth centuries saw a large and steady rise in the number of practitioners relative to the earlier period. At the same time, the proportion of priests and other religious involved in medical practice dropped dramatically, at least partly in response to a series of church decrees limiting the study and practice of medicine by clergy in major orders.[33]

The growth in the number of medical practitioners coincided with their increasing diversification. By the early fourteenth century, official sources clearly distinguished between physicians (who treated internal illnesses through diet and medication), surgeons, apothecaries, and barbers (who bled people and provided other minor surgical services).[34] Italy even

29 See Park, "Medicine and Society in Medieval Europe," pp. 74–5.

30 André Vauchez, *Sainthood in the Later Middle Ages*, trans. Jean Birrell (Cambridge: Cambridge University Press, 1997; orig. 1988), pp. 444–62. See also Sigal, *L'homme et le miracle dans la France médiévale*, pp. 312–13.

31 Cited in Vauchez, *Sainthood in the Later Middle Ages*, p. 466 n. 75.

32 See, for example, Monica H. Green, "Documenting Medieval Women's Medical Practice," in *Practical Medicine from Salerno to the Black Death*, ed. Luis García-Ballester et al. (Cambridge: Cambridge University Press, 1994), pp. 322–52.

33 Darrel W. Amundsen, "Medieval Canon Law on Medical and Surgical Practice by the Clergy," *Bulletin of the History of Medicine*, 52 (1978), 22–44; and André Goddu, "The Effect of Canonical Prohibitions on the Faculty of Medicine at the University of Paris in the Middle Ages," *Medizinhistorisches Journal*, 20 (1985), 342–62, especially pp. 347–8.

34 The literature on the relationships between these groups of practitioners is voluminous, for the period before 1350. See, for example, Cornelius O'Boyle, "Surgical Texts and Social Contexts: Physicians and Surgeons in Paris, c. 1270–1430," in García-Ballester et al., *Practical Medicine from Salerno to the Black Death*, pp. 156–85; Danielle Jacquart, "Medical Practice in Paris in the First Half of the Fourteenth Century," in García-Ballester et al., *Practical Medicine from Salerno to the*

saw the appearance of surgical subspecialties, like those mastered by Maria Gallicia, licensed in Naples in 1308 to treat "wounds, swellings, hernias, and conditions of the uterus."[35] These changes suggest the development of more complicated and specialized techniques, (re)learned from newly available Greek and Arabic surgical texts or elaborated in the course of apprenticeship and practice, and transmitted through original Latin surgical textbooks by contemporary authors.[36]

Indeed, the thirteenth and early fourteenth centuries saw a general expansion of text-based practice, which was no longer largely confined to clerics (of whom fewer and fewer studied and practiced medicine) but also included lay physicians and eventually surgeons. This development coincided with the appearance of academic training for medical practitioners alongside private study and apprenticeship, first in southern Italy (in the late eleventh or early twelfth century) and then (in the years around 1200) at the universities in Bologna, Paris, and Montpellier. With their increasingly formalized curricula and teaching methods, university medical faculties institutionalized the idea of medicine as a body of text-based knowledge, embracing both theory and practice, as opposed to a set of empirically acquired skills. At the same time, they supplied clear credentials for one who had acquired such knowledge, in the form of a university degree. Surgery could also be studied at the university – this was particularly the case in Italy – but the typical holder of a medical degree was the physician (*physicus*), whose knowledge of healing was underpinned and buttressed by an extensive education in the theoretical principles that underlay the body's anatomy and physiology, the nature of disease, and the physical order of the universe itself.[37]

This period also saw the first attempts at formal licensing of particular groups of healers. Given the diversity of healers and their training, there could

Black Death, pp. 186–210; and Jacquart, *Milieu médical en France du XIIe au XVe siècle*, pp. 27–55. The situation in other areas is discussed in Getz, *Medicine in the English Middle Ages*, chap. 1; and McVaugh, *Medicine before the Plague*, pp. 38–49.

35 Raffaele Calvanico, *Fonti per la storia della medicina e della chirurgia per il regno di Napoli nel periodo angioino (a. 1273–1410)* (Naples: L'Arte Tipografica, 1962), passim (quotation at p. 141). See in general Katharine Park, "Eyes, Bones, and Hernias: Surgical Specialists in Fourteenth- and Fifteenth-Century Italy," in *Medicine from the Black Death to the French Disease*, ed. Jon Arrizabalaga (London: Ashgate, 1998), pp. 110–30.

36 Siraisi, *Medieval and Early Renaissance Medicine*, chap. 6; and the following essays on literate surgery in García-Ballester et al., *Practical Medicine from Salerno to the Black Death*: Jole Agrimi and Chiara Crisciani, "The Science and Practice of Medicine in the Thirteenth Century According to Guglielmo da Saliceto, Italian Surgeon," pp. 30–87; Nancy G. Siraisi, "How to Write a Latin Book on Surgery: Organizing Principles and Authorial Devices in Guglielmo da Saliceto and Dino Del Garbo," pp. 88–109; Pedro Gil-Sotres, "Derivation and Revulsion: The Theory and Practice of Medieval Phlebotomy," pp. 110–55; and Michael R. McVaugh, "Royal Surgeons and the Value of Medical Learning: The Crown of Aragon, 1300–1350," pp. 211–36.

37 The literature on the rise of university education in medicine is vast; for an introduction, with appropriate references, see Siraisi, *Medieval and Early Renaissance Medicine*, chap. 3. The academic culture of *physica* is brilliantly analyzed in Agrimi and Crisciani, *Edocere Medicos*. On the development of a textual tradition in surgery, see Michael R. McVaugh, *The Rational Surgery of the Middle Ages* (Florence: SISMEL/Edizioni del Galluzzo, 2006).

be no single standard for medical qualification, which varied dramatically depending on the geographical region and the type of practice involved. In some areas (for example, England), licenses might be granted by local bishops, whereas in others (Spain and southern Italy) this was a matter for royal and municipal authorities and contingent on either proven competence or a medical degree.[38] In some university towns (most notably Paris), the faculty of medicine itself claimed this privilege, which in many northern Italian cities belonged to whatever craft guild incorporated medical practitioners – in the case of Florence, for example, the Guild of Doctors, Apothecaries, and Grocers (which will be discussed further).[39] With the possible exception of Paris and some Italian cities, however, there is little evidence that the pressure for licensing came largely or exclusively from medical practitioners. Rather, it was the work of patients and the public, who came increasingly to accept the knowledge claims of formally (though not necessarily university) educated doctors and as a result began to look for ways to distinguish those worthy of their trust.[40] This new willingness to submit to the authority of trained doctors also appears in the increasing public reliance on medical experts in situations where lay testimony previously would have sufficed: in legal judgments concerning impotence or cause of death, for example; in the official identification of lepers; and in canonization processes, where doctors were called on to certify the natural impossibility of miraculous cures.[41]

Despite the increasing number of licensed medical practitioners in this period, it is important not to overstate their importance. Until recently, historians of medicine have tended to overestimate the degree to which university-educated physicians and, to a lesser degree, formally trained and licensed general surgeons dominated the health care of medieval Europeans, often referring to them as lying at the "center," in opposition to a host of other "marginal" practitioners, and conferring on them an authority that they did not yet possess. This model fits poorly with the realities of medieval health care, where the extraordinary variety of practitioners, as well as the many customs and regulations that governed them, make it difficult to speak meaningfully of either margins or center. It is important to remember that only a small minority of thirteenth- and fourteenth-century Europeans lived in cities, where they might have had access to physicians and formally trained

[38] Shatzmiller, *Jews, Medicine, and Medieval Society*, p. 15; McVaugh, *Medicine before the Plague*, pp. 69–72, 95–103; and Luis García-Ballester, Michael R. McVaugh, and Agustín Rubio-Vela, "Medical Licensing and Learning in Fourteenth-Century Valencia," *Transactions of the American Philosophical Society*, 79, no. 6 (1989), 1–128.

[39] O'Boyle, "Surgical Texts and Social Contexts," pp. 171–5; Jacquart, "Medical Practice in Paris in the First Half of the Fourteenth Century," pp. 199–201; and Park, *Doctors and Medicine in Early Renaissance Florence*, chap. 1.

[40] McVaugh, *Medicine before the Plague*, chaps. 6 and 7.

[41] See, e.g., McVaugh, *Medicine before the Plague*, chap. 7; Shatzmiller, *Jews, Medicine, and Medieval Society*, pp. 3–4; and Joseph Ziegler, "Practitioners and Saints: Medical Men in Canonization Processes in the Thirteenth to Fifteenth Centuries," *Social History of Medicine*, 12 (1999), 191–225.

surgeons; country dwellers (and city dwellers of limited means) mostly made do with the healing skills of family members and neighbors – some of whom were known for their special abilities in this area. Furthermore, many regional authorities licensed not only physicians and general surgeons but empirically trained (or self-trained) surgeons skilled in the treatment of a single condition, such as cataracts, fractures, or hernias, and sometimes herbalists as well. In many cities, including some with well-defined licensing procedures, there is little evidence that prohibitions against unlicensed practice were consistently and effectively enforced – although prosecutions for unlicensed practice did become more common in the first decades of the fourteenth century.[42] In general, however, the marketplace for medical services remained relatively open and uncontrolled.

The nature and limits of medical authority in this period are clearly visible in the contracts that increasingly governed the relationship between patient and doctor. One typical agreement was drawn up in Genoa in 1244:

> I, Roger de Bruch of Bergamo, promise and agree with you, Bosso the wool carder, to return you to health and to make you improve from the illness that you have in your person, that is in your hand, foot, and mouth, in good faith, with the help of God, within the next month and a half, in such a way that you will be able to feed yourself with your hand and cut bread and wear shoes and walk and speak much better than you do now. I shall take care of all the expenses that will be necessary for this; and at that time, you shall pay me seven Genoese lire. . . . If I do not keep my promises to you, you will not have to give me anything. And I, the aforementioned Bosso, promise to you, Rogerio, to pay you seven Genoese lire within three days after my recovery and improvement.[43]

Such contracts (oral or written) seem to have shaped a significant portion of medical care by full-time, formally trained practitioners, and examples have been published from all parts of Western Europe – though their currency and the chronology of their spread, especially outside of the Mediterranean area, have yet to be explored.[44] The majority governed individual bouts of illness, though agreements also survive in which an individual or institution contracted for the services of a medical practitioner as needed over a particular period of time. Like the agreement just mentioned, they involved mutual promises by practitioners and patients – the former to cure his client, and the latter to follow the doctor's directions and to pay in the event of a cure. (This practice had clear echoes in the conditional vow that increasingly accompanied the petition for saintly healing, as described earlier.) Some

42 See McVaugh, *Medicine before the Plague*, pp. 71–2, 102–3.

43 Cited in Pomata, *Contracting a Cure*, p. 28.

44 Ibid., chap. 1; see especially references to published contracts on pp. 198–9 n. 8. For other examples, with discussion, see Shatzmiller, *Jews, Medicine, and Medieval Society*, pp. 124–31; McVaugh, *Medicine before the Plague*, pp. 174–85; and Getz, *Medicine in the English Middle Ages*, p. 77.

contracts further specified an initial payment at the beginning of treatment and installments at clearly indicated points along the way.

One of the most striking aspects of these contracts, as Gianna Pomata has recently argued, is that illness, improvement, and cure were defined from the point of view of the patient, in negotiation with his or her doctor, rather than according to some general standard established by medical practitioners, either individually or as a group.[45] This clearly reflects the differences between modern professional practice and medical care in the late Middle Ages, when malpractice was conceived as breach of contract rather than as negligent behavior that violated standards for treatment upon which physicians and surgeons generally agreed. Thus, even when medical experts were called in to adjudicate a legal disagreement concerning a course of treatment, they did so only by establishing whether the practitioner and patient had met the terms specified in the contract rather than whether the doctor's conduct (and the patient's health) met professionally established norms.[46]

The overall picture of health care in this period, then, is one of profound change, from a world of ill-defined and wholly unregulated practices and practitioners to an order that acknowledged, at least in theory, that particular types of healing required particular skills. There was a difference between competent and incompetent healers, and patients needed a way of distinguishing between them (before, rather than after, the fact), even if they thereby had to forfeit some freedom of choice. Here again, the issue came down to the doctor's authority and the patient's trust; clients were willing to give up a degree of autonomy – though many insisted on the safety of a contract – if they were convinced that they were putting their welfare in capable hands. Yet the novel authority claimed by medical practitioners was fragile and required continual reinforcement, as is clear not only from contemporary treatises on medical conduct, which advised physicians on matters ranging from manners to extracting payment,[47] but also from continued conflicts concerning who might practice, in what way, and on whom.

These issues were particularly urgent when it came to two groups of practitioners, women and Jews, whose claim to authority over patients (normatively, Christian male patients) raised obvious problems in the juridical and ideological context of medieval Christian Europe. Both groups had long been involved in medical practices of various sorts. Women treated a range of patients, both male and female, in a range of domestic and commercial

45 Pomata, *Contracting a Cure*, pp. 27–8.

46 McVaugh, *Medicine before the Plague*, pp. 184–5. For specific cases of this sort, see Pomata, *Contracting a Cure*, pp. 30–3, 199 n. 10, 244–5 n. 89; and Madeline Pelner Cosman, "Medieval Medical Malpractice: The Dicta and the Dockets," *Bulletin of the New York Academy of Medicine*, 49 (1973), 22–47.

47 McVaugh, "Bedside Manners in the Middle Ages"; Luis García-Ballester, "Medical Ethics in Transition in the Latin Medicine of the Thirteenth and Fourteenth Centuries: New Perspectives on the Physician–Patient Relationship and the Doctor's Fee," in *Doctors and Ethics: The Earlier Historical Setting of Professional Ethics*, ed. Andrew Wear et al. (Amsterdam: Rodopi, 1993), pp. 38–71.

settings, though their practice was often informal, and the very low levels of female literacy in this period meant that few learned or worked from written texts.[48] Jewish practitioners, in contrast, had a more highly developed textual tradition. Their numbers varied geographically, but in areas with well-established Jewish populations – Spain, Germany, Provence, and parts of Italy – they were well represented, even at the highest levels of formal practice.[49] Despite their importance (and their differences) as practitioners, however, both women and Jews found themselves subject to increasing limitations and restrictions, as well as intermittent prosecution for unlicensed practice. (The best-known case is that of Jacqueline Félicie, convicted in Paris in 1322 together with two men and three other women, one of them a Jew.)[50] Although clearly part of the attempts of better-established practitioners to restrict competition, such cases also spoke to the more general issue of medical authority.

Indeed, women and Jews seem to have had difficulty establishing and maintaining the authority commanded, at least in theory, by their Christian male counterparts. Both groups were barred in principle from earning university degrees, the gold standard of medical competence, though this had changed for Jews by the early fifteenth century. In addition, each was viewed as generally suspect, whether because of moral and intellectual weakness (in the case of women) or because of implacable hostility to Christians (in the case of Jews).[51] Stereotypes of this sort buttressed official attempts – on the whole unsuccessful – to prevent Jews from treating Christian patients and to limit women to female clients.[52]

The scarcity of women in the ranks of licensed full-time practitioners, especially physicians, was overdetermined – by their lack of formal education and their family responsibilities as well as by legal restrictions. But the peculiar

[48] See in general Monica H. Green, "Women's Medical Practice and Health Care in Medieval Europe," *Signs*, 14 (1989), 434–73; Green, "Documenting Medieval Women's Medical Practice"; Green, *Making Women's Medicine Masculine*, passim, especially chap. 3; Katharine Park, "Medicine and Magic: The Healing Arts," in *Gender and Society in Renaissance Italy*, ed. Judith C. Brown and Robert C. Davis (London: Longman, 1998), pp. 129–49, especially pp. 133–7; McVaugh, *Medicine before the Plague*, pp. 103–7; Montserrat Cabré, "Women or Healers? Household Practices and the Categories of Healthcare in Late Medieval Iberia," *Bulletin of the History of Medicine*, 82 (2008), 18–51; and Carole Rawcliffe, *Medicine and Society in Later Medieval England* (New York: Sutton, 1995), chap. 9.

[49] Shatzmiller, *Jews, Medicine, and Medieval Society*; McVaugh, *Medicine before the Plague*, passim; and Jacquart, *Milieu médical en France du XIIe au XVe siècle*, pp. 160–7.

[50] On the case of Jacqueline Félicie, see Monserrat Cabre and Fernando Salmon, "Poder académico *versus* autoridad femenina; la Facultad de Medicina de París contra Jacoba Félicié (1322)," *Dynamis*, 19 (1999), 55–78; Pearl Kibre, "The Faculty of Medicine at Paris, Charlatanism and Unlicensed Medical Practice in the Later Middle Ages," *Bulletin of the History of Medicine*, 27 (1953), 1–20. On the licensing of women and Jews in general, see Shatzmiller, *Jews, Medicine, and Medieval Society*, pp. 16–22; McVaugh, *Medicine before the Plague*, pp. 98–100, 106; and Green, "Women's Medical Practice and Health Care in Medieval Europe," pp. 446–7.

[51] McVaugh, *Medicine before the Plague*, pp. 184–6; Rawcliffe, *Medicine and Society in Later Medieval England*, pp. 171–8; and Shatzmiller, *Jews, Medicine, and Medieval Society*, pp. 32–3.

[52] Calvanico, *Fonti per la storia della medicina e della chirurgia per il regno di Napoli nel periodo angioino*, p. 277.

legal difficulties faced by Jews in the course of medical practice suggest that generalized prejudice could compromise the authority of women and religious minorities, and the trust of their patients, in more subtle ways. In particular, Jewish doctors were disproportionately targeted in malpractice suits and prosecuted for prescribing poisons or supplying poisons to clients. This was not only because such activities conformed to obvious stereotypes but also because, as McVaugh hypothesizes, Jewish doctors were less able to control their relationships with Christian patients, so that litigious patients "were in effect expressing the reservations and fears they felt when submitting themselves to treatment by a Jew."[53]

THE ELABORATION OF MEDICAL INSTITUTIONS, 1350–1500

In the period after 1350, medical practice continued to develop along the lines described in the previous section, as the continuing rise in prestige and currency of text-based medicine created an expanding market for medical services, together with public acknowledgment that safe and effective practice required a well-defined set of skills. These changes had important institutional implications. For one thing, potential patients felt an increasing need to be able to determine who possessed those skills and who did not, which spoke to the need for licensing procedures and other institutional markers of competence. For another, as communities and individuals became increasingly convinced of the public benefits of medical care and medical expertise, they elaborated institutions to increase access to both. At the same time, practitioners themselves began to organize in order to take advantage of the social and economic opportunities offered by this new situation. The result was a flowering of institutions devoted to the ordering and provision of medical care.

In this section, I will focus on two principal kinds of medical institutions elaborated in the fourteenth and fifteenth centuries: those concerned with regulating practice and defending the interests of practitioners, notably guilds and colleges of doctors, and those concerned with mobilizing medical expertise for the benefit of the larger community, notably hospitals for the sick poor and other institutions relating to charity and public health. Neither form of institution was new in the period after 1350; both had roots that went back a century or more. But they acquired a more clearly medical character beginning in the late thirteenth century – as the nature of medical practice itself acquired clearer definition – and accelerating into the fourteenth and fifteenth.

[53] McVaugh, *Medicine before the Plague*, p. 186. See also Jacquart, *Milieu médical en France du XIIe au XVe siècle*, p. 167; and Park, *Doctors and Medicine in Early Renaissance Florence*, p. 74.

The late-medieval corporations of medical practitioners, whether guilds, companies, colleges, or faculties, had two separate sets of functions. They provided (at least in theory) a framework for licensing and regulating medical practice and for helping practitioners resolve disputes in a formal and dignified manner. In this way, they buttressed the authority (and earnings) of their membership. At the same time, they served as fraternal organizations, providing mutual aid, solidarity, and protection, thereby promoting (again in theory) a sense of group identity. In practice, however, it proved difficult to reconcile these two different goals, given the enormous diversity of urban healers, which included not only physicians, master surgeons, surgical specialists, apothecaries, and barbers but a disparate range of informally trained healers. Any corporation that embraced all these groups could claim a relatively comprehensive monopoly and control of medical practice, as well as the wealth and influence that went with a large membership. By the same token, however, it possessed very little in the way of group identity and solidarity since its well-educated physicians and wealthy apothecaries had very little in common with its lower ranks. Conversely, organizations of more narrowly defined groups of practitioners (companies of barbers, guilds of apothecaries, colleges of physicians, and so forth) were more homogeneous, but their claim to monopolize and control medical practice was limited, and they inevitably squandered their resources and energies in turf battles with each other.

Fourteenth- and fifteenth-century Europe offers examples of both the inclusive and the exclusive models, as well as a spectrum of intermediate forms. The exclusive model tended to characterize cities with strong university medical faculties, such as Bologna or Paris, where the academic (or academically trained) physicians' strong sense of distinctiveness seems to have acted to dissolve any sense that medical practitioners were a group with shared economic or political interests.[54] In Paris, the medical faculty found itself in a long-term tug-of-war with separate companies of surgeons, barbers, and apothecaries, and seems to have been uniquely successful in this arena. In other cities, members of the college of physicians, although wealthy and socially prominent, were usually too few to exert effective control over other practitioners, despite their claims to intellectual authority.[55]

In towns without universities, however, medical guilds tended to conform to some version of the inclusive model, of which the Florentine Guild of

[54] On Paris, see O'Boyle, "Surgical Texts and Social Contexts"; Jacquart, "Medical Practice in Paris in the First Half of the Fourteenth Century"; and Goddu, "Effect of Canonical Prohibitions on the Faculty of Medicine at the University of Paris in the Middle Ages." On Bologna, see Nancy G. Siraisi, *Taddeo Alderotti and His Pupils: Two Generations of Italian Medical Learning* (Princeton, N.J.: Princeton University Press, 1981), pp. 19–23.

[55] For example, on the small size of Italian colleges of physicians, see Richard Palmer, "Physicians and the State in Post-Medieval Italy," in *The Town and State Physician in Europe from the Middle Ages to the Enlightenment*, ed. Andrew W. Russell (Wolfenbüttel: Herzog August Bibliothek, 1981), pp. 50–2.

Doctors, Apothecaries, and Grocers was a particularly dramatic example. Not only did this guild incorporate spice merchants and other sellers of dry goods, but the "doctors" (*medici*) of its title referred to a vast range of medical practitioners, from physicians with international reputations down to a woman of modest means who identified herself as a "ringworm doctor" and a shoemaker who couched cataracts on the side.[56] The Florentine Guild enjoyed a reasonably effective monopoly among full-time practitioners, but its statutes demonstrate inevitable problems in enforcing a sense of group identity and solidarity among medical practitioners, leading to the founding of a subcorporation of university-educated physicians (the College of Doctors) in 1392.[57]

In addition to bodies developed to license and regulate medical practitioners, the cities of the late Middle Ages produced a variety of institutions that drew on the expertise of practitioners who met the emergent criteria for medical competence and that aimed to broaden access to the particular forms of health care that they provided. These institutions were intended to meet the special needs of the urban population, which included growing numbers of the poor. Beginning in the early thirteenth century, for example, Italian communes began to employ salaried municipal doctors to ensure that quality health care was both available and affordable. The practice varied from city to city. Smaller towns usually wished only to ensure the presence of one competent practitioner, whereas larger ones might subsidize a whole stable; Venice employed thirty-one physicians and surgeons in 1324.[58] Whereas smaller communes tended to employ physicians – presumably on the grounds that they could treat a wider variety of illnesses – large cities, already well supplied with physicians, might hire only surgeons with highly specialized skills, such as treating wounds resulting from judicially prescribed torture and amputation.[59] Some cities required their municipal doctors simply to remain in residence, engaging in regular private practice, whereas others insisted that they treat the poor gratis or in accordance with a sliding scale of fees. In addition to providing doctors with ongoing positions, public authorities employed them as occasional consultants – on the nature of particularly serious epidemics, for example, or to offer evidence in judicial proceedings. When the Savoyard Pierre Gerbais was accused of poisoning in 1380, seven doctors and a barber testified at his trial.[60] As this example suggests, the public employment and consultation of doctors soon

56 Park, *Doctors and Medicine in Early Renaissance Florence*, especially pp. 24–34.

57 Ibid., pp. 17–27, 37–42.

58 Vivian Nutton, "Continuity or Rediscovery? The City Physician in Classical Antiquity and Mediaeval Italy," in Russell, *Town and State Physician in Europe from the Middle Ages to the Enlightenment*, especially pp. 24–34.

59 For example, Florence. See Park, *Doctors and Medicine in Early Renaissance Florence*, pp. 90–3.

60 Jacquart, *Milieu médical en France du XIIe au XVe siècle*, p. 289. Italian examples are in Katharine Park, "The Criminal and the Saintly Body: Autopsy and Dissection in Renaissance Italy," *The Renaissance Quarterly*, 47 (1994), 4–8.

spread from Italy to other parts of Europe.[61] Over the course of the late Middle Ages, therefore, doctors supplemented their contracts with private individuals by contracts with institutions, which increasingly included not only public authorities but hospitals as well.

Hospitals had existed in Europe since late antiquity, often associated with monasteries and cathedrals. The eleventh and twelfth centuries had seen an explosion of new foundations, often started by laypeople, at least partly in response to the increasing incidence of leprosy.[62] But these early hospitals were not properly medical institutions; for the most part small, rural houses, they catered to a miscellany of the needy – pilgrims, travelers, and the old, as well as the sick and the disabled.[63] Even those devoted to the care of lepers almost never employed staff identified as doctors. This situation changed in the thirteenth century, with the appearance of large institutions, sometimes with several hundred beds, founded to serve the growing urban population. Some of these hospitals hired formally trained doctors to visit the sick on a regular basis, as at two of the larger thirteenth-century foundations, St. Leonard at York and the Hôtel-Dieu at Paris.[64] The best-organized and best-documented institutions of this sort were in Italy, which saw the emergence of large specialized hospitals devoted to treating the acutely ill in addition to providing a range of other medical services to the city poor. Florence seems to have had the largest number of such institutions; by 1500, it boasted four large hospitals for the sick, including the hospital of Santa Maria Nuova, which employed ten physicians and surgeons, as well as a staff of pharmacists.[65]

As their names suggest, late-medieval hospitals were in the first instance religious institutions, established as Christian charities and concerned at least as much with the spiritual as with the physical needs of their dependents; in 1428, the Florentine hospital of San Matteo employed four priests, but only two doctors, to serve its forty-five patients. But these hospitals nonetheless reflected an order in which religious and natural healing had become defined

[61] Nutton, "Continuity or Rediscovery?"; Shatzmiller, *Jews, Medicine, and Medieval Society*, pp. 112–18; and McVaugh, *Medicine before the Plague*, chap. 7.

[62] See, for example, Kealey, *Medieval Medicus*, chaps. 4 and 5; Carole Rawcliffe, *Leprosy in Medieval England* (Woodbridge: Boydell Press, 2006); and Luke Demaitre, *Leprosy in Premodern Medicine: A Malady of the Whole Body* (Baltimore: Johns Hopkins University Press, 2007).

[63] See in general Miri Rubin, "Development and Change in English Hospitals, 1100–1500," in *The Hospital in History*, ed. Lindsay Grandshaw and Roy Porter (London: Routledge, 1989), p. 51. Recent overviews of this topic include Peregrine Horden, "A Discipline of Relevance: The Historiography of the Later Medieval Hospital," *Social History of Medicine*, 1 (1988), 359–74, with copious references, and Jetter, *Das europäische Hospital*.

[64] Nutton, "Medicine in Medieval Western Europe," p. 152.

[65] John Henderson, *The Renaissance Hospital: Healing the Body and Healing the Soul* (New Haven, Conn.: Yale University Press, 2006); Katharine Park, "Healing the Poor: Hospitals and Medical Assistance in Renaissance Florence," in *Medicine and Charity before the Welfare State*, ed. Jonathan Barry and Colin Jones (London: Routledge, 1991), pp. 26–45; and Katharine Park and John Henderson, "'The First Hospital among Christians': The Ospedale di Santa Maria Nuova in Early Sixteenth-Century Florence," *Medical History*, 35 (1991), 164–88, which includes a translation of the hospital's early-sixteenth-century statutes.

as complementary but autonomous. The staff of earlier medieval hospitals had engaged in both. Asked for "herbs and roots" to cure a tertian fever, the warden of the twelfth-century hospital of Castle Donnington dispensed holy water instead. By the end of the fifteenth century, however, priests had one well-defined set of functions and doctors another, and medical personnel rarely prescribed religious cures; a notebook of remedies "tried and tested in the hospital of Santa Maria Nuova," compiled in 1515 by a hospital doctor, contained hundreds of medical recipes using natural ingredients, and only a smattering of prayers and charms.[66]

One final index of the growing complementarity of the realms of medical and religious healing was the appearance in the late fifteenth century of a new form of hospital. The plague hospital, or *lazaretto*, was motivated exclusively by concerns relating to medicine and public health. Plague had appeared in medieval Europe in the winter of 1347–8 and returned in the form of periodic epidemics over the next three hundred years, generating enormous public concern. At first, local governments confined themselves to traditional preventive measures, principally prohibitions against dirt and odors. Over time, however, the authorities became increasingly convinced of the contagious nature of plague – in line with contemporary medical theory – and began to emphasize techniques for isolating the sick. As was the case with other medical institutions, the Italian cities were in the vanguard; by the end of the fifteenth century, they had begun to experiment with quarantine procedures and to establish special isolation hospitals, which, although they offered confessors for the dying, cannot be considered religious institutions at all.[67] Indeed, concerns about contagion increasingly set public health authorities at odds with the church. Whereas religious authorities responded to epidemics of plague with calls for religious processions, public preaching, and church attendance, public health officials tried to ban assemblies of all sorts. Preaching in Venice during the epidemic of 1497, one friar called the civic authorities to task:

> Gentlemen, you are closing the churches for fear of the plague, and you are wise to do so. But if God wishes, it will not suffice to close the churches. It will need a remedy for the causes of the plague, which are the horrendous sins committed, the schools of sodomy, the infinite usury contracts made at Rialto, and throughout the sale of justice and the favoring of the rich against the poor.[68]

66 Florence, Biblioteca Nazionale, MS. Magl. XV, 92, fol. 27r–192r. See Henderson, *Renaissance Hospital*, chap. 9.

67 See Carlo M. Cipolla, *Public Health and the Medical Profession in the Renaissance* (Cambridge: Cambridge University Press, 1976), chap. 1; Ann Gayton Carmichael, *Plague and the Poor in Renaissance Florence* (Cambridge: Cambridge University Press, 1986), chap. 5; Ann Gayton Carmichael, "Contagion Theory and Contagion Practice in Fifteenth-Century Milan," *Renaissance Quarterly*, 44 (1991), 213–56; and Richard Palmer, "The Church, Leprosy, and Plague," in Sheils, *Church and Healing*, pp. 79–100.

68 Marin Sanudo, *Diarii*, as cited in Palmer, "Church, Leprosy, and Plague," p. 96 (translation slightly edited); see in general pp. 96–9.

It would be wrong, however, to overemphasize the split between religious and naturalistic healing. In 1500, as in 1050, most Europeans (including most doctors) continued to subscribe to an ontology that accepted illness as having both spiritual and natural causes, and spiritual and natural cures; after all, God had created the natural world and used it to accomplish his purposes, such as punishing the wicked. People still engaged in charity, performed acts of penance, used religious charms, and went on pilgrimages when medical remedies had failed. But even the humble turned first to medical practitioners rather than the heavens. In the first place, medical practitioners lay closer at hand, at least in cities; not only had the period seen an enormous expansion in the number of doctors, but institutions such as hospitals rendered their services more accessible to the poor. Even more important, a public consensus had emerged that a relatively well-defined corpus of activities constituted medical practice – just as there was a relatively well-defined corpus of medical practitioners – and that those activities were the proper first line of defense against disease. This process, which some historians refer to as "medicalization,"[69] should not be confused with the emergence of medicine as a profession in any recognizable modern sense. Medical practice and medical practitioners were still too varied to qualify as such, and although the emergence of medicine as a learned discipline had created a generally shared set of assumptions and procedures, practitioners overall continued to lack standardized training and a sense of group identity. Furthermore, the contractual relationship between patient and practitioner meant that payment of the doctor depended ultimately on the patient's satisfaction rather than on adherence to a set of professional standards and norms. By the end of the Middle Ages, Europeans had granted considerable public authority to their medical practitioners: as forensic experts, interpreters of the natural order, and consultants on issues relating to public health. But they reserved to themselves the power to judge the effectiveness of their own doctors' care.

[69] McVaugh, *Medicine before the Plague*, p. 3; and Shatzmiller, *Jews, Medicine, and Medieval Society*, chap. 1.

27

TECHNOLOGY AND SCIENCE

George Ovitt

Henry Adams had the dizzying fortune – whether for good or ill – to witness the tumultuous birth of modernity. Adams related that he began life (in 1838) a "medieval, primitive, crawling infant" and that, by the turn of the century, he was an "automobiling maniac."[1] Of particular interest for the historian of technology are the great historian's ruminations on "the dynamo and the Virgin" contained in his masterpiece of detached autobiography, *The Education of Henry Adams*. Confronted by the "chaos" of the Great Exposition of 1900, Adams speculated on his own ignorance of machines, and, in recognizing that the world would be forever changed by these brute forces, he indulged in a nostalgic meditation on the lost world symbolized by the Virgin of Chartres.

Adams's use of this jarring image of dynamo and Virgin provides an excellent starting point for the study of medieval technology. Lynn White, among a handful of truly original scholars in this field, used Adams's imagery to argue, in a vein that was distinctly opposed to the spirit of the originator of the metaphor, that "the Virgin and the dynamo are not opposing principles . . . they are allies."[2] White's thesis was that medieval technology helped to liberate Western man from the drudgery of manual labor; unlike Adams, White believed that the history of medieval technology should be seen as part of the history of human freedom: "The labor saving power machines of the later Middle Ages were harmonious with the religious assumptions of the infinite worth of even the most seemingly degraded human personality and with an instinctive repugnance toward subjecting any man to a monotonous drudgery."[3] Although it seems clear that White has oversimplified the connection between medieval religion and the extraordinary

1 Henry Adams, *The Education of Henry Adams* (New York: The Modern Library, 1970), p. ix.

2 Lynn White, Jr., "Dynamo and the Virgin Reconsidered," in his *Dynamo and Virgin Reconsidered* (Cambridge, Mass.: MIT Press, 1968), p. 72.

3 White, "Dynamo and the Virgin Reconsidered," p. 73.

developments of medieval technology, he was correct in pursuing a connection between European culture – which was overwhelmingly religious – and its quest of technological innovation as a means to material and economic progress.

Although this chapter cannot hope to do justice to the whole history of medieval technology, it will accept the sage advice of Adams and "follow the track of the energy" by first looking at the construction of an intellectual framework for technological progress and offering a theory about this progress. Next there follows a sketch of six key sources of the energy of which Adams wrote – in agriculture, power generation, textile production, warfare, shipbuilding, and building construction. These brief surveys are intended to reveal the propensity of European inventors to make practical use of borrowed techniques while demonstrating the link between technological development and economic growth.

THE INTELLECTUAL CONTEXT OF MEDIEVAL EUROPEAN TECHNOLOGY

Historians attuned to the apparent paradox of a religious culture embracing technological change have examined the assumptions of the medieval church itself, looking for attitudes that might have fostered such worldly pursuits as invention, the assimilation of imported technologies, and economic growth. At the turn of the twentieth century, Max Weber examined this paradox; during the 1960s, the theologian Ernst Benz extended Weber's search into "the specifically Christian premises of our Western culture" for an interpretation of Western technological success.[4] Given the primary role played by the Church of Rome during the Middle Ages, such a search seems warranted, if problematic in its tendency toward historical determinism and reductionism.

Benz made two important points in his analysis of the theological foundations of Western technology. First, he argued that the medieval image of God as "artificer" (citing Isa. 45:9–12), and the well-known texts from Genesis enjoining Adam and his progeny to "subdue" the earth and its creatures, provided a deeply felt justification for the sanctity of labor and, by extension, of invention.[5] Humans, argues Benz, have obtained a divinely ordained license to dominate the earth, an idea that, in Benz's view, characterizes the

[4] Ernst Benz, "I fondamenti cristiani della tecnica occidentale," in *Tecnica e casistica*, ed. E. Castelli (Rome: Centro Internazionale di Studi Umanistici, 1964), pp. 243–4. See also Ernst Benz, "The Christian Expectation of the End of Time and the Idea of Technical Progress," in *Evolution and Christian Hope: Man's Concept of the Future from the Early Fathers to Teilhard de Chardin* (Garden City, N.Y.: Doubleday, 1966), pp. 121–42.

[5] Benz, "Christian Expectation of the End of Time and the Idea of Technical Progress," p. 122. See also Lynn White, Jr., "Cultural Climates and Technological Change," in his *Medieval Religion and Technology* (Berkeley: University of California Press, 1978), pp. 236–8.

Judeo-Christian worldview. In a series of essays written from the 1950s to the 1970s, Lynn White took up this idea and argued for the medieval idea of the "dualism of man and nature," a dualism that engendered the exploitation of nature for man's "proper ends."

This argument is problematic on several counts, in particular because medieval theological commentaries on the biblical verses cited by both Benz and White (e.g., Gen. 1:27 and 9:6) suggest a more complex interpretation of the divine call to "subdue" nature. In the hexaemeral literature – commentaries on the six days of Creation – composed between the Jewish writer Philo Judaeus's *On the Making of the World* (first century) and Robert Grosseteste's *Hexaemeron* (thirteenth century), the patient reader is more likely to encounter an ethic of stewardship and coöperative partnership between man and nature than an ethic of exploitation.[6]

Benz's second point had to do with the Christian conception of time. Linear time, including belief in the idea of progress and the conviction that the end of history in a great redemptive flourish is, even now, upon us (as in Matt. 24 and 25), led, according to Benz, to the construction of a theory of meaningful action within the limited scope of a finite linear history.[7] There is great intuitive appeal in this argument. Cyclic time, unending rounds of birth and rebirth, seems to preclude any urgency attached to the world of human affairs. The famous expression of the doctrine of *niskama karma* ("disinterested action") found in the great Sanskrit text of the *Bhagavadgita* would seem to argue for a detachment from this world, a deemphasis on working "to redeem the time," and a lack of incentive to seek out the tools of material progress.[8] And yet, as appealing as the dichotomy between a progressivist Christian West and an unworldly Hindu–Buddhist–Taoist East might be, the case cannot be made that belief in linear time and the end of the world did, in fact, propel medieval society to develop mouldboard plows, overshot waterwheels, and the crossbow. Therefore, in order to suggest a causal connection between Christianity and progress, Benz argues that in the Middle Ages, in the well-known formulation of St. Benedict, labor was worship, and "inactivity is the enemy of the soul."

Another argument for a unique western technological receptivity and creativity might begin by reversing the order of influence. Instead of a permanent ideology creating the conditions in which technology could thrive, perhaps one should notice instead that the church was famously resilient,

6 See George Ovitt, Jr., *The Restoration of Perfection* (New Brunswick, N.J.: Rutgers University Press, 1987), pp. 85–7, for a more detailed argument on this point.

7 Benz, "Christian Expectation of the End of Time and the Idea of Technical Progress," p. 126. For a general overview of the idea of progress during the Middle Ages, see M. D. Chenu, "Tradition and Progress," in *Nature, Man, and Society in the Twelfth Century*, ed. and trans. Jerome Taylor and Lester K. Little (Chicago: University of Chicago Press, 1968), pp. 310–30.

8 For the relevant texts, see *The Bhagavadgita in the Mahabharata*, trans. and ed. J. A. B. van Buitenen (Chicago: University of Chicago Press, 1981), pp. 93–5: "No one becomes a man of discipline without abandoning the intention of fruits."

adapted to changing material conditions, and constructed a "theology of labor" to celebrate economic and technological advances that had taken place independently of Christian theological concerns. The key changes in "cultural climate" traced by Weber, Benz, and White can all be dated to the twelfth century, when the church was developing a new theory of labor and progress to account for events that it did not control.[9] What seems clear in any case is that the medieval church did nothing to inhibit the development of technology. On the contrary, when coupled with the salutary example of economically self-sufficient, and therefore technologically progressive, monastic establishments, the church treated work in the world as a fitting preparation for the real work on the spiritual self that would free men from the burden of labor forever.

CLASSICAL AND ASIAN INFLUENCES ON MEDIEVAL TECHNOLOGY

A profound discontinuity existed between the technological achievement of the Greco-Roman period and that of the Latin Middle Ages. Roman society inherited a wide spectrum of tools and techniques from the early civilizations of Europe and the Middle East, including agricultural implements, mining technology, iron tools, techniques of textile production, the semicircular arch and barrel vault that made bridge and aqueduct construction possible, square-rigged sailing ships, and many other important technologies that were refined by Roman farmers, craftsmen, and engineers.[10] Yet, despite this rich tradition, the Romans displayed little interest in original invention. Columella's *De re rustica* and Vitruvius's extraordinary first-century B.C. treatise *De architectura* (a work of engineering as well as architectural design) are exceptions to the often-noted phenomenon of a great empire based on an inherited technical tradition that it did little to further. A reliance on slave labor and disdain for manual work led to the paradoxical situation of a technologically advanced society that had little use for invention. Roman society valued law, military superiority, and leisure, and it showed itself to be better adapted to the exploitation of inherited techniques than the pursuit of new ones.[11]

9 This argument is made by Jacques Le Goff in several essays, most notably in his "Labor, Techniques, and Craftsmen in the Value System of the Early Middle Ages (Fifth to Tenth Centuries)," in *Time, Work, and Culture in the Middle Ages*, trans. Arthur Goldhammer (Chicago: University of Chicago Press, 1980), pp. 86ff.

10 Paul-Marie Duval, "The Roman Contribution to Technology," in *A History of Technology and Invention: Progress Through the Ages*, ed. Maurice Daumas, trans. Eileen B. Hennessy, 5 vols. (New York: Crown Publishers, 1970), vol. 1, pp. 245ff. See also M. I. Finley, "Technical Innovation and Economic Progress in the Ancient World," *Economic History Review*, 2nd ser., 18 (1945), 29–45.

11 See Brian Stock, "Science, Technology, and Economic Progress in the Early Middle Ages," in *Science in the Middle Ages*, ed. David C. Lindberg (Chicago: University of Chicago Press, 1978), p. 4.

Even so, early-medieval Europe lost ground in many areas of technological development between the sixth and ninth centuries. In mining techniques, metallurgy, and textile manufacturing, there were clear losses of skill and technical knowledge. Even if most of these losses were temporary, they were significant, and it took the remarkable infusions of Asian – Chinese, Indian, and Islamic – inventions to stimulate Western inventiveness and the economy of Europe during the ninth and tenth centuries.

The magnetic compass was among the first Chinese inventions to reach the West; it is mentioned in Alexander Neckham's *De natura rerum* in 1190 (tenth century). Paper (twelfth century), the wheelbarrow (twelfth century), and cast iron (thirteenth century) would follow. Woodblock printing was used in China in the seventh century and wooden movable type by 1050. Gunpowder was known in China as early as the ninth century, and by the middle of the following century, the Chinese were using a mixture of salt-peter, sulfur, and carbonaceous materials to power rockets. Gradually, this invention made its way along the trade routes to the West. From Persia, Western Europe may have learned of the windmill, though differences in design suggest independent discovery. Design differences also suggest independent discovery for the escapement clock. India gave the West the churka, or cotton gin.[12]

Borrowing occurred also at a more abstract or theoretical level. Key scientific texts were translated from Greek and Arabic in the twelfth and thirteenth centuries, fueling the intellectually innovative period of the twelfth through fourteenth centuries. Of specific relevance to the development of Western technology were translations of works on mathematics and mathematical science. Included were texts that taught the use of Arabic numerals, practical geometry, and algebra, and also treatises on the astrolabe and mathematical astronomy. These materials were incorporated into the curricula of the schools as the "quadrivial sciences": arithmetic, geometry, astronomy, and music. Mathematical science also appeared in the university curriculum under the rubric of "mixed sciences" – scientific disciplines such as astronomy and optics conceived to be simultaneously mathematical and physical. The knowledge thus made available focused attention on the world of measurable entities and technological activity that required calculation, providing tools that would eventually come to define the modern character: in White's formulation, "precise, punctual, calculable, standard, bureaucratic, rigid, invariant, finely coordinated, and routine." By the fifteenth century, these traits would give the West its lead in both technological and economic development.[13]

12 For Chinese contributions to Western science and technology, see Joseph Needham, *Science and Civilization in China*, 7 vols. (Cambridge: Cambridge University Press, 1954), vol. 5, pt. 7, pp. 14ff. For gunpowder in particular, see Joseph Needham, *Science in Traditional China* (Cambridge, Mass.: Harvard University Press, 1981), pp. 27–56.

13 David C. Lindberg, "The Transmission of Greek and Arabic Learning to the West," in Lindberg, *Science in the Middle Ages*, pp. 62ff; A. I. Sabra, "Situating Arabic Science: Locality versus

AGRICULTURAL TECHNOLOGY

A full belly is, of course, the precondition for all higher forms of human culture. The first great technological revolution of the Middle Ages took place in the development of labor-saving devices for farming. Commencing in the first decades of the sixth century, a series of remarkable technological innovations completely transformed agriculture in Northern Europe.[14] Of greatest importance were the development of the carruca, or mouldboard plow – suitable for tilling the heavy soils of Northern Europe – and the gradual transition from oxen to horses as the primary draft animal. Properly harnessed, a horse is capable of pulling a given load fifty percent faster than an ox. Such an extraordinary gain in productivity led to the clearing of more wasteland and the transport of half again as many goods from the countryside to market towns and trade fairs. Crucial to the efficient use of horsepower were the breast-strap harness, invented in the sixth century, and the horse collar, introduced in the ninth century. These elegant inventions shifted the weight of the horse's load from the neck and windpipe to the shoulders and chest. Combined with the reinvented horseshoe (which had, inexplicably, disappeared from European history for three centuries), the whippletree (which increased plow-team traction by allowing for a variety of harnessing arrangements), and significant innovations in plow design (especially the mouldboard, which turned heavy soils), the shift in draft animals created a revolution in farming techniques and production. Splendid illuminations from the late Middle Ages provide iconographic documentation of the important changes in plow design, harnessing, and cart construction that led to extraordinary gains in agricultural productivity from the sixth to the twelfth centuries.[15] In addition to these key agricultural tools, the revolution in food production of the early Middle Ages saw innovations in planting and cultivation that would substantially increase per-acre yields. Especially important were shifts to three-field crop rotation, the use of fallowing to "rest" fields exhausted by nitrogen-intense cereal production, and, most important, the gradual expansion of manure use during the late Middle Ages to increase yields.[16]

Essence," *Isis*, 87 (1996), 654–70; and Ahmad Y. Al-Hassan and Donald R. Hill, *Islamic Technology* (Cambridge: Cambridge University Press, 1986).

[14] The following works deal with this transformation: B. H. Slicher van Bath, *The Agrarian History of Western Europe, A.D. 500–1850* (London: Blackwell, 1963); Marc Bloch, *Land and Work in Medieval Europe: Selected Papers*, trans. J. E. Anderson (Berkeley: University of California Press, 1967); Georges Duby, *Rural Economy and Country Life in the Medieval West*, trans. Cynthia Postan (Columbia: University of South Carolina Press, 1968); E. M. Jope, "Agricultural Implements," in *A History of Technology*, ed. Charles Singer, 8 vols. (New York: Oxford University Press, 1954–1978), vol. 2, pp. 81–102; and Del Sweeney, ed., *Agriculture in the Middle Ages: Technology, Practice, and Representations* (Philadelphia: University of Pennsylvania Press, 1986), p. 9.

[15] John Langdon, *Horses, Oxen, and Technological Innovation* (Cambridge: Cambridge University Press, 1986), pp. 9, 15, and passim, for illustrations.

[16] George Duby, *The Early Growth of the European Economy* (Ithaca, N.Y.: Cornell University Press, 1974), p. 192.

Other agricultural implements that came to be used during the great revolution in farming of the medieval period included the triangular harrow (used in France by the thirteenth century) and the bar-handled scythe – essentially the same design still in use today. Spread over four hundred years, the net effect of these improvements in agricultural tools and techniques included the creation of an interdependent food economy. Previously self-sufficient local production became more specialized and regional. Enhanced means of ground transport led to the movement of foodstuffs along regular trade routes in both the Mediterranean region and Northern Europe. The rise of market towns was also a by-product of agricultural surplus. Europe became more urbanized during the twelfth and thirteenth centuries as the population doubled or tripled.[17] None of these changes would have been possible without technological innovations in agriculture.

POWER TECHNOLOGIES

Closely related to developments in agriculture was the invention of labor-saving power technologies. The famous statistics of the Domesday Book (prepared under orders from William the Conqueror in 1086) indicate not only an extraordinary expansion in the use of water power but also a shift in diet – from porridge to baked bread – based upon the extension of cultivated areas.[18] The expansion of the use of water power during the tenth to twelfth centuries extended throughout Christian Europe. Both undershot and overshot waterwheels – named for the way the motive force of moving water was applied to the blades of the wheel – were used in the grinding of grain, the fulling of cloth, the driving of wine presses, and the powering of bellows for iron forges. The spread of new monastic orders like the Cistercians and Premonstratensians during the eleventh and twelfth centuries favored this expansion of the use of water power. Devoted as they were to economic self-sufficiency, these orders embraced labor-saving technologies. When, on monastic estates, the power of moving water was joined to the camshaft – invented in China and diffused through Islam to the West – the vertical overshot waterwheel could achieve forty to sixty horsepower of work, enough to drive a small-scale "industrial revolution" in the textile and building trades during the twelfth and thirteenth centuries.[19] On an even larger scale, water mill construction stimulated the building of dams

[17] J. C. Russell, "Population in Europe, 500–1500," in *The Middle Ages*, ed. Carlo Cipolla (The Fontana Economic History of Europe, 1) (New York: Fontana Press, 1976), p. 36.

[18] Duby, *Early Growth of the European Economy*, p. 187.

[19] For the history of waterwheels, see Terry Reynolds, *Stronger Than a Hundred Men: A History of the Vertical Water Wheel* (Baltimore: Johns Hopkins University Press, 1983), pp. 83ff. For the technological implications of the commercial revolution of the late Middle Ages, see Robert S. Lopez, *The Commercial Revolution of the Middle Ages, 950–1350* (Englewood Cliffs, N.J.: Prentice-Hall, 1971).

beginning in the twelfth century. These engineering projects were designed to divert the flow of streams to clusters of mills, and their design required the mastery of advanced engineering techniques.

A second important power technology during the late Middle Ages was the windmill. Neither the Greeks nor the Romans had developed technologies for harnessing wind power. It appears that the windmill was originally Persian, having been developed during the seventh century. However, the Western windmill of the twelfth century is different enough in design from the Persian version to suggest independent discovery. In particular, Western windmills utilized a greater "sail" area for wind capture than did Eastern windmills, thus obtaining a more consistent rotary motion. This change in design was an obvious advantage when wind power was used to pump water or to power a mill.[20] By the fourteenth century, the post-and-tower windmills of England and Northern Europe dominated the work of water-lifting and the powering of machinery. Iconographically, the first illustration of a post-mill dates from around 1270, in the "windmill Psalter" written at Canterbury. Post-mills turned on a central post axis so that the windmill sails could always face the wind. These efficient and flexible windmills provided a viable alternative to water-driven mill power in regions of England and Northern Europe where fast-moving streams were not available.

Although technological advances in the use of water and wind power had short-term positive effects on agriculture and manufacturing, the longer-term effects were often socially divisive. Peasant farmers were forced to pay a fee to use mills for the large-scale grinding of grain and fulling of cloth, and during the extended economic crisis of the fourteenth century, landlords collected substantial rents from their control of power machinery. Ironically, technological progress drove land rents higher. In the late Middle Ages, as in the modern era, "labor-saving" technology did not necessarily ease the financial burdens of workers.[21]

TEXTILE PRODUCTION

Economically, the most important direction in which the energy generated by wind and water would flow was toward the manufacture of cloth. During the early Middle Ages, cloth making was exclusively the province of women. With the expansion of agricultural productivity and the consequent freeing of labor time for other tasks, the manufacture of cloth became a domestic industry, with men weaving cloth while their wives prepared the yarn. Technological advances in loom design led to the gradual adoption of the horizontal, treadle-powered loom (perhaps based on a Chinese design),

[20] Rex Wailes, "A Note on Windmills," in Singer, *History of Technology*, vol. 2, pp. 617–18.

[21] Duby, *Early Growth of the European Economy*, pp. 221–4.

which was more efficient than the older vertical looms. By the twelfth century, cloth manufacture had evolved into a multinational enterprise, with England supplying raw wool and Flanders and Italy doing most of the fulling, dying, and weaving. Although domestic production of cloth remained in the hands of women, production for the market was taken over by men; indeed, among the earliest guilds were the cloth merchant guilds of tenth-century Italy.[22]

The mechanics of wool manufacture presents an interesting case study of the rationalization of production during the "first commercial revolution" of the Middle Ages; the development of this industry demonstrates the convergence of technological innovation with economic progress. Raw wool, often grown on monastic lands, first had to be sorted into damaged, coarse, medium-fine, and fine grades according to strict regulations enforced by the guilds. The divided bundles of wool were then washed with a variety of mild corrosives to remove grease and dirt. Next, the sun-dried wool was cleaned by hand, using fuller's tools like the forceps or the head of the thistle plant (*Dipsacus fullonum*), called a "teazel." The process of "carding" separated the fibers and prepared the wool for spinning; the carding machine essentially cross-combed the tangled strands of wool to separate them into individual fibers. A ball of carded (or combed) wool was then attached to the head of the distaff, from which it was spun into thread. In the ancient world, some spinning was done by hand, but for the most part a spindle and whorl (weight) were employed to convert the wool to thread. Until the thirteenth century, spinning in Europe used the spindle and whorl method; the earliest reference to the more energy-efficient spindle-wheel occurs in a Drapers' Guild regulation of the thirteenth century.[23] By the fifteenth century, refinements to the simple spinning wheel had given it the design that it would still have in the eighteenth century. Once the thread was spun, it had to be prepared for attachment to the loom by being reeled (unwound from the spindle) and prepared for the warp by being wound around the beams of the loom. When the cloth was woven on the horizontal loom, it was finished by being fulled; that is, the fibers were thickened to eliminate gaps in the weave. By the eleventh century, waterwheels drove the wooden mallets that beat (fulled) the cloth. Cleaning, brushing, and dying completed the process of preparing wool for the market.

By the thirteenth century, the profitability of cloth manufacture was great enough to allow some merchants to set up a putting-out system. Acting as a middleman, the merchant would purchase raw wool from a variety

[22] For a lucid account, see Francis Gies and Joseph Gies, *Cathedral, Forge, and Waterwheel: Technology and Invention in the Middle Ages* (New York: HarperCollins, 1994), pp. 183–4.

[23] My summary is based largely on R. Patterson, "Spinning and Weaving," in Singer, *History of Technology*, vol. 2, p. 202. Also excellent is John H. Munro, "Textile Technology," in *Dictionary of the Middle Ages*, ed. Joseph R. Strayer, 13 vols. (New York: Charles Scribner's Sons, 1988), vol. 11, pp. 693–711.

of sources – mostly monastic – and then sell it (at a profit) to various small producers (weaver, fuller, dyer), who would complete a step in the production process at a piecework rate and then sell the cloth back to the merchant. Finally, the finished cloth was taken to market and sold at a considerable markup. A logical development in this process of expanding and rationalizing the manufacture of textiles was the establishment of joint stock companies to raise capital for expanding production and for spreading out the risk of failure over several investors. As textile companies developed, primarily in Italy during the late thirteenth and fourteenth centuries, they naturally came to invest in shipping and to expand their dealings to other products – silks, linens, metals, and so forth. As business expanded, the need for accurate records became critical; hence the use of mathematical techniques to establish the first European account books. Bills of exchange and letters of credit followed the gradual, century-long process of developing double-entry bookkeeping. These instruments of credit made long-distance, transnational trade in a variety of commodities more efficient and may be seen as the foundation of the European banking system.[24] Even in crude outline, this long and quite complex history of the evolution of a modern capitalist enterprise out of a medieval handicraft reveals the connection between technological innovation and the economic expansion of European enterprise.

MILITARY TECHNOLOGY

The risk and reality of worker–peasant uprisings – such as that over mill fees at St. Albans (England) in 1330 – coupled with the normal course of dynastic and territorial conflict and the fear of invasion, contributed to continuously escalating military needs on the part of kings and noblemen. As the growth of powerful, centralized sources of coercion grew in the late Middle Ages – as nations took shape and feudal lords became divine monarchs – the technology of warfare became increasingly important. By the late eleventh century, the crossbow had evolved from a fairly primitive Roman prototype into the most terrible weapon of medieval warfare. So deadly was the crossbow that the Second Lateran Council (1139) had to ban it from use against any but the "infidel." This ban was not respected; indeed, over the next two hundred years, the crossbow became even more formidable. Forged in iron or primitive steel, this combination of gears, pulleys, and levers carried razor-sharp shafts farther and more accurately than any other medieval weapon.

The majority of other medieval weapons were developed in response to the wave of fortress-building that began in the tenth century and continued

[24] Gies and Gies, *Cathedral, Forge, and Waterwheel*, pp. 183–4.

through the high Middle Ages. The trebuchet, which had arrived in the West from China as early as the first century, became an important tool of siege warfare in the twelfth century. This weapon, which used the ingenious substitution of heavy counterweights for human dead weight, allowed military engineers to hurl destructive missiles with enormous momentum and energy against enemy fortifications.[25] Although cannon had entered the fray in Italy already by the fourteenth century, the trebuchet continued in use into the fifteenth century.

The best-known item of military technology of the medieval period is perhaps the stirrup. To this invention, Lynn White credited sweeping changes in the course of European history by ensuring the victory of Charles Martel over Muslim forces at Tours, leading to the eventual establishment of the feudal economic system.[26] Although the ancient stirrup – invented in China in the fourth century – was without question an aid to armed and armored riders, giving them greater stability as rider and horse were joined as a fighting machine of unprecedented effectiveness, there is some controversy about the extent of the impact of the stirrup on medieval warfare. Most tellingly, it appears that early-medieval tactical style favored the skirmish and siege over a set battle that would favor mounted "shock troops" of the sort described by White.[27] Nonetheless, whatever qualifications are added to White's analysis of the impact of the stirrup, the fact remains that a mounted man wielding a steel lance, battle-axe, and high-tempered sword was a more formidable weapon than Europe had ever seen. The definitive proof of its devastating power would come after the establishment of Spain's presence in the New World.

Of course, gunpowder would eventually make the armored knight an anachronism. Roger Bacon has been credited with the first European mention of gunpowder (ca. 1268), a description that seems clearly to have been based on (at least) second- or third-hand familiarity with Chinese fireworks.[28] Although the date of the first European firearms is unclear, unambiguous references to guns appear in Italy during the first quarter of the fourteenth century. Cannon were deployed in Germany in 1331 and by Edward III at Calais in 1346. These early cannon were made of bronze or iron and their cannonballs of lead or stone; by the fifteenth century, cannon were being cast in stronger iron and cannonballs were being made of iron as well. Although cannon were not necessarily as accurate or transportable as other siege weapons like the trebuchet, and poorly mixed gunpowder led to premature explosions,

25 Donald Hill, "Trebuchets," *Viator*, 4 (1973), 110–22.

26 Lynn White, Jr., *Medieval Technology and Social Change* (London: Oxford University Press, 1962), pp. 1–32.

27 See P. H. Sawyer and R. H. Hilton, "Technical Determinism: The Stirrup and the Plough" *[a review of Lynn White's Medieval Technology and Social Change], Past and Present*, 24 (1963), 90–5. See also Kelly De Vries, *Medieval Military Technology* (New York: Broadview Press, 1992), pp. 99–110.

28 Gies and Gies, *Cathedral, Forge, and Waterwheel*, p. 206.

ongoing work on this weapon of great potential destructiveness led to continuous improvements.[29] By the end of the fourteenth century, cannon were also being mounted on ships, allowing for the extension of war to the high seas.

MEDIEVAL SHIPS AND SHIPBUILDING

Medieval improvements in ship design made sea trade faster, safer, and far more profitable. In this area, as elsewhere during the late Middle Ages, a direct correspondence existed between the pursuit of profit and the development of technologies. In the early-medieval period, the hulls of Northern European ships were constructed with overlapping external planks riveted together in order to create a strong and flexible shell – a design suited to the rough seas of the North Atlantic. Mediterranean ships, operating in calmer seas, had less flexible hulls constructed of abutting planks joined with mortises and tenons – a tedious and difficult method of construction.[30]

The most significant advance in medieval shipbuilding technology took place in the gradual shift from shell construction, where the hull rather than the framing ribs provides the ship's strength, to skeleton construction, where an internal frame is built first and supplies the hull's resiliency.[31] During the late Middle Ages, Mediterranean shipbuilders converted to skeleton-built ships, reaping the advantages of shorter building times and lighter overall ship weight; northern shipbuilders continued to use shell construction.[32]

The standard northern merchant ship of the Middle Ages was the cog, which carried a single square sail amidships. Cogs were at first flat-bottomed and shell-built, but by the twelfth century the conversion to a rounded hull had made the ship more seaworthy. The stern-post rudder, long used in China, replaced the older lateral rudder in the Baltic Sea during the thirteenth century. By the fourteenth century, after being shortened by cutting a port in the stern to put the tiller directly in the helmsman's hand, the strengthened stern-post rudder was also used in the Mediterranean.

Although the single-masted cog was a vast improvement over the shallow-draft, multioared galley – in terms of seaworthiness, available cargo space, range of travel, and so forth – the fore-and-aft-rigged merchant vessel equipped with lateen sails was far better than any previous design. The lateen sail is essentially a modified square-rigged sail that has its leading edge

[29] Philippe Contamine, *War in the Middle Ages*, trans. Michael Jones (London: Blackwell, 1984), pp. 114ff. Also excellent is Bert S. Hall, *Weapons and Warfare in Renaissance Europe* (Baltimore: Johns Hopkins University Press, 1997).

[30] I have relied here on Richard Ungar, *The Art of Medieval Technology: Images of Noah the Shipbuilder* (New Brunswick, N.J.: Rutgers University Press, 1991), pp. 51ff.

[31] B. Greenhill, *Archaeology of the Boat* (London: A. and C. Black, 1976), pp. 60ff.

[32] Ungar, *Art of Medieval Technology*, p. 53.

cut short so that the ship can tack into the wind. It is unclear where the lateen sail was developed, but it was in common use throughout European waters by the twelfth century.[33]

During the last decade of the twelfth century, Alexander Neckham, in *On the Nature of Things*, made the first European reference to the mariner's compass. The assimilation of the compass – invented originally in China and independently in the West in the twelfth century – made all-weather travel possible in European waters. The doubling of the number of voyages that could be undertaken – from summer-only travel to winter travel as well – led, of course, to increased profits for the merchants of Europe and further demonstrated the value of technological innovation. In addition to the compass, the development of ships' charts, hourglasses (to assist in the calculation of approximate speed and distance traveled – not for the computation of longitude), and the astrolabe (for measuring approximate latitude) all made navigation safer and more profitable. The rise of seafaring city-states – preeminently Genoa – is not attributable solely to advances in technology, but these navigational tools did make the projection of European power possible on a far wider scale.

BUILDING CONSTRUCTION AND THE GOTHIC CATHEDRALS

Whereas ships, water mills, windmills, and siege machinery have to be visualized largely from the rich iconographic record of a book like *The Plan of St. Gall*, one can still marvel at that greatest surviving symbol of medieval engineering genius – the cathedral. It is also with the construction of the cathedrals that we learn something of the working life of medieval craftsmen. The master masons who supervised the building of cathedrals made sketches and kept notebooks, the best of which is the sketchbook of Villard de Honnecourt, a rich guide to late-medieval engineering, architecture, machine building, and practical mathematics.[34] Villard was especially adept at the use of geometrical figures as a means of estimating proportions for the construction of buildings, the fixing of sites, and the design of machines like waterscrews, water-powered saws, and hoisting equipment. Although one could not use Villard's plans as precise guides to the reproduction of tools or buildings, they do reveal a fertile thirteenth-century mind at work on what we would now regard as pure problems of engineering.

The skills that went into the construction of a cathedral represent a cross section of the technological achievements of the late Middle Ages and

33 For the development of the lateen sail, see John H. Pryor, *Technology and War: Studies in the Maritime History of the Mediterranean, 649–1571* (Cambridge: Cambridge University Press, 1988), pp. 30–8.

34 L. V. Salzman, *Building in England Down to 1540* (Oxford: Oxford University Press, 1952).

demonstrate both indigenous talent and the absorption of foreign innovations. Most notable, as Lynn White argues, was the importation of the pointed arch from India in the eleventh century. Once the load-bearing function of the outer wall was shifted to the arch and rib vault, the walls themselves could be freed to bear the rather different aesthetic burden of stained glass, the world of color and biblical narrative that Abbot Suger so famously embodied at St. Denis in the early twelfth century.[35] Contemporaries like Theophilus Presbyter and Gervase of Canterbury have left detailed accounts of the work of glaziers and metal smiths, which complemented the work of the master masons. With the addition of the functional and aesthetically pleasing flying buttresses at Notre Dame and Rheims (sketched by Villard de Honnecourt), the "Gothic" cathedral assumed the form that was to encapsulate the medieval combination of pious aesthetic and technological genius – soaring vertical spaces, articulated rib vaults rising toward heaven, cobalt-blue and copper-red glass demonstrating the proposition that "God is light," airy flying buttresses, and each niche filled with sculpture – by any measure the greatest demonstration of a wealthy, theocentric, technologically advanced society.[36]

CONCLUSION

Even this brief review of some of the key technological changes of the Middle Ages reveals something of the connection that developed, quite early in the history of the West, between technological progress and economic development. While it would be unhistorical to deny that technology, for some, eased the burden of labor, it would be equally false to assert that the West's interest in invention was spurred on by a humanistic desire to extend human freedom.

Indeed, a direct connection can be perceived between the development of the technologies of warfare and the success of agricultural and power technologies, on the one hand, and the creation of a profitable market economy on the other. It may be that medieval men were moved to invention out of some restless spirit of creativity and the desire to subdue nature. Some of the important architectural innovations of the late Middle Ages can be traced to purely creative activity – consider, for example, the great sketchbook of Villard de Honnecourt, which anticipated the playful creativity of Leonardo da Vinci. Yet, at least in Europe, the assimilation and enhancement

[35] The best "reading" of the philosophy of the great cathedrals is Otto von Simpson, *The Gothic Cathedral* (New York: Harper and Row, 1962). Also classic on this subject is Erwin Panofsky, *Gothic Architecture and Scholasticism* (New York: Meridian Books, 1957).

[36] Georges Duby, *The Age of Cathedrals: Art and Society, 980–1420*, trans. Eleanor Levieux and Barbara Thompson (Chicago: University of Chicago Press, 1981), puts the art and engineering of the cathedrals into the social context of the high Middle Ages.

of imported technologies favored the development of a gradually expanding capitalist economy. Whereas the Chinese, for example, abandoned any notions of a global or even a regional empire, Europeans used gunpowder, steel, and even, one might argue, microbes to project their economic desires and cultural convictions throughout the world.[37] Technological innovation helped to stimulate European economic growth during the late Middle Ages, and, in turn, the heady expansionism of the fifteenth century encouraged it further. As this expansion proceeded, innovations in the areas of shipbuilding, navigation, and warfare outpaced change in power technologies and agriculture.

37 Two popular accounts of European "exceptionalism" and the role played by technology in the rise of capitalism are Jared Diamond, *Guns, Germs, and Steel* (New York: W. W. Norton, 1997), and David S. Landes, *The Wealth and Poverty of Nations* (New York: W. W. Norton, 1998).

INDEX

N.B.: The index does not cover the notes.

Printed in the United States
By Bookmasters